Microbial Diversity and Functions

MICROBIAL DIVERSITY AND FUNCTIONS

Editors :

D.J. Bagyaraj

NASI Senior Scientist
Department of Microbiology
University of Agricultural Sciences
G.K.V.K. Campus, Bangaluru, Karnataka

K.V.B.R. Tilak

NASI Senior Scientist
Department of Microbiology
Osmania University, Hyderabad, Andhra Pradesh

H.K. Kehri

Reader
Department of Botany
University of Allahabad, Allahabad, Uttar Pradesh

2012

New India Publishing Agency
Pitam Pura, New Delhi-110 088

Published by
Sumit Pal Jain *for*
New India Publishing Agency
101, Vikas Surya Plaza, CU Block, L.S.C. Mkt.,
Pitam Pura, New Delhi- 110 088, (India)
Phone: 011-27341717, Fax: 011-27341616
E-mail: info@nipabooks.com
Web: www.nipabooks.com

ISBN : 978-93-81450-10-9

Typeset at: Harminder Singh Kharb *for* Typographiya
Printed at: Jai Bharat Printing Press, Delhi

Dedicated to

Prof. Sudhir Chandra

D. Phil., F.N.R.S., F.B.S., F.P.S.I., F.N.A.Sc.
Former Head & Emeritus Professor
Department of Botany, University of Allahabad
Allahabad

Preface

Since their very first appearance on this planet, the microorganisms have benefitted our society in many ways. They are supposed to be the simplest but most versatile and talented products of evolution. They thrive in habitats extremely hostile to human life and are infinitely more skilled than any human chemist in their synthetic activities. Their rich diversity and their functional aspects make them indispensable components of our ecosystems.

In view of their benefits, the microbiologists are trying to conserve their diversity on one hand and on other are trying to understand their functions in different ecosystems so that they may be exploited in a better way to increase the benefits manifold.

This book has been published on the eve of 75th Birth Day of Prof. Sudhir Chandra, Former Head and Emeritus Professor, Department of Botany, University of Allahabad, Allahabad, who is well known for his concern to the microbes and made significant contributions in exploring their beneficial activities in different ecosystems.

The book contains 31 articles written by distinguished scientists of the country having expertise in dealing with the microbes and exploiting their potential for the benefits of mankind. The articles included in the book are thought provocating and deals with the topics of Taxonomy, Diversity and Applications of VAM fungi in different Ecosystems, Applications of Microbial Technology for Treatment of effluents of a Gelatine Factory, Biodiversity of Mycotoxigenic Fungi and Trichoderma, Useful microbes of Mangrove Ecosystem, Extremophiles, PGPRs, Phytotoxins, Litter decomposition, Biopesticides, Botanical Pesticides, biofertilizers and so many others including major concerns about the Evolution and Conservation of Microbial Biodiversity. All the articles written by the authors are original, timely and appropriate.

We are very much thankful to the students of Prof. Sudhir Chandra and his colleagues of the Department for their encouragement without which it would have not been possible to bring out this book. We are also grateful to the authors for readily agreeing to contribute articles and adhering to the schedule fixed by us for the timely publication of this book. We are also thankful to the Staff of NIPA, New Delhi for taking keen interest and making all efforts to bring out this book in time.

In recognition to his outstanding contributions to science we have great pleasure to dedicate this book to Prof. Sudhir Chandra with all humility.

D.J. Bagyaraj
K.V.B.R. Tilak
H.K. Kehri

List of Contributors

A.B. ADE
Professor
Department of Botany
University of Pune
Pune-411007
avinashade@unipune.ac.in

AKHILESH KUMAR
Assistant Professor
Department of Botany
Banaras Hindu University
Varanasi-221005
akhilesh_100@hotmail.com

AJAY KUMAR MOHAPATRA
Chief Executive
Regional Plant Resource Centre
Bhubaneswar-751105
otelp@rediffmail.com

A.K. PANDEY
Chairman
Madhya Pradesh Universities
Regulatory Commission
Block-4, Flate No.-1-2
Bhoj Open University Campus
Kolar Road, Bhopal-462001
akpmycol@yahoo.in

A.K. ROY
Professor
Department of Botany
T. M. Bhagalpur University
Bhagalpur-812007
botanyakr@yahoo.co.in

ANUJ KUMAR SINGH
College of Forestry
Orissa University of Agriculture and
Technology, Bhubaneswar-751003

AMIT JAKHAL
Assistant Professor
Department of Botany
Post Graduate Government College
Sector 11, Chandigarh-160011
amit_jakhal@yahoo.co.nz

ANJU TANWAR
Research Fellow
Department of Botany
Kurukshetra University
Kurukshetra-136119
anjutanwarbotany@gmail.com

ANIL VYAS
Professor
Microbial Biotechnology and
Biofertilizer Laboratory
Department of Botany
J.N.V. University, Jodhpur-342003
dranilvyas@plantmicrobeinteraction.com

ANURADHA JHA
Post-Doctoral Fellow
National Research Centre for Agroforestry
Jhansi-284003

AVINASH PRATAP SINGH
Research Scholar
Department of Botany
University of Allahabad, Allahabad-211002
singh.avinash42@yahoo.co.in

ARCHANA SINGH
DST Women's Scientist
Laboratory of Herbal Pesticides
Centre of Advanced Study in Botany
Banaras Hindu University
Varanasi-221005
drarchanabhu@gmail.com

ASHA SINHA
Professor
Department of Mycology and
Plant Pathology, Institute of Agricultural
Sciences, B.H.U., Varanasi-221005
asinha_iasbhu@yahoo.co.in

ASHISH ANEJA
Medical Officer
Vaidyanath Health Care Complex and
Research Centre, 8/3, Urban Estate Market
Kurukshetra-136118

ASHOK AGGARWAL
Professor
Department of Botany
Kurukshetra University
Kurukshetra-136119
aggarwal_vibha@rediffmail.com

ASHOK SHUKLA
Post-Doctoral Fellow
Lab. of Microbial Technology and
Plant Pathology, School of Biological and
Chemical Sciences, Department of Botany
Dr HS Gour Central University
Sagar-470 003
ashokshukla06@gmail.com

BENDANGMENLA
Research Assistant
Department of Botany
Nagaland University, Lumami,
Mockokchong, Nagaland

BHANU PRAKASH
Junior Research Fellow
Laboratory of Herbal Pesticides, Centre
of Advanced Study in Botany, Banaras
Hindu University, Varanasi-221005
bhanubhu08@gmail.com

CHETAN SHARMA
Research Scholar
Department of Microbiology
Kurukshetra University
Kurukshetra-136119

C. MANOHARACHARY
Emeritus Professor
Department of Botany
Osmania University
Hyderabad-500 007
cmchary@rediffmail.com

DEEPMALA PANDEY
WOS-A, DST, New Delhi
Department of Botany
D.D.U.Gorakhpur University
Gorakhpur-273009

DEEPAK VYAS
Professor
Lab of Microbial Technology and
Plant Pathology
School of Biological and Chemical
Sciences, Department of Botany
Dr HS Gour Central University
Sagar-470003
dvyas64@yahoo.co.in

DIWAKAR BAHUKHANDI
Senior Scientist
Indian Grassland and Fodder Research
Institute, Jhansi-284003
diwakar14jan@yahoo.co.in

D.J. BAGYARAJ
Senior Scientist, NASI
Department of Microbiology
GKVK Campus, Bangalore

D.K. AGARWAL
Ex. Professor
Division of Plant Pathology
IARI, Pusa Road
New Delhi-110012
dk_agarwal@rediffmail.com

G.D. SHARMA
Pro Vice-Chancellor (STM) and Dean
School of Life Sciences
Department of life science
Assam University, Silchar-788011
gduttasharma@yahoo.co.in

G.L. TIWARI
Emeritus Professor
Department of Botany
University of Allahabad, Allahabad-211002

HARBANS KAUR KEHRI
Reader
Department of Botany
University of Allahabad
Allahabad-211002
kehrihk@rediffmail.com

H.N. GOUR
Former Professor and Head
Department of Plant Pathology, Maharana Pratap University of Agriculture and Technology, Udaipur-313001
hngour49@gmail.com

JAMALUDDIN
Scientist Emeritus (CSIR)
Department of Bioscience, Rani Durgawati University, Jabalpur-482001
jamaluddin_125@hotmail.com

K. JOSHI
Lecturer
Department of Biotechnology
Mewar Girls College, Chittorgarh-312001
drkkhushbu-joshi@rediffmail.com

KAMAL PRASAD
Scientist
Centre for Mycorrhiza Research
Biotechnology and Management of Bioresources Division
The Energy and Resources Institute, Darbari Seth Block, IHC, Lodhi Road, New Delhi-110003
kamalp@teri.res.in

KAMAL RAI ANEJA
Professor
Department of Microbiology
Kurukshetra University
Kurukshetra-136119
anejakr@yahoo.ca

K.V.B.R. TILAK
Emeritus Professor
Department of Botany
Osmania University, Hyderabad-500007
tilakkvbr@yahoo.com

L.V. GANGAWANE
Professor Emeritus (UGC)
Soil Microbiology and Pesticides Laboratory
Department of Botany, Dr. Babasaheb Ambedkar Marathwada University
Aurangabad-431 003,
Present Address: 5, Vinayshree, Nandanvan Colony, Aurangabad-431 004.

MANISHA SRIVASTAVA
Assistant Professor
Department of Botany
Harish Chandra P.G. College
Varanasi-221001

MEHBOOB CHOUHAN
Microbial Biotechnology and Biofertilizer Laboratory
Department of Botany
J.N.V. University, Jodhpur–342003
mehboob012001@gmail.com

MINAL TAMBOLI
Microbial Biotechnology and Biofertilizer Laboratory, Department of Botany
J.N.V. University, Jodhpur–342003
mt.bittu@gmail.com

M.N.KHARE
Ex-Dean and Professor Emeritus
Department of Plant Pathology
Jawaharlal Nehru Agricultural University
Jabalpur-482004
mnkhare8@gmail.com

MOHNISH VYAS
Microbial Biotechnology and Biofertilizer Laboratory, Department of Botany
J.N.V. University, Jodhpur-342003
mhnshvs@gmail.com

M.P. SHARMA
Retired Professor
Department of Botany, University of Chandigarh, Chandigarh-160047
Res.1545, Sector 49 B, Pushpac Complex
Chandigarh-160047
amit_jakhal@yahoo.co.nz

NAVNEET KAUR
26 A, Kanchanjunga Apartments
Sector 53, Noida (U.P.)

NEETU
Research Fellow
Department of Botany, Kurukshetra University, Kurukshetra-136119
n.khushi15@yahoo.in

NIBHA GUPTA
Senior Scientist
Regional Plant Resource Centre
Bhubaneswar-751105
nguc2003@yahoo.co.in

NISHA MISRA
Professor
Department of Botany, DDU Gorakhpur University, Gorakhpur-273009
nishamisra11@yahoo.in

NISHI MATHUR
Department of Biotechnology
Mahila P.G. Mahavidhyalaya
Jodhpur-342001
drnishimathur@scientist.com

N.K. DUBEY
Professor
Laboratory of Herbal Pesticides
Centre of Advanced Study in Botany
Banaras Hindu University
Varanasi-221005
nkdubey2@rediffmail.com

N.L. MANDAL
Professor of Botany
T.M. Bhagalpur University
Bhagalpur-812007

N.S. JAMIR
Professor
Department of Botany
Nagaland University, Lumami, Mockokchong, Nagaland

PALLAVI RAI
Research Scholar
Department of Botany
University of Allahabad
Allahabad-211002
pallvigoodan@gmail.com

PARVEEN SURAIN
Research Scholar
Department of Microbiology
Kurukshetra University
Kurukshetra-136119

PRADEEP SAXENA
Principal Scientist
Indian Grassland and Fodder Research Institute, Jhansi-284003
pradeepsax@yahoo.com

PRANITA BORAH
Research Fellow
Department of Life Science
Assam University
Silchar-788011

PRIYANKA SINGH
Senior Research Fellow
Laboratory of Herbal Pesticides
Centre of Advanced Study in Botany
Banaras Hindu University
Varanasi-221005
priyabhusingh@gmail.com

POOJA RAI
Research Scholar
Department of Botany
University of Allahabad, Allahabad-211002
raigreatpooja09@gmail.com

RAVINDRA SHUKLA
Senior Research Fellow
Laboratory of Herbal Pesticides
Centre of Advanced Study in Botany
Banaras Hindu University
Varanasi-221005
ravischolarbhu@gmail.com

RICHA TANDON
Research Scientist, DST, New Delhi
Department of Botany
University of Allahabad, Allahabad-211002
richatandon12@rediffmail.com

R.K. SHARMA
Senior Scientist
Division of Plant Pathology
IARI, Pusa Road, New Delhi-110012
rksharma35@yahoo.co.in

ROHIT SHARMA
Assistant Professor
Centre for Microbial Biotechnology
CIL Building, Panjab University
Chandigarh-160014
rohit28@yahoo.com

ROLLIE VARMA
Department of Botany
University of Allahabad
Allahabad-211002
rollie.verma@gmail.com

RAGHVENDRA PRATAP NARAYAN
Assistant Professor
Department of Botany
Dr Bhim Rao Ambedkar Government
College, Anaugi, Kannauj-209733
Uttar Pradesh
raghu707@rediffmail.com

RICHA RAGHUWANSHI
Assistant Professor
Department of Botany
Banaras Hindu University
Varanasi-221005
richabhu@yahoo.co.in

R.S. MEHROTRA
Professor and Former Head
Department of Botany
Kurukshetra University
Kurukshetra-136119
aggarwal_vibha@rediffmail.com

R.S. UPADHYAY
Professor
Department of Botany
Banaras Hindu University
Varanasi-221005
upadhyay_bhu@yahoo.co.uk

SANTOSH M. SHARMA
Research Fellow
Biotechnology Resource Centre
G/1, Adinath, Shaikh Misry Road,
Wadala (E)
Mumbai-400037
santoshatbrc@gmail.com

S. GIRISHAM
Professor
Department of Microbiology
Kakatiya University
Warangal-506001
sivasrigirisham@gmail.com

SHAGUN SHARMA
Centre for Microbial Biotechnology
CIL Building, Panjab University
Chandigarh-160014
drigu2905@gmail.com

SHARMILA ROY
Senior Scientist (Entomology)
Central Arid Zone Research Institute
Jodhpur-342003
roysharmila@yahoo.com

SHILPA YADAV
Microbial Biotechnology and Biofertilizer
Laboratory, Department of Botany
J.N.V. University, Jodhpur-342003
yadavshilpa14@gmail.com

S.M. PAUL KHURANA
Director
Amity Institute of Biotechnology
Amity University, Amity Education Valley, Gurgaon-122413 (Manesor)
smpkhurana@rediffmail.com

S.M. REDDY
Emeritus Professor
Department of Microbiology
Kakatiya University
Warangal-506001
profsmreddy@yahoo.com

S.N. KANADE
Assistant Professor
Soil Microbiology and Pesticides Laboratory, Department of Botany
Dr. Babasaheb Ambedkar Marathwada University, Aurangabad-431003
sarikakanade81@gmail.com

S.P. TIWARI
Professor
Department of Plant Pathology
Jawaharlal Nehru Agricultural University, Jabalpur-482 004
suresh274@gmail.com

S. ROY
Professor of Botany
T.M. Bhagalpur University
Bhagalpur-812007

SUDHIR D. GHATNEKAR
Managing Director
Biotechnology Resource Centre
G/1, Adinath, Shaikh Misry Road, Wadala (E), Mumbai-400037
brc_suvash@hotmail.com

TALI AJUNGLA
Assistant Professor
Department of Botany
Nagaland University, Lumami
Mockokchong, Nagaland

UDAY CHAND BASAK
Scientist
Regional Plant Resource Centre
Bhubaneswar-751105
uc_basak07@yahoo.co.in

VARUN KHARE
Sadasivan Mycopathology Laboratory
Department of Botany
University of Allahabad
Allahabad-211002
varunk_2k@rediffmail.com

V.C. KHILARE
Assistant Professor
Botany Research Centre
Vasantrao Naik College
Aurangabad-431003
vikramkhilare@gmail.com

V.K. DWIVEDI
Project Fellow
Department of Botany
University of Allahabad
Allahabad-211002
dwivedi668@gmail.com

V. KOTESWARA RAO
Project Fellow
Department of Microbiology
Kakatiya University
Warangal-506001
koti_micro08@yahoo.co.in

❑❑❑

Contents

Microbial Diversity and Functions, 2012
© *D.J. Bagyaraj, K.V.B.R. Tilak, H.K. Kehri (eds.), pp. 01-16*
New India Publishing Agency, New Delhi (India)
E-mail : info@nipabooks.com; Website : www.nipabooks.com

Chapter 1

Microbial Bio-Diversity, Evolution & Conservation: Major Concerns & Missing Links

H.N. Gour, S.M. Paul Khurana and K. Joshi

ABSTRACT

Biodiversity is often understood in terms of wide variety of plants, animals and micro-organisms existing on the earth. So far, about 1.75 million species have been identified, mostly small creatures such as insects. According to scientific estimates, there are actually about 13 million species, although range from 3 million to 100 millions. The present article deals with the importance and threats of biodiversity as well as tries to highlights the role of government and non-government sectors in conservation of biodiversity. The initial part of the article indicates the concept of evolution and biodiversity, wherein we have tried to identify the present scenario. We find that a wide range of microbial, plant and animal species are sources of various biological products for human use and welfare. These species are getting lost or extinct due to lack of conservation over a period of long time. Therefore, an effort has been made to highlight the conservation acts and important rules and regulations for protecting and conserving the bio-resources.

The article also throws light on the bio-safety programmes. The agriculture biotechnology sector has recently expanded in India and is concerned with bio-fertilizers, bio-pesticides, tissue culture techniques etc. These sectors work to produce different GM crops (cotton, tomato, potato, brinjal and golden rice etc.) of high value. In this way the MoEF, and DBT under the ministry of ST work for implementation of the regulations of bio safety. The Indian bio-safety regulatory frame work, comprising the 1989 rules and guidelines of DBT, cover the entire spectrum of activities relating to GMOs.

Keywords : Evolution, Biodiversity, Bio-safety, Conservation, GM crops.

Introduction

Al Gore mused about Biodiversity as "you look at that river gently flowing by. You notice the leaves rustling with the wind. You hear the birds, you hear the tree frogs. In the distance you hear a cow. You feel the grass. The mud gives a little bit on the river bank. It is quiet; it's peaceful. And all of a sudden, it's a gear shift inside you. And it's like taking a deep breath and going..... Oh' yeah, I forgot about this" (An inconvenient Truth). The former Vice-President of the United States had to reinvent himself to become nature's oracle (or Goracle). The rate of consumption of the earth's resources by the human race is posing a threat to the sustainability of life on the planet earth. Nevertheless, the global energy industry, along with its patrons and affiliates, has finally come around to accepting this verdict. However, the efforts of the scientists might not have made the world wakeup to the perils of global warming if a protagonist with a powerful voice had not stepped forward to convey the message. One may be reminded of the words uttered by the great leader of the Native American Squamish Tribe, Cheaf Seattle, who was a scholar with love for his land and people. The great leader said about a century and half ago, "Teach your children what we have taught ours, that the earth is our mother, whatever, befalls the earth befalls the sons of the earth. The earth does not belong to man; man belongs to the earth. Man did not weave the web of life, he is merely a strand in it. We do not inherit the earth from our ancestors; we borrow it from our children". These words spoken by Cheaf Seattle are more pertinent and most relevant in today's context of human activities on earth.

The five years (December 27, 1831 - October 2, 1836) Charles Darwin, the English naturalist spent on board HMS Beagle in around the world voyage gave him the opportunity to study and compare the fauna, flora, and geology of many distant lands. It led him to wonder about the diversity of life forms he found and why creatures occupying similar environments in places around the globe could be so vastly different. The idea that biological species were not immutable but were capable of change was in itself not new at the time. Darwin would have been familiar with the speculations of his own grandfather, Erasmus Darwin and the French Zoologist, Jean - Baptiste Lamarck. But within a couple of years following the Beagle Voyage, Darwin was going much farther. He was thinking about a common origin for all life on the planet when he sketched in his note book a tree of life, implying that all species had diversified from a common stalk.

However, Darwin was not the only one thinking along such lines. In 1858, he received a letter suggesting ideas remarkably like his own; it was from Alfred Russell Wallace, who was collecting biological specimens in south-east Asia. Papers putting forth both points of view were duly presented at a meeting of the Linnean Society of London. *The origin of species* (As Darwin's 1859 *Magnum opus* came to be titled in 1872, in the sixth edition) marshalled a vast body of evidence and presented his arguments in favour of evolution driven by a process of natural selection that allowed traits best suited to a particular environment to spread in a population. Evolution and a common origin for all life lie at the heart of biology. In an essay strikingly titled 'Nothing in biology makes sense except in the light of evolution', the geneticist and evolutionary biologist Theodosius Dobzhansky declared, "Without that light Biology becomes a pile of sundry facts - some of them interesting or curious but making no meaningful picture as a whole". The elucidation of the structure of DNA, the unravelling of genetic code, and the ability to sequence the entire genome of even complex organisms have served only to lay bare the processes that produce life, which all living organisms share, and show how evolutionary pressures act on those processes. As though this were not enough, Darwin's ideas have inspired, over the past century and a half, "Powerful images and insights in science, humanities and arts", as an essay in *Nature* reminds us.

In fact, what Darwin would have deciphered about the facts of biologicals, could have been traced in the ancient epic scriptures where it has been stated that all the worldly manifestations have single divine origin. Religiously may be but scientifically more we can have an analogy of DNA-the blue print of life, which all organisms share uniformly and diversity exists in each and every living object of the world because of laddered language of nucleotides in different modes every time. We are required, therefore, to respect and value the ecological system. In His prophetic teachings, Lord Krishna says that "I am the cause of whole world, I am there in fish, in vegetation, in animals and all human beings. All worldly affairs are regulated by me and thus the cycle of universe goes on". Consequently, the richness of biodiversity is a manifestation of power of DNA (as science would endorse). Whether it is life forms or life molecule, the origin of all species in ecology is from a single strand/source. What an amazing spectacle of nature? Biodiversity encompasses the variety of all life on earth. India is identified as one of the 12 mega biodiversity rich countries. With only 2.5% of the land area, India already accounts for 7.8% of the global recorded species. India is also rich in traditional and indigenous knowledge, both coded and formal.

The biodiversity of earth is astounding. Our planet supports between three and 30 million species of plants, animals, fungi, protozoa, nematodes, bacteria, viruses, etc. Despite two centuries of research, systematists have described only about 01.4 million species. The ecology or role of these species in ecosystems, has been studied far less than one per cent. We know more about large, economically important plants and animals than we do about fungi and bacteria, despite their important ecological roles. Microbial diversity has driven the evolution of life on earth as well as the nutrient cycle, which are keys to the operation of biosphere (Xu,2006). As we explore the microbial world fully, we must keep in mind that microbes evolve more quickly than we can study them, providing an ever-increasing diversity of function for industrial application (Handelsmen and Wackett, 2002). These rapid evolution events are predicted on the great diversity and apparent plasticity of microbial products. Microbial diversity is a great source for biotechnological exploration of novel organism products and processes. Novel methods and approaches enable us to explore this vast diversity (Woese,1987). There are plants of great variability possessing several different types of chemical constituents, biochemical processes and genetic variabilities such as tallest tree which is called as stratosphere Giant, which grows in the Rockefeller Forest, Humboldt Redwoods state Park, California, is 112. 32m. The General Sherman giant sequoia in sequoia National Park, California, USA, is the world's largest living thing. It is 83.3m tall and measures 2.53m round its mighty trunk. It weighs about 2000 tonnes including its huge root system. On the contrary, *Wolffia,* a kind of duckweed, is just 0.6 mm long and weighs about as much as two grains of salt. Its seeds are the tiniest known- they weigh only 70 micrograms, as much as a single grain of salt. It is the smallest flowering plant known on the earth. There are carnivorous plants too; plants that can grow luxuriously on marshy lands as well and thus we can see a large diversity of species in plant kingdom documented thus far.

Interestingly enough the researchers examining plants growing in the geothermal soils of Yellowstone National Park and Lassen Volcanic National Park have found evidence of symbiosis between fungi and plants that may hold clues to how plants adapt to and tolerate extreme environments. This research was published in the *Science.*

Biologists Regina Redman of the University of Washington and Joan Henson of Montana State University and their colleagues examined 200 samples of *Dichanthelium lanuginosum,* also called "Geyser's Dichanthelium", for fungal colonisation. They found what may be a new species of the fungus *Curvularia* that survives only in temperatures greater than 98 degrees C when it associates with plants. The researchers suggest that thermo tolerance may occur through symbiotic mechanisms like heat dissipation by pigment, such as melanin, or activation of a 'biological trigger' that tells the plant to react to temperature changes or strongly than plants that lack the fungus.

Haksworth (1991), estimated that on a world wide basis there are about 1.5 million species of fungi. To date, however, only about 80,000 species have been described. There is a tremendous discrepancy between the numbers of known versus estimated species appears to relate to the fact that there has been woefully inadequate sampling of fungi in many parts of the world, most notably tropical and subtropical regions. Currently there is lots of interest in and actually a sense of urgency about documenting the world's fungi. This is coming about at the same time that we are beginning to see reports concerning alarming decreases in both the total number of fungal species and the quantity of individual species in Europe. In the case of tropical and subtropical regions it would indeed be a tragedy to lose species to extinction before we even have determined that they exist. This is, however, more than just a philosophical or ethical concern. Fungi are extremely valuable sources of chemicals, including various antibiotics, and also have great potential as biological controls for many serious pests. As noted by Hawksworth (1991), "the world's undescribed fungi can be viewed as a massive potential resource which awaits realization".

The reproductive structures produced by some species and, in some cases, wood permeated by hyphae actually may give off visible light causing them to glow in the dark. This phosphorescent glow has long fascinated and even frightened humans, and much has been written about the subject. Observations on bioluminescent fungi can be traced as far back as Aristotle times. Apparently, people have long used pieces of bioluminescent wood to mark their paths at night, and there even are reports of soldiers attaching pieces of luminous rotten wood to their helmets in order to be visible to one another at night. In the United States the glow produced by bioluminescent fungi has been referred to as foxfire (Alexopolous *et al.*, 2001).

One of the better known species is the so-called Jack-0-Lantern mushroom, whose orange gills glow in the dark. Unfortunately for the discipline of mycology, few individuals realize how intimately our lives are linked with those of fungi. However, it truly can be said that scarcely a day passes during which all of us are not benefited or harmed directly or indirectly by these inhabitants of the microcosm. For example, if you are much under the age of 50, it is probably difficult for you to comprehend how many lives have been saved by penicillin.

It is possible that you too may have survived to study fungi because of the antibiotic. It is obvious that individuals who study fungi must do a better job of educating the general public about the importance of these organisms. The example of a recent event that probably has done more along these lines than anything else in many years was the report that *Armillaria bulbosa,* a fungal species that is a facultative parasite of tree roots, may be among the largest and oldest living organisms. This report not only was highlighted in the scientific press but also was reported widely in newspapers in North America and Great Britain. Basically, what scientists reported was that one clone of *Armillaria bulbosa – dubbed* "fungus humongous" by the popular press—occupies a minimum of 30 acres in a Michigan forest and that the thallus or body of the organism weighs in excess of 10 tons. The age of this thallus was estimated to be more than 1500 years.The majority of life found in environments are microbial (Satyanarayan *et.al.,* 2005). According to Satyanarayan *et. al.,* (2005) extreme environments include high temperature, pH, nutrient concentration and water availability, and also conditions having high levels of radiation, harmful heavy metals and toxic compounds (organic solvents). Culture-dependent and culture-independent (molecular) methods have been employed for understanding the diversity of microbes in these environments. Extremophiles are of interest to both basic and applied biology. In the basic sense, these organisms held many biological secrets, such as biochemical limits to macromolecular stability and genetic instruction for constructing macromolecule stable to one or another extreme, and in applied sense, these organisms have yielded an amazing array of enzymes capable of catalyzing specific biochemical reactions under extreme conditions. It is generally believed that enzymes from extremophiles mimic the stability characteristics of the source organism: that is heat, salt or acid resistant bacteria will produce heat, salt or acid resistant enzymes (Madigan and Marra, 1997 and Selleck and Choudhary 1999).

Extensive global research efforts have revealed the novel diversity of extremophilic microbes. These organisms have evolved several structural and chemical adaptations, which allow them to survive and grow in extreme environments. There are some types of microbes which are found in extreme conditions and following are their adaptations:

Psychrophiles : Various species within the genera *Alcaligenes, Alteromonas, Aquaspirillum, Arthrobacter, Bacillus, Bacteroides, Brevibacterium, Gelidibacter, Methanococcoides, Methanogenium, Methanosarcina, Microbacterium, Micrococcus, Moritella, Octandecabacter, Phormidium, Polaribacter, Polaromonas, Psychroserpens, Shewanella* and *Vibrio* have been reported to be psychrophilic (Morita, 2001). The genus *Moritella* appears to be composed of psychrophiles only. For the first time, some psycrophillic species, *viz. Leifsonia aurea* (Reddy *et al.,* 2003a), *Sporosarcina macmurdoensis* (Reddy *et al.,* 2003b) and *Kocuria polaris* (Reddy *et al.,* 2003c) have been reported from Antarctica .

Thermophiles : Microbes capable of growth at high temperatures are a wide variety of prokaryotes as well as eukaryotes.

A few thermophilic fungi belonging to Zygomycetes (*Rhizomucor miehei, R. pusillus*), Ascomycetes (*Chaetomium thermophile, Thermoascus aurantiacus, Dactylomyces thermophilus, Melanocarpus albomyces, Talaromyces thermophilus, T. emersonii, Thielavia terrestris*), Basidiomycetes (*Phanerochaete chrysosporium*) and Hyphomycetes (*Acremonium alabamensis, A. thermophilum, Myceliophthora thermophila, Thermomyces lanuginosis, Scytalidium thermophilum, Malbranchea cinnamomea)* have been isolated from composts, soil nesting materials of birds, wood chips and many other sources (Mouchacca, 1999; Satyanarayana, *et al.* 1977 and Tansey and Bock, 1978).

Several bacteria and archaebacteria, which are capable of growth at elevated temperatures, have been classified into moderate (*Bacillus caldolyticus, Geobacillus stearothermophilus, Thermoactinomyces vulgaris, Clostridium thermohydrosulfuricum, Thermoanaerobacter ethanolicus, Thermoplasma acidophilum*), extreme (*Thermus aquaticus, T. thermophilus, Thermodesulfobacterium commune, Sulfolobus acidocaldarius, Thermomicrobium roseum, Dictyoglomus thermophilum, Methanococcus vulcanicus, Sulfurococcus mirabilis, Thermotoga mritima*) and hyperthermophiles (*Methanococcus jannaschii, Acidianus infernos, Archaeoglobus profundus, Methanopyrus kandleri, Pyrobaculum islandicum, Pyrococcus furiosus, Pyrodictium occultum, Pyrolobus fumarii, Thermococcus littoralis, Ignicoccus islandicum, Nannoarchaeum equitans*)- based on their optimum temperature requirements (Reysenbach, *et al.*, 2002, Stetter, 1998; Gonzalez, 1999; Kristjansson, *et al.*, 2000 and Ghosh *et al.*, 2003). These have been isolated from composts, sun-heated soils, terrestrial hot springs, submarine hydrothermal vents and geothermally heated oil reserves and oil wells.

Value of Biodiversity

Bio-resources are important components for progress and economic activities of a nation. But bio-resources management and utilization for human welfare is very important for the optimum utilization of the bio-resources. Awareness of the importance and implications of bio-resources among common people as well as elite educated citizens for safeguarding and protecting the optimum and balanced way of using the bio-resources needs critical studies to focus the natural bio-resources wealth for the benefit of not only the present generation of people but also to the future generations for their better, healthy and peaceful living on the earth. The problems facing at present is the over exploitation of bio-resources which would not only have negative impact on the environment but also sometimes totally destroy and erode the important bio-resources which are now available at local level, regional level and national levels. Biological resources includes genetic resources, organisms or parts thereof, populations, or any other biotic component of ecosystems with actual or potential use or value for humanity. Therefore, handling bio-resources in a proper manner in an appropriate way is important for the optimum use without over exploitation of our bio-resources wealth. National Biological Diversity Act (NBDA) of 2002 is meant for research and development of biodiversity in India. Some of the provisions of the act have been amended on the basis of reaction to the Act (Ejnavarzala *et al.*, 2010). Enactment of NBDA in 2002 accomplished two things:

- It established national sovereignty over biological resources in the country, and
- It gave effect to the Convention on Biological Diversity (CBD) in 1993.

Threats to Biodiversity

Biodiversity is being lost as on today more rapidly than at any time in the past several million years. Some biologists believe that about 60,000 of the world's 2,40,000 plant species perhaps even higher proportions of vertebrate and insect species could have become extinct within the next thirty years if the same trends continue. But even a species at no risk of extinction can lose much of its potential through the loss of genetic material by reduction in range, numbers and varieties. Losses in biodiversity and the emergence of new infectious diseases are among the greatest threats to life on planet (Andrew & Pieters , 2010 and Jones *et al.*, 2010).The current losses to biodiversity can be attributed to direct causes including habitat loss and fragmentation, invasion of introduced species, over exploitation of living resources and modern agriculture and forestry practices. The basic problems of losses to biodiversity include:

- The unsustainably high rate of human population growth and natural resources consumption.
- The steadily narrowing selection of traded products from agriculture, forestry and fisheries.
- Economic systems that fail to value economic resources.
- Inequity in ownership, management and flow of benefits; from both the use and conservation of biodiversity.
- Deficiencies in knowledge and application.
- Legal and institutional systems that promote unsustainable exploitation.

Loss of species and genetic diversity presents a serious threat to the goal of sustainable agriculture. Species and genetic diversity provide sources of pest resistance and control, new domesticates and the genetic raw material for plant breeding and genetic engineering.

After an extensive and intensive consultation process involving the stakeholders, the Central Government of India has brought out a Biological Diversity Act, 2002, with the following features:

- To regulate access to biological resources of the country with the purpose of securing equitable share in benefits arising out of the use of biological resource and associated traditional knowledge relating to biological resources.

- To conserve and sustainable use of biodiversity and biological resources.
- To respect and protect traditional knowledge of local communities relating biodiversity.
- To secure sharing of benefits with local people as conservers of bio-resources and holder of knowledge and information relating to the use of bio-resources.
- Conservation and development of areas of importance from the status of biological diversity by declaring them as biological diversity heritage site for protection and rehabilitation of threatened species.
- Involvement of the Institutions and State Governments in the broad scheme of the implementation of the Biological Diversity Act through constitution of State Biodiversity Boards (SBB) and Biodiversity Management Committees (BMC).

Bio-safety

It is very important to protect and conserve the biodiversity for the welfare of the human beings and awareness on the importance of biodiversity and its economic value as well as threats to biodiversity among the common people, elite citizens, teachers, scientists, administrators, scholars, students and even school children needs to be created to save our national wealth of biodiversity.

The agricultural biotechnology sector has recently demonstrated a major expansion in India. For example, there has been a 55 per cent increase in the number of agricultural Biotechnology firms from 2001 to 2003. The 132 firms concerned deal with bio-fertilizers, bio-pesticides and tissue culture in the main, with relatively few (20) involved in GM crops. "The first approval for the commercial production of any GM crop in India occurred in March 2002 when the Indian competent authority approved three varieties of GM cotton (MECH 12, MECH 162 & MECH 184 expressing the *cry1Ac* gene) amid widespread protests by anti-GM activists". This was followed by a significant increase in the availability of approved *Bacillus thuringiensis* (Bt) cotton (currently 135) hybrid varieties better suited to Indian cultivation.

The goal of the Indian regulatory system is to ensure that approved GM crops pose no major risk to food safety, environmental safety or agricultural production. As such, the Government of India has adopted a policy of careful assessment of the benefits and risks of GMOs at various stages of their development and field release to ensure biosafety.

The existing regulatory framework takes the form of rules and guidelines and is based upon three specific provisions (*viz.* Sections 6, 8, and 25) of the Environment Protection Act of 1986 (EPA).

- *Section 6* of the Act empowers the central government to make rules on procedures, safeguards, prohibition and restrictions for handling of hazardous substances.
- *Section 8* of the Act prohibits a person from handling hazardous substances, except in accordance with procedures and after complying with safeguards.
- *Section 25* of the EPA empowers the central government to lay down rules regarding procedures and safeguards for handling hazardous substances. These provisions of the EPA led to the adoption of the 1989 rules for the manufacture, use, import, export and storage of hazardous microorganisms, genetically engineered organisms or cells.

In 1994, the Government of India revised its earlier guidelines of 1990, entitled "Revised Guidelines for Safety in Biotechnology". These revised guidelines aimed at regulating shipment and importation of GMOs for, laboratory research, as well as large-scale production and the deliberate release of GMOs, plants, animals and products into the environment. By 2002, an array of legislation had come into existence. This included the National Biodiversity Act, 2002 (NBA), and the Protection of Plant Varieties and Farmers' Rights Act, 2001 (PPVFR), the latter of which derived from a broad-based consultation with a view to incorporate a form of farmers' rights into the national plant variety rights legislation. The bio-safety rules have since been supplemented by the Biotechnology Safety Guidelines. These Biotechnology Safety Guidelines have been issued in pursuance of Rule 4(2) of the Bio-safety Rules (1989), which require manuals of guidelines to be brought out by the Review Committee on Genetic Manipulation (RCGM).

Therefore, the Indian bio-safety regulatory framework, comprising the 1989 Rules and the Guidelines issues under this in 1990, 1994 and 1998 DBT guidelines, covers the entire spectrum of activities relating to GMOs. This includes "research involving GMOs, as well as genetic transformation of green plants, recombinant DNA (rDNA) technology in vaccine development, and large-scale production and deliberate/accidental release into the environment of organisms, plants, animals and products derived from rDNA technology'. Production facilities such as distilleries and tanneries that use GMOs are also covered. In India, the risk assessment and regulatory approval for releases of GMOs and GM products are mandatory. The concept of 'bio-safety' used in the regulations is a broad one, covering the health safety of humans and livestock, environmental safety (ecology and biodiversity) and economic impact.

Two nodal agencies, the Ministry of Environment and Forests (MoEF) and the Department of Biotechnology (DBT) under the -Minisrry of Science and Technology (DST) are responsible for the implementation of the regulations. The life-cycle of a GM product features four domains, pre-research, research, release and post-release, and the approval mechanism involving six competent authorities. The Recombinant DNA Advisory Committee (RDAC) is in the pre-research domain as it triggers research through its initial approval mechanisms. The Review Committee on Genetic Manipulation (RCGM) resides in the DBT and functions in the research domain, closely monitoring the process of research and experimental releases. It requests food biosafety, environmental impact and agronomic data from applicants who wish to do research or conduct field trials and will give permits to import GM material for research. Pursuant to Rule 4(2) of the "1989 Rules", the RCGM is also required to produce manuals of guidelines.

The RCGM is primarily made up of scientists from various disciplines and can request experts with specialized knowledge to review cases. It has a Monitoring cum Evaluation Committee (MEC) that monitors limited and large-scale field trials of GM crops and is primarily made up of agricultural scientists. Commercial production of GM crops, large-scale field trials of GM crops, and the imports of GM commercial products and GM-derived products (for example foodstuffs, ingredients in foodstuffs, and additives including processing aids containing or consisting of GMOs) come under the authority of the Genetic Engineering Approval Committee (GEAC) at the MoEF. The committee members are scientific experts and bureaucrats representing different ministries and, like the RCGM, can request assistance from experts with specialized knowledge. Additional to these national committees are the State Biotechnology Coordination Committee (SBCC) and the District Level Biotechnology Committee (DLC), who, along with the MEC, basically occupy the post-release domain, although they also contribute to the research domain activities through data-provisioning to the RCGM. Completing the regulatory apparatus are the Institutional Biosafety Committees (IBSC) which undertake the monitoring and implementation of safeguards at the R&D sites, under the close supervision of the RCGM, the SBCC and the DLC. IBSCs must be established in any public or private institute using rDNA in their research and comprise scientists from their respective institutes and at least one member nominated by the DBT. There are more than 230 IBSCs in India, of which 70 deal with agricultural biotechnology. They can approve contained research at institutes unless the research uses a particularly hazardous gene or technique which will require specific approval from the RCGM. All the IBSC approved projects are sent to RCGM for further approvals. In general, these authorities are vested with non-overlapping responsibilities.

Genetically modified plants can break down pollutants. This may be an effective way to clean soil contaminated by industrial chemicals and explosives used by the military, defense scientists.

Tests on six-inch tall GM poplar cuttings which had a gene from a rabbit inserted into them showed that they could remove up to 91 per cent of a chemical called trichloroethylene from the water used in their feed. This chemical, used as an industrial degreaser and one of the most common contaminants of ground water, was broken down by the plants into harmless by-products more than 100 times faster than by unaltered plants.

In view of their large size and extensive root systems, these *transgenic poplars* may provide the means to effectively clean sites contaminated with a variety of pollutants at much faster rates and at power costs than can be achieved with current conventional techniques, (Sharon Doty, University of Washington, Seattle, Proceedings of the National Academy of Sciences,2003).

The GM poplars also broke down other common environmental pollutants such as chloroform, a by product of the disinfection of drinking water, the solvent carbon tetrachloride, and vinyl chloride, used to make plastics. Poplars use an enzyme called cytochrome-P450 to break down contaminants. trichloroethylene is turned into a harmless salt, water and carbon dioxide.

Another study, also published in the PNAS, demonstrated a way to break down the military explosive RDX. Widespread contamination of land and ground water has resulted from the use, manufacture, and storage of the military explosive RDX. This contamination has led to a requirement for a sustainable, low-cost method to remediate this problem. According to Neil Bruce, from University of York, "One of the biggest concerns of RDX as a pollutant is that it migrates readily through soil into the ground water and subsequently contaminates drinking water supplies". A team genetically modified *Arabidopsis* plants to express enzymes called XplA and XplB, which are known to break down RDX. At their best, the plants reduced RDX concentrations from soil by up to 97 per cent in one week.

Though the GM plants may be an effective way to treat pollutants, Dr. Doty acknowledged that people might have concerns at the thought of forests of GM trees. In the United States and Britain, such plants can currently be grown only for research purposes. Dr. Doty added that poplars were fast-growing and could grow for several years without flowering, so there was reduced risk of their genes being transferred into wild populations of the tree.

Conservation of Biodiversity

The science of protecting and restoring biodiversity and ecological health, the basic principles of which include ecology, evolutionary biology, genetics, and physiology, all applied to achieve the goals of maintaining and managing biological diversity as well as ecological health.

It is heartening to note that the world's largest population of Irrawady has been found recently in Bangaldesh's waters according to a five year wildlife study. Until now, it was believed that the small light-grey mammal, famed as an aquarium attraction, was threatened and the International Union of Conservation of Nature had put five of its South-East Asian populations on its list of critically endangered animals.

These require society at large, legislatures and governments to be involved and to pay special heed. Sadly, while the governments of about 190 nations have signed to honour the Declaration, many of them do not seem to look around us. The U.S. government wants to deface Alaska by digging for oil there, and also refuses to believe in global warming. Japan, Russia and some Scandinavian countries want the ban on whale-hunting lifted, knowing fully well that the ban has been imposed to save the majestic marine mammal from extinction. Then we learn about the steady depletion of the tropical rain forests of the Amazon.

Look closer to home. The government is pushing ahead with the *Sethusamudram* Project, even as environmentalists and several scientists warn us of the loss of precious coral reefs and associated marine life forms. Soon after the asbestos-laden ship Clemenceau was turned away from the Alang Shipwrecking yards, (thanks to environmentalists' pressure), yet another of its kind is permitted in.

The river Ganga, held sacred by millions, is polluted day in and day out by domestic and industrial wastes. Yet the government dithers about the modes of implementing its own Ganges Action Plan (started almost fifteen years ago).

The government's intentions may be genuine, the money has been set apart and yet the follow-through is taking years. The sanctity of the river, in the meanwhile, continues to be debased. One Chief Minister wanted to deface the ambience around the glorious Taj Mahal with bricks and cement, to build a shopping mall, before the courts stopped it.

Let us think what sort of legacy we wish to bequeath to our off spring.

Sadly, while just about every political leader and government around the world profess commitment to guard the environment and concern for the future generations, most of them have not lived up to their promise.

In this regard, what the Israeli Parliament (the Knesset) has attempted to do is admirable and worthy of emulation by India (and indeed many other nations). Called the Israeli Commission for Future Generations, it was created by law as an inner parliamentary entity. Its role is to overview each legislative process, with special regard to long-term issues, and attempt to prevent potentially damaging legislation from passing the Knesset. This commission is given the authority to initiate bills that advance the interests of the future generations.

It is also entitled to provide the parliament with recommendations, and the opinions. Recommendations of this commission have to carry a scientific character, be detailed and include comparative research.

In fact, the financial crisis for which we must now pay so heavily prefigures the real collapse, when humanity bumps against its ecological limits. On Friday, (October, 10). Pavan Sukhdev, the Deutsche Bank economist leading a European study on ecosystems, reported that we are losing natural capital worth between $ 2 trillion and $ 5 trillion as a result of deforestation alone. Whereas, the losses incurred so far by the financial sector amount to between $1 trillion and $1.5 trillion. Sukhdev arrived at his figure by estimating the value of the services - such as locking up carbon and providing fresh water - that forest perform, and calculating the cost of either replacing them or living without them. The credit crunch is petty when compared to the nature crunch.

The two crises have the same cause. In both cases, those who exploit the resource have demanded impossible rates of return and invoked debts that can never be repaid.

Ecology and economy are both derived from the Greek word *oikos* - a house or dwelling. Our survival depends on the rational management of this home: the space in which life can be sustained. The rules are the same in both cases. If you extract resources at a rate beyond the level of replenishment, your stock will collapse. These are the points to ponder seriously. It is surprising that economy so often predominates eco-logy.

The Biodiversity at Galapoga and Malay island was on account of conserved flora and fauna, that could be observed by Darwin and Wallace, respectively to formulate the theory of evolution. Modern biologists need to explore missing links between biodiversity and evolutionary forces, namely natural selection.

References

Alexopoulos, C. J., Mims, C. W. and Blackwell, M. (2001). *Introductory Mycology*. Wiley India Pvt. Ltd

Andrew, R.B. and Pieter, T.J.J. (2010). When an infection turns lethal. *Nature*, 465:881.

Ejnavarzala, H. (2010). DNA Barcoding access to biodiversity and benefit-sharing policy issues in the Indian context. *Curr. Sci.*, 99 (5): 594-599.

Ghosh, D., Bal, B., Kashyap, V. K. and Pal, S. (2003). Molecular phylogenetic exploration of bacterial diversity in a Bakreshwar (India) hot spring and culture of *Shewanella*-related thermophiles. *Appl. Environ. Microbial.*, 69: 4332–4336.

Gonzalez, J. M. (1999). Thermophiles. In: *Extremophiles in Deep-Sea Environments* (Eds.) K. Horikoshi and K. Tsujii), Springer Verlag, Berlin, 1999: 113–154.

Government of India, the Biological Diversity Act, 2002; http//www.nbaindia.org/act/act_english.htm.

Handelsman, J. and Wackett L.P. (2002). Ecology and Industrial microbiology, microbial diversity sustaining the earth and industry, *Curr. Opin. Microbiol.*, 5: 221.

Hawksworth, D.L. (1991). The fungal dimensions of biodiversity: magnitude, significance & conservation. *Mycol. Res.*, 95: 641-655.

Jones, K.E., Mitchell, C.E., Myers, S.S., Bogich, T. and Osfeld, R.S. (2010). Impacts of biodiversity on the emergence and transmission of infectious diseases. *Nature*, 468; 647-652.

Kristjansson, J. K., Hreggvidsson, G. O. and Grant, W. D. (2000). Environment and its exploitation. In: *Applied Microbial Systematics* (Eds.) F.G. Priest and M. Goodfellow, Kluwer, Dordrecht: 231–291.

Madigan, M. T. and Marrs, B. L. (1999). Extremophiles. *Sci. Am.*, 276-282.

Morita, R.Y. and Moyer, C.L. (2001). Psychrophiles, origin of. In: *Encyclopedia of Biodiversity*, Academic Press, New York, vol. 4:917-924.

Mouchacca, J. (1999). Thermophilic fungi: Present taxonomic concepts. In: *Thermophilic Moulds in Biotechnology* (Eds.) B. N. Johri, T . Satyanarayana and J. Olsen, Kluwer, Dordrecht: 43–83.

Reddy G.S.N., Prakash, J.S.S., Srinivas, R., Matsumoto, G.I. and Shivaji, S. (2003c). *Leifsonia rubra* sp. nov. and *Leifsonia aurea* sp nov. psychrophillic species isolated from a pond in Antarctica. *Int. J. Syst. Evol. Microbiol.*, 53:977-984.

Reddy, G.S.N., Matsumoto, G.I. and Shivaji, S. (2003a). *Sporosarcina macmurdoensis* sp. nov., from a *Cynobacterial* mat sample from a pond in Macmurdo Dry Valleys, Antarctic. *Int. J. Syst. Evol. Microbiol.*, 53: 1363-1367.

Reddy, G.S.N., Prakash, J.S.S., Prabahar, V., Matsumoto, G.I., Stackebrandt, E. and Shivaji, S. (2003b). *Kocuria polaris* sp. Nov., an orange pigmented psychrophilic bacterium isolated from an Antarctic Cynobacterial Mat sample. *Int. J. Syst. Evol. Microbiol.*, 53:183-187.

Reysenbach, A. L., Gotz, D. and Yernool, D. (2002). Microbial life of marine and terrestrial thermal springs. In: *Biodiversity of Microbioal Life* (Eds.) J. T. Stanley and A. L .Reysenbach, Wiley-Liss Inc., New York: 345–421.

Satyanarayan, T., Raghukumar, C. and Shivaji, S. (2005). Extremophilic microbes: Diversity & perspectives. *Curr. Sci.*, 89: 78-89.

Satyanarayana, T., Johri, B. N. and Saksena, S. B. (1977). Seasonal variation in mycoflora of nesting materials of birds with special reference to thermophilic fungi. *Trans. Br. Mycol. Soc.*, 68: 307–309.

Sellek, G.A. and Chaudhuri, J.B. (1999). Biocatalysis in organic media using enzymes from extremophiles, *Enzyme Microb. Technol.*, 25; 471.

Stetter, K. O. (1998). Hyperthermophiles: isolation, classification and properties. In: *Extremophiles: Microbial Life in Extreme Environments* (Eds.) K. Horikoshi and W. D. Grant, Wiley-Liss Inc., New York: pp 1–24.

Tansey, M. R. and Brock, T. D. (1978). Microbial life at high temperatures: ecological aspects. In: *Microbial Life in Extreme Environments* (Ed.) D. J.Kushner, Academic Press, London, pp. 159–216.

Woese, C.R. (1987). Bacterial evolution. *Microbiol. Rev.*, 51; 221.

Xu, J. (2006). Microbial Ecology in the age of genomics and metagenomics: concepts, tools and recent advances. *Mol. Ecol.*, 15: 1713.

❑❑❑

Microbial Diversity and Functions, 2012

New India Publishing Agency, New Delhi (India)
E-mail : info@nipabooks.com; Website : www.nipabooks.com

Chapter 2

Soil Organism : Key Factor to Plant and Soil Health

Raghvendra Pratap Narayan and Pallavi Rai

ABSTRACT

Microorganisms play a key role for the proper running of the ecosystem. They help in nitrogen fixation, which is unique and most necessary property found only in this group. They help in decompositions, mineralization, nutrient cycling, phosphate solubilization, soil aggregate formation, symbioses, parasitism and many other important works. A great emphasis has been given by many scientists to conserve the microbial diversity. Developing human civilization has a great negative impact not only on the above ground biodiversity but also on the belowground biodiversity. This article is an overview, what important and crucial role nitrogen fixers play for maintaining soil and plant health. Nitrogen fixer not only fixes atmospheric nitrogen but also produces growth promoting hormones which are very important for plant growth.

Keywords: Microbial diversity, nutrient cycling, ecosystem functioning

Introduction

Soil is one of the most important finite and non-renewable natural resource because it requires billions of year to be formed by chemical and biological processes. Soil functions are essential for the biosphere, but the soil is perhaps the most difficult and poorly understood component of environment. Soil performs wide range of ecosystem services that underpin the sustainability of both agriculturally productive and natural systems. The main ecological function of soil includes supply of nutrients to support plant growth; buffering and transformation of potentially harmful compounds; exchange of gases and water in ecosystem and biological habitat of gene and species. The loss of soil as well as soil quality is of great concern because most of the human population is dependent on it for the food. So, it is must for human race to maintain the quality and quantity of the soil. This requires sustainable use of soil. Soil is formed by complex chemical, physical and biological processes. The biological process is one of important factor which maintains the soil health. According to Doran and Saftey (1997) soil health can be defined as "The continued capacity of soil to function as a vital living system, within ecosystem and land-use boundaries, to sustain biological productivity, promote the quality of air and water environments, and maintain plant, animal and human health".

The soil organisms play a crucial role in maintaining the soil health by a diverse class of biological processes including nutrient acquisition (Smith and Read, 1997; Sprent, 2001), nitrogen cycling (Kowalchuk and Stephen, 2001), carbon cycling (Hogberg *et al.*, 2001) and soil formation (Rillig and Mummey, 2006). Moreover, soil microbes represent the unseen majority in soil and comprise a large portion of the genetic diversity on Earth (Whitman *et al.*, 1998). The biological activity is mostly concentrated in the topsoil, approximately upto 30 cm. The biological component occupy a tiny fraction (<0.5%) of the total soil volume and makes up less than 10% of total organic matter in soil. The biological component mainly consists of soil organisms such as algae, fungi, bacteria, protozoans and some meso and macrofauna as mites, collebola, micro-arthropods, earthworms, centipedes, larvae etc. It has been estimated that one gram of soil contains as many as 1010–1011 bacteria (Horner-Devine *et al.*, 2003), 6000–50, 000 bacterial species (Curtis *et al.*, 2002), and up to 200 m fungal hyphae (Leake *et al.*, 2004). Although biological component is small but it plays a crucial role in cycling of nutrients such as nitrogen, carbon, phosphorus, sulphur etc. (Pankhurst *et al.*, 1997). Soil microbes, including microbial pathogens, are also important regulators of plant community dynamics and plant diversity, determining plant abundance and, in some cases, facilitating invasion by exotic plants. Conservative estimates suggest that about 20 000 plant species are completely dependent on microbial symbionts for growth and survival pointing to the importance of soil microbes as regulators of plant species richness on Earth (Marcel *et al.*, 2008).

Soil Biodiversity

Biodiversity is defined as variability and variety of living organism and the ecosystem in which they occur. The biodiversity is divided into genetic, specific (taxonomic) and community and ecosystem (functional) diversity (Fig. 1). The total number of species that is specific biodiversity is controlled by the genetic diversity, which is greater than the specific diversity. In ecosystem many species perform same function. On other hand single function is performed by the number of species. The diversity at ecosystem level has three perspectives: alpha diversity (within community diversity); beta diversity (between community diversity) and gamma diversity (diversity at total geographical area). The above ground diversity, which is visible, gets more attention among the scientific community. There is community of living organisms which lives in the soil and this community constitute what is now referred as "below-ground biological diversity" (BGBD) or sometimes "soil biodiversity." Soil biodiversity is comprised of the organisms that spend all or a portion of their life cycles within the soil or on its immediate surface. The soil biodiversity encompasses plant, animals, invertebrates and microorganisms that are dependent on one another. This soil biodiversity is an important but poorly understood component of terrestrial ecosystems.

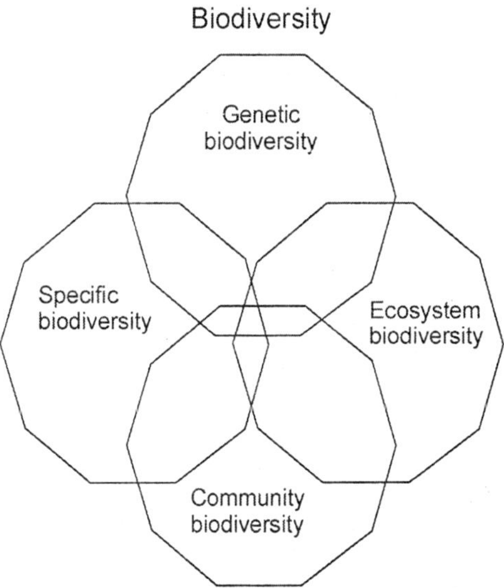

Fig. 1 : Different inter-related components of biodiversity.

Soil is actually one of the most diverse habitats on earth. It contains one of the most diverse assemblages of living organisms known to us such as fungi, algae, bacteria, protozoa, termites, earthworms etc. Giller *et al.* (1997) reported that a single gram of soil is estimated to contain 300,000 species of bacteria alone. Of the 1,500,000 species of fungi estimated to exist worldwide remarkably little is known about soil fungi, apart from the common fungal pathogens and the useful mycorrhizal species which improve crops' efficiency in taking up nutrients. Among the soil fauna some 100,000 species of protozoa 500,000 species of nematodes and 3,000 species of earthworms are estimated to exist, and several thousand species of the other invertebrate groups.

The easiest and most widely used system for classifying soil organisms is by using body size and dividing them into three main groups: macrofauna, mesofauna, microfauna and microbiota (Wallwork, 1970; Swift, *et al.*, 1979). *Microbiota:* these are the smallest organisms (< 0.1 mm in diameter) and are extremely abundant and diverse. They include algae, bacteria, cyanobacteria, fungi, yeasts, myxomycetes and actinomycetes that are able to decompose almost any existing natural material and can help in mineralization, nitrogen fixation, mineral absorption etc.

Microfauna: includes small collembola and mites, nematodes and protozoa that generally live in the soil and feed on microflora, plant roots, other microfauna and sometimes larger organisms. They are important to release nutrients immobilized by soil microorganisms. *Mesofauna*: (0.1–2 mm in diameter) includes mainly microarthropods, such as pseudoscorpions, springtails, mites, and the worm-like enchytraeids. Mesofauna have limited burrowing ability and generally live within soil pores, feeding on organic materials, microflora, microfauna and other invertebrates.

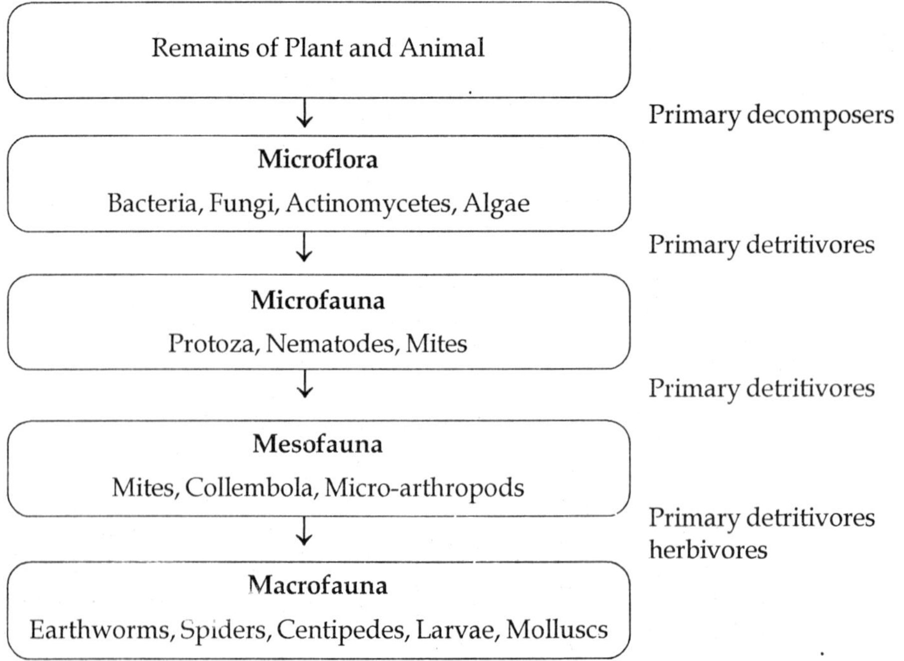

Fig. 2 : Soil Flora and Fauna involved in nutrient cycling.

Macrofauna: includes organisms visible to the naked eye (generally >2 mm in diameter). Such as vertebrates (snakes, lizards, mice, rabbits, moles, etc.) that primarily dig within the soil for food or shelter, and invertebrates (snails, earthworms and soil arthropods such as ants, termites, millipedes, centipedes, caterpillars, beetle larvae and adults, fly and wasp larvae, spiders, scorpions, crickets and cockroaches) that live in and feed in or upon the soil, the surface litter and their components. In both natural and agricultural systems, soil macrofauna are important regulators of decomposition, nutrient cycling, soil organic matter dynamics, and pathways of water movement as a consequence of their feeding and burrowing activities.

Contribution of Soil Organisms to the Soil Health

Invertebrates

The soil invertebrates which contribute in soil health maintenance has been classified by Swift and Bignell (1999) according to their feeding habits and distribution in the soil profile as (a) *Epigeic* species are biota which live and feed on the soil surface. These invertebrates reduce litter size and helps in mineralization (nutrient release), but do not actively redistribute plant materials. They are mainly arthropods e.g. ants, beetles, cockroaches, centipedes, millipedes, woodlice, orthopterans (grasshopper-type insects), together with gastropods (snails) and small, entirely pigmented (dark colored) earthworms. (b) *Anecic* species are biota which eat litter from the soil surface and transport it to the deeper soil layers. Through their feeding activities, a considerable amount of topsoil, minerals and organic materials become distributed through the soil profile; this is also accompanied by channel or structure formation and an increase in soil porosity. Fauna included in this group are earthworms (also called as ecosystem engineers), non-soil-feeding termites and arachnids (spiders). (c) *Endogeic* species are biota which live in the soil and feed on organic matter and dead roots, also ingesting large quantities of mineral materials. Fauna included in this group are microbes, non-pigmented earthworms and soil-feeding termites.

Bacteria

One of the most diverse groups of soil organism, and is found almost in every habitat. Thus they are known to perform a variety of function in ecosystem. They are known for cycling nutrients such as sulphur, nitrogen, phosphorus, carbon etc, mineralization of organic matter, decomposition, nitrogen fixation etc thus contribute up to a great extent to the soil health. Some of the functions of microbes in soil are discussed here.

Nitrogen fixation in soil

The biological nitrogen fixation that is by bacteria in soil is of two types associative or symbiotic that is done by the bacteria belongs to family rhizobiaceae that includes genera *Rhizobium, Bradyrhizobium, Sinorhizobium, Mesorhizobium* and *Azorhizobium* which forms nodules in legume plants. The Rhizobiaceae also have ability to form non-specific associative interaction with roots of other plants. (Reyes and Schimidt, 1979). These microbes associate with roots of non legume without forming nodule and stimulate their growth and yield (Shimshick and Herbert, 1979; Biswas *et al.*, 2000; Yanni *et al.*, 2001).

The other groups of nitrogen fixers are free living, in which some are partially associative. The first diazotroph was reported by Beijerinck in 1925 as *Spirillum lipoferum* (now *Azospirillum*). But the interest of the scientist in diazotrophs become intense after the discovery of high specific *Azotobacter paspali* and *Azospirillum* by Dobereiner *et al.*, 1972. The benefit of the diazotrophs over Rhizobiaceae is that they are associated with non legume crops especially with cereals. Several genera of diazotrophs has been discovered, which may include loosely or more intimately associated with plants such as *Acetobacter, Azocarpus, Azospirrilum, Beijerinkia, Herbaspirrilum, Klebsiella, Paenibacillus* and *Pseudomonas.* The biological nitrogen fixation(BNF) by these diazotrophs is highly variable in nature and depends on various environmental factors and variety of plants (Boddy *et al.*, 1991). The experiments have shown that BNF varies from 0-80% equivalent to 150-170 Kg N ha^{-1} Y^{-1} (Lima *et al.*, 1987; Boddy and Dobereiner, 1995; Garcia de Salamone, *et al.*, 1996; Malik *et al.*, 1997).

It is found that the nitrogen supplied by the root colonizing diazotrops is very low (Rao *et al.*, 1998). This is due to reason that the nitrogen fixed by these diazotrophs is not secreted directly into the host as in the case of symbiotic nitrogen fixation (Kleiner, 1984). In case of diazotrophs the nitrogen fixed remains in their cell and is available to the plants after death and decay of the bacterial hormones (Rao *et al.*, 1998). This process is inefficient and perhaps delayed, when compared with the active release of the immediate products of N_2 fixation by living bacteria as occurs in legume nodule (Mylona *et al.*, 1995). This is reason for this poor performance of free living nitrogen fixers. Another reason for this poor performance is due to insufficient release of the organic carbon in rhizosphere.

Production of plant growth promoting substances

Many bacteria are known to synthesise hormones that are called as phytohormones or plant growth regulators (PGRs). These phytohormones which, are organic substances by nature and influences the physiological processes of plants, and act at a very low concentrations. These phytohormones can lead to characteristic changes in plant growth and development. In 1979 production of auxin, cytokinin and gibberellins like substances were reported from *Azospirillum brasilense* (Tien *et al.*, 1979). Several other studies done by Okon and Kapulnik, 1986; Harari *et al.*, 1988 shows that the *Azospirrillum* inoculation have similar effect on the plants as that of phytohormones. Early seedling root growth of canola and lettuce was significantly promoted by the inoculation of certain strains of *R leguminosarum* (Noel *et al.*, 1996). The capacity to synthesise IAA (Indole Acetic Acid) ie Auxin is wide spread among soil and plant associated bacteria. It is estimated that 80% of bacteria isolated from soil could synthesise IAA (Patten and Glick, 1996). Some of the genera producing IAA are *Azotobacter* spp (Pati *et al.*, 1995); *Herbaspirrillum seropidecae* (Bastian *et al.*, 1998) *Klebsiella pneumonia* (El-Khawas and Adachi, 1999); *Acetobacter diazotrophicus* (Fuentes-Ramirez *et al.*, 1993) etc. The cytokines are known to produce by *Azotobacter* spp (Gonzalez-Lopez *et al.*, 1986); *Azospirillum* spp (Horemans *et al.*, 1986), *Rhizobium* spp (Upadhyaya *et al.*, 1997); *Paenibacillus polymixa* (Timmusk *et al.*, 1999). Gibberellins are produced by *Azotobacter* spp (Martinez-Toledo *et al.*, 1988), *Azospirillum brasilense* (Janzen *et al.*, 1992); *P. polymyxa* (Sattar and Gaur, 1987); *R. leguminosarum* (Atzorn *et al.*, 1988); *Azospirillum lipoferum* (Piccoli *et al.*, 1996); *Acetobacter diazotrophicus* and *Herbaspirrillum seropidecae* (Bastian *et al.*, 1998) and *Bacillus licheniformis* (Gutierrez-Manero *et al.*, 2001).

Nutrient uptake

Several studies show that many bacteria mostly PGPRs stimulate plant growth by facilitating the uptake of minerals such as NPK and other nutrients by plants. In experiments inoculated with *Azospirillum* spp have shown increased uptake of NPK from solutions at faster rate (Lin *et al.*, 1983). This uptake may be due to pectinolytic activity of *Azospirillum* cells (Okon, 1982) in which middle lamellae is hydrolysed, which may accelerate the water and nutrient uptake by roots (Sarig *et al.*, 1984; Bekri *et al.*, 1999). Another mechanism which helps in uptake of nutrients is production of siderophores. Siderophores are low molecular weight iron binding molecules that binds Fe^{3+} (Briat, 1992). The production of siderophore is reported in *Azospirillum lipoferum* (Shah *et al.*, 1992), *Azotobacter vinelandii* (Knosp *et al.*, 1984; Demange *et al.*, 1988).

Increased stress resistance

Many bacteria are known to increase the stress resistance power of the plants. Study show that sorghum plants inoculated with *Azospirillum* were less drought stressed and having more foliage water. Similar results were obtained with *Klebsiella* spp. In wet land rice together with increased nutrient uptake and proline content (Razi and Sen, 1996). Proline is good osmoregulator produced during water stress condition. The diazotrophic as well as many normal bacteria are known to minimize the oxidative stress caused by the reactive oxygen species such as super oxide anion radicals, hydrogen peroxide, hydroxyl radical and singlet oxygen (Sies 1997; Stajner *et al.*, 1995). *Azotobacter chroococcum* was reported to improve oxidative stress defense ability in sugar beet leaves (Stajner *et al.*, 1997). The inoculation with *A. chroococcum* increases chlorophyll and carotenoid content and increased super oxide dismutase, peroxidase and catalase activity.

Phosphate solubilisation

Phosphate is one of the most important nutrients for the plants. Most of the phosphate found in nature is in combined form, which is not readily available to the plants. Only 5% of total phosphate is available for the plants. In modern agriculture addition of soluble phosphatic fertilizer is common practice. However larger portion of the soluble inorganic phosphate applied to soil as fertilizer is rapidly immobilized by iron and aluminium in acid soils and by calcium in calcareous soil soon after application thus becoming unavailable to plants (Sanyal and De Datta, 1991; Holford, 1997). Soil microorganism are known to solubilize inorganic mineral phosphate and make them available to the plants (Illmer *et al.*, 1995; Jones, 1998). These microbes solubilize insoluble phosphate by using their enzyme phosphatase (Tarafdar and Junk, 1987; Garcia *et al.*, 1992). A large number of phosphate solubilizers have been isolated from the rhizosphere soil (Chabot *et al.*, 1993).

Vitamin production

Vitamins are organic compounds needed in small quantity and play a key role in growth and development of the living organism. Some of the vitamins are synthesized in plant body while others are supplied externally by soil microbes. Many bacteria such as *Azotobacter, Azospirillum, Bacillus, Rhizobium* etc are known to produce some or all water soluble B group vitamins such as niacin, pantothenic acid, thiamine, riboflavin etc (Martinez-Toledo *et al.*, 1996; Sierra *et al.*, 1999; Revillas *et al.*, 2000).

Control against pathogens

Many bacteria are known to control plant pathogens in one or other way. There are two main ways to control plant pathogens by these microorganisms, one is antagonism of pathogen and other is by changing the host plant's susceptibility *i.e.* by induced resistance. Bacteria can antagonize soil borne pathogens through various mechanisms such as competition, antibiosis, and parasitism (Handelsman and Stabb, 1996). Competition for nutrients is another mechanism to control pathogens. The best studied example is competition for iron. Many bacteria produce high affinity siderophore that sequester iron in the rhizoplane, making it less available to certain harmful rhizospheric microorganisms (Schippers *et al.*, 1987; O' Sullivan and O' Gara, 1992). In antibiosis bacteria are known to produce antimicrobial compounds such as antibiotics eg *Rhizobium leguminosarum* bv. trifolii produces an antibiotic peptide trifolitoxin (Breil *et al.*, 1996). *Azotobacter* have been found to be antagonistic against *Botrytis cinerea* Pers. These bacteria synthesise an antifungal compound of low molecular weight that inhibits the production of conidia by fungus (Doneche and Marcantoni, 1992). Similarly *Pseudomonas* spp (Whipps, 2001); *A. brasilense* and *A. lipoferum* (Tapia-Hermandez *et al.*, 1990); *Paenibacillus polymyxa* (Nielsen and Sorensen, 1997), *Bacillus* spp etc are known to produce antibacterial and antifungal compound.

Fungi

Fungi are one of the most diverse groups of living organisms, which play a great role in maintaining soil structure and function and supply nutrients to plants. In this section special attention has been given to the VAM fungi and their role for plant and soil health.

Supply of inorganic nutrients

VAM fungi are known for supplying inorganic nutrient from the soil to plants. Several experiments have shown that mycorrhizal fungi can overcome inorganic nutrient limitation by increasing acquisition of nutrients (Clark and Zeto, 2000). The mycorrhizae are especially known for absorbing P (Gianinazzi *et al.* 1995; Clark and Zeto, 2000). Improved uptake of phosphorus by mycorrhizal plants is attributed to the exploration of hyphae beyond "P" depletion zone (Beamer, 1992; Cui-Muyi and Nobel Park, 1992). It has been suggested that VAM fungi are able to take up phosphate more efficiently than the roots as well as they take it faster (Bolan *et al.*, 1984). VAM fungi have been reported to play a key role in improvement of uptake of nutrients other than phosphorus and nitrogen by altering acquisition mode of the absorbing system (Gildon and Tinker, 1983; Harley and Smith, 1983). They have been shown to be involved in the uptake of Zn, Cu and other trace elements having low mobility in soil. They have also been shown to increase Fe (Rai, 1988) and sulphate uptake (Harley and Smith, 1983). Uptake of Br, Cl, and others has been shown to be directly influenced by VAM fungi (Buwalda *et al.*, 1982).

Supply of organic nutrients

It is found that some litter inhabiting ECM fungi produce protease and distribute soluble amino compounds through hyphal network into the root (Read *et al.*, 1989). Recently, Glomus has been shown to transport the amino acids glycine and glutamine into wheat (Hawkins *et al.*, 2000; Dell *et al.*, 2000).

Supply of water

VAM plant interaction has been reported to increase absorption of water (Safir *et al.*, 1972), root hydraulic conductivity (Levy *et al.*, 1983) and plant water dynamics (Hardie, 1985). It has been proposed that in phosphate deficient situations, the mycorrhizal plants exhibit lower resistance to the flow of water and improve its absorption (Cooper, 1984). Several studies indicate that *Rhizopogon* an ECM fungi can help plants to tolerate and recover from water deficits and this can aid seedling establishment. Plants colonised by AM fungi may have increased tolerance to drought. In flax (*Linum usitatissimum*), AM mycorrhizal plants had lowered sensitivity to stress, higher assimilation combined with lower increase of transpiration, and enhanced root conductance (Vonreichenbach and Schonbeck, 1995). Unlike for AMs, large mycelia strands of ECM fungi may increase water flow by bridging the gap between the soil and the root (Lamhamedi *et al.*, 1992). It has been suggested that extramatrical hyphae bypass the dry zones surrounding the root during the drought period and maintain water continuity across the soil-root interface (Hardie, 1985).

Nutrient cycling and nutrient conservation

Soil microbes are of paramount importance in *cycling nutrients* such as carbon (c), nitrogen (n), phosphorus (P), and sulphur (S). Not only do they control the form of these elements [eg: specialized soil bacteria convert ammonium N (NH^+_4) to nitrate N (NO^-_3)], they can regulate the quantities of N available to plants. It is only through the action of soil microbes that the nutrients in organic fertilizers are liberated for plants and use by the other microorganisms. This process is termed as *mineralrsation* [the conversion of organic complexes of the elements to their inorganic forms. Eg. Conversion of proteins to carbon dioxide (CO_2) ammonium (NH^+_4) and sulphate (So_4^{2-})]. It is perhaps the single most important function of soil microbes as it recycles nutrients tied up in organic materials back into forms useable by plants and other microbes. It is through the process of mineralisation that crop. Residues, grass clippings, leaves, organic wastes etc are decomposed and converted to forms useable for plant growth as well as converted to stable soil organic matter called *humus*.

Soil structure maintenance

It is obvious from the examination of ECM mycelial mats, that mycorrhizal fungi have a big impact on soil structure. Yet, there is scant information in the literature regarding soils in tropical ecosystems. In agricultural soils, AM fungi increase the formation of soil aggregates (Bethlenfalvay *et al.*, 1999). Fungal filaments called hyphae, stabilizes soil structure because these threadlike structure ramify throughout the soil literally surrounding particles and aggregates like a hairnet. The fungi can be thought of as the threads of the soil fabric (Finlay and Read, 1986).

Biocontrol

VAM association with plants has been shown to provide them protection from pathogens (Zambolim, 1986). Many evidences, which show that plants inoculated with VAM symbiont exhibit an increased resistance to fungal root diseases, are available in literature (Roseldahl, 1985). Protection from *Olpidium brassicae* in lettuce and tobacco (Schoenbeck and Dehne, 1979), from *Thielaviopsis basicola* in citrus (Davis *et al.*, 1978), from *Pyrenochaeta terrestris* in onion (Becker, 1976), control of *Fusarium* wilt of cucumber Seedlings by AMF (Hao *et al.*, 2005), control of *Fusarium oxysporum f. sp. lycopersici* by AMF *Glomus intraradices* in tomato (Akkopru and Demir, 2005), from *Macrophomina phaseolina* in many crops (Kehri and Chandra, 1989) has been reported. Activities of other pathogens including nematodes have been shown to be suppressed by VAM fungi (Cooper and Grandison, 1987). Various morphological, physiological and biochemical mechanisms have been proposed to explain the resistance of mycorrhizal plants against pathogens, however, it is generally believed that the protection of host plant is due to accummulation of some metabolites, lignification of root cells and high chitinolytic activities in them or by siderophore production.

Improvement in hormone level

VAM fungi in symbiotic association with plants affect the hormone level; increase the level of cytokinin (Allen *et al.*, 1980) and alter level of gibberellins and abscisic acid (Allen *et al.*, 1982). They also affect the growth regulating substances, which are useful not only in root development and many basic processes of plant growth but also in mobilization of nutrients and their translocation (Letham *et al.*, 1978). The possible mechanism for the production of growth regulators by VAM symbiont or its host plant has not been investigated in detail; however, the production is possibly related to the physico-chemical changes in the host plant brought about by symbiont (Smith and Gianinazzi-Pearson, 1988).

Carbon transport

The fungal/plant interface provides a conduit for the movement of carbon from the plant to the fungus, and for movement between plants linked by mycelia (Simard *et al.*, 1997; Wu *et al.*, 2001). The nature of the interface and its mode of regulation are still being elucidated (Hall and Williams, 2000). It is generally believed that mycorrhizal plants direct more of their photosynthates into the soil than non-mycorrhizal plants. This extra carbon accumulates in patches and at the edge of hyphal mats (Finlay and Read, 1986), and boosts the energy supply to the detrital food web, benefiting saprohytic microbes and other soil organisms (Barea, 2000). Because the chemical (Dieffenbach and Matzner, 2000) and physical environment around mycorrhizas (the mycorrhizosphere) differs from nonmycorrhizas, presumably it provides microhabitats for soil biota that are not present in the rhizospere of nonmycorrhizal roots. Mycorrhizal fungi are estimated to consume from 15 to 50% of net primary production (Vogt *et al.*, 1982).

Food for animals

Long-distance dispersal of spores from ECM fungi with hypogeal (truffle-like) sporocarps depends largely on mammal mycophagy (Claridge and May, 1994). Mycophagy is widespread and has been demonstrated in Europe, Australasia and North America. Mycophagy serves to maintain populations of ECM fungi and provides nourishment to small mammals (Malajczuk *et al.*, 1987). Sporocarps are good sources of water, protein, carbohydrates and minerals (Johnson, 1994; Claridge *et al.*, 1999). The tripartite relationship between truffles/truffle-like fungi, vertebrates such as squirrels and many ground-dwelling marsupials, and the host trees, are well known. Less well known is the role that mycorrhizal fungi play as a food source for invertebrates and the role of invertebrates in dispersal of ECM and AM fungal spores.

References

Akkopru, A. and Demir, S. (2005). Biological Control of *Fusarium* Wilt in Tomato Caused by *Fusarium oxysporum f. sp. lycopersici* by AMF *Glomus intraradices* and some Rhizobacteria. *J. Phytopath.*, 153, 544–550.

Allen, M.F. (1982). Influence of vesicular arbuscular mycorrhizae on water movement through *Bouteloua gracilis* (H.B.J.) Lag. ex. Stend. *New Phytol.*, 91: 191-196.

Allen, M.F., Moore, T.S. and Christensen, M. (1980). Phytohormone changes in *Bouteloua gracilis* infected by vesicular arbuscular mycorrhiza. I. Cytokinin increases in the host plant. *Can. J. Bot.*, 58: 371-374.

Atzorn, R., Crozier, A., Wheeler, C.T., and Sandberg, G. (1988). Production of gibberellins and indole-3–acetic acid by *Rhizobium phaseoli* in relation to nodulation of *Phaseolus vulgaris* roots. *Planta*, 175: 532–538.

Barea, J.M. (2000). Rhizosphere and mycorrhiza of field crops. *Biol. Resour. Manag.*,: 81-92.

Bastián, F., Cohern, A., Piccoli, P., Luna, V., Baraldi, R., and Bottini, R. (1998). Production of indole-3–acetic acid and gibberellins A1 and A3 by *Acetobacter diazotrophicus* and *Herbaspirillum seropedicae* in chemically defined culture media. *Pl. Grow. Regul.*, 24: 7–11.

Beamer, E.E.J. (1992). Plant life span and response to inoculation with vesicular arbuscular mycorrhizal fungi. III. Responsiveness and residual soil P levels. *Mycorrhiza,* 1(4): 169-174.

Becker, W.N. (1976). *Quantification of onion Vesicular Arbuscular Mycorrhizae and their Resistance to Pyrenochaeta terrestris,* Ph.D. thesis, Univ. of Illinois, Urbana.

Beijerinck, M.W. (1925). Über ein *Spirillum* welches freien Stickstoff binden kann? *Zentbl. Bakt. Parasitkde, II,* 63: 353–359.

Bekri, M.A., Desair, J., Proost, P., Searle-van leeuwen, M., Vanderleyden, J., and Vande Broek, A. (1999). *Azospirillum irakense* produces a novel type of pectatelyase. *J. Bacteriol.,* 181: 2440–2447.

Bethlenfalvay, G.J., Cantrell, I.C., Mihara, K.L. and Schreiner, R.P. (1999). Relationships between soil aggregation and mycorrhizae as influenced by soil biota and nitrogen nutrition. *Biol. Fert. Soils,* 28: 356-363.

Biswas, J. C., Ladha, J. K., and Dazzo, F. B. (2000). Rhizobia inoculation improves nutrient uptake and growth of lowland rice. *Soil Sci. Soc. Am. J.,* 64: 1644–1650.

Boddey, R.M. and Döbereiner, J. (1995). Nitrogen fixation associated with grasses and cereals: recent progress and perspectives for the future. *Fertilizer Res.,* 42: 241–250.

Boddey, R.M., Urquiaga, S., Reis, V., and Döbereiner, J. (1991). Biological nitrogen fixation associated with sugar cane. *Plant Soil,* 137: 111–117.

Bolan, N.S., Robson, A.D. and Barrow, N.W. (1984). Increasing phosphorus supply can increase the infection of plant roots by vesicular-arbuscular mycorrhizal fungi. *Soil Biol. Biochem.,* 16: 419-420.

Breil, B. T., Borneman, J., and Triplett, E. W. (1996). A newly discovered gene, *tfuA,* involved in the production of the ribosomally synthesized peptide antibiotic trifolitoxin. *J. Bacteriol.,* 178: 4150–4156.

Briat, J. F. (1992). Iron assimilation and storage in prokaryotes. *J. Gen. Microbiol.* , 138: 2475–2483.

Buwalda, J.G., Ross, G.J.S., Stribley, D.P., and Tinker, P.B. (1982). The development of endomycorrhizal root system. III. Mathematical representation of the spread of vesicular arbuscular mycorrhizal infection in root system. *New Phytol.,* 9: 669.

Chabot, R., Antoun, H., and Cescas, M.P. (1993). Stimulation de la croissance du maïs et de la laitue romaine par des microorganismes dissolvant le phosphore inorganique. *Can. J. Microbiol.,* 39: 941–947.

Claridge, A.W. and May, T.W. (1994). Mycophagy among Australian mammals. *Austr. J. Ecol.,* 19: 251-275.

Claridge, A.W., and May. T.W. (1994). Mycophagy among Australian mammals. *Australian J. Ecol.* ,19: 251-275.

Claridge, A.W., Trappe, J.M., Cork, S.J. and. Claridge. D.L. (1999). Mycophagy by small mammals in the coniferous forests of North America: nutritional value of sporocarps of *Rhizopogon vinicolor*, a common hypogeous fungus. *J. Complete Physiol. Biol.*, 169: 172-178.

Clark, R.B. and Zeto, S.K. (2000). Mineral acquisition by arbuscular mycorrhizal plants. *J. Plant Nutrit.*, 23: 867-902.

Cooper, K.M. (1984). Physiology of VA mycorrhizal association. In: *VA mycorrhizae*. (Eds. C.LL. Powell and D.J. Bagyaraj), CRC Publication: 155-203.

Cooper, K.M. and Grandison, G.S. (1987). Effect of vesicular arbuscular mycorrhizal fungi on infection of tamarillo (*Cyphomandra betacea*) by *Meloidogyne incognita* in fumigated soil. *Plant Disease*, 71(12): 1101-1106.

Cui-Muyi and Nobel Park, S. (1992). Nutrient status, water uptake and gas exchange for three desert succulents infected with mycorrhizal fungi. *New Phytol.*, 122: 643-649.

Curtis, T.P., Sloan, W.T. and Scannell, J.W. (2002). Estimating prokaryotic diversity and its limits. *Proc. Natl Acad. Sci.* USA, 99, 10494–10499.

Davis, R.M., Menge, J.A. and Zentmeyer, G.A. (1978). Influence of vesicular arbuscular mycorrhizae on *Phytophthora* root rot of three crop plants. *Phytopath.*. 86: 1614.

Dell, B. and Malajczuk, N. (1994). Boron deficiency in eucalypt plantations in China. *Can. J. Forest Res.* 24:2409-2416.

Dell, B., Malajczuk, N., Dunstan, W., Gong, M.Q., Chen, Y.L., Lumyong, S., Lumyong, P., Supriyanto and Ekwey. L. (2000). *Edible forest fungi in SE Asia – Current practices and future management*. Proceedings of International Workshop BIOREFOR, Nepal, 1999. p. 123-130.

Demange, P., Bateman, A., Dell, A., and Abdallah, M. (1988). Structure of azotobactin D, a siderophore of *Azotobacter vinelandii* strain D (CCM 289). *Biochemistry* , 27: 2745–2752.

Dieffenbach, A., and Matzner, E. (2000). In situ soil solution chemistry in the rhizosphere of mature Norway spruce (*Picea abies* [L.] Karst.) trees. *Plant Soil*, 222: 149-161.

Dixon, R.K. (1990). Cytokinin activity in *Citrus jambhiri* seedlings colonized by mycorrhizal fungi. *Agric. Ecosys. Environ.*, 29 (1-4): 103-106.

Dobereiner, J., Day, J.M. and Dart, P.J. (1972). Nitrogenase activity and oxygen sensitivity of the *Paspalum notatum — Azotobacter paspali* association. *J. Gen. Microbiol.*, 71: 103–116.

Doneche, B. and Marcantoni, G. (1992). The inhibition of *Botrytis cinerea* by soil bacteria – A new opportunity for biological control of gray rot. *Comptes Rendus de L'Académie des Sciences Série III – Sciences de la Vie* 314: 279–283.

Doran, J. W. and Safley, M. (1997). Defining and assessing soil health and sustainable productivity. In: *Biological Indicators of Soil Health.* (Eds.) C.E., Pankhurst, B. M., Doube, and V. V. S. R., Gupta, CAB International, pp. 1-28.

El-Khawas, H. and Adachi, K. (1999). Identification and quantification of auxins in culture media of *Azospirillum* and *Klebsiella* and their effect on rice roots. *Biol. Fertil. Soils,* 28: 377–381.

Finlay, R.D. and Read, D.J. (1986). Hyphal mats. *New Phytol.,* 86: 199-212.

Fogel R. and Hunt, G. (1979). Fungal and arboreal biomass in a western Oregan Douglas fir ecosystem. Distribution pattern and turn over. *Can. J. Forest Res.,* 9: 256-265.

Francis, R. and Read, D.J. (1984). Direct transfer of carbon between plants connected by vesicular-arbuscular mycorrhizal mycelium. *Nature* (London),307: 53-56.

Fuentes-Ramirez, L.E., Jimenez-Salgado, T., Abarca-Ocampo, I.R., and Caballero-Mellado, J. (1993). *Acetobacter diazotrophicus,* an indoleacetic acid producing bacterium isolated from sugar cane cultivars of Mexico. *Plant Soil,* 154: 145–150.

Garcia de Salamone, *I.E.,* Döbereiner, J., Urquiaga, S., and Boddey, R.M. (1996). Biological nitrogen fixation in *Azospirillum* strain-maize genotype associations as evaluated by the 15N isotope dilution technique. *Biol. Fertil. Soils,* 23: 249–256.

Garcia, C., Fernandez, T., Costa, F., Cerranti, B., and Masciandaro, G. (1992). Kinetics of phosphatase activity in organic wastes. *Soil. Biol. Biochem.,* 25: 361–365.

Gianinazzi, S., Trouvelot, A., Lovato, P., Tuinen, D.Van., Franken, P. and Gianinazzi-Pearson, V. (1995). Arbuscular mycorrhizal fungi in plant production of temperate agroecosystems. In Special issue on Arbuscular Mycorrhizae: Biotechnological applications: an environmental sustainable biological agent. *Crit. Reviews Biotech.,* 15(3/4): 305-311.

Gildon, A. and Tinker, P.B. (1983). Interactions of vesicular-arbuscular mycorrhizal infection and heavy metal in plants. I. The effect of heavy metals on the development of vesicular-arbuscular mycorrhizas. *New Phytol.,* 95(2): 247-261.

Giller K. E., Barea, M.H., Havelle, P., Izac, A.M.N. and Swift, M.J. (1997). Agricultural intensification, soil biodiversity and agroecosystem function. In: Swift M J (Ed.), *Soil biodiversity, agricultural intensification and agroecosystem function. Appl. Soil Ecol.,* 6 (1): 3-16.

Gonzalez-Lopez, J., Salmeron, V., Martinez-Toledo, M.V., Ballesteros, F., and Ramos-Cormenzana, A. (1986). Production of auxins, gibberellins and cytokinins by *Azotobacter vinelandii* ATCC 12837 in chemically defined media and dialysed soil media. *Soil Biol. Biochem.,* 18: 119–120.

Gutierrez-Manero, F. J., Ramos, S. B., Probanza, A., Mehouachi, J., Tadeo, F. R., and Talon, M. (2001). The plant-growth-promoting rhizobacteria *Bacillus pumilus* and *Bacillus licheniformis* produce high amounts of physiologically active gibberellins. *Physiol. Plant,* 111: 206–211.

Hall, J.L. and Williams, L.E. (2000). Assimilate transport and partitioning in fungal biotrophic interactions. *Austr. J. Pl. Physiol.,* 27: 549-560.

Handelsman, J. and Stabb, E.V. (1996). Biocontrol of soilborne plant pathogens. *Plant Cell,* 8: 1855–1869.

Hao Z., Christie P., Qin, L. Wang C. and Li X. (2005). Control of *Fusarium* wilt of cucumber Seedlings by Inoculation with an Arbuscular Mycorrhical Fungus. *J. Plant Nutrit.*, 28: 1961–1974.

Harari, A., Kigel, J. and Okon, Y. (1988). Involvement of IAA in the interaction between *Azospirillum brasilense* and *Panicum miliaceum* roots. *Plant Soil*, 110: 275–282.

Hardie, K. (1985). The effect of removal of extraradical hyphae on water uptake by vesicular-arbuscular mycorrhizal plants. *New Phytol.*, 101: 677-684.

Harley, J.L. and Smith, S.E. (1983). *Mycorrhizal symbiosis*. Academic Press, New York: 483.

Haselwandter, K. and Bowen, G.D. (1996). Mycorrhizal relations in trees for agroforestry and land rehabilitation. *Forest Ecol. & Manage.*, 81: 1-17.

Hawkins, H.J., Johansen, A. and George, E. (2000). Uptake and transport of organic and inorganic nitrogen by arbuscular mycorrhizal fungi. *Plant Soil*, 226: 275-285.

Hogberg, P., Nordgren, A., Buchmann, N., Taylor, A.F.S., Ekblad, A., Hogberg, M.N. (2001). Large-scale forest girdling shows that current photosynthesis drives soil respiration. *Nature*, 411, 789–792.

Holford, I.C.R. (1997). Soil phophorus, its measurements and its uptake by plants. *Austr. J. Soil Res.*, 35: 227–239.

Horemans, S., De Koninck, K., Neuray, J., Hermans, R., and Vlassak, K. (1986). Production of plant growth substances by *Azospirillum* sp. and other rhizosphere bacteria. *Symbiosis*, 2: 341–346.

Horner-Devine, M.C., Carney, K.M. and Bohannan, B.J.M. (2004). An ecological perspective on bacterial biodiversity. *Proc. Royal. Soc. London, B.*, 271:113–122.

Illmer, P., Barbato, A., and Schinner, F. (1995). Solubilization of hardly soluble AlPO4 with P-solubilizing microorganisms. *Soil Biol. Biochem.*, 27: 265–270.

Janzen, R., Rood, S., Dormar, J. and McGill, W. (1992). *Azospirillum brasilense* produces gibberellins in pure culture and chemically medium and in co-culture on straw. *Soil Biol. Biochem.*, 24: 1061–1064.

Johnson, C.N. (1994). Mycophagy and spore dispersal by a rat-kangaroo: consumption of ectomycorrhizal taxa in relation to their abundance. *Functional Ecol.*, 8: 464- 468.

. Jones, D.L. (1998). Organic acids in the rhizosphere — a critical review. *Plant Soil*, 205: 25–44.

Kehri, H.K. and Chandra, S. (1989a). Interaction of *Glomus fasciculatum* and *Macrophomina phaseolina* in chickpea. *J.I.B.S.*, 68 : 363-365.

Kleiner, D. (1984). Bakterien und Ammonium. *Forum Mikrobiol.*, 7: 13–19.

Knosp, O., von Tigerstrom, M. and Page, J.P. (1984). Siderophore mediated uptake of iron in *Azotobacter vinelandii*. *J. Bacteriol.*, 159: 341–347.

Kowalchuk, G.A. and Stephen, J.R. (2001). Ammonia-oxidizing bacteria: a model for molecular microbial ecology. *Ann. Rev. Microbiol.*, 55, 485–529.

Lamhamedi, M.S., Bernier, P.Y. and Fortin, J.A. (1992). Hydraulic conductance and soil water potential at the soil-root interface of *Pinus pinaster* seedlings inoculated with different dikaryons of *Pisolithus* sp. *Tree Physiol.*, 10: 231-244.

Leake, J.R., Johnson, D., Donnelly, D.P., Muckle, G.E., Boddy, L. & Read, D.J. (2004). Networks of power and influence: the role of mycorrhizal mycelium in controlling plant communities and agroecosystem functioning. *Can. J. Bot.*, 82, 1016–1045.

Letham, D.S., Goodwin, P.B. and Higgins, T.J.W. (1978). *Phytohormone and related compounds: A comprehensive treatise*. Vol. 1 and 2, Elsevier, Amsterdam.

Levy, Y., Syvertsen, J.P. and Nemec, S. (1983). Effect of drought stress and vesicular-arbuscular mycorrhiza on citrus transplantation and hydraulic conductivity of roots. *New Phytol.*, 93(1): 61-66.

Lima, E., Boddey, R.M. and Döbereiner, J. (1987). Quantification of biological nitrogen fixation associated with sugar cane using 15N aided nitrogen balance. *Soil Biol. Biochem.*, 19: 165–170.

Lin, W., Okon, Y. and Hardy, R. W. F. (1983). Enhanced mineral uptake by *Zea mays* and *Sorghum bicolor* roots inoculated with *Azospirillum brasilense*. *Appl. Environ. Microbiol.*, 45: 1775–1779.

Malajczuk, N., Trappe, J.M. and Molina, R. (1987). Interrelationships among some ectomycorrhizal trees, hypogeous fungi and small mammals: Western Australian and northwestern American parallels. *Austr. J. Ecol.*, 12: 53-55.

Malik, K.A., Bilal, R., Mehnaz, S., Rasul, G., Mirza, M.S., and Ali, S. (1997). Association of nitrogen-fixing plant growth promoting rhizobacteria (PGPR) with kallar grass and rice. *Plant Soil*, 194: 37–44.

Marcel, G. A. , Heijden, van der, Bardgett, R. D. and Straalen, N. M. van (2008). The unseen majority: soil microbes as drivers of plant diversity and productivity in terrestrial ecosystems. *Ecol. Letters*, 11: 296–310.

Martinez-Toledo, M.V., de la Rubia, T., Moreno, J. and Gonzalez-Lopez, J. (1988). Root exudates of *Zea mays* and production of auxins, gibberellins and cytokinins by *Azotobacter chroococcum*. *Plant Soil* 110: 149–152.

Martinez-Toledo, M.V., Rodelas, B., Salmeron, V., Pozo, C. and Gonzalez-Lopez, J. (1996). Production of pantothenic acid and thiamine by *Azotobacter vinelandii* in a chemically defined medium and a dialysed soil medium. *Biol. Fertil. Soils*, 22: 131–135.

Mylona, P., Pawlowski, K. and Bisseling, T. (1995). Symbiotic nitrogen fixation. *Plant Cell*, 7: 869–885.

Nielsen, P. and Sorensen, J. (1997). Multi-target and medium independent fungal antagonism by hydrolytic enzymes in *Paenibacillus polymyxa* and *Bacillus pumilus* strains from barley rhizosphere. *FEMS Microbiol. Ecol.*, 22: 183–192.

Noel, T. C., Sheng, C., Yost, C. K., Pharis, R. P. and Hynes, M. F. (1996). *Rhizobium leguminosarum* as a plant growth-promoting rhizobacterium: direct growth promotion of canola and lettuce. *Can. J. Microbiol.*, 42: 279–283.

O'Sullivan, D.J. and O'Gara, F. (1992). Traits of fluorescent *Pseudomonas* spp. involved in suppression of plant root pathogens. *Microbiol. Rev.*, 56: 662–676.

Okon, Y. (1982). *Azospirillum*: physiological properties, mode of association with roots and its application for the benefit of cereal and forage grass crops. *Isr. J. Bot.*, 31: 214–220.

Okon, Y. and Kapulnik, Y. (1986). Development and function of *Azospirillum* inoculated roots. *Plant Soil,* 90: 3–16.

Pankhurst, C. E., Doube, B. M., and Gupta, V. V. S. R. (1997). Biological indicators of soil health: Synthesis. In: *Biological Indicators of Soil Health.* (Eds.) C. E., Pankhurst, B. M., Doube, and V. V. S. R., Gupta, CAB International, pp. 419-435.

Pati, B.R., Sengupta, S. and Chandra, A.K. (1995). Impact of selected phyllospheric diazotrophs on the growth of wheat seedlings and assay of the growth substances produced by the diazotrophs. *Microbiol. Res.,* 150: 121–127.

Patten, C.L. and Glick, B.R. (1996). Bacterial biosynthesis of indole-3–acetic acid. *Can. J. Microbiol.*, 42: 207–220.

Piccoli, P., Masciarelli, O. and Bottini, R. (1996). Metabolism of 17,17–[2H2]-gibberellins A4, A9 and A20 by *Azospirillum lipoferum* in chemically-defined culture medium. *Symbiosis,* 21: 263–274.

Rai, R. (1988). Interaction response of *Glomus albidus* and *Cicer Rhizobium* strains on iron uptake and symbiotic N2 fixation in calcareous soil. *J. Pl. Nutri.,* 11(6-11): 863-869.

Rao, V. R., Ramakrishnan, B., Adhya, T. K., Kanungo, P. K., and Nayak, D. N. (1998). Current status and future prospects of associative nitrogen fixation in rice. *World J. Microbiol. Biotechnol.,* 14: 621–633.

Razi, S.S. and Sen, S.P. (1996). Amelioration of water stress effects on wetland rice by urea-N, plant growth regulators, and foliar spray of a diazotrophic bacterium *Klebsiella* sp. *Biol. Fertil. Soils,* 23: 454–458.

Read, D.L., Leake, J.R. and Langdale, A.R. (1989). The nitrogen nutrition of mycorrhizal fungi and their host plants. In: *Nitrogen, Phosphorus and Sulphur Utilization by Fungi.* (Eds.) L. Boddy, R. Marchant and D.J. Read, Cambridge University Press, Cambridge. p 181-204.

Revillas, J. J., Rodelas, B., Pozo, C., Martinez-Toledo, M. V. and Gonzalez, L. J. (2000). Production of B-group vitamins by two *Azotobacter* strains with phenolic compounds as sole carbon source under diazotrophic and adiazotrophic conditions. *J. Appl. Microbiol.,* 89: 486–493.

Reyes, V.G. and Schimidt, E.L. (1979). Population densities of *Rhizobium japonicum* strain 123 estimated directly in soil and rhizosphere. *Appl. Environ. Microbiol.*, 37: 854–858.

Rillig, M.C. and Mummey, D.L. (2006). Mycorrhizas and soil structure. *New Phytol.,* 171, 41–53.

Roseldahl, S. (1985). Interactions between the vesicular-arbuscular mycorrhizal fungus *Glomus fasciculatum* and *Aphanomyces euteiches* root rot of peas. *J. Phytopath.*, 114 : 31-40.

Safir, G.R., Boyer, J.S. and Gerdemann, J.W. (1972). Nutrient status and mycorrhizal enhancement of water transport in soyabean. *Plant Physiol.*, 49: 700-703.

Sanyal, S.K. and De Datta, S.K. (1991). Chemistry of phosphorus transformations in soil. *Adv. Soil Sci.*, 16: 1–120.

Sarig, S., Kapulnik, Y., Nur, I. and Okon, Y. (1984). Response of non-irrigated *Sorghum bicolor* to *Azospirillum* inoculation. *Exp. Agric.*, 20: 59–66.

Sattar, M.A. and Gaur, A.C. (1987). Production of auxins and gibberellins by phosphate-dissolving microorganisms. *Zentralbl. Mikrobiol.*, 142: 393–395.

Schippers, B., Bakker, A.W. and Bakker, P.A.H.M. (1987). Interactions of deleterious and beneficial rhizosphere microorganisms and the effect of cropping practices. *Ann. Rev. Phytopathol.*, 25: 339–358.

Schoenbeck, F. and Dehne, H.W. (1979). Unterschungen zum pilzliche spores parasiten. *Olpidium brassicae*, T.M.V.Z. Pflenzenkrankh. *Pflanzenschutz*, 86: 103.

Shah, S., Karkhanis, V. and Desai, A. (1992). Isolation and characterization of siderophore, with antimicrobial activity, from *Azospirillum lipoferum* M. *Curr. Microbiol.*, 25: 347–351.

Shimshick, E.J. and Hebert, R.R. (1979). Binding characteristics of N2–fixing bacteria to cereal roots. *Appl. Environ. Microbiol.*, 38: 447–453.

Sierra, S., Rodelas, B., Martinez-Toledo, M.V., Pozo, C. and Gonzalez-Lopez, J. (1999). Production of B-group vitamins by two *Rhizobium* strains in chemically defined media. *J. Appl. Microbiol.*, 86: 851–858.

Sies, H. (1991). *Oxidative Stress: Oxidants and Antioxidants*, Academic Press, London.

Simard, S.W., Perry, D.A. Jones, M.D. Myrold, D.D. Durall, D.M. and Molina, R. (1997). Net transfer of carbon between ectomycorrhizal tree species in the field. *Nature* (London), 388: 579-582.

Smith, S.E. and Gianinazzi-Pearson, V. (1988). Physiological interactions between symbionts in vesicular-arbuscular mycorrhizal plants. *Ann. Rev. Plant Physiol. Molec. Biol.*, 39: 221-224.

Smith, S.E. and Read, D.J. (1997). *Mycorrhizal Symbiosis*, 2nd edn. Academic Press, London, UK.

Sprent, J.I. (2001). *Nodulation in Legumes.* Royal Bot. Gardens, Kew, UK.

Stajner, D., Gasaic, O., Matkovic, B., and Varga, Sz.I. (1995). Metolachlor effect on antioxidants enzyme activities and pigments content in seeds and young leaves of wheat (*Triticum aestivum* L.). *Agr. Med.*, 125: 267–273.

Stajner, D., Gasaic, O., Matkovic, B., and Varga, Sz.I. (1995). Metolachlor effect on antioxidants enzyme activities and pigments content in seeds and young leaves of wheat (*Triticum aestivum* L.). *Agr. Med.*, 125: 267–273.

Stajner, D., Kevreaan, S., Gasaic, O., Mimica-Dudic, N. and Zongli, H. (1997). Nitrogen and *Azotobacter chroococcum* enhance oxidative stress tolerance in sugar beet. *Biologia Plant,* 39: 441–445.

Swift, M.J. and Bignell, D. (1999). *Standard methods for assessment of soil biodiversity and land use practice.* Alternatives to Slash and Burn Project. ICRAF, Nairobi. 41 pp.

Swift, M.J., Heal, O.W. and Anderson, J.M. (1979). *Decomposition in terrestrial ecosystems.* Oxford, UK, Blackwell Scientific Publications.

Tapia-Hernandez, A., Mascarua-Esparza, M.A. and Caballero-Mellado, J. (1990). Production of bacteriocins and siderophore-like activity by *Azospirillum brasilense. Microbios,* 64: 73–83.

Tarafdar, J.C. and Junk, A. (1987). Phosphatase activity in the rhizosphere and its relation to the depletion of soil organic phosphorus. *Biol. Fertil. Soil,* 3: 199–204.

Tien, T.M., Gaskins, M.H. and Hubbell, D.H. (1979). Plant growth substances produced by *Azospirillum brasilense* and their effect on the growth of pearl millet (*Pennisetum americanum*). *Appl. Environ. Microbiol.*, 37: 1016–1024.

Timmusk, S., Nicander, B., Granhall, U. and Tillberg, E. (1999). Cytokinin production by *Paenibacillus polymyxa. Soil Biol. Biochem.* , 31: 1847–1852.

Upadhyaya, N.M., Letham, D.S., Parker, C.W., Hocart, C.H. and Dart, P.J. (1991). Do rhizobia produce cytokinins? *Biochem. Int.,* 24: 123–130.

Vogt, K.A., Grier, C.C., Meir C.E. and Edmonds, R.L. (1982). Mycorrhizal role in net production and nutrient cycling in *Abies amabilis* ecosystems in western Washington. *Ecology* 63: 370-380.

Vonreichenbach, H.G. and Schonbeck, F. (1995). Influence of VA-mycorrhiza on drought tolerance of flax (*Linum uaitissimum* L). 2. Effect of VA-mycorrhiza on stomatal gas exchange, shoot water potential, phosphorus nutrition and the accumulation of stress metabolites. *J. Appl. Bot.,* 69: 183-188.

Wallwork, J.A. (1970). *Ecology of soil animals.* London, UK, McGraw Hill.

Whipps, J. M. (2001). Microbial interactions and biocontrol in the rhizosphere. *J. Exp. Bot.,* 52: 487–511.

Whitman, W.B., Coleman, D.C. and Wiebe, W.J. (1998). Prokaryotes: the unseen majority. *Proc. Natl Acad. Sci.,* 95, 6578–6583.

Wu, B., Nara, K. and Hogetsu, T. (2001). Can 14C-labelled photosynthetic products move between *Pinus densiflora* seedlings linked by ectomycorrhizal mycelia? *New Phytol.,* 149: 137-146.

Xu, D. and Dell, B. (1998). Importance of micronutrients for productivity of plantation eucalyptus in east Asia. In: *Overcoming Impediments of Reforestation. Proceedings of the 6th International Workshop of* BIO-REFOR, Brisbane, Australia, December 2-5, 1997: 133-138.

Xu, D., Dell, B., Malajczuk, N. and Gong, M. (2001). Effects of P fertilization and ectomycorrhizal fungal inoculation on early growth of eucalypt plantations in Southern China. *Plant Soil*, 233: 35-44.

Yanni, Y.G., Rizk, R.Y., Abd El-Fattah, F.K., Squartini, A., Corich, V., Giacomini, A., de Bruijn, F., Rademaker, J., Maya-Flores, J., Ostrom, P., Vega-Hernandez, M., Hollingsworth, R.I., Martinez-Molina, E., Mateos, P., Velazquez, E., Wopereis, J., Triplett, E., Umali-Garcia, M., Anarna, J.A., Rolfe, B.G., Ladha, J.K., Hill, J., Mujoo, R., Ng, P.K., and Dazzo, F.B. (2001). The beneficial plant growth-promoting association of *Rhizobium leguminosarum* bv. trifolii with rice roots. *Aust. J. Plant Physiol.*, 28: 845–870.

Zambolim, L. (1986). In: *ANAIS da Reunais Brasieira Sobre Microrrhizas*, IInd A 14 de November de (1985). Larvas, M.G. (1986): 76-00.

□□□

Microbial Diversity and Functions, 2012
© D.J. Bagyaraj, K.V.B.R. Tilak, H.K. Kehri (eds.), pp. 39-94
New India Publishing Agency, New Delhi (India)
E-mail : info@nipabooks.com; Website : www.nipabooks.com

Chapter 3

Taxonomy of Arbuscular Mycorrhizal Fungi with Special Reference to *Glomus* Species

Ashok Aggarwal, Anju Tanwar, Neetu and R.S. Mehrotra

ABSTRACT

Mycorrhizal symbiosis is a highly evolved mutually beneficial relationship found between Arbuscular Mycorrhizal Fungi (AMF) and vascular plants. Knowledge of taxonomic and functional diversity of AM fungi is important in understanding the biology and ecology of such a widespread mycorrhizal symbiosis. Classification and identification of AM fungi remains in the formative stages of growth, with all the attendant turmoil. Since Glomus is known to be an important AM genera for providing multifacet benefit to several economically important plants. Hence, in this review article an attempt has been made to describe 120 Glomus species along with the line diagram of 63 Glomus species for their proper identification that can be used as an identification manual for mycorrhizal researchers.

Keywords: Taxonomy, Glomus species, Arbuscular mycorriza, AM Fungi.

Introduction

Mycorrhiza, dual organs of absorption is formed when symbiotic fungi inhabit tissues of most terrestrial plants (Trappe, 1996). They are of widespread occurrence among vascular plants and are increasingly believed to have played an important role in uptake of nutrients, production of growth hormones, salinity tolerance, disease resistance etc. Fossil records and molecular clock dating suggest that all land plants have arisen from an ancestral arbuscular mycorrhizal condition. Arbuscular mycorrhizas evolved concurrently with the first colonization of land plants some 450-500 million years ago and persist in most extant plant taxa (Cariney, 2000). During the last few decades there has been an increasing interest in the role of biodiversity and its importance in the sustainable development. Several studies have shown that plant species richness and composition affects plant productivity and ecosystem stability. While earlier studies focused on plant diversity, there is currently increased interest in the significance of microbial diversity. Microbes are essential component of soil. Since they provide unique and indispensable transformation in the biological cycle of the soil and provide plants with essential nutrients-arbuscular mycorrhizal fungi (AMF) fall into this group, associating with the majority of terrestrial plants, providing them with nutrients and protection from environmental stress (Smith and Read, 1997).

The term "Mycorrhiza" was coined by Frank (1885) who was fairly certain that these symbiotic plant fungus associations were required for the nutrition of both partners. According to Brundrett (2004) mycorrhizas occur in a specialized plant organ where intimate contact results from synchronized plant-fungus development. Vesicles and arbuscules are normally used to define VAM (Vesicular arbuscular mycorrhizas) association. There is disagreement about whether arbuscular mycorrhizal (AM) association or vesicular-arbuscular (VAM) association is the most appropriate term for these mycorrhizas. The term AM has gradually become more preferable because some fungi do not produce vesicles in roots.

Morphological Diversity of AM Fungi

Morphological characters that are stable and discrete are generally used to identify and classify the AM fungi. The different morphological characters are:

1. Hyphal characters

Based on their function, vegetative hyphae have been differentiated into infective, absorptive and runner hyphae. Fungal hyphae are filamentous network, which tend to form various shapes such as H-shaped parallel connections in *Glomus,* constricted hyphae near branch point as in *Acaulospora* and *Entrophospora* and coiled, swollen and irregular projections or knobs as in *Gigaspora* and *Scutellospora* (Mortan and Bentivenga, 1994).

Mycelia form specialised structures like arbuscules, vesicles and auxillary cells as described below:

2. Vesicles

Vesicles are spherical to oval, sac like globular to elongated terminal swellings of the hyphae or are intercalary aseptate which develop after the development of arbuscules. They have fat granules that serve as storage organs of the fungus. Vesicles in Glomaceae are sub-globose to elliptical, whereas in Acaulosporaceae they are pleomorphic and knobby (Abbott, 1992).

3. Arbuscules

Arbuscules are tree like, hyphae filled invaginations in cortical cells which provide intimate contact between plasmalemma of two symbiotic partners and the point of material exchange between host and fungus. Abrupt narrowing off branch hyphae reduction in hyphal width forms arbuscules in *Gigaspora* and *Glomus* (Brundrett and Kendrick, 1990). At later stage arbuscules are digested by the host cells.

4. Auxillary cells

Auxillary cells are clusters of thin walled cells which differ in shape, size and surface ornamentations. They are restricted to the sub order Gigasporineae (Morton, 1990). Functionally they act as temporary storage structures of carbon compounds.

5. Sporocarps

Sporocarps are formed by the aggregation of spores and fungal hyphae. They are formed in the soil and are surrounded by a wall layer known as peridium. AM fungal spores are either formed singly or in aggregates inside the sporocarps and most of the mycorrhizal members are being described and identified on the basis of morphology of their sporocarp and/or spores (Gerdemann and Trappe, 1974).

6. Zygospores/azygosopres

Spores may also be formed singly or in groups in sporocarps. Spore size varies from 10µm, as in *Glomus tenue*, to more than 1000µm, as in *Scutellospora* and *Gigaspora* species. Spore color varies from hyaline to black and the texture ranges from smooth to highly ornamented. Spores usually develop thick wall with more than one wall layer. They function as storage structures, resting stages and propagules.

7. Chlamydospores

The so called vesicles should be termed as chlamydospores, as suggested by Mehrotra (1993), because they are formed terminally, intercalary or laterally on an undifferentiated non-gametangial hyphae, secondly they are thin walled, but later develops smooth to ornamented thick layer (Mehrotra and Baijal, 1994; Wu *et al.*, 1995) and also because they can act as infective propagules and remain *via*ble for long periods (Biermann and Linderman, 1983).

Classification of AM Fungi

Knowledge of taxonomic and functional diversity of AM fungi is important in understanding the biology and ecology of such a widespread mycorrhizal symbiosis. Classification involves the arrangement of similar into taxonomic groups of different ranking according to selected characteristics or common developmental and evolutionary relationships. Formulation of a classification system for AM fungi has not been an easy task. The monograph of Gerdemann and Trappe (1974) gave a classification system and provided an orderly framework within which scientists could conduct research work. Since then most taxonomic efforts have concentrated more on the description of new species than on the evolution of taxonomic relationships.

AM fungi are hypothesized to have played a crucial role in the evolution of land plants as evidenced by mycorrhiza like structures in fossil tissues of *Rhynia* and *Asteroxylon* from the Devonian period (Pirozynski, 1981). Structures discovered in root fossils from the Triassic period also resemble vesicles and intraradical spores of extant AM fungi and arbuscules like structures in plants from Devonian period (Taylor *et. al.,* 1995).

Host specificity cannot be used in classification because AM fungal taxa do not show any evidence of being limited to association with particular host species. Anatomical features of AM fungi also have not shown distinction enough among taxa to promote their inclusion in classification.

Presently morphology and germination of anamorphs are used to classify AM fungi. Early history of classification of AM fungi is thoroughly discussed by Gerdemann and Trappe (1975), so only a brief summary will be presented here.

Link (1809, cited in Gerdemann, 1971) established the genus *Endogone.* Tulasne and Tulasne (1844) were the first to erect the genus *Glomus* and considered it to be closely related to *Endogone*. Fries (1849) established the family Endogonaceae but this family was placed in the order Mucorales by Bucholtz (1912).

Dangeard (1896) was the first to describe an arbuscular mycorrhiza which happened to have formed from poplar roots. The extraction of spores from soil was made possible by Wet Sieving and Decanting technique given by Gerdemann and Nicolson (1963). Nicolson and Gerdemann (1968) divided the fungi into two groups of Endogone, one forming extramatrical azygospore/ zygospore but producing no intramatrical vesicles and the other forming extramatrical chlaymydospore and producing intramatrical vesicles. Gerdemann and Trappe (1974) divided the old *Endogone sensulate* into seven genera including three nonmycorrhizal genera (*Endogone, Modicella and Glaziella*) and four mycorrhizal genera (*Glomus, Sclerocystis, Gigaspora* and

Acaulospora). They were all placed in Endogonaceae, Endogonales and Zygomycetes. Trappe and Schenck (1982) recognized another mycorrhizal genus *Entrophospora*. Walker (1987) added *Scutellospora* having dropped *Sclerocystis* from the total five mycorrhizal fungal genera.

Thaxter (1922) and Gerdemann & Trappe (1974) originally placed members of the Endogonaceae in the order Mucorales (Zygomycetes). Hypogenous associations with plants, aseptate hyphae, and reproductive structures which ranged from asexual chlamydospores and azygospores to sexual zygospores and a propensity for sporocarp development constituted the logic of this placement.

Morton and Benny (1990) placed the five genera into three families *i.e.*, Glomaceae, Acaulosporaceae, Gigasporaceae and two suborders *i.e.*, Glomineae and Gigasporaineae, both of them were placed in new order Glomales. Recently Morton and Benny (2001) recognized two new families with two new genera, the Archaeosporaceae having *Archaeospora* and Paraglomaceae having *Paraglomus*. The mycorrhizal fungi have now been raised to the rank of Glomeromycota =Glomeromycetes, Glomeromycete, Glomeromycotan), so the old name Glomales of Glomerales no longer represents the whole phylum (Schüßler *et al.*, 2001).

Several researchers have attempted to propose suitable classification system for arbuscular mycorrhizal fungi on the basis of morphological, biochemical and molecular characterization, some of which are listed below.

1. Gerdemann and Trappe, 1974

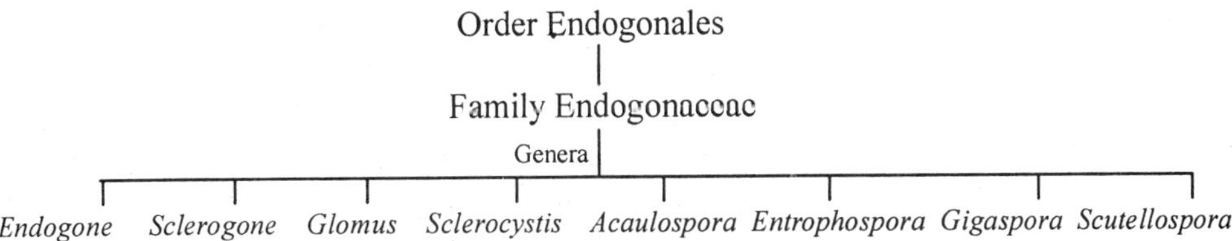

2. Morton and Benny, 1990

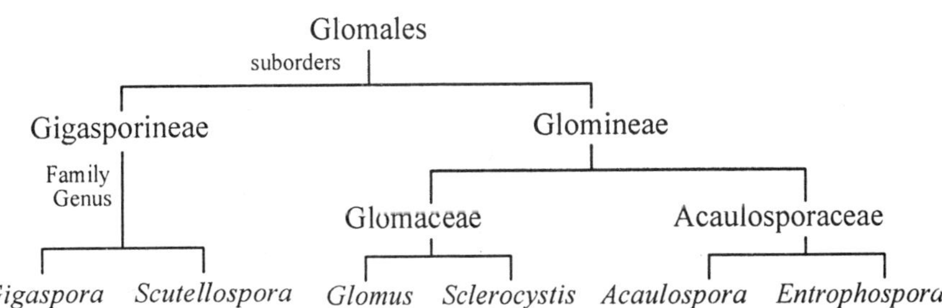

3. Morton and Redecker, 2002

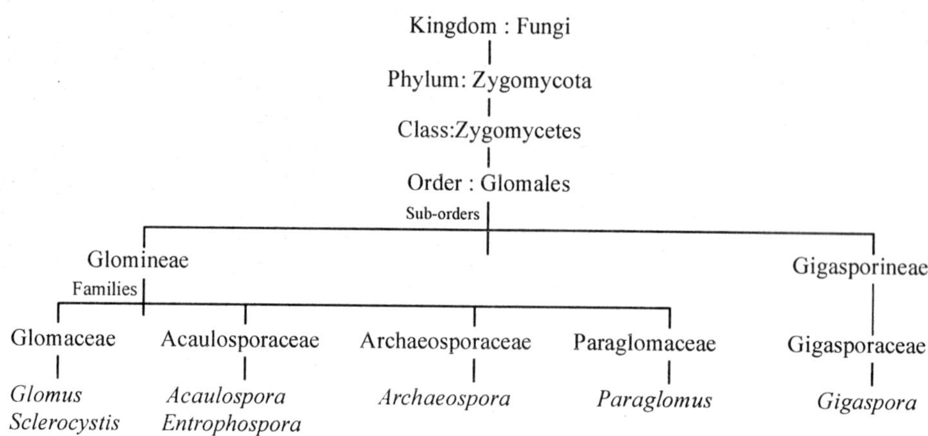

4. Link: www.lrz.de/schuessler/amphylo

The classification of AM fungi given here is that of Schüßler *et al.* (2001) with emendations of Oehl and Sieverding (2004), Walker and Schüßler (2004), Sieverding and Oehl (2006), Spain *et al.* (2006), Walker *et al.* (2007a,b), and Palenzuela *et al.* (2008).

Phylum : Glomeromycota
Class : Glomeromycetes

Orders	Families	Genera
Glomerales	Glomeraceae	*Glomus*
Diversisporales	Gigasporaceae	*Gigaspora* *Scutellospora* *Racocetra*
	Acaulosporaceae	*Acaulospora* *Kuklospora*
	Entrophosporaceae	*Entrophospora* (unclear phylogenetic affiliation)
	Pacisporaceae	*Pacispora*
	Diversisporaceae	*Diversispora* *Otospora* (unclear phylogenetic affiliation)
Paraglomerales	Paraglomeraceae	*Paraglomus*
Archaeosporales	Geosiphonaceae Ambisporaceae Archaeosporaeceae	*Geosiphon* *Ambispora* *Archaeospora* *Intraspora*

Schubler and Walker (2010) recognized under the phylum Glomeromycota one class Glomeromycetes with 4 orders, 11 families and 228 species.

Taxonomy of AM Fungi

Taxonomy is an essential subdivision of the biological sciences. Knowledge of taxonomic diversity of arbuscular mycorrhizal fungi is important in understanding the biology and ecology of this widespread root-fungus symbiosis. In the present paper a description of 131 species of *Glomus* is given to facilitate the researchers and others for their proper identification.

List of *Glomus* species worldwide

1. *Glomus achrum* Blaszk., Red., Koegel, Schuezek, Oehl & Kovacs
2. *Glomus africanum* (under publication)
3. *Glomus aggregatum* Schenck & Smith
4. *Glomus albidum* Walker & Rhodes
5. *Glomus ambisporum* Smith & Schenck
6. *Glomus antarcticum* Cabello = *Glomus intraradices*?
7. *Glomus arborense* McGee
8. *Glomus arenarium* Blaszk., Tadych & Madej
9. *Glomus atrouva* McGee & Pattinson = *Glomus botryoides*?
10. *Glomus aurantium* Blaszk., Blanke, Renker & Buscot
11. *Glomus aureum* Oehl & Sieverd.
12. *Glomus australe* (Berk.) Berch
13. *Glomus avelingiae* Sinclair
14. *Glomus badium* Oehl, Red. & Sieverd.
15. *Glomus bagyarajii* Mehrotra
16. *Glomus bistratum* Blaszk., Red., Koegel, Symanczik & Oehl
17. *Glomus boreale* (Thaxt.) Trappe & Gerd. = *Endogone borealis* Thaxt.
18. *Glomus botryoides* Rothwell & Victor
19. *Glomus caesaris* Sieverd. & Oehl
20. *Glomus caledonium* (Nicol. & Gerd.) Trappe & Gerd. = *Endogone macrocarpa* var. *caledonia* Nicol. & Gerd.
21. *Glomus canadense* = (Thaxt.) Trappe & Gerd. *Endogone canadensis* Thaxt.
22. *Glomus canum* McGee
23. *Glomus cerebriforme* McGee = *Glomus pallidum*?

24. *Glomus chimonobambusae* Wu & Liu
25. *Glomus citricola* Tang & Zang
26. *Glomus claroideum* Schenck & Smith = *Glomus maculosum* Mill & Walker = *Glomus multisubstensum* Mukerji, Bhattacharjee & Tewari = *Glomus fistulosum* Skou & Jakobsen
27. *Glomus clarum* Nicol. & Schenck
28. *Glomus clavisporum* (Trappe) Almeida & Schenck = *Sclerocystis clavispora* Trappe
29. *Glomus constrictum* Trappe
30. *Glomus convolutum* Gerd. & Trappe
31. *Glomus coremioides* (Berk. & Broome) Red. & Morton = *Sclerocystis coremiodes* Berk. & Broome = *S. alba, S. coccogena, S. dussii* (Pat.) Höhn. = *Xenomyces orchraceus* Ces.
32. *Glomus coronatum* Giovann.
33. *Glomus corymbiforme* Blaszk.
34. *Glomus cuneatum* McGee & Cooper
35. *Glomus custos* Cano & Dalpe
36. *Glomus delhiense* Mukerji, Bhattacharjee & Tewari
37. *Glomus deserticola* Trappe, Bloss & Menge
38. *Glomus diaphanum* Morton & Walker
39. *Glomus dimorphicum* Boyetchko & Tewari
40. *Glomus dolichosporum* Zhang & Wang
41. *Glomus dominikii* Blaszk.
42. *Glomus drummondii* Blaszk. & Renker
43. *Glomus eburneum* Kenn., Stutz. & Morton
44. *Glomus epigaeum* Daniels & Trappe
45. *Glomus etunicatum* Becker & Gerd.
46. *Glomus fasciculatum* (Thaxt.) Gerd. & Trappe = *Endogone fasciculata* Thaxt. = *E. macrocarpa* f. *media* = *E. arenaceae*
47. *Glomus fecundisporum* Schenck & Smith = *Endogone flavispora* Lange & Lund
48. *Glomus fistulosum* Skou & Jakobsen

49. *Glomus flavisporum* (Lange & Lund) Trappe & Gerd.
50. *Glomus formosanum* Wu & Chen
51. *Glomus fragile* (Berk. and Broome) Trappe & Gerd. = *Paurocotylis fragile* Berk. & Broome
52. *Glomus fragilistratum* Skou & Jakobsen
53. *Glomus fuegianum* (Spegazzini) Trappe & Gerd. = *Endogone fuegiana* Speg.
54. *Glomus fulvum* (Berk. & Broome) Trappe & Gerd. = *Paurocotylis fulvum* Berk & Broome = *Endogone fulvum* Berk. & Broome) Pat. = *Endogone moelleri* Henn. = *Endogone lignicola* Pat.
55. *Glomus geosporum* (Nicol. & Gerd.) Walker = *Glomus macrocarpum* var. *geosporum* Nicol. & Gerd. = *Endogone macrocarpa* var. *geospora*
56. *Glomus gerdemanni* Rose, Daniels & Trappe
57. *Glomus gibbosum* Blaszk.
58. *Glomus globiferum* Koske & Walker
59. *Glomus glomerulatum* Sieverd.
60. *Glomus halonatum* Rose & Trappe = *Glomus clarum* ?
61. *Glomus heterosporum* Smith & Schenck
62. *Glomus hoi* Berch & Trappe
63. *Glomus hyderabadensis* Swarupa, Kunwar, Prasad & Manoharachary
64. *Glomus indica* Manoharachary, Sharathbabu & Adholeya
65. *Glomus indicum* Blaszk., Wubet, HariKumar, Ryszka & Buscot
66. *Glomus iranicum* (under publication)
67. *Glomus insculptum* Blaszk. .
68. *Glomus intraradices* Schenck & Smith
69. *Glomus invermaium* Hall
70. *Glomus irregulare* Blaszk., Wubet, Renker & Buscot
71. *Glomus kerguelense* Dalpe & Strullu
72. *Glomus laccatum* Blaszk.
73. *Glomus lacteum* Rose & Trappe
74. *Glomus lamellosum* Dalpe, Koske & Tews
75. *Glomus leptotichum* Schenck & Smith
76. *Glomus liquidambaris* (Wu & Chen) Almeida & Schenck = *Sclerocystis liquidambaris* Wu & Chen = *S. cunninghamia*

77. *Glomus luteum* Kenn., Stutz & Morton
78. *Glomus macrocarpum* Tulasne & Tulasne = *Endogone macrocarpa* Tul. & Tul. = *Glomus macrocarpum* var. *macrocarpum* = *E. australis* = *Paurucotylis fulva* var. *zaelandica* = *E. pampaloniana* = *E. tenebrosa* = *E. guttulata* = *E. nuda*
79. *Glomus maculosum* Miller & Walker
80. *Glomus magnicaule* Hall
81. *Glomus manihotis* Howeler, Sieverd. & Schenck = *G. clarum* ?
82. *Glomus megalocarpum* Red.
83. *Glomus melanosporum* Gerd. & Trappe
84. *Glomus merredum* Porter & Hall
85. *Glomus microaggregatum* Koske, Gemma & Olexia
86. *Glomus microcarpum* Tulasne & Tulasne = *Endogone microcarpus* Tul. & Tul. = *E. neglecta*
87. *Glomus minutum* Blaszk. Tadych & Madej
88. *Glomus monosporum* Gerd. & Trappe
89. *Glomus mortonii* Bentiv. & Hetrick
90. *Glomus mosseae* (Nicol. & Gerd.) Gerd. & Trappe = *Endogone mosseae* Nicol. & Gerd.
91. *Glomus multicaule* Gerd. & Bakshi
92. *Glomus multiforum* Tadych & Blaszk.
93. *Glomus multisubstensum* Mukerji, Bhattacharjee & Tewari
94. *Glomus nanolumen* Koske & Gemma
95. *Glomus occultum* Walker
96. *Glomus pachycaulis* (Wu & Chen) Almeida & Schenck
97. *Glomus pallidum* Hall
98. *Glomus pansihalos* Berch & Koske
99. *Glomus pellucidum* McGee & Pattinson
100. *Glomus perpusillum* Blaszk. & Kovacs
101. *Glomus proliferum* Dalpe & Declerck
102. *Glomus przelewicensis* Blaszk.
103. *Glomus pubescens* (Sacc. & Ellis) Trappe & Gerd. = *Sphaerocreas pubescens* Sacc. & Ellis = *Sclerocystis pubescens* = *Endogone pubescens*

104. *Glomus pulvinatum* (Henn.) Trappe & Gerd. = *Endogone pulvinatus* Henn.
105. *Glomus pustulatum* Koske, Friese, Walker & Dalpe
106. *Glomus radiatum* (Thaxt.) Trappe & Gerd. = *Endogone radiata* Thaxt.
107. *Glomus reticulatum* Bhattacharjee & Mukerji
108. *Glomus rubiforme* (Gerd. & Trappe) Almeida & Schenck = *Sclerocystis rubiformis* Gerd & Trappe = *S. indicus* = *S. pachycaulis*
109. *Glomus scintillans* Rose & Trappe
110. *Glomus segmentatum* Trappe, Spooner & Ivory
111. *Glomus sinuosum* (Gerd. and Bakshi) Almeida & Schenck = *Sclerocystis sinuosa* Gerd. & Bakshi
112. *Glomus spinosum* Hu
113. *Glomus spinuliferum* Sieverd. & Oehl = *Glomus pansihalos*
114. *Glomus spurcum* Pfeiffer, Walker & Bloss
115. *Glomus sterilum* Mehrotra & Baijal (not validly published)
116. *Glomus taiwanense* (Wu & Chen) Almeida & Schenck = *Sclerocystis taiwanensis* Wu & Chen
117. *Glomus tenebrosum* (Thaxt.) Berch. = *Endogone tenebrosa* Thaxt.
118. *Glomus tenerum* Tandy
119. *Glomus tenue* (Greenall) Hall = *Rhizophagus tenuis* Greenall
120. *Glomus tortuosum* Schenck & Smith
121. *Glomus trimurales* Koske & Halvorson
122. *Glomus tubiforme* Tandy (originally *G. tubaeformis*)
123. *Glomus velum* Porter & Hall
124. *Glomus verruculosum* Blaszk.
125. *Glomus versiforme* (Karst.) Berch = *Endogone versiformis* Karst = *Glomus epigaeum* Daniels & Trappe
126. *Glomus vesiculiferum* (Thaxt.) Gerd. &Trappe = *Endogone versiculifera* Thaxt. = *E. tjibodensis*
127. *Glomus viscosum* Nicol.
128. *Glomus walkeri* Blaszk. & Renker
129. *Glomus warcupii* McGee
130. *Glomus xanthium* Blaszk., Blanke, Renker & Buscot
131. *Glomus zaozhuangianus* Wang & Liu

In this review article, emphasis has been made to understand the identification of different species of *Glomus*. A brief description of a total of 120 AM species has been given along with the line diagram of 63 AM species for their proper identification.

Glomus (Tulasne & Tulasne)

Glomus is a genus of endomycorrhizal fungi and all species are known to form symbiotic relationship with plant roots. *Glomus* is the largest genus of AM fungi having 136 species. *Glomus* is the only genus in the family Glomeraceae in the order Glomerales. *Glomus* includes both sporocarpic and non sporocarpic species. The spores develop terminally, on a single undifferentiated hypha. The spores are formed at the end of hyphae which may be constricted at the point of attachment to the spore, have parallel side walls become markely occluded at the point of attachment of spores. The spore wall can have one to many layers, without ornamentation.

Glomus achrum Blaszk, Red, Koegel, Schutz. & Kovacs

It produces small hyaline spores in aggregates in the soil and inside roots.

Spores are globose to subglobose, rarely egg shaped, oblong to irregular.

Wall consist of three hyaline layers a mucilaginous, short lived outermost layer I, a laminate middle layer II composed of loose sublayers and a flexible innermost layer III.

Glomus africanum (under publication)

Spore are formed in loose clusters and singly in soil, spores are pale yellow to brownish yellow, globose to subglobose, (60–) 87 (–125) μm diam, sometimes ovoid to irregular, 80–110 x 90–140 μm. The spore wall consists of a semi permanent, hyaline, outer layer I and a laminate, smooth, pale yellow to brownish yellow, inner layer II, which always is markedly thinner than the outer layer.

Glomus aggregatum Schenck & Smith (Fig. 1-A)

Chlamydospores formed in loose clusters or in sporocarps without a peridium, sporocarps are hyaline to pale yellow.

Chlamydospores are globose, subglobose, obovate, cylindrical to irregular, 50.4 - 91.2 μm in diameter, hyaline to yellow.

Spore wall is yellow to yellow brown, consisting of outer wall I which is slightly thicker and lighter in color than the inner wall II.

Hyphal attachment is straight or recurved.

Pore is open.

Glomus albidum Walker & Rhodes (Fig. 1-B)

Chlamydospores occur singly in soil. It appears hyaline to white when young and yellowish to brownish yellow when mature, globose to subglobose with only one subtending hypha.

Spore wall consist of two layers, outer layer I is hyaline but become roughened at maturity. Inner layer II is light yellow and finally laminated wall.

Subtending hypha has two wall layer continuous with the spore wall layer I & II. It is constricted at the spore base and is expended by thickening of outer wall to become slightly funnel shaped.

Glomus ambisporum Smith & Schenck (Fig. 1-C)

Spores produced are brown to black, from globose to subglobose. The wall layer consists of three layers. The outer layer (I) is subhyaline. The middle layer (II) consists of finely adherent sublayers, dark brown to black in color. The innermost layer (III) is thin and flexible.

The subtending hyphal wall consisting of two layers continuous with the outer two layers (I & II) of the spore wall.

The pore is occluded by a thickening of layer II together with III of the spore wall.

Glomus antarcticum Cabello (Fig. 1-D)

Sporocarp are globose or subglobose and without a peridium.

Chlamydospores are globose to subglobose, 50-70μm diameter and are yellow to brown in color.

Spore wall consists of three walls. Outer wall I is ephemeral, hyaline. The middle layer II is light brown, laminated. Inner wall III is hyaline.

Hyphal attachment is formed by the outer wall and the pore is occluded by the extension of inner spore wall.

Glomus arenarium Blaszk., Tadych & Madej (Fig. 1-E)

Spores are globose to subglobose, 55-120μm in diameter and orange to raw umber in color born singly in soil with single subtending hypha rarely two.

Spore consists of one wall with three layers. Layer I is evanescent, hyaline and closely adherent to layer II, smooth in juvenile spores, gradually deteriorate in mature spores. Layer II is flexible to semiflexible, hyaline to smooth and rarely present in mature spores.

Layer III is laminate, smooth, and orange to raw umber.

Subtending hypha is straight or recurvate, cylindrical or funnel shaped rarely constricted hyaline to yellowish white. Subtending hyphal layers are continuous with spore wall layer I and II.

Pore is occluded by a septum formed by inner laminae of wall layer III.

Glomus atrouva McGee & Pattinson (Fig. 1-F&G)

Sporocarps firm rounded lacking a peridium. Spores arising blastically from hyphal loops within a plexus of anastomosing hyphae.

Spores are globose to subglobose, 100-200 µm in diameter, tapering slightly toward the attached hypha.

Spore has two walls, outer wall I is brittle when mature, brown and inner wall II is laminated, brown and thickest at the spore base.

Subtending hypha is attached at right angle to this pore and is sometimes slightly constricted at the point of attachment and has two walls continuous with the spore walls.

Glomus aurantium Blaszk, Blanke, Renker & Buscot

Spores occur singly, 70-120µm, yellowish white when young, deep orange at maturity, globose to subglobose with single subtending hypha.

Wall consists of three layers, outermost layer I is permanent, flexible to semiflexible, smooth, hyaline and sometimes ballooning. Layer II is laminate, smooth, yellowish white to golden yellow. Layer III is flexible, smooth, hyaline and tightly adherent to layer II.

Glomus aureum Oehl & Sieverd

Sporocarp is light orange, irregular in shape, without a peridium having closely packed spores.

Spores are ovoid, 45-50µm in diameter, with single subtending hypha.

Spore has two layered walls, layer I is outer, evanescent, hyaline and ruptured in mature spores. Layer II is light orange, thick.

Hyphal attachment is thick at the point of attachment.

Glomus australe (Berkley) Berch (Fig. 1-H&I)

Spores are formed in loose clusters from central, branched, inflated hypha.

Spores are 120-180µm in diameter, hyaline to pale yellow in colour. Spore has two wall layers. Layer I is hyaline to pale yellow and layer II is light to dark brown.

Subtending hypha is broad at the point of attachment and continuous with the spore wall layer II.

Pore is open.

Glomus avelingiae Sinclair (Fig. 1-J)

Sporocarp yellowish brown to pale yellow in colour and irregularly elliptical and moderately convoluted. Peridium is absent.

Chlamydospores are globose to subglobose, 64-80µm in diameter.

Wall has four layers. Layer I is the innermost flexible layer, layer II is the middle pale yellowish brown, laminated layer and outer bilayer III and IV are subhyaline and appears as a single layer.

Subtending hypha is cylindrical having a septum.

Glomus badium Oehl, Red. & Sieverd. (Fig. 1-K)

Sporocarp is brownish orange to reddish brown, mainly ovoid to irregular, sometimes globose to subglobose radially originating from a hyphal plexus.

Spores are brownish orange to reddish brown, usually ovoid 35-65µm, rarely globose to subglobose with one subtending hypha.

Spore consists of spore wall with three layers. Layer I is evanescent, hyaline to pale orange and absent in mature spores. Layer II is laminated, brownish orange to reddish brown. Layer III is flexible, tightly adherent to the inner surface of layer II.

Subtending hypha is straight or recurved, cylindrical or slightly flared. Wall is thick at the spore base and composed of II layers continuous with the spore wall layers I and II.

Pore is occluded by spore wall layer III.

Glomus bagyarajii Mehrotra (Fig. 1-L)

Chlamydospores are formed singly, 100µm in diameter, yellow to yellow brown, globose, subglobose to irregular.

Spore wall is made up of four layers in two groups (A&B).

Group A: the first wall layer I is evanescent, hyaline. Wall layer II is laminated, yellow to yellow brown, with three laminae.

Group B: wall layer III is hyaline and IV is membranous and hyaline.

Subtending hyphae is hyaline to pale yellow, attached straight or recurved, cylindrical or funnel shaped. Hyphal wall layer is continuous with wall layer II & III.

A septum is present at the point of hyphal attachment formed by the innermost layer.

Glomus bistratum Blaszk., Red., Koegel, Symanczik & Oehl

Spores are small, globose to subglobose.

Wall composed of two permanent hyaline layers. The outer layer is unit, smooth and the inner one laminate.

Glomus boreale (Thaxt.) Trappe & Gerd.

Sporocarp is irregular, coherent, spongy, brown to almost chocolate brown, 8 mm in diameter.

The spores are yellow to reddish brown, ellipsoidal, 125 X 100μm diameter.

Spore wall is thick, reddish brown and about 8μm diameter.

Glomus botryoides Rothwell & Victor

Chlamydospores are born singly, or in tight clusters within loosely interwoven peridium.

Spores are reddish brown to black at maturity, globose, to subglobose, 145-250μm in diameter.

Spore wall consist up of two walls, the outer wall I is yellowish brown and with rough outer surface. The inner wall II is laminated, with fine projections upto 1μm long and over the outer surface.

Hyphal attachment is straight to recurved.

Glomus caesaris Sieverd. & Oehl (Fig. 2-A)

Spores are formed singly, 156-212μm diameter, light orange to dark orange, globose to subglobose, usually with one and rarely with two subtending hyphae.

Spore wall has five layer, Layer I is semi permanent, hyaline, roughened surface, mostly degraded in mature spores. Layer II is laminate and hyaline. Layer III is finally laminate, hyaline to pale yellow to pastel yellow and rarely

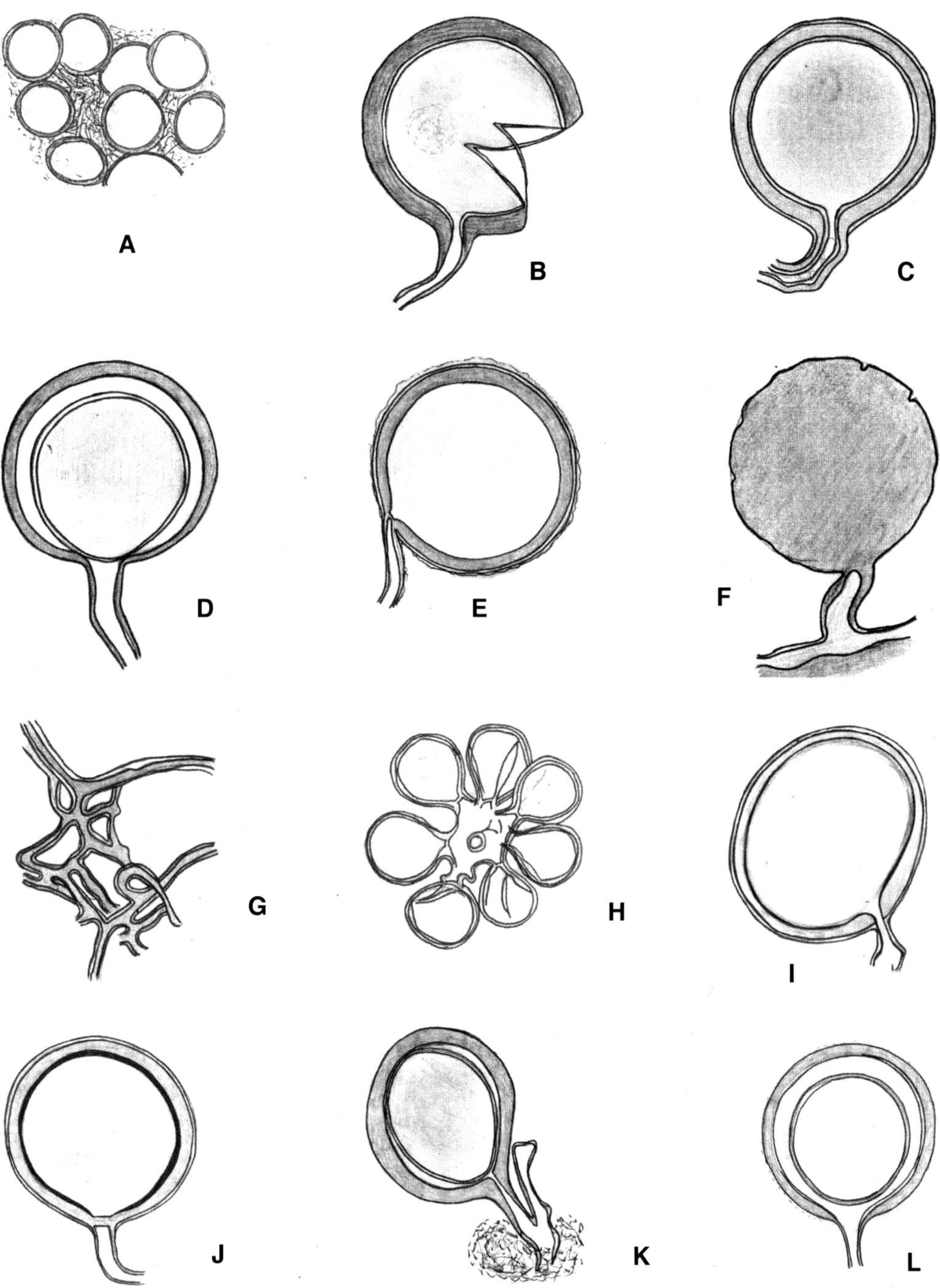

Fig. 1 : (A) *Glomus aggregatum*, (B) *Glomus albidum*, (C) *G. ambisporum*, (D) *G. antarcticum*, (E) *G. arenarium*, (F) & (G) *G. atrouva*, (H) & (I) *G. australe*, (J) *G. avelingiae*, (K) *G. badium*, (L), *G. bagyarajii*

separates from layer II. Layer IV is laminate, light orange to dark orange and tightly adherent to layer III. Layer V is semi flexible and is composed of two sublayers, concolours with layer IV.

Subtending hypha is straight or curved, cylindrical to flared, composed of four layers continuous with layer I, II, III and IV of spore wall layers.

Pore is closed by spore wall layer V.

Glomus caledonium (Nicol. & Gerd.) Trappe & Gerd. (Fig. 2-B)

Spores are pale orange yellow to yellow brown, globose, 180-320μm diameter to subglobose, rarely irregular.

Spore with four wall layers (I, II, III and IV). Outer layer I is hyaline, mucilaginous. Layer II is rigid hyaline layer, Layer III hyaline layer of granular consistency, usually attached to layer II. Layer IV is pale yellow brown, consist of sub layers.

Subtending hypha is cylindrical, occasionally slightly constricted and has three layers (I, II and III) continuous with spore wall layer.

A septum is formed by the innermost sublayer of spore wall layer IV which resembles a recurved septum.

Glomus canadense (Thaxt.) Trappe & Gerd

Sporocarp surface covered by a peridium, maximum breadth of spores more than 50μm but less than 70μm.

Glomus canum McGee (Fig. 2-C)

Sporocarp is rounded with a white cottony peridum.

Spores are globose to subglobose, 30-50μm in diameter, only single laminate yellow to pale brown wall is present.

Subtending hypha is yellow in colour, thin, attached at right angle to the spore.

A septum is present, formed by the innermost lamina of spore wall layer.

Glomus chimonobambusae Wu & Liu

Spores are borne singly in the soil, subhyaline or white, globose, subglobose, 110-180μm diameter, obovoid, triangular, or irregular.

Spore surface covered by tiny warts and coarse, clavate projections up to 12.5μm long. Spore wall distinguished by two groups, four layers. Wall I, ornamented by warts and projections, laminated. Wall layer II, a unit wall, closely adherent to wall 1. Wall III, a unit wall, closely attached to wall IV. Wall layer IV is a unit wall

Hyphal attachment is subhyaline or white, 8-16µm diameter, tubular, constricted at the attachment.

Attachment occluded by wall thickening.

Glomus claroideum Schenck & Smith (Fig. 2-D)

Chlamydospores are formed singly in soil, cream to light yellow, globose, to subglobose, 80-160µm.

Spore has four layers (I, II, III & IV). Layer I is hyaline, mucilaginous layer, tightly adherent, sloughing completely in mature spores. Layer II is hyaline, and it also degrades. Layer III is laminate, consisting of thin and tightly adherent pale yellow sublayers. Layer IV is concolorous with the laminate layer, producing folds when it is very thin.

Subtending hypha is cylindrical to slightly flared, three layers (I, II and III) continuous with the outer three layers of the spore wall, thick in the region of attachment, tapering distantly. Pore is occluded by a recurved septum formed either by a sublayer of the laminate layer III and IV or IV alone.

Glomus clarum Nicol. & Schenck (Fig. 2-E)

Spores are globose, to subglobose, 100-260µm diameter, sometimes elliptical, oblong, or irregular, colour varies from white to yellow-brown. Three layers (I, II and III). Layer I is hyaline mucilaginous layer. Layer II is a permanent hyaline layer consisting of sublayers that are not brittle and cracks are present. This layer is thicker in some regions so that the inner surface appears wavy. Layer III is also a permanent layer, usually consisting of 2-4 sublayers.

The subtending hypha is cylindrical to flared, sometimes slightly constricted also.

Glomus clavisporum (Trappe) Almeida & Schenck (Fig. 2-F)

Sporocarp brownish black, tightly packed around a central plexus of interwoven hyphae Peridium is absent.

Spores are dark brown, clavate, 90-160 X 35-45 µm diameter to subcylindric. Spore wall laminate, thickened at the apex. Subtending hypha is tapering to cylindric.

Glomus constrictum Trappe

Chlamydospore light brown to dark brown, shape oblong to ellipsoidal.

Outermost wall layer coloured, outer spore wall layer extending down subtending hyphae.

Wide subtending hyphae but narrow at the point of attachment.

Glomus convolutum Gerd. & Trappe

Chlamydospore bright yellow to dark brown or blackish, shape oval to globose.

Spores in group, number of spores 5-12 in cluster, spore attachment with hyphae straight or recurved at the point of attachment.

Glomus coremioides (Berk. & Broome) Red. & Morton

Sporocarp are 300-600 µm, pulvinate to subglobose,white when immature and tan to dull brown when found singly, flattened at the base, and are borne on a short stalk of upto 100 µm. peridium is thick and composed of thick walled, sparingly septate interwoven hyphae.

The chlamydospores are obvoid-ellipsoid, 65-100µm X 40-75µm diameter and brown in colour.

Chlamydospore wall is upto 4µm thick at the base and 2µm thick at the apex.

Subtending hypha has pore at the point of attachment.

Glomus coronatum Giovannetti (Fig. 2-G)

Spores are formed singly, greyish orange to brownish orange, globose to subglobose, 80-230µm in diameter with single subtending hyphae.

Spores have one wall with two layers, layer I is evanescent, hyaline and completely absent in mature spores. Layer II is laminated, greyish orange to brownish orange.

Subtending hypha is greyish orange to brownish orange and funnel shaped. Wall of subtending hypha is composed of two layers I & II continuous with the spore wall layer.

Pore is closed by a curved septum formed by innermost sublayer of spore wall layer II.

Glomus corymbiforme Blaszk. (Fig. 2-H)

Spore occurs in corymbiform sporocarps which are globose to subglobose, 180-490 μm diameter, sometimes ovoid, composed of 2-13 spores enveloped individually by hyphal mantle.

Spores are pastel yellow to orange coloured, globose to subglobose and sometimes ovoid or pyriform, with single subtending hypha.

The spore has three walls in one group (A). Wall layer I is smooth, hyaline to deep yellow closely attached to layer II. Layer II is laminated, pastel yellow to orange. Layer III is membranous, hyaline and usually tightly adherent to layer II.

Subtending hypha is cream to deep orange, straight or recurvate, having wall layers continuous with spore wall layers I & II.

Pore is occluded by growth of wall layer II.

Glomus cuneatum McGee & Cooper (Fig. 2-I)

Sporocarp is firm, rounded with a wide peridium formed by thin walled hyphae.

Spores are globose, to ovoid, clavate or irregular, 70 X 70 -120 X 180μm diameter.

Only a single laminated, black colored wall is present overlying a hyaline inner zone.

Subtending hypha is pale brown, attached at right angle to the spore.

A septum is present formed by the innermost laminae of spore wall layer.

Glomus custos Cano & Dalpe (Fig. 2-J)

Spores formed singly or in loose clusters of 2-5 spores in roots and soil.

Mature spores are pale yellow to brownish yellow, globose to subglobose, 110-172μm.

Spore wall is made up of four layers. Layer I is hyaline, mucilaginous, smooth but of irregular thickness. Layer II is hyaline, rigid, outer surface is smooth and inner surface is occasionally granular. Layer III is hyaline to pale yellow, semi flexible, smooth, easily separated from layer II. Layer IV is yellow to brownish yellow and laminated.

Subtending hypha is cylindrical or slightly flared. Pore is generally open.

Glomus deserticola Trappe, Boss & Menge (Fig. 2-K)

Spores are borne in soil singly or in loose aggregates without a peridium.

Spores are pale yellow to orange, globose to subglobose sometimes ovoid to pear shaped, 70-115μm in diameter with single subtending hypha.

Spore has one wall with two layers. Layer I is mucilaginous and hyaline. Layer II is laminated, smooth pale yellow to orange.

Subtending hypha is straight or curved or flared or funnel shaped, wall with two layers continuous with the spore wall layers I & II. Layer II frequently thickened at the spore base to form a collar.

Pore is opened or closed by a recurved septum formed by innermost lamina of spore wall layer II.

Glomus diaphanum Mor. & Walk. (Fig. 2-L & Fig. 3-A)

Azygospores borne in soil, spores mostly two in number, joined through a common hyphal attachment with mother hyphae.

Spores in variable shapes, yellow to yellowish brown, 100-250μm in diameter.

Subtending hyphae is hyaline and convoluted.

Glomus dominikii Blaszk

Chlamydospores are borne singly, white slightly pink when immersed in water and becoming light yellow to orange-yellow with age, ornamented, globose to subglobose, 75-165μm in diameter, rarely ovoid with single subtending hypha.

Spore wall has three layers (I, II & III) in two groups (A & B). Group A, has outer hyaline, unit wall layer I ornamented with fine warts. Group B has hyaline inner smooth, membranous layer II tightly adhering to a smooth, membranous innermost layer III.

Subtending hypha is hyaline, straight or slightly recurvate, 180-220μm long, with wall thicker at the spore base.

Pore is open.

Glomus drummondii Blaszk. & Renker (Fig. 3-B)

Spores occur singly, pastel yellow to maize yellow coloured, globose to subglobose, 58-85μm in diameter with single subtending hypha.

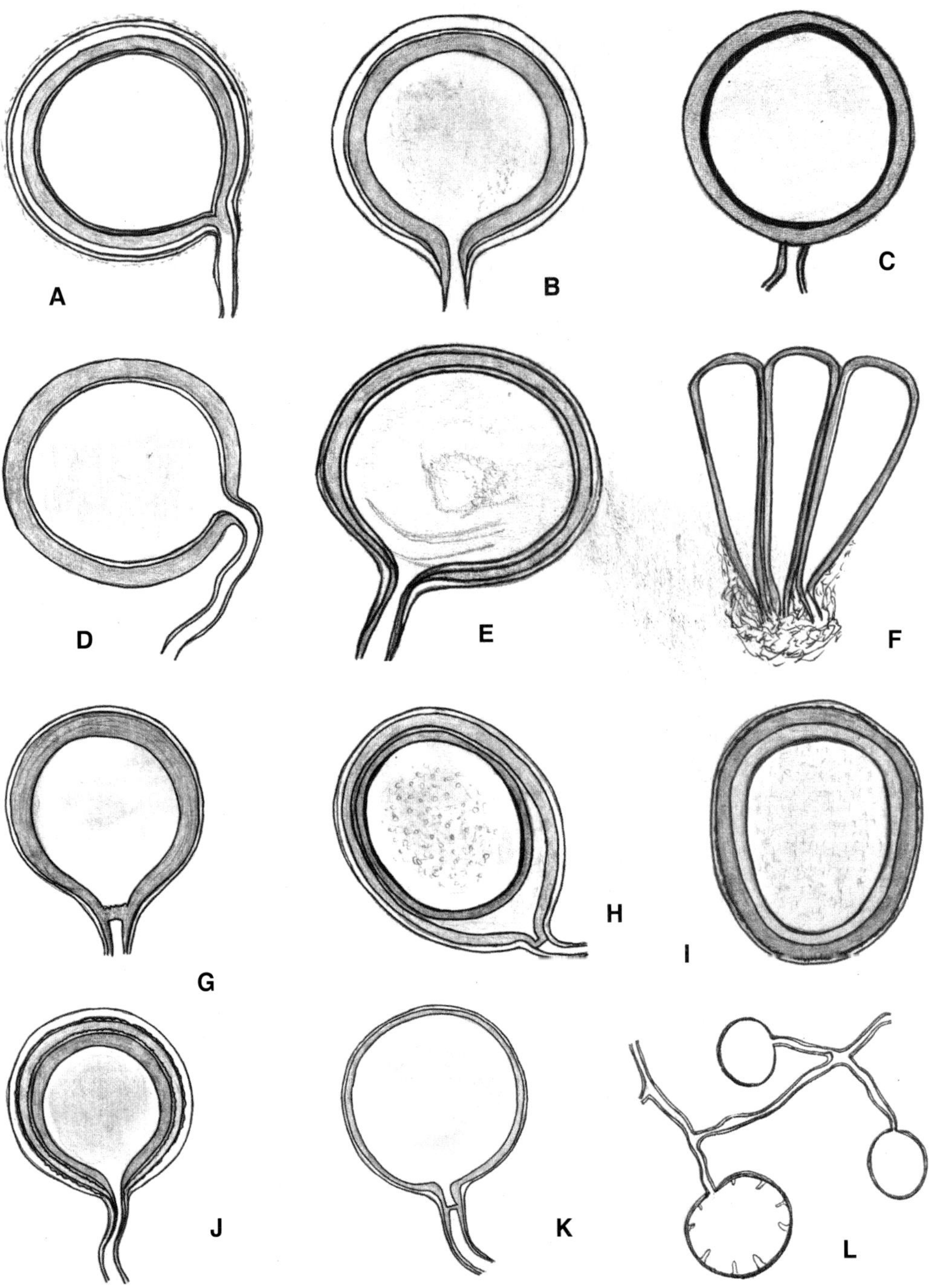

Fig. 2 : (A) *Glomus caesaris* (B) *G. caledonium*, (C) *G. canum*, (D) *G. claroideum*, (E) *G. clarum*, (F) *G. clavisporum*, (G) *G. coronatum*, (H) *G. corymbiforme* (I) *G. cuneatum*, (J) *G. custos*, (K) *G. deserticola*, (L) *G. diaphanum*

Spore has one wall with three layers, layer I is evanescent, hyaline and is completely absent in mature spores. Layer II is laminate, smooth and pastel yellow to maize yellow coloured. Layer III is flexible, smooth and hyaline.

Subtending hypha is hyaline to maize yellow, straight or recurvate, cylindrical or slightly flared. Wall is composed of two layers and continuous with the spore wall layers I & II.

Pore occluded by a septum continuous with spore wall layer III.

Glomus eburneum Kenn., Stutz. & Morton (Fig. 3-C)

Spores occur singly, yellowish white to butter yellow, globose to subglobose, 92-146μm diameter with one subtending hypha.

Spore has two wall layers, Layer I is semi-permanent, semi flexible and hyaline. Layer II is laminate, semiflexible, smooth, yellowish white to butter yellow.

Subtending hypha is straight or recurvate, cylindrical or slightly flared, with two wall layers continuous with spore wall layers I & II.

Pore is open or occluded by straight or recurved septum formed by innermost lamina of wall layer II.

Glomus epigaeum Daniels & Trappe

Spores dark brown to light blackish in colour, shape oval to oblong, subtending hyphae inserted into the spore wall.

Pore at point of attachment of subtending hyphae occluded by a aseptum like plug, spore borne singly.

Glomus etunicatum Becker & Gerd. (Fig. 3-D)

Spores are orange to red brown, globose, to subglobose, 60-160μm in diameter.

Spore wall consists of two layers (I & II). Layer I is outer layer, mucilaginous and degrades as spore matures. Layer II is thin, laminate, light orange brown to red brown.

Subtending hypha is cylindrical to slightly flared with two wall layers I & II continuous with the spore wall layer.

Pore is occluded by a septum formed by the innermost lamina of wall layer II.

Glomus fasciculatum (Thaxt.) Gerd. & Trappe (Fig. 3-E)

Spores globose or often mixed with sub globose, oval or ellipsoid or irregular spores, light yellow to yellowish brown, surface smooth to dull roughened, double wall with sporogenous hyphae.

Spore wall is 4.2μm thick, hyaline to yellow, the thicker walls often minutely perforated with thick inward projections, hyphal wall occluded at maturity.

Glomus fecundisporum Schenck & Smith

Chlamydospores are formed singly or in loose clusters.

Chlamydospores are globose, elongate to irregular, 83-150 X 107-207μm in diameter.

Spore wall is subhyaline to light yellow and becoming yellow brown to dark brown or black and laminate when mature, consisting of inner and outer wall of approximately equal thickness.

Subtending hypha has thin walls and pore is occluded in mature spores.

Glomus fistulosum Skou and Jakobsen (Fig. 3-F)

Spore is cream to light yellow, globose to subglobose, 80-140μm

Spore wall has four layers (I, II, III & IV). Layer I is hyaline and mucilaginous. Layer II is hyaline. Layers I & II are found in only young spores. Layer III is laminated and consist of thin and tightly adherent sublayers. Layer IV is thin and concolours with laminated layer.

Glomus flavisporum (Lange & Lund) Trappe & Gerd.

Sporocarp is firm, almost globose, somewhat lobed, 0.5 cm diameter. Peridium is whitish, brown.

Spores are irregularly arranged, ovate-oblong, 149-202 X 95-152μm diameter, often slightly constricted on the middle, rarely subglobose.

Wall is deep yellowish brown, outer thin layer is lamellate, hardly incrusted, bright yellowish and granular.

Glomus fragile (Berk. & Broome) Trappe & Gerd.

Chlamydospore yellowish brown, oval shaped.

Subtending hyphae simple, always hyaline and fragile.

Glomus fragilistratum Skou and Jakobsen

Spores are formed singly in soil, pale yellow, globose to ellipsoid or pyriform, 78-200μm in diameter.

Spore wall consists of five walls in two groups (A & B). Wall group A consists of three walls. Outermost layers I & II are thin, hyaline adhering walls. Wall I is evanescent and II is rigid unit wall. Wall III is yellow, laminated and fistular wall. Wall group B consists of two (IV & V) thin, hyaline, membranous wall.

Subtending hypha is hyaline.

Pore is open.

Glomus fuegianum (Spegazzini) Trappe & Gerd. (Fig. 3-G&H)

Sporocarp light brown, containing 7 spores radially arranged and is enveloped with peridium, which is hyaline.

Chlamydospore sessile or almost sessile in cluster of 2 or more, outer wall is much thicker than inner.

Spore colour varies from light yellowish brown to dark brown, ovoid in shape, 80-150μm diameter.

The wall composed of two layers, layer I evanescent, smooth, hyaline. Layer II laminated, yellowish brown.

Glomus fulvum (Berk. & Broome) Trappe & Gerd.

Yellow, brown coloured sporocarp surrounded by peridium.

Spore oblong to irregular, subtending hyphae with thick walls and occluded by spore wall.

Glomus geosporum (Nicol. & Gerd.) Walker (Fig. 3-I)

Spores are globose to subglobose, 120-240μm in diameter and yellow-brown to dark orange-brown in color.

Spore wall, three layers (I, II and III), layer I is a hyaline layer. Layer II is a rigid layer consisting of adherent sublayers yellow-brown to orange-brown in color, and Layer III is a semi-rigid to rigid layer always adherent to layer II.

Hyphal attachment is straight to somewhat recurved. A recurved septum forms from the point of attachment to the spore.

Glomus gerdemanni Rose, Daniels and Trappe (Fig. 3-J)

Spores are naked, formed singly or in loose clusters or small sporocarps in the soil.

Spores are globose to subglobose, 140-198 X 149-230μm in diameter or ellipsoid in shape, hyaline and smooth when young, becoming pale yellow brown and roughened at maturity.

Spore wall is formed by five layers. The outermost layer I is hyaline, smooth when young and become roughened and cracked in mature spores. Layer II is hyaline and fused, laminated when young and becomes pale yellow to brown and degrades on maturity. layer III is hyaline smooth, layer IV hyaline and adherent to layer V. The innermost layer V is thick and yellow.

Hyphal attachment is straight and occluded by the ingrowth of wall layer II.

Glomus gibbosum Blaszk. (Fig. 3-K)

Sporocarp globose to subglobose containing 2-8 spores enclosed by hyaline to pale yellow hyphal mantle.

Spores are hyaline to yellow, globose to subglobose, 93-115μm in diameter.

Single wall layer made up of five layers (I-V). Layer I is smooth or slightly roughened, hyaline. Layer II is rigid; hyaline to light yellow. Layer III is laminated, hyaline. Layer IV and V is tightly adherent, hyaline, flexible.

Subtending hypha is hyaline to yellowish white, straight or recurvate, cylindrical or sometime slightly funnel shaped. Wall layer is continuous with spore walls I-III.

Pore is occluded by septum formed by the innermost lamina of wall layer III.

Glomus globiferum Koske and Walker (Fig. 4-A)

Spores are formed singly or in pairs inside the peridial hyphae, orange brown to rich red brown to fuscous black, globose to subglobose, 150-260 X 150-270μm in diameter.

Spore wall has four layers, outer layer I is hyaline to pale yellow brown unit wall. Wall II is laminated, orange brown to red brown with up to nine sub equal laminae. Wall III and IV are hyaline, membranous.

Subtending hypha is thick walled, straight or recurved or funnel shaped.

Pore is closed partially by inner laminae or spore wall layer II and partially by a granular plug.

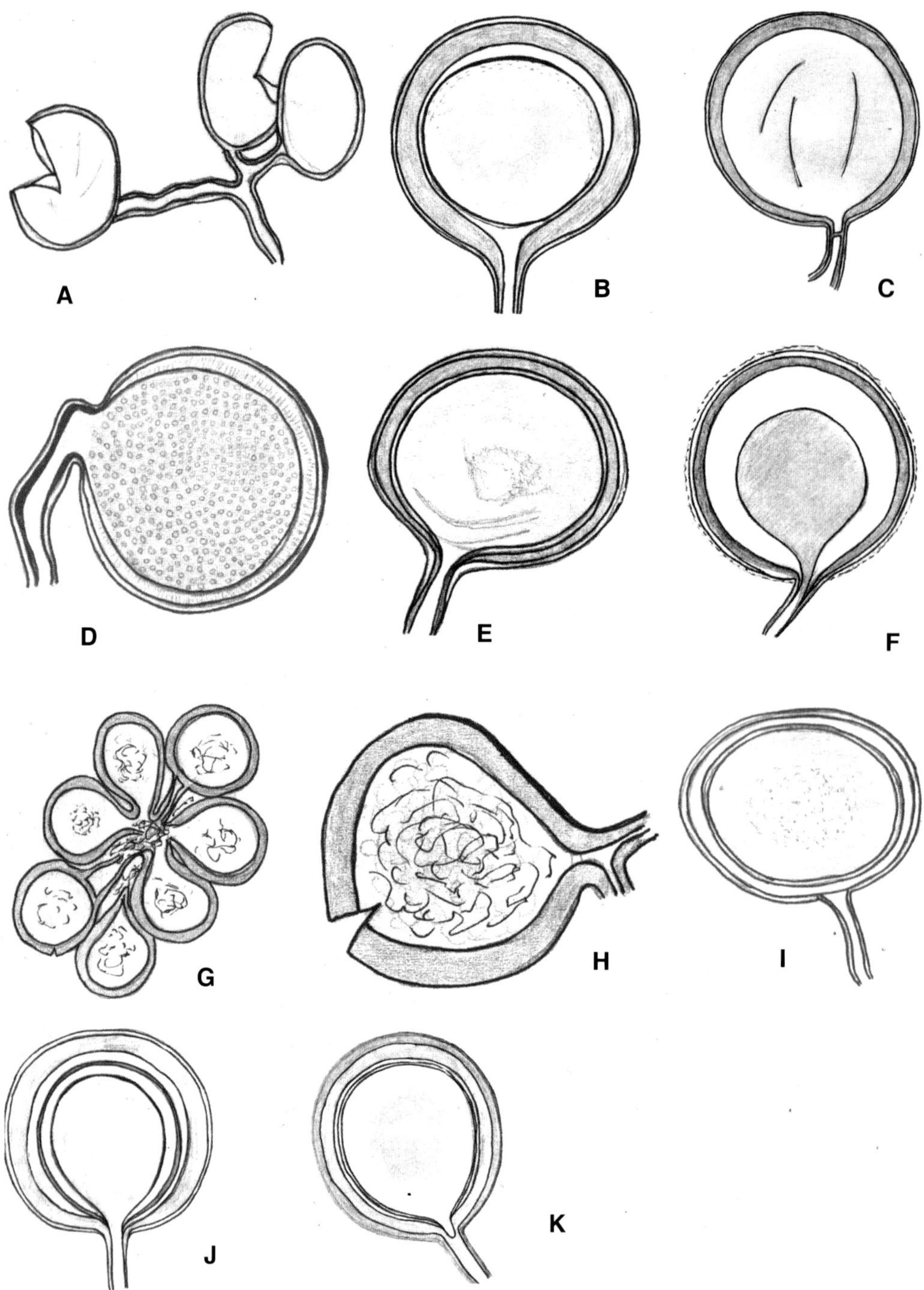

Fig. 3 : (A) *Glomus diaphanum*, (B) *G. drummondii*, (C) *G. eburneum*, (D) *G. etunicatum*, (E) *G. fasciculatum*, (F) *G. fistulosum*, (G) & (H) *G. fuegianum*, (I) *G. geosporum*, (J) *Glomus gerdemanni*, (K) *G. gibbosum*

Glomus glomerulatum Sieverd

Sporocarp dark brown, irregular and without a peridium.

Spores are globose to subglobose, 50-65μm in diameter. Wall has two layers I & II, layer II is light orange to golden yellow. Layer II is hyaline, flexible and tightly adherent to inner surface of layer I. Spores have two subtending hyphae.

Glomus halonatum Rose & Trappe (Fig. 4-B)

Spores are yellow to orange brown, globose, 202.5 X 200μm diameter.

Spore wall has three layers, layer I is evanescent, hyaline, mucilaginous and is smooth in young spores and with wide depressions in mature spores. Layer II is yellow to orange brown, laminated with 1.0-2.4μm tall spines on external surface. Layer III is hyaline and membranous.

Single subtending hypha is present with a septum near the insertion formed by layer III. The wall of subtending hypha is formed by wall layer I and II continuous with the spore wall layers.

Glomus heterosporum Smith & Schenck

Sporocarp is light to dark brown and formed in aggregates, having spores radially arranged around a central plexus. Peridium is absent.

Chlamydospores are light to dark brown, obovoid, 99-206 X 61-201μm diameter to ellipsoid, occasionally globose.

Spore with two distinct walls, inner wall laminate brown and outer wall is smooth, evanescent, hyaline, thick and is absent in the mature spores.

Spores have multiple attachments, which are branched.

Glomus hoi Berch & Trappe

Spores occur singly, light brown, globose to subglobose, ellipsoidal or irregular, 80-120 X 75-120μm in diameter.

Wall has two distinct layers. Outer layer I is orange yellow to light yellow, disintegrates in mature spores. Inner layer II is hyaline or light yellow, membranous.

Subtending hypha is single, cylindrical or slightly flared at the point of attachment with single wall.

Pore is occluded by a fine, curved septum.

Glomus hyderabadensis Swarupa, Kunwar, Prasad & Manoharachary (Fig. 4-C)

Spores are formed singly, honey colored to orange brown, globose, subglobose to ellipsoidal, 97-136 µm diameter.

Spore wall has three layers in a single group (group A). Outer wall I is smooth or roughened, dull yellow and perforated in older spores. Middle wall is single, non layered, rigid, unit and orange brown in colour. Inner wall is rigid, non layered and dull yellow in colour.

Subtending hypha is cylindrical to slightly flared having three layers, thick in the region of attachment and are continuous with the spore wall layers I, II & III.

The hyphal pore is occluded by a septum formed either by a sublayer of laminated layer III and IV or by wall layer IV alone.

Glomus indica Manoharachary, Sharathbabu & Adholeya

Chlamydospores naked, formed singly in the soil, sub-globose to globose, 117 X 129µm, dark brown, shiny smooth,

Spore wall is 5-10µm thick, one layered or occasionally two layered. The base of spore is straight or occasionally funnel shaped.

Subtending hypha is straight or recurved, constricted at the base, on one of the lateral side at base; a knob-like protuberance is present.

Glomus indicum Blaszk., Wubet, Harikumar, Ryszka & Buscot

It forms small, hyaline spores in hypogeous aggregates. The spores are globose to subglobose, (17-) 32 (-52) µm in diameter, rarely egg-shaped, oblong to irregular.

The single spore wall consists of two hyaline layers (I & II). A mucilaginous, short-lived, thin outer layer I and a laminate, smooth, permanent, thicker inner layer II.

Glomus insculptum Blaszk. (Fig. 4-D)

Spores are borne singly in the soil, yellowish white to golden yellow, globose to subglobse, 50-85µm diameter with single subtending hypha.

Spore consists of one wall with two layers. Layer I is hyaline continuous with the subtending hypha. Layer II is laminated with smooth outer surface and evenly pitted in a surface. Pits are round separated by ridges.

Subtending hypha is hyaline to pale yellow, straight to recurved, cylindrical or slightly flared. Wall consists up of two layers which is continuous with the spore wall layers.

Pore open or occluded by a recurved septum which is continuous with the spore wall layer II.

Glomus intraradices Schenck and Smith (Fig. 4-E)

Chalamydospores are pale cream to yellow brown, globose to subglobose irregular, 40-140μm in diameter.

Spore has three layers, layer I is the outermost layer, hyaline, muciliganous and degrades as the spore matures. Layer II is hyaline and adherent to layer I. layer III is pale yellow to brown, laminate.

Subtending hypha is cylindrical to slightly flared, occasionally slightly constricted. Hypha has three layers (I, II & III) continuous with the spore wall layers.

Pore is occluded by the innermost lamina of wall layer III.

Glomus invermaium Hall

Spores 50-75μm in diameter.

Outermost wall layer always hyaline extending down subtending hypha as a tightly fitting sleeve.

Pore is open.

Glomus iranicum (under publication)

It forms spores in loose clusters and singly in soil and sometimes inside roots. Spores are hyaline to pastel yellow, globose to subglobose, (13–) 40 (–56) μm diameter, rarely egg-shaped, prolate to irregular, 39–54 x 48–65 μm. The spore wall consists of three smooth layers: one mucilaginous, short-lived, hyaline, and outermost; one permanent, semi rigid, hyaline, and middle; and one laminate, hyaline to pastel yellow, innermost.

Glomus irregulare Blaszk., Wubet, Renker & Buscot

Spores occur mainly in aggregates, rarely singly, hyaline to pale yellow, mostly ovoid, sometimes oblong or irregular, rarely globose to subglobose, 60-130 X 80-240μm with single subtending hypha and occasionally with two.

Spore wall is composed of one wall with three layers, outermost layer I is semipermanant, smooth, hyaline before disintegration, and frequently thickened at the top to form a cap like swelling. Layer II is semipermanat, hyaline in closely adherent wall layer I. Layer III is laminate, smooth, hyaline to pale yellow.

Subtending hypha is straight or recurvate, mostly funnel shaped, rarely cylindrical, hyaline to pale yellow in color. Wall is continuous with the spore wall layers I - III.

Pore is open but occasionally occluded by a curved septum continuous with the innermost lamina of spore wall layer III.

Glomus kerguelense Dalpe & Strullu

Spores are formed singly or in clusters of 3 to 5 spores.

Mature spores are globose to subglobose, 153-216μm in diameter yellow and sometimes exhibits a granular appearance.

Mature spore wall consists of three layers. Layer I is hyaline, evanescent, mucilaginous, smooth to roughened, irregular, sometimes disappears in mature spores. Layer II is hyaline to pale yellow, finely laminated, smooth surface. Layer III is hyaline to pale yellow, finely laminated and the inside surface is ornamented with irregularly spaced warts.

Glomus laccatum Blaszk

Spores are found singly in the soil, hyaline, glistening, globose to subglobose, (50-) 87 (-130) μm diameter, sometimes ovoid with one subtending hypha.

Spore consists of one wall with two layers (I & II). Layer I is evanescent, hyaline, tightly adherent to layer II, usually absent in mature spores. Layer II is laminate, hyaline, smooth, composed of easily separating sublayers.

Subtending hypha hyaline, straight or curved, slightly funnel-shaped. Wall of subtending hypha is hyaline thick at the spore base and composed of two layers (I & II) continuous with spore wall layers.

Pore gradually narrows with age because of thickening of the inner layer of subtending hypha.

Glomus lacteum Rose & Trappe

Spores globose to subglobose, 150-220μm in diameter, milky white, shiny smooth, with two subtending hyphae.

The spore wall is simple, consisting of only one layer I, which is hyaline, 3-5μm thick.

Two subtending hyphae positioned parallel to each other for some distance before merging into one attachment. The hyphal wall is hyaline, 0.5-2μm thick.

Glomus lamellosum Dalpe, Koske & Tews (Fig. 4-F)

Spores are cream to pale yellow, globose to subglobose, 80-140μm. Spore wall has three layers, layer I is hyaline to light yellow and degrade on maturity to give a flaky appearance. Layer II is thin pale yellow to dark yellow, consists of sublayers. Layer III is thin concolours with laminated layer.

Subtending hypha is cylindrical to slightly flared, occasionally recurved slightly. It has two wall layers I & II continuous with the spore wall.

A recurved septum is formed by wall layer III.

Glomus leptotichum *Schenck & Smith*

Chlamydospores are formed singly or in loose clusters, globose to irregular, 48-262μm, white coloured.

Spore has single wall which is cream, to pale yellow-brown, loosely organized, with an indistinct alveolate reticulum of shallow ridges.

Subtending hypha is broad at the spore attachment and narrow gradually with distance, hyphal wall is continuous with the spore wall.

Pore is occluded by the spore wall in all spores.

Glomus liquidambaris (Wu & Chen) Almeida & Schenck (Fig. 4-G&H)

Sporocarps are golden brown to brown, globose to subglobose, 300-400 μm X 310-600μm diameter, consist of chlamydospores formed radially.

Chlamydospores are up to 200μm long, yellowish brown to reddish brown, ellipsoid to obovoid, sometimes with a septum at the base.

Spore wall is brown to reddish brown and is much thicker at the apex.

Glomus luteum Kenn, Schutz. & Morton (Fig. 4-I)

Spores formed are pale yellow to dark yellow in colour and globose to subglobose with size 60-180μm.

Spore has four layers (I, II, III & IV), all are adherent, layer I is hyaline mucilaginous layer, intact on young spores and then degrades as the spore matures. Layer II is hyaline and III is a rigid permanent layer, consisting of adherent sublayers. This layer extends into the subtending hypha. Layer IV is thin flexible layer concolorous with the laminate layer, very thin. Subtending hyphae is cylindrical to slightly flared and the spore wall occludes the hypha lumen, or a thin septum forms from layer III.

Glomus macrocarpum Tulasne & Tulasne

Chlamydospores are present in sporocarp, spores are yellowish brown to dark brown, formed singly, globose to subglobose, 100-115µm in diameter.

Spore wall has two layers, layer I hyaline, flexible. Layer II is smooth and yellow.

Subtending hyphae not inserted, pore without a plug.

Glomus maculosum Miller & Walker

Chlamydospores are formed singly in soil, 80-160µm, cream to light yellow, globose to subglobose.

Spore wall has four layers (I, II, III and IV). Layer I is hyaline, mucilagenous layer, difficult to distinguish from layer II because both layers are so tightly adherent, sloughing completely in mature spores. Layer II is hyaline, and degrades along with wall layer I. Layer III is laminate, consisting of thin and tightly adherent pale yellow sublayers. Layer IV is concolorous with the laminate layer, producing folds when it is very thin.

Subtending hypha is cylindrical to slightly flared, three layers (I, II and III) continuous with the outer three layers of the spore wall, thick in the region of attachment, tapering distantly.

Pore is occuluded by a recurved septum formed either by a sublayer of the laminate layer III and IV or IV alone.

Glomus magnicaule Hall

Outer spore wall layer much thicker than inner pore at point of attachment of spore and parent hyphae occluded or partially occluded by a plug and lateral thickening, outer spore wall layer not extending down subtending hyphae.

Glomus manihotis Howeler, Sieverd & Schenck (Fig. 4-J)

Spores are globose, 100-260µm, subglobose, sometimes elliptical, oblong, or irregular, colour varies from white to yellow-brown.

Spore wall has three layers (I, II & III). Layer I is hyaline mucilaginous layer. Layer II is a permanent hyaline layer consisting of sublayers that are not brittle and cracks are present. This layer is thicker in some regions so that the inner surface appears wavy. Layer III is also a permanent layer, usually consisting of 2-4 sublayers.

The subtending hypha is cylindrical to flared, sometimes slightly constricted also.

Glomus megalocarpum Red.

Sporocarp firm, irregularly lobate, 3.8 × 2 cm, and with a brownish rind, whitish-velvety on the outside. Peridial hyphae rust-colored at higher magnification. Composite spore mass in the interior pale tan to ochre. Spores obovate to pyriform or slightly irregular with dimensions of (56)65–95 (115) × 38–57(62) µm.

Spore wall single-layered, hyaline, and 1.7–4 µm thick.

Hyphal attachments is usually not persistent.

Spores occluded by a prominent septum, 7–8µm broad, continuous with an inner lamina of the spore wall, separating a triangular cavity from the remainder of the spore lumen.

Glomus melanosporum Gerd. &Trappe

Spores and sporocarp often exuding a sticky latex when cut.

Spores containing white latex, spore wall grading in colour from dark at outer surface to subhyaline near inner surface.

Glomus merredum Porter & Hall

Spore oblong to globose, light brown to dark brown, subtending hyphal wall II layered, inner wall of spore striated.

Maximum diameter of subtending hyphae at widest part 25-60µm.

Glomus microaggregatum Koske, Gemma & Olexia (Fig. 4-K)

Spores are formed singly or in clusters, hyaline to pale yellow to brownish yellow, globose, or subglobose to irregular, 30 X 30 µm in diameter

Spore wall has one or two walls in one group. Wall I is smooth, brittle, hyaline to pale yellow to brownish yellow. Wall II if present is membranous and concolorous with wall I.

Hyphal attachment is straight or infundibuliform; concolorous with wall I. Pore is usually open but sometimes closed by septum formed by wall layer II.

Glomus microcarpum Tulasne & Tulasne (Fig. 4-L)

Spores are formed singly or in clusters, hyaline to pale yellow to brownish yellow, globose, 30 X 30µm in diameter or subglobose to irregular.

Spore wall has one wall which is smooth, brittle, laminated.

Hyphal attachment is straight or infundibuliform; concolorous with wall I. Pore is usually open.

Glomus minutum Blaszk. Tadych & Madej (Fig. 5-A)

Spores formed singly or in loose aggregates. Spores are globose to subglobose, 18-65µm diameter with single subtending hypha.

Spore consist of one wall with two hyaline components (1&2). Component 1 is the outermost, smooth, semiflexible component adherent to component 2. Component 2 is laminated, smooth.

Subtending hypha is hyaline, straight or recurved, cylindrical or slightly flared. Wall is hyaline and continuous with both the wall components.

Pore is occluded by a septum formed by innermost lamina of wall component 2.

Glomus monosporum Gerd. & Trappe

Spores ranging in colour from yellowish to brown.

Sporocarps containing 1-3 spores, outer surface of inner wall ornamented with minute projections and thin outer wall not always obvious.

Delicate and branched subtending hyphae.

Spores often filled with hyphae, maximum diameter of subtending hyphae at widest part 8-26µm.

Glomus mortonii Bentiv. & Hetrick (Fig. 5-B)

Globose to irregular sporocarps are formed in the soil. Mantle is hyaline and envelops individual spore.

Spores of size 125 X 149µm are produced singly or in sporocarps, spores yellow to yellow brown in color.

Spore wall is made up of three layers in one group A. Outermost layer I is hyaline, mucilaginous. Middle layer II is laminated and is continuous with hyphal attachment. Innermost layer III is yellow to brown in colour, laminated with up to 9 laminae.

Hyphal attachment is highly variable, straight or occasionally recurved, cylindrical to flared to irregular.

Pore at point of hyphal attachment open in immature spores, becoming occluded by innermost lamina of wall III.

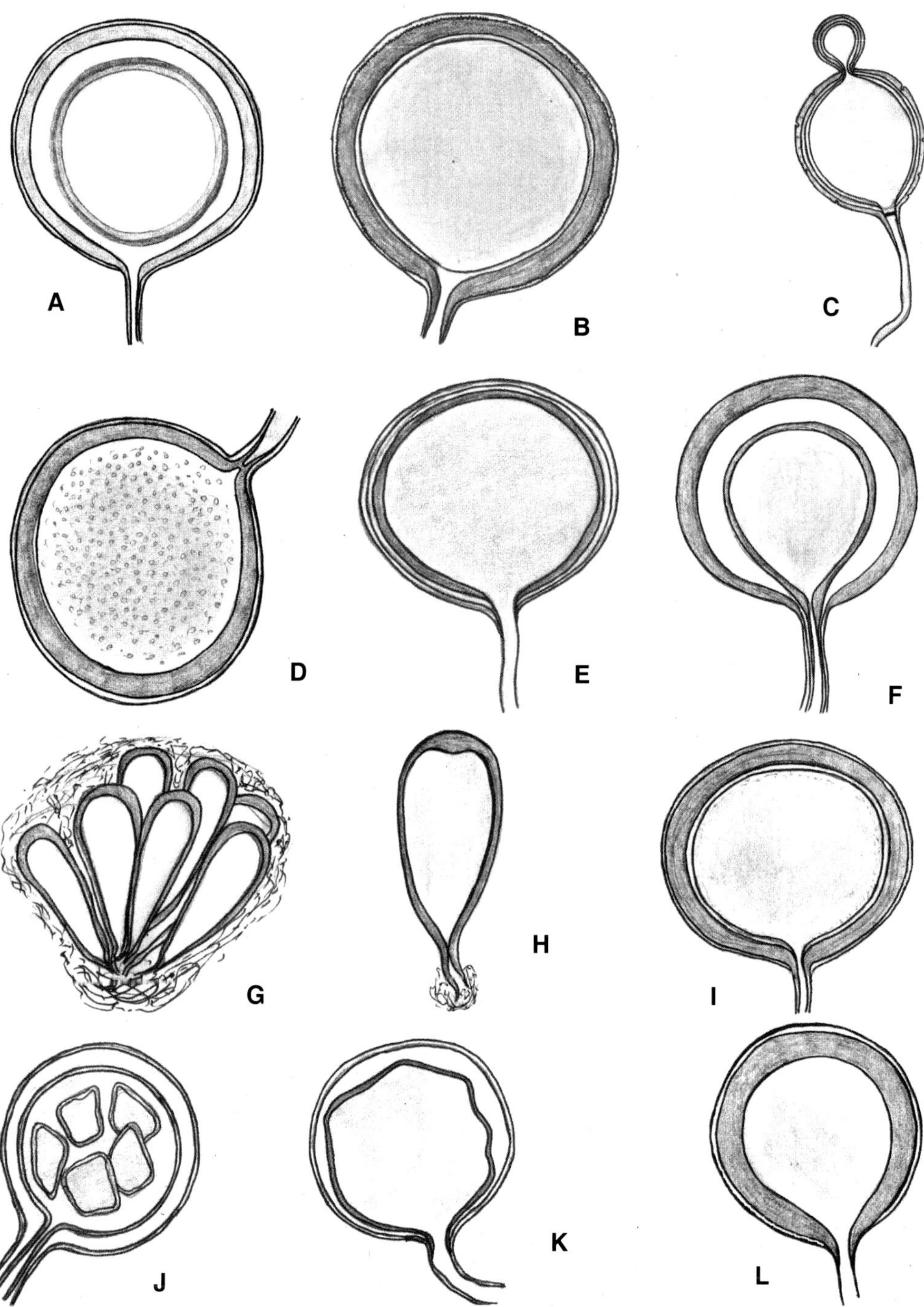

Fig. 4 : (A) *Glomus globiferum*, (B) *G. halonatum*, (C) *G. hyderabadensis*, (D) *G. insculptum*, (E) *G. intraradices* (F) *G. lamellosum*, (G) & (H) *G. liquidambaris*, (I) *G. luteum*, (J) *G. manihotis*, (K) *G. microaggregatum*, (L) *G. microcarpum*

Glomus mosseae (Nicol. & Gerd.) Gerd. & Trappe (Fig. 5-C)

Chlamydospore yellowish to brown, globose to subglobose, sometimes ellipsoid to irregular.

Outer surface of inner wall not ornamented, funnel shaped subtending hyphae.

Spores are more than 100µm in size, subtending hyphae is generally funnel shaped with cup shaped septum.

Glomus multicaule Gerd. & Bakshi (Fig. 5-D)

Chlamydospores are light brown to golden brown, shape oval to ellipsoidal.

Surface of spore ornamented with some projections, more than one subtending hyphal attachment.

Glomus multiforum Tadych & Blaszk. (Fig. 5-E)

Spores are borne singly, deep yellow to brown, globose to subglobose, 170-230µm diameter, with a single subtending hypha.

Spore consists of one wall with three layers. Outermost layer I is mucilaginous, smooth hyaline and closely adherent to wall layer II. Wall layer II is hyaline, ornamented with ingrowths filling the pits of wall layer III. Layer III is laminated, deep yellow to brown and evenly pitted with round or ovate deep depression separated by ridges.

Subtending hypha is straight or recurvate, funnel shaped, rarely constricted, deep yellow to brown. Wall is continuous with spore wall layer I - III.

Pore is occluded by a septum continuous with the innermost lamina wall layer III.

Glomus multisubstensum Mukerji, Bhattacharjee & Tewari

Chlamydospores are formed singly in soil, 80-160 µm, cream to light yellow, globose to subglobose.

Spore wall has four layers (I, II, III and IV). Layer I is hyaline, mucilaginous layer, difficult to distinguish from LII because both layers are so tightly adherent, sloughing completely in mature spores. Layer II is hyaline, and degrades along with wall layer I. Layer III is laminate, consisting of thin and tightly adherent pale yellow sublayers. Layer IV is concolorous with the laminate layer, producing folds when it is very thin.

Subtending hypha is cylindrical to slightly flared, three layers (I, II and III) continuous with the outer three layers of the spore wall, thick in the region of attachment, tapering distantly.

Pore is occluded by a recurved septum formed either by a sublayer of the laminate layer III and IV or IV alone.

Glomus nanolumen Koske & Gemma

Sporocarps are subglobose to irregular, composed of loosely to tightly packed spore and sporogenous hyphae.

Spores are sparking yellow, reddish yellow or rose pink in color, subglobose, 24-52 X 20-63µm diameter, pyriform, ovoid to irregular.

Spore wall has two walls (I & II) in one group A. Outer unit wall I is golden yellow brown. Wall II is laminated, pale yellow mainly hyaline. Spore lumen is small, connecting to sporogenous hyphae through an open channel.

Subtending hypha is funnel shaped.

Glomus occultum Walker

Chlamydospores borne singly or in aggregates in the soil, or in compact clusters in the root cortex, often broader than long, ovoid to obvoid, subangular to irregular, less frequently globose to subglobose, 15-100 X 20-120µm, hyaline to white.

Spore wall 1 to 2 layered with an additional rough outer deposit of granular material which sloughs with age. Outer wall when present less than 1µm thick, often indistinct.

Inner wall 1-5µm thick, usually of two, sometimes with indistinct laminations.

Subtending hypha is funnel shaped to simple, attached axially or eccentrically and recurved to straight, sometimes closed distally by a septum.

Glomus pachycaulis (Wu & Chen) Almeida & Schenck

Sporocarp yellow brown to dark brown, subglobose to somewhat irregular, 180 X 80-375 X 675µm, resembling a miniature blackberry consisting of a layer of chlamydospores surrounding a central plexus of hyphae. Peridium is absent.

Spores are yellow brown to dark brown, obvoid to elliptical, 27-125 X 29-87µm.

Inner wall is yellow brown to brown, laminate, often perforate and with thick perforated projections on inner surface. Outer wall is hyaline, evanescent and absent in mature spores.

Subtending hypha with small pore opening or occluded by wall thickening with two thin septa.

Glomus pallidum Hall

Spore white to pale yellow, with a slightly constricted subtending hypha with two layered walls.

Pore not occluded with a septum in mature spores, wall becoming laminated with age.

Glomus pansihalos Berch & Koske (Fig. 5-F)

Sporocarp is irregularly ellipsoid up to 15 X 12 X 7 mm light or dark brown in color. Peridium is absent.

Spores 60-120μm in diameter, hyaline to pale yellow in color, globose in shape and occasionally become subglobose, obovoid and ellipsoid, born singly or in loose clusters.

Spore wall distinguished by two layers. Wall I is hyaline to pale yellow ornamented by cerebriform folds and spins, laminated 2.5-5.0μm thick. Wall II is brown to reddish brown covered by spins and laminated.

Glomus pellucidum McGee & Pattinson (Fig. 5-G)

Sporocarp is hypogeous, rounded to irregular lacking a peridium.

Spores are globose to subglobose, 60-110μm in diameter.

Wall has three layers, outer wall layer I is hyaline, thick, mucilaginous when immature and becomes flaky and lost at maturity. Middle layer II is yellow to yellow brown, laminate. Inner layer III is hyaline and membranous.

Subtending hypha is broad at the point of attachment with three wall layers continuous with the spore wall layer.

A septum is formed by the inner wall.

Glomus perpusillum Blaszk. & Kovacs

Spores are small, hyaline hypogeous aggregates and occasionally inside roots, globose to subglobose, 10-30μm diameter, rarely egg-shaped, oblong to irregular.

Single spore wall consists of two permanent layers (I & II). A finely laminate, semiflexible to rigid outer layer I and a flexible to semiflexible inner layer II.

Glomus proliferum Dalpe & Declerck

Spores are hyaline, globose to subglobose, 22-76 μm diameter smooth cracking easily under pressure.

Mature spores with a wall comprising four smooth hyaline layers, layer I is thick, slightly laminated. Layer II is thick, tightly adherent to wall layer I. Layer III composed of tightly adherent sublayers. Layer IV is thick and membranous.

Subtending hypha is unique, hyaline to very pale yellow, cylindrical to slightly flared and wall is continuous with the spore wall layer.

Pore is usually open.

Glomus pubescens Saccardo & Ellis (Trappe & Gerd.) (Fig. 5-H)

Spores are formed singly or in clusters inside the well organized sporocarps with distinct peridial hyphae that form tuft on the surface of sporocarp.

Spores are hyaline to pale yellow to brownish yellow, globose, 30 X 30μm in diameter or subglobose to irregular.

Spore wall has two walls. The outer wall is thick to which an inner membranous, wrinkled wall is attached.

Hyphal attachment is straight or infundibuliform, concolorous. Pore is usually open.

Glomus pulvinatum (Henn.) Trappe & Gerd.

Pore partially occluded and always with a distinct septum.

Glomus pustulatum Koske, Friese, Walker & Dalpe (Fig. 5-I)

Spores are formed singly in soil, 86-140μm diameter, pale yellow to yellow brown or orange brown, globose to irregular.

Spore wall has three layers in one group (group A). Wall I is yellow brown to orange brown and has circular to irregular blister like area on the outer surface. Wall II is pale yellow to yellow brown, laminated. Wall III is thin hyaline, membranous and adherent to wall II.

Subtending hypha is straight or recurved pale yellow to yellow brown and continuous with spore wall layer II. Wall of subtending hypha sometimes slightly thickened at the junction of spores.

Pore is closed by a septum formed by ingrowth of spore wall layer II.

Glomus radiatum (Thaxt.) Trappe & Gerd.

Spores in sporocarps arranged in radial rows.

Glomus rubiforme (Gerd. & Trappe) Almeida & Schenck

Sporocarp yellow brown to dark brown, subglobose to somewhat irregular, 180 X 80-375 X 675 μm, resembling a miniature blackberry consisting of a layer of chlamydospores surrounding a central plexus of hyphae. Peridium is absent

Spores are yellow brown to dark brown, obovoid to elliptical, 27-125 X 29-87 μm.

Inner wall is yellow brown to brown, laminate, often perforate and with thick perforated projections on inner surface. Outer wall is hyaline, evanescent and absent in mature spores.

Subtending hypha with small pore opening or occluded by wall thickening with two thin septa.

Glomus scintillans Rose & Trappe

Chlamydospore formed singly, globose to subglobose, size is > 150μm, dark brown to black in colour.

Long subtending hyphae with hooked end.

Spore wall is not distinct with outer rough surface.

Glomus segmentatum Trappe, Spooner & Ivory (Fig. 5-J)

Spores are borne in epigeous sporocarp. Sporocarps are white, pulvinate and firm.

Spores are hyaline to yellowish white, globose to subglobose, 65-90μm in diameter with single subtending hypha.

Spore consists up of spore wall with three layers. Layer I is evanescent, hyaline. Layer II is finally laminated. Layer III is flexible, usually tightly adherent to the inner surface of layer II.

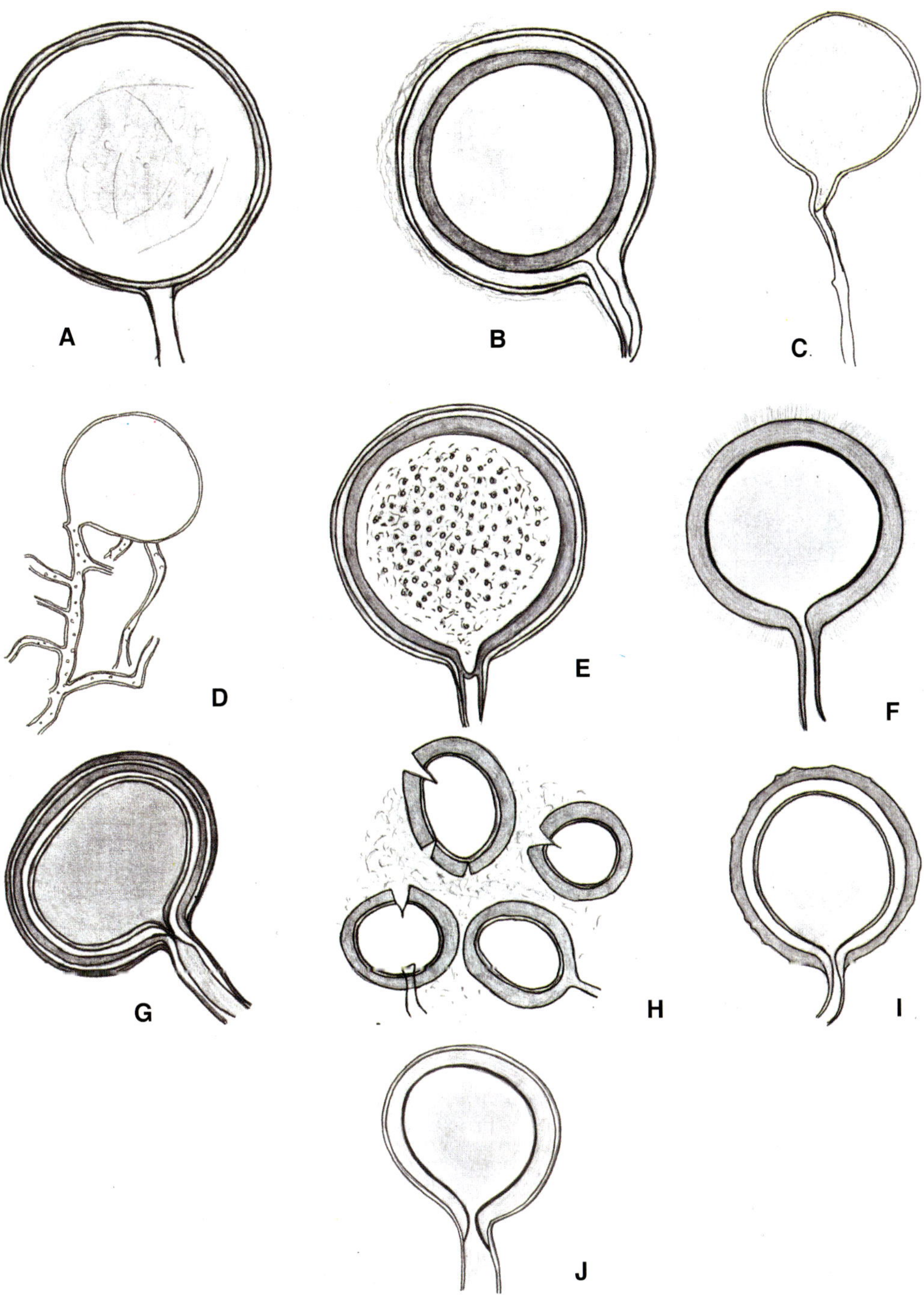

Fig. 5 : (A) *Glomus minutum*, (B) *G. mortonii*, (C) *G. mosseae*, (D) *G. multicaule*, (E) *G. multiforum*, (F) *G. pansihalos*, (G) *G. pellucidum*, (H) *G. pubescens*, (I) *G. pustulatum*, (J) *G. segmentatum*

Subtending hypha is straight or recurved, sometimes cylindrical. Wall composed of three layers continuous with the spore wall layer.

Pore is open or closed by a straight or slightly curved septum continuous with the innermost spore wall layer III.

Glomus sinuosum (Gerd. & Bakshi) Almeida & Schenck (Fig. 6-A&B)

Sporocarp is globose, 200-360μm diameter, subglobose, orange in colour when immature and orange-brown to dark orange-brown when mature.

Peridium is formed by a dense layer of tightly interwoven hyphae, 9-19 μm thick, covering all spores to keep them tightly packed.

Spores are 28-63 x 50-95μm, orange-brown in colour, clavate, sometimes obovate to elliptical and are organized in a single layer from a central plexus of hyphae.

Only single pale orange-brown wall layer which is surrounded by dense peridial hyphae.

Subtending hypha is cylindrical to slightly flared and pore is occluded by a thin septum formed by spore wall.

Glomus spinosum Hu

Sporocarp is pale brown, irregular, upto 5 mm in diam., formed by loose interwoven yellowish brown hypahe. Peridium is absent.

Chlamydospores are formed singly or inside the sporocarps, globose to subglobose, 40-90μm in diam., spores have roughened surface.

Spore wall has three wall layers in one group. Wall layer I is hyaline to subhyaline to pale yellow, unit wall. Wall II is orange brown, red brown to dark brown or nearly black, laminated, ornamented with subhyaline warts. Wall III is yellowish brown, membranous.

Subtending hypha is single, straight, or slightly recurved, and wall layer is continuous with the spore wall layer II.

Pore is occluded by a septum formed by innermost wall layer III.

Glomus spinuliferum Sieverd. & Oehl (Fig. 6-C)

Spores are formed singly, dark yellow to orange brown, globose, 120-140μm diameter to subglobose.

Spore wall has one wall group with four layers (layer I-IV). Outer layer I is hyaline to cream white, evanescent and absent in mature spores. Layer II is hyaline, covered with fine spines which grow into wall layer I .Wall layer III is dark yellow to orange brown and finally laminated and is often thick at the point of hyphal attachment. Layer IV is hyaline to concolorous with wall layer III, thin semiflexible and tightly adherent to wall layer III.

Subtending hypha is straight or curved and wall layers are continuous with spore wall layers I-III.

Pore is usually closed by wall layer IV.

Glomus spurcum Pfeiffer, Walker & Bloss (Fig. 6-D)

Spores are subhyaline to pale yellow, globose, 60-120µm in diameter to subglobose.

Spore wall consist of two layers. Layer I is hyaline to pale yellow, thin, often wrinkled and produces many folds. Layer II is thin, hyaline to subhyaline sublayers. The innermost sublayer often produces numerous fine wrinkles.

Subtending hypha is cylindrical to slightly flared, very fragile and consists of single layer I.

Pore is occluded by thickening of layer II of spore wall.

Glomus sterilum Mehrotra & Baijal

Sporacarp hypogeous, 256-576 X 320-640µm, pale yellow, yellow brown to orange brown, with peridium composed by loose mass of hyphae.

Chlamydospores formed singly, pale yellow, orange brown to reddish brown, globose, subglobose to irregular.

Single wall group composed of two sublayers I and II. Wall I is single, brittle, unit layer, yellow to reddish brown. Wall II when present is membranous, yellow to yellow brown.

Subtending hypha is straight or infundibuliform.

Pore is closed by a septum formed by inner spore wall layer.

Glomus taiwanense (Wu & Chen) Almeida & Schenck

Sporocarp has only four chlamydospores, formed radially around a central plexus of hyphae.

Chlamydospores are 200µm in diameter, with two wall layers I & II. Outer layer I is reddish brown and inner layer II is light brown.

Glomus tenebrosum (Thaxt.) Berch. (Fig. 6-E)

Spores are 24X230μm in diameter, globose to subglobose . Spore wall is single, thick and yellow to dark brown and appears laminated near the hyphal attachment .The outer surface is smooth or may bear flattened tubercles.

Subtending hypha is dark coloured near the attachment and fades from dark brown to yellow or hyaline away from attachment.

Pore remains open.

Glomus tener Tandy

Subtending hyphae constricted at point of attachment.

Innermost wall layer hyaline.

Glomus tenue (Greenall) Hall

Spores dark brown with subtending hyphae ellipsoidal to oblong and outer wall thick.

Subtending hyphae extending on outer wall.

Glomus tortuosum Schenck & Smith

Spores are formed singly with a peridium covering the spore wall; many spores appear to develop randomly within a dense hyphal peridium. Peridium consisting of thin-walled hyphae. Each sporocarp contains between 2-6 spores and ranges from 290-540μm in diameter.

Spores are light yellow-brown to orange-brown, but most are pale orange-brown, globose, to subglobose; 120-220μm in diameter occasionally irregular. Spore wall has a single layer I surrounded by dense peridial hyphae. Layer I is pale yellow-brown with sublayers that usually remain adherent.

Subtending hypha is 12.5-15μm, cylindrical to slightly flared, with a single wall layer less than 1μm thick.

Glomus trimurales Koske & Halvorson (Fig. 6-F)

Spores are formed singly pale yellow to pale brownish yellow, globose to subglobose, ellipsoid, pyriform or irregular, 121 X 118μm in diameter.

Spore wall has three layers, layer I is laminated, pale yellow to yellow brown. Layer II is unit wall and pale yellow to brownish yellow. Wall III is laminated, hyaline to pale yellow and brownish in mature spores.

Hyphal attachment is straight or slightly constricted at the point of attachment, hyaline to pale yellow.

Pore is usually closed by a granular plug.

Glomus tubiforme Tandy

Sporocarp is compact, white, epigeous containing large number of spores. Small sporocarps may be globose but larger ones are usually flattened cushions, occasionally even plate like. Sporocarp has ribbed hollow trumpet-shaped cells which is an unusual character to this species.

Glomus velum Porter & Hall (Fig. 6-G)

Two layered spore wall, spores yellow, shape slightly irregular.

Pits on outer surface of inner layer but outer layer very thin.

Glomus verruculosum Blaszk. (Fig. 6-H&I)

Spores are formed singly, yellow to orange, globose to subglobose, 150-165μm in diameter with single subtending hyphae.

Spore has one wall with two layers. Layer I is semi flexible, hyaline. Layer II is laminated, yellow to orange, ornamented with evenly distributed warts projecting inwards from the innermost lamina.

Subtending hypha is straight or recurved, funnel shaped to almost cylindrical. Hyphal wall is yellow to orange and continuous with the spore wall layer I and II.

Pore is closed by a curved septum continuous with the innermost lamina of spore wall layer II.

Glomus versiforme (Karst.) Berch (Fig. 6-J)

Spores are orange to red brown, globose, 60-160 μm, subglobose and sometimes ovoid.

Wall layers consisting of two layers. L1 is the outer layer semi-permanent, subyhyaline, adherent to L2. L2 has many thin adherent sublayers, pale yellow-brown to orange-brown in colour.

Subtending hypha is cylindrical to slightly flared.

Glomus vesiculiferum (Thaxt.) Gerd. & Trappe

Spores in sporocarp, pore at point of attachment of spore and attached hypha never or rarely with a distinct septum.

Sporocarps are more than 1 mm in diameter.

Glomus viscosum Nicol. (Fig. 6-K)

Spores are globose to subglobose, 50-120μm and subhyaline to pale straw coloured. Spore wall with three layers (I, II & III). Layer I is a semi-flexible hyaline layer. Layer II is a thin semi-flexible hyaline layer with discrete edges so that it appears as a raised outer edge on layer III when tightly adherent. Layer III is a hyaline, thick layer consisting of sublayers. This layer is more rigid than layers I or II. Hyphal attachment is cylindrical to slightly flared, sometimes slightly constricted also.

Glomus walkeri Blaszk. & Renker

Spores are found singly in the soil, white to pale yellow, globose to subglobose, sometimes ovoid, 55-95µm in diameter with a single subtending hypha.

Spore has one wall with three layers. Layer I is semi-permanent, hyaline, deteriorating with age. Layer II laminate, smooth, white to pale yellow. Layer III flexible, smooth, hyaline,

Subtending hypha is white to pale yellow, straight or recurvate, cylindrical or slightly flared, rarely funnel-shaped or constricted. Wall of subtending hypha white to pale yellow, composed of two layers continuous with spore wall layers I and II.

Pore occluded by a septum, continuous with spore wall layer III.

Glomus xanthium Blaszk., Blanke, Renker & Buscot (Fig. 6-L)

Spores occur singly, light yellow to yellow ochre, globose, to subglobose, 70µm in diameter with single subtending hypha.

Spore has one wall with three layers. Layer I is the outermost, rigid, smooth, hyaline layer which deteriorates with age. Layer II is rigid, smooth hyaline and closely adherent to wall layer I. layer III is laminate, smooth, light yellow to yellow ochre. Subtending hypha is straight or recurvate, cylindrical or flared, rarely constricted. Wall is light yellow to yellow ochre and is continuous with spore wall layers I- III.

Pore is occluded by a septum continuous with the innermost lamina of layer III.

Conclusion

According to Dkhar *et al.* (2007) knowledege of microbial diversity is important because of their role in regulating the population of other organisms and ecosystem processes and of equal importance of microbial diversity itself are variations in their own population that affect host plant diversity. Microbial communities particularly AM fungi constitute an essential component of biological characteristics in soil ecosystems. In response to the increased concern for environmental quality, innovative approaches need to be incorporated

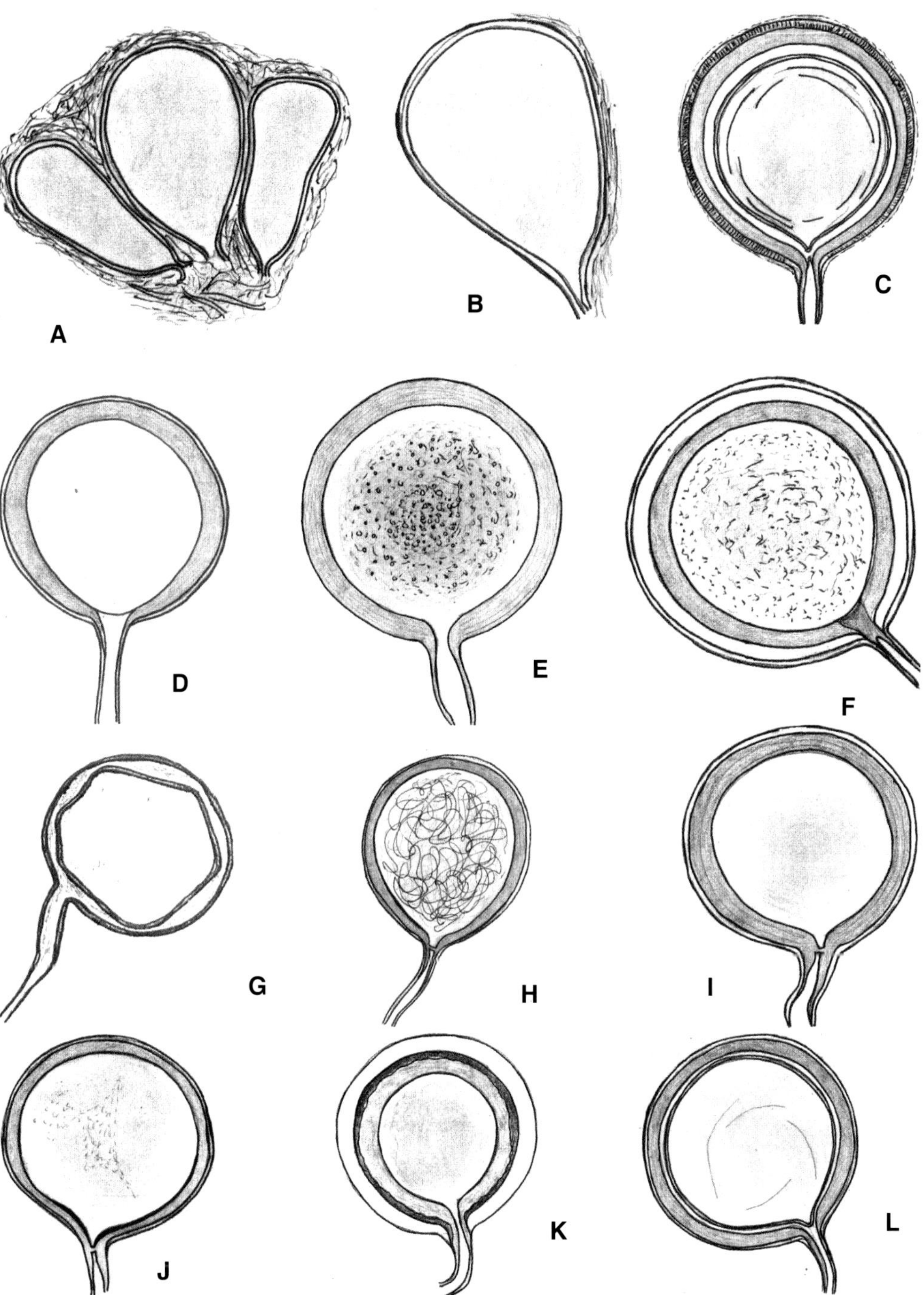

Fig 6 : (A) & (B) *Glomus sinuosum*, (C) *G. spinuliferum*, (D) *G. spurcum*, (E) *G. tenebrosum*, (F) *G. trimurales* (G) *G. velum*, (H) & (I) *G. verruculosum*, (J) *G. versiforme*, (K) *G. viscosum*, (L) *G. xanthium*

into agricultural systems. Management of AM fungi is one important aspect of such an approach. However, the obligate symbiotic nature of AM fungi continues to impede research on improving our understanding of these fungi. Furthermore, limited supply of AM inocula has restricted field testing. High production costs as well as problems associated with soil based pot cultures have dampened the enthusiasm for commercial development. Much is unknown about the storage of AM fungal inocula. Further research on the storage properties of AM inocula will definitely spur commercial production and the development of new formulations. An inoculum production and inoculation technologies of AM fungi undergo further development, involvement of all those concerned with crop production and plant health will ensure AM inocula of consistent high quality and availability.

Classification and identification of AM fungi remains in the formative stages of growth, with all the attendant turmoil. Still much more is being learned about the extent of species diversity occurring in nature. Present taxonomy of AM fungi is based on spore morphology and other parameters. These characteristics may be essential for fungal identity but this may be unrelated to any benefits that are given to the plant. Routine identification of AM fungi will probably continue to be based primarily on structural characters and thus an increased appreciation of the relationship between anatomy and ribosomal gene sequences will be important. The ability of properly naming of fungi, avoid duplication of names and relate the species to one another depends on collections such as those held by INVAM, the international culture collection of arbuscular fungi (http:// invam. caf. wvu. edu/) and BEG/ IBG International Bank for the Glomeromycota (http:// www. kent. ac. uk/bio/ beg/).

Glomus is the most important mycorrhizal fungus for increasing the biomass of crop plants, in agriculture system and forest management. A detailed description of *Glomus* species has been compiled to facilitate the researchers, field workers and academicians in the proper identification of *Glomus* species which will help in utilization of mycorrhizal fungi for increased productivity and sustainable development.

References

Abbott, L.K. (1992). Comparative anatomy of vesicular arbuscular mycorrhizal fungi: advances and future prospects. *Afr. J. Biotechnol.*, 30: 485-495.

Aggarwal, A., Parkash, V. and Mehrotra, R.S. (2006). VAM fungi as biocontrol agent for soil borne pathogens. In: *Plant Protection in New Millennium* (Eds.) A.V. Gadewa and B.P. Singh, Satish Serial Pub., New Delhi, India: 137-162.

Amerian, M. R., Stewart, W.S. and Griffiths, H. (2001). Effect of two species of AM fungi on growth assimilation and leaf water relation in maiza (*Zea mays*). *Asp. Appl. Biol.*, 63: 73-76.

Ames, R.N., Reid, C.P.P., Porter, K.K. and Cambardella, C. (1983). Hyphal uptake and transport of nitrogen from two ^{15}N labelled source by *Glomus mosseae* arbuscular mycorrhizal fungus. *New Phytol.*, 95: 381-396.

Azcon, R. and Ocampo, J.A. (1981). Factors affecting vesicular-arbuscular infection and mycorrhhizal dependency of thirteen cultivars. *New Phytol.*, 87: 677-685.

Barea, J.F. and Azcon-Anguilar, (1982). Production of plant growth regulating substances by the vesicular-arbuscular mycorrhizal fungus *Glomus mosseae*. *Appl. Environ. Microbiol.*, 43: 810-813.

Barea, J.F. and Jeffries, P. (1995). Arbuscular mycorrhizas in sustainable soil-plant systems. In: *Mycorrhiza structure, function and physiology* (Eds.) A.Varma and B. Hock, Springer-Verlag, Berlin.

Barna, T. (2002). Mycorrhizal role in afforestation. *Acta Microbiologica et Immunologica Hungarica*, 49(2/3): 215-226.

Bidartondo, M.I., Burhgardt, B., Gebaucer, G., Bruns, T.D. and Read, D.J. (2004). Changing partner in the dark isotopic and molecular evidence of Ectomycorrhizal liaison between forest orchids and trees. *Proct. R. Soc. Lond., B*. 271: 1799-1806.

Biermann, B. and Linderman, R.G. (1983). Use of vesicular-arbuscular mycorrhizal roots, intraradical vesicles and extraradical vesicles as inoculum. *New Phytol.*, 95: 97-105.

Brundrett, M. (2004). Diversity and classification of mycorrhizal associations. *Biol. Rev.*, 79: 473-495.

Brundrett, M.C. and Kendrick, B. (1990). The roots and mycorrhizas of herbaceous woodland plants I. Quantitative aspects of morphology. *New Phytol.*, 114: 457-468.

Bucholtz, F. (1912). Beiträge zur Kentnis der Gattung *Endogone* Link., *Beih. Bot. Zbl.*, 29: 147-225.

Busse, M.D. and Ellis, J.R. (1985). Vesicular-arbuscular mycorrhizal (*Glomus fasiculatum*) influence on soybean drought tolerance in high phosphorus soil. *Can. J. Bot.*, 63: 2290-2294.

Cariney, J.W.G. (2000). Evolution of mycorrhiza systems. *Naturwissenchaften*, 87: 467-475.

Caris, C., Horett, W., Jawkins, H.J., Ramheld, V. and George, E. (1998). Studies on the iron transport by arbuscular mycorrhizal hyphae from soil to peanut and sorghum plants. *Mycorrhiza*, 8: 35-39.

Chilvers, G.A., Lapeyrie, F.F and Horan, D.P. (1987). Ectomycorrhizal vs. Endomycorrhizal fungi within the same root systems. *New Phytol.*, 97: 441-448.

Clarke, R.B. and Zeto, S.K. (2000). Mineral acquisition by arbuscular mycorrhizal plants. *J. Plant Nutr.* 23: 867-902.

Dangeard, P.A. (1896). Une maladie du peuplier dans l'ouest de la France. *Botaniste*, 58: 38-43.

Dkhar. (2007). Microbial Diversity-A review. In: *Biodiversity and its significance* (Eds.) P. Tandon, Y.P. Abrol and S. Kumaria, Internatioanal Publ. House Pvt. Ltd., New Delhi: 192-203.

Feber, B. A., Zasoki, R.J., Burau, R.G and Urio, K. (1990). Zinc uptake by corn as affected by vesicular arbuscular mycorrhizae. *Plant Soil,* 129: 121-130.

Frank, A.B. (1885). Uberdie auf Wurzelsytouse Beruhende Ernahrung Gewisser Banice durch unteridische. *Pilze Ber. Duet. Bot. Ges.,* 3: 128-145.

Friberg, S. (2001). Distribution and diversity of arbuscular mycorrhizal fungi in traditional agriculture on Niger Inland delta, Mali, West Africa. *CBN Skriftserie,* 3:53-80.

Fries, E.M. (1849). *Summa Vegetabilium Scandinaveae,* 2: 261-572.

Gaur, A. and Adholeya, A. (2004). Prospects of arbuscular mycorrhizal fungi in phytoremediation of heavy metal contaminated soils. *Curr. Sci.,* 86(4): 528-534.

Gerdemann, J.W. (1968). Vesicular arbuscular mycorrhiza and plant growth. *Ann. Rev. Phytopathol.,* 6: 397-418.

Gerdemann, J.W. (1971). Fungi that form the vesicular-arbuscular type of endomycorrhiza. In: *Mycorrhizae* (Ed.) E. Hacskayla, Proceedings of the first North American conference on mycorrhizae. USDA Misc. Publ., 1189: 9-18.

Gerdemann, J.W. (1975). Vesicular-arbuscular mycorrhizae. In: *The Development and Functions of Roots,* (Eds.) J.G. Torrey and D.T. Clarkson, Academic Press, London: 575-591.

Gerdemann, J.W. and Nicolson, Y.H. (1963). Spores of mycorrhiza *Endogone* species extracted from soil by wet sieving and decanting. *Trans. Brit. Mycol. Soc,.* 46: 235-244.

Gerdemann, J.W. and Trappe, J.M. (1974). Endogonaceae in the Pacific Northwest. *Mycol. Mem.,* 5: 1-76.

Gerdemann, J.W. and Trappe, J.M. (1975). Taxonomy of the Endogonaceae. In: *Endomycorrhizas* (Eds.) F.E. Sanders, B. Mosse, and R.B. Tinker, Academic Press, London: 35-51

Hadley, G. (1975). Organisation and fine structure of orchid Mycorrhiza. In: *Endomycorrhizas* (Eds.) F.E. Sanders, B. Mosse, and R.B. Tinker, Academic Press, London: 373-389.

Harley, J.L. (1989). The significance of mycorrhizae. *Mycol. Res.,* 92: 129-139.

Harley, J.L. and Smith, S.E. (1983). *Mycorrhizal symbiosis.* Academic press, London.

Jackson, R.M. and Mason, P.A. (1984). *Mycorrhiza.* Edward Arnold, Ltd., London, ISBN 0-7131-2876-3: 60.

Jamal, A., Ayub, N., Usam, M. and Khan, A.G. (2002). Arbuscular mycorrizal fungi enhance zinc and nickel uptake from contaminated soil by soybean and lentil. *Int. J. Phytoremed.,* 4(3): 202-221.

Jeffries, P., Gianinazzi, S., Perotto, S., Turnau, K. and Barea, J.M. (2003). The contribution of arbuscular mycorrhizal fungi in sustainable maintainance of plant health and soil fertility. *Biol. Fert. Soil*, 37: 1-16.

Johansson, J., Paul, L. and Finlay, R.D. (2004). Microbial interactions in the mycorrhizosphere and their significance for sustainable agriculture. *Microbial Ecol.*, 18: 1-13.

Joshee, N., Mentreddy, S.R. and Yadav, A.K. (2007). Mycorrhizal fungi and growth and development of micropropagated *Scutellaria integrifolia* plants. *Ind. Crop. Prod.*, 25(2): 169-177.

Karthikeyan, B., Jaleel, C.A., Changxing, Z., Zoe, M.M., Srimannarayan, J. and Deiveekansundaram, M. (2008). The effect of AM fungi and phosphorus level on the biomass yield and ajmalicine production in *Catharanthus roseus*. *Eur.Asia J. Bio. Sci.*, 2: 26-33.

Khade, S.W. and Adholeya, A. (2007). Feasibility bioremediation through arbuscular mycorrhizal fungi imparting heavy metal tolerance: a retrospective. *Bioremediation J.*, 11(1): 33-43.

Koide, R.T. (1991). Nutrient supply, nutrient demand and plant response to mycorrhizal infection. *New Phytol.*, 117: 365-386.

Kothari, S.K., Marschner, H. and George, E. (1990). Response of *Citronella java* to VA mycorrhizal fungi and soil compaction in relation to phosphorus supply. *Plant Soil*, 178(2): 231-237.

Link, H.F. (1809). Obrevationes in ordines plantarum naturals. Die Gesellschaft naturforschender Freunde zu Berlin: Magazine für die neuesten Entdeckungen in der gesammten. *Naturkunde*, 3:33.

Link: www.lrz.de/schuessler/amphylo

Liu, R., Li, M., Liu, X.M.X. and Li, X. (2000). Effects of AM fungi on endogenous hormones in corn and cotton plants. *Mycosystema*. 19(1): 91-96.

Mali, B.L., Shah, R. and Bhatnagar, M.K. (2009). Effect of VAM fungi on nutrient uptake and plant growth performance of Soyabean. *Indian Phytopath.*, 62(2): 171-177.

Manoharachary, C., Kunwar, I.K., Reddy, S.V. and Adholeya, A. (2009). Ecological implications and ectomycorrhiza. *Mycol. News*, 21(1): 2-8.

Manoharachary, C., Laxami, A.N. and Kunwar, I.K. (2000). Microbial ecology of polluted soil: some aspects. In: *Glimpses in Botany* (Eds.) K.G. Mukerji, B.P. Chamola and A.K. Sharma, APH Pub. Corp., New Delhi: 314-325.

Mehrotra, V.S. (1993). The so called vesicles in vesicular-arbuscular mycorrhiza are chlamydospores. *Philip. J. Sci.* 122: 377-395.

Mehrotra, V.S. and Baijal, U. (1994). Advances in Taxonomy of vesicular-arbuscular mycorrhizal fungi. In: *Biotechnology in India* (Eds.) B.K. Dwivedi and G. Pandey, Bioved Research Society, Allahabad, India: 227-286.

Melin, E. (1923). *Mykologische unterschungen und Berichte*, 2: 73-331.

Meney, K.A., Dixon, K.W., Scheltema, M. and Pate, J.S. (1993). Occurrence of vesicular arbuscular mycorrhizal fungi in dryland species of Restionaceae and Cyperaceae from south western Australia. *Aust. J. Bot.*, 41(6): 733-737.

Morton, J.B. and Benny, G.L. (1990). Revised classification of Arbuscular Mycorrhizal fungi (Zygomycetes). New order Glomales, two new sub orders Glomineae and Gigasporineae and two new families Acaulosporaceae and Gigasporaceae with emendation of Glomaceae. *Mycotaxon*, 37: 471-491.

Morton, J.B. and Bentivenga, S.P. (1994). Levels of diversity in endomycorrhizal fungi (Glomales, zygomycetes) and their role in defining taxonomic and non-taxonomic groups. *Plant Soil*, 159: 47-59.

Morton, J.B. (1990). Evolutionary relationship among arbuscular mycorrhizal fungi in the Endogonaceae. *Mycologia*, 82: 192-207.

Morton, J.B. and Benny, G.L. (2001). Two new families of Glomales, Archaeosporaceae and Paraglomaceae, with two new genera *Archaeospora* and *Paraglomus*, based on concordant molecular and morphological characters. *Mycologia*, 93: 181-195.

Morton, J.B. and Redecker, D. (2002). Two new families of Glomales, Archaeosporaceae and Paraglomaceae with two new genera, *Archaeospora* and *Paraglomus*, based on concordant molecular and morphological characters. *Mycologia*, 93: 181-195.

Mukherji, K.G. and Dixon, R.K. (1992). Mycorrhizae in forestation. In: *Proc. Int. Symposium on rehabilitation of tropical rain forest ecosystem* (Eds.) N.M. Mahid, I.A.A.Malak, M.Z. Hamzad, and K. Jusoff, Univ. Pert. Malaysia: 66-82.

Nehra, S., Pandey, S. and Trivedi, P.C. (2003). Interaction of arbuscular mycorrhizal fungi and different levels of root-knot nematode on ginger. *Indian Phytopath.*, 56(3): 297-299.

Nicolson, T.H. and Gerdemann, J.W. (1968). Mycorrhiza *Endogone* species. *Mycologia*, 60: 313-325.

Oehl, F. and Sieverding E. 2004. *Pacispora*, a new vesicular arbuscular mycorrhizal fungal genus in the Glomeromycetes. *J. Appl. Bot.*, 78: 72-82.

Ojha, S., Chakraborty, M.R., Dutta, S. and Chatterjee, N.C. (2008). Influence of VAM on nutrient uptake and growth of custard apple. *Asian J. Exp. Sci.*, 22(3): 221-224.

Ortas, I., Harris, P.J. and Rowell, D.L. (1996). Enhanced uptake of phosphorus by mycorrhizal sorghum plants as influenced by forms of nitrogen. *Plant Soil*, 184: 255-264.

Palenzuela, J., Ferrol, N., Boller, T., Azcón-Aguilar C. and Oehl, F. (2008). *Otospora bareai*, a new fungal species in the Glomeromycetes from a dolomitic shrubland in the Natural Park of Sierra de Baza (Granada, Spain). *Mycologia* (in press).

Parkash, V., Sharma, S., Kaushish, S. and Aggarwal, A. (2009). Endomycorrhizal association of khair (*Acacia catecheu* Willd.): a traditionally important tree in Himachal Pradesh. *Mycorrhiza News*, 21(1): 15-18.

Pirozynski, K.A. (1981). Interactions between fungi and plants through the ages. *Can. J. Bot.*, 59: 1824-1827.

Puthur, J.T., Prasad, K.V.S.K., Sharmila, P. and Saradhi, P. (1998). Vesicular arbuscular mycorrhizal fungi improves establishment of micropropagated *Leucaena leucocephala* plantlets. *Plant Cell Tiss. Org.*, 53(1): 41-47.

Rai, M.K. (2001). Current advances in mycorrhization in micropropagation. *In Vitro Cell Dev. Biol.-Plant*, 37(2): 158-167.

Rani, P. Aggarwal, A. and Mehrotra, R.S. (1998). Establishment of nursery technology through *Glomus mosseae*, *Rhizobium* sp and *Trichoderma harzianum* on better biomass yield of *Prosopis cineraria* Linn. *Nat. Acad. Sci., Allahabad*, 68(B), III and IV: 301-305.

Schüßler, A. and Walker, C. (2010). http://www.lrz.de/~schuessler/amphylo/.

Schüßler, A., Schwarzott, D. and Walker, C. (2001). A new fungal phylum, the Glomeromycota: phylogeny and evolution. *Mycol. Res.*, 105(12): 1413-1421.

Selvaraj, T. (1998). *Studies on mycorrhizal and rhizobial symbiosis on tolerance of tannery effluent treated Prosopis juliflora*. Ph. D. Thesis, University of Madras, Chennai, India: 209.

Sharma, S., Aggarwal, A. and Kumar, A. (2008). Biocontrol of wilt of *Albezzia lebbeck* by using AM fungi and *Trichoderma viride*. *Ann. Pl. Protec. Sci.*, 16(2): 422-424.

Sieverding, E. and Oehl, F. (2006). Revision of *Entrophospora* and description of *Kuklospora* and *Intraspora*, two new genera in the arbuscular mycorrhizal Glomeromycetes. *J. Appl. Bot. Food Qual.*, 80: 69-81.

Smith, F.A. and Smith, S.E. (1997). Structural diversity in vesicular-arbuscular mycorrhizal symbiosis. *New Phytol.*, 118: 87-93.

Smith, S.E. (1995). Discoveries, discussion and directions in mycorrhizal research. In: *Mycorrhiza* (Eds.) A. Verma and B. Hock, Springer-Verlag, Berlin: 3-24.

Smith, S.E. and Read, D.J. (1997). Vesicular-arbuscular mycorrhizas. In: *Mycorrhizal symbiosis* 2nd Edn. (Eds.) S.E. Smith and D.J. Read, Academic Press, London: 9-160.

Spain, J.L., Sieverding, E. and Oehl, F. (2006). *Appendicispora*: a new genus in the arbuscular mycorrhiza-forming Glomeromycetes, with a discussion of the genus *Archaeospora*. *Mycotaxon*, 97: 163-182.

Taylor, T.N., Remy, W., Haas, H. and Kerp, H. (1995). Fossil arbuscular mycorrhizae from the early Deovonian. *Mycologia*, 87: 560-573.

Tester, M., Smith, S.E. and Smith, F.A. (1987). The phenomenon of "non-mycorrhizal" plants. *Can. J. Bot.*, 65: 419-431.

Thaxter, R. (1922). A revision of the Endogoneae. *Proc. Am. Acad. Arts Sci.*, 57: 292-348.

Trappe, J.M. (1996). What is Mycorrhiza? In: *Mycorrhizas in integrated systems from genes to plant development*. Proceeding of the 4th European symposium on mycorrhizas.

European Comission, Directorate-General XII, Science, Research and Develpoment, Brussels: 3-6.

Trappe, J.M. and Schenck, N.C. (1982). Taxonomy of the fungi endomycorrhizae. A. Vesicular arbuscular mycorrhizal fungi (Endogonsles). In: *Methods and principles of mycorhizal research* (Ed.) N.C. Schenck, American Phytopathological Socirty, St. Paul, Minn.: 1-10.

Tulasne, L.R. and Tulasne, C. (1844). Fungi nonnulli hipogaei, novi v. Minus cogniti auct. *G. Bot. Ital.*, 2: 55-63.

Van der Heijden, M.G.A., Klironomos, J.N., Ursic, M., Moutoglis, P., Streitwolf-Engel, R., Boller, T., Wiemken, A. and Sanders, I.R. (1998). Mycorrhizal fungal diversity determines plant biodiversity, ecosystem variability and productivity. *Nature*, 396: 69-72.

Walker, C. (1987). Current concepts in the taxonomy of the Endogonaceae. *Procedings of the 7th NACOM. IFAS*, University of Florida, Gainesville, Fla.

Walker, C. (1995). AM or VAM: What's in a word? In: *Mycorrhiza* (Eds.) A. Varma and B. Hock, Springer-Velag, Berlin: 25-26.

Walker, C. and Schüßler, A. (2004). Nomenclatural clarifications and new taxa in the Glomeromycota. *Mycol. Res.*, 108: 979-982.

Walker, C., Vestberg, M. and Schüßler, A. (2007b). Nomenclatural clarifications in Glomeromycota. *Mycol. Res.*, 111: 253-255.

Walker, C., Vestberg, M., Demircik, F., Stockinger, H., Saito, M., Sawaki, H., Nishmura, I., and Schüßler, A. (2007a). Molecular phylogeny and new taxa in the Archaeosporales (Glomeromycota): *Ambispora fennica* gen. sp. nov., Ambisporaceae fam. nov., and emendation of *Archaeospora* and Archaeosporaceae. *Mycol. Res.*, 111: 137-13.

Wang, F., Lin, X. and Yin, R. (2005). Heavy metal uptake by arbuscular mycorrhizas of *Elsholtzia splendens* and the potential for phytoremediation of contaminated soil. *Plant Soil*, 269: 225-232.

Wright, S.F. and Upadhyaya, A. (1996). Extraction of an abundant and unusual protein from soil and comparision with hyphae protein from arbuscular mycorrhizal fungi. *Soil Sci.*, 161: 575-586.

Wright, S.F. and Upadhyaya, A. (1998). A survey of soils for aggregate stability and glomalin, a glycoprotein produced by hyphae of arbuscular mycorrhizal fungi. *Plant Soil*, 198: 97-107.

Wu, Chi-Guang, Liu Yen-Sher, Huang Ling-Ling, Wang-Yin Po and Chao Cheng Ching. (1995). Glomales of Taiwan: V. *Glomus chimnobambusae* and *Entrophospora kentinensis* spp. Nov. *Mycotaxon*, 53: 289-294.

❑❑❑

Microbial Diversity and Functions, 2012

New India Publishing Agency, New Delhi (India)
E-mail : info@nipabooks.com; Website : www.nipabooks.com

Chapter 4

Diversity of PGPR Microbes in Pesticides Degradation

V.C. Khilare, S.N. Kanade, A.B. Ade and L.V. Gangawane

ABSTRACT

Pesticide degradation is the breaking down of toxic pesticides into a non-toxic compounds and, in some cases, down to the original elements from which they were derived. The most common type of degradation is carried out in the soil by microorganisms, especially the fungi and bacteria. Microbial degradation depends not only on the presence of microbes with the appropriate degradative enzymes, but also on a wide range of environmental parameters. Pesticides which are rapidly degraded are called non-persistent while those which resist degradation are termed as recalcitrant (persistent). Microorganisms are able to degrade a large variety of compounds, including pesticides under laboratory conditions. However, methods have yet to be developed to decontaminate the environment from pesticide residues. Continuous efforts are required in this direction, and at present several PGPR capable of degrading pesticides have been isolated from the natural environment. Catabolic genes responsible for the degradation of several xenobiotics, including pesticides, have been identified, isolated, and cloned into various other organisms such as Streptomyces, algae, fungi, etc. In addition, recombinant DNA studies have made it possible to develop DNA probes that are being used to identify microbes from diverse environmental communities with a unique ability to degrade pesticides. Bioremediation treatment still requires more intense research through specific microbial degradation of broad range of organic pollutants. This review describes recent advances in biodegradation of pesticides by addressing the microbes involved in pesticide degradation.

Keywords: Pesticide degradation, Bioremediation, PGPR.

Rhizosphere is the zone of soil where microorganisms are stimulated by plant roots. The rhizosphere bacteria those help plant growth in various ways are designated as Plant Growth Promoting Rhizobacteria (PGPR). They support seed germination, increase plant growth, produce nodules on legumes, enhance nitrogen fixation and also suppress crop diseases. Simultaneously the use of pesticides has also become an integral and economically essential part of agriculture. It is essential when resistance fails in the host plant and disease infection spoils the crop in the field. At this moment one has to advocate the use of pesticide for the management of disease or pest. Pesticide production in India was 85338 MT in 2008-09. But consumption has been declined up to 42378 MT during 2005-2006. Cotton and paddy sharing 44.5 % and 22.8 % consumption of pesticides respectively in India. India's consumption of pesticides per hectare is low when compared with world averages - 0.380 kg/ha against Korea's 16.56 kg/ha and Japan's 10.80 kg/ha. India spends $3 /ha on pesticides compared with $24/ha spent by Philippines, $225/ha by South Korea and $633 by Japan. Based on the data on consumption of pesticides for the year 2007-08, five States *viz.*, Uttar Pradesh, Punjab, Haryana, Maharashtra and Tamil Nadu consumed more than 3,000 MT pesticides annually. Presently a total of 221 pesticides belonging to different groups are registered under section 9(3) of the insecticide act 1968 (Dureja and Gupta, 2009).

Pesticides are indispensable to the modern agriculture. Currently among various groups that are being used world over, organophosphate form a major and most widely used group *i.e.* more than 36% of the total world market. Malathion, phorate, parathion, monocrotophos, dimethioate are some of the widely used organophosphorus pesticides (Kanekar *et al.*, 2004). Organophosphate pesticides account for about one half of the insecticides used. These are active against a broad spectrum of insects and are used for food crops as well as in residential and commercial buildings and on ornamental plants and lawns (National Report on Human Exposure to Environmental Chemicals, U.S.). There are almost 900 different bug killers that have been used in the U.S. out of which 37 belongs to class insecticides (organophosphates).

Hazardous Effects of Organophosphates

There are different hazardous effects of organophosphates, if these are consumed or come in contact with human body. Different harmful symptoms have been quoted like mild poison such as runny nose, chest tightness, shortness of breathing, stomach cramp, muscle twitching, confusion etc. (Antonella *et al.*, 2001). The excessive use of pesticide leads to an accumulation of huge amount of residues in the environment which causes the substantial health hazard for the current and future generation due to uptake and accumulation of these toxic chemicals in food chain and drinking water (Mohamed, 2009). A recent study of University of Washington, 2002, found that 109 out of 110 urban and suburban children have pesticide in their urine samples. Pesticides are suspected to cause the neurological defects in children through damage to the developing fetus. Human Brest milk and amniotic fluid are contaminated in many women due to wide spread use of pesticide and herbicide in all over the world. Young children are particularly susceptible to damage from these potent chemicals due to exposure. In adults also pesticides are found to be dangerous. Out of the 26, 12 pesticides are classified as carcinogen by EPA. Pesticides are responsible for certain types of cancer, Parkinson's disease, birth defects, miscarriages and sperm abnormality (Elizabeth *et al.*, 1998; Hayes, 1982). The worldwide death and chronic illness due to pesticide poisoning number is about 1million/ year (EHP, 1999).

Degradation by Microbes

A review of Bollen (1961) suggests the interaction between pesticide and soil microorganisms disturbs the balance of soil microorganisms, however, several effects to the microbes are not observed. Soil microorganisms show an early warning about soil disturbances by frozen chemicals hence they serve as pollution indicators of the soil. This may be used as one of the parameters of soil pollution (Choudhary *et al.*, 2008). In order to study the residual pesticide in the agricultural field, the multi residual analytical method was developed especially for organophosphates (Consalter and Guzzo, 1991). Kannan and Job (2007) also projected their studies on the residual level of pesticide pollution in the water reservoir. To maintain the appropriate residual level of pesticide in the soil, the pesticide degradation practices were found to be important. The mechanism of pesticide degradation in compost was described by Craig (1999, 2000). In his findings, he involved some steps such as composting, adsorption, humification, biological transformation and volatilization. Application of microbes in pesticide degradation was widely studied by many workers by using diversified microbes (Table 1). Reed *et al.* (1987) has given an account for characterization of microorganisms in soil which accelerates the process of pesticide degradation. With modified soil condition, pesticide degradation carried out using microbes where glyphosate, atrazine, acetochlor and metachlor were tasted under laboratory condition. The pesticide degradation in the soil can also be carried out by PGPR. These bacteria are capable of promoting plant growth by colonizing the plant roots which were mainly used for the uptake of the nutrients or preventing plant disease. Myresiotis *et al.* (2009) have highlighted the degradation of pesticide by PGPR and their effects on bacterial growth where *Bacillus amyloliquifaciens* and *B. pumilus* were characterised for the degradation of broad range of pesticide.

Microbes select a wide range of pesticides during degradation. Chacko *et al.* (1966) showed that, the microbes especially the actinomycetes and the filamentous fungi are responsible for degradation of chlorinated pesticides such as DDT. Felsot *et al.* (2003) have given the review about the disposal and degradation of the pesticide waste, where they have described that pesticide can be degraded by oxidation process. They have focused on the composting and land forming activity for the same. Panagiotis, *et al.* (2005) have highlighted the environmental degradation of two pesticides *i.e.* azinphos methyl and parathion methyl where they have stated that temperature and relative humidity are responsible for the degradation of these pesticides. Alexander (1981) has described the microbial degradation of dichlorvous (organophosphorous insecticide) in detail. It was carried out by *Pseudomonas* and *Bacillus*. TLC, GLC and mass spectral analysis was done to characterize the metabolites released in degradation. Pesticide degradation in soil was described by Blum Horst (1996) which provides information on its persistence, degradation, volatilization and mineralization. The organophosphate pesticide ethion was subjected to microbial degradation by using species of *Pseudomonas* and *Azospirillum* by Australian workers (Foster *et al.*, 2006). Microbial degradation of chorphyrifos was described by Bhagobaty *et al.* (2007). In their review, they have cited the work of many authors regarding microbiological transformation of chlorpyrifos and their metabolites to various systems. Degradation of methyl by the novel bacterial strain *Stenotrophomonas maltophila* M1 was described by Mohamed (2009).

Degradation of Malathion

Matsumura and Bousch (1966) revealed the Malathion degradation property of *Trichoderma viride* and *Pseudomonas* spp. highlighting the activity of carboxyesterase enzyme. Soil degradation of Malathion was studied by Konrad *et al.* (1969) where they have shown that Malathion degradation was rapid in both sterile and non sterile soil system and no lag phase was found prior to degradation. Malathion degradation by *Arthrobacter* spp. was described by Walker and Stojonovic (1974) where, bacterium is most efficient for Malathion utilization and it degrades in to several derivatives. This was proved using GLC, TLC and IR spectroscopy. In water the kinetic mechanism of Malathion degradation was worked out by Wolfe *et al.* (1975). Degradation of Malathion by salt marsh microorganisms was described by Bourquin (1977) in which Malathion was used as sole source of carbon due to the phosphatase activity. Degradation of Malathion on wheat and corn moisture content was seen by Kadoum and La Hue (1979) who have obtained residue data during a 12 month period from corn and wheat of various moisture levels. White and Nowicki (1985) studied the effect of temperature and duration of Malathion residue in dry rape seed. Degradation of Malathion along with other insecticides in sea water was studied by Cothan and Bidleman (1989). Similarly, Malathion degradation in the grain drying system was observed by Wintersteen and Foster (1992), where they have revealed that the malathion was found to be degraded on stored grain with respect to specific moisture content and temperature. Some other researchers showed that the degradation of malathion was possible in thermally generated aerosols (Brown *et al.*, 1992).The decay of the pesticides including malathion in apple samples was worked out by Barrio *et al.* (1995) using gas chromatography. The experimental biofilm of microbes was prepared for the Malathion degradation which involved mixed bacterial culture of three different strains of *Micrococcus* spp. and two strains of *Pseudomonas* spp. by Kumari *et al.* (1998). In wheat germ the degradation of Malathion and phenthoate by glutathione reductase was studied by Yoshi *et al.* (2000). The comparative degradation study of Malathion in a biosimulator with the utilization of *Pseudomonas* spp. and indigenous microbes with varied dissolved oxygen concentration was studied by Hashmi and Kim (2003). The reaction products were analyzed by gas chromatography with various parameters such as pH, DO, COD etc. Hashmi *et al.* (2004) have given the account of Malathion degradation by *Pseudomonas* using activated sludge treatment system. This degradation was measured by the estimation of dissolved oxygen and by HPLC techniques. Biodegradation by *Fusarium oxysporum* was described by Kim *et al.* (2005) where citation of the functions of two lipolytic enzymes fungal cutinase and yeast esterase were pointed out. The degradation of Malathion on stored maize and wheat grains was worked out by Rowlands (2006) where, degraded products of Malathion from wheat and maize grains for six months and degradation products were identified by TLC. Xuliang Zhuang *et al.* (2007) put forth new advances in PGPR for bioremediation. The progress of PGPR remediation of soils contaminated with pesticides. Role of various micronutrients Malathion degradation was worked out by Francis and Gangawane (2009). Biodegradation and detoxification of Malathion by *Bacillus thuringiensis* was noted by Zeinat Kamal *et al.* (2008) where the efficiency of *B. thuringiensis* MOS-5 (Bt) isolated from agricultural waste water contaminated with Malathion in Egypt.

Degradation of Phorate

Degradation of phorate in laboratory incubation studies with soils of varying physico-chemical characteristics. Phorate was more persistence in non-flooded soils than in flooded soil. Phorate sulphoxide was recovered as the only metabolite of phorate in non-flooded soils while three metabolites (diethyl, dimethiophosphate, triethyl dithiophosphate and unidentified metabolites) were formed in flooded soils. The study indicated that, non-flooded soil phorate was degraded *via* oxidation while in flooded soils hydrolysis is the major degradation process. Photodegradation of phorate was also studied by Sharma and Gupta (1994). The photolysis of phorate has been studied as a thin film on a glass surface and in a solution of methanol, water (60:40) by UV light. The rate of disappearance of phorate in the solution was characterized by NMR and MS. The degradation of phorate due to temperature was worked out by Patterson and Rawlins (1968). The persistence, degradation and bioactivity of phorate was studied by Getzin and Shanks (1970) by using TLC, gas liquid chromatography and insect bioassay. Degradation of phorate was accelerated by an increase in temperature by repeated free exposure of soil and application of microorganisms in their transformation (Singh *et al.*, 2003). The degradation pattern of soil applied insecticides was described season wise by Bhuvaneshwari *et al.* (2008). The sorption and degradation of phorate was examined using 14 soil samples collected at different depth from two soil types which differed in their organic matter and clay mineralogy. The sorption of phorate as measured by the distribution coefficient was greater for the Egmant than for the Tokomaru soil and decreased with depth for both soils (Bolan and Baskaran, 1997).

The environmental fate of phorate and its metabolites, phorate sulphone and phorate sulphoxide was examined in mesocosmos placed in South Dacota Wetlands. The metabolites of phorate increased significantly in the water during the study (Dieter *et al.*, 1995). Persistence and degradation of phorate along with other organophosphates in the soil and their uptake by carrots was worked out by Suett (1999). Decomposition of phorate in aqueous solution by photolytic ozonation was carried out by Ku and Lin (2002) using various experimental conditions. The initial step of photolytic decomposition of phorate was considered to be the breakage and subsequent oxidation of various organic intermediates. The other chemicals were also found active for the degradation of phorate. Doong and Chang (1997) showed that degradation of phorate can be done by hydrogen peroxide which was mediated by photo assisted titanium dioxide. A study by Burns (1971) in California, USA confirms that organic soil adsorbs more phorate than clay soil and clay soils more than sand soils. Another study by Fuhremann and Licstenstein (1980) conducted in Wisconsin, USA and summarized in the review and showed that phorate was more persistent in a silt clay loam of organic matter 4.7 % than a sand 6 % organic matter. Along with this, the degradation of phorate was reported due to the soil microbes by Kadam and Gangawane (2005). The effect of environmental factors and soil types on the phorate degradation was also studied by Menzer *et al.* (1970).

Pesticide degradation is analysed by various methods. These include the pesticide degradation in the liquid (water) and solid (soil). To implement the pesticide degradation naturally, application of pesticide degrading microbes was described by number of authors. The PGPR properties were also utilized for pesticide degradation. Response of *Azospirillum brasilience* to the pesticides, bromopropionate and methidion on chemically defined media and dialized soil media was characterized by Gomez *et al.* (1998). Degradation of oxyfluorfen by *Azotobacter chroococcum* was shown by Chakraborty *et al.* (2002). Mrkovaaki *et al.* (2002) have highlighted the pesticides effects on *Azotobacter chroococcum* using four different pesticides, inhibitory effects were observed at higher concentration. Biodegradation of certain organic floatation reagents such as dodecyl amine, diamine, sodium isopropyl xanthate and sodium oleate from alkaline solutions was done by *Bacillus polymyxa* (Dev and Natarajan, 1998). Langlois *et al.* (1970) have studied some factors affecting degradation of organochlorine pesticides by number of bacteria such as *Bacillus cereus, B. coagulans, B. subtilis, Escherichia coli* and *Enterobacter aerogenes.* The bacteria, *Rhizobium* and *Bradyrhizobium* showed the degradation of organo-phhosphorous pesticides due to the enzymes phophodiesterase and phosphotriesterase. It is essential to emphasize on plant growth promoting rhizobacteria and their interactions with the plants. (Abd- Alla, 1994). Nezarat and Gholami (2009) screened plant growth promoting bacteria, *Pseudomonas putida, P. fluorescens, Azospirillum lipoferum* and *A. brasiliense* for enhancing seed germination, seedling growth and yield of maize plant.

Vanloon (2007) highlighted the plant responses to plant growth promoting rhizobacteria. PGPR were also found beneficial for phyto-stabilization of mine tailings where the re-vegetation of an extremely acidic high metal content tailing sample require 15 % compost amendment for normal plant growth. It has been minimized due to PGPR (Christopher *et al.*, 2008). The antioxidant states, photosynthesis enhancement, mineral uptake and growth under saline soil was enhanced in lettuce plant PGPR (Han and Lee, 2005). There are several PGPR such as nitrogen fixers like, *Rhizobium, Azotobacter, Azospirillum, Pseudomonas* and *Bacillus* are known for the production of hormones for the growth of plants as well as they are also acting as biocontrol agents against soil pathogens (Yang *et al.*, 2009). *Rhizobium* is a nitrogen fixer associated with the legumes is also involved in the pesticide degradation was isolated and characterised by many workers (Ade 2004). *Azotobacter* and *Azospirilllum* was isolated for the pesticide degradation by Kadam (2002). *Pseudomonas* spp. (Ade, 2004), *Bacillus polymyxa* (Stansly and Schlosser, 1947).

Isolation and characterization of *Pseudomonas oliovorans* degrading the chloroacetamide herbicide ocitochlor was done by Zun Xu *et al.* (2006).

Isolation and characterization of *Azotobacter* and *Azospirillum* strains from the sugarcane rhizosphere was done by Tejera *et al*, (2005). Taxonomic characterization and plant colonizing ability of some bacteria related to *Bacillus* was done by Reva *et al*, (2004) in which phylogenetic relationship of 17 *Bacillus* strains isolated from plant and soil were determined on the basis of 16s rRNA, gyrase A (gyrA) and the cheA histidin kinase. Isolation and functional-structural characteristics of *Bacillus polymyxa* were worked out by Iashenko (2001). Isolation and characterization of phorate degrading bacteria were isolated from phorate contaminated sites. *Ralstonia eutropha* strain AAJ1 isolated from soil was found to degrade phorate up to 85% in 10 days in liquid medium Rani et *al*, (2009).

Reed *et al*, (1987) characterized the microorganisms in soil exhibiting accelerated pesticide degradation. Nesmith and Jenkins (1978) have given new selective medium for monitoring population of *Pseudomonas solanacearum* in naturally and artificially infested soils. This medium was derived by modification of a standard triphenyl tetrazolium chloride medium. The characterization of *P. fluorescens* by isolating its secondary metabolites was carried out by Reddy and Reddy (2009). They have isolated 20 *P. fluorescens* strains rice growing soil samples and characterized one of the isolate was identified from the dual culture test. It was fermented for secondary metabolite in a small scale and extracted with ethyl acetate. The structure of the compound was eluicited by high resolution NMR spectroscopy. *Rhizobium* was isolated in rhizosphere flora of *Cerasus sachaliensis* by Yu *et al*. (2007) along with other bacterial genera such as *Bacillus, Pseudomonas* and *Flavobacterium*. These were also associated with actinomycetes, *Mucor, Aspergillus* and *Penicillum*. Identification of Rhizobial strains was also carried out by Johnstan and Beringer (1975) from pea root nodules using genetic markers. The *Rhizobium tropici,* nodulating *Faciolatus vulgaris* beans and *Leucaena* sp. tree was isolating using multilocus enzyme electrophoresis, DNA-RNA hybridization, 16s rDNA analysis and phenotypic characteristics by Romera *et al*, (1991). Molecular biodiversity and identification of free living Rhizobial strains from diverse Egyptian soil by direct isolation without trap hosts was carried out by 16s rDNA analysis by Shamseldin *et al*. (2008). Isolation and characterization of *Azotobacter* species for the production of polybetahydroxy alkanoate was done by Quagliano *et al*. (1994).

PGPR are able to degrade the pesticide both in water and soil because of the enzyme system they have. For organophosphate pesticides two well-characterized enzymes were found to be responsible. The organophosphate biodegradation in anaerobic bacteria was described using immobilized enzymatic activity by Haddane *et al*, (1991) in which two organophosphorus compounds, parathion and paraxon were studied in a biphasic reactor stimulating an anaerobic media highly charged in organic matter. The first hydrolytic steps were rapid and hydrolysing factors (phosphatase) was highly bound to the soil phase. The phosphatase was chromatographied to apparent purity and its kinetic constant and amino acid composition was determined. Spatial distribution of soil phosphatase activity within a riparian forest was worked out by Amador *et al.*, (1997) where phosphatase activity was considered as indicator of soil quality in which it was co-related with physical and chemical properties of the soil. Phosphatase activities in pesticide treated, growing and developing cells of *Dictyostelium discoideum* were described by Gayatri and Chattarjee (2006) where the effects of BHC (Benzene Hexa Chloride) were seen for the phosphatase activity, which had been co related with cytotoxicity with alkaline phosphotages. Influence of pesticide presence on phosphatase activity and selected microbial biota of a natural lake system was worked out by Lopez *et al*, (2006) where they have studied the phosphatase degradation status of aldrin, lindane, dimethoate and methyl parathion along with herbicide altrazine and fungicide captan significantly increased phosphatase activity after 28 days of incubation.Another phosphatase activity was highlighted by Bhalerao and Puranik (2009) during microbial degradation of monocrotophos by *Aspergillus oryzae*, where the phosphatase activity was verified by analysing degradation process through HPTLC and FTIR methods.

Along with the phosphatase, another enzyme, carboxyesterase was also characterized for the degradation of organophosphate insecticides. Degradation of Thifensulfuron methyl in soil by carboxyesterase activity was studied by Brown *et al.*, (1977) which was found to degrade rapidly in diverse non-sterile agricultural soil. Carboxyesterase activity was used for bioremediation of organophosphorus pesticides was determined by Zhang *et al*, (2006) in which the carboxyesterase enzyme was detected from mosquito surface. Zhang *et al*, (2006) have given the account of decontamination of the vegetables spread organophosphate by organophosphorus hydrolase and carboxyesterase (B1). For that a genetically engineered *Escherichia coli* cells was used for the expression for these enzymes. The product of the pesticide hydrolysis with the treatment of these enzyme extracts were determined to be virtually non-toxic. Characterization of carboxyesterase from Malathion degrading bacterium *Pseudomonas* was carried out by Singh *et al.* (1989).

In order to apply the activity of these enzymes for contaminated sites, several workers have given their ideas in various publications (Hutchinson *et al.*, 1993, Lee *et al.*, 1998). Enzyme activity and degradation of organic micromolecules by neustonic and planktonic bacteria in Estuarine Lake was worked out by Mudryk and Skorczewski (2006). Co-expression of two detoxifying pesticide degrading enzymes was studied in a genetically engineered bacterium. The vector pETDuet was designed for the co-expression of the two target genes. This enzyme was characterized as organophosphate hydrolase and carboxyesterase (Lan *et al.,* 2006). The genes governing the expression of pesticide degrading enzymes were characterized by several authors. In most of the cases these were found on the bacterial plasmids. Fisher *et al.* (1978) discovered the plasmid pJP-1 from *Alcaligenes paradoxus* was responsible for pesticide degrading activity.

Kumar *et al.* (1996) have given the critical review in microbiology article regarding molecular aspects of pesticide degradation by microorganisms. They reviewed pesticidal degradative genes located on plasmids, transposones, and/ or on chromosomes. Expression of the 2,4-D degrading plasmids pJP-4 of *Alcaligenes eutrophus* in *Rhizobium trifoli* which was transferred by conjugation, the transferred *Rhizobium* could show 2,4-D degradation and degradation products were confirmed by Gas Chromatographic analysis (Feng *et al.,* 2004). Cloning and expression of plasmid gene, encoding resistance to chromate and cobalt in *Alcaligenes eutrophus* was worked out by Nies *et al,* (1989). *Rhizobium,* carbaryl was found to have carboryl hydrolase genes which encode for the enzyme esterase (Hashimoto *et al.,* 2002). Cloning of the organophosphorus pesticides hydrolase gene clusters of seven degradative bacteria isolated from a methyl parathion contaminated site and the evidence of their horizontal gene transfer was worked out by Zhang *et al,* (2006).

The bacterium was isolated from the intestine of the fresh water fish *Labeo rohita* by an enrichment technique and plasmid mediated dimethioate degradation by *Bacillus licheniformis* was worked out by Mandal *et al,* (2005). The bacterial genes were transferred regarding pesticide degradation in common soil fungus *Gliocladium virens* by Xu *et al,* (1996) successfully. Molecular cloning and expression of the 3 chlorobenzoate degrading gene from *Pseudomonas* was done by Weisshaar *et al,* (1987). The hydrolase genes were recorded from so many bacterial genera such as, *Pseudoaminobacter, Acromobacter, Brucella, Ochrobacterium, Plesiomonas,* etc. by Zhang *et al,* (2005). Genes encoding triazine degradation were reported as plasmid born in *Klebsiella pneumoniae* by Karns and Eaton (1997). As pesticide degradation was shown by many bacterial strains as reported by several workers, the applications of these strains were also formulated in the form of sequential steps in the bioreactor technology.

A model describing pesticide by availability and biodegradation in the soil which account for sorption to the soil surfaces, diffusion into the internal matrix of soil organic matter for aggregates and microbial growth was proposed by Shelton and Doherty (1997). The effect of initial concentration, co-application and separated application on pesticide degradation in biobed mixture was studied by conducting the experiments and for the degradation of chlorpyriphos and metalaxyl by Vischetti *et al.* (2008).

Table 1 : Diversified microbes involved in pesticides degradation

Pesticide	Microorganisms involved	Reference
Aldrin	*Bacillus* sp.	Tu *et al.*, 1968
	Corynebacterium sp.	Tu *et al.*, 1968
	Micrococcus sp.	Patil *et al.*, 1970
	Micrococcus sp.	Tu *et al.*, 1968
	Nocardia sp.	Tu *et al.*, 1968
	Pseudomonas sp.	Patil *et al.*,1970
	Streptomyces sp.	Tu *et al.*, 1968
	Thermoactinomyces sp.	Tu *et al.*, 1968
γ- BHC (Lindane)	*Aerobacter aerogenes*	
	Bacillus cereus	Meksongsee and Guthrie, 1965
	B. megaterium	Meksongsee and Guthrie, 1965
	Citrobacter freundii	Jagnow *et al.*, 1977
	Clostridium sp.	Raghu and MacRae, 1966
	C. rectum	Jagnow *et al.*,1977; Ohisa and Yamguchi 1978; Ohisa *et al.*,1980
	Escherichia coli	Francis *et al.*, 1975; Vonk and Quirijins 1979
	Mixed microbial culture	Engst *et al.*, 1979
	Pseudomonas fluorescens	Mekosongsee and Guthrie, 1973
	P. putida	Benezet and Matsumura,1973
	Soil micro organism	Tu, 1975
DDT	Actinomycetes	Chacko *et al.*, 1966
	Aerobacter aerogenes	Wedemeyer, 1966, Johnson *et al.*, 1967, Plummer *et al.*, 1968
	A. tumifaciens	Johnson *et al.*, 1967
	Arthrobacter sp.	Patil *et al.*, 1970
	Bacillus sp.	Patil *et al.*, 1970
	B. cereus	Johnson *et al.*, 1967, Plummer *et al.*, 1968, Langlois *et al.*, 1970
	B. coagulans	Langlois *et al.*, 1970
	B. megaterium	Plummer *et al.*, 1968
	B. subtilis	Johnson *et al.*, 1967, Langlois *et al*, 1970
	Clostridium pasteurianum	Johnson *et al*, 1967
	C. michiganense	Johnson *et al.*, 1967
	Enterobacter aerogenes	Langlois *et al*, 1970
	Erwinia amylovora	Johnson *et al.*, 1967
	E. ananas	Johnson *et al.*, 1967
	E. carotovora	Johnson *et al.*, 1967
	E. chrysanthemi	Johnson *et al*, 1967
	Escherichia coli	French and Hoppingarner, 1970, Langlois *et al.*, 1970
	Hydrogenomonas sp.	Focht and Alexander, 1970, 1971

	Klebsiella pneumonie	Wedemeyar, 1966
	Kurthia zopfii	Johnson *et al.*, 1967, Langlois *et al.*, 1970
	Nocardia sp.	Chako *et al.*, 1966,
	Proteus vulgaris	Barker *et al.*, 1965
	Pseudomonas sp	Patil *et al.*, 1970
	P. fluorescens	Meksongsee and Guthrie, 1965
	Sewage sludge microorganisms	Albone *et al.*, 1972
	Streptococcus sp.	Ledford and Chen, 1969
	S. anamoneus	Chacko *et al*, 1966
	S. auerofaciens	Chako *et al.*, 1966
	S. viridochromogenes	Chako *et al.*, 1966
	Xanthomonas pruni	Johnson *et al.*, 1967
	X. stewarti	Johnson *et al.*, 1967
	X. uredororus	Johnson *et al.*, 1967
	X. vesicatoria	Johnson *et al.*, 1967
Dieldrin	*Aerobacter aerogenes*	Wedemeyar, 1968
	Arthrobacter sp.	Jognow and Hyder, 1972
	Corynebacterium sp.	Jognow and Hyder, 1972
	Mocrococcus sp.	Jagnow and Hyder, 1972
	*Microbacterium*sp.	Jagnow and Hyder, 1972
	Nocardia sp.	Jagnow and Hyder, 1972
	Pseudomonas sp.	Jagnow and Hyder, 1972
	P. melophthora	Bousch and Matsumura, 1967
	Soil microorganisms	Patil *et al.*, 1970
Endosulfan	*Rhodococcus* sp.	Gaur and Saraswat, 1995
	Soil bacteria	Sutherland *et al.*, 2000
	Azotobacter chroococcum	Balaji and Mahadevan, 1999
	Rhizobium sp.	Gangawane and Francis, 1997
Endrin	*Aerobacter aerogenes*	Meksongsee and Guthrie, 1965
	Arthrobacter sp.	Patil *et al.*, 1970
	Bacillus sp.	Meksongsee and Guthrie, 1965
	B. megaterium	Meksongsee and Guthrie, 1965
	Micrococcus sp.	Patil *et al.*, 1970
	Pseudomonas sp.	Patil, 1970
	P. fluorescens	Meksongsee and Guthrie, 1965
Heptachlor	*Arthrobacter* sp.	Miles *et al.*, 1969
	Bacillus sp.	Miles*et al.*, 1969
	Coryneobacterium sp.	Miles *et al.*, 1969
	Mocromonopora sp.	Miles *et al.*, 1969
	Nocardia sp.	Miles *et al.*, 1969
	Soil microorganisms	Carter and Sringer, 1970

	Streptomyces sp.	Miles *et al.*, 1969
	Thermoactinomyces sp.	Miles *et al.*, 1969
Toxaphene	*Bacillus cereus*	Meksongsee and Guthrie, 1965
	B. megaterium	Meksongsee and Guthrie, 1965
	Pseudomonas fluorescens	Meksongsee and Guthrie, 1965
	Aerobacter aerogenes	Meksongsee and Guthrie, 1965
Bromophos	Soil microorganisms	Stenesen, 1969
Diazinon	*Arthrobacter* sp.	Gunner and Zuckerman, 1968 Sethunathan and Pathak, 1972
	Flavobacterium sp.	Sethunathan and Yoshida, 1973
	Mixed microbial culture	Adhya *et al.*, 1981, Munnecke, 1976
	Pseudomonas sp.	Rosenberg and Alexander, 1979
	P. melophthora	Bousch and Matsumura, 1967
	Soil microorganisms	Getzin, 1967, Bro-Rasmussen *et al.*, 1968, Gunner and Zukerman, 1968, Hsu and Bartha, 1979
	Streptomyces griseus	Gunner and Zuckerman, 1968
Dichlorvos	*Psedomonas melophthora*	Bousch and Matsumura, 1967
Dimethoate	*Aspergillus niger ZHY 256*	Liu *et al.*, 2001
	Rhizobium sp.	Gangawane and Francis, 1997
	Soil microorganisms	Congregado *et al.*, 1979
Fenitrothin	*Bacillus subtilis*	Miyamoto *et al.*, 1966
	Mixed microbial culture	Munnecke, 1976
	Soil microorganisms	Spillner *et al.*, 1979
Malathion	*Azotobacter chroococcum*	Athar *et al.*, 1998
	Esturine microorganisms	Munnecke, 1976
	Flavobacterium sp.	Sethunathan and Yoshida, 1973
	Pseudomonas sp.	Matsumura and Bousch, 1966, Rosenberg and Alexander, 1979
	Rhizobium sp.	Mostafa *et al.*, 1972
	Soil microorganisms	Spillner *et al.*, 1979
	Trichoderma viride	Matsumura and Boush, 1966
Methyl parathion	*Bacillus subtilis*	Miyamoto *et al.*, 1966
	Flavobacterium sp.	Adhya *et al.*, 1981
	F. balsutinm	Manavati *et al.*, 1999
Parathion	*Flavobacterium* sp.	Sethunathan and Yoshida, 1973; Adhya *et al.*, 1981
	Mixed microbial culture	Munnecke, 1976, Munnecke and Hsieh, 1974
	Pseudomonas sp.	Sidramappa *et al.*, 1973; Munnecke and Fischer, 1979; Adhya *et al.*, 1981; Talbot *et al.*, 1982

	P. aeruginosa	Gibson and Brown, 1974
	P. diminuta	Serdar *et al.*, 1982
	P. melophthora	Boush and Matsumura, 1967
	P. stutzai	Daughton and Hsiegh, 1977
	Rhizobium japonicum	Mick and Dahm, 1970
	R. meliloti	Mick and Dahm, 1970
	Soil microorganisms	Barik and Sethunathan, 1978, Hsu and Bartha, 1979; Barik *et al.*, 1979; Ferris and Lichtenstein, 1980.
Phorate	*Chlorella pyrenoidosa*	Ahmed and Casida, 1958
	Pseudomonas fluorescens	Ahmed and Casida, 1958
	Soil microorganisms	Getzin and Shanks, 1970
	Thiobacillus thiooxidans	Ahmed and Casida, 1958
	Torulopsis utilis	Ahmed and Casida, 1958
	Azotobacter chroococcum	Kadam, 2002
Carbaryl	*Aspergillus flavus*	Liu and Bollag, 1971
	A. terreus	Liu and Bollag, 1971
	Gliocladium roseum	Liu and Bollag, 1971
	Micrococcus sp.	Ninnekar and Doddmani, 2001
	Pseudomonas cepacia	Venkateshwarlu *et al.*, 1980
	P. melophthora	Boush and Matsumura, 1967
	Pseudomonas sp.	Larkin and Day, 1986
	Rhizobium sp.	Gangawane and Francis, 1997
	Soil microorganisms	Liu and Bollag, 1971, Rodriguez and Dorough, 1977
Carbofuran	*Pseudomonas cepacia*	Venkateswarlu *et al.*, 1980, Caro *et al.*, 1973, Venkateswarlu and Sethunathan, 1978
Mancozeb	*Azotobacter chroococcum*	Athar *et al.*, 1998
Metalaxyl	*Soil bacteria*	Mahopatra and Awasthi, 1995
Metalachlor	*Bacillus circulans*	Sexena *et al.*, 1987
Zectran	*Soil microorganisms*	Benezet andMatsumura, 1973
Anilophos	*Micrococcus* sp.	Gowrisanker *et al.*, 2000
Atrazine	*Bacillus* sp.	Gowrisankar *et al.*, 2000
	Proteus sp.	Gowrisankar *et al.*, 2000
	Pseudomonas sp.	Gorisankar *et al.*, 2000, Topp *et al.*, 2000, Behki and Khan, 1986
2,4-D	*Alcaligenes eutrophus*	Perkins and Lurquin, 1988
	Azotobacter chroococcum	Gahlot and Narula, 1996, Balajee and Mahadevan, 1990, Athar *et al.*, 1998
á -2,6-dichlorobenzalamine	*Pseudomonas putrefaciens*	Milbarrow, 1964

3-Chorobenzoate	*Alkaligenes faecalis B 16*	Grover *et al.*, 1993
	Pseudomonas putida	Chatterjee *et al.*, 1981, Grover *et al.*, 1993
	Pseudomonas sp.*DP 5*	Grover *et al.*,1993
	P. paucimobilis	Grover *et al.*, 1993
4-Chlorobenzoate	*Pseudomonas* sp.	Sahashrabudhe and Modi, 1991
5-Chlorosalicylate	*Bacillus brevis*	Crawford *et al.*, 1979
4-Chlorobenzoic acid	*Arthrobacter* sp.	Marks *et al.*, 1984
4-Chloro-phenoxyacetate (4-CPA)	*Pseudomonas* sp.	Evans *et al.*, 1971b
	Aspergillus niger	Faulkner and Woodcock 1965
	Arthrobacter sp.	Loss *et al.*, 1967
	Achromobacter sp.	Steensen and Walker 1957
	Flavobacteium sp.	Steensen and Walker 1957
	Pseudomonas sp.	Fernley and Evans, 1958, Evans *et al.*, 1971b
Chlordemeform	*Streptomyces grseus*	Johnson and Knowles,1970
	Sarratia marcescens	Johnson and Knowles,1970
Chlorpropharm (CIPC)	*Pseudomonas* sp.	Kaufman and Kearney, 1965
	Agrobacterium sp.	Kaufman and Kearney, 1965
	Flavobacterium sp.	Kaufman and Kearney, 1965
	Achromobacter sp.	Kaufman and Kearney, 1965
2,4-Dichlorobenzoate	*Alcaligenes denitrificans*	Van DenTwell *et al.*, 1987
3-5-Dichlorobenzoate	*Pseudomonas* sp.	Reinike and Anackmuss, 1980
1,3-Dichlorobenzene	*Alcaligens* sp.	DeBont *et al.*, 1986
Dimethyl tetra-phthalate	*Pseudomonas acidovorons*, D_4	Desai *et al.*, 1995
2,3-Dichloro-1-propanol	*Agrobacterium* sp.	Dancer, 2000
2-(2,4) Diclorophenoxy propionic acid	*Flavobacterium* sp.	Strichbier *et al.*, 1990,
Dalapon	*Nocardia, Pseudomonas* sp.	Hirsch and Alexander, 1960
	Pseudomonas dehalogenaes	Jensen, 1960
	Arthrobacter sp.	Jensen, 1960
	Agrobacterium sp.	Jensen, 1960
	Bacillus sp.	Kaufman, 1974
	Alkaligenes sp.	Kaufman, 1974
	Arthrobacter sp.	Kerney *et al* 1964
4-(2,4-Dichlorophenoxyl) butyric acid (2,4-DB)	*Flavobacterium* sp.	MacRae *et al.*, 1963
	Nocardia opaca	Webley *et al.*, 1957

Dichloroacetate	*Trichoderma* sp.	Jensen, 1960
	Penicillium sp.	Jensen, 1960
	Fusarium sp.	Jensen, 1960
	Agrobacterium sp	Jensen 1960
	Pseudomonas sp.	Hirsch and Alexander, 1960
	Nocardia sp.	Hirsch and Alexander, 1960
Endothall	*Arthrobacter globiformis*	Sikka and Saxena, 1973
2-methyl-6-dinitrophenol (DLOC)	*Corynebacterium simplex*	Gundersen and Jensen, 1956
Maleic hydrazide	*Alcaligens faecalis*	Limbeck and Colmer, 1957
	Flavobacterium biffisum	Limbeck and Colmer, 1957
2-methyl-chloro-phenoxyacetate (MCPA)	*Pseudomonas* sp.	Evans *et al.*, 1971; Gaunut and Evans, 1971
	Corynebacterium sp.	Jensen and Petersen, 1952
	Aspergillus niger	Clifford and Woodcock, 1964
4-(4-chloro-2-methylphenoxyl) butyric acid (MCPB)	*Nocardia opaca*	Webley *et al.*, 1958
Monochloro-acetate	*Pseudomonas* sp.	Hirch and Alexander, 1960
	Nocardia sp. *Trichoderma* sp., *Penicillium* sp., *Fusarium* sp.	Hirch and Alexander, 1960; Jensen, 1960
Monuron	*Pseudomonas* sp.	Suzuki and Nose, 1972
Naphthlene	*P. putida DP 99*	Desai *et al.*, 1995
Pentachlorophenol	*Arthrobacter* sp., *Flavobacterium* sp.	Steirt *et al.*, 1988
	Bacillus	Suzuki and Nose, 1972
	Basidiomycetes	Lyr, 1963
Sodium trichoroacetate (TCA)	*Micromonospora*	Erikson,1941
	Pseudomonas	Hirsch and Alexander, 1960
	Agrobacterium	Jensen, 1960 ; Kearney *et al.*, 1964
2,4,5-T	*P. cepasia*	Chatterjee *et al.*, 1982

The biodegradation behaviour of agricultural pesticides in anaerobic batch reactors was studied by Elefsinotis and Li (2008) in which biodegradation potential of two agricultural pesticides (2,4-D and isoproturon) was assessed as well as their effect on the performance on the aerobic digestion process was studied. Biodegradation of mixed pesticides by mixed pesticide enriched cultures was carried out by Ramakrishna and Phillip (2009) in which the pesticide mixture of lindane, methylparathion and carbofuran was used mixed microbial consortium including *Pseudomonas aeruginosa, Bacillus* sp and *Chryseobacterium foostei.* Development of *in situ* biodegradation technology was proposed by Erikson (1992) in his project, which was objected for a model based simulation for biodegradation.

Fogg *et al.* (2003) performed a study to determine whether biobeds can degrade relatively complex pesticide mixtures when applied repeatedly. A pesticide mixture containing isoproturon, pendimethalin, chlorpyrifos, chlorothalonil, epoxiconazole and dimethoate was incubated in biomix and topsoil at concentrations to simulate pesticide disposal. The effect of sorption on degradation of pesticide, 2,4-D was studied in a soil, amended with various amount of activated carbon. The relationship between sorption and decay of 2, 4-D was analyzed using analytical solutions for equilibrium sorption and to a two site non equilibrium adsorption model coupled with two first order degradation term for the dissolved and sorbed pesticide respectively (Guo *et al.*, 2000). Grundman *et al.* (2007) described the application of microbial hot spots enhanced pesticide degradation in soils of various types under laboratory and outdooe conditions. The microbes extracted from a soil having high native ability to mineralize the pesticides were established on extended clay particles and distributed to various soils in the form of hotspots.

For critical analysis of degradation product of the pesticides, several techniques are recommended. Aguera and Fernandez (2006) provided an overview of the application of GC-MS and LC-MS techniques for carrying out pesticide degradation studies by means of advanced oxidation processes. Attention is focused on the crucial role of these techniques in the evaluation degradation products. GCMS technique was used for revealing products of pesticide degradation by Kiss *et al.* (2007). The degradation of S-triazine type pesticides were allowed to degrade by UV light source. The degradation processes were followed by TLC, GC and GCMS techniques. Residue analysis of 500 high priority pesticides was done using GCMS as well as LCMS comparatively by Alder *et al.* (2006). Khilare and Kanade (2010) unpublished work showed *Pseudomonas fluorescence* and *Bacillus polymyxa* degrade phorate more than 80% in medium analysed by GCMS (Table 2).

Biodegradation is a natural process, where the degradation of a pesticide by an organism is primarily a strategy for their own survival. Most of these microbes work in natural environment but some modifications can be brought about to encourage the organisms to degrade the pesticide at a faster rate in a limited time frame. This capability of microbe is sometimes utilized as technology for removal of contaminant from actual site. Knowledge of physiology, biochemistry and genetics of the desired microbe may further enhance the microbial process to achieve bioremediation with precision and with limited or no scope for uncertainty and variability in microbe functioning (Singh, 2008). Gene encoding for enzyme has been identified for several pesticides, which will provide a new inputs in understanding the microbial capability to degrade a pesticide and develop a super strain to achieve the desired result of bioremediation in a short time.

Table 2 : Degradation of phorate by different PGPR microbes analysed by GC

PGPR	Phorate added in Medium (μg/ml)	Residue of Phorate from Medium (μg/ml)	Phorate degraded (μg/ml)	% Degradation
Azotobacter chroococcum	100	26.54	73.46	73.46
Azosprillum lipoferum	200	50.18	149.82	74.91
Bacillus polymyxa	200	35.5	164.5	82.25
Pseudomonas fluorescence	200	33.12	166.88	83.44
Rhizobium meliloti	50	23.12	26.88	53.76

Due to the diversity of chemistries used in pesticides, the biochemistry of pesticide bioremediation requires a detailed research aim to wide range of catalytic mechanisms and enzymes. Recently research and technology development have found ways to stimulate biological activity and thus decrease the length of time of treatment which can reduce the extent of environmental hazards. In general, bioremediation treatment still requires more intense research through specific microbial degradation of broad range of organic pollutants. In comparison to other remediation processes *i.e.* incineration, thermal disposition, land farming etc., the bioremediation has a better future in development of technology for removal of contaminants from actual site. There is huge potential for industrial applications of pesticide degrading bacteria. Bioremediation has been proven to work effectively under laboratory conditions. Short-term studies show that it also works under several field conditions. In spite of its limitations, bioremediation is benefiting from the rush to use biotechnology to solve public health problems. The EPA and the DOE are even investigating the feasibility of bioremediation to remove radioactive wastes from contaminated soil and water. Bioremediation's popularity is further enhanced because it is perceived as being more "green" than other remediation technologies. Companies and individuals are investing in biotechnology futures in spite of the high risks. As a result bioremediation companies have a *via*ble future regardless of its long-term effectiveness.

References

Ade, A.B. (2004). *Studies on mutation breeding of Rhizobium and other microbes for integrated management of certain soil borne diseases of groundnut.* (Ph. D. thesis), Dr. Babasaheb Ambedkar Marathwada University, Aurangabad, India.

Adhya, T.K., Barik, S. and Sethunathan, N. (1981). Stability of commercial formulations of fenitrothion, methyl parathion and parathion in anaerobic soils. *J. Agric. Food. Chem.*, 29: 90-93.

Aguera, A. and Fernandez, A. R. (2006). *GC-MS and LC-MS evaluation of pesticide degradation products generated through advanced oxidation processes*: An Overview. CAB Abstracts.

Ahmed, M. K. and Casida, J. E. (1958). Metabolism of some organophosphorus insecticides by microorganism. *J. Econ. Entomol.*, 51(1): 59-63.

Albone, E. S., Eglington, G. Evans, N.C. and Rhed, M.M. (1972). Formation of bis-(p-chlorophenyl) acetonitrile (pp-DDCN) from pp-DDT in anaerobic sludge. *Nature*, Lond. 140: 420-422.

Alder, L., Kerstin, G., Gunther, K. and Barbel, V. (2006). Residue analysis of 500 high priority pesticides: Better by GCMS/ LCMS/MS? *Mass Spectrom. Rev.*, 25(6): 838-865.

Alexander, M. (1981). *Microbial degradation of pesticides.* Final report, October 77-sep 81 : 1-18.

Amador, J. A., Gluckman, A. M., Lyons, J. B. and Gorres, J. H. (1997). Spatial distribution of soil phosphatase activity within a Riparian forest. *Soil Sci.*, 162(11): 808-825.

Antonella, F., Bent, I., Manuela, T., Sara, V., Marko, M. and Fengesheng, H. (2001). Preventing health risks from use of pesticides in agriculture. *WHO*, 9-12

Athar, R., Khan, S.T. and Ahmed, M. (1998). Isolation of pesticides and heavy metal tolerant strain of *Azotobacter* from the rhizosphere region of wheat crop. In: *Biofertilizer and Biopesticide.* Ed. A.M. Deshmukh, Technoscience Publ. Jaipur, India, 145-149.

Balaji, S. and Mahadevan, A. (1990). Dissimilation of 2,4-dichlorophenoxyacetic acid by *Azotobacter chroococcum. Xenobiotica.* 6: 607-617.

Barik, S. and Sethunathan, N. (1978). Biological hydrolysis of parathion in natural ecosystems. *J. Env.. Qual.*, 7: 346-348.

Barik, S., Wahid, P.A., Ramakrishna, C. and Sethunathan, N. (1979). A change in the degradation pathway of parathion after repeated application to flooded soil. *J. Agric. Food Chem.*, 27: 1391-1392.

Barker, P.S., Morrison, F.Q. and Whittaker, R.S. (1965). Conversion of DDT to DDD by *Proteus vulgaris*, a bacteria isolated from intestinal flora of a mouse. *Nature* (London), 205: 621-622.

Barrio, C. Sonez, A., Plaza Medina, J.M. Perez Clavijo, M. and Galbon B. J. (1995). Evaluation of the decay of malathion dichlofluanid and furnitrothion pesticides in apple samples using gas chromatography. *Food Chem.*, 52(3): 305-309.

Benezet, H.J. and Matsumura, F. (1973). Isomerization of ã- BHC to á-BHC in the environment. *Nature,* 243: 480-481.

Bhagobaty, Kumar, R., Joshi, S. and Malik, A. (2007). Microbial degradation of organophosphate pesticdes: chlorpyrophos (Minireview). *The Internat J. Microbiol.,* 4(1).

Bhalerao, T. S. and Puranik, P.R. (2009). Microbial degradation of monocrotophos by *Aspergillus oryzae. Int. Biodeterior. Biodegrad.,* 63(4): 503-508.

Bhuvaneswari, K., Regupathy, A. and Raghvan, G.S.V. (2008). The degradation pattern of soil applied insecticides. *Am. Soc. Agril. Biol. Engineers.*

Blum, H. and Michael R. (1996). Experimental parameters used to study pesticide degradation in soil. *Weed Technol.,* 10: 169-173.

Bolan, N. S. and Baskaran, S. (1997). Sorption and degradation of phorate as influenced by soil depth. *Austr. J. Soil Res.,* 35(4): 763-776.

Bollen, W. B. (1961). Interactions between pesticides and soil microorganisms. *Annl. Rev.Microbiol.,* 15:69-92.

Bourquin, A. W. (1977). Degradation of malathion by salt marsh microorganisms. *Appl. Environ. Microbiol.,* 33(2): 356-362.

Bousch, G.M. and Matsumura, F. (1967). Inseticidal degradation by *Pseudomonas melophthora,* the bacterial symbiont of the apple maggot. *J. Econ. Entomol.,* 60: 918-920.

Bro-Rasmussen, F., Noddegaard, E. and Voldun-Clausen, K. (1968), Degradation of diazinon in soil. *J. Agri. Food Chem.,* 17: 138.

Brown, H. M., Joshi, M.M. Ailen, T. V., Theodore, H. C., Dulka, J. J. Patrik, M. C., Robert, W. R., Robert S. L. and James, D. (1977). Degradation of thifensulphuron methyl in soil: Role of microbial carboxyesterase activity. *J. Agric. Food Chem.,* 45(3): 955-961.

Brown, J.R., Melson, R.O. and Breaud, T.P. (1992). Degradation of malathion in thermally generated aerosols. *JAM Mosq. Control Assoc.,* 8(2): 191-192

Burns, R.G. (1971). The loss of phosdrin and phorate insecticides from a range of soil types. *Bull. Environ. Contamin. Toxicol.,* 6 (4): 316-321.

Carter, F.L. and Sringer, C.A. (1970). Residue and degradation products of technical heptachlor in various soil types. *J. Econ. Entomol.,* 63: 625-628.

Chacko, C.I., Lockwood, J.L. and Matthew, Z., (1966). Chlorinated hydrocarbon pesticides degradation of microbes. *Science* , 154 (3751): 893-895.

Chakraborty, S.K., Bhattacharya, A. and Chowdhary, A. (2002). Degradation of oxyfluorfen by *Azotobacter chrococcum* (Beijerink). *Bull. Environ. Contamin. Toxicol.,* 69 (2): 203-209.

Chatterjee, D.K., Kilbane, J.J and Chakrabarty, A.M. (1982). Biodegradation of 2,4,5-trichlorophenoxyacetic acid in soil by a pure culture of *Pseudomonas cepacia*. *Appl. Environ. Microbial.*, 44:524:516.

Chatterjee, D.K., Kellogg, S.T., Hamada, S. and Chakrabarti, A.M. (1981). Plasmid specifying total degradation of 3-chlorobenzoate by a modified ortho pathway. *J. Bacterial.*, 146: 639-646.

Choudhary, A., Pradhan, S., Saha, M. and Sanyal, N. (2008). Impact of pesticides on soil microbiological parameters and possible bioremediation strategies. *Ind. J. Microbiol.*, 48 (1): 114-127.

Christopher, J., Grandic M. O., Mendez, J. C., Machada B., and Maier R M. (2008). Plant growth promoting bacteria for phytostabilization of mine filings. *Environ. Sci. Technol.*, 42(6):2079-2084.

Clifford, D.W. and Woodcock, D. (1964). Metabolism of phenoxy acetic acid by *Aspergillus niger* van Tiegh. *Nature*, London. 203:763.

Congregado, F., Simon-Pujol, D. and Juarez, A. (1979). Effect of two organophosphates insecticides on the phosphate dissolving soil bacteria. *Appl. Environ. Microbiol.*, 37: 169-171.

Consalter, A. and Guzzo, V. (1991). Multiresidue analytical method for organophosphate pesticides in vegetables. *Fresenious J. Anal. Chem.* 339: 390.

Craig, C. (1999). Mechanisms of pesticide degradation in compost. *Compost Sci. Utiliz.* : 7.

Craig, C. (2000). Mechanisms of pesticide degradation in compost. *Compost Sci. Utiliz.* : 8.

Crawford, R.L., Olsen, P.E., and Frick, T.D. (1979). Co-metabolism of 5-chlorosalicylate by a *Bacillus* isolated from the Mississippi river. *Appl. Environ. Microbiol.*, 38:379-384.

Dancer, B.N., Effendi, A.J. and Greenaway, S.D. (2000). Isolation and characterisation of 2,3-dichloro-1-propanol-degrading Rhizobia. *Appl. Environ. Microbial.*, 66(7):2882-2287.

Daughton, C.G. and Hsiegh, D.P.H. (1977). Parathion utilization by bacterial symbionts in chemostat. *Appl. Environ. Microbiol.*, 35: 175-184.

DeBont, J.M., Vorage M.W., Hartmans, S. and Van Den Tweel, M.J. (1986). Microbial degradation of 1, 3-dichlorobenzene. *Appl.Environ. Microbial.*, 52:677-680.

Desai, A.J., Patel, D.S., Desai, J.D. and Ramakrishna, C. (1995).Biodegradation of Dimethyl-tetrapthalate by *Pseudomonas acidovarans* D 4. Proc. Natl. *Symp. Frontiers in Appl. Environ. Microbial.*, 11-13 Dec 1995

Desai, A.J., Hanson, K.G. and Bhagata, S.W. (1995). Influence of some parameter on naphthalene degradation by *Pseudomonas putida* DP 99. Proc. Natl. *Symp. Frontiers in App. Environ. Microbiol.*, 11-13 Dec. 1995.

Dev, N. and Natarajan, K.A. (1998). Biodegradation of some organic flotation creagent by *Bacillus polymyxa*. *Bioremediation J.*, 2 (3&4): 205-214.

Dieter, Charles, D., Walter G. D., and Lister D. F. (1995). Environmental fate of phorate and its metabolites in Northern Prarie wetlands. *J. Freshwater Ecol.,* 10(2): 103-109.

Doong, R.A. and Chang W.H. (1997). Photoassessed titanium dioxide mediated degradation of organophosphorus pesticides by hydrogen peroxide. *J. Photochem. Photobiol. Chemistry, Elsevier,* 107(1-3):239-244.

Dureja, P. and Gupta, R.L. (2009). Status of pesticides in India. *Pesticide Res. J.,* 21 (2): 202-210.

Elefsiniotis, P. and Li, W. (2008). Biodegradation behaviour of agricultural pesticides in anaerobic batch reactors. *J. Environ. Sci. Health B.,* 43(2):172-179.

Elizabeth, G. A., Meza, M. M., Aqilar, M. G., Soto A. D., and Garcia E. I. (1998), An anthropological approach to the evolution of preschool children exposed to pesticides in Mexico. *Environ.Health Prospect.,* 106 (6): 347-353.

Engst, R., Fritche, W., Klell, R., Kujawa, M., Macholz, R.M., and Straube, G. (1979). Interim Results of studies of microbial isomerization of gamma hexachlorocyclohexane. *Bull. Environ. Contam. Toxicol.,* 22: 699-707.

EHP (1999). Environmental Health Perspectives. Forum: Killer Environment. 107 (2): A 62. http://ehp.niehs.nih.gov/docs/1999/107-2/forum.html#rev

Erikson, L. E. (1992). *Development of insitu biodegradation technology.* A Project of Kansas State University.

Erikson, R. (1941). Studies on some lake-mud strains of *Micromonospora. J.Bacterial.* 41: 227-300.

Evans, W.C., Smith, B.S.W., Fernely, H.N. and Davies, J.I. (1971a). Bacterial metabolism of 2, 4-dichlorophenoxyacetate. *Biochem. J.,* 122: 543-551.

Evans, W.C., Smith B.S.W., Moss P. and Fernley, H.N. (1971b). Bacterial metabolism of 4-chlorophenoxyacetate. *Biochem. J.,* 122:509-517.

Faulkner, J.K. and Woodcock, D. (1965). Fungal detoxification Part. VII. metabolism of 2,4-chloro-phenoxy acetic acid and 4-chloro-2methyl phenoxy acetic acid by *Aspergillus niger. J. Chem. Soc.,* London: 1187.

Felsot, A. S. Racke, K.D. and Hamilton, D.J. (2003). Disposal and degradation of pesticide waste. *Rev. Environ. Contam. Toxicol.,* 177: 123-200.

Feng, L., Zwieten, L.V., Kenedy, I.R., Rolfe B.G., and Gartner, E. (2004). Expression of the 2,4-D degrading plasmid pJP4 of *Alcaligenes eutrophus* in *Rhizobium trifoli. Acta Biotech.,* 14(2): 119-129.

Fernley, H.N. and Evans, W.C. (1958). Oxidative metabolism of polycyclic hydrocarbons by soil Pseudomonads. *Nature,* 182:373-375.

Ferris, I. G. and Lichtenstein, E. P. (1980). Interactions between agricultural chemicals and soil microflora and their effects on the degradation of parathion in a cranberry soil. *J. Agric. Food. Chem.,* 28: 1011-1019.

Fisher, P.R., Appleton. J., and Pemberton J. M. (1978). Isolation and characterization of the pesticide degrading plasmid pJP 1 from *Alcaliegenes paradoxus. J. Bacteriol.,* 135(3): 798-804.

Focht, D.D. and Alexander, M. (1970). DDT metabolites and analogous ring fission by *Hydrogenomonas. Science,* 170: 91-92.

Focht, D.D. and Alexander, M. (1971). Aerobic cometabolism of DDT analogs by *Hydrogenomonas* sp. *J. Agric. Food Chem.,* 19: 20-22.

Fogg, P., Alistaic, B.A., Walker, B. A. and Jukes, A. (2003). Pesticide degradation in a biobed composting substrate. *Pest Manage. Sci.,* 59(5): 527-537.

Foster, L., John, R., Kwan, B. H. and Vancov, T. (2006). Microbial degradation of the organophosphate pesticide ethion. *FEMS Microbial. Letters,* 240(1): 49-53.

Francis, A.J., Spanggord, R.J., and Ouchie, G. I. (1975). Degradation of lindane by *Escherichia coli. Appl. Microbiol.,* 29 : 567-568.

Francis, R.P. and Gangawane, L.V. (2009). Effect of micronutrients on malathion degradation on *Rhizobium. Bioinfolet.,* 6(1). 57-59

French, A.L. and Hoppingarner, R.A. (1970). Dechlorination of DDT by membranes isolated from *Escherichia coli. J. Econ. Entomol.,* 63: 756-759.

Fuhremann, T.W. and Lichtenstein, E.P. (1980). A comparative study of the persistence, movement and metabolism of six carbon -14 insecticides in soils and plants. *J. Agric. Food Chem.,* 28 (2): 447-452

Gahlot, R. and Narula, N. (1996). Degradation of 2,4-dichlorophenoxy acetic acid by resistant strain of *Azotobacter chroococcum. Ind. J. Micribiol.,* 36:141-143.

Gangawane, L.V. and Francis, R.P. (1997). *Rhizobium* from wild legumes and pesticide degradation. Two in one. In: *Microbial Biotech.* (Eds.) S.M. Reddy *et al. Scientific Publ.,* Jodhpur, India: 59-63.

Gaunut, J.K., and Evans, W.C. (1971). Metabolism of 4-chloro-methyl-phenoxyacetate by a 'soil *Pseudomonas.* Preliminary evidence for the metabolic pathway. *Biochem. J.,* 122:519-526.

Gaur, A.K. and Saraswat, R. (1995). Bioremediation of â-Endosulfan by *Rhodococcus sp. Ind. J. Microbiol.,* 33 (3) : 249-253.

Gayatri, R. and Chatterjee S. (2006). Phosphatase activity in pesticides treated growing and developing cells of *Dictyliostellium discoidium . J.Appli. Toxicol.,* 13(4): 297-300.

Getzin, L.W. (1967). Metabolism of diazinon and zinophos in soil. *J. Econ. Entomol.,* 60: 505-508.

Getzin, L.W. and Shanks, C.H. (1970). Persistance, degradation and bioactivity of phorate and its oxidative analogs in soil. *J. Econ. Entomol.,* 63: 52-58.

Getzin, L.W. and Shanks, J.R.C.H. (1970). Persistence, degradation and bioactivity of phorate and its oxidative analogues in soil. *J. Econ. Entomol.,* 61(1): 52-58(7).

Gibson, W.L. and Brown, L.R. (1974). The metabolism of parathion by *Pseudomonas aeruginosa. Dev. Ind. J. Microbiol.*, 16: 17-87.

Gomez, F, V., Rodelas, S.B.,Told M.V. M., and Lopez ,J. G.(1998). Response of *Azospirillum brazilense* to the pesticides bromoprophylate and malathion on chemically defined media and dialysed soil media. *Ecotoxicol.*,7(1): 43-47.

Gowrisankar, R., Murugesan, P., Palanippan, R. and Sundaram, C.M. (2000). Growth pattern and herbicide utilization profile of atrazine and anilofos resistant autochthonous bacterial isolates. *Poll. Res.*, 19(4): 591-595.

Grover, N., Johl, P.P., Singh, G. and Kahlon, R.S. (1993). Degradation of chlorobenzoates and effect of induction of chlorobenzoate utilization in *Pseudomonas sp. Ind. J. Microbial.*, 33(2):105-110.

Grundman, S., Schmid, R. F. M., Laschinger, M., Ruth, B., Schulin. R., Munch, J.C. and Schroll, R. (2007). Application of microbial hotspots enhances pesticide degradation in soil. *Chemosphere*, 68(3): 511-517.

Gundersen, K. and Jensen, H.L. (1956). A soil bacterium decomposing organic nitro compounds. *Acta. Agric. Scand.*, 67:100-114.

Gunner, H.B. and Zuckerman, B.M. (1968). Degradation of diazinon by synergistic microbial action. *Nature*, (London). 217: 1181-1184.

Guo, L.,. Jury, W. A., Wagenet R. J. and Flury M. (2000). Dependence of pesticide degradation on sorption: Non equilibrium model and application to soil reactors. *J.Contam. Hydrol.*, 43(1): 45-62.

Haddane, M.A., Rambaud M.A., and Previero C. (1991). Organophosphates biodegradation in anaerobic media by immobilized enzymatic activity. *Environ. Technol.*, 12(10):887-896.

Han, H.S. and Lee K. D. (2005). Plant growth promoting rhizobacteria effect on antioxidant status photosynthesis, mineral update and growth of lettuce coder soil salinity. *Res. J. Agri. and Biol. Sciences*, 1(3): 210-215.

Hashimoto, M., Fukui, M., Hayano, K. and Hayatsu .M. (2002). Nucleotide sequence and genetic structure of a novel carbaryl hydrolase gene (Ceh A) from *Rhizobium* sp. strain AC100. *Appli. and Environ. Microbiol.*, 68(3): 1220-1227.

Hashmi, I., and Kim, J. G. (2003).Comparison of degradation studies in abiosimulators by continuous cultivation of *Pseudomonas* and indigenous microorganisms with varied dissolved oxygen concentrations. *Int. J. Environ.and Poll.*, 19(2): 123-138.

Hashmi, I., Khan, M. A. and Jong-G.K. (2004). Malathion degradation by *Pseudomonas* using activated sludge treatment system. *(Biosimulator) Biotechnol.*, 3(1): 82-89.

Hayes, W. (1982). Pesticide Studies in Man: a thorough review of toxic effect and metabolism of pesticides. *Baltimore Williams and Wilkins*.

Hirsch P. and Alexander, M. (1960). Microbial degradation of halogenated propionic and acetic acids. *Can. J. Microbial.*, 6:241-249.

Hsu, T.S. and Bartha, R. (1979). Accelerated mineralization of two organophosphate insecticides in the rhizosphere. *Appl. Environ. Microbiol.*, 37: 36-41.

Hutchinson, P.J.,Shea,P.J., Takacs .J. M., and Staswick ,P. E. (1993).Immunoassay to detect enhanced carbomothioate degradation in soil. *Weed Technol.*,7: 396-403.

Iashenko, B.N. (2001). Isolation and functional structural characteristics of *Bacillus polymyxa* RP4:Mucts 62 transcipients. *Microbiol .Virusol.*, 2(2): 8-13.

Jagnow, G. and Hyder, K. (1972). Evolution of CO2 from soil incubated with dieldrin and the action of soil bacteria on labelled dieldrin. Soil. *Biol. Biochem.*, 4 : 43-49.

Jagnow, G., Haider, K. and Ellwardt, P.C. (1977). Anaerobic dechlorination and degradation of hexachlorocyclohexane by anaerobic and facultative anaerobic bacteria. *Arch. Microbiol.*, 115:285-292.

Jensen,H.L. (1960). Decomposition of chloroacetates and chloropropionates by bacteria. *Acta. Agric. Scand.*, 2:215-231.

Jensen,H.L. (1960). Decomposition of chloroacetates and choropropionates by bacteria. *Acta. Agric. Scand.*, 10:83-103.

Jensen, H. L. and Peterson, H.J. (1952). Detoxification of herbicide by soil bacteria. *Acta. Agric. Scand.*, 2: 215-31.

Johnson, B.T. and Knowles, C.O. (1970). Microbial degradation of acaricides N-(4-chloro-o-tolyl)-N,N-dimethylformidine. Bull. *Environ. Contam. Toxicol.*, 5: 158-163.

Johnson, B.T., Goodman, R.N. and Goldberg, H.S. (1967). Conversion of DDT to DDD by pathogenic and saprophytic bacteria associated with plants. *Science*, 157: 560-261.

Johnstan, A.W. and Beringer, J.E. (1975). Identification of the Rhizobial strain in pea root nodule using genetic markers. *J. Gen. Microbiol.*, 87(2): 343-350.

Kadam, T.A. and Gangawane L.V. (2005). Degradation of phorate by *Azotobacter* isolates. *Ind. J. Biotechnol.*, 4: 153-155.

Kadam, T.A. (2002). *Studies on pesticide degradation by free living nitrogen fixing microorganisms from Marathwada soils.* (*Ph.D. Thesis*) Dr. B. A. M. U. Aurangabad.

Kadoum, A. M. and La Hue, Delmon, W. (1979). Degradation of malathion on wheat and corn of various moisture content. *J. Econ. Entomol.*, 72(2): 228-229 (2).

Kanekar, P. P., Bhadbhade, B. J., Deshpande ,N. M. and Sarnaik S. S. (2004), Biodegradation of organophosphorus pesticides. *Proc. Ind. Nat. Sci. Acad.* B70 No. : 57-70.

Kannan, V. and S.V. Job (2007). Studies on residual levels of pesticide pollution in Sathiar reservoir. *J. Radioanal. Nuclear Chem.*, 53 (1-3): 247-253.

Karns, J. S. and Eaton, R. W. (1997). Genes encoding s-threonine degradation are plasmid born in *Klebsiella pneumoniae* strain 99. *J. Agric. Food Chem.*, 45(3): 1017-1022.

Kaufman, D.D. (1974). Degradation of pesticide by soil microorganisms. In: Guenzi, W.D. (ed.) pesticides in soil and water. *Soil Sci. Soc. Am. Madison, Wiscosin* : 133-202.

Kaufman, D.D. and Kearney, P.C. (1965). Microbial degradation of isopropyl-N-3-chlorophenyl carbamate and 2-chloroethyl-N-3-chlorophenyl carbamate. *Appl. Microbiol.*, 13(3):443-446

Khilare, V.C. and Kanade, S.N. (2010). Degradation of phorate by certain PGPR microbes. (Unpublished).

Kim, Y.H., Ahn, J.Y,. Moon, S.H., and Lee J. (2005). Biodegradation and detoxification of organophosphate insecticide malathion by *Fusarium oxysporum* sp. *pisi cutinase*. *Chemosphere,* 16 (10): 1349-1355.

Kiss, A., Rapi. S., and Sutoras, C.S. (2007). GCMS studies on revealing products and reaction mechanism of photodegradation of pesticides. *Microchemical J.,* 85(1): 13-20.

Konrad, J.G., Chesters, G. and Armstrong D.E. (1969). Soil degradation of malathion,a phosphodithioate insecticide. *Soil. Sci. Soc. Am J.*, 33: 259-260.

Ku, Y. and Lin H. S. (2002). Decomposition of phorate in aqueous solution by photolytic ozonation, *Water Res.*, 36(16): 4155-4159.

Kumar, S., Mukerji ,K.G., and Lal R. (1996). Molecular aspects of pesticide degradation by microorganisms. *Crit. Rev. Microbiol.*, 22(1): 1-26.

Kumari, B.,A., Guha, M.G., Pathak, T.C., Bora, and Roy, M. K. (1998). Experimental biofilm and its application in malathion degradation. *Folia Microbiol.*, 43:27-30.

Lan, W.S., Gu, J.D. , Zhang, J.L., Shen, B.C., Liang, H., Mulchandani, A., Chen W.and Quiao, C.L. (2006). Coexpression of 2-deoxyfying pesticide degradative enzymes in a genetically engineered bacterium. *Int. Biodeterio. and Biodegrad.*, 58: 70-76.

Langlois, B.E., Collins, J. A.,and Sides, K.G. (1970). Some factors affecting Degradation of organochlorine pesticides by bacteria. *J. Dairy Scien.*, 53(12): 1671-1675.

Larkin, M.J. and Day, M.J. (1986). The metabolism of carbaryl by three bacterial isolates. *Pseudomonas* sp. (NCIB 12042 and 12043) and *Rhodococcus* sp. (NCIB 12038) from garden soil. *J. Appl. Bacteriol.*, 60: 233-242.

Ledford, R.A. and Chen, J.H. (1969). Degradation of DDT to DDE by cheese microorganisms. *J. Food Sci.*, 34: 386-388.

Lee, R. P., Parkinson. A., and Forket, P. G. (1998). Isozyme selective metabolism of ethyl carbamate by cyclochrome p450 (CyP2E1) and carboxyesterase (Hydrolase A) enzymes in marine liver microsomes. *Drug Metabol. and Disposition.*, 26(1):60-65.

Lembeck, W.L. and Colmer, A.R. (1957). Aspect of decomposition and utilization of maleic hydrazine by bacteria. *Weeds*, 5(1): 34-39.

Liu, S.Y. and Bollag, J.M. (1971). Metabolism of carbaryl by a soil fungus. *J. Agric. Food Chem.*, 19:487-490.

Liu, Y.H., Chung, Y.C. and Xiong, Y. (2001). Purification and characterization of a dimethoate degrading enzyme of *Aspergillus niger* ZHY 256, isolated from sewage. *Appl. Environ. Microbiol.* , 67(8): 3746-3749.

Lopez, L., Pozo, C., Rodelas, B.,Calvo .C., and Lopez J. G. (2006). Influence of pesticides and herbicides presence on phosphatase activity and selected bacterial microflora of a natural lake system. *Ecotoxi.,* 15: 487-489.

Loss, M.A., Roberts, R.N. and Alexander M. (1967). Phenols as insecticides in the decomposition of phenoxyacetates by an *Arthrobacter* species. *Can. J. Microbial.,* 13:691-699.

Lyr, H. (1963). Enzymatic detoxification of pentachlorophenol. *Phytopathol.,* 47:73- 83.

Mahopatra, S. and Awasthi, M.D. (1995). Degradation of metalaxyl in soil and by soil microbial cultures. *Proc. Natl. Symp. Frontiers in Appl. Environ. Microbiol.* 11-13 Dec. 1995.

Manavati, B., Somara, S. and Siddavatam (1999). Indiginous organophosphorus pesticide degrading plasmid, pBC9 of *Flavobacterium balustirnum* construction of plasmid library and detection of clone, *Ind. J. Microbiol.,* 39: 105-108.

Mandal, M. D., Mandal, S., and Pal, N.K. (2005), Plasmid mediated dimethoate degradation by *Bacillus licheniformis* isolated from a fresh water fish, *Labeo rohita. J. Biomed. Biotechnol.,* 2005(3): 280-286.

Marks, T. S., Smith, A.R., and Quirk, A.V. (1984).Degradation of 4-chloro-benzoic acid by *Arthrobacter sp. Appl. Environ. Microbial.,* 48:1020-1025.

Matsumura, F. and Boush G. M. (1966). Malathion degradation by *Trichoderma viride* and a *Pseudomonas* sp. *Science,* 153(3741): 1278-1280.

McRae, I.C., Celo. and Jovenia S. (1974). The effects of organo-phosphorus pesticides on the respiration of *Azotobacter vinelandii. Soil Biol. and Biochem.,* 6(2): 109-111.

MacRae, I.C., Alexander, M. and Rovira, A.D. (1963). The decomposition of 4-(2,4-dichlorophenoxy) butyric acid by *Flavobacter* sp. *J.Gen. Microbiol.* 32:69-76.

Meksongsee, H. and Guthrie, F.E. (1965). Degradation of chlorinated hydrocarbon insecticides by certain soil bacteria in broth culture. *J. Elisha Michel. Sci. Soc.,* 81: 81.

Menzer, R.E., Fontanilla, E. L. and Ditman, L.P. (1970). Degradation of disulfoton and phorate in soil influenced by environmental factors and soil types. *Bull. Environm.Contam.Toxicol.,* 5 (1): 1-5.

Milbarrow, B.V. (1964). 2,6-Dichlorobenzonitrate degradation. *J. Expt. Bot.,* 15:515-524.

Miles, J.R.W., Tu, C.M. and Harris, C.R. (1969). Metabolism of heptachlor and its degradation products by soil microorganisms. *J. Econ. Entomol.,* 62: 1334-1337.

Mohamed, M. S. (2009). Degradation of methyl by the novel bacterial strain *Stenotrophomonas maltophilia* M1. *Environm. Biotechnol.,* 12(4): 1-6

Mrkovacki, N. B., Nikola, A. and Vera, M.M. (2002) Effects of pesticides on *Azotobacter chroococcum. Proceedings for Natural Sciences, Matica Srpska Novi Sad.,* 102: 23-28.

Mudryk, Z. Jan., and Piotr, S. (2006). Enzymatic activity and degradation of organic macromolecule by neustonic and planktonic *bacteria* in an eustuarine lake. Pol. *J. Ecol.,* 54(1): 3-14.

Munnecke, D.M. (1976). Enzymatic hydrolysis of organophosphate insecticides, a possible pesticide disposal method. *Appl. Environ. Microbiol.,* 32 : 7-13.

Munnecke, D.M. and Fischer, H.F. (1979). Production of parathion hydrolase activity. *Eur. J. Appl. Microbiol. Biotechnol.,* 8: 103-112.

Munnecke, D.M. and Hsieh, D.P.H. (1974). Microbial decontamination of parathion and p-nitrophenol I aqueous media. *Appl. Microbiol.,* 28: 212-217.

Myresiotis, C., Viryzas, Z. and Papadopoulou, E. (2009). Degradation of pesticide by plant growth-promoting rhizobacteria (PGPR) and their effect on bacterial growth. Abstract (Poster Session-B) In: Confr. '*Pesticide behaviour in soil, water and air*' York, UK 14-16 Sept. 2009.

Nezarat, S. and Gholami.A. (2009), Screening plant growth promoting rhizobacteria for improving seed germination, seedling growth and yield of maize. *Pakistan J. Biol. Sci.,* 12(1): 26-32.

Nies, A., Nies, D. H. and Silver ,S. (1989). Cloning and expression of plasmid genes encoding resistance to chromate and cobalt in *Alcaligenes eutrophus. J. Biotech.,* 171(9):5065-5070.

Ninnekar, H.Z. and Doddmani, H.P. (2001). Biodegradation of carbaryl by a *Micrococcus* sp. *Curr. Sci.,* 43: 69-73.

Ohisa, M. and Yamaguchi, M. (1978). Gamma-BHC degradation accompanied the growth of *Clostridium* isolated from paddy field. *Agri. Biol. Chem.,* 42: 1819-1823.

Ohisa, M., Yamaguchi, M. and Kurihara, M. (1980). Lindane degradation by cell free extracts of *Clostridium rectum. Arch. Microbiol.* ,125 (3): 221-225.

Panagiotis, A., Kyriakisis, N.V. and Panagiotis S. (2005). A study on the environmental degradation of pesticides Azinphos methyl and parathion methyl. *J. Environ. Sci. Health* Part B 39(2): 297-309.

Patil, K.C., Matsumura, F. and Boush, G.N. (1970). Degradation of endrin, aldrin and DDT by soil microorganisms. *Appl. Microbiol.,* 19:879-881.

Patterson, R.S. and Rawlins W.A. (1968). Loss of phorate from a granular formulation applied in the soil. *The Florida Entomol.,* 51(3): 143-150.

Perkins, E.J. and Lurguin, P.F. (1988). Duplication of 2,4-dichloro-phenyxyacetic acid mono oxigenase gene in *Alcaligenes eutrophus* JMP 134. *J.Bacteriol.,* 170:5669-5672.

Plummer, J.R., Kearney, P.C., Endt, D.W., and Von (1968). Mechanism of conversion of DDT to DDD by *Aerobacter aerogenes. J. Agric.Food Chem..* 16: 594-597.

Quagliano, J.C., Alegre, P., and Miyazaki, S.S. (1994). Isolation and characterization of *Azotobacter* sp. for the production of poly-beta hydroxyl karoate. *Rev. Argent. Microbiol.,* 26(1): 21-27.

Raghu, K. and MacRae, I.C. (1966). Biodegradation of gamma isomer of BHC in submerged soils. *Science,* 184 : 263-264.

Rama, K. K. and Ligy, P. (2009). Biodegradation of mixed pesticides by mixed pesticide enriched culture. *J. Environ. Sci. Health Part B* 44(1): 18-30.

Rani, R., Lal, R., Kanade, G.S. and Juwarkar, A (2009) Isolation and characterization of a phorate degrading bacterium. *Letters in Appl. Biol.,* 49(1): 112-116.

Reddy, B.P. and Reddy, M.S. (2009). Isolation of secondary metabolites from Pseudomonas fluorescens and its characterization. *Asian J. Res. Chem.,* 2 : 6-29.

Reed, J.P., Kremer, R.J. and Keaster A.J. (1987) Characterization of microorganism in soils exhibiting accelerated pesticide degradation. *Bull. Environ. Contam. Toxicol.,* 39: 776-782.

Reinike, W. and Knackmuss, H.J. (1980). Hybrid pathway for chlorobenzoate metabolism in *Pseudomonas* sp. B13. *J. Bacteriol.,* 142:467-473.

Reva, O. N., Dixelius, C., Meijec, J. and Priest, F. G. (2004).Taxonomic characterization and plant colonizing ability of some bacteria related to *Bacillus amyloliquifaciens* and *Bacillus subtilis. FEMS. Microbiol Ecol.,* 48(2): 249-259.

Rodriguez, I.D. and Dorough, H.W. (1977).Degradation of carbaryl by soil microorganisms. *Arch. Environ. Contam. Toxocol.,* 6: 47-56.

Romera, E., Segovia, M.L., Mercante, F.M., Franco, A.A., Graham, P. and Pardo, M.A. (1991). *Rhizobium trapici,* a novel species nodulating *Phaseolus vulgaris* L. Beans and *Leucena* sp tree. *Int. J. Syst. Bacteriol.,* 41:417-426.

Rosenberg, A. and Alexander, M. (1979). Microbial cleavage of various organophosphorus insecticides. *Appl. Environ. Contam. Toxicol.,* 6:47-56.

Rowlands, D.G. (2006). The degradation of malathion on stored maize and wheat grains. *J. Sci., Food and Agri.,* 15(12): 824-829.

Sahashrabudhe, S. R. and Modi, V.A. (1991). Degradation of 4-chlorobenzoate and mode of dehalogenation in *Pseudomonas* sp. *Ind. J. Expt. Biol.* , 20:252-255.

Serdar, D.M., Gibson, D.T., Munnecke, D.M. and Lancaster, J.H. (1982). Plasmid involvement in parathion hydrolysis by *Pseudomonas diminuta. Appl. Environ. Microbiol.,* 44: 246-249.

Sethunathan, N. and Yoshida, T. (1973). A *Flavobacterium* sp. that degrades diazinon and parathion. *Can J. Microbiol.,* 19: 873-875.

Sexena, A., Zang, R. and Bolla, J.M. (1987). Microorganisms capable of metabolising the herbicide, metolachlor. *Appl. Environ. Microbiol.,* 53:390-395.

Shamseldin, A., Sadowsky M.J., Saadani, M. E. and Chung S. A. (2008). Molecular biodiversity and identification of free living *Rhizobium* strains from diverse Egyptian soils as assessed by direct isolation without trap hosts. *Am. Eurasian J. Agric. Environ. Sci.,* 4(5): 549-549

Sharma, B. K. and Gupta, N. (1994). Photodegradation of organophosphate insecticide, phorate. *Toxicol. Environ. Chem.,* 41(3 and 4): 249-254.

Shelton, D. R. and Doherty M. A. (1997). A model describing pesticide bioavailability and biodegradation in soil. *Soil Sci. Soc. Am. J.*, 61: 1078-1084.

Sidramappa, R., Rajaram, K.P. and Sethunathan, N. (1973). Degradation of parathion by bacteria isolated from flooded soil. *Appl. Microbiol.*, 26: 846-849.

Sikka, H.C. and Saxena, J. (1973). Metabolism of Endothal by bacteria isolated from flooded soil. *Appl. Microbiol.*, 26:846-849.

Singh, A. K., Srikanth, N.S., Malhotra O.P., and Seth P.K. (1989). Characterization of carbonyl esterase from malathion degrading bacterium, *Pseudomonas* SPM-3. *Bull. Environ. Contamin. and Toxicol,*. 42(6):860-867.

Singh, D.K. (2008). Biodegradation and bioremediation of pesticide in soil: concept, method and recent developments. *Ind. J. Microbiol.*, 48 (1): 35-40.

Singh, N., Singh, B., Dureja, P. and Sethunathan N. (2003), Persistance of phorate in soil, role of moisture, temperature, free exposure of microorganisms. *J. Environ. Sci. Health,* Part B 38(6): 723-735.

Spillner, C.J., Debaun, J.R. and Menn, J.J. (1979). Degradation of fenitrothion in forest soil and effects on forest soil microbes. *J. Agric. Food Chem.*, 27: 1054-1060.

Stansly, P.G. and Schlosser M. E. (1947). Studies on polymixin: Isolation and identification of *B. polymyxa* and differentiation of polymixin from certain known antibiotics. *J. Bacteriol.*, 54(5):549-556.

Steensen, T.I. and Walker, N. (1957). The pathway of breakdown of 2,4-dichloro, 2-methyl-phenoxyacetic acid by bacteria. *J. Gen. Microbiol.*, 16: 146-155.

Steensen, T.I. and Walker, N. (1957). The pathway of breakdown of 2,4-dichloro and 4-chloro-2 methyl-phenoxyacetic acid by bacteria. *J. Gen. Microbiol.*, 16:146-155

Steirt, J.G., Thoma, W.J.,Ugurbic, K. and Cowford R.L. (1988). 31P nuclear magnetic resonance studies of effect of some chlorophenols on *Escherichia coli* and a pentachlorophenol degrading bacterium. *J. Bacteriol.*, 170: 4954-4957.

Stenersen, J. (1969). Degradation of p-bromophos by microorganisms and seedlings. *Bull. Environ. Contam. Toxicol.*, 4: 104-112.

Streichsbier, F., Horvath, M., Ditzemiller, G. and Loidi, M. (1990). Isolation and characterisation of a 2-(2,4-dichlorophenoxy propionic acid) degrading soil bacterium. *Appl. Microbial. Biotechnol.*, 33:213-216.

Suett, D.L. (1999). Persistence and degradation of chlorfenvinphos, diagram, fonofos and phorate in soils and these uptake by carrots, *Pest Manag. Sci.*, 2(3): 105-112.

Sutherland, T.D., Horne, H., Lacey, M.J., Harcourt, R. L., Russel, R.J. and Oakeshott, R.G. (2000). Enrichment of an endosulfan degrading mixed bacterial culture. *Appl. Environ. Microbiol.*, 66(7): 2822-2828.

Suzuki, T. and Nose, K. (1972). Decomposition of pentachlotophenol in farm soil part 2: PCP metabolism by a microorganism isolated from soil. *Noyaka Sesian Gijutsu*, 26:21.

Talbot, H.W., Johnson, L.M., Barik, S. and Williams, D. (1982). Properties of a *Pseudomonas* sp. derived parathion hydrolase immobilized to porous glass and activated alumina. *Biotechnol. Lett.*, 4: 209-214.

Tejera, N. C., Llunch, M. V., Toledo ,M. and Lopez J. G. (2005). Isolation and characterization of *Azotobacter* and *Azospirillum* strains from the sugarcane rhizosphere. *Plant Soil,* 270:223-232.

Topp, E. H., Zhu, S., Lewis, M. and Cuppels, D. (2000). Characterisation of an Atrazine degrading *Pseudominobacter* sp. isolated from Canadian and French agriculture soils. *Appl. Environ. Microbial.*, 66 (7): 2773-2782.

Tu, C.M. (1975). Interaction between lindane and microbes in soils. *Arch. Microbiol.,* 105 : 131-134.

Tu, C.M., Miles, J.R.W. and Harris, C.R. (1968). Soil microbial degradation of aldrin. *Life Sci,.* 7: 311-322

Van, D., Twell, W.J., Kok,J.B. and DeBont, J.A.(1987). Reductive dechlorination of 2,4-dichlorobenzoate to 4-chlorobenzoate and hydrolytic dehalogenation of 4-chloro-4-bromo and 4-iodobenzoate by *Alcaligenes denitrificans* NTB-1. *Appl. Environ. Microbiol.,* 53:810-815.

Van Loon L. C. (2007). Plant responses to plant growth promoting rhizobacteria. *Eur. J. Pl. Pathol.,* 119(3): 243-254.

Venkateshwarlu, K., Chenrayan, K. and Sethunathan, N. (1980). Persistence and biodegradation of carbaryl in soils. *J. Environ. Sci. Health* ,B15: 421-429.

Venkateswarlu, K., and Sethunathan, N. (1978). Degradation of carbofuran in rice soils as influenced by repeated application and exposure to aerobic conditions following anaerobiosis. *J. Agric. Food Chem.*, 26:1148-1151.

Vischetti, C., Monaci, E. Cardinali, A., Casucci, C and Perucci, P. (2008). The effect of initial concentration co-application and repeated application on pesticides degradation in a biobed mixture. *Chemosphere* 72(11): 1739-1743.

Vonk, J.W. and Quirijins, J.K. (1979). Anaerobic formation of a hexachloro cyclohexane in soil and by *Escherichia coli. Pestic. Biochem. Physiol.*, 12: 68-74.

Walker, W.W. and Sojanovic B.J. (1974). Malathion degradation by an *Arthrobacter* sp. *J. Environ .Qual.*, 3: 4-10.

Webley, D.M., Duff R.S. and Farmer V.C. (1957). Formation of β-hydroxy acetic acid is an intermediate in the microbiological conversion of monochlorophenoxybutyric acids to the corresponding substituted acetic acids. *Nature* (London), 179: 1130-1131.

Wedemeyer, G. (1966). Dechlorination of DDT by *Aerobacter aerogenes. Science,* 152 : 647.

Weisshaer, M. P., Franklin F.C. and Reineke, W. (1987). Molecular cloning and expression of 3-chlorobenzoate degrading genes from *Pseudomonas* sp. strain B13. *J. Bacteriol.,* 169(1): 394-402.

White, N.D.G. and Nowinki .T.W. (1985). Effects of temperature and duration of storage on the degradation of Malathion residues on dry rape seeds. *J. Stored Products Res.,* 21(3): 111-114.

Wintersteen, W. K. and Foster, D. E. (1992). Degradation of malathion on wheat and corn of various moisture content. *J. Econo. Entomol.,* 85(3): 1015-1022 (8).

Wolfe, N.L., Zepp, R.G., Baughman,G.L. and Gardon J.A. (1975). Kinetic investigation of malathion degradation in water. *Bull. Environ. Contamin. and Toxicol.,* 30 (6):707-713.

Xu, B., Wild J. R. and Kenedy C. M. (1996). Enhanced expression of a bacterial gene for pesticide degradation in a common soil fungus. *J. Ferment. Bioeng.,* 81(6): 473-481.

Xuliang, Z., Chain.J., Hojaushin and Bai, Z. (2007). New advances in plant growth promoting rhizobacteria for bioremediation. *Environ. Int.,* 33(3): 406-413.

Yang, J., Kloepper, J.W. and Ryu, C.M. (2009). Rhizosphere bacteria help plants tolerate abiotic stress. *Trends Plant Sci.,* 14(1):1-4.

Yoshi, K., Tsumura, Y., Ishimitsu S., Tonogi, Y. and Nakamura, K. (2000).Degradation of malathion and phenthoate by glutathione reductase. *J. Agri. Food Chem.,* 48(6): 2502-2505.

Yu, C., Lu D.G., Quin S J., Du G.D., and Liu G.C. (2007). Microbial flora in *Cerasus sachaliensis* rhizosphere. *Ying Yong Sheng Tai Xue Bao* 18(10): 2277-2281.

Zeinat, K. M., Nashwa, A.H., Fetyan, A., Ibrahim M.A. and Nagogy S. E. (2008). Biodegradation and detoxification of malathion by *Bacillus thuringiensis* MOS-5. *Austr. J. Basic Appl. Science,* 2(3): 724-732.

Zhang, R., Cui, Z., He, J. D. J., Gu, X .and Li, S. (2005). Diversity of organophosphates pesticide degrading bacteria in a polluted soil and conservation of their organophosphorus hydrolase genes. *Can. J. Microbiol.,* 51(4): 337-343.

Zhang, R., Cui, Z., Zhang, X., Jiang, J., Gu, J. D. and Li, S. (2006). Cloning of the organophosphorus pesticide hydrolase gene cluster of seven degradative bacteria isolated from a methyl parathion contaminated sites and evidence of their horizontal gene transfer. *Biodegr.,* 17(5): 465-472.

Zun, X., Qui, X., Dai, J., Cao, H., Yang, M., Zhang, J. and Xu M. (2006). Isolation and characterization of *Pseudomonas oliovarans* degrading the chloroacetaminde herbicide, acetochlor. *Biodegr.,* 17(3):219-225.

❑❑❑

Microbial Diversity and Functions, 2012
© D.J. Bagyaraj, K.V.B.R. Tilak, H.K. Kehri (eds.), pp. 127-148
New India Publishing Agency, New Delhi (India)
E-mail : info@nipabooks.com; Website : www.nipabooks.com

Chapter **5**

Application of State-of-the-Art Microbial Biotechnology Coupled with Phycoremedial Measures for the Treatment of Secondary Liquid Effluents of a Gelatine Manufacturing Industry

Sudhir D. Ghatnekar and Santosh M. Sharma

ABSTRACT

The present paper describes application of synergistic action of microalgae viz., Spirulina platensis, Chlorella vulgaris, and Oscillatoria princeps coupled with state-of-the-art microbial biotechnology (that involved strains of Bacillus subtilis, Nitrobacter winogradskyi, Aspergillus flavus, and A. niger) developed by Biotechnology Resource Centre (BRC), Mumbai for treatment of secondary liquid effluents from a gelatine manufacturing industry. The growth of mixed cultures of selected microbial and algal species was measured respectively in terms of Colony Forming Units (CFUs) and Optical density (O.D.) of the culture. The results were indicative of efficient degradation potential of the proteinaceous liquid effluents. Moreover their BOD and COD loads reduced by 30.75 % and 84.10 % respectively after the treatment. Furthermore, the bio-safety of algal biomass cultured in selected effluents as feed supplement was tested on Swiss albino mice (Mus musculus). Mice in treated group didn't exhibit any observable adverse effect on their growth or on their reproductive performance. In fine, the liquid effluents of gelatine manufacturing industries were polished along with production of probiotic food supplement.

Keywords : bio-safety, microbial biotechnology, phycoremedial measures, probiotic, gelatine.

Introduction

According to the United Nations Special Rapporteur on the Right to Food, Olivier de Schutter, one billion people sleep without food across the world, and every six seconds a child dies of malnutrition. The global food crisis is far from being abated. To add to the woes of food crisis, competition for water resources among individuals, regions, and countries and associated human activities is already occurring with the current world population. About 40 % of the world's people live in regions that directly compete for shared water resources (Pimentel *et al.,* 1996).

There is a direct interrelation between the water crisis and food security. A recent report titled 'Charting our Water Future', released by a Water Resources Group in Washington DC stated that, by 2030, the global water demand will overshoot current supply by 40%, assuming average economic growth and no water efficiency gains. The agriculture, which claims 71% of global water resources, is a prime driver of this gap that may intensify the tradeoff between food security and water security.

On analyzing the gravity of above mentioned situation, it becomes crystal clear that human beings need to tackle two urgent problems of global proportion and magnitude: food and water crisis. The researchers all over the world are constantly striving to find an alternative solution to tackle this foreseen scenario in coming decade.

The increasing scarcity of water in the world in relation to rapid population increase gives reason for concern. Besides it also points to the urgent need for appropriate water management practices. Very little investment has been made in the past on treatment of liquid effluents, which contains water as its major constituent. However, due to the trends in urban and industrial development, wastewater treatment deserves greater emphasis (Sarah, 2003). It shall enable recovery of misplaced resources in form of nutrients as well as a large quantity of water from them.

On the food crisis front, the probable solution to embark upon it is to look out for a safe naturally existing food sources such as algae. Interestingly the algal population constitutes about two-thirds of the Earth's biomass. Thousands of algal species covering the Earth are now being identified as source of food, pharmaceuticals, biochemicals and fertilizers. Algae thus represent one of the potential solutions as we need to produce food while restoring our planet.

In this context, Microalgae such as *Chlorella, Spirulina* etc. are acclaimed as non-conventional food besides their multidimensional applications.

Environment *vis-à-vis* Economy

Man correlates his development with rapid industrialization and urbanization. Increasing human population needs a large number of industries to fulfill its diverse necessities. Then again, industrial activities generate a huge amount of obnoxious wastes which have to be released into the environment. However, if these wastes are not treated suitably before discharging, they can cause detrimental effects to the environment and its components (Ghatnekar *et al.*, 2009 a). These effects have already manifested in the form of hazardous phenomena like acid rain, global warming and deterioration of water resources.

Wastewater: an invaluable resource of water

A considerable proportion of the liquid effluents generated by industries, consist of water, a valuable resource. In the industrial towns millions of gallons of water in the form of these effluents are thrown into the gutters everyday while we raise a hue and cry about the water crisis (Ghatnekar, 1983). Moreover, if we fail to recover organic wastewater from urban areas it would mean a huge loss of life-supporting resources. These resources instead of being used fruitfully shall fill rivers or sea with undesirable pollutants (Rose, 1999).

Water supply and treatment often received more priority than wastewater collection and treatment. According to the World Bank, "The greatest challenge in the water and sanitation sector over the next two decades will be the implementation of low cost sewage treatment that will at the same time permit selective reuse of treated effluents for agricultural and industrial purposes" (Looker, 1998).

Nevertheless, with the growing concern for the environment, stricter regulations are now being enforced on industries for safer waste disposal. In the recent past, environmental regulations have undergone vast changes. As a result, conventional treatment technologies have been further refined and new technologies for wastewater treatment are being implemented (Kumar *et al.*, 2008).

Bioremedial measures for treatment of wastes are looked upon as more sustainable alternatives in relation to conventional waste treatment technologies. Biological methods such as, constructed wetlands (Babatunde *et al.*, 2008), vermiculture treatment (Ghatnekar *et al.*, 2009 b and 2010 b), phytoremediation (Meagher, 2000) etc. are attracting considerable attention.

Phycoremedial measures for treatment of wastewater

Phycoremedial measure may be defined in a broad sense as the use of macro algae or micro algae for the removal or biotransformation of pollutants, including nutrients and xenobiotics from wastewater and CO_2 from waste air (Olguín, 2003). This field has evolved from the earlier works done by Oswald and Gotaas (1957) that involved the application of micro algae for tertiary treatment of municipal wastewater to many other applications.

Algal blooms occurring naturally in wastewater are considered as important bio-indicators of pollution. In natural water bodies and wetlands, natural forces (chemical, physical, and solar) act to purify the water. Then again, multiple forms of bacteria, planktons, microalgae and macrophytes act together or in succession, thereby achieving wastewater treatment (Ghatnekar *et al.,* 1979).

This phenomenon provides an insight upon ability of these organisms to flourish in abundant organic and inorganic matters that are present in the wastewater. Simultaneously, it also points out at nature's method of self purification of its resources.

Algal members known to occur in sewage are Chlorophycean forms like *Chlorella, Chlamydomonas, Ankistrodesmus, Spondylomorum, Pandorina, Eudorina, Scenedesmus, Oocystis;* Myxophycean forms like *Anabaenopsis, Oscillatoria, Merismorpedia;* Diatoms like *Navicula, Nitzschia, Cymbella, Cyclotella* and the Euglenophycean forms like *Euglena* and *Phacus* (Venkataraman, 1969).

The fundamental principles involved in treatment of domestic sewage and effluent by these photoautotrophic organisms are: Wastewater consists of organic and inorganic matter rich in Carbon (C), Nitrogen (N), Phosphorus (P), Sulphur (S), Potassium (K) etc. The oxidation and degradation of complex matter in the wastewater is accomplished by symbiotic action of algae and microorganisms. The oxygen liberated by the algae during photosynthesis is utilized by the microorganisms which oxidize the organic compounds in the sewage into forms that are suitable for utilization by algae and other phototrophic organisms as nutrients for their growth.

The treatment of raw and anaerobically digested animal waste by algal cultures (Ayala and Bravo, 1984) and suspended algae in high rate pond systems (Olguin, 2003) has been reported earlier.

This present paper describes the application of commercially important microalgae *viz., Spirulina platensis, Chlorella vulgaris,* and *Oscillatoria princeps* coupled with state-of-the-art microbial biotechnology developed by the Biotechnology Resource Centre (BRC), Mumbai for treatment of secondary liquid effluents generated by a gelatine manufacturing industry.

Thus, the present study might be of specific interest for biotreatment of effluents from gelatine manufacturing industries and also for cultivation of commercially important microalgae using liquid effluents as resource of water and nutrients (Plate 4).

Biomanagement of Secondary Liquid Effluents Generated by Gelatine Manufacturing Industry

Gelatine industry

Gelatine, a heterogeneous mixture of water-soluble proteins derived from the collagen of animal hide or bone is composed of various essential amino acids required for human nutrition. It has a diversity of uses in food science and industry, including wine fining, confectionary, matches and photography (Maree *et al.*, 1990); therefore, the gelatine manufacturing industries are of great importance to the society. Figure 1, displays various steps involved in manufacture of gelatine and its useful applications.

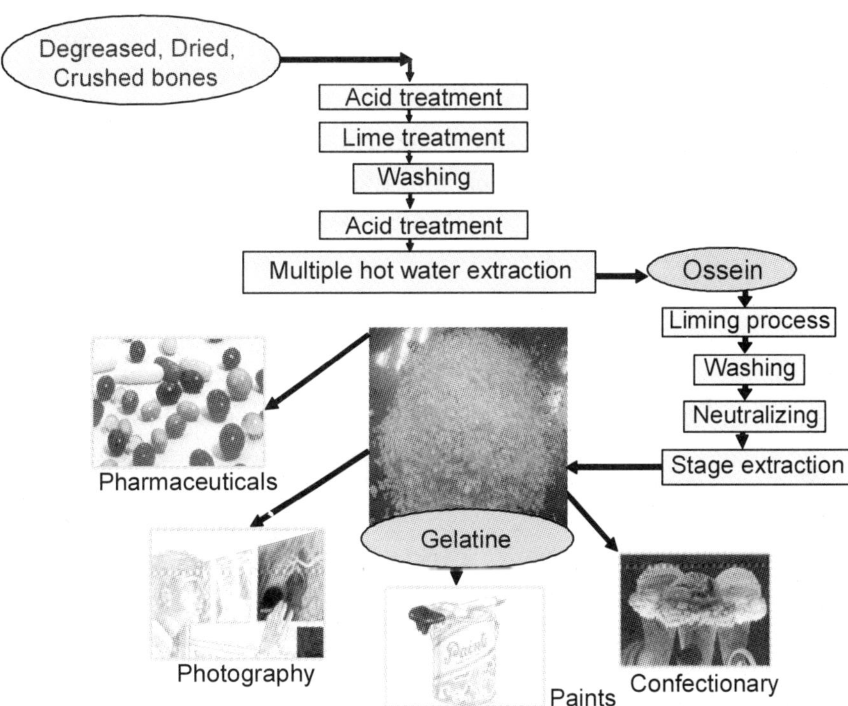

Fig. 1 : Gelatine manufacturing process and It's applications

Characteristics of liquid effluents generated by gelatine manufacturing industry

The major consideration during the manufacture of all forms of gelatine is the generation of large amounts of process waste effluents. These waste effluents contain mineral components and lipid material, which create a high biological oxygen demand (BOD), and can be alkaline or acidic (NOSB TAP Review Compiled by OMRI for the USDA National Organic Program, 2002).

Maree *et al.* (1990) reported that the pit effluents of a gelatine manufacturing unit are high in pH, COD and calcium content, which constitute a major source of pollution. The bone waste collected from a discharging outlet of the ossein industry seems to be fibrous and greases with a pasty matter containing little pieces of bones and hairs. It is composed of a higher quantity of proteins and a small amount of lipids. This is thought to be a major contributor of the odour associated with the ossein industry (Pualchamy *et al.,* 2008).

Conventionally, these effluents are treated in Effluent Treatment Plants (ETP) using the activated sludge method (Plate: 1a) (Buyukgungor and Gurel, 2009). Later, the active sludge is dehydrated to form a sludge cake, which is subsequently disposed. However, the disposal of organically-rich sludge generated from conventional ETPs requires a large dump yard. Furthermore, it may potentially cause environmental pollution by leaching or harbouring pathogens. Discharge of effluents without suitable treatment can cause environmental damage. Sharma and Rao (1996) reported adverse effects of gelatine factory effluents discharged into the Narmada River. These included a negative impact on seed germination and seedling growth of important crop plants irrigated with river water.

In India, Pollution Control Boards in some states are now imposing additional conditions, which read "concerned Industry shall utilize the whole quantity of treated effluent within their premises for plantation and horticulture etc. Effluent shall not be discharged outside the factory premises in any circumstances. Zero discharge conditions shall be maintained" (MPPCB, 2009).

Effluent treatment protocol adopted in the present study

IGCL, Vapi, Gujarat produces food and pharma grade gelatine. During gelatine manufacturing processes, the plant generates 3000 m^3 of liquid effluents every day that amounts to an average of around 30 tonnes of proteinaceous bio-solid sludge (Ghatnekar *et al.,* 2009 b). The protocol adopted for Pilot scale treatment of selected liquid effluents was as follows.

1). Analysis of physico-chemical characteristics of selected liquid effluents

The selected liquid effluents were analyzed to determine the major physico-chemical parameters (pH, BOD, COD, TS, and TDS) before treatment (Trivedi and Goel, 1986). The results display its acidic nature. It exhibited high BOD (452.2 ± 5.0) and COD (1818 ± 8.46) values. The organic and inorganic content of the effluent mainly constituted of the suspended solids (TS) and dissolved solid content (TDS). The analysis results are presented in

Table 2. Ghatnekar *et al.* (2009 a) reported that the gelatine industry wastes are rich in nitrogen and other minerals like calcium, phosphorus and silica.

2. Isolation and characterization of microorganisms from selected waste and selection of degrading brigade for the effluent treatment.

The microbial populations in the selected waste were isolated and characterized (Plate 2 a). Characterization studies revealed dominance of strains of bacteria *viz., Bacillus sp., Cellulomonas sp., Nitrobacter sp.,* and fungi *viz., Trichoderma sp., Aspergillus sp.* The selected effluents also exhibited presence of algae *viz., Oscillatoria sp., Anabaena sp.* etc (Plate 1 b). Presence of *Bacillus, Aspergillus and Trichoderma* species in gelatine waste was earlier reported by Abrusci *et al.* (2005).

Based on characterization of potential degrading microbial species from the selected wastes, the following microbial and algal species were finally selected for the degradation experiments.

- Bacterial sp.- *B. subtilis, N. winogradskiyi.*
- Fungal sp. - *A. flavus, A. niger.*
- Algal sp. - *S. platensis, O. princeps and C. vulgaris.*

3. Acclimatization of selected microbial and algal species in liquid effluents

The pure cultures of selected microbial species were procured from the culture bank of BRC. Later, the effluent resistant isolates of selected microbial species *viz., B. subtilis* (BRC-4), *N. winogradskyi* (BRC-5), *A. flavus* (BRC-27), and *A. niger* (BRC-28) were cultured in flasks and maintained as standard type strains at 30 °C and 90 rpm in rotary shakers (Orbitek). Later these isolates were multiplied in semi-automated fermentors (BRC-Bio-Boom).

Axenic cultures of the selected algal species were procured from BRC. The waste acclimatized algal strains *S. platensis* (BRC-45), *O. princeps* (BRC-46), *C. vulgaris* (BRC-47) were multiplied in flask culture and used for further experiments. Polyculture of selected algal species was initially maintained in laboratory conditions (*i.e.* artificial light (12 hours/day) provided by milky fluorescent tubes (25 W) and temperature 32 (± 3) °C during day time and 27 (± 3) °C during the night). Later, the cultures were gradually transferred to external conditions. During these steps the effluent treatment protocol was optimized and finally scaled up in Green house condition (Plates: 3a, b).

3. Enzymes

The selected liquid effluents are rich in proteinaceous and cellulosic content hence protease and cellulase enzymes (procured from Egenix Biotech Pvt. Ltd., Bangalore) were selected for the pretreatment of complex organic matter (Ghatnekar *et al.,* 2009 a).

4. Treatment of selected liquid effluents

The pilot scale experiments were conducted in 5 circular ferro- cement tanks each 1.2m in diameter and 0.7m height (capacity, 700 L). The experiments constituted of five sets (Plate 3b). In Set I (control) the selected algal species (polyculture system) were cultured in BRC-Spiro medium (100%). Sets II (T1), III (T2), IV (T3), and V (T4) comprised of treated sets in which the effluents treatment were carried. The effluents at concentrations 25% (T1), 50% (T2), and 75% (T3) mixed with BRC-Spiro medium were selected for treatment process. BRC-Spiro medium is an organic medium developed by BRC for the commercial cultivation of micro algae such as *Spirulina* (Tamhane and Ghatnekar, 2004). The experimental set T4 consisted of 100% effluents.

The details of the experimental components contained in each set are presented in Table 1.

Table 1 : Concentration of liquid effluents used in combination with BRC-Spiro medium for each experimental sets.

Treatments	Culture media	Inoculant
Set I (C)	BRC – Spiro media (100 %)	Polyculture of selected algal spp.
Set II (T1)	BRC – Spiro media (75 %) + liquid effluent (25 %)	Waste acclimatized enzymes, selected microbial and algal spp.
Set III (T2)	BRC – Spiro media (50 %) + liquid effluent (50 %)	Waste acclimatized enzymes, selected microbial and algal spp.
Set IV (T3)	BRC – Spiro media (25 %) + liquid effluent (75 %)	Waste acclimatized enzymes, se lected microbial and algal spp.
Set V (T4)	Liquid effluent (100 %)	Polyculture of selected algal spp.

C, Control; T, Treated;

The cultures in the experimental tanks were initiated with volume of 150 L and gradually scaled up to 500 L. The mixture of selected concentrations of liquid effluents and BRC-Spiro medium were introduced into the tanks. The combination of protease and cellulase enzymes (2:1) and mixed culture of selected microbial species was inoculated (at 5% concentration) in treated sets, T1, T2, and T3. Control and set T4 were not subjected to the enzymatic or microbial treatment. Algal cultures (at 10% concentration) were then inoculated in all the sets.

The cultures in treatment tanks were maintained in semi sterile conditions under ambient light and temperature regime throughout the experimental period. The cultures were initially aerated for 45 minutes, 2 to 3 times per day. However, as the density of cultures increased the aeration rate was also increased (5 to 7 times per day).

The effluents in each experimental set were monitored for selected parameters. The physico-chemical changes in the effluents recorded after a period of 45 days are presented in Table 2. In all the experiments total Colony Forming Units (CFUs) and CFUs of selected microbial species were enumerated as indicators of biodegradation potential (Table 2). The status of microalgal cultures were regularly monitored under compound microscope (Model CXL, Labo, India) (Plate 2b). The O.D. of cultures in experimental tanks recorded using 215 D Visible Spectrophotometer (Chemito) at 560 nm was indicative of growth rate of algal species (Table 3).

Table 2 : Chemical and Microbiological analysis of the Liquid effluents before and after treatment

Parameters	Before treatment	After treatment (45 days)				
Chemical Analysis		Control	T_1	T_2	T_3	T_4
pH	6.14 ± 0.103	7.52 ± 0.10	6.9 ± 0.03	7.26 ± 0.04	7.44 ± 0.04	7.9 ± 0.03
Temperature	Ambient	Ambient	Ambient	Ambient	Ambient	Ambient
Total Solids (TS) %	0.316 ± 0.004	0.342 ± 0.008	0.386 ± 0.004	0.424 ± 0.004	0.468 ± 0.004	0.484 ± 0.008
Total Dissolved Solids (TDS) %	0.268 ± 0.007	0.272 ± 0.004	0.182 ± 0.006	0.154 ± 0.004	0.238 ± 0.002	0.284 ± 0.002
Biological oxygen demand (BOD) - ppm	452.2 ± 5.0	407 ± 5.61	341.8 ± 2.19	313.0 ± 3.00	363 ± 1.22	381 ± 2.45
Chemical oxygen demand (COD) - ppm	1818 ± 8.46	667 ± 4.36	550 ± 5.48	289 ± 10.54	612 ± 8.0	870.6 ± 13.36
Microbiological analysis						
(Total CFU)	1.86×10^5	1.46×10^5	$2.07 \times 10^{5.5}$	2.03×10^7	$2.36 \times 10^{6.5}$	2×10^5
B. subtilis	-	-	1.74×10^1	$0.44 \times 10^{1.1}$	0.46×10^1	-
N. winogradskyi	-	-	3.21×10^1	$0.69 \times 10^{1.25}$	0.93×10^1	-
A. flavus	-	-	$1.96 \times 10^{1.5}$	$2.1 \times 10^{1.7}$	$1.78 \times 10^{1.5}$	-
A. niger	-	-	$2.43 \times 10^{1.7}$	$2.79 \times 10^{1.85}$	$0.74 \times 10^{1.7}$	-

C, Control; T, Treated; TS, Total Solids ; TDS, Total Dissolved Solids ; BOD, Biological oxygen demand ; COD, Chemical oxygen demand

Data collected were analyzed as per the methods used by Ghatnekar *et al.* (2009 c).

Degradation potential of the selected liquid effluents

Enzymatic pre-treatment of the liquid effluents rendered them as nutrient rich medium for growth of selected microorganisms and algal species. Ossein waste from gelatine industry also promoted production of protease enzyme by the microbial population in it (Jegannathan and Viruthagiri, 2009).

Table 3 : Effect of different concentrations of selected effluents on Growth rate of algal cultures in terms of Optical Density (O.D.) of the culture.

Number of days	O.D. (560 nm) C	% I	T1	% I	T2	% I	T3	% I	T4	% I
0	0.47	-	0.51	-	0.54	-	0.59	-	0.67	-
15	0.68	44.68	0.62	21.56	0.70	29.62	0.68	15.25	0.71	5.97
30	0.82	74.46	0.75	47.05	0.82	51.85	0.75	27.11	0.74	10.44
45	0.91	93.61	0.81	58.82	0.89	64.81	0.78	32.20	0.79	17.91

C, Control; T, Treated; % I – percentage increase in absorbance in relation to zero reading

The mixed cultures of selected microbial and algal species exhibited efficient degradation potential. Enzymes and microbes accomplished degradation of the organic contents in the liquid effluents. It further initiated a series of alternate aerobic and anaerobic microbial reactions as the result the microbial population exhibited exponential increase.

Within 45 days, the total CFU increased from 2×10^5 to 2×10^7 in set T_2. Likewise, the inoculated microbial species displayed prolific growth and the highest CFU count of *B. subtilis* 1.80 (± 0.0081) × $10^{1.25}$, *A. flavus* 1.77 (± 0.031) X $10^{1.7}$, *A. niger* 1.87 (± 0.033) X $10^{1.85}$, and *N. winogradskyi* 1.74 (± 0.013) X $10^{1.5}$ in set T_2 manifested conducive probiotic factor (Table 2).

Amongst the other sets, set T3 exhibited better degradation potential in relation to set T1. The total CFU in set T3 increased from 2.77 ± 0.037 × 10^{5} (0 days) to 2.02 ± 0.1× $10^{6.5}$ (45 days). However, after 45 days, the CFU of selected microbial species was higher in set T1 [*B. subtilis* 4.31 (± 0.149) × 10, *N. winogradskyi* 6.25 (± 0.172) × 10, *A. flavus* 2.23 (± 0.189) × $10^{1.5}$, and *A. niger* 2.10 (± 0.172) × $10^{1.7}$ in relation to set T3 [*B. subtilis* 1.51 ± (0.05) × 10, *N. winogradskyi* 3.67 (± 0.28) × 10, *A. flavus* 2.12 (± 0.191) × $10^{1.5}$, *A. niger* 2.2 (± 0.19) × $10^{1.7}$.

The Control set exhibited least total CFU count [1.992 (± 0.028) × 10^5] amongst all the sets. However, the occurrence of microorganisms in Control set was due to natural succession process as the dead cell of algal population increase in the culture tanks. This phenomenon poses as difficulty in continuous cultivation of algae.

Set T4 (100% effluents untreated with enzymes and microorganisms) exhibited very slow increase in microbial population its total CFU count was the least [2.56 (± 0.157) × 10^{5}] amongst the treated sets at the end of 45 days this suggests the need for suitable treatment of the liquid effluent.

The results indicated steady growth in the microbial population in the effluents in all the sets. However, the highest increase in Total CFU and CFU of selected microbial species was observed in set T2.

Characteristics of liquid effluents after treatment

The results of physico-chemical analysis of selected effluents after treatment are presented in Table 2. After 45 days of treatment, the effluents at 50% concentration (T_2) exhibited significant reduction in BOD (30.75 %) and COD (84.10 %) values. Set T_2 demonstrated the lowest BOD and COD values amongst all the four sets (BOD 313 ± 3.0 ppm and COD 289 ± 10.54 ppm). On the other hand, 100% concentration of effluents exhibited least reduction in BOD (15.70%) and COD (52%) value indicating need for suitable treatment. The photosynthetic machinery of the algae provided partial aerobic environment besides constant aeration also reduced the BOD.

The decrease in TDS content after treatment was indicative of degradation potential of the effluents. Also it indicated the ability of algae to assimilate the dissolved contents as their nutrients. The set T_2 displays 42.5 % reduction in TDS that was due to excellent growth of algal population.

In present study, the growth of *Spirulina* increased the alkalinity of the culture medium which in accordance with study conducted by Van Hille *et al.* (1999) on heavy metal precipitation from acidic water streams. The acidic nature of the selected effluents can thus be raised to neutrality due to algal growth.

Growth of algal species in treated effluents

Percentage increase in O.D. (% I) with respect to O.D. at the beginning of experiment was indicative of favourable growth of algal population (Table 3).

The control set (100% BRC-Spiro medium) exhibited the highest % I in O.D. (93.61%). Among the treated group, Set T2 exhibited the highest % I in O.D. (64.81 %) followed by set T1 (58.82 %) then T3 (32.20 %) and least % I in O.D. was recorded in set T4 (17.9 %).

Highest biomass harvested from the experimental sets was in Control (7.85 Kg) while amongst the treated sets, set T2 yielded higher harvest efficiency

(6.34 Kg) than other sets. Set T4 yielded lowest yield (1.85 Kg). Lower concentration of effluents 25 % and 50 % supported better microbial and algal growth than 75 % and 100 % concentrations. The highest algal growth in set T2 was indicative of optimal concentration of nutrients in relation to other treated sets (Table 4).

Table 4 : Fresh weight of biomass harvested from circular ferro-cement tanks at the end of 45 days.

Number of days	Fresh weight of biomass harvested (kg)				
	C	T1	T2	T3	T4
45	7.85	5.86	6.34	5.87	1.85

C, Control ; T, Treated;

In the present study, the microorganisms and algal population exhibited excellent synergistic action in bio-treatment of selected proteinaceous liquid effluents. The microorganisms selected in present study justified their role in waste treatment process.

Application of microbial biotechnology in waste treatment process

Fungi are known to be an important degrader in activated sludge. These multicellular organisms can metabolize organic compounds and can successfully compete with bacteria under certain environmental conditions in a mixed culture (Water Environment Association, 1987). In present study higher CFU content of selected fungal species in relation to bacterial species can be argued on above finding.

Kavian and Ghatnekar (1998) utilized the fungal species *viz., A. flavus* and *A. niger* in the treatment of the waste from pharmaceutical industry. Ghatnekar *et al.* (2009 a, b) used combination of *A. flavus, A. niger* and *Bacillus licheniformis* in the "three-tier vermiculture biotechnology" for conversion of bio-solid waste from the gelatine manufacturing industry into useful plant probiotics.

In another study, Ghatnekar *et al.* (2009 d) reported the bio-management of liquid effluents discharged after the secondary treatment from gelatine manufacturing industry using combination of *A. flavus* and *A. niger*. Furthermore, authors successfully cultured *S. platensis* on these pre-treated effluents.

Nitrifying bacteria, *Nitrobacter* is reported to efficiently convert ammonia to nitrate ($2NO_2 + O_2 \rightarrow 2NO^{3-}$ + energy) in activated sludge (Water Environment Association, 1987). In present study, the increase in *N. winogradskyi* was conspicuous after period of 30 days. It is because the longer mean cell residence time, allows an adequate population of nitrifying bacteria to be built up. Moreover, the oxygen demand for complete nitrification is high and the rapid growth of algal population provides the microbes with necessary oxygen supply during their rapid photosynthesis.

Bacillus sp are important denitrifying (facultative anaerobic) bacteria (Claus and Berkeley, 1986) that use Nitrate from the wastes and return N_2 and N_2O to the surrounding. These components are absorbed by the algal cultures for their growth.

Application of microalgae in treatment of effluents

Microalgae are known to play a fundamental role in primary and secondary wastewater treatment (Shushu and Chipeta, 2002). Cultivating microalgae by using waste water as a nutrient source can be considered as a heterotrophic or more precisely mixotrophic cultivation (Lodi *et al.*, 2005). As heterotrophic metabolism is faster than autotrophic (Chen *et al.*, 1996), simple C sources in form of treated effluents can be employed to sustain the algal growth.

In recent past, intensive algal cultures is being employed in outdoor ponds as a tertiary treatment for removal of residual organic compounds (Kawai *et al.*, 1984). Moreover, cultivation of commercially *via*ble microalgae in wastewater offers multiple benefits of which few can be listed as; cost effective and sustainable method for bio-treatment of domestic and industrial wastewater, reducing input cost of microalgal production by utilizing the nutrients available in wastewater, recovery of invaluable resource *viz.*, water, and the most important of all would be, production of protein rich microalgal biomass that finds tremendous application in food industry, pharmaceutical and cosmetic industry, animal husbandry etc. Thus, the wastes of the industries can be polished along with production of nutritional food supplement (Ghatnekar and Ka*vian*, 1992; Vonshak, 1997).

The different species of microalgae selected in present study have also been tested for their potential role in efficient wastewater treatment. Several attempts at utilization of these species can be cited as follows;

Siew - Moi (1987) and Tanticharoen *et al.* (1993) suggested growth of *S. platensis* in anaerobically treated effluents from palm oil mill and starch factories respectively. Van Hille *et al.* (1999) used *Spirulina* in treating either a tannery or sewage effluent as part of an integrated process.

Chen *et al.* (1996) reported heterotrophic and mixotrophic growth of *Arthrospira (Spirulina)* and *Chlorella*. Soni *et al.* (2006) reported satisfactory growth of *Oscillatoria sp.* on secondary activated sludge effluent supplemented with 1% $NaHCO_3$. Shushu and Chipeta, (2002) reported use of *Spirulina* and *Oscillatoria* isolated from Gaborone oxidation ponds for reduction of excess nutrients from wastewater effluents before discharge into the receiving streams. They observed that the growth rate of algae cultured in wastewater was similar to algae cultured in specific media.

Ghatnekar *et al.* (2010 a) demonstrated the use of aquatic eco-web for perpetual treatment of secondary liquid effluents obtained from a gelatine manufacturing industry without disturbing its biodiversity. The objective of the above study was to monitor the co-existence of these biotic components and their synergistic role in the biomanagement of selected liquid effluents. In addition it focused perpetual zero discharge process.

Economic importance of algal species selected for present study

Commercially important microalgae such as *Chlorella, Spirulina, Oscillatoria* etc. are well known for their nutraceutical properties as well as abundance of commercially important bioactive compounds. The microalgae *viz., Spirulina* and *Chlorella* have been acclaimed in the health food industry as single cell protein (Ghatnekar and Sharma 2009). Sharma *et al.* (2008) reported application of organically cultured *S. platensis* as a feed supplement for Mozambique Tilapia and Indian major carps.

Ghatnekar *et al.* (2009 e) suggested that the microalgae *Spirulina,* can be rightly called as a probiotic nutraceutical food supplement of 21st century due to its exceptional nutritional and therapeutic values. The authors further proposed the integration of *Spirulina* cultivation with conventional agriculture for the multiple benefits of farmer.

The amino acid composition of the *Oscillatoria* algal protein compares favorably with the tentative standard for ideal protein defined by FAO/WHO (Hashimoto and Furukawa, 1989).

Besides, *Spirulina* and *Oscillatoria* are reported to form a source of commercially important phycobiliprotein *viz.,* Phycocyanin (Soni *et al.,* 2006; Silveira *et al.,* 2007).

In addition, their ability to grow in wastewater is being tapped by researchers for biomanagement of industrial and domestic wastewaters (Becker, 1994).

Bio-safety evaluation of microalgae cultured in selected liquid effluents

The algal biomass harvested in the present study consisted of nutritional algae such as *Spirulina*, *Chlorella* and *Oscillatoria*. The biomass was dried and further tested for evaluation of its bio-safety on Swiss albino mice (*Mus musculus*). The experiments were conducted in 3 sets. Set I (Control) consisted the mice fed with only standard mice feed (SMF) while Set II (Treated set - T1) consisted of mice fed with SMF + organically cultured *Spirulina* and Set III (Treated set - T2) consisted of mice fed with SMF + microalgal biomass cultured *in effluents.*

The effect of inclusion of microalgae cultured in effluents as a feed supplement was monitored in terms of growth and reproductive performance of the mice (Plate 5). The nutraceutical efficacy of such algal biomass was compared with standard mice feed (Control set).

In trial period of 3 months, feeding microalgae cultured in effluents did not exhibit any adverse effect on the mice. The increase in weight of mice in Control and Treated groups is presented in Table 5.

Table 5 : Average weight of mice administered with E-*Spirulina* supplement in relation to mice administered with O-*Spirulina* and SMF.

Parameter	Average weight of 3 mice (g) in Control and Treated groups					
Days	Set I (C)	% I	Set II (T 1)	% I	Set III (T 2)	% I
0	13.2	-	15.8	-	11.4	-
15	16.4	24.2	17.6	11.39	11.9	4.38
30	20.63	56.28	19.8	25.31	14.47	26.92
45	20.86	58.03	26.42	67.21	17.84	56.49
60	22.49	70.37	28.44	80.0	19.0	66.6
75	23.02	74.39	27.13	71.7	20.35	78.5
90	23.96	81.53	29.8	88.60	21.98	84.08

% I – Percentage increase in relation to zero day weight, C - Control, T - Treated
SMF - Standard Mice feed; O – *Spirulina* – Microalgae cultured in Organic medium; E – *Spirulina* – Microalgae cultured in effluents
Set I – Mice fed with SMF; Set II – Mice fed with SMF + O – *Spirulina* powder (10% concentration); Set III - Mice fed with SMF + E – *Spirulina* powder (10% concentration)

Inclusion of protein supplements in treated set is evident right from the beginning of the trials as mice in both the treated groups exhibited higher percentage increase (88.60 % and 84.08 %) in weight in relation to mice in control group fed with only standard mice feed (81.53 %).

Plate 1a : Primary treatment of liquid effluents at the ETP of gelatine industry.

Plate 1b : Secondary liquid effluents at the ETP of gelatine industry exhibiting natural growth of algae.

Plate 2a : Colony Forming Units of selected microorganisms flourishing in the liquid effluents at different dilution.

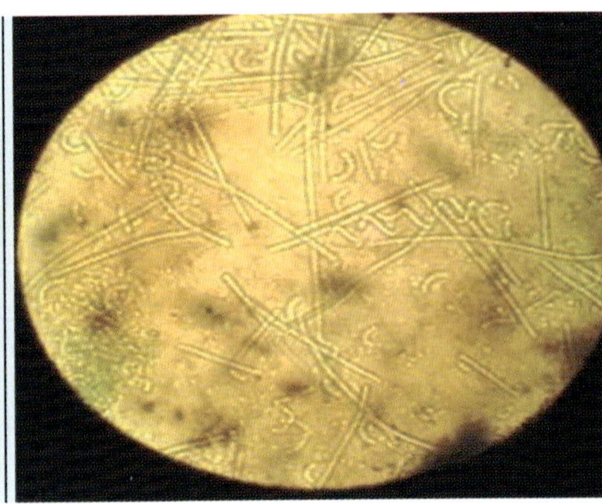

Plate 2b : Microscopic view of selected micro algae cultured in the gelatine industry effluents (10 x)

Plate 3a : Effluent treatment studies (Laboratory scale)

Plate 3b : Effluent treatment studies (Pilot scale)

Plate 4 : *Spirulina* cultured in Secondary liquid effluents (50% concentration) mixed with BRC-Spiro medium

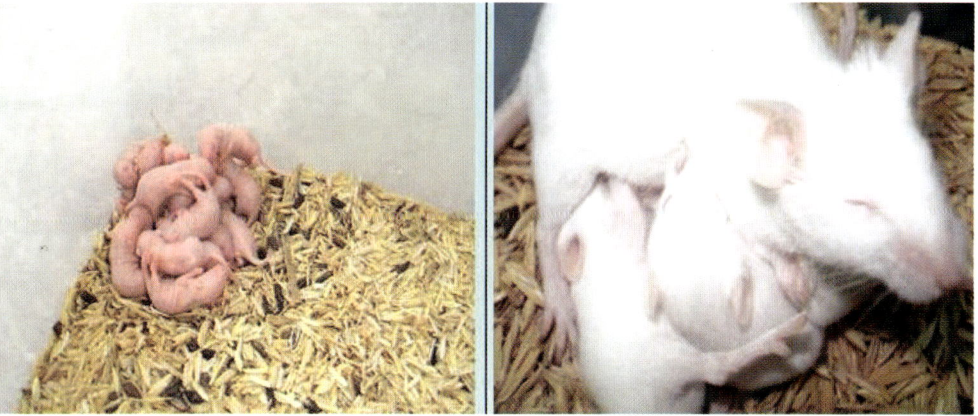

Plate 5 : Evaluation of bio-safety of S.Platansis cultured in liquid effluents on growth rate and reproductive performance of Swiss albino mice.

Becker and Venkataraman (1982) reported that *Spirulina* cultured in waste such as bone meal and human or cow urine can be incorporated safely at 5% level in chicken rations. Potential application of *Spirulina* cultured in waste water as animal feed had been suggested by Ghatnekar and Ka*via*n (1995).

Conclusion

Microalgal cultivation coupled with the state-of-the-art microbial biotechnology that involves selected microbes and waste specific enzymes can be used as an effective tool for biomanagement of secondary liquid effluents from gelatine manufacturing industry. *S. platensis*, *O. princeps*, and *C. vulgaris* can synergistically accelerate the degradation of selected liquid effluent. Though algae can grow in untreated liquid effluents in association with naturally occurring microbes, their growth rate can be relatively enhanced by the pretreatment with waste specific enzymes and microbes.

In the present study, the secondary liquid effluents generated by gelatine manufacturing industry at 50 % concentration mixed with 50 % BRC-Spiro medium exhibited optimum degradation potential and relative suitability for the growth of selected algal cultures in relation to other concentrations. The technology used here can therefore be applied for recovery of valuable resource: water from an undesirable nuisance: wastewater.

Furthermore, the microalgae cultured in effluents were biosafe and could be used as a nutritional supplement in mice feed. The findings from the present study thus indicate that such biomass can be suitably applied even in aquaculture, poultry and animal husbandry. These animals are the well established protein source for the human beings.

In fine, optimization of the microalgal mass cultivation using liquid effluents as a nutrient and water resource in addition to potential application of algal biomass cultured in effluents as a nutraceutical food or feed supplement offers an excellent measure to grapple the concerning foreseen situation of probable water as well as food crisis.

Acknowledgements

Authors like to acknowledge Mr. Viren Mirani (Executive Director) and Mr. D. R. Vora (C.E.O.) of IGCL, Vapi for providing the required effluent samples.

References

Abrusci, C., Martín-González, A., Amo, A. Del, Catalina, F., Colladod, J. and Platas, G. (2005). Isolation and identification of bacteria and fungi from cinematographic films. *Internat. Biodeterior. & Biodegrad.*, 56, (1) : 58-68.

Ayala F.J. and Bravo R.B. (1984). Animal wastes media for *Spirulina* production. *Arch. Hydrobiol,*. 67(3) : 349–355.

Babatunde, A. O., Zhao, Y.Q., O'Neill M. and O'Sullivan B. (2008). Constructed wetlands for environmental pollution control: A review of developments, research and practice in Ireland. *Environ. Internat.*, 34 (1):116-126.

Becker, E. W. (1994). *Microalgae: Biotechnology and Microbiology*, Cambridge Studies in Biotechnology 10, Cambridge: Cambridge University Press.

Becker, E. W. and Venkataraman, L. V. (1982). Biotechnology and Exploitation of Algae: The Indian Approach, *Deutsche Gesellschaft fur Technische Zusammenarbeit (GTZ)*, Germany.

Buyukgungor, H. and Gurel, L. (2009). The role of biotechnology on the treatment of wastes. *Afr. J. Biotechn.*, 8 (25) : 7253-7262

Claus, D., and Berkeley, R. C. W. (1986). Genus *Bacillus, Bergey's Manual of Systematic Bacteriology*, 2. Williams & Wilkens, Baltimore, MD: 1105-1139.

Chen, F., Zhong, Y. and Guo, S. (1996). Growth and phycocyanin formation of *S. platensis* in photo-heterotrophic culture. *Biotechnol. Lett.*, 18 (5): 603-608.

Ghatnekar, S. D. (1983). Wealth from Waste, *Soc. Sci.*, 6 (3-4) : 44 – 56.

Ghatnekar, S. D., Pandey, R. P., Kulkarni, U. B. and Iyer, S. P. (1979). Biomanagement of Malathion Effluent in Excel Industries Limited, Paper presented at the seminar - cum exhibition on "*Environmental Management and Pollution Control*" organized by Indian Chemical Manufacturer at Bombay, Feb 16 – 17.

Ghatnekar, S. D. and Kavian, M. F. (1992). New Spins-offs of Biotech after the Rio-Earth Summit, Proceedings of inaugural session of Biotech India '92, *World Trade Review*, Oct. 28-31 : 10-14.

Ghatnekar, S. D. and Kavian, M. F. (1995). Application of vermiculture in Indian industries: In Chandra S, Khanna KK, Kehri HK (Eds) *Microbes and Man*, BSMPS, Dehradun, India : 137-149.

Ghatnekar, S. D. and Sharma, S. M. (2009). Therapeutic applications of *Spirulina. Pharma. Buzz.*, 4(8) : 20-24.

Ghatnekar S. D., Ghalsasi D. S. and Tamhane B. M. (2009 a). The novel three-tier biotechnology to convert solid waste of gelatine manufacturing unit into useful plant probiotics. *Ind. J. Environ. Protec.*, 29 (9) : 767-774.

Ghatnekar, S. D., Ghatnekar, S. S. and Ghalsasi, D. S., (2009 b). Three-Tier vermiculture biotechnology to treat bio-solid wastes into bio-clean probiotics for agriculture. *Proceedings of 14th European biosolids and organic resources conference, seminar and exhibition*, organized by Aqua Enviro Technology Transfer, The Royal Armouries, Leeds, UK. 9th to 11th November : 1-13.

Ghatnekar, S. D., Kavian, M. F. and Ghalsasi, D. S. (2009 c). Activities of *Lumbricus rubellus* and *Eudrilus euginiae* and fate of *Acacia auriculiformis, Casurina equisetifolia* and *Dalbergia sisoo* seeds. In: Vermitechnology I. Dynamic Soil, Dynamic Plant 3 (Special Issue 2) (Eds.) N. Karmegam : 82-86.

Ghatnekar, S. D., Tamhane, B. M., Sawant, S. A. and Sharma, S. M., (2009 d). Biomanagement of liquid effluents from gelatine manufacturing industry using combination of *Spirulina platensis, Aspergillus flavus* and *Aspergillus niger*. *J. Ind. Bot. Soc.*, 88 (3-4) : 165–169.

Ghatnekar, S. D., Sharma, S. M., Ghatnekar, S. S. and Karnik, D. A. (2009 e) *Spirulina* (Blue-green micro alga) – A probiotic nutraceutical food supplement for 21st century. *Green Farming,* 2 (5) : 335 – 338.

Ghatnekar, S. D., Sharma, S. M. and Ghalsasi (Dighe), D. S. (2010 a) Prototype of aquatic eco-web for the biomanagement of liquid effluents. *J.I.B.S.,* 89 (3 & 4) : (In Press).

Ghatnekar, S. D., Kavian, M. F., Sharma, S. M., Ghatnekar, S. S. Ghatnekar, G. S. and Ghatnekar, A. G. (2010 b) Application of Vermi-filter-based Effluent Treatment Plant (Pilot scale) for Biomanagement of Liquid Effluents from the Gelatine Industry. In: Karmegam N (Ed) *Dynamic Soil, Dynamic Plant* 4: (Special issue 1) (In Press) http://www.globalsciencebooks.info/Journals/images/InPressList.pdf.

Hashimoto, S. and Furukawa K., (1989) Nutrient removal from secondary effluent by filamentous algae. *J. Ferment. and Bioengin.,* 67(1) : 62-69.

Jegannathan, K. R. and Viruthagiri, T., (2009) Ossein Waste: A Potential Raw Material for Protease Production. *Internat J. Chem Reactor Engin.* : 7.

Kavian, M. F. and Ghatnekar, S. D. (1998). Biotechnological innovation to treat waste from pharmaceutical industry. *The Sixth Congress of Pharmaceutical Sciences of Iran.,* August 26-27 : 28.

Kawai, H., Grieco, V. M. and Jureidini, P. (1984) A study of the treatability of pollutants in high rate photosynthetic ponds and the utilization of proteic potential which proliferate in the ponds. *Environ. Technol. Letters,* 5: 505-516.

Kumar, M., Rawat, S. and Singh G. P. (2008) Impact of different culture conditions on growth and pigment contents of *Spirulina platensis* (Jal – Mahal Isolate) *J. I. B. S.,* 87 (3-4) : 267 – 271.

Lodi, A., Binaghi, L., De Faveri, D., M. Carvalho, J. C., Converti, A. and Delborghi, M. (2005) Fed-Batch Mixotrophic cultivation of *Arthrospira (Spirulina) platensis* (Cyanophyceae) with Carbon source pulse feeding. *Annals Microbiol.,* 55 (3): 181-185.

Looker, N. (1998) Municipal Wastewater Management in Latin America and the Caribbean, R.J. Burnside International Limited, Published for *Roundtable on Municipal Water for the Canadian Environment Industry Association.*

Maree J. P, Cole, C. G. B., Gerber, A. and Barnard, J. L. (1990) Treatment of gelatine factory effluent. *Water SA,* 16 (4): 265 – 268.

Olguin, E. J. (2003) Phycoremediation: Key issues for cost-effective nutrient removal processes. *Biotech. Adv.,* 22: 81–91.

Oswald W. J. and Gotas, H. B. (1957). Light conversion efficiency in photosynthetic oxygenation. *Sixth Progress Report, Institute of Engineering Research,* Series 44(6): University of California, Berkeley, U.S.A. : 38,123.

Meagher, R. B. (2000). Phytoremediation of toxic elemental and organic pollutants. *Curr. Opinion Plant Biol.,* 3(2): 153-162.

Madhya Pradesh Pollution Control Board (2009). Available online: http://www.mppcb.nic.in/pdf/Disp_Consent/DC_122009/Narmada_Gelatine.pdf.

NOSB TAP Review Compiled by OMRI for the USDA National Organic Program (2002) Available online: http://www.scribd.com/doc/20812834/Gelatin.

Pimentel, D., Huang, X., Cordova, A. and Pimentel, M. (1996). Impact of population growth on food supplies and environment Presented at *AAAS Annual Meeting*, Baltimore, MD, 9 February.

Pualchamy, C., Dharmaraj, P. and Laxmanan, U. (2008). A preliminary study on co-digestion of ossein industry waste for methane production. *EurAsian J. of BioSci.*, 2:110-114.

Rose, G.D. (1999). Community-Based Technologies for Domestic Wastewater Treatment and Reuse: Options for Urban Agriculture, N.C. *Division of Pollution Prevention and Environmental Assistance, CFP Report Series:* Report 27.

Sarah, V. (2003). Sustainable Wastewater Treatment and Reuse in Urban Areas of the Developing World *Written April 2003 for the requirements of CE 5993 Field Engineering in the Developing World.*

Sharma, A. and Rao, C. L. S. N. (1996). Effect of gelatin factory effluents on seed germination and seedling growth of some important crop plants. *The J. Environ. Biol.*, 17 (2) : 143-148.

Sharma, S. M., Tamhane, B. M., Sawant, S. A. and Ghatnekar, S. D. (2008) Feeding *Spirulina* to Mozambiq Tilapia and Indian Major Carps. *J. I. B. S.* 87 (3-4) : 248 – 251.

Shushu, D. D. and Chipeta, P. (2002). Secondary wastewater treatment by microalgae isolated from Gaborone oxidation ponds. *Botswana J. Technol.*, 11(1) : 39-43.

Siew – Moi, P. (1987). Agro – Industrial waste water reclamation in Peninsular Malaysia, *Archiv für Hydrobiol.*, 28: 77.

Silveira, S.T., Burkert, J.F.M., Costa, J.A.V., Burkert, C.A.V. and Kalil S.J. (2007). Optimization of phycocyanin extraction from *Spirulina platensis* using factorial design. *Bioresource Technol.*, 98(8) : 1629-1634.

Soni, B., Kalavadia, B., Trivedi, U. and Madamwar, D. (2006). Extraction, purification and characterization of phycocyanin from *Oscillatoria quadripunctulata* – Isolated from the rocky shores of Bet-Dwarka, Gujarat, India. *Process Biochem.*, 41(9): 2017-2023.

Trivedy, R. K. and Goel, P. K. (1986). Chemical and biological methods for water pollution studies. *Environmental Publications*, Karad, India: 209.

Tamhane, B. M. and Ghatnekar, S. D. (2004). Effect of chemicals *vis-à-vis* natural ingredients in the culture media of *Spirulina platensis*. In: *Environmental Science and Technology in India,* (Eds.) Arvind Kumar and R.K. Somashekhar, Daya Publishing house: 540.

Trivedy, R. K. and Goel, P. K. (1986). Chemical and biological methods for water pollution studies. *Environmental Publications*, Karad, India: 209.

Tanticharoen, M., Bunnag, B. and Vonshak, A. (1993). Cultivation of *Spirulina* using secondary treated starch waste water, *Austr. Biotech.*, 3: p223.

Van Hille R. P., Boshoff, G. A., Rose P. D. and Duncan, J. R. (1999). A continuous process for the biological treatment of heavy metal contaminated acid mine water. *Resources, conservation and recycling, International Symposium of the International Society for Environmental Biotechnology* No4, Belfast, ROYAUME-UNI: 27(1-2): 216.

Venkataraman, G. S. (1969). *The cultivation of Algae*. Monograph on Algae (9th Series). Indian Council of Agricultural Research, New Delhi: 190-197.

Vonshak, A. (1997). *Spirulina*: Growth, Physiology and Biochemistry, In: *Spirulina platensis (Arthrospira): physiology, cell biology and biotechnology,* (Ed.) A. Vonshak, Taylor & Francis Publishers, London, United Kingdom : 67 – 77.

Water Environment Association (1987). *Activated Sludge, Manual of Practice* #9.

□□□

Microbial Diversity and Functions, 2012
© D.J. Bagyaraj, K.V.B.R. Tilak, H.K. Kehri (eds.), pp. 149-183
New India Publishing Agency, New Delhi (India)
E-mail : info@nipabooks.com; Website : www.nipabooks.com

Chapter 6

Biodiversity in Mycotoxigenic Fungi

S.M. Reddy, S. Girisham and V. Koteswara Rao

ABSTRACT

Elaboration of mycotoxins by several moulds in different agricultural commodities has been reported to be widespread. The mycotoxins are reported to cause variety of health hazards in different animals including Chicks, cow, horse, rat, Mouse, pig, rabbit, dog, including man. More than 150 species of moulds are reported to elaborate chemically and biologically diverse toxic metabolites. These mycotoxins are known to induce variety of diseases commonly known as mycotoxicoses.The fungi which elaborate mycotoxins are diverse belonging to different groups of fungi.Mycotoxigenicity of these moulds not only differ among different genera but also among different species. Some fungi such as Aspergillus flavus, A. terreus, Penicillium griseofulvum, P.expansum, P.nordicum, P.verrucosum, Fusarium acuminatum, F.culmorum, F. graminearum, F.moniliforme, F.oxysporum, F.solani, F. tricinctum and Cheatomium globosum are known to produce more one type of mycotoxin which differ in their chemical structure and biological activity. On the other hand, some of mycotoxins like sterigmatocystin, cytochalasins, gliotoxin and citrinin are elaborated by more than one taxonomically unrelated fungus. In general mycotoxigenic fungi are spread over among ascomycetes and duteriomycetes. Interestingly no basidiomycete or phycomycetes are reported to elaborate mycotoxin. The relationship between systemic position and mycotoxins producing ability are discussed. The nature of fungi and type of mycotoxins produced and their ecological significance are discussed.

Keywords: *Biodiversity, Mycotoxins, Aspergillus, Penicillium, Fusarium, Conidia, Conidiophores.*

As a sequel to the report of aflatoxin production by *Aspergillus flavus* in 1960 (Sargeant *et al.*, 1961) a great interest has been arose to know the harmful effects of fungi associated with stored foods and feeds. Large assembly of fungi are now known to elaborate wide variety of metabolites which exhibit varying toxicity ranging from chronic to acute toxicity. This fungus mainly exists in anamorphic state and has ascomycete's connection. These fungi which are comparatively xerophilic and potential to produce conidial mass in abundance and need warmer conditions to eloborate these toxic metabolites on nutritionally rich substrates making them unsuitable and hazardous to health of a man and animal are popularly called moulds. More than fifty genera of fungi are known to contain mycotoxigenic species, most of them remain in asexual stage abscuring their phylogenetic relationships and makes prediction of toxigenic potential difficult. Most of these moulds produce their sexual states from groups of ascomycetous orders eurotiales and hypocreales with a smaller number distributed among other orders of ascomycetes. The genera which are reported to contain mycotoxigenic species are precised in table 1.

Table 1 : Genera of fungi containing Mycotoxigenic species

Acremoniium	*Dichotomomyces**	*Myrothecium*	*Rosellinia**
Aspergillus	*Diplodia*	*Microdochium*	*Sclerotinia**
Bipolaris	*Drechslera*	*Monographella* *	*Sphacelia*
Lasiodiplodia	*Epichloe**	*Nigrosabulum**	*Stachbotrys*
*Byssochlamys**	*Epicoccum*	*Nigrospora*	*Talaromyces**
*Ceratocystis**	*Fusarium*	*Paecilomyces*	*Thielavia**
*Cheatomium**	*Gibberella**	*Penicillum*	*Trichoderma*
Cladosporium	*Gliocladium*	*Periconia*	*Trichhothecium*
*Claviceps**	*Gloeotinia**	*Phoma*	*Verticillium*
Colletotrichum	*Khuskia**	*Phomopsis*	*Verticimonosporium*
Curvularia	*Metarrihizium*	*Pithomyces*	*Zygosporium*

* - Teleomorphic stage

Perusal of table 1 leads to the conclusion that toxigenic ability is sporadically distributed that any fungus is as likely as any other to be mycotoxigenic. However, it is logical to suggest that these fungi are phylogenically related as these mycotoxins are synthesized at the end of long enzyme mediated reaction sequences. Majority and most significant mycotoxigenic species share same basic biosynthetic pathways but specific toxins are not shared between the orders. Species of other orders of ascomycetes produce various toxins but these toxins are also produced by species of either eurotiales or hypocreales. However, claviceptales only produce ergot alkaloids.

Members of eurotiales dominate toxigenic species represented by *Aspergillus* and *Penicillium*. Other toxigenic members of this group are species of *Byssochlamys* and *Paecilomyces*.

Toxins produced by members of this order are diverse representing atleast six of the main biosynthetic categories. These include among others, (1) polyketides-patulin, citrinin, ochratoxins, maltoryzine, viriditoxin cytochalasins, rugulosin, ergochromes, aflatoxins and sterigmatocystins, (2) tetramic acid - cyclopiazonic acid; (3) diketopiperzines - aspergillic acid, fumitrogens and oxaline (4) neurotrophic peptide-tryptoquivaliene; the C_6 C_3 product - xanthocilin and (6) nonadride - rubratoxin a compound of mixed fatty acid – TCA cycle origin. However, a few of them are produced by non-eurotialean genera such as a ergochromes are also formed by *Claviceps, Phoma* and lichen. Cyclopiazonic acid is produced by species of *Alternaria, Pyricularia* and *Phoma*. Similarly, sterigmatocystins are produced by species of *Bipolaris* and *Chaetomium*.

Toxigenic species are found in all of the six subdivisions of *Penicillium* and out of 18 groups of Penicillia, 10 contained toxigenic. Penicillia are found in four subgenera but the production of specific toxins is not localized to any subgenus. For example, patulin is produced by teleomorphic genera *Byssochlamys* and *Eupenicillium* and by two sections each of *Penicillium* and *Aspergillus*. On the other hand, aflatoxins are produced by only two species of *A. flavus* group, but structurally similar sterigmatocystins are produced by species belonging to four groups including the *A. flavus* group, one species of *Penicillium* and other non-eurotialean fungi. The ability to produce particular toxin by species is not necessarily to produce toxins by other closely related species. Variation in the ability to produce a particular toxin also rests among different strains of the same species. For instance *Aspergillus flavus, A. parasticus, A. sojae* and *A. oryzae* are difficult to distinguish morphologically but *A. flavus* and *A. parasticus* produce aflatoxins, while no strain of *A. oryzae* and *A. sojae* which are used in production of food are not known to produce toxins. Wicklow and Cole (1984) feel that these species evolved from *A. flavus* and *A. parasiticus* and lost toxigenicity under domestication.

Over 40 species of anamoprhs of hyocrealean fungi are toxigenic and belong to *Acremonium, Fusarium, Gliocladium, Myrothecium, Trichoderma* and *Verticillium*. Teleomorphs are not known for *Metarrhizium* or *Verticimonosporium* but their morphology and biology suggests their hypocrealean nature. These fungi are known to produce comparatively less number of toxins which include zearalenone (polyketide), trichothecenes (sesquiterpenes) and butenolides whose biosynthetic pathway is as yet unknown when butenolides are produced by only these fungi, zearalenone is also produced by species of *Curvularia*, a genus of pleosporales. On the other hand, trichothecenes are produced exclusively by species of *Fusarium, Myrothecium, Acremonium, Trichothecium* and *Trichoderma*.However; *Stachybotrys* which has affiliation to sordariales are also produce macrocyclic trichothecenes. Similarly *Microdochium nivale* which was formerly known as *Fusarium nivale* now it is incuded in hyphonectriaceae of polystigmales and by the flowering plant genus *Baccharis* also produce trichothecenes.

Claviceps with anomorphic state *sphacelia* is the only genus of Claviceptales positively identified as toxigenic. The pharmacologically most active components of ergots are alkaloids which are amides of D-lysergic acid. No lysergic acid is found outside the *Claviceps*. The other two alkaloids produced by *Claviceps* are clavine alkaloids and ergolines. Clavine alkaloids are also formed by species of *Aspergillus* and *Penicillium*, while ergolines are found in *Aspergillus, Penicillium* and *Phoma* which is a genus of pleosporiales and lichens.

Acremonium typhium, the anomorph of *Epichloe typhina,* a pasture grass endophyte, has been linked to the fescue foot toxic syndrome. Butanolides produced by species of *Fusarium* have also been implicated in this condition.

Some of the anamorphic genera like *Phoma, Curvularia, Bipolaris, Drechslera, Alternaria, Diplodia* and *Periconia* are reported to have their teleomorphs in bitunicate ascomycetes belonging to pleosporales. Of these fungi, *Alternaria* is pharmacologically important and produces polyketides alternarioles and related altertoxins by its one of the species, *A. alternata*. This species also produces the tetramic acid, tenuazonic acid, in common with *Pyricularia* and *Phoma*. The other tetramic acid, the cyclopiazonic acid is elaborated by species of *Aspergillus* and *Penicillium*.

The other toxigenic fungi belong to four unrelated orders of ascomycetes and represented by one or two species. *Chaetomium* belonging to order sordariales is reported to produce cytochalasins and Sterigmatocystins with several other unrelated fungi. Several species of *Cladosporium* whose teleomorph stage is in the dothideales, produces epi- and fagicladosporic acids, fatty acid derivatives which can induce alimentary toxic aleukia in common with *Fusarium. Stachybotrys* an anamorph of the sordariales produces trichothecenes in common with *Monographella* and flowering plant, *Baccharis*.

Some anamorphic fungi such as *Pithomyces, Zygosporium* and *Epicoccum* are reported to be toxigenic. Teleomorphs are not known and it is not possible to speculate their affinities. Though conidial ontogeny in *Pithomyces* and *Epicoccum* are similar, they are not related. *P. chartarum* produces sporidesmins, the cause of photosensitivity, liver damage and death in sheep, cattle and deer. These toxins include gliotoxins produced by hypocrealean genus *Gliocladium* and flavipine produced by *Epicoccum purpurescens* and species of

Aspergillus, Zygosporium in common with many unrelated fungi produce cytochalasins.

Most of the known mycotoxins are derived either from polyketides or from sesquiterpenes, two basically different categories of secondary metabolites both of which are common in ascomycetes and fungi imperfecti. Thus, the basic pathways to most mycotoxins are generally resident within the ascomycetes and their anamorphs.

Key for the identification of these mycotoxigenic fungi

1. Sexual spores present in life cycle — 2

   1a. Sexual spores unknown, asexual reproduction by budding — *Torulopsis* (1)

2. Ascospores borne in a ascus, the ascus naked or enclosed within a ascocarp. The asexual state may have a separate name — 39

   2a. Ascospres not present — 3

3. Conidia formed and borne on simple or branched conidiophores or arising directly from vegetative mycelium — 7

   3a. Conidia absent, sclerotia present — 4

4. Sclerotia produced in the mature inflorescence of grasses *Claviceps* (2)

   4a. Sclerotia not in the inflorescence of grasses — 5

5. Sclerotia homogeneous in structure — 6

   5a. Sclerotia not homogeneous; — *Sclerotium* (3)

6. Sclerotia black, mycelium dark-green, becoming black, plant parasitic on cloves and other legumes — *Rhizoctonia* (4)

   6a. Sclerotia not black, very small, consisting of regular to irregular groups of cells. Sclerotia referred to as bulbils — *Papulospora* (5)

7. Conidia formed in pycnidia — 8

   7a. Conidia not formed in pycnidia — 12

8. Pycnidia separate, wall distinct, globose to subglobose with a circular short protruding opening. Immersed in host tissue — 9

   8a. Pycnidia without a distinct wall, but reduced to cavities in a stroma. Spores, hyaline, fusiform, large in bark — *Fusicoccum* (6)

9. Conidia one-celled — 10

   9a. Conidia two-celled — 11

10. Conidia elliptical, oval, short, cylindrical, hyaline *Phoma* (7)

    10a. Conidia oval to lanceolate, hyaline. Conidiophores long, tapering toward apex. Beta conidia (curved, needle like) present *Phomopsis* (8)

11. Conidia brown, 2-celled, cylindrical or slightly curved, with rounded apex. On roots, stalk, ear of corn *Diplodia* (9)

    11a. Conidia 2-celled with longitudinal striations *Lasidiplodia* (10)

12. Conidia formed in an acervulus 13

    12a. Conidia not formed in an acervulus 14

13. Conidia hyaline, one-celled. Dark setae within and around the acervulus present or absent *Colletorichum* (11)

    13a. Conidia with several cells. Central cells dark, end cells hyaline. Hyaline appendages arising from end cells hyaline *Pestalotiopsis* (12)

14. Conidia formed in sporodochia or naked conidiophore 15

    14a. Conidia not formed in sporodochia; conidiophores free on the host or Substrate 18

15. Sporodochia disk-, cup-, or cushion-shaped 16

    15a. Sporodochia not disk-, cup-, or cushion-shaped but thin, irregular, and consisting mainly of discrete aggregations of conidiophores 17

16. Sporodochia cushion-shaped to irregular.Conidiophores tightly packed, irregularly branched. Conidia fusiform, pale-green.*Dendrodochium* (13)

    16a. Sporodochia disk- or cup-shaped, with or without setae, sessile, or with short stalk.Conidiophore branched, terminated by an apical whorl of phialides. Conidia one-celled *Myrothecium* (14)

17. Macroconida hyaline, three – to many-celled, fusoid. Basal cell somewhat resembling a foot (foot cell). Conidiophores simple or brahnched, aggregated into slimy sporodochia *Fusarium* (15)

    17a. Conidia dark, large, globose, septa in various planes *Epicoccum* (16)

18. Conidiophore simple (*i.e.,* unbranched) 19

    18a. Conidiophore branched or unbranched 30

19. Conidia hyaline or light-colored 20

    19a. Conidia dark 25

20. Conidia one-celled 21

    20a. Conidia two-celled 24

21. Conidiophore mostly straight tapering gradually towards the apex 22

21a. Conidiophore not straight but modified by a swelling or crooks 23

22. Conidia aggregated in a slimy mass at the tip of the conidiophore, globose, oval to short cylindric *Acremonium* (17)

22a. Conidia arising from a convoluted mass of compact hyphae (stroma) in the young ovaries of certain grasses.Conidial state of *Claviceps*. *Sphacelia* (18)

23. Conidiophores (=conidiogenous cell) somewhat swollen at the base, borne directly on the hyphae, tapering to a long slender tube, conidia in chains. *Paecilomyces* (19)

23a. Conidiophore terminated by a swelling (vesicle). Sterigmata arising in one or two series from upper part or all of the surface of the vesicle *Aspergillus* (20)

24. Conidia with long thin rod like appendapes arising from apical cell *Clavaria* (21)

24a. Conidia without appendages, formed in chains; the latter composed of conidia oriented sideways to the long axis of the conidiophore. Colony pink. *Trichothecium* (22)

25. Conidia one-celled, brown to black, subglobose or ovoid. Conidiophore inflated at base, tapering toward apex *Nigrospora* (23)

25a. Conidia two-to many-celled, condiophore not inflated 26

26. Conidia only with transverse septa 27

26a. Conidia with septa in two or more planes (muriform) 29

27. Conidia cylindrical (some species taper in apical part), smooth; formed through a pore, germinating laterally, conidiophore straight, bent, or wavy in spore-producing zone . Mostly parasitic on grasses, *Pyrenophora* perfect state. *Drechslera* (24)

27a. Conidia not cylindric, perfect state not in *Pyrnophora* 28

28. Conidia fusiform or ellipsoid, straight or curved. If predominantly curved, end cells pigmented. Germination only from end cells. Conidia formed through a pore, perfect states in *Trichometasphaeria* and *Cochliobolus* *Bipolaris* (25)

28a. Conidia fusiform, dark, with light-colored end cells curved.Perfect state where known in *Cochliobolus*. *Curvularia* (26)

29. Conidia ovoid, obclavate, muroform, solitary or in chains, with a beak conidia formed through a pore. *Alternaria* (27)

 29a. Conidia oval, dark-brown, muriform, roughened; conidia holoblastic on dead plant parts *Pithomyces* (28)

30. Condiophore unbranched 38

 30a. Condiophore branched 31

31. Conidia formed in chains 32

 31a. Conidia not in chains 36

32. Conidophore with vesicle at the apex. One or two layers of phialides over surface of vesicle; the outer phialides produce conidia in chains *Aspergillus* (29)

 32a. Conidiophore without a swelling at the apex 33

33. Conidiophore brush like, composed of phialides which produce condia. 34
 33a. Conidiophore not brush like 35

34. Conidia ovoid, globose, cylindrical,produced in long chains of dry conidia arıse from phialides brus like complexly branched conidiophore. *Penicillium* (30)
 34a. Conidiogenous cell swollen at the base tapering to a long, slender, widely divergent tube. Conidia hyaline in chains. *Paecilomyces* (31)

35. Conidia developing blastically from the conidiophore branches or mostly from other conidia to form a complex of branched, conidial chains (a tree like system). Conidia dark, elliptical or lemon-shaped, one- or two-celled. *Cladosporium* (32)
 35a. Conidiophore dark with few, very short, apical branches; curved downward at the apex. Conidia dark, globose, developing successively in chains from globose conidiogenous cells. *Periconia* (33)

36. Conidiogenous cells in whorls; Conidia one-celled, ovoid, aggregating in slimy balls. Irregular branching sometimes presents *Verticillium* (34)
 36a. Conidiophore branches not in whorls 37

37. Conidiophore profusely and irregularly branched, conidiogenous cells, bottle-shaped, disposed irregularly, opposite or alternate. Irregular or false whorls of phialides common. Conidia mostly subglobose to short obovoid, in loose heads, culture green. *Trichoderma* (35)

 37a. Conidiophore branched irregular or simple. Each branch terminated by a cluster of phialides. Conidia one-celled, formed in a slimy mass around the phialides *Stachybotrys* (35)

38. Conidiophore irregularly branched or simple . conidiogenous cell flask-shaped or globose, terminated by a thread like zigzag apex. *Beauveria* (37)

38a. Conidiophore penicillately branched.Conidia light-colored or hyaline, ovoid, elliptical, or reniform. Conidia aggregate at the tip of phialides in slimy masses . Perfect states in *Nectria, Hypocrea, Hypomyces, Thuemenella* and *Lilliputia.* *Gliocladium* (38)

39. Ascocarp present 40

39a. Ascocarp lacking, asci arising in clusters on the mycelium. Ascospores hyaline. *Byssochlamys* (39)

40. Ascocarp a perithecium,Perithecia free or in a stroma 41

40a. Ascocarp not a perithecium 44

41. Ascospores one-celled 42

41a. Ascospores two- to many celled 43

42. Ascospores small, hyaline, hat-shaped. Asci globose, borne irregularly within the perithecium. Ascospores collect in slime at the tip of the long perithecial neck. Conidia present *Ceratocystis* (40)

42a. Perithecia superficial, globose, oval, vase-shaped, variously shaped hair like hyphae of diagnostic value arise from the upper part. Spores mostly lemon-shaped and olive-brown but shape and color may vary. Asci clavate, disappering early *Chaetomium* (41)

43. Ascospores with three or more cells, hyaline, elliptical to fusiform Asci elliptical to clavate, perithecia superficial or on an erumpent stroma. *Gibberella* (42)

43a. Stroma globose terminating in a stout or slender stipe. Asci very narrow with a thick apex transversed with a pore. Ascospores needle like. *Claviceps* (43)

1. *Aspergillus*

Ever since the reports of carcinogenic aflatoxins produced by *A. flavus*, the genus *Aspergillus* has gained importance. Extensive and intensive search for toxigenic moulds associated with foods and feeds revealed quite a large number of species of *Aspergillus* are mycotoxigenic and are important from the health point of view (table 2). In allergic reactions often of a serious nature. Correct identification of the species is, there is fore, of paramount importance.

Aspergillus is a large genus containing more than 100 species. There are number of teleomorphic genera which have *Aspergillus* conidial states but only two genera *Eurotium* and *Neosartoria* produce heat resistant ascospores and are mycotoxigenic and sporulate heavily in stored products.

Table 2 : Important mycotoxins producing species of *Aspergillus*

Mycotoxin	*Aspergillus* Species
Aflatoxin	*A. flavus, A. parasiticus, A. nomius*
Aflatrem	*A. flavus*
Citrinin	*A. terreus, Aspergillus carneus,*
Citreoviridin	*A. terreus,*
Austdiol, Austamide, Austocystin, Brevianamide	*A. ustus*
Cyclopiazonic acid	*A. flavus, A. tamari, A. versicolor*
Fumitoxins	*A. fumigatus*
Kojic acid	*A. flavus, A. nomius, A. parasiticus, A. tamarii*
Nidulotoxin	*A. versicolor, Emericella nidulans*
Ochratoxin A, Destruxin B	*A. ochraceus*
Patulin	*A. clavatus, A. terreus*
Physcions	*Eurotium amestelodani, E. chevalieri, E. repens, E. rubrum*
Echinulins	*E. chevalieri, E. amestelodai*
Sterigmatocystin	*A. versicolor, E. nidulans*
Terretrems	*A. terreus*
Tryptoqualins	*A. clavatus*
Cytochalasins	*A. clavatus*
Fumagilin, Gliotoxin	*A. fumigatus*

Gams *et al.* (1985) proposed nomenclaturally correct classification for species of genus *Aspergillus* by grouping into six sub-genera which are subclubbed with sections. In addition to traditional morphological taxonomic characters, chemical characteristics such as isozyme patterns, secondary metabolites, ubiquinone system and molecular characterstics have been used to clarify relationships with in the genus *Aspergillus*.

Raper and Fennell (1965) assigned species of *Aspergillus* into three groups based on conidiogeneous apparatus (table 3). They are

1. Those with one series of conidiogenous cells (uniseriate)
2. Those with two series (conidiogenous cells plus supporting cells) (biseriate).
3. Those in which both conditions are found. However, recent changes in the taxonomy of *Aspergillus* have divided the genus into six subgenera

(Gams *et al.*, 1985), each containing one or more sections. Sections correspond to the groups of Raper and Fennel (1965). Only those sections that have mycotoxiological importance will be discussed here.

Table 3 : *Aspergillus* sections according to structure of conidiogenous cells

Uniseriate section	Biseriate section	Both sections
Clavati	*Nidulans*	*Flavi*
Restricti	*Versicolores*	*Wentii*
Fumigati	*Flavipedes*	*Nigri*
Aspergillus (*A. glaucus group*)	*Terrei*	*Candidi Circumdati* (*A. ochraceus* group)

For identification of species of *Aspergillus* primary emphasis was on arrangement of conidiogenous cells, secondary emphasis on shape and pigmentation of conidial heads and finally on the presence or absence of ascomata, Hulle cells and pigmentation of the conidiophore. A key to 14 sections and mycotoxigenic species is given below.

Key to sections and species of *Aspergillus*

1. Conidiogenous cells strictly uniseriate 2
   Conidiogenous cells biseriate or uniseriate 5

2. (1) Conidial heads clavate Clavati
   Conidial structures not exceeding 4.0mm length *A. clavatus* (1)
   Conidial heads not clavate 3

3. (2) Conidial heads radiate, ascomata present *Aspergillus*

   Conidial heads large, radiate to loosely columnar, borne above the surface layer of cleistothecium and enveloping hyphae, ascospores lenticular,6 m or less in long axis, equatorial ridges facing furrow absent or showing only as a trace *A.repens* (2)

   Equatorial ridges low or rounded, furrow broad and *A.rubrum* (3) shallow, colonies strong red, orange red pigment

   Equatorial ridges thin, and flexuous, crest like ascospores resembling a pulley *A.chevalieri* (4)

   Conidial heads not radiate, ascomata absent 4

4. (3) Conidial heads columnar, conidia echinulate Restricti

   (4) Conidial heads columnar, conidia spinose Fumigati

Cultures strictly conidial, varying from velvety to floccose

A. fumigatus (5)

Cultures producing claistothecia conidial development generally limited

Vesicles often borne at an angle to the condiphore, thin *A. fischeri* (6) walled, vesicle, hyaline, often strongly noded *A. viridi-nutans* conidia in pale-blue green shades

5. Conidigenous cells strictly biseriate 6

Condiogenous cells uniseriate and biseriate 10

6. (5) Ascomata present Nidulantes

Ascospores present, orange-red in colour equtorial crests two in number, convex wall smooth, conidial stage dark yellow-green, crest entire

A. nidulans (8)

Ascomata absent 7

7. Conidial heads columnar 8

Conidial heads not columnar 9

8. (6) Conidial heads white to brown, conidiophores brown Flavipedes Conidiophores definitely pigmented in yellow to light-brownshades, conidial heads usually whiteto very plae-buff *A. flavipes* (9)
(7) Conidial heads cinnamon, conidiophores not-brown Terrei

Colonies velvety, conidial heads long compactly columnar in Cinnaman to orange-brown to brown shades borne on short conidiophores sclerotium like masses of swollen, relatively heavy walled cells lacking

A. terreus (10)

9. (8) Conidial heads radiate, dull-brown, conidiophore brown Usti

(9) Conidial heads olive- gray to drab or red-brown, hulle cells Versicolores scattered thoughout the colony or forming irregular masses not associated with pigmented mycelium. *A. ustus* (11)

Conidial head radiate, green, conidiophores not brown

Vesicles globose to somewhat elongate, fertile over most of the vesicular surface, globose to sub-globose, hulle cells often present.

A. versicolor (12)

Conidial heads variable in colour, light yellow-green buff orange-yellow

A. sydowi (13)

Conidial heads always blue-green when young

10. (10) Conidial heads yellow-green Flavi

    Sterigmata single or double, heads loosely columnar Heads radiate, stergmata uniseriate *A. flavus* (15)

    Colonies shifting to light-brownish-green, conidia large elliptical to globose to subglobose, irregularly *A.oryzae* (15)

    Conidial heads not yellow-green

11. (11) Conidial heads ginger-brown Wentii

    Conidial heads large upto 500 mm on smooth or slightly granular stalks which may reach several mm in length *A.wentii* (16)

    Conidial heads not ginger-brown 12

12. (12) Conidial heads globose, black Nigri

    Sterigmata biseriate, conidia globose, roughened

    Sterigmata uniseriate, conspicuously echinulate *A.niger* (17)

    Conidial heads globose, not black *A.japanicus* (18)

13. (13) Conidial heads globose, white Candidi

    (14) Conidial heads globose, yellow to ochraceus Circumdati Sclerotia abundant, small, pure-yellow then brown conidia globose to sub-globose to ovate *A.melleus* (19)

    Sclerotia large, pink to venaceous, purple, when mature ovate to cylindrical, conidia globose to sub-globose *A.ochraceous* (20)

    Number in ( ) is the number of section

2. Penicillium

Demonstration of Sakaki (1891) that the moldy unpolished yellow rice was fatal to dogs, rabbits and guinea pigs due to toxic metabolites resulting banning of the yellow rice in Japan in 1910. Another incident that moldy corn infested by *Penicillium puberulum* extract was toxic to animals in Nebraska in 1913. In 1940 Miyake *et al.* (1940) have isolated *P. toxicarium* from yellowed rice which produced highly toxic metabolite, citreoviridin. Cole and Cox (1981) have recorded no less than 85 species of penicillia which are known to be mycotoxigenic and about 42 mycotoxins are known to be produced by one or other species of penicillia. The literature on this aspect has accumulated mainly on the chemistry and toxicology and most of the literature lacks specificity and resulted in misidentification of species of *Penicillium*. For instance citrinin was reported to be produced by atleast 22 species which may be attributed to misidentification. Misidentifications even by experts have also been common. Of 12 *Penicillium* species reported in the literature to produce tremorgenic toxin only two were accurate. Pitt (1979) has developed a classification of species of *Penicillium* based on broader range of standardized characters that used previously. Classification based on secondary metabolites, electrophoretic finger printing of certain isozymes was also proposed. Samson and Pitt (1990) have developed more convenient guide to the identification of toxigenic penicillia.

Penicillium is a large genus with 150 species recognized which probably is the most conservative. The *Penicillium* is divided into sub-genera based on number and arrangement of phialides and metulae and rami on the main stalk cells (stipes). Pitt (1989) classified *Penicillum* into four subgenera Aspergilloides in which phialides are borne directly on the stipe without intervening supporting elements, while *furcatum* and *biverticillatum* in which phialides are supported by metulae. On the other hand, *Penicillium* in which both metulae and rami are usually present. Majority of important mycotoxigenic and food spoilage species are found in the subgenus *Penicillium.*

Good computer assisted key for identification of species of *Penicillium* is provided by Pitt (1988), while Samson *et al.* (1995) and Pitt (1996) classified food-borne species of *Penicillium* based on secondary metabolites. Correct identification of species of *Penicillium* can be made by observations under standard conditions such as media, incubation period and temperature, in addition to microscopic morphology, gross physiological features including colony diameter, nature of conidiophore, metulae and sterigmata, colour of conidia and colony pigments. Culture media, colony characters (colours), exudates, odour and growth rate are other characters which are taken into consideration.

About 14 species of *Penicillium* are reported to be principal mycotoxigenic. Majority of them produce single toxin, while some species such as *P. expansum* (citrinin and patulin), *P. crustosum* (CPA, penetrim A and roquisfortin), *P. roquiforti* (Patulin, PR toxin, Roquifortine) and *P. aurantiogrisem* (penicillic acid, vermcosidin, viomellein and xanthomegnin) produces more than one mycotoxins. Similarly toxins like penetrim A and patulin are elaborated by four species, while CPA by seven species. Toxins like citrinin and patulin are simple molecules, while others like penetrims are very complex. Thus *Penicillium* toxins are of great complex and diverse. Only ochratoxin A is carcinogenic, while citreoivirdin, citrinin, patulin and penetrim A exhibit wide range of toxicity from acute to chronic. The toxins produced by *penicillium* can be broadly divided into 2 groups. Those affect liver and kidney are asymptomatic or cause generalized debility in humans or animals, while others are neurotoxins which are characterized by sustained trembling. With the low levels of *Penicillium* in nature animal health is affected resulting reduced fertility, increased susceptability to infectious diseases, reduced feed conversion and ill thrift. Many species produce several compounds which are known to be toxic and the possibility of synergy also cannot be ignored (table 4). The potential involvement of *Penicillium* toxins in low level diseases in humans undoubledly warrants continued research. In assessing the claim of toxigenicity of *Penicillium,* several points need to be kept in mind.

Table 4 : Different spesies of *Penicillium* producing mycotoxins

Name of the species	Name of the mycotoxin
Penicillium aethiopicum	Griseofulvin
P. allii	Roquefortine -C, Cyclopenin
P. brevicompactum	Mycophenolic acid
P. camemberti	Cyclopiazonic acid
P. caseifulvum	Rugulovasine, Cyclopenin
P. chrysogenum	Roquefortine -C, PR.Toxin
P. citrinum	Citrinin
P. commune	Cyclopiazonic acid , Patulin
P. crustosum	Cyclopenin
P. digitatum	Tryptoquilanins
P.discolor	Territrems
P.expansum	Patulins, Citrinin, Roquefortine-C
P. griseofulvum	Roquefortine -C, Patulin, Cyclopiazonic acid Griseofulvin
P. italicum	Italinic acid, Brevinamide
P. nalgiovense	Penicillin -F
P. nordicum	Ochra toxin -A&B, Citrinin
P. olsonii	Verrucolone, Breviones
P. roqueforti	Rubra toxin A & B, P.R. toxin
P. rubrum	Rubratoxin A & B
P. tricolor	Xanthomegnin, Viomellein
P. verrucosum	Ochra toxin -A&B, Citrinin

1. Many of the identifications reported in the literature are incorrect and thus has resulted in confusion.
2. A number of genuinely toxic compounds are produced by species not usually occurring in foods or feeds. Such toxins are of little practical importance.
3. Some compounds often cited as mycotoxins in the literature are of low toxicity.
4. Growth of moulds does not always mean production of toxin.

Key to species

1. Conidiophore heads symmetrical, with sterigmata, usually metulae, but no branches 2

   1a Conidiophore asymmetrical with sterigmata, metulae and branches 7

2. Heads, with sterigmata only *P. citreo-viride* (1)

   2a. Heads, with sterigmata and metulae 3

3. Sterigmata flask-shaped, metulae loosely packed *P. citrinum* (2)

   3a. Sterigmata lanceolate, metulae closely packed 4

4. Colony reverse on CZ deep red, pigment diffusing into surrounding agar. 5

   4a. Colony reverse on CZ, yellow-brown to orange-brown, pigment not diffusing into surrounding agar 6

5. Conidia rough-walled, colony producing a fragrant, fruity odor *P. purpurogenum* (3)

   5a. Conidia smooth-walled, colony odorless *P. rubrum* (4)

6. Conidia smooth-walled, colony diameter exceeding 2 cm in 2 weeks *P. islandicum* (5)

   6a. Conidia rough-walled, colony diameter not exceeding 2 cm in 2 Weeks *P. rugulosum* (6)

7. Conidia strongly elliptical and greater than 5 in diameter *P. oxalicum* (7)

   7a. Conidia not strongly elliptical and less than 5 in diameter 8

8. Colonies blue 9

   8a. Colonies other than blue

9. Fascicles regularly produced on M *P. expansum* (8)

 9a Fasicles not regulary produced on M 10

10. Colony growth and conidium production restricted on M 11

 10a. Colony growth and conidium production not restricted on M

11. Colony reverse on CZ, reddish-maroon, pigment diffusing into surrounding agar *P. martensii* (9)

 11a. Colony reverse on CZ, colourless to purple-brown, pigment not diffusing into surrounding agar 12

12. Fascicles regularly produced on CZB *P. cyclopium* (10)

 12a. Fascicles rarely produced on CZB *P. puberulum* (11)

13. Colony light gray and producing a fragrant, fruity odor *P. urticae* (12)

 13a. Colony green, producing a strong musty odour 14

14. Colony granular to fasciculate, bright-green to yellow-green *P. viridicatum* (13)

 14a Colony velvety, dull-green to grey-green 15

15. Conidia delicately rough-walled, rarely exceeding 4 mm in diameter *P. olivio-viride* (14)

 15a. Conidia smooth-walled, regularly exceeding 4 mm in diameter *P. palitans* (15)

CZB = CZ plus 20 ppm Botran; CZ = Czapek solution agar; M = malt extract agar.

3. Fusarium

Importance of *Fusarium* species with reference to mycotoxins infections may sometimes occur in developing seeds especially in cereals and also in maturing fruits and vegetables having potential to produce mycotoxins in foods.

Eever since it was proved that species of this genus incited grave disorders in man and animals (Joffe, 1978) by the production of toxins, determination of the exact identity of these species and strain has become a matter of acute importance. The taxonomy of the sporotrichella, section is a matter of some dispute. The various species in this section are bunched by Snyder and Toussom (1965) under *F. tricinctum*. This section includes species which are not of particular phytopathological importance and therefore, the snyder-Hansen system is not acceptable.

The fusaria were reported to be the cause of alimentary toxic aleukea (ATA), human mycotoxicosis epedimic in the USSR which killed about one lakh people between 1942 and 1948 (Joffe, 1978). It also occurred in Russia during 1913, 1932 and also earlier. ATA also occurred in other countries including England during 16th to 18th Centuries. Recent researches have shown that *Fusarium* species are capable of producing a bewildering array of mycotoxins. Formost among these are atleast about 140 trichlothecenes are known to be produced by fusaria and suspected to be responsible for acute and chronic diseases of man and animals.

The taxonomy of *Fusarium* is complex because many of species, even when grown from monospore cultures, exhibit extreme variability in their morphological characteristics. Hence it has undergone a phenomenal change in species concept due to multitude problems during last 60 years. Wollenweber and Reinking (1935) recognized over 100 species, while Snyder and Hansen (1945) and Tousoun and Nelson (1968) have accepted only nine species each with many formae species. On the other hand, Booth (1971) and Gerlach and Nirenberg (1982) have recognized about 75 species. Identification of *Fusarium* species is ideally carried from the growth on carnation leaf agar, an effective medium for macroconidium production (Nelson *et al.*, 1985).

Determined attack on the taxonomy of these fungi by International collaborative group led by Nelson has resolved most of the conflicts and developed most acceptable classification with 30 valid species (Nelson *et al.*, 1985). An attempt is made to present precisely the characteristics of those *Fusarium* species that are known to produce mycotoxins (table 5).

Table 5 : Different species of *Fusarium* producing mycotoxins

Name of the species	Name of the mycotoxin
Fusarium acuminatum	Diacetoxyscirpenol, HT-2 toxin, Neosolaniol, T-2 toxin,
F.anguioides	Diacetoxyscirpenol
F.avenaceum	Acetyl T-2 toxin, Acetyldeoxynivalenol, Acetylneosolaniol Diacetoxyscirpenol, Acetoxyscirpenediol , Fumonisin B1, HT-2 toxin, Moniliformin, Monoacetoxyscirpenol, Neosolaniol, Nivalenol,
F.bacteridioides	Diacetoxyscirpenol
F. crookwellense	Zearalenone
F. culmorum,	Acetoxyscirpenediol, Acetyl T-2 toxin, Acetyldeoxynivalenol, Acetylneosolaniol, Deoxynivalenol, Diacetoxyscirpenol, Fumonisin B1, HT-2 toxin, Moniliformin, Monoacetoxyscirpenol, Neosolaniol, Nivalenol, Zearalenone

F.diversisporum	Diacetoxyscirpenol
F.episphaeria	Diacetoxyscirpenol, Fusarenon-X, Zearalenone
F. equiseti	Acetoxyscirpenediol, Acetyldeoxynivalenol, Acetylneosolaniol Acetyl T-2 toxin, Diacetoxyscirpenol, Fusarenon-X, Moniliformin, Monoacetoxyscirpenol, Neosolaniol, Nivalenol, Zearalenone,
F. graminearum	Deoxynivalenol, Nivalenol, Zearalenone, Diacetoxyscirpenol
F.lateritium	Diacetoxyscirpenol,
F. moniliforme	Acetoxyscirpenediol, Acetyl T-2 toxin, Acetyldeoxynivalenol, Acetylneosolaniol , Deoxynivalenol diacetate, Deoxynivalenol, Diacetoxyscirpenol, Fumonisin B1, Fusaric acid, HT-2 toxin, Moniliformin, Monoacetoxyscirpenol, Neosolaniol, Nivalenol, T-2 toxin, Zearalenone,
F. nivale	Acetoxyscirpenediol, Acetyldeoxynivalenol, Acetylneosolaniol, Acetyl T-2 toxin, Deoxynivalenol diacetate, Fumonisin B1, HT-2 toxin, Moniliformin, Monoacetoxyscirpenol, Deoxynivalenol, Nivalenol, Diacetoxyscirpenol, T-2 toxin, Fusarenon-X, Zearalenone
F. oxysporum	Acetoxyscirpenediol, Acetyl T-2 toxin, Acetyldeoxynivalenol, Acetylneosolaniol, Diacetoxyscirpenol, Fusarenon-X,Moniliformin, Monoacetoxyscirpenol, Nivalenol, T-2 toxin, Zearalenone
F.poae	Diacetoxyscirpenol, HT-2 toxin, Neosolaniol, T-2 toxin,
F.rigidiusculum	Diacetoxyscirpenol, Neosolaniol, T-2 toxin
F. roseum	Acetoxyscirpenediol, Acetyl T-2 toxin, Acetyldeoxynivalenol, Acetylneosolaniol, Diacetoxyscirpenol, Fusarenon-X, HT-2 toxin, Moniliformin, Monoacetoxyscirpenol, Neosolaniol, Nivalenol, T-2 toxin, Zearalenone
F.sambucinum	Diacetoxyscirpenol
F.scirpi	Diacetoxyscirpenol, HT-2 toxin,T-2 toxin, Zearalenone
F. semitectum	Deoxynivalenol, Fusarenon-X, Neosolaniol, Zearalenone,
F. solani	Diacetoxyscirpenol, HT-2 toxin, Neosolaniol, T-2 toxin,
F.sporotrichioides	Diacetoxyscirpenol, HT-2 toxin,T-2 toxin
F.sulphurium	Diacetoxyscirpenol, HT-2 toxin,T-2 toxin
F. tricinctum	Deoxynivalenol, HT-2 toxin, Neosolaniol, T-2 toxin, Zearalenone

The most prominent of these are *F. poae* and *F. sporotrichioides,* which grow in the Soviet Union on over wintered cereals and cause the often fatal disease, alimentary toxic aleukia in man and *F. tricinctum,* which grows in the United States on mouldy corn, wheat and fescue hay and produces toxic symptoms in many animals. Metabolites of *Fusarium* have received a great deal of attention from research scientists because of their mycotoxic potential.

A direct consequence of confusion in taxonomy is the confusion in species mycotoxin associations. *Fusarium* isolates producing a particular toxin have been given different names as a result of the different taxonomic systems used or simply as a result of misidentification. Marasas *et al.* (1984) after intensive study of about 200 toxigenic *Fusarium* isolates provided accurate information for species identification and toxins produced by them. They listed 24 species of which they considered *F. sporotrichioides, F. equiseti, F. graminearum* and *F. moniliforme* to be most important from human health point of view.

Of the sixteen sections of *Fusarium* accepted by Gerlach and Nirenbreg (1982), 12 have one or more toxigenic species. But three sections Sporotrichella, Liseola and Roseae are rich in toxigenic species. Species belonging to section Sporotrichella produce trichothecene and butanelides, while species of section Liseola produce zearalenone.Species placed in Rosea produce both trichothecenes and zearalenone.

The taxonomy of species of the Sporotrichiella section is a matter of some dispute. The various species in this section are bunched together by Snyder and Toussoun (1965) under *F. tricinctum.* This section includes species which are not of particular phytopathological importance and therefore, the Snyder-Hansen system is not acceptable.

Most significant change in the species composition is of *F. moniliforme* along with two other species which formed a Liseola section of Wollenweber is reduced to one species *F. moniliforme* and one sub-species *F. moniliforme* var. *subglutinans* by Booth (1971). Another interesting nomenclatural change is that of *F. roseum* which was used by Snyder and Hansen (1945) and commonly used in phytopathological literature has been applied to five species and one variety in three sections of *Fusarium* (Gerlach and Nirenberg, 1982). Nirenberg (1976) replaced *F. moniliforme* with the older name *F. verticilloides* and expanded the old concept of *F. moniliforme* to include in addition to *F. verticilloides, F. proliferatum* and *F. lactus*. She divided the old *F. moniliforme* var *subglutinans* among *F. sacchari, F. sacchari* var. *subglutinans, F. neoceras* and *F. anthophilum.*

Another problem with regard to *Fusarium nivale* is reported to have teleomorph in *Micronectriella nivalis* and *Calenectria nivalis*. The teleomorph of this is also reported to be *Monographella* and anamorph superficially resembles

Fusarium and now it is placed in *Gerlachia* (Games and Muller ,1980) and subsequently in *Microdochia* (Samuels and Hallett, 1983) which produces trichothecenes.

Gams and Muller (1980) feel that *Fusarium* species be typified by material producing only microconidia, while other argue that macroconidia should be taken as a character for classifying species.

Another group of *Fusarium* of considerable toxicological interest is the *Discolor* section, comprising such species as *F. graminearum (Gibberella zeae), F. sambucinum (G. pulicaris)* and *F. culmorum*. Snyder and Hansen (1945) devoted much attention to these species, which cause well-known cereal diseases. Largely on the basis that they all attack cereals, species as disparate as these *Discolor* species were grouped together with *F. equiseti (G. intricans)* of the *Gibbosum* section, and *F. avenaceum (G. avenancea)* of the *Roseum* section. All these species belonged to different sections, but were included by Snyder and Hansen (1945) in their Roseum section. In view of this, a grouping that disregards important morphological and physiological features is completely unacceptable. On the other hand, Snyder and Hansen's treatment of other sections of interest to toxicologists, such as the Arachnites *(F. nivale)*, Liseola *(F. moniliforme)*, and Martiella *(F. solani)* sections, de*via*tes only in less important details from the generally accepted schemes based to some degree on Wollenweber and Reinking (1935). A toxicologist or other scientist who wishes to identify a species of *Fusarium* for toxicological purposes may have difficulty utilizing Snyder and Hansen's (1945) system, since it is oversimplified and tailored for use in plant pathology.

According to Nelson *et al.* (1985) there are 12 sections or groups in *Fusarium*. Only four sections contain most common toxic species. They are (1) Sporotrichella (*F. sporotrichoides, F. poae*), (2) Gbbosum (*F. equiseti*) (3) Discolor (*F. graminearum, F. culmorum*) and (4) Liseola (*F. moniliforme, F. proliferatum* and *F. subglutinans*). Samson *et al.* (1995) have provided excellent keys for identification of species of *Fusarium*.

The toxigenicity of *F. graminearum, F. sambucinum, F. culmorum, F. equiseti,* and *F. avenaceum* has been studied fairly extensively, especially with regard to the estrogenic and emetic effects observed in swine and some other animals in the United States and toxicoses in humans, animals, and plants. *F. nivale,* which is highly lethal to animals, induced pathological changes, degenerations, and necroses in various organs and produced cytotoxic and strongly irritant effects on rabbit skin. *F. moniliforme* is the most prevalent fungus on corn kernels and causes acute toxicity in different animals. *F. moniliforme* produced a cytotoxic compound with antitumor activity. Neurotoxic syndrome (Wilson and Maronpot, 1971) and necrosis in the brain (leukoencephamalacia) in horses in South Africa was reported (Kellerman *et al.*, 1990). *F. solani*, a soil fungus, causes numerous destructive plant diseases. *F. solani* also cause intoxication of horses in Japan, characterized by disturbances of the central nervous system, and hemorrhages and crust formation on rabbit skin (Ueno *et al.*, 1972). It was responsible in some cases for corneal disease (keratitis) in man in Spain (Guarro, *et al.*, 2003). Characteristic features of different sections are given below.

1. Section Arachnites Wr. (*F. nivale, F. larvarum*) Macroconidia on aerial mycelium, rarely in sporodochia or pionnotes, curved, apedicellate. Chlamydospores and sclerotia absent.
2. Section Sporotrichiella Wr. Joffe (*F. poae, F. sporotrichioides, F. sporotrichioides* var. *tricinctum, F. sporotrichioides* var. *chlamydosporum*): The basis for classification in this section is (a) shape of microconidia, whether lemon or pear shaped, globose, ellipsoid, or elongate, dispersed in aerial mycelium or formed in false heads; (b) the relative frequency of micro - and macroconidia. Macroconidia sparse, small, oblong, narrowly fusoid to falcate, pedicellate, formed in aerial mycelium or in sporodochia. Chlamydospores intercalary, terminal, in chains or knots, occasionally with plectenchymatous sclerotia. Perithecial states absent.
3. Section Roseum Wr. (*F. avenaceum, F. arthrosporioides*) : Cultures yellow, ochre, carmine, purple or red. Microconidia absent or sparse, Chlamydospores absent, Macroconidia in sporodochia, pionnotes, or on aerial mycelium, sublunate, slender almost filiform, falcate with thin walls, narrowing at both ends, pedicellate. Sclerotia white, yellow, purple to brown. Stroma yellow-red.
4. Section Liseola Wr. (*F. moniliforme, F. moniliforme* var. *subglutinans, F. moniliforme* var. *anthophilum*): Cultures white-cream-brown, organge, violet. Microconidia on aerial mycelium, usually in long chains or small false heads, oval, fusiform, oblong, rarely pyriform. Macroconidia thin walled in sporodochia, pionnotes, or on aerial mycelium, sublunate, spindle shaped to cylindrical, straight or curved with narrow apex and base cells, typically 3-septate.Chalmydospores absent.
5. Section Gibbosum Wr. em. Joffe (*F. equiseti* and *F. equiseti* var. *acuminatum, F. equiseti* var. *compactum, F. equiseti* var. *caudatum, F. equiseti* var. *longipes*): Culture white-pale pink, pale ocher, olive, carmine-red. Microconidia absent or sparse in aerial mycelium. Macroconidia in pionnotes and sporodochia, falcate, narrowing at both ends with elongated apical cell and well-developed pedicellate foot cell, dorsiventral, parabolic or hyperbolic, typically 5- rarely 3-septate. Chlamydospores intercalary, abundant, smooth or rough walled, single or in chains and knots, yellow-brown.

6. Section Discolor Wr. (*F. heterosporum, F. graminearum, F. sambucinum, F. sambucinum* var. *coeruleum, F. sambucinum* var. *trichothecioides, F. cumorum, F. tumidum):* Cultures white-rose, peach, grayish, rose, red to brown. Microconidia absent; Macroconidia thick-walled, in aerial mycelium, sporodochia and pionnotes, either broad, falcate with short apical cell, and well developed foot cell, spindle or sickle-shaped with elongated gradually narrowing apical cell, well-marked basal cell, typically 5-septate. Chlamydospores intercalary, sometimes terminal, often in knots and chains. Sclerotia purple-blue, brown to dark. Stroma yellow.

7. Section Martiella Wr. em. Joffe Palti (*F. solani, F. solani* var. *coeruleum, F. solani* var. *ventricosum, F. javanicum*): Cultures white, cream, orange-blue to brown. Microconidia abundant, oval or oblong, hyaline. Macroconidia in aerial mycelium, sporodochia or pionnotes, fusoid, cylindrical, curved or elongate, with thick walls and short, rounded apical cells and foot cells. Chlamydospores globose, oval, smooth or rough-walled, terminal and intercalary, single or in pairs, short chains or knots.

Key to species

1. Microconidia very heterogeneous, singly or in false heads or in short chains. 2

   1a. Microconidia abundant, broadly oval, elongated, singly or in false heads. 7

2. Microconidia elongated in chains usually with a broad and rounded apex 3

   2b. Microconida absent or rare 11

3. Microconidia pyriform, cylindricate, clavate, lemon shaped or oval 8

   3a. Microconidia globose, spherical with basal papilla, rarely pear-shaped 11

4. Microconidia pyriform to ellipsoid, globose or elongate 5

   4a. Microconidia narrowly clavate with rounded apex, pyriform to fusiform, not spherical or globose 6

5. Microconidia abundant. Macroconidia very sparse, small, curved, falcate, usually without foot cell, typically 3-septate. Sporodochia and pionnotes absent. *F. poae* (1)

   5a. Microconidia often as numerous as macroconidia. Macroconidia formed in curved, sporodochia, falcate, narrowly fusoid,typically 3-to 5-septate. *F. sporotrichioides* (2)

6. Microconidia more abundant than macroconidia. The latter formed in sporodochia, rarely in aerial mycelium, curved, spindly ellipsoid, falcate, 3-to 5-septate, typically 3-septate. *F. tricinectum* (3)

   6a. Microconidia more or less numerous than macroconidia, the latter curved, fusoid with narrowly pointed apex and marked foot cell. Sporodochia absent or very rare, 3- to 5-septate, typically 3-septate. *F. chlamydosporum* (4)

   6b. Macroconidia elliptical, sausage-shaped, elongated, slightly curved, with round apical and round basal cells or marked foot cell, 3 to 5-septate, typically 3-septate *F. solani*(5)

7. Culture peach to violet. Macroconidia spindly, elongated, cylindrical, straight, or slightly falcate with narrow curved apical cell and marked foot cell, 3-to 5-septate, typically 3-septate, Microconidia monophialidic *F. moniliforme* (6)

   7a. Microconidia are produced from polyphialidic conidiophores *F. proliferatum* (7)

8. Culture light-cream, brown. Microconidia rare, short, oval, spindle-shaped or clavate, 0 to 2-septate. Macroconidia falcate to parabolic with more or less gradually or sharply narrowing elongated, curved or non-curved apical cell and well-marked foot cell, 3 to 7-septate,typically 5-septate *F. equiseti* (8)

   8a. Cultures from below sulphureus conidia 3-4 septate. *F.sulphureum* (9)

9. Culture sometimes whitish, mostly strongly colored, rose, red, dark red, or crimson, Microconidia rare, produced short elliptical to spindle-shaped. Macroconidia elongated filiform or slightly hyperbolic, narrowing apical cell and well-marked foot cell, 3 to 7-septate, typically 5-septate *F. avenaceum* (10)

   9a. Microconidia absent. Macroconidia usually in aerial mycelium, small, often sparse, 1 to 3-septate, curved, sometimes aseptate

10. Microconida with an ovate or hooked apical cell, macroconidia curved, beaked or elongated apical cells *F. lateritium* (11)

    10a. Culture light-colored, slow-growing. Macroconidia short, elliptical, curved, sickle-shaped with pointed apex and wedge-shaped or round basal cell, 0 to 3-septate. *F. nivale* (12)

11. Microconidia absent. Macroconidia typically 5-septate, strongly dorsiventral, with a short and wide pointed apex and well-marked foot cell 12

    11a. Macroconidia 3-to 7-septate, broad, *F. culmorum* (13)

12. Macroconidia cylindrical, curved, strongly dorsiventral, falcate, elliptic or sickle-shaped with short beaked apical cell or elongated apical cell, and well-marked foot cell. 13

    12a. Macroconidia cylindrical, curved, strongly dorsiventral, fusoid with short beaked apical cell, 3-to 7-septate. *F. sambucinum* (14)

13. Macroconidia elongated up to 78 mm, spindle-or sickle-shaped, slightly dorsiventral, curved with elongated apical cell, 3 to 7-septate. *F. graminearum* (15)

    13a. Apical cell of macroconidia elongated narrowing eventy to a point with chlamydospores formed in knots or chains. *F. acuminatum* (16)

    13b. Microconidia present or absent macroconidiate usually present macroconidia with wedge shaped foot cell, culture from below beige-brown becoming dark-brown *F. semitectum* (17)

4. Alternaria

Colonies effuse usually grey, dark blackish-brown or black. Conidiophores mcaronematous, mononematous, simple or irregualry and loosely branched, pale-brown or brown, solitary or in conidiogenous cell, integrate terminal, becoming, intecalary, polytrete, sympoidal sometimes monotretic cicatrized. Conidia catenate or solitary dry typically ovoid or obclavate, often rostrate, pale to mid olivaceous brown or brown smooth or verrucose with trans verrucose and frequently also oblique and long longitudinal septa. Species of *Alternaria* produce Alteraniol, Alternarion methylester, Alternuene, isoaltneuene, ATX-1, ATX-II and Tenuazonic acid.

Key to species

1. Conidia solitary or in short chains 2

    1a. Conidia in long chain 3

2. Conidia solitary or in short chains obclavate, beak of the length of the conidium, golden brown, 20-95mM *A. tenuisima* (1)

    2a. Conidia in short chain, shape variable *A. solani* (2)

3. Conidia mostly in chains often or more chains branched. *A.alternata* (3)

    3a. Conidia irregularly shaped, beak short not exceeding 1/3 of conidial length, medium golden brown *A.kukichiana* (4)

5. Stachybotrys

Stachybotrys is a saporophytic and worldwide in distribution. Soviet scientists have described this fungus under the name *S. alternans* Bonorden var. *jateli* with two different variants, one toxic and other nontoxic, while in the literature of other parts of the world used the name *S. atra* Cords and in a few cases as *S. charatarum* (Ehrenberg ex Link) Hughes. *S. atra* is a typical cellulose-decomposing fungus that grows well on cellulose-rich substrates such as straw and hay, but it has also been isolated from cereal grains, plant debris, soil, peas, cotton, and sugar cane roots. It produces satratoxins, macrocyclic trichothecene.

Key to species

1. Conidia mostly more than 20 mm long *S.theobromae* (1)
   1a. Conidia always less than 17 mm 2
2. Conidia reniform 3
   2a. Conidia not reniform 4
3. Phialides brone in 1 or 2 rings around the swollen apex of the Conidiophore *S. oenanthes* (2)
   3a. Phialides borne in one ring, simple, dark-green and verrucose apex of the conidiophore *S.nephrospora* (3)
4. Conidia with dark longitudinal striations *S.cylindrospora* (4)
   4a. Conidia without striations 5
5. Conidia when mature smooth 6
   5a. Conidia when mature verrucose 8
6. Conidia ellipsoidal or boat shaped truncate at the base *S. Sanseveriae*(5)
   6a. Conidia ellipsoidal not truncate at the base 7
7. Conidia dark-brown, 3.0-5.5 mm long. *S. parvispora* (6)
   7a. Conidia greenish or grayish-brown 6.0-11.0 mm long *Melanopsamma pomformis* (7)
8. Conidia often obliquely attenuated at the base, 8-12 x 4-6 mm *S. dichroa* (8)
   8a. Conidia not obliquely attenuated at the base 9
9. Conidia 11-15 x 6-8 mm *S.kampalensis* (9)
   9a. Conidia 8-11 x 5-11 mm *S. atra* (10)
   9b. Conidia 6-8 x 4-5 mm. *S.atra var. microspora* (11)

6. Myrothecium

The genus *Myrothecium* includes those imperfect molds in which the fructification is a sporodochium of hyaline, compacted, irregularly branched conidiophores bearing unicellular slimy phialospores which are green to black in mass. Two other genera, *Metarrhizium* (Metsch.) Sorokin and *Dendrodochium* Bonorden, include species in which unicellular spores, green to black in mass, are borne on branched conidiophores grouped in sporodochia. *Metarrhizium* isolates can be distinguished from those of *Myrothecium* by their dry spores which adhere in tall columns, but green-spored *Dendrodochium* spp., including *D. toxicum* are very similar to *Myrothecium*. Tulloch (1972) considers that it is necessary to extend the present concept of *Myrothecium* to include them.

Three of the 13 species included in *Myrothecium* by Tulloch (1972) have been reported to produce material toxic to mammals. *M. roridum, M. verrucaria* and *M. leucotrichum* (synonyms: *M. jollymannii* Preston, *M. indicum* Rama Rao). *M. leucotrichum* is distinguished from the other two toxin-producing species by the hyaline, thin-walled, several-celled setae which surround its sporodochia. *M. verrucaria* by fusiform spores, 6.5 to 8.0 mm in length, and *M. roridum* by rod-shaped or narrowly ellipsoid spores, 5.5 to 7.0 mm in length.

Key to species

1. Conidiophores in synnemata or sporodochium — *M. masonii* (1)
   1a. Conidiophores single — 2
2. Conidia not more than 4 mm long on agarics — *M.inundatum* (2)
   2a. Conidia more than 4 mm — 3
3. Setae present, conspicuous mostly non-septate — *M.gramineum* (3)
   3a. Setae absent or rarely seen — 4
4. Conidia with longitudinal striations — *M. cinctum* (4)
   4a. Conidia without longitudinal striations — 5
5. Conidia slightly curved, sporodochia stalked. Nectria bacteridoides (5)
   5a. Conidia straight, sporodochia usually sessile — 6
6. Conidia cylindrical or narrowly ellipsoidal — 7
   6a. Conidia navicular, limoniform or broadly ellipsoidal — 8
7. Conidia 5.7 x 1.5-2.5mm, usually on Coffea — *M. advena* (6)
   7a. Conidia mostly 6-8 x 1.5-2.5 mm — *M. roridum* (7)
   7b. Conidia mostly 10-11 x 1.3 mm — *M. carmichaelii* (8)
8. Conidia 6-10 x 1.5-2 mm — *M. verrucaria* (9)
   8a. Conidia 12-17 x 7-9 mm — *Nectria ralfsii* (10)

7. Trichoderma

Colonies spread, repeatedly branched, hyphae hyaline, septate, conidiophore in tufts with divergent often irregularly branched and flask shaped phialides. Conidiophores may end in a sterile appendage with phialides borne on lateral branches in some species. Conidia hyaline or usually green, o-septate, globose to spherical. Hyaline chlamydospores are usually present in the mycelium of older cultures.

Ishikawa *et al.* (1973) have reported production of trichodermin and trichoderminol by *T. hamatum and T. harzianum. Hypocrea* is the perfect stage of some species belonging to *Trichoderma* and *Trichothecium roseum.*

Key to species

1. Conidiophore and its branches short and stout often with sterile hyphal elongation bearing crowded phialides Colonies white, whitish-green or green 2

   1a. Conidiophore and its brnahces strong and slender without sterile hyphae elongation, phialides not crowded. Colonies yellowish, bright-dull to dark-green 5

2. Sterile hyphal elongation absent, conidia lobose, hyaline *T. piluliferum* (1)

   2b. Sterile hyphal elangation present or modified or rarely absent, conidia not globose 3

3. Conidia green, short, ellipsoidal, surrounded by a wide irregular veil *T. saturnisporum* (2)

   3a. Conidia smooth walled or finally punctuated 4

4. Conidia hyaline, small, 2.4-3.8 x 1.8-2.2 mm *T. polysporum* (3)

   4b. Conidia green, small to large, 3.8-6.0 x 2.2-2.8 mm *T. homatum* (4)

5. Conidia rough, 3.6-4.8 x 3.5-4.5 mm *T. viride* (6)

   5a. Conidia smooth walled 6

6. Conidiophore with complicated dendroid branching, Phialides regularly disposed in no. f 3 or more 7

   6a. Conidiophore with simple branching system, phialides irregularly laterally disposed often arising singly 9

7. Conidia ellipsoidal or oblong often appearing angular 3.0-4.8 x 1.9-2.8 mm *T. koningii* (6)

   7a. Conidia shorter with a length: width ratio of less than 1.5 mm 8

8. Conidia obovoid with truncate *T. aureoviride* (7)
   8a. Conidia shorter with a length: width ratio of less short obovoid *T. harzianum (8)*
9. Conidia subglobose to ovoid *T. reesei* (9)
   9a. conidia ellipsoidal 10
10. Phialides usually distinctly attenuate at the base, conidia smaller, pale-green, 2.8-4.8 mm, mostly oblong ellipsoidal *T. pseudokoningii* (10)
    10a. Phialides usually only slightly attenuate at the base, conidia large, partly dark-green upto 7.0 mm long, mostly ellipsoidal *T. longibrachiatum* (11)

8. Trichotheicum

Colonies fast growing at first whit later pinkish, culminate velvety, powdery form conidial formation zonate diurnal rhythm conidiophores erect of sub erect, single or in grooves ,unbranched or sparingly branched, hyaline, septate, the tip bear basipetal imbricate, zigzag chains of conidia held together by mucilage conidia ellipsoidal, two celled, thick walled, hyaline with well-marked truncate attachment point. This is monotypic genus. It produces trichothecene mycotoxin.

9. Pithomyces

Pithomyces chartarum is the causative agent of pithomycotoxicosis, commonly called facial eczema, which occurs in sheep and cattle grazing pasture in summer and fall in New Zealand and less frequently in Australia and South Africa. *P. chartarum grows* saprophytically on dead vegetable matter and is cosmopolitan in distribution. It has been reported more often from the tropical warm temperate zones than from the cool temperate zones. It occurs in sufficient numbers to cause animal disease virtually only in pasture and under rather restricted regimes of moisture and temperature.

P. chartarum is an imperfect mold which forms rough, dark, several-celled, barrel-shaped aleurlospores, 8 to 20 x 10 to 30 mm, borne singly on short simple conidiophores. A small piece of denticle remains attached to the base of free spores. Septa in the spores are both transverse and longitudinal and their number, none to five (usually three) transverse, one to three (usually two) longitudinal, distinguishes the spores of *P. chartarum* from other species.

10. Phoma

Freshly isolated strains grow rapidly on standard nutrient media at 20 to 25°C, producing dark mycelium and forming pycnidia within a few days of inoculation; old cultures may form pycnidia less freely. The black pycnidia are globose or ovoid with ostioles at the tips of small papillae. Conidiophores are inconspicuous. The very numerous pycnospores are ovoid, colorless, guttulate, and either unicellular or with a single septum. The presence of both one-and two-celled pycnospores has caused the mold to be assigned to both *Phoma* and *Ascochyta*. Macroscopically, cultures of *P. herbarum* var. *medicaginis* are brown-black, moist and somewhat wrinkled with little aerial mycelium.

11. Phomopsis

Lupin plants are infected by *P. leptostromiformis* and cause field outbreaks of a mycotoxicosis known as lupinosis in sheep, cattle, horses and pigs.

The fungus produces black stromatic pycnidia on the stems, pods, and seeds of infected lupin plants. Black, stromatic pycnidia are produced in rows or irregularly scattered on stem lesions. Dark-brown, water-soaked spots develop on infected pods; these pods then become dark-brown to black with pycnidia frequently produced in concentric rings, A conspicuous, coarse, white mycelium is visible in such discolored pods and infected seeds become shriveled and brown. The interior of the pycnidial stroma consists of white pseudoparenchymatous cells surrounded by a black sclerotial plectenchyma. A flat fertile locule eventually develops in the upper part of the stroma under the black outermost layer .Along the inner wall of the locule, a dense pallisade of slender, simple or branched, tapering conidiophores are produced. Conidia are hyaline, one-celled, biguttulate, cylindrical with tapered ends, straight or very slightly curved, 5 to 12 x 1.5 to 2.5 mm . b-conidia are produced in separate pycnidia on overwintered stems, and these are hyaline, filiform, and curved to strongly hooked

Conidia are extruded through the ostioles of mature pycnidia in cream-colored to salmon-pink droplets after 14 to 21 days. Only a-conidia have been observed on artificial media and they are similar in shape and size to those produced on the host. The ascigerous state of *P. leptostromiformis* is presumed to be *Diaporthe lupini*.

12. Diplodia

Diplodia maydis (Berk.) Sacc. [=*D. zeae* (Schw.) Lev. occurs in nature as an important pathogen of maize (*Zea mays* L.). Maize ears infected by *D. maydis* cause field outbreaks of a mycotoxicosis known as diplodiosis in cattle and sheep.

The fungus causes a seedling blight, stalk-rot and ear-rot of maize wherever the crop is grown intensively. The ear rot usually progresses upward from the base of the ear and is visible as a conspicuous, coarse, and white to grayish-brown mycelial growth over the husk and kernels. The husks are often glued to the kernels by the white mycelium and infected kernels have a lusterless appearance and a dull-gray to light-brown color. Black pycnidia are produced late in the season on the invaded husks and kernels as well as on the rotted stalks. Pycnidia are subglobose or flask-shaped, 150 to 350 mm in diameter, completely immersed with a distinct, slightly protruding ostiole. Conidiophores are phialidic and formed from the cells of the inner pycnidial wall and are elongate, straight or curved, hyaline, simple, aseptate, 10 to 20 x 2 to 3 mm. Conidia are formed from the apex of the conidiophores and are cylindrical or ellipsoidal, narrowing at both ends to a rounded apex and truncate base, straight or curved, brown, smooth-walled, 0 to 2 but usually 1-septate, slightly or not constricted, 15 to 45 x 3 to 8 mm.

In addition to *D. maydis,* at least two other species of *Diplodia* are known to cause ear rot of maize, *i.e., D. macrospora* and *D. frumenti*. The three species of *Diplodia* on *Zea mays* can be identified with the help of following key.

Key to species

1. Conidia cylindrical or ellipsoidal, 0- to 3-septate, light-brown, smooth-walled, mycelium white 2

   1a. Conidia ellipsoidal, 0- to 1-septate, dark-brown, longitudinally striate, 15 to 31 x 10 to 15 mm, mycelium becoming brown. *D. frumenti*

2. Conidia cylindrical or ellipsoidal, 0- to 2-septate, 15 to 45 x 6 to 8 m *D. maydis*

   2a. Conidia cylindrical, 0- to 3-septate, 43 to 98 x 6 to 14 mm *D. macrospora*

13. Chaetomium

Chaetomium is one of the familiar genra of ascomycetes associated with agricultural commodities due to their cellulolytic activity. Species of this genus are characterized by dark-coloured perithecia with short neck which are covered with irregularly coiled or tightly spiraled hairs or with tuff simple or branched hairs which are of taxonomic importance. However, recent *Chaetomium* taxonomists have selected ascospore morphology as a major taxonomic character. These are world wide in occurrence and are found as contaminants of straw, roots, tubers, seeds, grains and may cause detereioration of foods and food stuffs.

The toxicity of *C. globosum* to roots was first demonstrated by Christenisen *et al.* (1966). Subsequently two toxins cochliodinol and chaetomin were isolated from cultures of *Chaetomium*. Subsiquantly group of new cytochalasins and chaetoglobosins from rice cultures of *C. globosum* were isolated. Oosporin was isolated from cultures of *C. aurum* and *C. trilaterale* by Cole *et al.* (1974). Brewer and Taylor (1978) reported chaetoglobins production by 3 strains of *C. globosum* and chaetomin production by *C. cochliodes, C. fimicola, C. globosum* and *C. umbonatum.* Mollicellins A-H is reported to be produced by *C. mollicellum.* Carter (1982) divided the genus into four subgenera *Chaetomium, Murora, Bomerella* and *Achaetomium.* The Subgenus *Chaetomium* consists of six sections (*Chaetomium, Bostrychodes, Torulosa, Brasiliensia, Crispata* and *Indica*), while *Murora* has five sections (Murora, Aura, Atrobruno, Spiralia and Cuniculosa). The mycotoxins produced by some of the *Chaemium* species are précised in table 6.

Table 6 : Different species of *Chaetomium* producing mycotoxins

Name of the species	Name of the mycotoxin
Chaetomium abuense	Cochliodin
C. coarctum	Chaetoglobosins
C. cochliodes	Chaetoglobosins, Chaetomin, Cochliodin
C. elatum	Cochliodin
C. fimeti (Chaetomidium)	Cochliodin
C. fumicola	Chaetomin
C. globosum	Chaetoglobosins, Cochliodin, Chaetomin
C. globosum var. *rectum*	Chaetoglobosins
C. megalocarpum	Chaetoglobosins
C. subuffine	Chaetoglobosins
C. subglobosum	Chaetomin
C. terissimum	Chaetochromin, Chaetomin
C. tetraspomum	Chaetochromin
C. umbonatum	Chaetochromin, Chaetomin
C. thielavioidarum	Chaetochromin, Sterigmatocystin Chaetomin,
Achaetomella virescens	Chaetochromin, Sterigmatocystin
A. udgawae	Chaetochromin, Sterigmatocystin
C. caprinum	Chaetochromin

At present much more work is still required to indicate how frequently toxigenic *Chaetomium* infested material are used as foods and feeds and more information on different factors contributing to human and animal mycotoxicoses is requered. In *Chaetomium,* a very well defined correlation between particular species and the mycotoxins they produce so that secondary metabolism could readily find a place as an adjunct to taxonomy. Further, identification of *Chaetomium* to species level will provide a good basis for recognization of the toxins produced.

Acknowledgements

Thanks are due to Head, Department of Microbiology, Kakatiya University, Warangal for providing necessary facilities.

References

Booth, C. (1971). The genus "*Fusarium*". Commonwealth Mycological Institute, Kew, Surrey : 237

Brewer, D. and Taylor, A. (1978). The production of toxic metabolites by *Chaetomium* spp. isolated from soil of permanent pasture. *Can. J. Microbiol.,* 24: 1082–1086.

Carter, A. (1982). *A taxonomic study of the ascomycete genus Chaetomium* Kunze. Ph.D. thesis. University of Toronto, Toronto.

Cole, R. J., and Cox, R. H. (1981). *Sterigmatocystins.* In : *Handbook of toxic fungal metabolites* (Eds.) R. J. Cole and R. H. Cox, Academic Press, New York : 67-93.

Cole, R. J., Kriksey, J.W. and Morgan-Joneswells, G. (1974). A new tremorgenic metabolite from *Penicillium paxilli* authors. *Can. J. Microbiol.,* 20: 1159-1162.

Gams, W. Christensen, M., Onions, A.H., Pitt, J.I. and Samson, R.A. (1985). Infrageneric taxa of Aspergillus. In: *Advances in Penicillium and Aspergillus Systematics.* E.ds. R.A. Samson and J.I. Pitt, Plenum Press, New York : 55-62.

Gams, W.J. and Müller, E. (1980). Conidiogensis of *Fusarium nivale* and *Rhynchosporium oryzae* and its taxonomic implications. *Neth. J. Plant Pathol.,* 86: 45-53.

Gerlach, W. and Nirenberg, H. I. (1982). The genus *Fusarium* - a pictorial atlas. Mitt. Biol. Bundesanst. Land- u. Forstwirtsch. *Berlin-Dahlem,* 209: 1-406.

Guarro J., Carmen R., Josepa G., Josep C., Joaquina G.,Rafael B.M., Jose, M. and Enrique, M. (2003). Case of Keratitis Caused by an Uncommon *Fusarium* Species. *J. Clinical Microbiol.,* 41: 58-5826.

Ishikawa, M., Oki,T. and Sawada,F.(1973).On the antifungal compounds formed by Hypocrea species. *Rept. Tottori.Mycol. Inst.,* 10:637-645.

Joffe, A.Z. (1978). *Fusarium poae* and *F. sporotrichioides* as principal causal agents of alimentary toxic aleukia. In: *Mycotoxic fungi,Mycotoxins, Mycotoxicoses.*An Encyclopedia Handbook, Vol. 3, (Eds. T.D.Wyllic and L.G.Morehouse), Marcel Dekker, New York : 21-86.

Kellerman, T. S., Marasas, W. F. O., Thiel, P. G., Gelderblom,W. C. A., Cawood, M. and Coetzer, J. A. W. (1990). Leucoencephalomalacia in two horses induced by oral dosing of fumonisin B1. *Onderstepoort. J. Vet. Res.,* 57: 269–275.

Marasass, W. F. O., Nelson, P. E. and Toussoum, T.A. (1984). *Toxigenic Fusarium Species Identity and Mycotoxicology*, Pennsylvania State Univ., University Park, Pennsylvania : 328.

Miyake, I., Naito, H. and Sumeda, H. (1940). *Report of the Research Institute for Rice Improvement* 1: 1.

Nelson, P.E., Toussoun, T.A. and Marasas, W.F.O. (1985). *Fusarium species, an illustrated Manual for Identification*, Penn. State. Univ. Press, Univ. Park.

Nirenberg, H. (1976). Untersuchungen uber die morphologische und biologischi Differenzierung in der Fusarium Sektion Liseeola. Mitteilungen aus der Biologischen Bundesant Salt purlandund Forstwirtschaft, *Berlin-dehlena*, 169: 1-117..

Pitt, J.I. (1979). *The genus Penicillium and its teleomorphic states Eupenicillium and Talaromyces*, Academic Press, London : 634.

Pitt, J. I. (1988). *A laboratory guide to common Penicillium species*. (2 ed.) North Ryde: Commonwealth Scientific and Industrial Research Organization (CSIRO), Division of Food Processing. North Ryde, N.S.W., Australia.

Pitt, J. I. (1989). Recent developments in the study of *Penicillium* and *Aspergillus* systematics. Symposium Supplement, *J.Appl. Bacter.*, 18: 37S-45S.

Raper, K. B. and Fennell, D. I. (1965). *The genus Aspergilllus*.Williams and Wilkins, Rober E.Kriegger Publ. Co. Inc. Baltimore : 686.

Pitt. J.I. (1996). What are mycotoxins? *Aust. Mycotoxin Newsle.*, 7: 1

Sakaki, J. (1891). Toxicological study on moldy rice. *Z.Tokyo-med. Ges.*, 5:1097-1104.

Samson, R.A. and Pitt, J.I. (1990).*Modern concepts in Penicillium and Aspergillus* Classification. Plenum Press, New York : 478.

Samson, R.A., Hockstra, E.S., Frisvod, J.C. and Filtenborg, O. (1995) *Introduction to food-Borne fungi*. Centralbureau voor schimmel cultures, Baarn, The Netherlands.

Snyder, W. C., and Hansen, H. N. (1945). The species concept in *Fusarium* with reference to Discolor and other sections. *Am. J. Bot.*, 32:657-66.

Snyder, W. C. and Toussoun, T. A. (1965). Current status of taxonomy in *Fusarium* species and their perfect stages. *Phytopath., 55:833-837.*

Sargeant, K., Sheridan, A. Kelley, J.O. and Garnagham, R.B.A. (1961) Toxic products with certain samples of groundnuts. *Nature* (London) 192: 1096-1097.

Samuels, G.J. and Hallett, I.C. (1983). *Microdochium stoveria* and *Monographella stoveri*. *Trans. Brit. Mycol. Soc.*, 81:473-483.

Toussoun, T.A. and Nelson, P.E. (1968). *A pictorial Guide to the identification of Fusarium species According to the taxonomic system of Synder and Hansen*. Pennsylvania State University Pres, University Park. Pennsylvania.

Tulloch, M. (1972). The genus *Myrothecium* Tode ex Fr. *Mycological Papers. Commonwealth Mycological Institute* Kew, Surrey, England. No. I 30.

Ueno, Y., Ishii, K., Sakai, K., Kanaeda, S., Tsunoda, H., Tanaka, T. and Enomoto, M. (1972). Toxicological approaches to the metabolites of Fusaria. IV. Microbial survey on, Bean-hulls poisning of horses'with the isolation of toxic trichothecenes, neosolaniol and T-2 toxin, from *Fusarium solani* M 1-1. *Japan. J. Med.*, 42: 187-203.

Wicklow, D. T. and Cole, R. J. (1984). Citreoviridin in standing corn infested by *Eupenicilliulll ochrosallllloneulll. Mycologia,* 76:959-961.

Wilson, B. J. and Maronpot, R.R. (1971). Causative fungus agent of leucoencephalomalacia in equine animals. *Vet. Rec.*, 88: 484-486.

Wollenweber, H.W. and Reinking, O.A. (1935). *Die Fusarien, inhre Beschreibung, Schadwirkung und Bekämpfung.* Paul Parey Verlag, Berlin, Deutschland: 35-53.

□□□

Microbial Diversity and Functions, 2012
© D.J. Bagyaraj, K.V.B.R. Tilak, H.K. Kehri (eds.), pp. 185-200
New India Publishing Agency, New Delhi (India)
E-mail : info@nipabooks.com; Website : www.nipabooks.com

Chapter 7

Diversity of Nematodes and their Problems in Agroecosystem

S.P. Tiwari and M.N. Khare

ABSTRACT

Nematodes are non-segmented worms typically 50 µm in diameter and 1 mm in length. Those few species responsible for plant diseases have received a lot of attention, but far less is known about the majority of the nematode community that plays beneficial roles in soil. Nematode diversity tends to be greatest in ecosystems with the least disturbance. Species of Aphelenchoides, Ditylenchus spp, Helicotylenchus, Meloidogyne spp., Pratylenchus, Scutellonema and Rotylenchulus are obnoxious plant parasitic nematode beside Heterodera and Globodera. Soil temperature, soil structure, soil pH, soil moisture and soil texture play major role in movement of nematode, infestation and completion of life cycle under abiotic and biotic stresses.

Keywords: Nematode, Soil properties, interaction, geographical distribution, parasites

Introduction

Nematodes are a component of the soil ecosystem that interacts with biotic and abiotic soil factors. Because of this interaction, nematodes are excellent bio-indicators of soil health, because they form a dominant group of organisms in all soil types, have high abundance, high biodiversity and play an important role in recycling within the soil (Neher, 2001, Schloter *et al.*, 2003). Nematodes are heterotrophs, higher in the food chain than micro-organisms and so serve as integrators of soil properties related to their food source, predators and parasites (Ferris *et al.*, 2001). Nematode diversity tends to be greatest in ecosystems with the least disturbance. The disturbance to the soil by environmental or land management practices changes the composition of nematode population (Bongers, 1990). There are a number of indices derived from nematode community analysis that can be used to determine the impact of management changes on the soil ecosystem. Further, nematodes are also aquatic organisms that depend on thin water films to live and move within existing pathways of soil pores of 25–100 μm diameters. Nematodes live in soil and are widely distributed or spread and persist as soil plant pest for indefinite period. Nematode abundance is highly dependent on the presence of the host plants. However, they are also influenced, to a greater or lesser extent, by most of the biological, chemical and physical components of the soil environment. Soil nematodes can be a tool for testing ecological hypotheses and understanding biological mechanisms in soil because of their central role in the soil food web and linkage to ecological processes. Ecological succession is one of the most tested community ecology concepts, and a variety of nematode community indices have been proposed for purposes of environmental monitoring. In contrast, theories of biogeography, colonization, optimal foraging, and niche partitioning by nematodes are poorly understood. Ecological hypotheses related to strategies of coexistence of nematode species sharing the same resource have potential uses for more effective biological control and use of organic amendments to foster disease suppression (Neher, 2010).

Nematodes in soil

Nematodes are non-segmented worms typically 50 μm in diameter and 1 mm in length. Those few species responsible for plant diseases have received a lot of attention, but far less is known about the majority of the nematode community that plays beneficial roles in soil. An incredible variety of nematodes function at several trophic levels of the soil food web. Some feed on the plants and algae; others are grazers that feed on bacteria and fungi; and some feed on other nematodes. Free-living nematodes can be divided into five broad groups based on their diet. Bacterial-feeders consume bacteria. Fungal-feeders feed by puncturing the cell wall of fungi and sucking out the internal contents. Predatory nematodes consume all types of nematodes and protozoa. They engulf smaller organism's whole, or attach themselves to the cuticle of larger nematodes, scraping away until the prey's internal body parts are extracted. Omnivores eat a variety of organisms or may have a different diet at each life stage. Root-feeders are plant parasites, and thus are not free-living in the soil.

Plant parasitic nematodes remain a major challenge to crop production. The diversity of nematode genera and species associated with cereal crops indicates the possibility of nematode population build up due to production intensification especially in soils with high sand content. Twenty-two nematode species from ten genera of plant parasitic nematodes were recovered in root samples of five cereal crops barley, maize, millet, sorghum and wheat Of these,12 nematode species were encountered in maize roots, namely *Aphelenchoides arachides, Aphelenchoides eltaybi,Ditylenchus* spp, *Helicotylenchus dihystera, Meloidogyne* spp., *Pratylenchus brachyurus, Pratylenchus goodeyi,Pratylenchus zeae, Scutellonema brachyurus, Scutellonema paralabiatum, Scutellonema clathricaudatum* and *Rotylenchulus borealis* (Talwana *et al.*,2008).

Nematodes and soil quality

Nematodes may be useful indicators of soil quality because of their tremendous diversity and their participation in many functions at different levels of the soil food web. Several researchers have proposed approaches to assessing the status of soil quality by counting the number of nematodes belonging to different families. In addition to their diversity, nematodes may be useful indicators due to their populations either relatively stable or change in response to moisture and temperature, yet nematode populations respond to land management changes in predictable ways. Nematodes are quite small and live in water films; changes in nematode populations reflect alternation in soil microenvironments. Soil texture also affects nematode population densities. In general sandy soils are nematode-loving soils. However, nematodes exist in soils of all textures some even being favoured by a more fine texture.

Since virtually all plant-parasitic nematodes spend some time in the soil and it is the medium in which plants grow, various edaphic factors have their influence on nematodes and plants. Of the three major nutrients N P K, potassium seems to be the most critical in reducing the expression of symptoms due to nematode feeding. Potassium plays many roles in the plant, but two important ones are its affect on water relations and cell walls. With adequate K, cell walls are thicker and provide more tissue stability. This impact on cell growth normally improves resistance to lodging, pests and diseases. Potassium is important to minimize stress in plants. Feeding by plant-parasitic nematodes definitely results in stress. Adequate potassium levels are required to achieve soybean yields in soybean cyst nematode-infested fields (Warner, 2009).

Soybean cyst nematode populations increased much more rapidly in sand (90 % sand) than in loamy sand (83 % sand) or sandy loam (76 % sand). For example, during the first year of the study in one plot, the soybean cyst nematode population density reached from 250 eggs per 100 cm^3 at planting to 90,000 eggs per 100 cm^3 at harvest in sand on a soybean cyst nematode-susceptible soybean variety. By the third to fourth year, the soybean cyst nematode population densities may be roughly equal over the three soil types.

The observations on population dynamics of soybean cyst nematodes appeared to be successfully surviving during winter and seem to be better in the more fine-textured as compared to the more coarse-textured soil (Tylka *et al.*, 2010).

Nematodes are also important in mineralizing, or releasing, nutrients in plant-available forms like protozoa. When nematodes consume bacteria or fungi, ammonium (NH4+) is released because bacteria and fungi contain much more nitrogen than the requirement of nematodes.

Geographical distribution of nematodes

Distribution patterns of different nematode species vary in soil. The factors controlling geographical and vertical distribution of nematodes are quite different. Nematode population levels fluctuate in an annual crop from harvest of one crop to planting and harvest of the crop; fluctuate throughout the year on an established perennial crop. The distribution of a particular nematode species can be the result of many differing factors. Distribution may reflect a lengthy association with natural vegetation or a relatively recent introduction to a new region. A nematode may not be found naturally within a particular area or it may not establish itself in an area because the soil and climatic environment are unsuitable *viz.*, the climatic variations in valleys, cool High Mountain and coastal regions are diverse enough to influence the pattern of nematode distribution. Some nematodes may have adapted to strict climatic requirements, others may be broadly adapted to survival and proliferation within any of these climatic extremes. Within a geographic area, distribution often largely depends on farming factors, but also may be determined by additional environmental requirements, such as availability of host plants as agricultural crops or weed species and particular soil physical and chemical characteristics. The soil texture, pH, water potential, organic matter, cation exchange capacity etc are also most consistently highly correlated with nematode population (Burn, 1971). Correlation of nematodes with soil factors is stronger in wet than in dry years. The highest number of nematodes is usually found in the lighter soils, except loamy sand where moisture probably was limiting factor. In general soil moisture levels below 20% saturation were limiting factor of the nematodes, except Dorylaims which survived in large numbers in soils with less than 20% saturation (Norton *et al.*, 1971).

1. Stubby root nematodes (*Trichodorus* spp.) predominate in soils with a high sand content (usually 70% or more sand) and, therefore, there are limits based on soil textural preferences within any one geographic area.
2. At the generic level lesion nematode species *Pratylenchus* are widely distributed and do not appear to be restricted by climate, soil characteristics, or host crop preferences. Thus, root lesion nematodes

are encountered in any soil sample. However, various species of *Pratylenchus* show preferences to a variety of characters that influence their geographic distribution. For example, both *P. scribneri* and *P. thornei* are found throughout in all geographic and climatic agricultural regions and have wide host ranges on annual crop field and vegetable crops. However, *P. scribneri* shows a textural preference for sandy loam and coarser textured soils, whereas *P. thornei* shows a preference for the clay and loam soils(Dickerson,1979).

Pratylenchus neglectus is the most widely distributed root lesion nematode. It is commonly associated with annuals like alfalfa and other forage crops that may support grassy weeds. *Pratylenchus brachyurus* appears to be adapted to the warmer soil temperatures found in cotton and bean areas. *Pratylenchus vulnus* is the common root lesion nematode of perennial crops, although other species such as *P. neglectus* can be found among grassy orchard and vineyard floors. They are not restricted by soil types. *Pratylenchus penetrans* is associated with cool regions, especially where cherries or apples are grown. A significant host preference between the two species that attack perennials is that *P. vulnus* is a severe pathogen of figs, walnuts, citrus, and grapes, whereas *P. penetrans* is not considered an important pathogen on these crops and is seldom associated with them. The difference in host preferences and pathogenicity among species of root lesion nematodes demonstrate the critical importance in determining which lesion nematode species is present within a given field.

3. The stem and bulb nematode (*Ditylenchus dipsaci*) reflects its spread on infested planting material as in garlic cloves. This is a good example where agricultural and cultural practices are primary determinants of nematode distribution.

   The distribution of two foliar nematode species, *Aphelenchoides fragariae* and *A. ritzemabosi* reflects the movement of infested propagative plant material. Both are most commonly found where their ornamental hosts and strawberry hosts are grown. The infection process depends on moist conditions.

4. The Citrus nematode (*Tylenchulus semipenetrans*) is a warm, temperate and subtropical species with a wide distribution in the citrus growing areas. The host requirement appears to be the most important distribution determinant; its host range is confined to citrus species, olives, persimmons, grapes and several ornamental crops. Its wide distribution reflects movement of infested planting stocks during the earlier years of the citrus industry worldwide. It appears in all soil textures in which

citrus is planted, from gravelly sand and sandy soils to clay loams and clays. Ecological conditions in medium textured soils favour citrus nematode.

5. The distribution of sugar beet cyst nematode (*Heterodera schachtii*) is primarily related to sugar beet production. This nematode is not restricted by soil temperature. It is also adapted to the complete range of soil types, from the sandiest soils to the heavy clay soils to peat and muck soils.

6. Root-knot nematodes (*Meloidogyne* spp.) are distributed throughout agricultural areas in the world. The different species show some regional preferences, however. *Meloidogyne incognita, M. arenaria,* and *M. javanica* have similar distributions and are concentrated in the tropical and subtropical regions. *Meloidogyne hapla* is widely distributed in temperate regions. This species has temperature optima in different phases of the life cycle that are about 4 to 5°C lower than for *M. incognita* and *M. javanica,* as shown by the wider distribution of *M. hapla* in the cooler areas.

   The extensive host range is in both annual and perennial crops. The most common root knot nematodes and the periodic movement of infested rootstocks and planting material have no doubt expedited spread of these important pathogens. *Meloidogyne* species occur in a wide range of soil textures, but they appear to predominate in coarse textured sandy and sandy loam soils. Plant damage is often accentuated in sandy fields or in sandy patches or streaks within a field (McKenry and Roberts, 1985).

7. In soybean cyst nematode infested sites, failure to maintain potassium levels exhibited more serious symptoms and yield loss although K levels do not have impact on soybean cyst nematode population densities but found reducing symptoms. Since root-knot nematodes disrupt the xylem and phloem of their hosts, the transport of water and nutrients deficiency often results in wilting of plants.

8. The effect of soil type on population densities of plant parasitic nematode species in four different soil types in corn, soybean, wheat, and forage mixtures on dark-colored silty clay loams harboured high densities of *Helicotylenchus pseudorobustus* as compared to Tylenchorhynchus *acutus* which was highest in such soils. Light-colored silt loams favoured development of *Paratylenchus projectus,* however this nematode developed poorly in the darker soils. Comparable densities of *Xiphinema americanum* were found in all soils and on all crops, regardless of soil type. *Tylenchorhynchus martini,* although present but did not build up in any of the soils. Populations of *Pratylenchus* species were generally low in the rotated blocks of all soil types (Ferris and Bernard, 1971).

9. Studies on correlation of total nematodes, non-stylet nematodes, *X. americanum*, *Helicotylenchus pseudorobustus*, *Tylenchus* spp., *Aphelenchus avenae*, and other groupings of nematodes with pH; percentage sand, silt, and clay; percentage organic matter; cation exchange capacity and saturation percentage showed the most consistently significant correlations with the soil factors with *H. pseudorobustus*.

Nematode population

Nematode population fluctuates during the year and in different environments. Many factors determine the rate of increase and eventual size of population, such as fecundity which reflects the number of potential propagules produced eggs, any other means of multiplication, fertility number of variable unit produced, duration of life cycle, longevity, and substance available for parasitism and physical, chemical and biological environments have direct role in nematode community structure and population. Nutrition, aeration or other environmental factors can affect fecundity. Nematode density can change suddenly due to soil environment and phases of nematode life cycle. Eggs are the part of population but usually not studied due to extraction difficulties and sometimes even have a greater difficulty in distinguishing an egg of one species from that of another.

Natility and mortality also have direct relation with the population aggregation in both rhizosphere and rhizoplane of a crop in a particular field or locality. Natility is the rate of production of new individuals and is of three types, ecological natility, actual rate of production of new individuals in a specific environment and absolute or maximum natility which is limited only to inherent physiology of the nematodes such as *Aphelenchides ritzemabosi* which produces 2 eggs per day while *Aphelenchus avenae* laid 207 when the temperature remains 25C within 37 days, whereas *Ditylenchus dipsaci* 210 to 498 eggs/ 45-73 days on onion seedlings. Similarly, *Meloidogyne* spp. 2882eggs/ female between 24-121 days depending on the parasitic behaviour , *Pratylenchus penetrans* maximum 68 eggs *i.e.* 1.1/day which varied with the soil temperature, *Radopholus similis* 2.6 eggs/female *i.e.* 1.8 /day but excess water decline the reproduction factor quickly, *Rotylenchulus reniformis* produced 196 eggs/egg-mass and eggs+empty shells.

Damage by nematodes

The nematode, the host and the environment are the three interacting variables influencing the extent of yield loss in infested soils. An understanding of the mechanisms and principles involved in these interacting relationships is basic to being able to predict yield reductions from estimates of pre-planting nematode population densities (Pi). Damage is proportional to the nematode population density and the degree of damage is influenced by environmental factors. The yield harvested is determined by the amount of light intercepted by the crop, by how efficiently the intercepted light is converted into dry matter, and finally how that dry matter is partitioned into non-harvested and harvested yield. For some crops significant variations in moisture content will also affect final yield. Damage may be proportional to the nematode population density and the relationship is usually curvilinear. There is some evidence that at low densities the host plant can repair the damage and that growth may even be slightly stimulated. Seinhorst (1965) termed the population density (Pi) at which damage first became apparent as the tolerance limit (T). Equally, at very high values of Pi, increasing numbers of nematodes may not further reduce dry matter productivity. Seinhorst (1965) termed this the minimum yield (m). There are various reasons why it (m) may occur; there may be some growth before attack or after feeding, and a significant biomass may be planted. However, m applies to total dry matter and because of effects on partitioning, the harvest value of m may be greater or less than that for total dry matter. The third parameter in the Seinhorst equation is z, a constant slightly less than one. The equation is:

For Pi>Ty=1 *where* Pi? T

where y is the yield.

An important qualification is that y is expressed as a proportion of the nematode free yield. Hence, according to Seinhorst, greater the yield potential, greater the loss in tones per hectare for any value of Pi.. The Seinhorst equation is usually plotted with Pi on a logarithmic scale, producing a sigmoid curve .In practice T is usually small and the Pi value at which m is reached is so large that it is only the central part of the curve that is of practical use. Oostenbrink (1966) suggested that this approximated to a straight line. The equation for such a line is:

$$y = y\ (\text{max}) - \text{slope constant} \times \log \text{Pi}$$

The slope of the regression varies for several factors. These include differences in pathogenicity between species, e.g. *Meloidogyne* spp. may be inherently more damaging than *Tylenchus* spp. but we have no measure of their relative pathogenicities. An important consideration, often overlooked, is the basis of measuring Pi. Usually it is given as numbers per gram of soil. A more appropriate measure is per unit volume of soil as this allows for bulk density differences. Numbers per gram of root is probably the most appropriate, but is difficult to measure because it is always changing. This later aspect becomes important when trying to relate results from experiments where root densities are very different as in pot and field trials. A problem may be encountered when considering damage by nematodes that have two or more generations in the lifetime of a crop. Usually the Pi is measured at planting, but on a good host population of, for example, *Meloidogyne* spp., they can increase from below the value of T to a level in mid-season where they cause significant damage. Even so, it is a race between increasing Pi and increasing plant size that brings with it increasing tolerance in Seinhorst terms, increasing m. In such situations suitability increases host susceptibility and tolerance can have a marked effect on the degree of damage (Olabiyi, 2009).

Damage is proportional to the intensity of attack; this is often proportionally greater in sandy soils where nematodes can move more freely, than in heavier soils where movement is impeded. Adequate soil moisture is essential for free movement so attack is often limited as soils dry out later in the season. Temperature also influences the rate of nematode movement, but plant growth is usually equally affected.

With *Meloidogyne* spp., impaired water relations appear to contribute substantially to reduced rates of top growth. This is probably because the developing giant cell systems interfere with and disrupt the developing xylem. Clearly, with such damage, affect on growth and yield is likely to be greater where the plants are on the threshold of becoming moisture stressed. Other affect includes reduced photosynthetic efficiency and it has been reviewed by Trudgill (1992).

There is a good correlation in many crops between percent ground cover *i.e.* the percentage of ground occupied by a plant or a crop, when viewed from above that is covered by green leaves and percent light interception. Most annual crops start as individual, separate plants and a reduction in growth rate is directly reflected in ground cover and hence light interception. As they grow the leaves of neighbouring plants merge to form a continuous canopy. Nematode damage that only delays the production of a continuous canopy, and hence 100% light interception, will have a smaller effect on final yield than damage which prevents the crop achieving such full cover. Premature crop death will also proportionally reduce yield.

Soil environment

Soil is the home of most nematode and deserves more attention.Soil is a complex system consisting of minerals, organic substances, gaseous exchange and aqueous phase with mineral portions.

In soil, the nematode moves through a labyrinth of pore spaces, size and distribution of which depend on the size of soil crumbs. The moisture characteristic is an invaluable tool in assessing pore size and water distribution in soil. Nematodes are so small that soil pore size is probably insignificant in determining their distribution (Nelsen, 1949). Nematodes appear to be well adapted to living in soil pores and allow them to move through most of the soil cavities due to undulatory type of locomotion along with sinusoidal path of the same width as the nematode body. Extreme moisture in a soil does not favour nematode locomotion and survival. Dry conditions also depress nematode activity especially hatching of *Heterodera rostochiensis, Meloidogyne arenaria and Ditylenchus dipsaci* eggs. However in some species of *Heterodera* hatching is stimulated by exudates from the host plant root. Production of such hatching factors is greater when the soil pores are emptying.

The thickness of soil water film between the nematode cuticle and air-water interface is therefore important especially when nematode occur within soil crumbs. The content of O_2 and CO_2 in the soil varied with wide limits. There are no change of O_2 consumption when the O_2 consumption is decreased by 10% from the atmosphere (Nelsen,1949). He further concluded that within the variation of O_2 tensions occurring of normal soils, as O_2 consumption of nematodes is independent of O_2 tension. Triffitt (1930) showed that J_2 of *Heterodera rostochiensis* fail to emerge from cyst in the absence of O_2. Twenty hours exposure to CO_2 appeared sufficient to kill the nematode.

Soil temperature

The influence of soil temperature on plant nematodes can be divided into following arbitrary phases-

(i) Optimum temperature which favours the development of nematodes;

(ii) Non-lethal low and high temperature at which activity of nematodes is inhibited and

(iii) Lethal low and high temperature which affect reproduction, development and hatching of nematodes (Table 1).

When vertical soil section is examined 3 to 5 feet deep it was observed to be divided into different horizons or layers of different colours, texture and thickness from top to bottom. These horizons are designated as O - organic, A -top most minerals, B-accumulated layer of clay and iron + alluminium oxide,C - frequently dense and generally outside the area of major biological activity of the field crops and R -bed-rock. Mineral component of soil is mainly sand, silt and clay in composition (Norton, 1978).

Soil type has its affect on nematode movement as well as nutrient and water supply to the host. It also influences the nematode survival during periods of stress and species composition of nematode communities. The affect of fertilizer application and of water availability has interaction with host genotype and husbandry factors such as spacing and time of planting (Trudgill, 1987).

Table 1 : Influence of soil temperature on infestation of nematode activity.

Nematodes	Temperature (°C)	Inhibition of nematode activity	Reference
D.dipsaci, Meloidogyne spp	7-13	Infestation	Cairns,1954
H.glycines and *H rostochiensis*	10-11	Development	Ichinohe (1955), Chitwood and Buhrer (1946)
Aphelenchoides ritzemabosi, D dipsaci	8	Reproduction activity	Dolliver *et al.*, 1962; Courtney and Latta, 1934
H.glycines, H.schachtii	10 and 16	Emergence	Slack and Hamble, 1959; Wallace 1955
Meloidogyne spp	9	Reproduction	Tyler, 1933
Pratylenchus projectus	10	Moulting	Rhoades and Linford, 1959
H.glycines	38	Emergence	Courtney and Latta, 1934
H rostochiensis	32	Development	Slack and Hamble, 1959
	32	Emergence	Fenwick, 1951
	30-35	Development activity	Kampfe, 1958
	29		Mai and Harrison, 1959
Trichodorus	35	Development	Rohde and Jenkins, 1957

pH

Most plant-parasitic nematodes are soil-borne, and at least part of their life cycle occurs in the soil. Consequently, edaphic factors may directly affect their distribution and population dynamics (Table, 2) Pedersen *et al.*, (2010) and Olabiyi (2009).

Table 2 : Role of soil texture and pH on plant parasitic nematodes.

Soil texture (%) sand	silt	clay	pH	Plant parasitic nematodes	Soil series (class)
58	32	10	4.5	Most of the migratory plant parasitic nematodes	Loamy textural
80	6	14	5.3	Species of *Meloidogyne, Heterodera* and *Hemicycliophora*	Sandy loam textural
86-95	4-5	10	4.9-6	Species of *Pratylenchus, Hirschmaniella, Meloidogyne, Tylenchulus, Hemicycliophora* and *Heterodera*	Sand textural
76	10	4	5.3	Most of sedentary plant parasitic nematodes	Loamy sand textural

Damage caused

Damages due to the nematode *Pratylenchus brachyurus* on pineapples are very serious under high acidity (pH 4.2 - 4.5) which is favorable to the development of *P. brachyurus*. Soil pH has a significant effect on the development of *P. brachyurus* in pineapple roots. As pH increases, *P. brachyurus* populations drop. The lowest populations were observed when pH was increased up to 6.0. Consequently, plant growth, yield and sucker production were improved. This study shows that liming to maintain soil pH between 5.5 and 6.0 is recommended as one of the cheapest ways to control nematode population in small farms, where chemical nematicides are unaffordable (Osseni *et al.*, 2008).

Yield loss estimation methods

Pot study determines basic information on yield-loss relationships, but due to environmental differences and interactions, field studies are required. There are two approaches one is to use nematicides at relatively uniformly infested sites and the aspect to work at sites with a range of population densities .The first approach yield practical information on the effectiveness and potential value of a particular treatment but little about the nature of the relationship. The second one involves benefit of generating information on the relationship between Pi and yield. Pi estimates have large errors; accuracy must be improved by increasing plot size and by taking and processing multiple samples from each plot. The plot size must be large enough to obtain a realistic yield and adequate guard plants. Another option is to establish many small plots in large uniform fields. The plots can be split and a nematicide applied to one half. For each plot the Pi and yield are determined. The results will exhibit decreased yields with increasing Pi.

The most striking feature of nematode distribution and damage within a field is the irregularity of infested area. Damaged crops will appear as irregular patches or streaks that may vary in size, shape, and number. These variations usually reflect the compounding of nematode stress as a function of population density, feeding behavior, damage potential, host response etc. on a plant by such other factors as physical soil differences, irrigation, drainage patterns, soil texture, cropping history, and management practices. Cysts containing *via*ble eggs are introduced by contaminated harvesting or cultivation equipment, tractor wheels, human or animal feet, irrigation water or by wind blown from an adjacent infested area (Moreno *et al.*, 2006).

Soil sampling is a critical problem in nematode studies. At least uniform sampling and nematode processing is necessary to obtain a mean population density per 0.01 hectare plot of true mean with 95% confidence.

At seeding or planting in kharif, plant parasitic nematode populations typically are at a low level and increase towards harvest as long as roots are healthy enough to support nematode feeding. *Meloidogyne chitwoodi* on potatoes required, initial population of juveniles to monitored at weekly intervals from planting until harvest. Following harvest, nematode numbers decrease until another susceptible crop is planted or weeds become available to support reproduction. Though a susceptible host is one of the most important factor governing populations of obligate parasites however, a resistant plant might maintain a population and serve as Time Bridge until more susceptible hosts become available (Wallace, 1963).

Population dynamics of plant parasitic nematodes in perennial crops typically changes during the year with the timing of the fluctuations varying by crop and location.

Nematodes have various reproductive strategies. Some grow large and have long life cycles with low rates of population increase (K strategists), others are relatively small, have short life cycles and potentially higher reproductive rates (r strategists). An endoparasitic habit with induction of giant cells is continuously available as a food source, reduces exposure to predation and other stresses and further increases reproductive potential. A reduction in the number of J2 further decreases development time, thereby reducing generation time and increasing the potential for multiple generations in a season. A wide host range completes the adaptation of *Meloidogyne* spp., which can be regarded as the ultimate plant-parasitic nematode r strategists. Many *Longidorus* spp. are examples of K strategists. It is a characteristic of K strategists that they do best in stable environments where populations are usually close to the equilibrium density. In contrast r strategists increase rapidly where the environment is favourable, often overshooting the equilibrium density. Nematode multiplication can be modelled in different ways. For migratory nematode species that multiply continuously, Seinhorst (1966) proposed the following formula derived from a logistic equation:

$$Pf = aEPi/(a - 1)Pi + E$$

where a = the maximum rate of increase and E = the equilibrium density at which Pf = Pi. For sedentary nematodes with one generation at a time, e.g. potato-cyst nematodes, Seinhorst (1967) proposed an alternative model. Jones and Perry (1978) also proposed a model for sedentary nematodes with a logistic basis derived from the observation that sex determination is density dependent (Nicholson, 1933).

Differences can be modelled in terms of the maximum multiplication rate or the space required for successful multiplication (Seinhorst models), or in terms of fecundity or effects on the sex ratio (Jones and Perry model). An important method of expressing and comparing the effects of different cultivars or cropping regimes is to consider the equilibrium density, *i.e.* the point at which Pf = Pi. This density is usually observed at a Pi which is larger than that which gives the largest Pf (Warner, 2009).

Future thrust

Further research is needed on nematodes in natural and agricultural soils to synchronize nutrient release and availability relative to plant needs, to test ecological hypotheses, to apply optimal foraging and niche partitioning strategies for more effective biological control, to blend organic amendments to foster disease suppression, to monitor environmental and restoration status, and to develop better predictive models for land-use decisions.

References

Bongers, T. (1990). The maturity index: an ecological measure of environmental disturbance based on nematode species composition. *Oecologia*, 83: 14-19.

Burn, N. C. (1971). Soil pH Effects on Nematode Populations Associated with Soybeans *J. Nematol.*, 3(3): 238–245.

Cairns, E.J. (1954). Effect of temperature upon pathogenicity of mushroom spawn nematode, *Ditylenchus* sp. *Proceedings of International Conference in Science aspects of mushroom growing* 11. Gembloux, June , 1953,164-167.

Chitwood, B.C. and Buhrer, E.M. (1945). The life history of golden nematode of potatoes, *Heterodera rostochiensis* Wollenweber, under Long island, New York, conditions. *Phytopath.*, 36:180-189.

Courtney, W.D. and Latta, R. (1934) Some experiments concerning the revival of quiescent *Anguillulina dipsaci. Proc. Helminthological Soc.*, Washington 1:20

Dickerson, O.J. (1979). The Effects of Temperature on *Pratylenchus scribneri* and *P. alleni* Populations on Soybeans and Tomatoes. *J. Nematol.*, 11(1):23-6.

Dolliver, J.S., Hildebrandt,A.C. and Riker, A.J. (1962). Studies of reproduction of *Aphelenchoides ritzemabosi* (Schwartz) on plant tissues in culture. *Nematogogica*, 7:294-300.

Fenwick, D.W. (1951). Investigations on emergence of larvae fron the cyst of potato root eelworm, *Heterodera rostochiensis*.4.Physical conditions and their influence on larval emergence in the laboratory. J. *Helminthol.*, 25:37-48.

Ferris, V.R and Bernard, R.L. (1971). Effect of soil type on population densities of nematodes in soybean rotation fields. *J. Nematcl.*, 3:123-8.

Ferris, H., Bongers, T. and de Goede, R.G.M. (2001). A framework for soil food web diagnostics: extension of the nematode faunal analysis concept. *Appl. Soil Ecol.*, 18:13-29.

Ichinohe, M. (1955). Studies on the morphology and ecology of the soybean nematode. *Hokkaido Agricul. Exper. Station Report* 48:1-64.

Kampfe, L. (1958). Physiologische befunde zur Artrennung und zum Herkunfts nachweis in der Gattung Heterodera Scmidt (Nematodes). *Varh. Deutsch. Zool. Ges. Frankfurt*, 25:383-391.

Mai, W.F. and Harrison, M.B. (1959). The golden nematode. *Cornell Experimental Bulletin* : 870.

Moreno, S. S., Hideomi , M. Ferris, H. and Jackson, L. E. (2006). Linking soil properties and nematode community composition:effects of soil management on soil food webs. *Nematology*, 8:703-715

McKenry, M.V. and Roberts. P. A. (1985). *Phytonematology Study Guide*. University of California, Division of Agriculture and Natural Resources. pp 4045.

Neher, D.A. (2001). Role of nematodes in soil health and their use as indicators. *J. Nematol.*, 33:161-168.

Neher, D.A. (2010). Ecology of Plant and Free-Living Nematodes in Natural and Agricultural Soil. *Ann. Rev. Phytopath.*, 48: 371-394.

Nelsen, C. O. (1949). Studies on soil microfauna.11. The soil inhabiting nematodes. *Natura Futlandica*, 2:1-131

Nicholson, A.J. (1933). The balance of animal populations. *J. Animal Ecol.*, 2: 132-178.

Norton D.C., Frederick L.R., Ponchillia, P.E. and Nyhan, J.W. (1971). Correlations of nematodes and soil properties in soybean fields. *J. Nemat.*, 3:154-63.

Norton, D.C. (1978). *Ecology of plant parasitic nematodes*, John Wiley & Sons Inc., Canada. pp 266.

Olabiyi, T.I. (2009). Influence of Soil Textures on Distribution of Phytonematodes in the South Western Nigeria.*World J. Agricul. Sci.*, 5: 557-560.

Usseni, B., Sarah, J. and Hugon, R. (2008). Effect of soil pH on the plant population development of *Pratylenchus brachyurus* (Godfrey) in pineapple roots and on the growth and yield of the plant. *ISHS Acta Horticul* :425.

Pedersen, P., Tylka, G.L. Mallarino, A., Macguidwin, A.E. Koval, N.C and Grau, C.R. (2010). Correlation between soil pH, *Heterodera glycines* population densities, and soybean yield. *Crop Sci.*, 50:1458–1464.

Rhoades, H.L. and Linford, M.B. (1959). Molting of pre-adult nematodes of the genus *Paratylenchus* stimulated by root diffusates. *Science*, 130:1476-1477.

Rohde, R.A. and Jenkins, W.R. (1957). Effect of temperature on the life cycle of stubby root nematodes. *Phytopath.*, 47:29.

Schloter, M., Dilly, O. and Munch, J.C.(2003). Indicators for evaluating soil quality. *Agricul., Ecosyst. Environ.*, 98:255-262

Seinhorst, J.W. (1965). The relation between nematode density and damage to plants. *Nematologica*, 11: 137-154.

Seinhorst, J.W. (1966). The relationships between population increase and population density in plant-parasitic nematodes. I. Introduction and migratory nematodes. *Nematologica*, 12: 157-169.

Seinhorst, J.W. (1967). The relationships between population increase and population density in plant-parasitic nematodes. II. Sedentary nematodes. *Nematologica*, 13: 157-171.

Slack, D.A. and Hamblen, M.L. (1959). Factors influencing emergence of larvae from cysts of *Heterodera glycines* Ichinohe. Influence of constsnt temperature. *Phytopath.*, 49:319-320.

Talwana, H. L., Butseya, M. M. and Tusiime, G. (2008). Occurrence of plant parasitic nematodes and factors that enhance population build up in cereal based cropping system in Uganda. *Afr. Crop Sci. J.*, 16: 119 – 131.

Triffitt, M.G. (1930). Observations on the life cycle of *Hereodera schachtii*. *J. Helminthol.*, 8:185-196.

Trudgill, D.L. (1987). Effects of rates of nematicide and of fertilizer on growth and yield of cultivars of potato which differ in their tolerance of damage by potato-cyst nematodes (*G. rostochiensis* and *G. pallida*). *Plant Soil*, 104:185-193.

Trudgill, D.L. (1992). Resistance to and tolerance of plant-parasitic nematodes in plants. *Ann. Rev. Phytopath.*, 29: 167-192.

Tyler, J. (1933). Development of root knot nematode as affected by temperature. *Hilgardia* 7:392-415.

Tylka, P.G.L., Mallarino, A. Macguidwin, A.E., Koval, N.C. and Grau, C.R. (2010). Correlation between soil pH, *Heterodera glycines* population densities, and soybean yield. *Crop Sci.*, 50:1458–1464.

Wallace, H.R. (1955). Factors influencing the emergence of larvae from cysts of the beet eelworm, *Heterodera schachtii* Schmidt. *J. Helminthol.*, 29:3-16.

Wallace, H.R. (1963). *Biology of plant parasitic nematodes*. Edward Arnold (Publishers) Ltd, London pp 63-64.

Warner , F. (2009). Soil fertility, pH, texture and nematodes. *Diagnostic Services*, 13:21.

□□□

Microbial Diversity and Functions, 2012
© D.J. Bagyaraj, K.V.B.R. Tilak, H.K. Kehri (eds.), pp. 201-229
New India Publishing Agency, New Delhi (India)
E-mail : info@nipabooks.com; Website : www.nipabooks.com

Chapter 8

Potentials of *Trichoderma* as Biocontrol Agent, its Ecology and Biodiversity

D.K. Agarwal and R.K. Sharma

ABSTRACT

Trichoderma species play a vital role in the management of plant diseases in association with different commercial bioformutlations of the fungus. Its role as a successful pollution free environment and friendly mechanism of action has been highly exploited. The diversity of species based on location, type of agri-land and association with different fauna and flora has been instrumental in large upsurge of biological methods of Plant disease management. Pollution of agri-land soils with chemicals, heavy metals and development of resistance to diseases has become an emerging problem coupled with the high costs of Infrastructure. Thus, the interest in alternative biologically based systems is the need for bioremediation, bioecology and biocontrol with Trichoderma.

Keywords: Trichoderma, bioecology, biodiversity, management potentials.

Introduction

Trichoderma Pers. ex Fr., a genus under Deuteromycotina, Hyphomycetes, Phialosporae, Hyphales, and Dematiaceae has gained immense importance since last few decades due to its biological management ability against several plant pathogens (Kubicek and Harman, 1998; Mukhopadhyay and Mukherjee, 1996 and Papavizas, 1985). The researchers are interested on this genus because of its novel biological properties and biotechnological applications. Benefits of *Trichoderma* include the usage in fabric detergent, animal feed and fuel production, alternative to conventional bleaching, effluent treatment, and degradation of organochlorine pesticides and more importantly as biocontrol agent (Kuhls *et al.*, 1997 and Samuels, 1996). By recognizing the importance of this fungus as biocontrol agent against several plant pathogens, it is sold under different trade names for example: *Trichoderma viride* as Ecofit - Hectz; Niprot - Pest Control India Ltd., Bioderma - Biotech International Ltd., Bhasderma - Basarass Biocontrol Research labs in India, *Trichoderma harzianum* as F. Stop - Cornell University, USA, Trichodex and Binab - T in USA and as Trichodex, M.T.R. - 35 and T.39 from Israel.

Trichoderma was first described more than two hundred years ago by Persoon (1794) and was later envisaged into four genera. The identification and characterization of *Trichoderma* was worked out into a monograph (Rifai, 1969), in which this genus was classified into nine aggregates. Rifai recognized several biological species under each aggregate. He was unable to define limits of individual biological species (Samuels *et al.*, 1998). Bisset (1991 a,b and 1992) elevated Rifai's species aggregates to species level and recognized two to several species within each of five sections of the genus. Several articles have been published to decide whether these artificial approaches have a molecular or physiological basis (Samuels, 1996; Samuels *et al.*, 1998). Kiffer and Morelet (2000) have recognized a total of thirty - six species under the genus *Trichoderma*. Recently another species, *T. stromaticum* has been added to the list with molecular identification (Samuels and Padra - Schultheiss, 2000). The teleomorph of this genus is *Hypocrea* Fr. under Hypocreales, Ascomycotina (Rifai and Webster, 1966a, b; Webster, 1964).

Different species of *Hypocrea* producing *Trichoderma* anamorphs are identified (Bisset, 1991a,b and 1992; Samuels *et al.*, 1998; Rifai and Webster, 1966a, b; Webster, 1964) under different species names. The species concepts and interspecific relationships between the *Trichoderma* isolates was reviewed along with macromolecular and teleomorphic status (Samuels, 1996).

Among several fungi, the genus *Trichoderma* is a predominant cosmopolitan fungus distributed in soils of different agroclimatic zones. It can grow rapidly on diversified substracts and can resist the noxious chemicals and the toxic heavy metals. The fungus with the well-developed enzyme system plays a vital role in improving both plant and soil health apart from its industrial applications. Benefits of *Trichoderma* include the usage in fabric detergent, animal food and fuel production, alternative to conventional bleaching, affluent treatment, and degradation of organochloride pesticides more importantly as biological agent. Besides, the antibiotics and lytic enzymes are used for the, management of various plant pathogenic microbes infecting plants. Based on the efficacy of the genus *Trichoderma,* several products have been registered both at national and international levels. For soil borne plant pathogens, chemical control is arduos, uneconomical and not advisable due to risk of ground water pollution, death of non-target beneficial flora & fauna and evolution of fungicide resistant pathogen variants.

Trichoderma is easily identified in culture media, which produces large number of small green or white conidia from phialides present on the profusely or meagerly branched conidiophores. However, the identification of isolates to species level is difficult and confusing due to the complexity and closely related characters to the species. Species concepts within *Trichoderma* are very wide and this has resulted in the establishment of many specific and sub specific taxa (Samuels, 1996).

In India this genus was isolated by Thakur and Norris (1928) from soils of Madras. Thereupon it was reported from various substrates and locations and the identifications were based on the morphological characters. Various papers have been published in India on the biocontrol efficiency of *T. harzianum, T. koningi, T. longibrachiatum, T. virens* and *T. viride.* Even though the species names are mentioned in the research papers, invalid names are still used, which gives misconception regarding the identification of the potentials of *Trichoderma* as biocontrol agent.

The decades of laboratory experiments have led to the explosion in the field of biocontrol and introduced more than 50 commercial formulations comprising fungi and bacteria (Whipps, 1997). Among the antagonists, *Trichoderma* plays a vital role in plant disease management and has opened new vistas for the commercialization. *Trichoderma* species are one of the predominant components among the mycoflora in various soils in all climatic zones. *Trichoderma* (Hypocreales, Ascomycota) species are frequently associated with decaying wood produce more enzymes of industrial value (*Trichoderma reesei* = *Hypocrea jecorina*) (Kubicek and Penttila, 1998) and antibiotics (Sivasithamparam and Ghisalberti, 1998), or have application as biocontrol agents against plant pathogens (*i.e. T. harzianum* = *H. lixii; T. atroviride* = *H. atroviridis; T. asperellum*) (Hjeljord and Tronsmo, 1998). More recently, *T. longibrachiatum* was found to be an opportunistic pathogen of immuno compromised mammals including humans (Kredics *et al.*, 2003). *Trichoderma* has been recognized to comprise a significant amount of fungal biomass in soil (Nelson, 1982) and is frequently present as an indoor contaminant (Thrane *et al.*, 2001). These diverse implications of *Trichoderma / Hypocrea* with human society necessitate an accurate species and strain identification. However, classical approaches based on the use of morophological criteria are difficult to apply to *Trichoderma* due to the plasticity of characters. However, the knowledge on the diversity and distribution pattern of *Trichoderma* is highly essential to explore the best among the diverse species of *Trichoderma*. The analysis of the *Trichoderma* products registered with Central Insecticides Board (CIB), Faridabad, Haryana revealed that only *T. viride, T. harzianum* and *T. hamatum* have been explored to a maximum extent.

Eco-biology of *Trichoderma*

The genetic variation is a key factor to evaluate biodiversity. The genetic variations in *Trichoderma* are mainly attributed to their ability to adapt to different ecological conditions. *Trichoderma* occurs in all soils and other natural habitats that are rich in organic matter. It is also found on root surfaces of various plants, decayed bark, mainly after the decay by other fungi (Elisa Esposito and Manuela da Silva, 1988). The population of *Trichoderma* decreases, when it is exposed to dry conditions for a long period of time. Some strains of *T. hamatum* and *T. pseudokoningii* are adapted to excessive soil moisture and those of *T. viride* and *T. polysporum* are found to be restricted to areas of low temperature. *T. harzianum* is predominant in warm climatic regions, while *T. hamatum* and *T. koningii* are widely distributed in areas of diverse climatic conditions (Papavizas, 1985; Grondona *et al.*, 1997). It is also distributed in different soils including agricultural, forest, prairie, salt marsh and desert soils scattered in different agro-climatic zones. The propagules of *Trichoderma* were abundant in forest soils and they proliferated well in the forest soils rich in carbon and nitrogen (Danielson and Davey, 1973). Some of the species of *Trichoderma* survive as decomposers of woody and herbaceous materials and are also necrotrophic against the primary wood decomposers (Rossam, 1996).

The population dynamics of *Trichoderma* is also altered due to the variations in the iron content of the soil (Xu, 1996). Certain strains of *Trichoderma* have evolved and occupied the niches of higher elevations. Analysis of soil samples from elevations of about 2700m in the Himalayas reflected the presence of section *Longibrachyatum*. Based on the parsimony analysis section *Longibrachyatum* was grouped under *H. audinensis*. It was also isolated from 2300m elevation of Venezuelan Andes (Samuels *et al.*, 1998). Another member of the section *Longibrachyatum* namely, *T. konilangbra* was isolated from locations between 1750 and 3400m in Uganda.

pH and *Trichoderma* diversity

Hydrogen ion concentration played a major role in the adaptability of biocontrol agents to their nutrient environment. The activities of *Trichoderma* spp. are more pronounced at low pH and the efficacy as influenced by acidic conditions (Chet and Baker, 1980, 1981). The activity of *T. lignorum* was poor in neutral and alkaline soils (Dhingra and Khare, 1973). Acidic pH favoured disease suppression induced by *T. harzianum* (Chet and Baker, 1980; Liu and

Baker, 1980; Marshall, 1982; Harman and Taylor, 1988) and *T. hamatum* (Chet and Baker, 1981). The spore germination (Chet and Baker, 1980), mycelial growth, conidiophore production (Chet and Baker, 1980), antibiotic production (Dennis and Webster, 1971) or the activity of lytic enzymes (Chet and Baker, 1980) were enhanced under acidic conditions. Jeyarajan *et al.*, (1994) found that the isolates of *T. viride* perfomed well on a wide range of pH varying from 5.0 to 9.0. Hence a composite, compatible formulation of *T. viride* could easily be developed and used in disease management programs.

Moisture and *Trichoderma* diversity

The soil moisture content is very important than soil reactions in altering the population dynamics of the soil. It affects the growth of antagonist. The bacterial population dropped rapidly as the soil moisture failed below 15 to 20 percent while fungi and actinomycetes population is not affected. The maximum relative humidity required for the growth of *Trichoderma* is 91 per cent. The maximum growth of *Trichoderma* on ground pine needles occurs at 144 and 201 per cent moisture. It grows well at 103 per cent but is depressed at 46 per cent moisture. Growth of *T. viride* is more at 40 to 60 per cent moisture than all the isolates of *T. viride* and *T. harzianum* survived uniformly well at 40 per cent followed by 60 per cent moisture holding capacity.

Ecology and the diversity of *Trichoderma*

The diversity of the species of *Trichoderma* varies based on locations and with the microsites. The soils of Taiwan were rich in *T. asperellum*. But, the tree barks associated with lichen in Chitan forest in Taiwan was colonized by *Hypocrea jecorina, T. viride, T. atroviride* and *T. harzianum*. On the contrary, the tree bark, which was not having association with any of the fauna was dominated by the association of *Trichoderma sp. novum*. Hence, based on the association of different fauna, the diversity of *Trichoderma* spp. also differs to a greater extent.

Similarly the tropical forest soil of Cambodia was rich in *T. harzianum*, which indicates that it prefers warm soil temperature for its survival and multiplication. Analysis of the samples from the ruined walls of the temple indicates that they were predominant with *T. harzianum* and to certain extent the ruined walls were also found to be associated with *T. virens*. The soils near seashore of Thailand at Control Island and Koh Samet Island were colonized with *T. sp. novum* and *T. harzianum*. The wet soils of Kauchaburi province of Burma was colonized by *T. harzianum* and *T. ghanense*. The tropical rain forest soils of Malaysia had *T. virens, T. spirale, T. harzianum, T. asperellum* and *T. sp. novum*. The lake garden soil of Kuala Lumpur was associated with *T. virens, T. spirale, T. harzianum, T. sp. novum, T. koningii and T. atroviride.*

The forest soil of Singapore was colonized with *T. koningii, T. asperellum, T. virens,. T. spirale* and *T. harzianum.* Though there was a huge variation in the distribution pattern of the *Trichoderma* spp., the cultivated garden soils of South East Asia were predominant with *T. virens* which is being widely used as biocontrol fungi for the management of soil borne pathogens. This vividly explains that the ecological requirements are different for the species colonizing tree bark, forest soils and cultivated garden soils (Kubicek *et al.,* 2003).

Isolates of *T. spirale* were found to be more in South East Asia than the cosmopolitan *T. koningii.* Similarly several isolates of *T. spirale* were observed in soils of Canada. However, the abundance of the same was more in Central America, Columbia and South Africa which suggests that these species are more abundant in tropical climates (Kubicek *et al.,* 2003). *H. jecorina,* a pantropical ascomycete in abundance in South East Asia, The Pacific Island South America. But the same was previously reported to be restricted to the regions around equator (Samuels *et al.,* 1998). Soils from Central Russia, The Urals, Siberia and the Himalayan mountains were predominant with *T. atroviride, T. hamatum, T. virens, T. asperellum, T. koningii, T. harzianum and T. oblongisporum* (Kullinig *et al.,* 2002).

Biodiversity of *Trichoderma*

Trichoderma being cosmopolitan fungus is found to be widely distributed in different agroclimatic zones and habitats in both tropical and subtropical countries. However, the biogeographical distribution of *Trichoderma* has not been extensively studied except in European countries, Russia and China. Since *Trichoderma* has wide scope and potential to improve human life in different angles, attention must be focused to identify the diversity and distribution pattern of *Trichoderma.* About 35 *Trichoderma* species are recognized on the basis of morophological and molecular characters. The diversity of *Trichoderma* spp. in Russia, Siberia and the Himalayan mountains were identified through molecular approaches using ITS 1 and 2 sequences of the rDNA cluster of the 75 isolates obtained. The comparison of sequences confirmed the presence of *T. atroviride, T. virens, T. hamatum, T. asperellum, T. koningii* and *T. oblongisporum, T. harzianum* and *T.hamatun.* Sections *Longibrachiatum* and *T. stromaticum* were found distributed at an elevation of 2700 m in the Himalayas (Kullnig *et al.,* 2002).

Parsimony and distance analysis of DNA sequences from multiple genetic loci of *Trichoderma* with 18S rDNA sequences analysis suggested that the genus *Trichoderma* might be evolved 110 million years ago as *Hypomyces* and *Fusarium.* 28S rDNA sequences analysis shows that the genus *Trichoderma* is part of a monophyletic branch within the Hypocreaceae. Phylogenic trees inferred from a combined analysis of the nuclear ribosomal internal transcribed spacer provided strong statistical support for a consistent existence of four clades. Clade A comprises of species of Bissett's (1991 a, b) sect. *Trichoderma* in addition of *T. hamatum, T. pubescens, T. asperellum* and *T. strigosum;* clade C comprises all the species contained in section *Longibrachiatum* as revised by Samuels *et al., (*1998*)* and clade D contains only *T. aureoviride* which is genetically most distant to all other species. Clade B, on the other hand, contains a large and taxonomically heterogeneous mixture of species, among which several subclades could be identified: subclade B1 containing *H. lacteal, H. citrine, H. citrine* var. *americana, H. lutea* and an unnamed *T.* sp. 1; subclade B2 containing *T. stromaticum* and an unnamed *T.* sp. PPR13559; subclade B3 containing *T. fertile, T. oblongisporum* and *H. hunua;* subclade B5 containing *T. polysporum, T. roceum, Hypocrea pilulifera* and *T. minutisporum* and a larger subclade (B4), in which three strain clusters could be distinguished: One comprising *T. harzianum, T. hamatum, H. vinosa* and *T. aggressivum,* another one containing *T. fasciculatum, T. longipile, and T. strictipile* and a third containing *T. virens, T. avofuscum,* and *T. crassum*. The position of the remaining species of subclade B4 (*T. spirale, T. cfr aureoviride, H. tawa,* and *T. tomentosum*) was not resolved (Kullnig Gradinger *et al., 2002*).

The variations of *Trichoderma* in South East Asia (Taiwan and Indonesia) were identified through morphological, biochemical and through sequences analysis. Ninety six strains were isolated. Seventy eight isolates were identified as *T. harzianum/T. hamatum* (37 strains), *T. virens (*16 strains*), T. spirale (*8 strains*), T. koningii (*3 strains*), T. atroviride (*3 strains*), T. asperellum (*4 strains*), Hypocrea jecorina* (anamorph: *T. reesei;* 2 strains), *T. viride* (2 strains), *T. hamatum* (1 strain) and *T. ghanense* (1 strain). Analysis of biochemical characters revealed that *T. virens, T. spirale, T. asperellum, T. koningii, H. jecorina,* and *T. ghanense* formed well defined clusters, thus exhibiting species specific metabolic properties. In addition, two new taxa of *Trichoderma* were identified (Kubicek *et al.,* 2003).

The biodiversity and biogeography of *Trichoderma* in four areas of China *viz.,* North (Hebei province), South-East (Zhejiang province), West (Himalayan, Tibet) and South-West (Yunnan province) include one hundred and thirty five isolates. They were grouped tentatively based on morophological characters which showed the existence of 64 isolates. But, all were grouped into eleven species as *Trichoderma asperellum, T. koningii, T. atroviride, T. viride, T. velutinum, T. cerinum, T. virens, T. harzianum, T. sinesis, T. citrinoviride, T. longibrachiatum* and two putative new species (*Trichoderma* sp. C1, and *Trichoderma* sp. C2) by both morophological characters and phylogenetic analysis. *T. harzianum* accounted for almost half of the biodiversity and was predominant in North China. In addition, a unique haplotype of *T. harzianum* which is rarely found elsewhere was also identified (Zhang *et al.,* 2005).

Advantages of *Trichoderma*

- *Trichoderma* spp. are saprophytic, quickly growing and easy to culture and produce large amount of conidia.
- It grows on the rhizosphere and on the surface of the roots, thereby provides disease control and enhances root growth.
- It multiplies on its own and protects the roots for whole growing season.
- Its spores survive in the soil but the food it uses is mostly secreted from the root surface.
- It is commonly found in almost any soils and other natural habitats consisting of organic matter.
- It can be used along with organic matter.
- Low cost and longer efficiency than conventional fungicides.
- Does not allow to develop any resistance in the plant growth.
- It has shown a little host specificity because of preference to root exudates.
- It has the capacity to minimize and protect the host plants in a symbiotic relationship similar to that between nitrogen fixing bacteria and legume roots or that between mycorrhizae and root systems.
- Advantages of *Trichoderma* - rDNA

Ecology significance of *Trichoderma*

Trichoderma plays an immense role in creating a pollution free environment through multi farious actions. *Trichoderma* could be used for:

- Plant growth promotion and disease control
- Improving soil health
- Bioremediatioin of polluted soils
- Industrial applications
- Phytobial remediation

Biotechnological applications of *Trichoderma*

Biological control of plant pathogens

Trichoderma is the most widely used single genus as a potential biocontrol agent all over the world. *T. viride, T. harziaum, T. virens* and *T. hamatum* are the most exploited species for the management of various seed and soil borne pathogens of agricultural and horticultural crops (Table-1). Diseases due to *Rhizctonia solani, Macrophomina phaseolina* and *Fusarium* spp. cause high percentage of mortality of the plants and consequent yield losses in rain-fed

Table 1 : Successful application of *Trichoderma* species used against plant pathogens

Trichoderma species	Pathogen	Crop	Delivery system
Trichoderma sp. C62	*Sclerotium cepivorum*	Onion	Soil application
Trichoderma spp.	*P graminis*	Wheat	Spores added to soil
	var. *trittci R. solani*	Lettuce	Soil
		Lettuce	
T. harzianum	*Fusarium graminearum*	Radish	Soil
T. harzianum T951	*Pythium ultimum*	Cucumber	Soil
T. koningii	*S. rolfsii*	Tomato	Soil
T. koningii	*Pythium* spp.	Chickpea	Seed
	Pythium spp.	Pea	Seed and Soil
T. harzianum	*P. ultimum*	Cotton	Seed
		Maize	Seed
		Pea	Seed
T. harzianum+Rhizobium	*S. rolfsii*	Groundnut	Seed and Soil
T. viride, T. harzianum, M. phaseolina, T. longibrachiatum		*Sesamum*	Seed
T. viride	*M. phaseolina*	Groundnut	Seed
		Sesamum	
		Blackgram	
		Greengram	
		Pigeon pea	
		Sunflower	
T. hamatum	*R. solani*	Cotton	Soil-as mycelial
		Sugar beet	preparations
		Radish	
T. viride	*M. phaseolina*	Blackgram	Seed
T. viride	*M. phaseolina*	Greengram	Furrow application
T. viride	*F. udum, M. phaseolina*	Pigeon pea	Seed and Soil
T. viride	*F. oxysporum f. sp. cubense*	Banana	*Trichoderma* capsules
	R. *solani*	Cotton	Seed
	M. phaseolina	Green gram	Seed
		Cowpea	Seed + Soil
	Pythium ultimum	Chilies,	Solid Matrix priming
		Tomato	
Trichoderma	*M. phaseolina*	*Sesamum*	Seed
		Blackgran	Seed and Soil
T. viride	*R. solani*	Cotton	Seed
	S. rolfsii	Brinjal	
		Sunflower	
		Bhendi	

crops especially in pulses and oilseeds. Soil incorporation of *T. harzianum* in the field reduced the inoculum density of *R.solani* by 85 per cent and the fruit rot of tomato by 27-51 per cent. Soil application of *T. harzianum* was found effective in reducing the incidence of *Macrophomina* root rot of beans by 37 to 74 per cent. Coating of melon seeds with conidia of *T. harzianum* reduced the disease caused by *Macrophomina* to the extent of 46.3 per cent. Seed coating of green gram and cowpea @ 10g of *T. viride* isolate MG6 combined with soil application reduced dry root rot incidence and increased the yield (Nakkeeran and Sabitha Doraisamy, 2001). In addition to disease reduction, dry matter production was also enhanced which was attributed to the production of growth hormones. Treatment of sunflower seeds with *T. viride* @ 4g/kg reduced the root rot incidence to 18.15 as against 37.1 per cent in control. It increased the yield by 21 per cent. Root rot disease of blackgram could be managed by seed and soil application with *Trichoderma* (Dinakaran and Marimuthu, 2001).

The root rot incidence of *Sesamum* was reduced to 12.8 per cent when the seeds were treated with *T. viride* @ 4g/kg and increased the root, shoot length and oil content. It recorded an yield of 677 kg/ha as against 301 kg/ha in control. Results of the field demonstrations conducted in the farmer's fields for the control of root rot in groundnut, sesame, urdbean and chickpea against dry root rot through *T. viride* were found to be promising. *T. harzianum* was found effective in the management of *Fusarium* wilt diseases. Strain T35 checked *Fusarium* wilt of cotton and melon under natural soil conditions. Incorporation of *T. harzianum* as wheat bran-peat preparation into soil under green house conditions reduced the incidence of crown rot of tomato up to 80 per cent. Soil application increased the yield of tomato by 18.8 per cent (Table-2).

Soil application of *T.koningii* @ 2 g per vine of black pepper at the time of transplanting was highly effective compared to metalaxyl compounds against *Phytophthora* foot rot. The disease incidence was 23 to 35 per cent lower and yield was 30 to 40 per cent higher than control. Pre and post planting application of *T. viride* and *T. harzianum* was effective against foot rot disease of betel vine. Amendment of nursery soils with *T. harzianum* prior to the colonizaticn and damping off pathogen in nursery beds resulted in the reduction of pre and post emergence damping off of cardamom. Delivery of *Trichoderma* strains through solid matrix priming in tomato and chillies reduced the incidence of damping off under field conditions than dry seed treatments (Nakkeeran and Sabitha Doraisamy, 2001); Kannaiyan *et al.*, 2002). Bio-manure developed with coir pith and *T. viride* contained damping off disease of Eucalyptus seedlings (Kumar and Marimuthu, 1993) and turmeric rhizome rot (Ushamalini *et al.*, 2006). Seedlings (Kumar and Marimuthu, 1993) and turmeric rhizome rot (Ushamalini *et al.*, 2006) (Table -2).

Table 2 : Examples of successful control of plant diseases by *Trichoderma* spp.

Pathogen	Crop	Diseases	Antagonists	Methods and rate of application
Sclerotium rolfsii	Tomato, beans, groundnut, sugarbeet, chickpea	Collar rot	*Trichoderma viride, T. harzianum, T. hamatum*	Seed treatment 4g/kg. Furrow application 130-160 kg/ha.
Rhizoctinia solani	Groundnut, bean, cucumber, cotton, rice	Root/collar rot	*Trichoderma viride, T. harzianum, T. hamatum*	Seed treatment 4g/kg. Furrow application 130-160 kg/ha.
Pythium aphanidermatum	Tobacco, tomato, peas, chili, citrus, Cucumber	Damping - off	*T. harzianum T. hamatum*	Soil application to nurseries
Macrophomina phaseolina	Beans	Root-rot	*T. harzianum*	Furrow application 130-160
Fusarium oxysporum	Chickpea, Chrysanthemum. Cotton, melons, pigeonpeas	Wilt	*T. harzianum*	Soil application 160 kg/ha
Phytophthora citrophthora	Citrus	Gummosis	*Trichoderma spp.*	Filling holes with inoculum
Different wood rotting fungi	Wood trees logs	Root-rot	*T. viride*	Dusting on cut wounds
Botrytis cinerea	Grapes	Grey mold	*T. viride*	Foliar spray
F. oxysporum f. sp. ciceri + R. solani, S. rolfsii	Chickpea	Wilt	*T. harzianum*	Seed treatment 4g/kg + Soil application @ 160 kg/ha
Ceratocystis fimbriata	Sugarcane	Pineapple Disease	*T. harzianum*	Sett treatment @ 10g/kg + Soil application
Pythium aphanidermatum and Fusarium solani	Ginger and turmeric	Soft rot and ginger yellows	*T. harzianum*	Rhizome treatment @ 10g/kg + Soil application @ 130kg/ha.
M. phaseolina	Safflower	Root rot	*T. viride*	Soil application 160 kg/ha

Phoma sorghina	Mulberry	Collar rot	*T. harzianum* + *T. viride*	Cutting and soil treatment
Ganoderma lucidum	Arecanut	Foot rot/ Anabe	*T. harazianum*	Soil application/ plant @ 50g/ plant
Phytophthora spp.	Betelvine and Black Pepper	Foot rot/ root rot	*T. viride*	Soil application / plant @ 50g/ plant
Pellicularia koleroga, Fomes noxius F. oxysporum f. sp.	Coffee	Root diseases	*T. virens*	Soil application @ 160 kg/ ha

Commercial production and formulations of Talc based *Trichoderma* biofungicide

It is the formulation that can transform a potential biological control agent from a laboratory agent to a laboratory curiosity to a commercially successful venture in terms of disease management (Connick *et al.,* 1990). A successful bio-control formulation is one, which is economical to produce, safe, stable in the environment, easily delivered and should enable the bio-control agent to act effectively and consistently under varied environmental conditions. The nutrient status of the substrate or additives is one of the important factors influencing the efficacy of the formulation. Introduction of antagonists through organic carriers alle*via*te the competition from autochthonous microorganisms, since the organic carriers serve as protectant and a food base during the establishment of the antagonist (Deacon, 1988; Steinmetz and Schonbeck, 1992). Commercial formulations of *Trichoderma* are being widely used for disease management in agricultural and horticultural crops (Table - 3). Delivering of *Trichoderma* spp. to the site of infection *via* the organic food bases like diatomaceous earth granule with molasses (Backman and Rodriguez Kabana, 1975), wheat bran (Henis *et al.,* 1978), wheat bran + saw dust (Elad *et al.,* 1980), wheat bran + peat (Sivan *et al.,* 1984), molasses yeast (Papa*v*izae *et al.,* 1984), alginate pellets (Fravel *et al.,* 1985), pyrax (Papa*v*izas and Lewis, 1989), vermiculite + wheat bran (Lewis *et al.,* 1991, Nakkeeran *et al.,* 1997) and processed manure pellets (Kok *et al., 1996*) allow establishment and multiplication of the antagonist in the soil resulting in high population levels during an extended period of time.

Many of these companies are small, privately owned firms with a limited product-line. In India, *Trichoderma* based biofungicides as biocontrol formulations are marketed as wettable powder (WP) based on single strain of *Trichoderma.* The list of registered products available in the market are given as below. (Kulkarni and Sagar, 2007).

Table 3 : Commercial products from *Trichoderma* species for disease management

Product	Target pathogens	Crops recommended	Manufacturer
Bio-fungus-*Trichoderma* spp.	*Sclerotina, Phytophthora, R. solani* , *Pythium* spp. *Fusa rium, Verticillium*	Flowers, Strawberries, Trees, Vegetable	De Cuester, Belgium
Binab T-*T. harzianum* and *T. polysporum*	Pathogenscausingwilt, take all, root rot, and internaldecay of wood products	Flower, fruits, ornamentals, turf grass and vegetables	Bio-innovation Lab-UK
Root-Pro-*T. harzianum*	*R. solani, Pythium* spp, *Fusarium spp*. and *S. rolfsii*	Flowers and vegetables	Mycontrol Ltd., Israel
Root shield-*T. harzianum*	*R. solani, Pythium* spp, *Fusarium* spp	Trees, shrubs, & vegetables	Bioworks, Inc, USA
Supresivit-*T. harzianum*	Several fungalpathogens	Broad spectrum	Borregaard Bio Plant,
T22G, T22 planter box -*T. harzianum*	*R. solani, Pythium* spp. *Fusarium* spp, and *Sclertotinia homeocarpa*	Bean, cabbage, corn, cotton, cucumber, peanut, potato, Sorghum soybean, tomatoandsugarbeet.	Bioweorker
Trieco- *T. viride*	*R. solani, Pythium* spp *Fusarium* spp.,	Cotton, chilies, tomato sugarcane, citrus grapes sunflower, cereals and vegetables	Ecosense labs (1) Pvt. Ltd., Mumbai, India
Trichodex-*T. harzianum*	*B. cinerea, Colletotrichum spp., Fulvia fulva, Monilia Laxa, Plasmopara viticola, Pseudoperonospora Cubensis*	Cucumber, grapes nectarine, soybean	Makhteshim Chemical Works Ltd., New York
Trichpel, Trichoject, Trichdowel, Trichoseal *T. harzianum and T. viride*	*Armillaria, Botryosphaeria , Chondrostereum, Fusarium, Nectria , Phytophthora, Pythium , Rhizoctonia*	Broad spectrum	Agrimm Technologies Ltd., New Zealand
Trichoderma 2000-Trichodema spp.	*R. solani, Pythium* spp *Fusarium* spp, *and S. rolfsii*	Nurseryand field crops	Mycontrol Ltd., Israel

Kannaiyan et al., 2002

Talc based formulations

Tamil Nadu Agricultural University, Coimbatore, has developed the technology of mass production of talc-based formulation of *T. viride* and *Pseudomonas fluorescens* for seed treatment and soil application. Several private industries in Coimbatore, Bangalore, Chethali, Delhi and Chennai produce the same in large quantities. However, the demand exceeds the present supply. The annual requirement of *Trichoderma* has been estimated as 5,000 tones to cover 50 per cent area in India. Indirectly it also creates self-employment opportunities to the unemployed youths. The first commercial seed treatment formulation of bio-control agent in India was developed by Jeyarajan *et al.,* (1994). *T, viride* was grown in molasses yeast medium for 10 days in flasks. Then 2.51 of the culture were inoculated to 50L. of sterilized molasses yeast medium in fermentor. It was incubated for 10 days with 4-8 h aeration / day. The fungal biomass and the broth were mixed with 100 kg talc (super white) powder and 500 g of carboxy methyl cellulose as a sticker. It was dried in shade for 72 hr and packed in alkathene bags. The initial population of *Trichoderma* in the produce was 300 x 10^6 cfu/g. The product should contain a minimum of 20 x 10^6 cfu/g at the time of use. This is applied @ 4 g per kg of seed. The shelf life was 4 months. Angappan (1992) found that chickpea treated with this product maintained the rhizosphere population at 11 to 13 x 10^3 cfu/g of soil throughout crop growth.

Other formulations

Ranganathan *et al.,* (1995) found that gypsum was a good and cheaper substitute for talc. Nakkeeran and Jeyarajan (1996) tested two industrial waste namely precipitated silica and calcium silicate in the place of talc and found them to give a population of 99, 104 x 10^6 cfu/g respectively compared to 143 x 10^6 cfu/g in talc formulation after 4 months of storage. These two substrates were also much cheaper than talc.

Diatomaceous earth granules were added into a broth consisting 100ml black grape molasses, 900 ml of water, 3 g each of KNO_3 and KH_2PO_4 till the level of saturation. It was autoclaved at 121°C for 15 min and spread in shallow pans to a height of 3-5 cm and autoclaved again. *T. harzianum* (3 day old culture) was homogenized in a blender for 3 seconds and mixed with sterilized granules in shallow pans and incubated at 25°C for 4-7 days. Clumps were broken up and granules air dried with frequent stirring and used as an inoculum. The inoculum was mixed at 1:1 (v/v) with sterilized diatomaceous earth granules impregnated with 10 per cent molasses solution. It was applied to peanut at 140 kg/ha on 70 and 100 days after sowing to control *S. rolfsii.* The disease was reduced by 42 per cent over control and yield increased by 13.5 per cent (Backman and Rodriguez Kabana, 1975).

Table 4 : Commercial development of talc based *Trichoderma* biofungicide

Name	Bioagent	Manufacturers
SK_1-UASD	*Trichoderma harzianum*	Department of Plant Pathology, University of Agricultural Sciences, Dharwad, Karnataka
TRICHODERMA WP	*T. viride, T. harzianum*	Sun-Agro Industries Ltd., Porur, Madras
ANTAGON COMBI WP	*T. viride* + *Pseudomonas Fluorescens*	Green Tech Agro Plants Pvt. Ltd., Coimbatore
KADUR'S TRIJAIVA WP	*T. harzianum*	Kadur's Agro Malasandra, Banagalore
MONITOR WP	*Trichoderma* spp.	Agricultural & Biotech Pvt. Ltd., Gujrat
BINAB-T WP	*T. harzianum* + *T. polysporum*	Bio Innovations AB Toreboda Sweeden
BIOCURE-F WP	*Trichoderma* sp.	T. Stane's Coimbatore
BIOCURE (F + B) WP	*Trichoderma* + *P. fluorescens*	T. Stane's Coimbatore
TRICHOGUARD™	*T. viride*	Anu Biotech International, Delhi
TRICHODERMA SPP. WP	*Trichoderma* spp.	Innovative Pestcontrol laboratory, Bangalore
BIOFUNGICIDE WP	*T. harzianum*	Project, DBT, UAS, Dharwad
MBF TRICHO-ACTION WP	*Trichoderma viride*	Maharashtra Bio Fertilizers India Pvt. Ltd., Latur, Maharashtra
BIODERMA WP	*Trichoderma viride*	Biotech International Ltd., New Delhi
AGRODERMA WP	*Trichoderma harzianum*	Nuestras oficinas y bodega estan ubicadas por carretera Culiacan, Culican Sinaloa.
MTR35 WP	*Trichoderma harzianum*	Makhteshim Chemical, Works Ltd., Israel
ECODERMA WP	*Trichoderma viride*	Margo Biocontrols Pvt. Ltd., Banagalore

Wheat bran: Sawdust: tap water mixture (3:14 v/v) is taken in polypropylene bags and autoclaved for 1 h at 121°C for two successive days. The bags were inoculated with *T. harzianum* and incubated in illuminated chambers for 14 days at 30°C. It was applied at the time of sowing and mixed with the soil to a depth of 7-10 cm with a rotary hoe. It increased yield of beans (1500 kg/ha), tomato (300 kg/ha), cotton (500 kg/ha) and potato (40-600 kg/ha) and controlled *S. rolfsii* and *R. solani* (Elad *et al.*, 1986).

Wheat bran: Peat mixture (1:1 v/v) was autoclaved for 1h. Substrate moisture was adjusted to 50 per cent (w/w) with sterile water, medium was inoculated with 0.1 ml of a conidial suspension containing 2 x 10^4 conidia / ml and incubated for seven days at 30⁰C. It was mixed with rooting mixture for tomato at 10 per cent v/v to control crown rot (*Fusarium oxysporum f. sp. radicis lycopersici*). The disease was reduced by 29.7 per cent over control and yield increased by 7 per cent (Sivan *et al.*, 1984).

Trichoderma was multiplied in molasses - yeast medium for 10 days. Vermiculite (Grade 4) and milled wheat bran (250 mesh) were heated by placing over metal pans at 70⁰C in hot air oven for three days. Vermiculite (100g), wheat bran (3.3 g), liquid culture (14ml) and 0.05N HC1 (17.5ml) were mixed and packed. This was immediately used for soil application (Lewis *et al.*, 1991).

Sodium alginate (20g) was dissolved in 750 ml water at 40⁰C on a stirring hot plate. Wheat bran ground to pass through a 0.425 mm mesh screen was placed in a glass blender container with distilled water (50g/250ml). Kaolin can also be used in place of bran. It was autoclaved for 30 min and cooled. Fermentor biomass (16-21 g/I) was added to provide 7 x 10^6 conidia and chlamydospores. The mixture containing fungus, alginate and bran or kaolin was added drop wise into 500ml gallant solution (0.25 M $CaCl_3$, pH 5.4). As it entered, each droplet gelled and a distinct spherical bead formed. After 20 min in the gallant, beds were separated from the solution by gentle filtration, washed and dried for 24 h in a stream of air at 25⁰C (Lewis and Papavizas, 1986).

Shelf life

Shelf life of a bio-control agent plays a crucial role in adoption of the technology. In general the antagonists multiplied in a organic food base has greater shelf life than the inert or inorganic food bases. Viability of the encapsulated alginate pellets of *T.harzianum* isolate (Thr IDI) remained viable for at least 6 months when stored at 5⁰C and were reduced significantly when stored at 2⁰C (Dandurand and Knudsen, 1993). Jeyarajan *ei al.*, (1994) developed talc, peat lignite and kaolin based formulation of *T. viride*, which had a shelf life of 4 months. Shelf life of gypsum-based formulation was also four months, but was cheaper than talc based formulations (Ranganathan *et al.*, 1995). Vermiculite bran acid fermentor biomass (VBA-FB) of *T. viride* recorded the highest mean population of 20.5x10^7 cfu/g up to 75 days of storage at room temperature (Nakkeeran *et al.*, 1997).

Delivery system

Biocontrol agents can be applied to various plant parts like seed, root, foliage, inflorescence and in to the soil. Several effective methods of delivery system have been developed which enhanced the yield of crops considerable (Nakkeeran *et al.*, 1995; Nakkeeran and Renukadevi, 1997).

Seed treatment

Seed treatment is one of the methods for introducing bio-control agents into soil-plant environment. The efficacy of seed treatment depends on the ability of the antagonist to multiply in the rhizosphere region. Seed treatment with *T. harzianum* reduced the disease caused by *R. soalni* on cotton under field conditions (Elad *et al.*, 1982). Treatment of black gram, green gram, pigeon-pea, cowpea, groundnut, sunflower, sesame and cotton with *T. viride* @ 4g/kg reduced the incidence of root rot and wilt (Dinakaran *et al.*, 1995: Jeyarajan *et al.*, 1994: Jeyarajan and Nakkeeran, 1996; Muthamilan and Jeyarajan, 1996; Nakkeeran and Renukadevi, 1997).

Solid Matrix Priming

It is a process in which moistened seeds are mixed with an organic carrier and the moisture content of the mixture is brought to a level just below the required for seed sprouting. Solid matrix priming can further enhance the efficacy of antagonist in treated seeds as it grows on seed surface during priming process and thus the propagule numbers increase. Slurry of *Trichoderma* is first added to seeds followed by addition of lignite and water. *Trichoderma* required less moisture for growth. The primed seeds were incubated for 4 days at 80-100 per cent RH. During this process *Trichoderma* profusely sporulate on seed surface. In *Pythium ultimum* infested soil the stand was increased to 70-80 per cent as against 10 per cent in control. In tomato seed the population of *Trichoderma* increased by ten fold due to solid matrix priming (Harman *et al.*, 1989).

Soil application

High population of the antagonist in the soil is necessary for effective management of soil borne pathogens. Adams (1990) defined efficiency of biocontrol agents as the ratio of number of mycoparasites required to obtain disease control to the typical inoculum density of a plant pathogen. Therefore, attempts were made to develop suitable delivery system to soil. For controlling *R. solani, Trichoderma* population of 5 x 10^6 was required for each propagule of *R. solani.* Addition of wheat bran based inoculum to soil gave 80 per cent reduction of root rot over control in chickpea and bran (Elad *et al.*, 1986). Incidence of urdbean root rot was reduced by 91.3 per cent by adding *T. viride* + *T. harzianum* soil (Kousalya and Jeyarajan, 1988). Delivering *T. harzianum* through soil during sowing increased the survival of peanut (90 per cent), while in control none of the plants survived (Muthamilan and Jeyarajan, 1996).

Cut stump application

Richard (1981) developed a special shears to apply *T. viride* simultaneously with pruning to protect fruit trees against silver leaf disease caused by *Chonostereum purpureum.* The shears have a reservoir for suspension of *T. viride.* Each pruning cut was covered by spore suspension.

Hive insert

Thomson *et al., (*1992*)* developed an innovative method of application of bio-control agent right in infection court at the exact time of susceptibility. *Botrytis cinerea* (fruit rot of straw berry) was effectively controlled by *T. virens* through honey bees which acted as vehicle to carry the spores of *Trichoderma.* Peng *et al., (*1992*)* developed a novel and more efficient method of applying *T. virens* on talc and corn meal (5x10^8 cfug/g), which was placed inside the beehive in a dispenser. Bees acquired the conidia on their legs and bodies as they crawled. Each bee had a spore load of 88 to 1800 x 10^3. Each flower got 1.6 to 27 x 10^3 conidia of antagonist. It suppressed *B. cinerea* on stamens (54 per cent) and petals (47 per cent) and controlled fruit rot.

Population dynamics and its importance

Soil

The soil ecosystem is a hostile environment for introduced microorganisms, when often cannot establish and eventually disappear. The bio-control efficacy of *Trichoderma* isolates seems to depend on the population size, reached after introduction into soil. A successful antagonist, to survive better in soil and rhizosphere, should have the ability to proliferate well in the introduced environment (Baker and Cook, 1974). Population densities of *T. viride* (T-1-R4) and *T. harzianum* (WT-6-24) increased to 10^4 and 10^3 fold respectively in natural soil during the first 3 weeks of incubation (Lewis and Papa*v*izas, 1986). The decline in the population of *T. viride* in the soil might be due to lysis and disintegration of conidia. The stable population densities of *T. viride* from 9-36 weeks of incubation may be due to the survival of chlamydospores. In no case did conidia of *Trichoderma* and *Gliocladium* proliferated in soil without the addition of a food base (Papa*v*izas, 1985). Pyrax (Clay powder), alginate pellets, vermiculite-bran mixture based formulations allowed establishment and multiplication of the antagonist in soil, resulting in high population levels during the extended period of time. The activity and proliferation of *Trichoderma* spp. in the protection of radish seeds against *R. solani* was effective only in acidic soils (Chet and Baker, 1980;1981). A food base such as ground wheat bran enables the fungal propagule in a given preparation to germinate and to colonize the soil. In sterile conditions the propagation of *T. harzianum* was much greater than under un-sterile soils. The incorporation of *T. harzianum* conidia through processed manure pellets helped the colonization of non-sterile soils indicated the necessity of food base of establishment. The soil pH also influenced the population dynamics (Kok *et al.*, 1996).

Seeds

Bio-control agents must proliferate rapidly and become established on the place where they are applied. The first 12-24 h after planting is a crucial period for the bio-protectant applied as seed treatment against soil-borne pathogens and they need to be grown and established on the seed coat (Hubbard *et al.*, 1983; Taylor and Harman, 1990). Solid matrix priming is the process, in which seeds are moistened and mixed with a finely ground organic carrier, sufficient water was added to achieve the appropriate moisture potential for priming and then incubated for a given duration at constant temperature. Organic carriers have been shown to enhance the efficacy of *Trichoderma* spp. proliferate during the incubation period and become established, more active or have colonized the spermosphere thoroughly before planting (Harman *et al.*, 1981). Prime seeds consistently emerge more rapidly than the untreated seeds (Harman and Taylor, 1988). Solid matrix priming of tomato seeds increased the number of *Trichoderma* propagules by one order of magnitude in the spermosphere. The effectiveness of bio-control agents may depend partially on their ability to proliferate during a short period of favorable environmental conditions before they encounter plant pathogens. The number of colony forming units (cfu) of *Trichoderma* increased on seeds during the matrix priming in Agro-lig by about 10 fold. Number of *Trichoderma* on matrix priming were 10^4 cfu/seed after four days of incubation, while on seeds treated with *Trichoderma* alone were about 10^3 cfu/seed. Hence the increase in the cfu of *Trichoderma* provided better emergence of tomato (Harman and Taylor, 1988).

Industrial application of *Trichoderma*

Trichoderma, a wood degrading fungus of the forest has well defined extracellular enzyme production and it is widely used in textile and food industries. It is used commercially for the production of cellulases and other enzymes that degrade complex polysaccharides. The cellulases are also used in "biostoning" of denim fabrics to give rise to the soft, whitened fabric stone-washed denim. In addition the enzymes are used in poultry feed to increase the digestibility of hemicelluloses from barely or other crops. Apart from it, *Trichoderma* is also used in industries to increase ethanol production (Stevenson and Weimer, 2002). Hence, the concept of transforming renewable plant biomass to fuels attracts attention. In semi solid fermentation, use of cellulose and 8% cellulose (from *T. reesei*) increased ethanol production.

Trichoderma, a soil healh promoter

Lignocelluloses being the major source of carbon to ecosystems and the principal constituent of agricultural, forestry and municipal wastes, cellobiohydrolases and endoglucanases produced by *Trichoderma* spp. act on the lignocellulose and aids in the recycling nutrients by decaying them. It not only cleans the environment, but also improves the mineral pool in the soil and there by increase the soil health and plant health. *T. harzianum* and solubilized manganese dioxide, Zn metal and rock phosphate by chelation and reduction. In addition, it also aids in recycling carbon and other elements by decaying organic matter and dead trees in forests (Altomare *et al.*, 1999).

Bioremeidiation by *Trichoderma*

The variety of organic pollutants released by the ever increasing number of industries is the direct cause of environmental and health-related problems that have detrimental effects on living beings. Conventional approaches (*e.g.* land-filling, recycling, pyrolysis and incineration) to the remediation of contaminated sites are inefficient and costly and can also lead to the formation of toxic intermediates. Thus, biological decontamination methods are preferable to conventional approaches because, in general, microorganisms degrade numerous environmental pollutants without producing toxic intermediates.

Different organic pollutants that are present in the environment persist for varying periods of time. Depending on their half-life, some pollutants either disappear within a short period of time or transformed into non-toxic end products. Many others,, such as polychlorinated dibenzodioxfurans(PCDDF/Fs), dioxins, dichlorodiphenyltrichloroethane (DDT), chlordane and hexachlorocyclohexane (HCH) have long half lives in the environment and tend to enter the food web, where they are subsequently 'biomagnified'. On entering mammalian systems, most of these pollutants are often metabolized to generate more toxic and harmful intermediates owing to the nonspecific activity of different mammalian enzymes. Bioremediation is the process by which living organisms degrade or transform hazardous organic contaminants. *Trichoderma* plays a vital role in bioremediation of heavy metals, toxic chemicals such as xenobiotic comounds, cobalt, lignite, cyanide, nitrocellulose etc.

Trichoderma spp. are recognized as "keystone or controller organisms" in forest soil due to their involvement in decomposition, nutrient cycling, and their regulation of associated mycobiota. Recent evidences have proved their potential to decompose xenobiotics. *T. harzianum* and *T. viride* have been shown to be capable of degrading organochlorides under *in vitro*. The predominant association of *T. harzianum* and *T. viride* may influence the persistence of organochloride contaminants in forest soil (Smith, 1995). *T. harzianum* was capable of degrading chlorophenols and adsorbable organic halogens. It also had the capability to degrade ciliatine, glyphosate, and also phosphonic acid (Lupicka *et al.*, 1997).

Biodegradation of noxious chemicals

Trichoderma species also have the potential to degrade noxious chemicals like nitrocellulose, cobalt 60, cyanides and heavy metals. *T. reesei* had the capacity to degrade nitrocellulose which is commonly used in explosives and pharmaceutical industries. The fungus was able to degrade nitrocellulose completely after 22 days of application (Auer *et al.*, 2005). Cotton, beans or corn-grown on a polluted site (*i.e.* heavy metals, arsenic, cyanide and metallocyanides, nitrates, etc.) in association with *T. harzianum* increased the uptake of nutrients or toxicants by the plants due to the production of enzymes like rhodanase, formamide hydro lyase and cyanide hydrates (Ezzi and Lynch, 2005).

Biodegradation of pesticide

T. harzianum degraded DDT, dieldrin, endosulfan, PCNB, and pentachlorophenol. *T. harzianum* CCT-4790 degraded 60 per cent of the herbicide diuron in soil and opens a good potential for bioremediation of herbicide-contaminated soils. *T. viride* degraded pesticide photodieldrin into water soluble and non-insecticidal compounds after 4 to 5 weeks. Thus, *Trichoderma* strains play a crucial role in bioremediation process. It colonized plant rhizosphere and metabolize several pesticides. Pentachlorophenol (PCP) was metabolized to pentachloroanisole (Rigout and Matsumura, 2002). *Trichoderma* is a predominant microflora in the rhizosphere of oil polluted soils of Kuwait. It encouraged remediation of oil in the rhizosphere region of polluted soil (Lynch and Moffat, 2005).

Phytobiation

Physical and chemical techniques are the most commonly used methods for remediation of pollutants. The high costs of these methods have stimulated interest in alternative biologically-based systems, in particular, bioremediation and phytoremediation. Bioremediation consists of adding microorganisms to a polluted system, in the presence of nutrients, for degrading toxic compounds. Phytoremediation uses plants to degrade or remove toxicants from soil lacking a defined or managed root-microbial population. Combined application of both remediation is termed as phytobiation. Cultivation of cotton, beans or corn as well as ferns on a polluted site rich in heavy metals, arsenic, cyanide and metallocyanides, nitrates, etc. in association with *Trichoderma* spp. increased rapid uptake of the toxic elements by the plants. In addition there was strong reduction of toxicity caused by polycyclic aromatic hydrocarbons and phenolic compounds in water and soil by the use of *Trichoderma* (Vinale *et al.,* 2003).

Future thrust

The investigation of the biogeographical distribution of fungi is a challenging task. Firstly, it is generally regarded that only a small proportion of fungi (approximately 70,000 species) have been identified to date, with estimates of the numbers of fungi at 1.5 million. This estimate is not universally accepted as it is based on calculations from temperate regions indicating an approximately 6:1 ratio of fungal species to plant species. However, for tropical regions there are conflicting estimates, with this ratio either considered too high or too low or proportions of up 33:1 is considered more realististic. Consequently, the incomplete knowledge of the diversity and distribution of fungi has to be taken in to account in any fungal biogeographic study. Owing to the significant contribution of *Trichoderma* to human welfare and due to its rich biodiversity in India, the following issues may be addressed to conserve and to claim the IPR rights of our microbial wealth. They are:

- Adoption of polyphasic systematic to map the distribution of *Trichoderma* from different Agro-climatic zones.
- Identifying the habitat preference of individual taxa.
- Exploring the potential of all the strains of *Trichoderma* for human welfare.
- Research on bioremediation, phytoremediation, phytobiation has to be strengthened to select potential strains of *Trichoderma.*
- Transformation and gene manipulation strategies have to be strengthened for exploring the potential of *Trichoderma* other than *T. reesei.*
- Strengthening proteomic and genomic research of *Trichoderma* for better pest and disease control of plants.
- Improving the knowledge on the ecology of *Trichoderma* - plant-pathogen interaction.
- Assessment of human and plant health risks.
- Human Resource Development on bio-informatics, systems biology, digital biology with molecular techniques for the documentation and exploitation of the full potential of *Trichoderma* for the welfare of the country and the mankind as a whole.

Conclusion

The genus *Trichoderma* is very important in several commercial and biotechnical applications. The useful, harmful or deleterious species/strain have to be identified correctly for the use of apposite strain in the industry as well as in agriculture. The various isolates applied and exploited in India were generally obtained from certain source or isolated from the biotic or abiotic substrates. Mostly they belong to the species *viz., T. aureoviride, T. harzianum, T. koningii, T. psuedekoningii, T. virens* and *T. viride.* The additions to the list as new species or new reports are very few from India. Furthermore, the identification of species or strains till to date are based mostly on morphological characters only. The reasons for these are copious hence an exhaustive survey and identification studies are mandatory to avoid erroneous names in a variety of applications.

The studies of extreme habitats are one of the crucial areas where efficient strains of *Trichoderma* can be isolated and identified by means of traditional and macro-molecular techniques.

References

Adams, P.B. (1990). The potential of microparasites for biological control of plant diseases. *Ann. Rev. Phytopath.,* 28:59-72.

Altomare, C., Norvell, W.A., Bjorkmann, T. and Harman, G.E. (1999). Solubilization of phosphates and micronutrients by the Plant Growth Promoting and Biocontrol fungus, *Trichoderma harzianum* Rifai 1295-22. *Appl. Environ. Microbiol.,* 65 : 2926-2933.

Anagappan, K. (1992). *Biological control of chickpea root rot caused by Macrophomina phaseolina (Tassi) Goid.* M.Sc. (Ag.) Thesis, TNAU, Coimbatore, p.114.

Askew, D.J. and Lang, M.D. (1993). An adapted selective medium for the quantitative isolation of *Trichoderma* species. *Plant Pathol.,* 42:686-690.

Auer, N., Hedger, J.N. and Evans, C.S. (2005). Degradation of nitrocellulose by fungi. *Biodegradation,* 16 : 229-236.

Backman, P.A. and Rodrigue-Kabana, R. (1975). A system for the growth and delivery of biological control agents to the soil. *Phytopath.,* 65:819-821.

Baker, K.F. and Cook, R. J. (1974). *Biological control of plant pathogens.* W.H. Freeman and Co., San Francisco, C.A.: 433.

Bissett, J. (1991a). A revision of the genus *Trichodermna.* III. Sect. *Pachybasium. Can. J. Bot.,* 69: 2373-2417.

Bissett, J. (1991b). A revision of the genus *Trichodermna.* IV. Additional notes on section *Longibrachiatum. Can. J. Bot.,* 69: 2418-2420.

Bisset, J. (1992). *Trichoderma atroviride. Can. J. Bot.* 70:639-641.

Chet, I. and Baker, R. (1980). Induction of suppressiveness to *Rhizoctonia solani* in soil. *Phytopath.,* 70:994-998.

Chet, I. and Baker, R. (1981). Isolation and biocontrol potential of *Trichoderma hamatum* from soil naturally suppressive to *Rhizoctonia solani. Phytopath.,*71:286-290.

Connick , Jr., W.J., Lewis, J.A. and Quimby Jr., P.C (1990). *Formulation of biocontrol agents for use in Plant Pathology, New Directions in Biological control. Alternatives for Suppressing Agricultural Pests and Diseases* (Eds. R.R. Baker and P.E. Dunn). Alan R. Liss, Inc. New York, pp. 345-372.

Dandurand, L.M. and Knudsen, G. R. (1993). Influence of *Pseudomonas fluorescens* on hyphal growth and biocontrol efficacy of *Trichoderma harzianum* in the spermosphere and rhizosphere of pea. *Phytopath.,* 83: 265-270.

Danielson, R.M. and Davey, C.B. (1973). The abundance of *Trichoderma* propagules and the distribution of species in forest soils. *Soils Biol. Biochem.,* 5:486-494.

Deacon, J.W. (1988). Biocontrol of soil-borne plant pathogens with introduced inocula. *Phil. Trans. R. Soc. Lond. Ser.B.,* 318:249-264.

Dennis, C. and Webster, J. (1971). Antagonistic properties of species groups of *Trichoderma* I. Production of non-volatile antibiotics: *Trans. Br. Mycol., Soc.* 57:25-39.

Dhingra, O.D. and Khare, M.N. (1973). Biological control of *Rhizoctonia bataticola* on Urdbean. *Phytopath. Z., 76*:23-29.

Dinakaran, D. and Marimuthu, T. (2001). Effect of *Trichoderma viride* mutants on the management of root rot of blackgram. *Natl. Symp. On Pulses and Oilseeds for sustainable agriculture, July 29-31, 2001,* Directorate of Research, Tamil Nadu Agricultural University, Coimbatore, India, pp. 181 (Abst.).

Dinakaran, D. Ramakrishan, G. Sridar, R. and Jeyrajan, R. (1995). Management of sesamum root rot with biocontrol agents. *J. Oilseeds Res.,* 12: 262-263.

Elad, Y., Chet, I. and Henis, Y. (1982). Degradation of plant pathogenic fungi by *Trichoderma harzianum. Can J. Microbiol.,* 28:719-725.

Elad, Y. Chet, I. and Katan, P. (1980). *Trichoderma harzianum.* A biocontrol agent effective against *Sclerotium rolfsii* and *Rhizoctonia solani. Phytopath.,* 70:119-121.

Elad, Y., Zuiel, Y. and Chet, I. (1986). Biological control of *Macrophomina phaseolina* (Tassi) Goid. by *Trichoderma harzianum, Crop Prot.,* 5:288-292.

Elisa Esposito and Manuela da Silva. (1988). Systematics and environmental application of the genus *Trichoderma. Criti. Rev. Microbiol.,* 24:89-98.

Ezzi, M.I. and Lynch, J.M. (2005). Cyanide catabolizing enzymes in *Trichoderma* spp. *J. Enzymes Microbial. Technol.,* 31:1042-1047.

Fravel, D.R., Marois, J.J., Lumsden, R.D. and Connick Jr., W.J. (1985). Encapsulation of potential biocontrol agents in an alginate - clay matrix. *Phytopath.,* 75:774-777.

Grondona, I., Hermosa, R., Tejada, M., Gomis, M.D., Mateos, P.F., Bridge, P.D., Monte, E and Garcia-Acha, I. (1997). Physiological and biochemical characterization of *Trichoderma harzianum,* a biological control agent against soilborne fungal plant pathogens. *Appl. Environ. Microbiol.,* 63:3189.

Harman, G.E. and Taylor, A.G. (1988). Improved seedling performance by integration of biological control agents at favourable pH levels with solid matrix priming. *Phytopath.,* 78: 520-525.

Harman, G.E., Chet, I. and Baker, R. (1981). Factors affecting *Trichoderma hamatum* applied to seed as a biocontrol agent. *Phytopath.,* 71:569-572.

Harman, G.E., Taylor, A.G. and Stasz, T.E. (1989). Combining effective strains of *Trichoderma harzianum* and soil matrix priming to improve biological seed treatment. *Plant Dis.,* 73:631-637.

Henis, Y., Ghaggar, A. and Baker, R. (1978). Integrated control of *Rhizoctonia solani* damping-off of radish. Effect of successive plantings. PCNB and *Trichoderma harzianum,* pathogen and disease. *Phytopath.,* 68:900-907.

Hjeljord, L. and Tronsmo, A. (1998). *Trichoderma* and *Glicladiumin* Biological Control; An Overview. In: *Harman, G.E., Kubicek, C.P. (Eds.), Trichoderma and Gliocladium. Vol. 2, Enzymes, Biological Control and Commercial Applications.* Taylor and Francis Ltd., London, pp. 131-151.

Hubbard, J.P., Harman, G.E. and Hadar, Y. (1983). Effect of soil-borne *Pseudomonas* spp. on the biological control agent, *Trichoderma hamatum,* on pea seeds. *Phytopath.,* 73: 655-659.

Jeyarajan, R. and Nakkeeran, S. (1996). *Exploitation of biocontrol potential of Trichoderma for field use In: Current trend in life sciences Vol. XXI* (Eds.) K. Manibhushan Rao and A. Mahadevan), Today and Tomorrow's Printers and publishers, New Delhi:61-66.

Jeyarajan, R.G., Ramakrishnan, Dinakaran, D. and Sridar, R. (1994). Development of products of *Trichoderma viride* and *Bacillus subtilis* for biocontrol of root rot diseases. In: *Biotechnology in India* (Ed.) B.K. Dwivedi *Bioved Research Society,* Allahabad : 25-36.

Kannaiyan, S., Sabitha Doraiswamy, Ramakrishnan, G., Rabindra, R.J., Samiyappan, R., Udayasurian, V., Gunasekaran, K. Kennedy, J.S., Nakkeeran, S. and Sathiah, N. (2002). *Recent Advances in Eco-friendly Approaches for the Management of Pests and Diseases in Crops.* Tamil Nadu Agricultural University, Coimbatore : 1-389.

Kiffer, E. and Morelet, E. (2000). *The Deutermycetes Mitosporic Fungi Classification and Generic Keys.* Science Publishers, Inc. USA.

Kiffer, E. M. and Orelet, E. (2000). Processed manure as carrier to introduce *Trichoderma harzianum.* Population dynamics and biocontrol effect on *Rhizoctonia solani. Biocontrol Sci. and Tech.,* 6:147-161.

Kok, C. J., Hageman, P.E.J, Maas, P.W.T., Postma, J., Roozen, N.J.M. and Van Vuurde, J.W.L. (1996). Processed manure as carrier to introduce *Trichoderma harzianun.* Population Dynamics & Bio-control effect on *Rhizoctonia solani. Bio-control Sci. Technol.* 6: 117-161.

Kubicek, C. P. and Harman, G.E. (1998). *Trichoderma* and *Gliocladium, Vol* I & II. Taylor and Francis Ltd., London, U.K.

Kousalya, G. and Jeyarajan, R. (1988). Techniques for mass multiplication of *Trichoderma viride* Pers. Fr and *T. harzianum* Rifai. *National Seminar on management of crop diseases with plant products/biological agents.* 10-12 January, Agricultural College and Research Institute, Madurai : 31-33.

Kredics, L., Antal, Z., Doczi, I., Manczinger, L., Kevei, F. and Nagy, E. (2003). Clinical importance of the genus *Trichoderma.* A review. *Acta Microbiol. Immunom. Hung.,* 50:105-17.

Kubicek, C.P., Bissett, J. Druzhinina, I., Kullnig-Gradinger, C.M. and Szakacs, G. (2003). Genetics and metabolic diversity of *Trichoderma:* a case study on South East Asian isolates. *Fungal Genet. Biol.,* 38:310-319.

Kubicek, C.P. and Herman, G.E. (1998). Regulation of Production of Plant Polysaccharide Degrading Enzymes by *Trichoderma. In: Eds. G.E. Harman, C.P. Kubicek, Trichoderma Gliocladium. Vol. 2. Enzymes, Biological Control and Commercial Applications.* Taylor and Francis Ltd., London, pp. 49-71.

Kuhls, K., Lieckfeldt, E., Samuels, G.J., Meyer, W., Kubicek, C. and T. Borner, (1997). Revision of *Trichoderma* sect. *Longibrachiatum* including related teleomorph based on analysis of ribosomal DNA internal transcribed spacer sequences. *Mycologia* 89 (3): 442-460.

Kulkarni, Srikant and Sagar, S. (2007). *Trichoderma potential biofungicide of the millennium. Tech. Bull. 5*: spp. 1-19. Department of Plant Pathology, College of Agricultural Sciences, Dharwad.

Kullinig-Gradinger, C.M., Szakacs, G. and Kubicek, C.P. (2002). Phylogeny and evolution of the fungal genus *Trichoderma"* a multigene approach. *Mycol. Res.,* 106:757-767.

Kumar, A. and Marimuthu, T. (1993). Biological control of damping-off of *Eucalyptus camaldulensis* caused by *Rhizctonia solani* with *Trichoderma viride* and decomposed coconut coir pith. *Pl. Dis. Res.* 9: 116-121.

Lewis, J.A. and Papa*v*izas, G.C. (1986). Reduced incidence of *Rhizoctonia* damping off of cotton seedlings with preparations of biocontrol fungi. *Bio. Cult. Tests,* 1:48.

Lewis, J.A. and Papa*v*izas, G.C. and Lumsden, R.D. (1991). A new formulation system for the application of biocontrol fungi to soil. *Biocontrol Sci. Technol.,* 1:59-69.

Liu, S. and Baker, R. (1980). Mechanism of biological control in soil suppressive to *Rhizoctonia solani. Phytopath.,* 70:404-412.

Lupicka, K., Strof, W., Kubs, K., Skorupa, M., Wieczorek, P. Lejczak, B. and Kafarski, P. (1997). The ability of soil - borne fungi to degrade organophosphate carbon to phosphorus bounds. *Appl. Microbiol. Biotechnol.,* 48:549.

Lynch, N.M. and Moffat, A.J. (2005). Bioremediation-prospects for the future application of innovative applied biological research. *Annals Appd. Biol.,* 146(2):217-221.

Marshall, D.S. (1982). Effect of *Trichoderma harzianum* seed treatment and *Rhizoctonia solani* inoculum concentration on damping off of snapbean in acidic soil. *Pl. Dis.,* 66:788-789.

Mukhopadhya, A.N. and Mukherjee, P.K. (1996). Fungi as fungicides. *Internal. J. Trop. Pl. Dis* 14 : 1-17.

Muthamilan, M.R. and Jeyarajan, R. (1996). Integrated management of *Sclerotium* root rot of groundnut involving *Trichoderma harzinaum, Rhizobium* and Carbendazim. *Indian J. Mycol. Pl. Pathol.,* 26:204-209.

Nakkeeran, S. and Jeyarajan, R. (1996). *Exploitation of antagonistic potential of Trichoderma for field use. Paper presented in National Symposium on disease of plantation crops and their management of Institute of Agriculture, Sriniketan held during February,* pp. 18-20, 1996.

Nakkeeran, S, and Renukadevi, P. (1997). Seed borne microflora of pigeonpea and their management. *Pl. Dis. Res.,* 12:103-107.

Nakkeeran, S. and Doraisamy, Sabitha (2001). Genetic improvement of *Trichoderma viride* and assessment of its efficacy against soil borne pathogens. *In: National Seminar on Emerging Trends in Pests and Disease Management.* Center for Plant Protection Studies, TNAU, Coimbatore : 130.

Nakkeeran, S., Gangadharan, Kousalya, and Renukadevi, P. (1995). Efficacy of organic amendments and antagonists on root and wilt pigeonpea. *National Symposium on organic farming,* TNAU, Madurai. Oct. 27-28: 141-142.

Nakkeeran, S. Sankar, P. and Jeyarajan, R. (1997). Standardization of storage conditions to increase shelf life of Trichoderma formulations. *J. Mycol. Pl. Pathol.,* 27:60-63.

Nelson, E.E. (1982). Occurrence of *Trichoderma* in a Douglas-fir soil. *Mycologia,* 74:280-284.

Papavizas, G.C. (1985). *Trichoderma* and *Gliocladium*:Biology, ecology and potential for biocontrol. *Ann. Rev. Phytopathol.,* 23:23-54.

Papavizas, G.C. and Lewis, J.A. (1989). Effect of *Trichoderma* and *Gliocladium* on damping-off and blight of snapbean caused by *Sclerotium rolfsii. Pl. Pathol.,* 38:277-286.

Papavizas, G.C., Dunn, M.T., Lewis, J.A. and Beagle-Ristanio, J. (1984). Liquid fermentation technology for experimental production of biocontrol fungi. *Phytopath.,* 74 : 1171-1175.

Peng, G., Sutton, J.C. and Kevan, P.G. (1992). Effectiveness of honey bees for applying the biocontrol agents *Gliocladium rodeum* to strawberry flowers to suppress *Botrytis cinerea. Can. J. Pathol.,* 14:117-129.

Persoon, C.H. (1794). Dispositio methodica fungorum..... *Romer's neues Mag. Bot.,* 1: 81-128.

Persoon, C.H. (1794) . Effectiveness of honey bees for applying the biocontrol agents *Gliocladium roseum* to strawberry flowers to suppress *Botrytis cinerea. Can. J. Pathol.,* 14: 117-129.

Ranganathan, K., Sridar, R. and Jeyarajan, R. (1995). Evaluation of gypsum as a carrier in the formulation of *Trichoderma viride*. *J. Biol. Control.*, 9:61-62.

Richard, J.L. (1981). Commercialization of *Trichoderma* based myco-fungicide. Some problems and solutions. *Biocontrol News and Information*, 2:95-98.

Rigout, J. and Matsumura, F. (2002). Assessment of the rhizosphere competency and pentachlorophenol-metabolizing activity of a pesticide-degrading strain of *Trichoderma harzianum* introduced into the root zone of corn seedlings. *J. Env. Sci and Health*, 37:201-210.

Rifai, M.A. (1969). Revision of the Genus *Trichoderma*. *Mycological Paper* 116: 1-56.

Rifai, M.A. and Webster, J. (1966a). Culture studies on *Hypocera and Trichoderma*. II. *H. qureovvirids* and *H. rufa*, *F. sterilis* F. nov. *Trans. Brit. Mycol. Soc.* 49 : 289-296.

Rifai, M.A. and Webster, J. (1966b). Culture studies on *Hypocrea* and *Trichoderma*. III. *H. lacteal* (=*H. citrina* and *H. pulvinata*. *Trans. Brit. Mycol. Soc.* 49 : 306-309.

Rossam, A.Y. (1996). morphological and molecular perspectives on systematics of Hypocreales. *Mycologia*, 88:1-19.

Samuels, G.J. (1996). *Trichoderma:* a review of biology and systematics of the genus. *Mycol Res.*, 100 :923-935.

Samuels, G.J. and Padra-Schultheiss, R. (2000). *Trichoderma stromaticum* sp. nov. a parasite of the Cacao witches broom pathogen. *Mycol. Res.*, 104: 760-764.

Samuels, G.J., Petrini, O., Kuhls, K., Lieckfeldt, E. and Kubicek, P. (1998). The *Hypocrea schweinitzii* I complex and *Trichoderma* sect. *Longibrachiatum*. *Studies in Mycology* 41:1-54. Centraalbureau voor schimmelcultures, Baarn/Deift. The Netherlands.

Sarbhoy, A.K., Agarwal, D.K. and Varshney, J.C. (1992). *Fungi of India* 1977-81 (1986) Assoicated Publishing Company, New Delhi.

Sivan, A., Elad, Y and Chet, I. (1984). Biological control. Effect of new isolates of *Trichoderma harzianum* on *Phytihum aphanidermatum*. *Phytopath.*, 74:498-501.

Sivasithamparam, K. Ghisalberti, E.L. (1998). Secondary Metabolism in *Trichoderma* and *Gliocladioum*. *In: Eds. C.P. Kubicek, G.E. Harman, Trichoderma* and *Gliocladium*. Vol. 1. *Basic Biology, Txonomy and Genetics*. Taylor and Francis Ltd., London, pp. 139-191.

Smith, W.H. (1995). Forest occurrence of *Trichoderma* species: emphasis on potential organochloride (xenobiotic) degradation. *Ecotoxicol. Environ. Sdf.*, 32:179.

Steinmetz, J. and Schonbeck, F. 1992. Applicability of different formulations of fungal antagonists for the control of soil-borne diseases. *Bulletin OILB-SROP*, 15:206-208.

Stevenson, D.M. and Weimer, P.J. (2002). Isolation and characterization of a *Trichoderma* strain capable of fermenting cellulose to ethanol., *Appl. Microbiol. Biotechnol.*, 59:721-726.

Taylor, A.G. and Harman, G.E. (1990). Concepts and technologies of selected treatements. *Ann. Rev. Phytopathol.*, 28:321-329.

Thakur, A.K. and Norris, R.V. (1928). A biochemical study of some soil fungi withj special reference to ammonia production. *J. Indian Inst. Sci.* XIA, 12: 141-160.

Thomson, S.V., Hansen, D.R., Flint, K. and Vandenberg, S.D. (1992). Dissemination of bacteria antagonistic to *Erwinia amylovora* by honey bees. *Plant Dis.*, 76:1052-1056.

Thrane, U., Poulsen, S.B., Nirenberg, H.I. and Lieckfeldt, E. (2001). Identification of *Trichoderma* strains by image analysis of HPLC chromatograms. *FEMS Microbiol Let.*, 203:249-255.

Ushamalini, C., Nakkeeran, S. and Marimuthu, T. (2006). Development of Biomanure for the management of Turmeric rhizome rot., *Archives Phytopath Plant Protection, (*Online*)* : 1-12.

Vinale, F., Abadi, K., Ruocco, M., Marra, R., Scala, F., Zonia, A., Woo, S. and Lorito, M. (2003). Remediation of pollution by using biological systems based on beneficial plant-microgan-ISMS interactions. *J. Plant Pathol.*, 85:301.

Webster, J. (1964). Culture studies on *Hypocrea* and *Trichoderma* 1. Comparison of perfect and imperfect states of *H. gelatinosa, H. rufa and Hypocrea* sp. 1. *Trans. Brit. Mycol. Soc.*, 47: 75-96.

Whipps, J.M. (1997). Developments in the biological control of soil-borne plant pathogens. *Bot. Res.*, 26:1-134.

Xu, T. (1996). Advances of molecular biology of *Trichoderma* spp. *Zhenjun Xuebao,* 15 : 143.

Zhang, C.L., Druzhinina, I.S., Kubicek, P.C. and Xu, T. (2005). *Trichoderma* biodiversity in China: Evidence for a North to South distribution of species in East Asia. *FEMS Microbiol. Lett.*, 251 : 251-257.

□□□

Microbial Diversity and Functions, 2012
© D.J. Bagyaraj, K.V.B.R. Tilak, H.K. Kehri (eds.), pp. 231-256
New India Publishing Agency, New Delhi (India)
E-mail : info@nipabooks.com; Website : www.nipabooks.com

Chapter 9

Litter Decomposition in Tropical Humid Forest

Pranita Borah and G.D. Sharma

ABSTRACT

Decomposition of litter is a dynamic process. It regulates cycling of nutrients in ecosystem. The productivity of an ecosystem depends on the turnover of carbon and other nutrients. Microbes play an important role in releasing of minerals locked up in organic complexes. The release of nutrients, however, is regulated by various climatic factors. The microbial succession at different stages of degradation of litter is controlled by the litter quality and species composition and biological activity. The present paper reviews the status of litter production, decomposition, species diversity and Biological activities as affected by the climatic and edaphic factors in tropical ecosystem.

Keywords : Litter decomposition, microbes, species diversity, biological activities, edaphic factors, tropical ecosystem.

Introduction

Decomposition is the process by which tissues of a dead organism break down into simpler forms of matter. The process is essential for new growth and development of living organisms because it recycles the finite matter that occupies physical space in the biome.

Litter decomposition has been widely studied in various tropical and subtropical (Heneghan *et al.*, 1998, Pandey, *et al.*, 2007), semiarid (Tateno *et al.*, 2007), temperate, (Lensing and Wilse, 2007) and Mediterranean climatic conditions (Martins *et al.*, 2006; Sirulink *et al.*, 2007).

Loss of litter in mixture is greater than in monoculture litter (Gartner and Cardon, 2004), suggesting that interaction among different litter species affect the rate of decomposition (Hattenschwiler *et al.,* 2005).

Process of litter decomposition begins with leaching of the most easily soluble carbon compounds, physical fragmentation into smaller bits which have greater surface area for microbial colonization and attack. This process is largely carried out by the soil invertebrate fauna, insects and fungi. Following this, the plant detritus consisting of cellulose, hemicellulose, and lignin, amino acids, sugars etc. undergo chemical alteration by microbial enzymes. Lignin is relatively resistant to decomposition but is decomposed by certain fungi , such as the black-rot fungi.

In most grassland ecosystems, natural damage from fire, insects that feed on decaying matter, termites, grazing mammals, and the physical movement of animals through the grass are the primary agents of breakdown and nutrient cycling, while bacteria and fungi play the main roles in further decomposition.

Significance of Plant Litterfall

In the biogeochemical cycle of organic matter and mineral elements, litterfall plays an important role in the relations among soil, vegetation and surrounding environment, constituting one of the essential ecological phenomena in the wooded ecosystems (Vitousek *et al.,* 1995). The litter production is an important process in the tropical forest nutrient cycle (Veneklaas, 1995), since the litterfall dynamics to the soil, is one of the most important determinants in the renovation of these forests. Besides it is the most important pathway for the return and transfer of carbon and other nutrients from the aerial part of vegetation the soil (Spain, 1984).

Forest constitutes an open system with chemical elements going in and out of it, or moving internally within it. The nutrients remain stored in the vegetation, constituting one of the greatest reserves of nutrients in the ecosystem (Vitousek *et al.,* 1995). Later, the nutrients move until the soil by means of falling leaves and rainwater flows. Another proportion of nutrients is stored in litter and in the other organic matter entering into the forest, from where they are gradually released by its decomposition. The circulation of these nutrients in the ecosystem depends on the organic matter and its rate of decomposition (Sundarapandian and Swamy1999). In the case of litter production and availability; this is determined by seasonal fluctuations. However, factors such as topography, edaphic conditions, vegetal species, age, and density of the forest are relevant as well (Hemandez *et al.,* 1992). In any type of forest, the massive litterfall takes place every year in a determined time causing a great accumulation of organic matter, and producing variations in the dynamics of the nutrient cycling (Jenny, 1980).

Litterfall is the major pathway for the return of organic matter and nutrients from aerial parts of the plant community to the soil surface and fertility. Litter production and decomposition rates have great importance in maintaining the fertility of the soil. A substantial portion of nutrients accumulated by plants is returned to the soil as litterfall followed by its decomposition. In tropical ecosystem maintenance of soil organic pool is achieved by the high and rapid circulation of nutrients through the fall and decomposition of litter. Standing crop of litter acts as an input-output system of nutrients and the rates at which forest litterfalls, the subsequently decay; regulate energy flow primary productivity and nutrient cycling in the forest ecosystem. Nutrient cycling rates in forests are usually inferred from a comparison of nutrient concentrations and amounts in litterfall, forest floor litter and crown drips, however, due to variation in canopy architecture and, amounts of rates of litterfall and decomposition show considerable spatial variation. This spatial heterogeneity allows co-existence of species and contributes to niche differentiation for seedlings and forest floor organisms.

Litter production

Evaluation of litter production is important for understanding nutrient cycling, forest growth, successional pathways and interactions with environmental variables in forest ecosystems. Litter production varies with climate, season, substrate quality and type of vegetation (Melillo *et al.*,1982;Upadhyay *et al.*,1989;Vitousek *et al.*, 1994).Chemical composition of litter which changes with type of plant community, influences structure and activity of microbial communities inhabiting soils and biological and physicochemical properties of topsoil (Heal and Dighton,1986).Release of nutrients not only depends upon litter composition but also upon soil type, microbial communities and soil properties. (Scholes and Walker, 1993; Rawat *et al.*, 2009). The production of litter depends primarily on the site productivity, but other properties of the environment as well, may introduce important variation.

Pattern of litter production In forest ecosystem

Forest litter is important to know the pattern of litter fall, whether distinctly seasonal or more or less continuous. Litter production depends primarily on the productivity of the plant community in a site, climate, rainfall and growing season, soil fertility, soil water retention and species composition within the same climatic range.

The pattern of litter fall varies greatly in individual species of tropical rain forest, but is not tied to any sort of annual cycle. It is extremely variable and dependent upon both external and internal factors. In the warm temperate forests litter fall goes on through out the year, but maximum in spring and early summer. In the Eucalyptus megnans forests of Victoria, Australia leaf fall in summer as well as in winter while in Western Australia the young and old forests leaf litter mainly is from January to March. In New Zealand the peak leaf-fall is connected with the development of new leaves in the spring months of October and November.

In cool temperate forests, seasonal pattern of leaf fall is often striking with autumn cooling, leading to more or less complete leaf fall in deciduous species. Among Gymnosperms the pattern of deposition ranges from distinctly seasonal to irregular throughout the year.

Individual species in equatorial forests may also show seasonal variation to a lesser degree than in most temperate forests .The longevity of evergreen gymnospermous leaves depends mainly on both internal and external conditions.

Litter production shows definite seasonal pattern with the variation of vegetation type and latitude (Bray and Gorham, 1964).Litter production in Equitorial rain forests throughout the year, while in temperate evergreen forests there are peaks related to leaf production may be either at the beginning or near the end of the dry period.

Environmental factors affect the seasonal pattern. In two riverine forests of Belgium, woody litter increased during the winter due to the strong winds (Hermy, 1987). Spain (1984) reported positive correlation between litter fall, temperature and rainfall. Factors, like strong wind, soil characters, mineral deficiency, rain and mechanical effects, have been found to affect leaf defoliation (Hopkins, 1966: Jonn, 1973; Vitousek, 1984). Litter production is strongly correlated with the densities, basal areas and important value index of the tree species. In many studies peak leaf fall has been related to a period of water stress during the dry season.

In natural sal stand maximum leaf fall was observed in mid February to mid May. In *Tectona* grandis plantation maximum annual leaf fall was recorded in mid November to mid February. A greater portion of annual leaf litter in popular plantation is shed during the winter months (Das, 1998).

Rate of Litterfall

The decomposition study becomes more relevant in tropical regions where soil nutrients are scarce and the growth of plant depends on the release of nutrients from the litter. When compared under different environmental condition, the rate of decomposition of forest litter varies with forest types. Slower decomposition has been reported for the subtropical mountain forest than for tropical deciduous forest (Singh, 1990, Kshattriya, 1990).

Comparing the litter decomposition of same plant species suggests three times more time at higher altitude than at lower altitude. Decomposition is faster in tropical forests than in sub-tropical forests.

Soil moisture and temperature were responsible for faster decay in tropical forest. Canopy is also a regulating factor, for temperature and wetting cycle (Singh,1990). Low pH lowers the decomposition of litter by influencing the activity of the decomposer community (Sanchez,1976). In a deciduous forest ecosystem, the litter is decomposed in the soil by a variety of microorganisms, particularly the fungal species which are the chief decomposers. Break down of the litter follows the rate of CO_2 evolution as a measure of metabolic activity in the soil. Plant litter in coniferous forests is decomposed mainly through microbial activity. The quantitative contribution of soil animal to decomposition is considered to be minimal. Environmental factors are most important in regulating the turnover rate of litter tend to be those that regulate the activity of microorganisms. In addition to climate, the litters chemical composition regulates mass-loss rates. Nitrogen and phosphorus in litter have been found favourable for, where as lignin has been suggested as a rate retarding compound (Foegel and Cromack, 1977). In some cases an initially high decomposition rate can decrease so strongly that only after a few months the mass loss rate is almost too low to be measured (Berg *et al.*, 1982b). The decomposition rate of leaf litter varies depending on its biochemical composition of the material. The mineralization cycles in forest ecosystem are related to the nature of the litter layer and its chemical composition. The soil organic layer consists of both the litter and the well humified product of microbial catabolism of the litter. Microbial decomposition of its structure is controlled primarily by the leaf bacterial flora and the physical and chemical prosperities of the leaf. The actual rate of decomposition is related at least in part to micro-macro faunal shattering of the over all leaf structure.

Factors Responsible For Regulating Litter Decay

Litter decomposition process represents an essential phase in the organic matter and nutrient cycle. The litter decay rate determines forest soil fertility and plays an important role in ecosystem functioning (Swift *et al.*,1979). It is influenced by litter quality(Bernhard *et al.*, 2003),decomposer organisms (Lavelle *et al.*,1993; Dilly *et al.*, 2004) and environmental condition (Dayer *et al.*, 1990; Kurzatkowski *et al.*, 2004). The relative importance of these factors varies according to the characteristics of the ecosystems (Ibrahim *et al.*, 2010).

Litter parameters such as toughness and lignin content including cellulose and hemicellulose have been reported as factors which affecting the nutrient release patterns. The most common factor that may regulate the litter decay is related to litter quality including N elemental concentration and ratios such C:P (Berg *et al.*,1982a; Berg and Mc Claugherty, 1989); organic matter fractions such as lignin (Meentemeyer, 1978;), lignocellulose index lignin; N index, alkyl C content of waxes and cutin (Trofymow *et al.*, 1995) elevated CO_2 concentration (De Angelis *et al.*, 2000) and tanin contents (Mesquita *et al.*, 1998). These considerations are more important under litter diversity conditions. However, when the substrate is the same, the chemical composition cannot be correlated to the decomposition rate.

Release of nutrients not only depends upon litter composition but also upon soil type, microbial communities and soil properties (Kutsch and Dilly,1999;Scholes and Walker,1993). Plant chemical composition significantly impacts on (eg. microbial immobilization and nitrification)nutrient cycling, as these ecosystem functions improve with increased plant diversity (Hooper and Vitousek,1998). Also, high stocking rates lead to reduced litter production and root biomass (Cantarutti *et al.*, 2002; Christie, 1979). An ecological perspective was explained with examples that poor soils support vegetation communities which are adapted to it. There is a two-way relationship between structure or vegetation communities and type of soils, but it is still not clear which plays a greater role in determining the other (Rawat *et al.*, 2009).

Effect of Climate on Decomposition

Climate directly influences litter decomposition through temperature and moisture; however, climate can also have an indirect effect on litter chemistry through influence on plant community composition and litter quality (Perez *et al.*, 2007).

On the other hand direct effect on litter decomposition due to the effect of temperature and moisture, have been exhibited by Aerts *et al.* (1997). Other climatic index, such as Actual Evaporation (AET) have been used to estimate the relative contribution of climate to litter decomposition rates (Aerts *et al.*,1997). The climatic factors most often accounted for are of litter decomposition was more widely spread because the effect of either was not constant in time and space and they interact in complicated ways. (Taylor *et al.*,1988). Moisture is considered as the major factor controlling litter decomposition(Woods *et al.*,1983). Nevertheless, some studies have shown that temperature has a dominant effect (Anderson *et al.*, 1973; Reinke *et al.*,1981). Other studies have demonstrated that temperature and moisture content are so interdependent that their interaction is a key point ,individual effects being barely relevant (Andersonn *et al.*,1991; Esser *et al.*,1992; Ibrahim et *al.*, 2010).

Temperature and moisture effects on decomposition

In most models which simulate soil organic matter turnover decomposition is calculated using a function such as:

Decomposition=K× Tm× Wm× other factors. where,

K=Potential decomposition rate.

Tm=Mathematical function which demonstrate the relative rate of microbial activity with increasing soil temperature, usually between a value of 0 and1.

Wm=Mathematical function which demonstrates the relative rate of microbial activity with increasing water availability, usually between a value of 0 and 1.

Other factors=C:N ratio factor, usually between a value of 0 and1.

Microbial activity increases rapidly up to 30°c with optimal temperature between 35°c and 45°c.

It is often suggested that sensitivity of microbial communities generally decreases with increasing mean annual temperature and rainfall (Kirschbaum,1995). Indeed, both the population and metabolism of the microbial population changes with water content (Sulkava *et al.*, 1996) and microbes may adopt to the variation in the frequency of wetting and drying cycle.

In addition to microbial activity, several physical processes that can affect microbial activity vary with soil water content, particularly water movement, and gas and solute diffusion. As a consequence, the relationship between soil water content and microbial processes in soil is complex because this relationship varies between soils, depending on the soil moisture retention curve, porosity, concentration of organic matter, pH and soil depth (Goncalves and Carlyle ,1994;Rodrigo *et al.*, 1997;Leiros *et al.*, 1999).

The Tm and Wm function may be dependent on substrate quality. More easily decomposable substrate is likely to be influenced by the soil temperature and water (eg.Vigil and Kissel 1995). Difference in substrate quality may explain why microbial activity in litter layer and soil are not similarly affected by the temperature and water (Quemada and Cabrera,1997)and why the effects of temperature and water on microbial activity was greater in the soil surface layer than the underlying mineral soil layers(Leiros *et al.*,1999).

Composition and activities of soil communities

In terms of biomass and species numbers of soil organisms are involved in organic matter turnover, particularly the bacteria and fungi. Recycling of carbon and nutrients during decomposition is a fundamentally important ecosystm process (Swift *et al.*,1979) that has major control over the carbon cycle, nutrient availability and consequently, plant growth and community structure (Wardle 2002, Bardgett, 2005). Plant species composition in turn significantly affects ecosystem nutrient cycling through plant nutrient uptake and use, rhizosphere interactions, production of litter of specific quality and microenvironmental changes (Eviner and Chapin, 2003).

The role of litter diversity and activity of soil microbial communities during decomposition has rarely been studied. This circumstances is surprising because litter quality as the overriding determinant for decomposition within a given climate varies tremendously among species (Perez-Harguindeguy *et al.*, 2000, Hattenschwiler, 2005). Similarly, the ecosystem consequences of the diversity of soil organisms are little understood, except for some keystone species or ecosystem engineers such as earthworms, termites and ants (Jones et al 1994,Anderson 1995). Despite the reasonable expectation that the diversity and composition of functional groups or feeding groups are important for ecosystem processes (Setälä 2002, Heemsbergen *et al.*, 2004), the existence and the significance of species diversity within functional groups is puzzling (Hattenschwiler *et al.*,2005).

Forest canopy in organic matter decomposition

Rates of key soil processes involved in recycling of nutrients in forests are governed by temperature and moisture, and the chemical and physical nature of the litter. The canopy of forest influences these factors and thus has a large influence on nutrient cycling. The increased availability of nutrient s in soil in clearcuts illustrates how the canopy retains nutrients (especially N) on site, both by storing nutrients in foliage and through the steady input of available C in litter. The idea that faster decomposition is responsible for the flush of nitrate has not been supported by experimental evidence. Soil N availability increases in canopy gaps as small as 0.1kg per ha, so natural disturbances or partial harvesting practices that increase the complexity of the canopy by creating gaps will similarly increase the spatial variability in soil N cycling and availability within the forest.

Canopy characteristics affect the amount and composition of leaf litter produced, which largely determines the amount of nutrients to be recycled and the resulting nutrient availability were thought to be brought about largely through differences in the decomposition rate of their foliar litter. Studies indicate that the effect of tree species can be better predicted from the mass and nutrient content of litter produced, hence total nutrient return, than from litter decay rate. The greater canopy complexity in mixed species forests creates similar heterogeneity in nutritional characteristics of the forest floor.

Microbial succession

The decomposition of leaf litter by microbes is an important process as it involves changes and release of many nutrients including N, P, K thus playing an important role in improving soil nutrient status besides helping in nutrient cycling and sustenance of vegetation.

The chemical breakdown of plant litter is mainly due to microbes though the fauna under favourable conditions may also consume the litter. Elemental loss was recorded from angiospermic litter during decomposition by fungi, bacteria and actinomycetes while few other elements still remained in the host litter even after two years. It has been observed that plants with more nitrogen show a decline in organic nitrogen as net mineralization occurs.

Neely *et al.* (1991) reported that the change in the quality of the organic matter induces a succession of microbial communities with some dominating fungi at specific stages in the decomposition process. Wardle *et al.* (1999) and Robinson *et al* (1999) suggested decline in bacterial proliferation during the decomposition process due to the release phenolic compounds from the plant residues (Hoorens and stroetenga, 2003).

Nutrients release *via* litter decomposition

In terms of chemical composition and quantity of organic matter, three main factors can be distinguished; the first is the easily soluble fraction, which can be quickly lost; the second is a non-soluble but easily degradable fraction and is composed mainly of hemicellulose and cellulose; and the third, which lasts much longer, is composed of lipids, lignin and lignified carbohydrates (Heal *et al*, 1997). Relationship between initial litter quality characteristics and decomposition rates for a large number of plant species related to the carbon-nitrogen ratio(C: N) has been demonstrated to be a good index of the of litter degradation (Berg *et al*,.1982; Taylor *et a*l.,1989).In general litter with a low C:N ratio is decomposed faster than litter with a high C:N ratio (Adama and Atwill, 1982). However, when C: N ratios exceed 75-100, other indexes such as lignin: N may be better (Heal *et al.*, 1997).

Dynamics in N content are usually characterized by a net immobilization (net increase in content due to incorporation of N into the litter from the surroundings) and net mobilization (release).The immobilization of N during decomposition often occurs in temperate ecosystem (Hasegawa and Takeda, 1996, Enoki and Hawaguchi, 2000).

When N is a limiting factor during litter decomposition, microbes and fungi not only immobilize N but may import N from the surrounding litter substrates (Pleguezuelo *et al.*, 2009).

Increase in atmospheric concentration of Co_2 and O_3 can have some effects on ecosystem productivity, yet their independent and interactive effects on other ecosystem process rates remain poorly understood. In temperate deciduous forests, about 41%of above ground net primary productivity is allocated to leaf production and litterfall amount can be altered by both elevated Co_2 and O_3 (Liu *et al,*, 2009). Further, these trace gases can impact the concentrations of nutrients and carbon based constituents including sugars, starch, organic acids, tannins, phenolics, lipids, hemicellulose, cellulose, pectin, and lignin. High C:N, and lignin:N or concentration of secondary compounds such as phenolics, tannins,and lignin,can reduce litter decomposition and nutrient mineralization rates (Berg and Laskowski, 2006), trace gas related alteration of litter chemistry can alter nutrient supply to plants and microbes. For example, elevated CO_2 can lower litter N concentration (Cotrufo *et al.*, 2005). Similarly, O_3 has detrimental effects on plant growth and development (EPA, 2006) and elevated O_3 can trigger antioxidant defense responses including increase foliar and litter concentrations of phenolic acids (Liu *et al.*, 2005) tannins (Booker *et al.*,1996; Liu *et al.*, 2005). Despite these findings, understanding of the effects of elevated CO_2 and O_3 on chemical decomposition dynamics has been constrained incubation studies (Parsons *et al.*, 2004; Chapman *et al.*, 2005).

In addition to changes in chemical composition, elevated CO_2 and O_3 can alter litter production, which may be important to biochemical cycling (Liu *et al.*, 2009). A meta analysis by Curtis and Wang (1998) indicated that leaf production increased by 31% under elevated CO_2, whereas elevated O_3 can significantly reduce forest productivity (King *et al.*, 2005). Furthermore, atmospheric changes in CO_2 and O_3 may alter plant community composition and quality and quantity of litterfall. Elevated level of CO_2 and O_3 also reduced the competitive ability of beech assemblage. As foliar chemistry varies across tree species, canopy compositional changes are likely to have consequences for litter decomposition and nutrient cycling (Luo *et al.*, 2004; Bradley and Pregitzer, 2007). Changes in litter quality and quantity may also influence below ground process rates, with profound effects on substrate availability for microbial metabolism (Giardina *et al.*, 2005), forest nutrient availability, carbon storage, and ultimately ecosystem productivity (Strain and Bazaz, 1983).

Elevated CO_2 and O_3 have the potential to alter litter decomposition not only by changing the quality and quantity of litter, but also by modifying forest-floor environmental conditions such as soil moisture and temperature

(Pendall *et al.*, 2003; 2007). Changes such as these in the forest environment would further affect biogeochemical process rates. Reciprocal transport studies with common litter substrate across several FACE experiments indicate d that environmental changes under elevated CO_2 had no significant influence on litter decomposition rates (Knops *et al.*, 2007). This suggests that any differences in litter decomposition rate observed under elevated CO_2would be derived from a change in litter quality and /or quantity, rather than changes in microenvironment or even microbial and faunal community composition.

The litter layer helps in to maintain favourable condition for decomposition by regulating the microclimate (Sayer, 2006)and creating habitats for arthropods. Litter removal and addition treatments in temperate forests not only caused substantial changes to the microclimate on the forest floor, but also affected the number of decomposer organisms such as arthropods and fungi. Furthermore, changes in litterfall modify the amounts of available nutrients in the soil (Sayer, 2006),which may also affect decomposition rates (Rothstein *et al.*, 2004). The amount of leaf litter on the forest floor influences its decomposition (Sayer *et al.*, 2006).

Leaf litter is a major component in the top layer of natural soils. At least 50-60% of this is incorporated as lignocellulosic substances (Kirk, 1983).This recalcitrant material has to be broken down by microorganisms to maintain the carbon cycle (Hammel, 1997). Microbial decomposition is in fact mediated by enzymes.

Cellulose is a major constituent of plant litter and its degradation in situ is exclusively a biological process (Chidthaisong and Conrad, 2000). Cellulolytic enzymes, such as Carboxymethylcellulase (CMCase) and filter paper activity (FPA), are primarily involved in cellulose degradation (Takashima *et al.*, 1998). While lignin, an abundant aromatic compound in terrestrial biomass, usually combines closely with cellulose. Cellulose is not available as a carbon source for other microbial decomposers (Boominathan and Reddy, 1992). Many enzymes are involved in the oxidative degradation of lignin, including lignin peroxidase (LiP), manganese peroxidase (MnP), and laccase (Sugiura *et al.*, 2003).To date, many studies have addressed the changes in functional enzymes during various kinds of litter decomposition (Carreiro *et al.*, 2000).

Soil organisms constitute the basic consumer trophic level of the decomposer subsystem (Wardle *et al.*, 1997). As such they control the breakdown of organic matter and hence the release of nutrients and their availability to other organisms, thus contributing to maintain long-term agricultural sustainability through net productivity of the agro-ecosystem (Robertson *et al.*, 1994).

The mineral cycling in ecosystem occurs as a function of time. Therefore, annual as well as seasonal cycles are important .The biological cycle involves the cyclic circulation of nutrient between the forest soil and plant and animal communities. It includes the phenomenon of uptake (by roots) retention (in biomass) and restitution of losses (By leaf litter, organic debris etc.). A quantitative estimation of mineral nutrients in the plant soil is helpful in evaluating the primary productivity of the ecosystem (Ovington and Madgwick, 1959b).It is seen that there is a tendency of the minerals in the plant during the process of succession which is facilitated by the internal cycle of elements such as nitrogen, phosphorus and potassium etc. remain immobilized in the organic part of the ecosystem. (Nye and Greenland, 1960).

In considering the important effects of organic matter on soil phenomenon, it is closely linked with the action of diverse forms of organic substances and the microorganisms participation in the natural circulation of element like iron, sulphur, calcium, silica, phosphorus etc.

Besides being a source of nutrients for the plants, organic matter has a fundamental effect on the physical properties of the soil and determines a large degree of physic o-chemical properties as the exchange capacity and buffering properties. These properties not only control the uptake of nutrients by the plants but also their retention in the soil and in suppressing the deleterious effect of soil acidity. The simplest forms of the action of organic substances on the mineral part of the soil are the dissolution of phosphate, calcium and magnesium carbonates and other compounds by root exudates. Microbes are responsible for bringing about the conversion of a number of chemical elements into available to the plants.

Plants utilize the elements of carbon, nitrogen and mineral nutrients under condition in which substances are cycled by biological and geological processes. The elements scattered through the atmosphere are accumulated by autotrophic and heterotrophic organisms in the form of living substance and are subsequently liberated during living processes and after the death of the organisms. This is a biological cycle of immense state and significance occurring within the trajectory of the geological cycle.

The admission of nutrients from the geological cycle into the small biological cycle does not have the characteristic of a closed system. The decomposition of plants and animals, after death do not proceed completely into final products of mineralization of organic substances.

A large reserve of CO_2 and carbon in inorganic and organic forms are present in the atmosphere, water, sea and soil crust. However, only a part of these nutrients is returned to the soil in the form of leaf litter and root residue.

The course of return of plant nutrient is not limited only to the masses left behind by the plants, it is also necessary to take into account the return of nitrogen, phosphorous and other elements during the growth of the plants. An item of income in the balance is provision of plant elements from soil material and of nitrogen from the atmospheric precipitation .An essential role in the nitrogen balance is played by the free living bacteria, which fix atmospheric nitrogen.

Forests conserve nutrients under natural conditions. Significantly higher amount of litter fall on the forest floor needs special attention in almost all nutrient cycling studies (Bray and Gorham, 1964). The functioning of most of the forest ecosystem is influenced by the availability of nutrients, their distribution and rates of cycling. Singh and Mishra (1978); Singh and Singh (1986) and (Chaturvedi and Singh, 1987) have summarized nutrients status and cycling in certain Tropical forests of India.

The level of nutrients in the soil is the net outcome of the inputs and outputs of the system. Generally a great proportion of the nutrients are found on the surface of the soil, due to massive input of nutrients to the soil through litter fall. The distribution of nutrients in different components differs considerably on the account of variation in biomass and nutrients were found to reside on the foliage subsystem, amongst other above ground tree components. But in the evergreen oak forest, the position of foliage was found to be intermediate with regard to nutrient storage patterns. Comparative study of soils in different forests showed great variation in the nutrient content (Johnson and Risser, 1974, Foster and Morrison, 1976, Maclean and Wein, 1978, Vancleave *et al.*, 1981).These differences in nutrient contents are due to nutrient concentration, soil depth and bulk density. Leaf fall generally accounts for 78-85% and wood litter fall for 10-35% (Bernhard and Reversal, 1972), Pandey and Singh (1981) estimated about 80-85% of the nutrient input through leaf litter fall and 17% to 20% through wood and miscellaneous fall for an Oak and conifer forests.

The differences in annual return of nutrient in various forests are numerous. The nutrient content of the leaf varies considerably within the same species or among the species in different localities. The quantity of the litter fall, which depends on density or age of the forests, habitat also affecting the nutrient returns is affected by the differences in soil nutrient level (Ovington,1956). The edaphic factors have a considerable influence on nutrient return. The soil receives nutrients consequent to release of decomposition of litter. This release from litter amounts to 60-75% of the nutrients contained in annual litter fall.

The amounts of nutrients and their distribution in various soil are related to growth and development of the plantation as vegetation reacts with the environment and develops specific soil properties for particular tree species.

Depending on climatic condition, community characteristic, nutrient dynamics and soil properties, wide variety of forests types have been recognized .Organic matters enter forest soil predominantly *via* two cycles, *i.e.*, root exudates and decaying roots and as forest litter. In temperate forests and in tropical forests, both root and litter input are seasonally pulsed,*i.e.* the most biomass enters the soil during the growing season. Litter layer decomposition occurs throughout the year during those times when temperature and moisture levels are conducive to microbial decomposition. There is considerable variation in the rate of defoliation in plants which depends on a number of factors like rainfall, temperature, wind velocity, soil moisture, physiological condition of the plant and species involved. Fresh organic matter enters forests primarily during autumn and spring. Above ground biomass inputs into the litter layer include leaf and needle litter, insect debris plus any organic matter produced by under growth like fern, mosses and annuals growing in the system. Litter layer thickness depends on the balance between input and decomposition .The litter decomposition is influenced by complex of factors like environmental factors including physic- chemical properties of the substrate. The conifer litters are more bio-decomposition resistant than deciduous litter. The litter and associated organic decomposition products layer thickness varies from several inches to a foot thick. These organic layers serve as reservoir for plant nutrient, effect soil physical properties and subsequent nutrient transfers between the microbial community and plants and control plant root and overall plant development.

The litter plays a great role as a pool of plant nutrients. The primary procedure for estimation of nutrient cycling in forest ecosystem follows nutrient release and cycling in the litter layer. Many excellent studies on nutrient cycling in forest system have been reported (Vitousek, 1984, Chaturvedi and Singh, 1987). The prime conclusion from each of these studies is that the litter provides a major source of nutrients for the growing plants. The mineral nutrients supplied by the litter is less than that found in fresh plant tissue because a major portion of this constituent is translocated to perennial tissues before leaf fall. With the more available nutrients, such as nitrogen, the process involved in nutrients, transformation, is primarily biological, although some chemical immobilization into humic substance can occur (Stevenson, 1982). The distribution of those nutrients which are likely to be precipitated by soil ions are controlled partially by the geological process. A more difficult problem to asses is the distribution of nitrogen by the microbial process of immobilization and mineralization. This is important for ecosystem function because nutrients immobilized by the microbial community are not immediately available for plant growth, they become part of the soil nutrient pool as leaf litter decomposes organic nitrogen which is mineralized .But when the litter contains excess carbon, a major portion of this nitrogen is assimilated by the microbial community, into new cell biomass. Hence, this nitrogen becomes unavailable to the plant community until death and decay of the microbial cells. Although these nutrients are not available for immediate utilization by the plant, they form a pool or reserve of nutrients for future biomass production. Once the litter is decomposed more mineral nitrogen is produced by decomposition than, is required by the microbes. Some of the nitrogen becomes directly available to the plant as it is mineralized to facilitate the compact mineralization of nutrient cycle in forest soils. Litter was defined as that material exhibiting net nitrogen immobilization, where as soil exhibits net nitrogen mineralization. Thus, in the forest floor profile, where the litter material has been decomposed into the soil, in that position nitrogen becomes available to the plant roots. Generally the association of roots and litter decomposers would become most prominent in the profile. A small portion of the water soluble litter nutrients are lost from the soil by leaching (Berg and Wessen, 1984). The physical structure of the soil ecosystem, considering the quantities of nutrients mobilized by the microbial community, allows fewer losses much smaller. The nutrients that escape the plants and microorganisms in the upper most surface layer of the soil organic layer will usually be trapped and metabolized by those organisms living deeper in the soil profile.

Litter production, ecology of decomposition and mineral cycling of the forest trees have been studied by various workers. Many excellent studies of nutrient cycling in the forest system have been reported Vitousek (1984).

Very little work has been done on litter dynamics and nutrient dynamics in various forest systems in this part of the country. Assam which falls in Eastern Himalayan bio-ecological zone has rich and varied forest ecosystem attributed to the diverse geographical area, varied topography, climate and soil variability, marked seasonal periodicity. Forest ecosystem, which is more or less self managed, self regulated require, balanced litter dynamics and nutrient cycling for growth and productivity, especially in the plantation forests. The structure, function and ecosystem services of tropical forest depend on its species richness diversity, dominance, and the patterns of changes in the assemblages of the tree population over time. Therefore a detailed study on litter dynamics and nutrient cycling of the plantation forests will go a long way in the proper management of forests.

Conclusion

Teak, (*Tectona grandis* Linn., family:Verbenaceae), is one of the most valuable timber species of the world. It occurs naturally between 9'to 26'N latitude and 73′to140′E longitude in the tropical and sub-tropical region of south and south-east Asia which include penninsular India, Myanmar ,Thailand and Laos. Total area of natural teak forests in India has been estimated to be 9.77m ha. which is about 13%of the total forest area of the country.

Large scale plantation of teak has been raised both within and outside its range of natural distribution due to ever-increasing demand of its timber. Globally, teak ranks third among the tropical hardwood species in plantation area.

In India, the history of teak planting dates back the year 1842 when the first teak plantation was attempted at Nilamber in Kerala state. Presently, there is more than 1.5m ha area under teak plantation in India. Teak plantations are being raised at an annual rate of about 50,000 ha. (Subramanian *et al.*, 2000, Shukla, 2009).

Therfore, *Tectona grandis* which has much economic value as timber plant is extensively grown as plantation tree species and need an investigation with regards to their productivity, litter dynamics and nutrient budget.

References

Aber, J. D. and Melillo, J. M. (1982). Nitrogen immobilization in decaying hard wood leaf litter as a function of initial nitrogen content. *Can.J.Bot.*, 60:2263-2269.

Adams, M.A., Atwill, P.M.(1982). Nitrogen mineralization and nitrate reduction in forests. *Soil Biol.Biochem.*, 14:197-202.

Aerts, R. (1997). Climate, leaf litter chemistry and leaf litter decomposition in terrestrial ecosystem:a triangular relationship.*Oökos*,79:439-449.

Anderson, J.M. (1995). Soil organisms as engineers microsite modulation of macroscale processes. In: *Linking Species to Ecosystem*, (Ed.) C.G. Jones, JH Lawton,pp.94-106.New York:Chapman and Hall.

Anderson, J.M. (1973).The breakdown and decomposition of sweet chestnut (*Castanea sativa* Mill.) and beech (*Fagus saylvaticus* L.). Leaf litter in two deciduous woodland soils.*Oecologia*, 12:251-274.

Andersonn, J.M. (1991). The effects of climate change on decomposition processes in grassland and coniferous forests. *Ecol. Applic.*, 1:326-347.

Anitha, K., Joseph, S., John, R., Chandran, E. V., Ramasamy, S. Prasad, N. (2010). Ecological Complexity, In Press, Corrected Proof, Available online 3 March.

Bardgett, R.D. (2005). *The Biology of Soil: A Community and Ecosystem Approach*.oxford: Oxford Univ.Press.253 pp.

Berg, B. and Mc Claugherty, C. (1989). Nitrogen and phosphorus release from decomposing litter in relation to the disappearence of lignin. *Can. J. Bot.*, 67:1148-1156.

Berg, B., Ekbohm, G. and Mc Caugherty, C. (1984). Lignin and holocellulose relations during longterm decomposition of some forest litters.*Can. J. Bot., 62:2540-2550.*

Berg, B., Staaf, H.,Wessen, B. and Ekbohm, G. (1982a). Nitrogen level and decomposition in Scots pine needle litter. *Oikos*, 38:291-396.

Berg, B., Hannus, K., Popoff, T. and Theander, O. (1982b).Changes in organic chemical components of needle litter during decomposition.Long term decomposition in a scots pine forest. *Can .J. Bot.*, 60:1310-1319.

Berg, B., Laskowsk, R. (2006). Litter decomposition: a guide to carbon and nutrient turnover. *Adv. Ecol. Res.*, 38:448.Amsterdam: Elsevier.

Bernhard-Reversal, F. (1972). Decomposition de la litiere de fevilles enforet ombrophile de basse coted ivoire d' Ivoire. *Oecol. Planta.*, 7:279-300.

Bernhard-Riversat, F., Main, G. Holl, K., Loumeto, J. and Ngoo, J. (2003). Fast disappearance of the water-soluble phenolic fraction in Eucalyptus leaf litter during laboratory and field experiments. *Appl. Soil Ecol.*, 23:273-278.

Booker, F.L., Prior, S.A., Torbert, H.A., Fiscus, E.L., Pursley, W.A. and Hu, S.J. (2005). Decomposition of soybean grown under elevated concentrations of CO_2 and O_3.*Glob. Chang. Biol.*, 11:685-98.

Boominathan, K. and Reedy, C.A. (1992). Fungal degradation of lignin .In: *Handbook of Applied Mycology. Vol.4.*(Eds.) D.K. Arora, R.P. Elander, and K.G. Mukherji, Morcel Dekker, New York:763-782.

Bradley, K.L., Pregitzer, K.S. (2007). Ecosystem assembly and terrestrial carbon balance under elevated CO_2.*Trends Ecol.*, 22:538-47.

Bray, J.R. and Gorham, E. (1964). Litter production in forest of the world. *Adv. Ecol. Res.*, 2:101-175.

Burns, R.G. (1978). Enzymes activity in soil: some theoretical practical consider-actions. In: *Soil Enzymes* (Ed.) R.G. Burns, Academic Press, London UK. : 295-340.

Cantarutti, R.B., Tarr'e, R., Macedo, R., Cadisch, G., Rezemde, C. P.D., Pereira, J. M., Braga, J. M., Gomide, J.A, Ferreira, E, Alves, B.J.R.,Urqeiaga, S. and Boddey, R. M. (2002).The effect of grazing intensity dynamics in Brachiaria pastures in the Atlantic forest region of the south of Bahia,Brazil.*Nutri. Cycl. in Agroeco.*, 64:257-271.

Carney, K.M., Hungate, B.A., Drake, B.G. and Megonigal, J.P. (2007). Altered soil microbial community at elevated CO_2 leads to loss of soil carbon. *Proc. Nat. Acad. Sci., USA* 104:4990-5.

Carreiro, M.M., Sinsabaugh, R.L., Repert, D.A. and Parkhurst, D.E. (2000). Microbial enzyme shifts explain litter decay response to simulated nitrogen decomposition. *Ecology*, 81:2359-2365.

Chapman, J.A., King, J. S., Pregitzer, K. S. and Zak, D.R. (2005). Effects of elevated CO_2 and tropospheric O_3 on trii fine root decomposition. *Tree Physiol.*, 25:1501-10.

Chaturvedi, O.P. and Singh, J.S. (1987).The structure and function of pine forest in Central Himalaya.I.Dry matter dynamics. *Ann. Bot.*, 60:237-252.

Chidthaisong, A., and Conrad, R. (2000). Pattern of non-methanogenic and methanogenic degradation of cellulose in anoxic rice field soil. *FEM Microbial Ecol.*, 31:87-94.

Christie, E.K. (1979). Ecosystem processes in semiarid grasslands. ii. Litter production, decomposition and nutrient dynamics. *Austr. J. Agricul. Res.*, 30:29-42.

Cotrufo, M.F., De Angelis, P. and Polle, A. (2005). Leaf litter production and decomposition in a poplar short-rotation coppice exposed to free air CO_2 enrichment (POPFACE). *Glob. Chang. Biol.*, 11:971-82.

Curtis, P.S. and Wang, X.Z. (1998). A metanalysis of elevated CO_2 effects on woody plant mass form and physiology.*Oecologia*, 113:299-313.

Das, M. (1998). *Studies of litter production and Nutrient cycling in Teak forest*. Ph.D. Thesis, Gauhati University, Guwahati, India.

David, Ponge, J.F., Arpin, P. and Vannier, G. (1995). Experimental modifications of litter supplies in a forest mull and reaction of the nematode fauna. Fund. *Appl.Nematol.*, 18:371-389.

De Angelis P., Chigwerewe, K.S. and Mugnozza, G.E.S. (2000). Litter quality and decomposition in a CO2 enriched Mediterranean forest ecosystem.*Plant Soil*, 224: 31-41.

Dickinson, C.H. and Pugh, G. J. F. (1974). *Biology of Plant Litter Decomposition*, 2nd Volume Academic Press, New York.

Dilly, O. and Munch, J.C. (2004). Litter decomposition and microbial characteristics in agricultural soils in northern, central and southern Germany. *Soil Sci. Plant Nurt.*, 50:843-853.

Dyer, M.L., Meentemeyer, V. and Berg, B. (1990). Appearent controls of mass loss rate of leaf litter on regional scale. *Scand.J.For.Res.*, 5:311-323.

Ebermayer, E. (1876). "*Diegesamte Lahreder Waldstreu mil Riicksicht aufdie chemische statikdes waldbaues,*" 116 pp. Berlin:Julius springer.

Enoki, T., Hawaguchi, H. (2000). Initial nitrogen and topographic moisture effect on the decomposition of pine needles. *Ecol. Res.*, 15: 425-434.

Esser, G. (1992). Implications of climate changes for production and decomposition in grasslands and coniferous forests. *Ecol. Applic.*, 2:47-54.

Eviner, V.T, Chapin, F.S. (2003). Functional matrix: a conceptual framework for predicting multiple plant effects on ecosystem processes. *Ann. Rev. Ecol. Evol. Syst.*, 34: 487-515.

Foegel, R. and Cromak, K. (1977). Effect of habitat and substrate quantity on Douglas fir litter decomposition in Western Oregon. *Can. J. Bot.*, 55:1632-1640.

Foster, N.W., Morrison, I.K. (1976). Distribution and cycling of nutrients in a natural *Pinus banksiana* eco-system. *Ecology*, 57:110-120.

Gartner, T.B., Cardon, Z.G. (2004). Decomposition dynamics in mixed species leaf litter, *Oikos*, 104:230-246.

Giardina, C.P., Coleman, M.D., Binkley, D., Hancock, J.E., King, J.S., Lilleskov, E.A., Loya, W.M., Pregitzer, K.S., Ryan, M.G. and Trettin, C.C. (2005). The response of belowground carbon allocation in forests to global change. In: *The Impacts of Global Climate Change on Plant-Soil Interaction.* : Binkley D,Menyailo.

Goncalves, J.L.M. Carlyle, J.C. (1994). Modelling the influence of moisture and temperature on net nitrogen mineralization in a forested sandy soil. *Soil Biol. Biochem.*, 26:1557-1564.

Gosz, J.R., Likens, G.E. and Bormann, F.H. (1973). Nutrient release from decomposing leaf and branch litter in the Hubbard Brook Forest, New Hampshire. *Ecol. Monogr.*, 43 : 173-91.

Guoyi Zhou, Lili Guan, Xiaohua Wei, Dequiang Zhang, Qianmei Zhang, Junhua Yan, Dazhi Wen. Juxiu Liu, Shuguang Liu, Zhongliang Huang, Guohui Kong, Jiangming MO,Qingfa Yu (2007). Litter fall production along successional and altitudinal gradients of sub-tropical monsoon evergreen broadleaved forests in Guangdong, China. *Plant Ecol.*, 188:77-89.

Hammel, K.E. (1997). Fungal degradation of lignin. In: *Diven by Nature: Plant Litter Quality and Decomposition,* (Ed.) *I.E.* Giller , Wallinford, U.K., CABI International : 33-45.

Hasegawa, M., Takeda, H., Takeda, H. (1996). Carbon and nutrient dynamics in decomposing pine needle letter in relation to fungal and faunal abundances. *Pedobiologia,* 40:171-184.

Hattenschwiler, S., Gasser, P. (2005). Soil animals alter plant litter diversity effect on decomposition. *Proc .Nat. Acad. Sci., USA,* 102 : 1519-24.

Hattenschwiler, Stephan, Tiunov, Alexei, V. and Scheu Stefan (2005). Biodiversity and Litter Decomposition in Terrestrial Ecosystems. *Ann. Rev. Ecol. Syst.*, 36:191-218.

Heal O W and Dighton J (1986) Nutrient cycling and decomposition in natural terrestrial ecosystem. M.J. Mitchell and J.P. Naka (Eds). In : *Microflora and Faunal Interaction in Natural and Agroecosystem*. Nijhoff and Junk, Dordrecht : 14-73.

Heal, O.W. Anderson, J. M., Swift, M.J. (1997). Plant litter quantity and decomposition an historical overview. In: *Driven by nature: Plant Litter Quality and Decomposition.* (Eds.) G. Cadish and K.E. Giller, CAB International, Wallinford, Oxon.

Heemsbergen, D.A., Berg, M.P., Loreau, M., Van Haj, J.R., Faber, J.H., Verhoef, H.A. (2004). Biodiversity effects on soil processes explained by interspecific functional dissimilarity. *Science,* 306:1019-20.

Hemandez, I., Santa-Regina, M.I., Gallardo, J.F. (1992). Dinamica de la dccomposition forest ald em bisques de la Cuenca del Duero(Provincia de Zamora):Modelization de la perlida de peso. *Soil Res. Rehabil.*, 6:339-355.

Heneghan, L., Colemn, D.C., Zou, X., Crosley, J.D.A., Haines, B.L. (1998). Soil microarthropod community structure and litter decomposition dynamics; a study of tropical and temperate sites. *Appl. Soil Ecol.*, 9:33-38.

Hermy, M. (1987). Path analysis of standing crop and environmental variables in the field layer of two Belgian rivervine forests. *Vegetation*, 70:127-133.

Hobbie, S. E., Vitousek, P.M. (2000). Nutrient limitation of decomposition in Hawaiian forests. *Ecology*, 81.1867-1877.

Hooper, D.U. and Vitousek, P. M. (1998). Effects of plant composition and diversity on nutritional cycling. *Ecol. Monogr.*, 68:121-149.

Hooper, D. U. (1996). *The effects of plant functional group diversity on nutrient cycling in a California serpentine grassland*. Thesis, Department of Biological Science, Standford University, California, USA.

Hoorens, B., Aerts, R. and Stroetenga, M. (2003). Does initial litter chemistry explain litter mixture effects on decomposition? *Oecologia*, 137:578-586.

Hopkins, B. (1966).Vegetation of the Olokemeji Forest Reserve, Nigesia, vi. The litter and soil, with special reference to their seasonal changes. *J. Ecol.*, 54:687-703.

Ibrahim, A., Gilon, D. and Joffre, R. (2010). Leaf Litter Decomposition of Mediterranean Tree Species in Relation to Temperature and Initial Water Imbibition under Microcosm Experiments. *Res. J. Agricul. Biol. Sci.*, 6(1):32-39.

Jenny, H. (1980). *Soil genesis with ecological perspectives*, Spinger-Verlog. New York 560pp.

Jones, C. G., Lawton, J. H., Shachak, M. (1994). Organisms as ecosystem engineers. *Oikos*, 69:373-86.

Jonn, D.M. (1973). Accumulation and decay of litter and net production of forest in tropical West Africa. *Oikos*, 24:430-435.

Kainulainen, P., Holopainen, T., Holopainen, J. K. (2003). Decomposition of secondary compounds from needle litter of Scots pine grown under elevated CO_2 and O_3. *Glob. Chang. Biol.*, 9:295-304.

Kirk, T. K. (1983). Degradation and conversion of lignocellulose. In: *The Filamentous Fungal Technology*, (Eds.) J. E. Smith, D. R. Berry and B. Kristiansen, Vol 4. Edward Arnold, London:266-295.

Kirschbaum, M.U.F. (1995). The temperature dependence of soil organic matter decomposition and the effect of global warming on soil organic carbon storage. *Soil Biol. Biochem.*, 27:753-760.

Knops, J. M. H., Naeemw, S. and Reich, P. B. (2007). The impact of elevated CO_2, increased nitrogen availability and biodiversity on plant tissue quality and decomposition. *Glob. Chang. Biol.*, 13:1360-71.

Kozonits, A. R., Matyssek, R., Blaschke, H., Gottlein, A. and Grams, T.E.E. (2005). Competition increasingly dominates the receptiveness of juvenile beech and spruce to elevated $CO_2$and / or O_3concentrations through two subsequent growing seasons. *Glob. Chang. Biol.*, 11:1387-401.

Kshattriya, S. (1990). *Ecological studies on the microbial and biochemical aspects of forest litter decomposition at higher and lower altitudes of Meghalaya*. Ph. D. Thesis. North Eastern Hill University, Shillong , India.

Kubisek, M. E., Quinn, V.S., Marquardt, P. E. and Karnosky, D.F. (2007). Effects of elevated atmospheric CO_2 and or O_3 on intra and interspecific competitive ability of aspen. *Plant Biol.*, 9:342-55.

Kurzatkowski, D., Martins, C., Höfer, H., Garcia, M., Förster, B., Beck, L. and Vlek, P. (2004). Litter decomposition microbial biomass and activity of soil organic matter in three agroforestry Sites in central Amazonia. *Nutr. Cycl. Agroecosyst.*, 69:257-267.

Lamb, R. J. (1985). Litter fall and nutrient turnover in two eucalyptus woodland. *Austr. J. Bot.*, 33:1-14

Lavelle, P., Blanchart, E., Martin, A., Spain, A., Toutain, F., Barois, I., Schaefer, R. (1993). A hierarchical model for decomposition in terrestrial ecosystems; application to soil of the humid tropics. *Biotropica*, 25:130-150.

Lavelle, P., Blanchart, E., Martin, A. and Martin, S. (1993). A hierarchial model for decomposition in terrestrial ecosystem: Application to soils of the humid tropics. *Biotropica*, 25:130-150.

Leiros, M. C., Trasar-Cepeda, C., Seoane, S., Gil-Sotres, F. (1999). Dependence of mineralization of soil organic matter on temperature and moisture. *Soil Biol. Biochem.*, 31: 327-335.

Lensing, J. R., Wilse, D.H. (2007). Impact of changes in rainfall amounts predicted by climate changing models on decomposition in a deciduous forest. *Appl. Soil Ecol.*, 35:523-53f4.

Liu L. L., King, J. S., Giardina, C.P. (2005). Effects of elevated concentrations of atmospheric CO_2 and tropospheric O_3 on leaf litter production and chemistry in trembling aspen and popper birch communities. *Tree Physiol.*, 25:1511-22.

Liu, L. L., King, J. S., Booker, F. L., Giardina, C. P., Allen, L. H., Hu, S. J. (2009). Enhanced litter input rather than changes in litter chemistry drive soil carbon and nitrogen cycles under elevated CO_2: a microcosm study. *Glob. Chang. Boil.*, 15:441-53.

Lugo, A. E. and Duggar, K. (1978). Structure, productivity and transpiration of a sub-tropical dry forest in Puerto Rico. *Biotropica*, 10(4):278-291.

Luo, Y., Su, B., Currie, W. S., Dukes, J. S., Finzi, A., Hartwig, U., Hungate, B., Mc Murtric, R. E., Oren, R., Parton, W. J., Pataki, D. E., Shaw, M. R., Zak, D. R., Field, C. B. (2004). Progressive nitrogen termitation of ecosystem responses to rising atmospheric carbon dioxide. *Bioscience*, 54:731.

Maclean, D. A. and Wein, R.W. (1978). Weight loss and nutrient changes in decomposing litter and forest floor material in new Brunswick forest stands. *Can. J. Bot.*, 56:2730-2749.

Martins, A., Azevedo, S., Raimundo, F. and Madeira, M. (2006). Decomposicao e evolucaoda composicao estructural e do teor em nutrientes. In: *II congresso International de ciecia do Solo, Livrode resumos, Huelva.*

Meentemeyer, V., Box, E.O. and Thompson, R. (1982). World patterns and amounts of terrestrial plant litter production. *Bioscience*, 32:125-128.

Mellilo, J. M., Aber, J. D. and Muratore, J. F. (1982). Nitrogen and lignin control of hardwood leaf litter decomposition dynamics. *Ecology*, 63:621-626.

Mesquita, R. C. G., Workman, S. W. and Neely, C. L. (1998). Slow litter decomposition in a Cecropia dominated secondary forest of central Amazonia. *Soil Biol. Biochem.* 30:167-175.

Muller, P.E. (1887). "*Studies iiber die naturlichen Humusformen and deren Einwirkung auf vegetation and Boden*", 324 pp.Berlin :Julius Springer.

Neely, C. L., Beare, M. H., Hargrove, W. L., Coleman, D. C. (1991). Relationship between fungal and bacterial substrate-induced respiration, biomass and plant residue decomposition. *Soil Biol. Biochem.*, 23:947-995.

Nye, P. H. and Greenland, B. J. (1960). The soil under shifting cultivation, ech. *Commum. Bur. Soil Sci.* : 51.

Okoh, I. A., Badejo, M.A., Nathanie., I.T. and Tian,G. (1999b). Studies on the bacteria, fungi and springtails (Collembola) of an agro forestry arboretum in Nigeria. *Pedobiologia,* 43: 18-27.

Ovington, J.D. (1956). The composition of tree leaves. *Forestry*, 29.22-28.

Ovington, J. D. and Madgwick, H.A.I. (1959b). Distribution of organic matter and plant nutrients in plantation of scots pine. *Forest Sci.*, 5:344-55.

Pandey, U. and Singh, J. S. (1981). A quantitative study of the forest floor, litter fall and nutrient return in an oak-conifer forest in Himalaya. I composition and dynamics of forest. *Oecol. General.*, 2:49-61.

Pandey, R.R., Sharma, G., Tripathi, S. K. and Singh, A.K. (2007). Litter fall, litter decomposition and nutrient dynamics in a subtropical oak forest and managed plantation in north eastern India. *Forest Ecol. Manage.*, 249:96-104.

Parsons, W. F. J., Bockheim, J. G. and Lindroth, R. L. (2008). Independent, interactive and species-specific responses of leaf litter decomposition to elevated $CO_2$and O_3 in a northern hard wood forest. *Ecosystems*, 11:505-519.

Pendall, E., Del Grosso, S., King, J.Y., Cain, D.R., Milchunas, D.G., Morgan, J.A., Mosier, A.R., Ojima, D.S., Parton, W. A., Tans, P. and white, J.W.C. (2003). Elevated atmospheric CO_2 effects and soil water feedback on soil respiration components in a Colorado grassland. *Global Biogeochem. Cycles*, 17:10.1029/2001 GB 001821.

Pçrez, H. N., Díaz, S., Vendramini, F., Guruich, D.E., Clingoloni, A.M., Giorgis, M., Cabido,M. (2007). Direct and indirect effects of climate on decomposition in native ecosystems from central Argentina. *Aust.Ecol.*, 32:749-757.

Perez-Harguvindeguy, N., Diaz, S., Cornelissen, J.H.C., Venramini, F., Cabido, M. and Castellanos, A. (2000). Chemistry and toughness predict leaf litter decomposition rates over a wide spectrum of functional types and taxa in central Argentina. *Plant Soil,* 218:21-30.

Person, H. (1980). Fine root dynamics in a scot pine stand with and without near optimum nutrient and water regims. *Acta Phytogeogr. Suecia,* 68:101-110.

Prescott Cindy, E. (2002). The influence of the forest canopy on nutrient cycling. *Tree Physiology,* 22:1193-1200 @2002 Heron Publishing -Victoria, Canada.

Quemada, M. and Cabrera, M.L. (1997). Temperature and moisture effects on C and N mineralization from surface applied clover residue. *Plant Soil,* 189:127-137.

Rawat, Neelam, Nautiyal, B.P. and Natiyal, M.C. (2009). Litter Production pattern and nutrients discharge from decomposing litter in an Himalayan alpine ecosystem. *New York Sci. J.,* 2(6): 1554-2000.

Reinke, J. J., Adriano, D.C. and Mc Leod, K.W. (1981). Effects o f litter alteration on carbon dioxide evolution from a south carolina pine forest floor. *Soil Sci. Soc. Am.J.,* 45: 620-623.

Robertson, L.N., Kettle, B.A., Simpson, G. B. (1994).The influence of tillage practices on soil microfauna in a semi arid agroecosystem in North Eastern Australia. *Agric. Ecosyst. Environ.,* 48(2):149-156.

Robinson, C. H., Kirkham, J.B. and Litterwood, R. (1999). Decomposition of root mixtures from high arctic plants:a microcosm study. *Soil Biol. Biochem.,* 31:1101-1108.

Rodrigo, A., Recous, S., Neel, C. and Mary, B. (1997). Modelling temperature and moisture effects on C-N transformation in soils:comparison of nine models. *Ecol. Model.,* 102:325-339.

Rodriguez Pleguezuelo, C.R., Duran Zuazo, V.H., Muriel Fernandez, J.L., Peinado, F. J., Martin, T. and Franco, D. (2009). Litter decomposition and nitrogen release in a sloping Mediterranean subtropical agroecosystem on the coat of Granada (SE, Spain):Effects of floristic and topographic alteration on the slope. *Agriculture, Ecosys. Environ.,* 134:79-88.

Rothstein, D.E., Vitousek, P.M. and Simmons, B.L. (2004). An exotic tree alters decomposition and nutrient cycling in a Hawaiian montane forest. *Ecosystem,* 7: 805-814.

Sanchez, P.A. (1976). *Properties and management of soils in the Tropics.* Wiley, London.

Sariyildiz, T. (2008). Effects of tree canopy on litter decomposition rates of *Abies nordmanniana, Picea orientalis and Pinus sylvestris. Scandinavian J. Forest Res.,* 23: 330-338.

Satchell, J.E. (1974). Litter Interface of animate in animate matter. In: *Biology of Plant Litter Decomposition* (Eds.) C.H. Dickson and G.J.F. Pugh, Academic press, London and New York.

Sayer, E.J., Janner, E.V.J. and Lacey A.L. (2006). Effect of litter manipulation on early stage abundance in a tropical moist forest. *Forest Ecol. Manage.*, 229:285-293.

Sayer, E. J., Janet, E.V.J. and Cheesman, A.W. (2006). Increased litter fall changes fine root distribution in a moist tropical forest. *Plant Soil*, 281(in press).

Scholes, R. J. and Walker, B. H. (1993). *An African savana-synthesis of Nylsvleystudy* Cambridge University Press,Cambridge, 306pp.

Setälä, H. (2002). Sensitivity of ecosystem functioning to changes in tropic structure, functional group composition and species diversity in belowground food webs. *Ecol. Res.*, 17:207-15.

Seth, S.K., Kaul, O.N and Gupta, A.C. (1963). Some observation on Nutrient cycle are return of nutrients in plantation at New Forest. *Ind. Forester*, 89:90-98.

Shukla, P K. (2009). XIII World forestry congress. Buenos Aires, Argentina, 18-23 October 2009.

Singh, A.K. (1990). *Litter production, nutrient and release during decomposition in degraded moist deciduous and sub-tropical wet hill forests of Meghalaya.* Ph.D.thesis.North Eastern Hill University, Shillong, India.

Singh, K.P. and Misra, R. (1978). Structure and functioning of natural modified and silvicultural ecosystem of eastern Uttar Pradesh, 155 pp MAB. *Technical Report, Banaras Hindu University*, Varanasi, India.

Singh, K.P. and Singh (1986). Structure and function of Central Himalayan of Forest. *Proc. Ind. Acad. of Sci. (Plant Sciences)*, 96:159-89.

Sirulnik, A.G., Allen, E.B., Meixner, T. and Allen, M.F. (2007). Impacts of anthropogenic N addition on nitrogen mineralization from plant litter in exotic animal grass lands. *Soil Biol.Biochem.*, 39.24-32.

Spain, A.V. (1984). Litter fall and the standing crops of litter production and leaf-litter decomposition of selected tree-species in tropical forest at Kodayar in Western Ghat, India. *Forest Ecol. Managem.*, 23:231-244.

Stevenson,F.J. (1982). *Humus Chemistry*. John Wiley and Sons, New York:443.

Strain, B.R., Bazzaz, F.A. (1983).Terrestrial plant communities in London ER, Ed. CO_2 and plants. *AAAS Selected Symposium*. Westview, Boulder, Co., USA: 177-222.

Suiguira, M., Hirai, H. and Nishida,T. (2003). Purification and characterization of a novel lignin peroxidase from the white rot fungus *Phanerochaete sordid* YK-624. *FEMS Microbiol.Lett.*, 224:285-290.

Sulkava, P., Huhta,V. and Laakso, J. (1996). Impact of soil faunal structure on decomposition and N-mineralisation in relation to temperature and moisture in forest soil. *Pedologia*, 40:505-513.

Sundarapandian, S.M. and Swamy, P.S. (1999). Litter production and leaf-litter decomposition of selected tree species in tropical forest at Kodayar in western Ghats,India. *Forest Ecol. Manage.*, 123:231-244.

Swift, M.J., Heal, O. and Anderson, J.1 (979). Decomposition in terrestrial ecosystem, Blackwell scientific publication. *Studies Ecol.* (5),Oxford London:372.

Takashima, S., Likura, H., Nakamura, A., Hidaka, M., Masaki, H. and Uozumi,T. (1998). Overproduction of recombinant *Trichoderma reesei* cellulases by *Aspergillus oryzae* and their enzymatic properties. *J.Biotechnol.*, 65:163-171.

Tateno, R., Tokuchi, N., Yamanaka, N. D.U.,S., Otsuki, K, Xue, Z., Wang, S. and Hou, Q. (2007). Comparison of litter fall production and leaf litter decomposition between an exotic black locust plantation and an indigenous oak forest near Yanán on the Loess Plateau. *China Forest Ecol .Manage.*, 241:84-90.

Taylor, B. and Parkinson, D. (1988b). Aspen and pine leaf litter decomposition in laboratory microcosm. ii. Interaction of temperature and moisture level. *Can. J. Bot.*, 66:1966-1973.

Trofymow J.A, Moore, T. R., Titus, B., Prescott, C., Morrison, I., Siltanen, M., Smith, S., Fyles, J., Wein, R., Camire, C., Duschene, L., Kozak, L., Kranobetter, M. and Visser, S. (2002). Rates of litter decomposition over 6 years in Canadian forests; influence of litter quality and climate. *Can J. For. Res.*, 32:789-804.

Upadhyay V.P., Singh, J. and Meentemeyer, V. (1989). Dynamics and weight loss of leaf litter in central Himalayan forest ecosystem. *Trop. Eco.*, 34(1):44-50.

Upadhyay, V.P and Singh, J. S. (1989). Pattern of nutrient immobilization and release in decomposing forest litter in Central Himalaya India. *J. Ecol.*, 77:127-146.

Upadhyay,V.P and Singh, J. S. and Meentemeyer, V. (1989). Dynamics and weight loss of leaf litter in Central Himalayan Forests.Abiotic versus litter quality influences. *J. Ecol.*, 77:147-161.

Vancleve, K. and Noonan, I. (1975). Litterfall and nutrient cycles in the forest floor of birch and aspen stands in interior Alaska. *Can. J. For. Res.*, 5:626-639.

Veneklaas, E.J. (1995). Water and nutrient in two montane rain forest canopies, Central cordillera, Columbia chapter 14. In: *Studies on Tropical Andean Ecosystem.* Volume 4. Ecoandes.

Vigil, M.F. Kissel, D.E. (1995). Rate of nitrogen mineralization from incorporated crop residues as influenced by temperature. *Soil Sci. Soc. Ame. J.*, 59:1636-1644.

Vinokurov, M.A. and Tyurmenko, A.N. (1958). Materials in the forests, biological cycle of nitrogen and phosphorus. *Pochvovedenie*, 7:787-791.

Vitousek, P. M., Turner, D. R., Parton, W. J. and Sanford, R. L. (1994). Litter decomposition on the Mauna Loa environmental matrix, Hawaii: Patterns, mechanism and models. *Ecology*, 75:418-429.

Vitousek, P., Gerrish, G., Tuener, D.R., Walker, L. and Mueller-Dombois, D. (1995). Litterfall and nutrient cycling in four Hawaira Montane rain forest. *J. Trop. Ecol.*, 11:189-203.

Vitousek, P.M. (1984). Litter fall, Nutrient cycling and nutrient limitation in tropical forests. *Ecology*, 65(1):285-298.

Wardle, D.A. (2002). *Communities and Ecosystems:Linking the Aboveground and Belowground Components*. Princeton, NJ: Princeton Univer.Press:392.

Wardle, D.A., Bonner, K.I., Barker, G.M., Yeates, G.W., Nicholson, K.S., Badgett, R.D., Watson, K.N. and Ghani, A. (1999). Plant removals in perennial grassland: vegetation dynamics, decomposers, soil biodiversity and ecosystem properties. *Ecol. Monogr.*, 69(4):535-568.

Wardle, D.A., Bonner, K.I., Nicholson, K.S. (1997). Biodiversity and plant litter; experimental evidence which does not support the view that enhanced species richness improves ecosystem function. *Oikos*, 79:247-258.

Woods, P. and Raison, R.J. (1983).Decomposition of litter In sub-alpine forests of *Eucalyptus delegatensis, E .pauciflore and E. divers. Austr. J.Ecol.*, 8:287-299.

❑❑❑

Microbial Diversity and Functions, 2012
© D.J. Bagyaraj, K.V.B.R. Tilak, H.K. Kehri (eds.), pp. 257-272
New India Publishing Agency, New Delhi (India)
E-mail : info@nipabooks.com; Website : www.nipabooks.com

Chapter 10

Microbes and Green Technology : Soil Plant Microbe Interaction Perspective

Jamaluddin and Anuj Kumar Singh

ABSTRACT

Microbes are integral part of the terrestrial ecosystem. Their function ranges from nutrients mineralization, stress alle*via*tion to bioremediation in soils .There is a great possibility of using microbial inoculants as a biological tool for sustainable forestry and agriculture. This review gives an overview on the benefits of microbial technology and their contribution in development of green technology.

Keywords: Green technology, Microbes, AMF, PGPRs, Microbial consortium, Carbon sequestration.

Introduction

Microbes have a unique role in the soil system. Soil microbial biomass, living part of the soil organic matter, is an agent of transformation of organic matter and source of available nutrients. The potential of soil microorganisms has been recognized widely in improvement of soil quality, soil formation, aggregation and revegetation through their activities in litter decomposition and nutrient cycling. Microbial activities such as phosphate solubilization, nitrogen fixation, oxidation of various inorganic components of soil or mineralization of inorganic components and mycorrhizal symbiosis are major beneficial activities that play a very important role in soil system functioning. Implications of plants and their symbionts like mycorrhizal fungi, nitrogen fixing bacteria and free-living rhizosphere population of bacteria promote plant's establishment and growth. An active soil microbial biomass is an essential factor in the long-term fertility of soils. Microorganisms improve the nutrient status and texture by addition of organic matter (Palaniappan and Natarajan, 1993). Application of microbial technology has become an integral element of forestry and agriculture. It has substantially contributed in forest and agricultural productivity. This article discusses some of the important aspect of microbial applications in forestry and agriculture.

Soil-Plant-Microbe Interactions: A Review

Microbial activity in the rhizosphere affects rooting patterns and the supply of available nutrients to plants thereby modifying the quality and quantity of root exudates (Bowen and Rovira, 1999; Gryndler, 2000; Barea, 2000). Carbon fluxes are crucial determinants of rhizosphere function. The release of root exudates and decaying plant material provide sources of carbon compound for the heterotrophic soil biota as either growth substrates or structural material for root associated microbiota (Werner, 1998). Nitrogen-fixing microbes can exist as free-living organisms in associations of different degrees of complexity with other microbes and plants. The most abundant element in the atmosphere (N_2) is very often the limiting element for the growth of most organisms. Many soil organisms interact with each other to overcome of the limitation. Positive effects of *Rhizobium* sp. inoculation in combination with *Azotobacter* sp. or *Azospirillum* sp. inoculants have been reported for different forage and grain legumes. The role of microorganisms in nutrient cycling is unique. An active biomass is an essential factor in the long-term fertility of restored soil. It is therefore essential that microbes beneficial for plant growth have to be introduced to the mine spoils. Among the different microbes arbuscular mycorrhizal fungi, nitrogen fixers and phosphate solubilizers are very important for any plant. Phosphorus is an important plant nutrient next to nitrogen for growth. The most important aspect of phosphorus cycle is microbial mineralization, solubilization and immobilization besides chemical fixation of phosphorus in the soil. Phosphorus solubilising microorganisms convert insoluble inorganic phosphate compounds into soluble form. A considerable higher concentration of phosphate solubilising bacteria is commonly found in the rhizosphere soil. Also the fungal genera with this capacity are *Penicillium* and *Aspergillus* (Suh *et al.*, 1995; Whitelaw *et al.*, 1999). *Pseudomonas* is a typical PGPR and their interactions with AM fungi mutually enhance each other's colonization and achieve additive plant growth enhancement (Singh and Jamaluddin, 2008). Another mechanism of action of PGPR on plant growth is the production of siderophores. The siderophores are produced by most fungi and bacteria including *Pseudomonas, Rhizobium* and *Azotobacter* (Meyer and Linderman, 1986). Arbuscular mycorrhizal, fungi (AMF) which are an important group of soil-borne microorganisms; contribute substantially to the establishment, productivity, and longevity of natural or man-made ecosystems. These fungi form symbiotic association with most terrestrial plant families. Due to the extensive network of external hyphae which function as plant rootlets and increase phosphorus uptake, AMF are considered as beneficial microorganisms. The species of *Pseudomonas, Bacillus, Aspergillus, Penicillium* etc. have been reported to be active in the bioconversion of insoluble phosphorus. These organisms produce organic acids like citric, glutamic, succinic, lactic and tartaric acids which are responsible for solubilization of insoluble forms of phosphorus. Phosphorus solubilising microorganisms synergistically interact with nitrogen fixing microorganisms. These PGPRs are widely being exploited as biofertilizers in agriculture, horticulture, forestry and agro-forestry (Gaur, 1990).

Microbes and Phosphorus Nutrition of Plant

Phosphatase enzyme activity

Phosphorus is an important plant nutrient next to nitrogen for plant's growth. The most important aspect of phosphorus cycle is microbial mineralization, solubilization and immobilization besides chemical fixation of phosphorus in the soil. Phosphorus solubilising microorganisms convert insoluble inorganic phosphate compounds into soluble form. A considerable higher concentration of phosphate solubilising bacteria is commonly found in the rhizosphere soil. A significant amount of phosphorus is bound in organic forms in the rhizosphere. Phosphatase enzymes produced by plants and microbes are presumed to convert organic phosphorus into available phosphate, which is absorbed by plants. Phosphate availability is one of the major growth limiting factors for plants in many natural ecosystems (Antibus and Dighton, 1993). Plants absorb phosphorus from the soil as inorganic phosphate ions, but their availability is severely restricted by reactions of inorganic and organic phosphates with soil constituents. Phosphatase enzymes can significantly import the availability and recycling of phosphorus in and around the rhizosphere. Mine spoils are infertile strata for the plant growth (Bradshaw, *et al.*, 1975). Since calcareous mine spoils present low level of key nutrients like N,P and K and also lack sufficient level of organic matter. An appropriate level of nutrients and microbial system is a requisite for the growth and establishment of plants. Phosphorus is one of the major constituents of the soil nutrient system, which often becomes inadequate in mine spoils. In such unfavorable conditions phosphatase enzyme activity of plants may prove significant in the phosphorus liberation from organically bound phosphorus to available inorganic form. This liberated and available form of phosphorus is mineralized by rhizosphere microbes and is made available for plant uptake.

A study on effect of AMF inoculation on phosphatase enzyme activity conducted by Singh and Jamaluddin (2008) reported that the inoculation of AM fungi influenced both the phosphatases (acid and alkaline) activity. By inoculation of AM fungi; activity of both acid and alkaline phosphatases was found to be increased in the rhizosphere soil of *Jatropha curcas*. There was significant variation ($p<0.05$) in the acid phosphatase activity in the rhizosphere soil of inoculated and un-inoculated plants. Alkaline phosphatase activity was also found to be enhanced in the rhizosphere soil of inoculated plants. From the above records, it is clearly evident that AMF affected the activity of both the phosphatases. Mutualistic association thus may improve the phosphorus mineralization and its acquisition by plants. In this study, acid phosphatase activity was found greater than alkaline phosphatase activity in rhizosphere soil of both inoculated and uninoculated plants. Acid phosphatases contribute significantly to the mineralization of organic phosphorus compounds in soil, thus enhancing the biological availability of released inorganic phosphorus. Kennedy and Rangarajan (2001) observed significant increase in both the acid and alkaline phosphatase activity in three varieties of papaya and observed comparatively higher acid phosphatase activity than alkaline phosphatase activity. Mohandas (1990) observed significant enhancement in the activity of acid and alkaline phosphatase in the root surface of papaya inoculated with AM fungi, also observed that *Glomus mosseae* inoculation resulted in a higher increase compared to *G. fasciculatum* and acid phosphatase activity was higher than the alkaline phosphatase activity. Tisserant *et al.* (1993) suggested that alkaline phosphatase activity was induced by colonization of host roots and this enzyme could be used for analyzing the symbiotic efficiency of arbuscular mycorrhizal infection. Thiagarajan and Ahmad (1994) reported increased acid and alkaline phosphatase activity in AM inoculated cowpea plants as compared to uninoculated plants. Joner and Johansen (2000) studied comparative effect of different AMF on phosphatase activity in soil and found that *G. intraradices* showed the highest external phosphatases activity in two experiments and *G. claroides* had the same activity in one experiment. Dodd *et al.* (1987) observed comparatively enhanced acid and alkaline phosphatase activity in the rhizosphere of AM inoculated plants than uninoculated ones. Studies conducted at the International Crops Research Institute for the Semi Arid Tropics (ICRISAT), Andhra Pradesh, India showed that in sterilized soil, acid and alkaline phosphatase activities in the rhizospheres of mycorrhizal and non-mycorrhizal plants were higher than in non- rhizosphere soil. With acid phosphatase, the activity was highest in the rhizosphere of mycorrhizal plants followed by rhizosphere of non-mycorrhizal plants and non-mycorrhizal soil (Krishna and Bagyaraj, 1985). These studies reflect significance of microbes in the phosphorus nutrition of plants.

Microbial solubilization of calcite rocks

Soil microorganisms inhabiting the rhizosphere environment interact with plant roots and mediate nutrient cycling. Solubilization of rock materials by microorganisms contribute to a great extent and play very important role in maintaining supplies of inorganic nutrients for plants. Limestone mine overburden dumps are composed of large amount of infertile soil along with waste rocks. These over burden dumps are deficient in nutrients and are unable to support vegetation on mined dumps. In such alkaline soil, the activity of calcium and its minerals become high. As these minerals are rock bound, their release into the soil is very slow. The solubilization of these rock bound minerals is very important for the enrichment of the fertility of the mined spoils. The soil microorganisms play an active role in the solubilization of calcite rocks and the minerals released into the soil may be utilized by the plants grown in calcite medium of limestone mined spoils. Singh and Jamaluddin (2010) reported an increase in the phosphate solubilization efficiency of *Pseudomonas fluorescens* and *Azospirillum sp.* when the isolates of both the bacterial strains were repeatedly exposed in calcium rich medium. The *in vitro* assay showed the bacterial inoculation solubilized the added rock powder more efficiently when they were repeatedly inoculated in calcium rich medium. It may be an indication of development of more tolerant and more efficient strain of the same bacterial isolates. Such efficient strains may be developed and multiplied for use in the field poor in nutrients phosphorus in particular. Leyval *et al.* (1990) also reported bacterial solubilization of Mica mineral which led to the release of unavailable nutrients to pine and beech. Watteau and Berthelin (1994) emphasized that the organic acids produced by rock dissolving bacteria enhanced the weathering of primary rock minerals by chelating cations from minerals. The rhizosphere of the plant which is rich in microbial diversity especially of bacteria, contribute significantly to the supply of nutrients to plants (Hinsinger and Gilkes,1993).These bacterial population can be developed in more capable strains and can be multiplied for application in the field. Chang and Li (1998) demonstrated induced weathering processes by combined effect of plant root and soil microbes which may be responsible for transformation of mineral structure, removing cations from the minerals for rapid uptake by plants. Seshadri *et al.* (2000) tested three strains of *Azospirillum sp.* to study their capability to solubilize inorganic phosphate *in- vitro*. Gaur (1990) reported solubilization of tricalcium phosphate in liquid medium by using bacterial inoculants and demonstrated phosphorus solubilising bacteria as a significant bacterial biofertilizer. Agnihotri (1970) tested some fungal isolates for efficiency of solubilization of insoluble phosphate and observed the solubilization process to a maximum extent. The periodic solubilization of ferric and aluminium phosphate was investigated in liquid medium by Gaur and Gaind (1983). Ahmad and Jha

(1968) reported the solubilization of rock phosphate and hydroxyapatite by some bacteria and fungi. Wani *et al.* (1979) also confirmed the rock solubilization ability of selected strains of *Pseudomonas striata* and *Bacillus polymyxa* on both solid and liquid media. The investigations carried out in the present study are an indication of the development of new strains of the test organisms. These improved strains can be further multiplied in bulk and can be used for biological solubilization of rock bound minerals in nutrient deficient soils like limestone mined overburden dumps.

Mycorrhizal Technology and Soil Fertility

Arbuscular mycorrhizal associations are important in natural and managed ecosystems due to their nutritional and non- nutritional benefits to their symbiotic partners. They can alter plant productivity because AMF acts as biofertilizers, bioprotectants or biodegraders and are known to improve plant growth and health by improving mineral nutrition or increasing resistance or tolerance to biotic and abiotic stresses.

Mycorrhizal fungi are species of fungi that intimately associate with plant roots forming a symbiotic relationship, with the plant providing carbohydrate for fungi and the fungi providing nutrients such as phosphorus, to the plants. Mycorrhizal fungi can absorb, accumulate and transport large quantities of phosphate within their hyphae and release to plant cells in root tissue. Mycorrhizal symbiosis is very important in agriculture and forestry. Plants inoculated with arbuscular mycorrhiza have shown to be more resistant to some root diseases. It is now generally recognized that they improve not only the phosphorus nutrition of the host plant but also its growth, which may result in an increase in resistance to drought stress and some diseases. Therefore, AM fungi offer a great potential for sustainable agriculture, and the application of AM fungi to agriculture has been developed.

AM fungi modify the quality and abundance of micro flora and alter overall microbial activity. Following host root colonization the AMF induces changes in the host root exudation pattern, which alters the microbial equilibrium in the rhizosphere. These interactions enhance plant growth, health and productivity. Applications of biofertilizers have great potential in preventing soil degradation and restoring soil fertility of drastically disturbed lands. Plant succession following drastic disturbance due to mining is a subject of both practical and ecological interest. An adequate understanding of natural revegetation processes should be included in all rehabilitation efforts without which no desired plant cover will be possible. In some countries the AM fungal inocula have been commercialized. Since it is laborious and cost-consuming for production of AM fungal inocula because of their obligate biotrophic nature, the ways to increase the function of the indigenous AM fungi in soil have also been developed. AM fungal association enhances nitrogen gain in ecosystems by increasing the nitrogen fixation rates of plant and nitrogen fixing bacterial associations. In many ecosystems a large fraction of the available nitrogen in the soil is ammonium and mycorrhizal fungi readily transport ammonium from soil to plant (Ames *et al.*, 1983). Nitrogen fixed in association with one plant is transported to an adjacent plant *via* hyphal connections. Nitrogen fixed by soybeans was transported to maize *via* the mycorrhizae and significantly increased the growth and nitrogen status of maize plants (Van Kessel *et al.*, 1985). So, arbuscular mycorrhizal sysmbiosis has a very important role in nutrient transportation in soil-plant system.

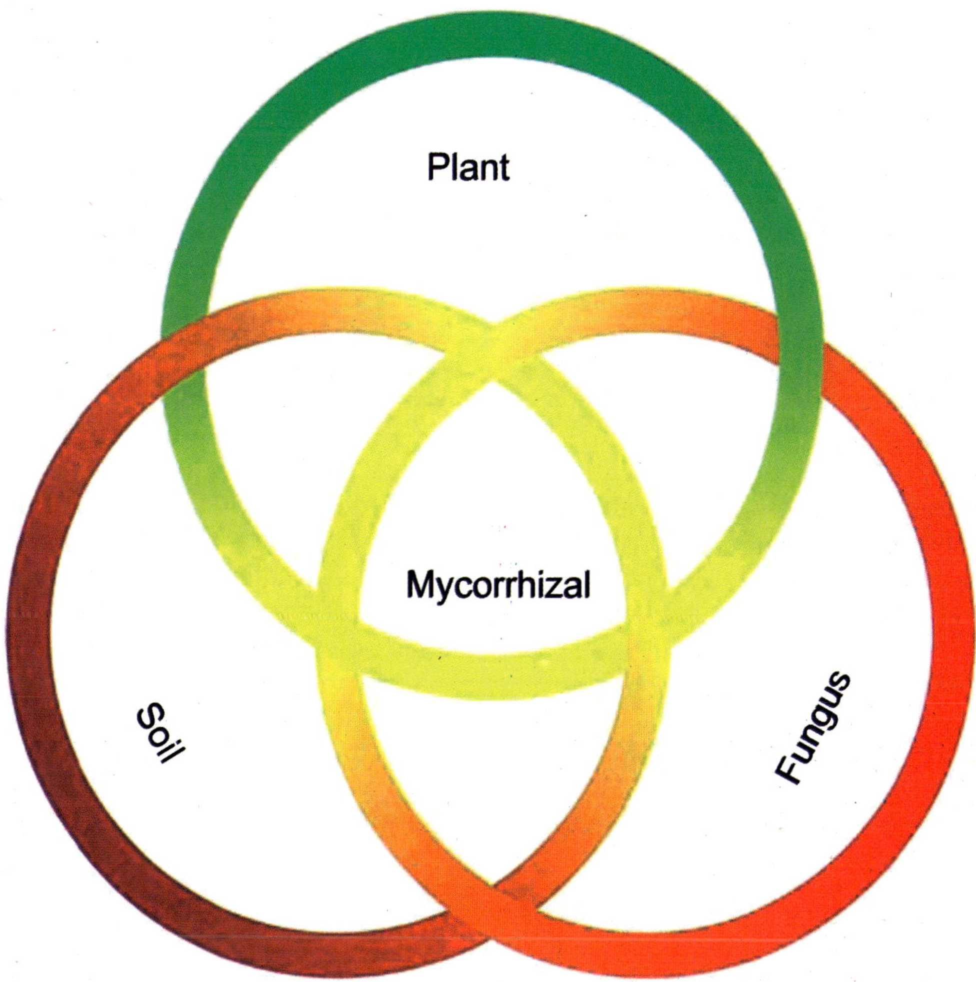

Fig. 1 : Mycorrhizal association, showing the interactions between fungus, plant and soil. (Source: Brundrett *et al.*, 1996).

The benefits of optimization (*via* inoculation) of AMF in production systems with low indigenous inoculum or efficacy have included:

- Increased plant nutrient uptake *via* the AM fungus e.g. phosphorus
- Increased tolerance to root pathogens by the plant system
- Increased tolerance to water stress and adverse environmental conditions (e.g. heavy metal pollutants)
- Increased efficacy of N-fixation by *Rhizobium* and other nitrogen fixing bacteria
- Increased natural succession
- Increased plant biodiversity in restored ecosystems
- Increased stability of soils

Green Technology in Mines Restoration

Numerous studies have demonstrated that land restoration benefits from microbial inoculation in plantations on mines spoil. Plantation allows jump start succession. The catalytic effects of plantations are due to changes in under story microclimatic conditions (increased soil moisture, reduced temperature etc.), increased vegetational structure complexity, development of litter and humus layers and the soil physical and chemical environment and accelerating development of diversity on degraded sites. Plantations have an important role in protecting the soil surface from erosion and altering the accumulation of fine particles. They can reverse degradation process by stabilizing soil through development of extensive root systems. Plantation of suitable species speed up succession that fulfills revegetation goal. Besides controlling leaching of nutrients through soil erosion increases plant diversity. Earlier studies indicated that well adapted plant species could be recommended to establish self-sustaining cover, which require little maintenance activities. In restoration, emphasis is given first to build soil organic matter, nutrients and vegetation cover to accelerate natural recovery process. Plantation can be used as a tool for mine spoil restoration as well as carbon sequestration. Once plantation is established, plants increase soil organic matter, lower soil bulk density, moderate soil pH and bring mineral nutrients to the surface and accumulate them in available form. The plants accumulate these nutrients and re-deposit them on the soil surface in organic matter, from which nutrients are much more readily available for microbial breakdown. Once the soil characteristics have been restored, it is not difficult to form the full suit of self sustained plantation on mined lands. Some of the plant species *viz.*, *Jatropha curcas*, *Pongamia pinnata*, *Ailanthus excelsa* and *Withania somnifera* have been successfully experimented on limestone mined out areas of Madhya Pradesh in India. These important biodiesel and medicinal plant species are suitably surviving and also have attracted different other shrubs, grasses and tree species to grow. A number of restoration ecologists have suggested many approaches for restoration of mined land; however afforestation approach is uniformly accepted.

The major aims of restoration of mined spoils should be (Soni *et al.*, 1989)

- Speedy development of vegetal cover capable of reducing erosion and pollution.
- To provide ecological site stability in terms of favourable soil environment to support colonization of diverse flora and fauna
- Enrichment of soil nutrient levels, weathering of overburden materials and humification of organic matter.
- Bio-rejuvenation of soil system. Growth and survival of above ground and under ground flora.
- Creation of self sustaining ecosystem

Sequence of activities under afforestation approach

I. Studies on mined lands, overburden dumps and spoils

a) Ecological survey of major vegetative association and natural succession

b) Characterization of mine spoils and dumps and identification of limiting factors for plant growth

II. Stabilization of sites through mechanical measures

III. Selection of site specific species suitable for the site

IV. Planting technique

V. Use of amendments - Application of biofertilizers, mulching, manuring and fertilization

VI. Conservation of moisture and water harvesting

VII. Protection, monitoring and evaluation

Selection of site-specific species suitable for the site

For sustainable stability of the ecosystem selection of most suitable species is important. The basis of species selection is:

- Indigenous species of the particular eco-climatic or agro-climatic zone
- Ecological survey for identifying pioneering species of grasses, herbs, shrubs and trees
- Species trial in nursery

The selected species should be

- Capable of colonizing degraded areas
- Fixing atmospheric N_2 as well as conserving soil
- Capable of attracting birds and other faunal population
- Fast growing species should be given preference
- Preference should be given to indigenous one over the exotic

Mycorrhizal Technology in Mines Restoration

AMF – PGPRs Interaction and plantation establishment

Soil acts as a critical controlling component in the development of any ecosystem. Mine spoils are not suitable for both plant and microbial growth because of low organic matter content, unfavorable pH, drought arising from coarse texture or oxygen deficiency due to compaction. The other factors for revegetation of mine spoil may be salinity, alkalinity, poor water holding capacity, inadequate supply of plant nutrients and accelerated rate of erosion. The soils of disturbed sites are frequently low in available nutrients and lack the nitrogen-fixing bacteria and mycorrhizal fungi usually associated with root rhizospheres. As such, land restoration in semi-arid areas faces a number of constrains related to soil degradation and water shortage. As mycorrhizae may enhance the ability of the plant to cope with water stress situtations associated to nutrient deficiency and drought. The use of AMF could reduce the amount of fertilizer needed for the establishment of vegetation and could also increase the rate at which the desired vegetation becomes established by stimulating the development of beneficial microorganisms in the rhizosphere. Degraded soils are common targets of revegetation efforts in the tropics, but they often exhibit low densities of AMF fungi . This may limit the degree of mycorrhizal colonization in transplanted seedlings and consequently hamper their seedling establishment and growth in those areas. Soil inoculation with consortium of AMF and PGPRs has significantly enhanced plant growth and biomass production in limestone mine spoils. Inoculation of native and well adapted microbial flora may prove a proficient tool for restoration of heavily degraded limestone mine spoils. Native benefifical microbial flora like AMF along with phosphorus solubilising bacteria(PSB) and nitrogen fixing bacteria were isolated, multiplied and re-inoculated in different important plant species *viz., Pongamia pinnata, Jatropha curcas, Withania somnifera* and *Ailanthus excelsa*. All the inoculated plants exhibited enhanced growth and development as compared to uninoculated ones. Moreover, inoculation with beneficial plant growth promoting rhizobacteria (PGPRs) changed the soil characteristics and also allowed increased invasion and natural succession on planted spoil as compared to unplanted sites (Singh and Jamaluddin, 2009a).

Green technology and carbon sequestration

Many research organizations and industries have been developing technologies to mitigate carbon dioxide concentrations. The available options include separation and capture of CO_2 from the energy system and sequestration in the deep ocean, sequestration in the geologic formations, sequestration in the depleted oil reserves, sequestration in the explored coal seam, and terrestrial sequestration in plants and soil. Out of many available technologies, afforestation approach is universally accepted due to its economic feasibility and practicability. Terrestrial ecosystems which consist of vegetation and soils are considered to be a major sink for carbon at present time. Carbon sequestration by plants is a natural process and does not require any specific sophisticated technological inputs. The process of photosynthesis combines atmospheric carbon dioxide with water, subsequently releasing oxygen into the atmosphere and incorporating the carbon atoms into the cells of plants. Additionally, forest soils capture carbon. Trees, unlike annual plants that die and decompose yearly, are long-lived plants that develop a large biomass, thereby capturing large amounts of carbon over a growth cycle of many decades. Thus, a forest ecosystem can capture and retain large volumes of carbon over long periods. Forests operate both as vehicles for capturing additional carbon and as carbon reservoirs. A young forest, when growing rapidly can sequester relatively large volumes of additional carbon. An old-growth forest acts as a reservoir, holding large volumes of carbon even if it is not experiencing net growth. Thus, a young forest holds less carbon, but it is sequestering additional carbon over time. An old forest may not be capturing any new carbon but can continue to hold large volumes of carbon as biomass over long periods of time. Managed forests offer the opportunity for influencing forest growth rates and providing for full stocking, both of which allow for more carbon sequestration. Forest systems operate on a cycle of many decades and centuries, rather than annually or over a few years as would be the case with most crops and non-tree vegetation. As forest biomass expands, the amount of carbon contained in plant increases. As the biomass contract, the forest holds less carbon. In an unmanaged state, forests ebb and flow in response to disturbances in the natural system. Forest disturbance regimes are part of the natural ecological system, with wind, disease, fire and other natural, *i.e.*, non-anthropogenic, events causing forest destruction and death. These events result in the release of carbon into the atmosphere but also are typically followed by the regrowth of the forest, which, in turn, begins a new process of carbon buildup in the forest. In some cases, these disturbances are catastrophic in that large areas of the forest landscape are disturbed, as with large wildfires such as are common in many pine and boreal forests. In other cases, the disturbances are highly localized, as with an occasional tree death due to disease or old age such as is common in many tropical forests. Carbon release is occasioned by the disturbance and often in the decay and decomposition of dead matter that follows. However, most natural forests have provisions for natural regeneration and regrowth, which, once again, captures carbon. Thus carbon is recycled in the forest ecosystem.

AMF technology in bioremediation

By definition, bioremediation is the use of living organisms, primarily microorganisms, to degrade the environmental contaminants into less toxic forms. It uses naturally occurring bacteria and fungi or plants to degrade or detoxify substances hazardous to human health and/or the environment. The microorganisms may be indigenous to a contaminated area or they may be isolated from elsewhere and brought to the contaminated site. Contaminant compounds are transformed by living organisms through reactions that take place as a part of their metabolic processes. Biodegradation of a compound is often a result of the actions of multiple organisms. When microorganisms are imported to a contaminated site to enhance degradation we have a process known as bioaugmentation. For bioremediation to be effective, microorganisms must enzymatically attack the pollutants and convert them to harmless products. As bioremediation can be effective only where environmental conditions permit microbial growth and activity, its application often involves the manipulation of environmental parameters to allow microbial growth and degradation to proceed at a faster rate.

Work with AMF strains tolerant to heavy-metal has provided evidence for their rapid adaptation to contaminated soils. Copper (Cu) was absorbed and accumulated in the extraradical mycelium of three AMF isolates, as observed in a study with *Glomus spp.* (Gonzalez-Chávez *et al.,* 2002). Other references indicated resistance of arbuscular mycorrhizal fungi to aluminium. Soil aluminum normally causes significant reduction in tissue calcium and magnesium concentrations (Cumming and Ning, 2003). *Glomus caledonicum* seems to be a promising mycorrhizal fungus for bioremediation of heavy metal contaminated soil (Liao *et al.,* 2003). Rufykiri *et al.* (2002) found that AM fungus could take and translocate uranium towards the roots. At varying zinc levels, mycorrhizal colonization increases zinc absorption and accumulation in the roots. This may help to explain the alle*via*tion of zinc toxicity at high concentrations (Chen *et al.,* 2003). Mycorrhizae were found to ameliorate the toxicity of trace metals in polluted soils growing in soybean and lentil plants (Jamal *et al.,* 2002). In *Cynara cardunculus* mycorrhiza survived to pesticide employed in commercial nursery and enhanced wild carrdoon plant productivity (Marin *et al.,* 2002). Mycorrhizal colonization, however, was reduced in field plots through applications of the fungicide benomyl as a soil drenche (O'Connor *et al.,* 2002). The goal in bioremediation is to stimulate microorganisms with nutrients and other chemicals that will enable them to destroy the contaminants. The bioremediation systems in operation today rely on microorganisms native to the contaminated sites, encouraging them to work by supplying them with the optimum levels of nutrients and other chemicals essential for their metabolism. Researchers are currently investigating ways to augment contaminated sites with nonnative microbes including genetically engineered microorganisms specially suited to degrading the contaminants of concern at particular sites. It is possible that this process, known as bioaugmentation, could expand the range of possibilities for future bioremediation systems.

References

Agnihotri, U.P. (1970). Solubilization of insoluble phosphates by some soil fungi isolated from nursery seedbeds. *Cana. J. Microbiol.*, 16:877-880.

Ahmed, N. and Jha. K.K. (1968). Solubilization of rock phosphate by microorganisms isolated from Bihar soils. *J. Gen. Appl. Microbiol.*, 14:89-95.

Ames, R.N., Reic, C.P.P., Proter, L.K. and Cambardella, C. (1983). Hyphal uptake and transport of nitrogen from two N-labeled Sources by *Glomus mosseae* a vesicular arbuscular mycorrhizal fungus. *New Phytol.*, 95: 381-396.

Antibus, R.K. and Dighton, J. (1993). Acid phosphatase activities and oxygen uptake in mycorrhizas of selected deciduous forest trees : *Proceedings of the 9th North American Conference on Mycorrhizae*, 8-12 August, 1993,University of Guelph, Guelph, Ontario, Canada : 65.

Barea, J.M. (2000). Rhizosphere and mycorrhizae of field crops, In: *Biological Resource Management, Connecting Science and Policy (OECD)*, (Eds.) P. Toutant, E. Balazs and E. Galante, INRA and Springer, Berlin, Heidelberg, New York.

Bowen, G.D. and Rovira, A.D. (1999). The rhizosphere and its management to improve plant growth. *Adv. in Agron.*, 66:1-102.

Bradshaw, A.D., Dancer, W.S., Handley, J.F. and Sheldon, J.C. (1975). Biology of land revegetation and reclamation of china-clay wastes, 363-384. In: *The Ecology of Resource Degradation and Renewal*, (Eds.) M.J. Chadwick, and G.T Goodman, Blackwell Scientific Publication, Oxford, England.

Brundrett, M., Beegher, N., Dell, B., Groove, T. and Malajczuk, N. (1996). *Working with mycorrhizas in Forestry and Agriculture.* ACIAR Monograph 32.374+xp.ISBN186320 1815

Chang, T.T. and Li, C.Y. (1998). Weathering of limestone, marble and calcium phosphate by ectomycorrhizal fungi and associated microorganisms. *Taiwan J. Forestry Sci.*, 13(2):85-90.

Chen, B.D., Li, X.L., Tao, H.Q., Christie, P., Wong, M.H. (2003). The role of arbuscular mycorrhiza in zinc uptake by red clover growing in calcareous soil spiked with various quantities of zinc. *Chemosphere*, 50(6): 839-846

Cook, C.W., Hyde, R.M. and Sims, P.L. (1974).Guidelines for revegetation and stabilization of surface mined areas in the western states. *Colorado State University, Range Science Series* 16: 70.

Cumming, Jr. and Ning, J. (2003). Arbuscular mycorrhizal fungi enhance aluminium resistance of broomsedge (*Andrpongon virginicus* L). *J.Exp. Bot.*, 54 (386):1477-1459.

Dodd, J.C., Burton, C.C., Burns, R.G. and Jeffries, P. (1987). Phosphatase activity associated with the roots and the rhizosphere of plants infected with vesicular-arbuscular mycorrhizal fungi. *New Phytol.*, 107(1): 163-172.

Gaur, A.C. and Gaind, S. (1983). Microbial solubilization of phosphates with particular reference to iron and aluminium phosphate. *Sci. Cult.*, 49:110-112.

Gaur, A. C. (1970). Role of phosphate solubilizing microorganisms and organic matter in soil productivity. *National symposium on Soil productivity*, National Academy of Sciences : 259-268.

Gaur, A. C. (1990). Phosphate solubilizing bacteria as biofertilizer. In: *Phosphate Solubilizing Microorganisms as Biofertilizers*. Omega Scientific Publishers, New Delhi : 160-176

Gonzalez-Chavez, M.C., Carrilo-Gonzalez, R., Wright, S.F. and Nichols, K. (2004). The role of Glomalian, a protein produced by arbuscular mycorrhizal fungi, in sequestering potentially toxic elements. *Environ.Pollution*,130:317-323.

Gryndler, M. (2000). Interactions of arbuscular mycorrhizal fungi with other soil organisms. In: *Arbuscular mycorrhizas: Physiology and function*, (Eds.) Y. Kapulnik, and D.D. Douds Jr,., Dordrecht Kluwer academic publishers, The Netherlands : 239-262.

Hinsinger, P. and Gilkes, R.J. (1993). Root-induced dissolution of phosphate rock in the rhizosphere of lupins grown in alkaline soil. *Austr. J. Soil Res.* 33: 477-489.

Jamal A, Ayub, N., Usman, M. and Khan, A. G. (2002). Arbuscular mycorrhizal fungi enhance zinc and nickel uptake from contaminated soil by soybean and lentil. *Inter. J. Phytoremediation*, 4(3):203-221.

Joner, E.J. and Johansen, A. (2000). Phosphatase activity of external hyphae of two arbuscular mycorrhizal fungi. *Mycol. Res.*, 104:81-86.

Kennedy,Z.J. and Rangarajan, M. (2001).Biomass production, root colonization and phosphatase activity by six mycorrhizal fungi in papaya. *Ind. Phytopath.*, 54 (1): 72-77.

Krishna, K.R. and Bagyaraj, D.J. (1985). Phosphatases in the rhizospheres of mycorrhizal and non-mycorrhizal groundnut. *J. Soil Biol. and Ecol.*, 5(2): 81-85.

Leyval, C., Laheurte, F., Belgy, G. and Berthelin, J. (1990). Weathering of micas in the rhizospheres of maize, pine and beech seedlings influenced by mycorrhizal and bacterial inoculation. *Symbiosis*, 9:105-109.

Liao, J.P., Lin, X.G., Cao, Z.H., Shi, Y.Q., Wong, M.H. (2003). Interactions between arbuscular mycorrhizae and heavy metals under sand culture experiment. *Chemosphere,* 50 (6): 847-853.

Marin, M., Ybarra, M., Fe, A., Garcia-Ferriz, L. (2002). Effect of arbuscular mycorrhizal fungi and pesticides on *Cyanara cardunculatus* growth.*Agricul. Food Sci. Finland,* 11 (3): 245-251.

Meyer, R.J. and Linderman, R.G.(1986). Response of subterranean clover to dual inoculation with vesicular arbuscular mycorrhizal fungi and plant growth promoting rhizobacterium Pseudomonas putida. *Soil Biol.Biochem.,* 18 (2):185-190.

Mohandas, S. (1990). Enhanced phosphatase activity in mycorrhizal Papaya (*Carica papaya* cv. Coorg honey dew) roots : In: *Current Trends in Mycorrhizal Research,* (eds.) B.L. Jalali. and H. Chand, Proceedings of the National Conference on Mycorrhiza, Haryana agriculture University, Hisar, Haryana : 55-56.

O'Connor, P.J., Smith, S.E. and Smith, F. (2002). Arbuscular Mycorrhizas influence plant diversity and community structure in a semi arid herb land. *New Phytol.,* 154 (1):209-218.

Palaniappan, S.P. and Natarajan, K. (1993). Practical aspects of organic mater maintenance in soils : In: *Organics in soil health and crop production,* (Ed.) P.K.Thompson, Tree crops development foundation, India : 23-41.

Sheshadri, S., Muthukumarasamy, Lakshminarsimhan, C. and Ignacimuthu, S. (2000). Solubilization of inorganic phosphates by *Azospirillum halopraeferans. Curr. Sci.,* 79(5): 565-567.

Rao, A.V., Tak, R. (2002). Growth of different tree species and their nutrient uptake in limestone mine spoil as influenced by arbuscular mycorrhizal (AM)-fungi in Indian arid zone. *J. Arid Environ.,* 51 (1):113-119.

Rufykiri, G., Thiry, Y., Wang, L., Delvaux, B. and Declerck, S. (2002). Uranium uptake and translocation by the arbuscular fungus, Glomus intraradices, under root-organ culture conditions. *New Phytol.,* 156(2):275-281.

Singh, A. K. and Jamaluddin (2008). Phosphatase activity in the rhizosphere of medicinal plants inoculated with Arbuscular Mycorrhizal Fungi. *Mycorrhiza News,* 19(3): 11-12.

Singh, A. K. and Jamaluddin (2009a). Effect of bio-inoculants and stone mulch on performance of biodiesel and medicinal plant species on limestone mined spoil. *Mycorrhiza News,* 21(3):31-33

Singh, A. K. and Jamaluddin (2009b).Effect of microbial inoculants on performance of *Pongamia pinnata* in calcareous mined spoils. *Ind. Forestry,* March: 1-5.

Singh, A. K.and Jamaluddin (2010). Synthesis of resistance and enhanced mineral solubilization capacity of Plant Growth Promoting Rhizobacteria (PGPRs) native to calcareous soils of limestone mined spoils. *Ind. Forester* (In press).

Soni, P., Vasistha, H.B., Kumar, O. and Bhatt, V. (1989). Successional trends in mined ecosystems. *J. Nat. Cons.*, 1:115-122.

Suh, J. S., Lee, S. K., Kim, K.S. and Seong, K. Y. (1995). Solubilization of insoluble phosphates by *Pseudomonas putida*, *Penicillium* sp. and *Aspergillus niger* isolated from Korean Soils. *JKSSF*, 28(3):278-286.

Thiagarajan, T.R. and Ahmad, M.H. (1994). Phosphatase activity and cytokinin content in cowpeas (*Vigna unguiculata*) inoculated with a vesicular arbuscular mycorrhizal fungus. *Biol. Fertil. Soils*, 17(1):51-56.

Tisserant, B., Gianinazzi-Pearson, V., Gianinazzi, S. and Gollett, A. (1993). Plants histochemical staining of fungal alkaline phosphatase activity for analysis of efficient arbuscular mycorrhizal infections. *Mycol. Res.*, 972: 245-250.

Vankessel, C., Singleton, P.W. and Hoben, H.J. (1985). Enhanced nitrogen transfer from a soybean to maize by vesicular-arbuscular mycorrhizal fungi. *Pl. Physiol.*, 79: 562-563.

Wani, P.V.,More, B.B. and Patil, P.L. (1979). Physiological studies on the activity of phosphorus solubilizing microorganisms. *Ind. J. Microbiol.*, 19:23-25.

Watteau, F. and Berthelin, J. (1994). Microbial dissolution of iron and aluminium from soil minerals: efficiency and specificity of hydroxamate siderophores compared to aliphatic acids. *European J. Soil Biol.*, 30:1-9.

Werner, D. (1998). Organic signals between plants and micro-organisms. In: *The Rhizosphere : Biochemistry and organic substances at the soil-plant interfaces*, (Eds.) R. Pinton, , Z.Varanini and P. Nannipieri, , New York, Marcel Dekker : 197-222.

Whitelaw, M.A. (2000). Growth promotion of plants inoculated with phosphate-solubilizing fungi. *Adv. Agron.*, 69:99-151.

□□□

Microbial Diversity and Functions, 2012
© D.J. Bagyaraj, K.V.B.R. Tilak, H.K. Kehri (eds.), pp. 273-290
New India Publishing Agency, New Delhi (India)
E-mail : info@nipabooks.com; Website : www.nipabooks.com

Chapter 11

Mangrove Ecosystem : A Marine Resource for Useful Microorganisms

Nibha Gupta, Uday Chand Basak and Ajay Kumar Mohapatra

ABSTRACT

Microorganisms are potential tool for the production of large amount of potent and effective health care, food products and biofertilising agents. Recently, the marine microorganisms are considered as good sources of such products which may possess novel properties. There is a high scope for the exploration of marine microbes in order to search new product of economic importance from marine resources. Mangrove habitat is also part of marine habitat with the plant community having unique adaptation in their morpho- physiological and reproductive behavior. The high level of organic content in mangrove rhizosphre also leads to the colonization of degradative microorganisms and make them responsible for the creation of diverse and potential microbial community. Hence, mangroves are also considered as untapped resources of microbial metabolites and products. Several microorganisms including bacteria, fungi, algae, cyanobacteria, lichens are documented from mangrove habitats. These inhabitant microbes are the good source of secondary metabolites, therapeutic enzymes, food protective agents and mineral solubilisers. However, information on these aspects is very meager and thus the present review is an attempt to compile informations on the sources, properties and application of important microbes from mangrove ecosystems.

Keywords: Mangroves, marine, microbes

Mangrove ecosystems are forest vegetation growing in saline environments found along the coastlines of tropical and subtropical regions in the inter tidal zones of river deltas (Ravishankar *et al.*, 2004). According to Fisher and

Spalding (1993) the total area of the world mangroves is about 198, 818 sq. km and some 60 species of trees and shrubs are exclusive to the mangrove habitat. The mangrove habitat, according to Walsh (1974) has two broad distributions; 1) the indo pacific regions and (2) Western African and American regions. According to forest survey of India (1999), out of 487100 ha of Indian mangroves, nearly 56.7% (275,800 ha) is present along the east coast, where as 23.5% (114,700 ha) is found along the west coast, and the remaining 19.8% (96600 ha) is located in the Andaman and Nicobar islands (Ravishankar *et al.*, 2004).

Plants and microbes colonizing such habitats exhibit remarkable adaptation in their morphology, physiology and reproductive behaviour. The composition, richness and diversity of mangrove species is impacted by degree of inundation, seasonal rainfall, salinity gradient, soil character and silt-clay-sand proportion. The colonizers of mangrove ecosystem creates excellent habitat for diverse mangrove dependent organisms to flourish, along the successional pathway.

Microorganisms present in mangrove habitat is linked to degradation and mineralization of mangrove leaves, wood, sea grasses and other lignocellulosic materials, which become significant primary sources in the food web of the tropical marine ecosystem. Those microbes survive on sediments as well as in the intertidal region of mangrove stands (Alongi *et al.*, 2005).

Mangrove ecosystem are highly productive, biologically diversified and support other biological system by supplying them various organic matter. These materials are ideal substances for wood destroying fungi, bacteria and animals. Compared to terrestrial ecosystem, information on microbes of marine environment in general and mangrove ecosystem in particular is scarce. However, a number of studies have recently reported the occurrence of bacteria and fungi from mangrove ecosystem of Asia-Pacific region (Holguin Bashan, 1996; Benka and Olmungin, 1996; Klekonski *et al.*, 1994; Alongi, 1994; Mann and Steinke, 1993; Namsaraev *et al.*, 1992; Kristensen *et al.*,1998; Hartung *et al.*, 1998; Widiastuti *et al.*, 1996; Abdelmonen *et al.*, 1997, Woitchik *et al.*,1997; Jones *et al.*, 2005; Jones *et al.*, 1999; Poonyth *et al.*, 1999).

Several fungi and bacteria are reported for their occurrence and beneficial activity in mangrove ecosystem (Ananda and Sridhar, 2004; Kathiresan and Selvam, 2006; Suryanarayan and Kumaresan 2000). A vertical distribution of marine fungi associated with Avicennia marina is reported by Abdelwahab *et al.*, (2001). This study contributed two new intertidal swampomyces species from red sea mangroves in Egypt. A new species of cryptovalsa from Mai Po mangrove in Hongkong was reported by Inderbitzein and Desjardin (1999).

Ascomauritian, a lignicolous, isolated from submerged wood in Marititus (Poonyth *et al.* 1999). Baker *et al.*, (2001) studied the ultrastructure of ascus and ascospore appendages of the mangrove fungi *Halosarpheia ratnagiriensis.* Biodiversity of intertidal estuarine fungi on phragmites at Mai po marshes Hong Kong was studied well.

Occurrence of fungi from Indian mangroves have been reported by several scientist (Purkayastha and Pal, 1996; Sengupta and Choudhuri, 2002; Chinaraj, 1992; Patil and Borse, 1985; Prasannarai and Sridhar, 2001). Maximum number of fungal species was recorded in rhizopshere of *Rhizophora apiculata* follwed by *Avicennia officinalis, A. marina.* (Sarma *et al.*, 2001). This study resulted in identification of 88 fungi from east coast of India. These include 65 ascomycetes, one basidiomycetes, 22 microsporic fungi. Vishwakiran *et al.* (2001) studied the spatial and temporal distribution of fungi in the coastal tropical water of Goa , India. Gupta *et al.* (2005) isolated the fungi from different mangrove plants of Bhitarkanika mangroves of Orissa and analysed them for their extracelluar activity.

Manglicolous fungi from India were reported by Chinnaraj (1992). Higher marine fungi of Lakshdeep islands, Maldeeves were reported by Chinnaraj (1992, 1993). Kumarsan and Suryanarayan (2002) obtained 36 species of fungi from some halophytes from estuarine mangrove forests. In this study *Camarosporum* species showed some degree of host preference as they were the dominant endophytes of the halophytes of Chenopodiaceae. These are reported to be very good for the litter degradation. Higher leafy decomposition after leaf fall lead to better endophytic population of *Cladosporium cladopsoroides, Phyllostica* was reported. Kumarsan and Suryanaryan (2001) Studied 7 dominant mangrove species of an estuarine mangrove forest in South India for the occurrence of foliar endophytes. Microsporic , ascomycetes and sterile mycelia were isolated from the leaves of the mangroves .

The diversity and abundance of higher marine fungi on woody substratum along the west coast of India was also studied well (Prasanna rai and Sridhar, 2001). Out of 3327 wood samples scanned 72% possesed sporulaing marine fungi. Altogether 88 species belonging to 47 were encounters. Anada and Sridhar (2002) reported the *Aniptodera indica, a* new species of mangrove inhabiting ascomyctes from west cost of India.

Mangroves ecosystem of different countries of world were analysed for the different aspects of bacterial population (Holguin and Bashan, 1996; Benka and Olmungin, 1996; Klekowaski *et al.*, 1994; Alongi, 1994; Mann and Seinke, 1993; Namsaracv *et al.*, 1992; Kristensen *et al.*, 1998; Hartung *et al.*, 1998; Widiastuli *et al.*, 1996; Abdelmonen *et al.*, 1997; Wortchick *et al.*, 1997; Ando *et al.*, 2001). There is sufficient published evidences are available on the occurrence of bacteria from mangrove ecosystem of abroad (Abdolmonem *et al.*, 1997; Jones *et al.*, 2005). Wong Wu Rong (1993) studied on the microbial flora of Rensui river mangroves and reported 215 stains of bacteria belonging to *E. coli, Salmonella, Shigella Psudomonas, Monococcus, Bacillus, Sarcina, Staphylococcus and Streptococcus.*

There are so many strains of bacteria reported from mangrove ecosystem. Sengupta and Choudhari (1991) studied the ecology of heterotrophic dinitrogen *fixation* in the rhizosphere of mangrove plants community at the Ganges River estuary in India. They isolated nine isolates of N-fixing bacteria. They differ in their Holotolerance and N-fixation capacity. Venthanayagam (1991) isolated the purple photosynthetic bacteria from mangrove. These were *Chromatium* species and *Rhodopseudomonas*. Various species of different nutritional categories was isolated from the Indian mangroves system. Agate *et al.* (1993) isolated the salt tolerant bacteria from the rhizosphere of mangroves, *Rhizophora mucronata* and *Avicennia officinalis* of the Goa and Konkan region. The samples were analyzed for the sediment analysis and bacteria and they found the *Bacillus, Pseudomonas, Staphylococcus, Micrococcus* species were highly salt tolerant. Panchandikar (1993) reported the microbial activity of mangrove ecosystem of Goa and Konkan, soil was analyzed for physiochemical properties, and bacterial population was also analyzed and it was found that *Bacillus, Pseudomonas, Staphylococcus* and *Micrococcus* species were widely distributed there. In the same region of mangrove chemo-heterotrophy was demonstrated in vitro by Rao and Krishnamurthy (1994). Isolation and characterization of methanogenic bacteria from mangrove sediments of Pichavaram forest was done by Mohanraju *et al.* (1997) and they reported the presence of nonspore forming methanogenic bacteria from this habitat. Besides the isolation of specific bacteria from the mangrove area several workers have reported that the microbial activities in the mangroves ecosystem in its vegetation and sediments. Sardessi and Bhosle (2002) worked the organic solvent tolerant bacteria from mangroves. Bacterial density of Azospirillum was studied in the root of *Avicennia marina* (Ravikumar *et al.*, 2002). The growth production of IAA and the role of N_2 fixation in the *Azospirillum brasilense* were found to be maximum at pH 7. 0 and at 3 % NaCl.

In bacterial group Cyanobacteria is very important for nitrogen fixation and nutrient cycling. Toledo *et al.*, (1995) studied the cyanobacteria and black mangroves in northwest Mexico. They were also studied *in vitro* the colonization and increase in nitrogen fixation of seedlings roots of black mangroves inoculated by a filamentous cyanobacteia. (Toledo *et al.*, 1995). Holguin and Bashan (1996) studied the N-fixation by *Azospirillum brasilense*. The nitrogen *fixing* ability of A. *Brasilense* increased significantly when grown in mixed culture with the non-nitrogen-fixing bacteria *Staphylococcus* epidermis but not with *micrococcus lyre,* though, both were isolated from mangroves. Holguin *et al.* (1992) isolated *Staphylococcus* species isolated from mangrove trees. They stated the importance of rhizosphere community for the performance of nitroge*N-fixing* bacteria. Transformation and transport of inorganic N in sediment of a south East Asian mangrove forest studied by Kristensen *et al.* (1998) they elaborated the significance of nitrifying bacteria. The nutrient cycling is depended upon the bacterial population. Few studies have been made on this direction. Alongi (1994) studied the role of bacteria in recycling in tropical and other coastal benthic ecosystem here the role of bacteria as mineraliseres of organic detritus and recycles of essential nutrient is discussed in relation to tropical and coastal sediments. In mangrove water and sediments, proteolysis bacteria are also isolated from this environment, (Namsaraev *et al.,* 1992).

The existence of taxonomically diverse populations of phylogenetically distinct Actinomycetes is residing in the marine environment. The mangrove environment is a potent source for the isolation of antibiotic-producing Actinomycetes (Sivakumar *et al.,* 2005). The distribution of Actinomycetes in the aquatic environment remained largely undescribed for many years and most of the workers suspected the indigenous nature of aquatic Actinomycetes because these bacteria produce resistant spores that are known to be transported from land to sea and other aquatic bodies where they remain dormant for many years. The research work on mangrove Actinomycetes has been done only very few institutions in the world like Taiwan and Japan. Hatono (1997) was studied the Actinomycetes population from 6 types of mangrove forest of different islands of Japan. They were represented as *Micromonosperma, Streptomyces, Actinobacter, Norcardia, Rhodococcus, Actinomadura, Microtetraspore,* etc. *Gordania rhizophera* species isolated form the mangrove ecosystem and analysed their phylogenetic analysis (Takeuchi and Hatano, 1998, 1999).

Streptomyces may be considered a major part of the actinomycetes communities in soil *and* widely distributed in coastal environment Gupta *et al.,* 2009. They resemble fungi in their structure. Streptomycetaceae is well filled by the prolific genus *Streptomyces.* Their branching filamentous arrangement of cells form a network called mycelium. They are found worldwide in soil and are largely responsible through secretion of chemicals called geosmins (practical streptomyces genetics). The genus remains a focus of systematic research not only being the promising source of commercially significant compounds but also because current molecular biological methods are having an increasing impact on conventional streptomyces systematics that are based on phenotypic characteristics .

Recent studies have shown that *Streptomyces* have been isolated from mangrove swamps and other aquatic habitats (Mishra, 2009, Gupta *et al.*, 2009). Guan *et al.* (2005) isolated *Streptomyces* from a mangrove plant *Kandelia candel*. The organism produced p-aminoacetophenonic acids. A strain of Streptomyces nodusus (NPS007994) isolated from marine sediment in Scripps Canyon, La Jolla, California was found to produce lajollamycin antibiotic. *Streptomyces* extracted from marine and mangrove proved to be good antibacterial activity against human pathogens like *Bacillus subtilis*, *Pseudomonas vulgaris*, *Shigella flexneri*, *Klebsiella pneumoniae*, *Vibrio cholerae* and *Staphylococcus aureus* (Sivakumar *et al.*, 2005; Dhanasekaran *et al.*, 2005).

A few workers have also studied association of phytopathogenic bacteria with mangroves. Hartung *et al.* (1998) has reported of a novel plant disease called sundri which was noticed in the mangrove belt of Bangladesh at the Bay of Bengal, which represents the largest mangroves of the world. Widiastuli *et al* (1996) isolated *Bacillus thuringiensis* from the mangroves and characterized completely. Bacterial species of different nutritional categories was isolated from the Indian mangroves system. Agate *et al.* (1993) isolated the salt tolerant bacteria from the rhizosphere of *Rhizophora mucronata* and *Avicennia officinalis* of the Goa and Konkan region. The samples were analyzed for the sediment analysis and bacteria and they established that *Bacillus, Pseudomonas, Staphylococcus, Micrococcus* species were highly salt tolerant. Venthanayagam (1991) isolated the purple photosynthetic bacteria from mangrove. These were *Chromatium* species and *Rhodopseudomonas*. Isolation and characterization of methanogenic bacteria from mangrove sediments of Pichavaram forest was done by Mohanraju *et al.* (1997) and they reported the presence of nonspore forming methanogenic bacteria from this habitat. Antibiotic resistant *Vibriochorelra* was also isolated from the Palangi pettal coastal environment of South India. Besides the isolation of specific bacteria from the mangrove area several workers have reported on the microbial activities in the mangroves ecosystem in its vegetation and sediments. Panchandikar (1993) reported the microbial activity in mangrove ecosystem of Goa and Konkan. Soil was analyzed for physiochemical properties along with bacterial population and it was found that *Bacillus, Pseudomonas, Staphylococcus* and *Micrococcus* species were widely distributed there. In the same region of mangrove chemoheterotrophy was demonstrated in vitro by Rao and Krishnamurthy, 1994). Sengupta and Choudhari studied (1991) the ecology of heterotrophic dinitrogen-fixation in the rhizosphere of mangrove plants community at the Ganges River estuary in India. They isolated nine isolates of *N-fixing* bacteria belonging to all known O_2 responses groups. They differ in their halo tolerance and N_2-fixation capacity.

Plant growth promoting bacteria are the potential tool for arid mangrove reforestation. The role of sediment microorganisms in the productivity, conservation and rehabilitation of mangrove ecosystem are very well reported (Holugin *et al.*, 1992; Bashan and Holguin, 2002). Ando *et al.* (2001) isolated a bacterium useful for the degradation of sea sludge. Mangrove rhizosphere bacteria are reported to be beneficial in terms of plant growth promotion. Bashan *et al.* (2001) studied the effect of halo tolerant bacteria and *Azospirillum* spp on the growth of seawater-irrigated oilseed halophytes *Salicornia bigelobii.*

Greibler *et al.* (2001) has given the DAP I and SYBER green II method for the enumeration of total bacterial numbers in aquatic sediments including mangroves. Work on the bioremediation effect of the microbial community in oiled mangrove sediments was also conducted. Ando *et al.* (2001) used mangrove soil to promote initial fermentation of the sludge constituents. The result suggested that certain microorganisms that were thought to inhabit the subtropical mangrove soil had the potential and significant role in the fermentation and that the use of the microorganism in the mangrove soil might be useful for composting sea sludge.

There are so many strains of bacteria reported from mangrove ecosystem. Sengupta and Choudhari (1991) studied the ecology of heterotrophic dinitrogen fixation in the rhizosphere of mangrove plants community at the Ganges River estuary in India. They isolated nine isolates of N fixing bacteria. They differ in their Holotolerance and N fixation capacity. Venthanayagam (1991) isolated the purple photosynthetic bacteria from mangrove. These were *Chromatium* species and *Rhodopseudomonas*. Various species of different nutritional categories was isolated from the Indian mangroves system. Agate *et al.* (1993) isolated the salt tolerant bacteria from the rhizosphere of mangroves, *Rhizophora mucronata* and *Avicennia officinalis* of the Goa and Kokan region. The samples were analyzed for the sediment analysis and bacteria and they found the Bacillus, Pseudomonas, Staphylococcus, Micrococcus species were highly salt tolerant. Panchandikar (1993) reported the microbial activity of mangrove ecosystem of Goa and Kokan, soil was analysed for physiochemical properties, and bacterial population was also analysed and it was found that Bacillus, Pseudomonas, Staphylococcus and Micrococcus species were widely distributed there. In the same region of mangrove chemoheterotrophy was demonstrated in vitro by Rao and Krishnamurthy (1994). Isolation and characterization of methanogenic bacteria from mangrove sediments of Pichavaram forest was done by Mohanraju *et al.* (1997) and they reported the presence of nonspore forming methanogenic bacteria from this habitat. Antibiotic resistant *Vibrio chorelra* was also isolated from the coastal environment of South India. Besides the isolation of specific bacteria from the mangrove area several workers have reported that the microbial activities in the mangroves ecosystem in its vegetation and sediments. Sardessi and Bhosle (2002) reported the organic solvent tolerant bacteria from mangroves. Bacterial density of *Azospirillum* was studied in the root of *Avicennia marina* (Ravikumar *et al.*, 2002). The growth and production of IAA and the role of N_2 fixation in the *Azospirillum brasilense* were found to be maximum at pH 7. 0 and at 3 % NaCl.

Mangrove rhizosphere was very well studied in abroad as far as biological interaction is concerned. Rojas *et al.* (2001) studied the synergism between *Phyllobacterium* sp. and *Bacillus licheniformis* in response to microbial system participate in bio-mineralization of organic matter and bio-transformation of minerals. Several Nitrogen-fixing enzyme producers and mineral solubilisers are reported from mangrove rhizosphere..

Now it is very clearly understood that mangrove rhizosphere bacteria can be used as a tool to enhance reforestation of mangrove seedlings. This can be done by inoculating seedlings with plant growth-promoting bacteria participating in one or more of the microbial cycles of the ecosystem. Kathiresan *et al.* (2000) described the potential of beneficial bacteria in promoting mangrove seedling growth. Forty-eight bacterial strains were isolated from rhizosphere soil of a mangrove species (*Rhizophora mucronata*). Two strains which doubled the growth of mangrove seedlings were identified as *Azotobacter vinelandii* and *Bacillus megaterium*. This raises the possibility of using the bacterial species to enhance the growth of mangrove seedlings.

A study was also made on the microbiological processes of production and destruction of organic matter in mangroves. It was reported that significant amount of organic matter and high temperature promote activity of aerobic and nonaerobic bacteria. In mangrove water and sediments, protyolytic bacteria were also isolated from this environment, (Namsaraev *et al.*, 1992). Woltchik *et al.*, (1997) studied the *in situ* de-composition of senescent leaves of two abundant mangrove species. Enrichment and activity of dinitrogen-fixing bacteria during decomposition were also investigated. Benka and Olumagin (1996) studied the effects of waste drilling fluid on bacterial isolate from a mangrove swamp oilfield located in Nigeria. Similarly, Klenkowaski *et al.* (1994) studied the effect of mutagens to tropical marine environment. The best treatment for reducing seed rot seedling disease and mortality was evaluated by Abdelmonem *et al.* (1997).

Asparaginases are found in diverse sources in nature including bacteria, yeasts, mold, plant and vertebrates. The enzymes derived from microorganisms are the major source for practical clinical use. Intensive research for L-asparaginase has been carried out among microorganisms and other higher forms during past decades (Imada *et al.*, 1973). Microbial L-asparaginase produced from mangroves of Andaman Islands was optimized for growth temperature of 37°C and pH of 7.2 (Shome and Shome, 2001). *Streptomyces* isolated from prawn *Penaeus indicus* from Veli estuarine lake along Kerala coast has been found to show maximun L-asparaginase activity at pH 7, 37°C and 2-5% NaCl levels. Marine organisms have been reported as potential sources for anti cancer compounds. Sahu *et al.* (2007) studied the *L-asparaginase* enzyme of actinomyceters isolated from esturarine fishes. *L-asparaginase* activity of Streptomycetes isolated from fish, shelfish and sediment of estuarine lake along Kerala coast was studied by Dhevendaran and Annie (1999). L asparagines activity in growing conditions of Steptomyces spp. associated with *Therapon jarbua* and Villorita cyprinoids of Veli Lake was studied by Dhanasekharan *et al.* (2005). Pradhan and Gupta (2008) reported the production and properties of L asparaginase from *Streptomyces* species associated with mangrove. However, very rare reports are available on the asparaginases producing microbes from mangrove origin (Das *et al.*, 2006; Shome and Shome, 2001, Kathirasan, 2000).

Lipases are important due to their ability to catalyze with wide spectrum of lipid substrates and stability under varied temperature and pH. Most of these enzymes are commercially available, being mostly from microbial source produced in extracellular condition (Sztajer *et al.*, 1988, Rapp and Backhaus, 1992; Sharma *et al.*, 2004). Microbial lipases have biotechnological impact due their practical properties, like stability in organic solvents, independence of cofactors, wide spectrum substrate specificity (Lescic *et al.*, 2004).They have been used as an additive in food processing, pesticide, drug formulations, etc. (Li *et al.*, 2001). Mangrove ecosystem is also one of the important habitats for microbial activity but not studied systematically as far as lipase producers are concerned (Mishra, 2009).

Arbuscular mycorrhizal fungi are of immense value for being a transporter and carrier of various minerals like P, Zn, Mn, Mg, Cu, and Al etc. from soil to the host plants helping in enhancement of growth and productivity of plants. A wide spectrum of studies has been carried out on their ecology and distribution, their effects on host physiology, biochemistry and genetics in the national and international laboratories. Their biofertilising potential has been very well documented with respect to various agricultural, horticultural and forest plants. Several aspects of studies have been made nationally and internationally on the AM fungi. Variation in the habitat may certainly change the micro habitat and their behavior too. Mangrove ecosystem are saline habitat which have also been studied for the occurrence of various fungi and bacteria including AM fungi (Gupta *et al.*, 2002). Several mangrove plants species have been found to be mycorrhzal. Gupta *et al.* 2002 reported on the AM mycorrhization in mangrove plant species in non saline and saline natural soil. Recently, few more reports have been noted regarding the mycorrhzation of mangrove tree and other plant species. The association of AM fungi with mangrove has also been reported from Pichavaram forest and Ganges river estuary by Chaudhuri and Sengupta (1990). It is evident from the study of Gupta *et al.* (2002) that arbuscular mycorrhizal fungi are present in the saline soils of Bhitarkanika which confirmed the occurrence of AM fungi.

Mangrove ecosystem is known to be highly rich due to high amount of dissolved and particulate organic matter due to various microbial activities. Microbes from this area play an important role in biodegradation of plant material through their various enzymatic and metabolic activities. A large number of heterotrophic and autotrophic soil microorganims now known to have the capacity to solubilize inorganic phosphate through their metabolic activities directly or indirectly. The occurrence of phosphate solubilising bacteria has been reported in mangrove habitat. Phosphate solubilising bacteria of the rhizosphere of the mangrove ecosystem of Great Nicobar island of India is recently reported (Kothamsi *et al.*, 2006). The phosphate solubilizing potantial of the rhizosphere microbial community in mangroves was demonstrated when culture media supplemented with insoluble tribasic calcium phosphate, and incubated with roots of black *Avicennia germinus* and white *Laguncularia racemosa* (L.) Gaertn. Mangrove became transparent after few days of incubation (Vazquez *et al.*, 2000). Degradation of phosphonate by Streptomyces was also reported well (Obojska *et al.*, 1999). In mangrove ecosystem, phosphorus is present in insoluble inorganic forms. However, mangrove plants don't exhibit P deficiency due the presence of P solubilisers in rhizosphere. Seshadri *et al.* (2002) studied the variation in population of PSM in southeast coast of India.

A considerable amount of work has been done on microbial diversity of Bhitarkanika mangrove ecoysystem also. Bhitarkanika Wildlife Sanctuary (BWLS) is the second largest contiguous mangrove forest in India. It is located on the east coast of India between the latitudes 20 °30′-20° 50′ N and the longitudes 86 ° 45′ -87 ° 10 E′ and lies in the northeast of the Mahanadi delta in Kendrapara district. Preliminary studies related with occurrence and distribution of fungi and bacteria have been done (Gupta *et al.*, 2006). Gupta *et al.* (2001) and (2009) studied the Microbial population in phyllosphere of mangroves grown in different salinity zones of Bhitarkanika. A good number of fungi have been isolated and studied in respect of their morphology and characteristics, and salt tolerance behaviour (Gupta *et al.*, 2005, 2007a,b,c and d). Arbuscular mycorrhizal association of mangroves in saline and non saline soils has also been studied (Gupta *et al.*, 2002a and b). A good account of occurrence of *Streptomyces* from Bhitarkanika magnroves is documented and reported in various journal (Gupta *et al.*, 2007a and b and 2009b,c and e Mishra *et al.*, 2009).

Antifungal microbes from Bhitarkanika mangrove ecosystem has also been isolated and worked out in details. Sabat and Gupta (2007, 2009b, and 2010) worked out the nutritional factors affecting antifungal acitivity of Penicillium steckii from mangrove origin. Antifungal activity of mangrove fungi against phyllosphere pathogen of rose is also reported (Sabat and Gupta, 2006). Selection and evaluation of antifungal metabolite from mangrove bacteria active against *Verticillium albo-atrum* was done (Gupta *et al.*, 2006a). Antifungal activity of bacteria associated with mangrove trees is also studied (Gupta *et al.*, 2006a,b).

Many microbial cultures isolated from Bhitarkanika mangrove were found to be mineral solublisers (Gupta and Das, 2008). Gupta *et al.* (2010) studied the TCP and rock phosphate solubilzation by mangrove fungi grown under different pH and temperature regime in liquid culture. Evaluation of in vitro solubilization potential of phosphate solubilising *Streptomyces* isolated from phyllosphere of *Heritiera fomes* (mangrove) was done by Gupta *et al.* (2010). Sabat and Gupta (2009a) studied the iron ore solubilzation by *Penicillium restrictum*. Sabat *et al.*, (2008) evaluated the difference in mineral solubilising activity of two strains of *Penicillium steckii* isolated from *Avicennia marina* and *Tamarix troupii* inmangrove ecosystem of Bhitrakanika . Useful enzyme activity of microbes isolated from Bhitarkanika mangrove ecosystem of Orissa coast has also been studied well (Gupta *et al.*, 2007b).

It is clearly evident that mangrove environment is endowed with great microbial diversity. Though, the cultural and physiological phenomenon of mangrove microbes have been reported well, their ecological distribution, functional dynamics and exploratory potential still seems to be properly known and defined.

References

Abdelmonem, A.M., Rasmy, M.R., Prochazkova, Z. and Sutherland, J.R. (1997). *Proceedings of the ISTA Tree Seed Pathology Meeting*, Openo, Czech republic, 9-11 October 1996, 67-73.

AbdelWahab, M. A., Sharouney, H.E. and Jones, E.B.G. (2001). Two new intertidal lignicolous Swampomyces species from Red Sea mangroves in Egypt. *Fungal Diversity*, 8: 35-40.

Agate A.D., Lieth, H. and Al-Masoom, A.A. (1993). Salt tolerant bacteria from the rhizosphere of mangrove plants. Towards the rational use of high salinity tolerant plants Vol: 1. Deliberations about high salinity tolerant plants and ecosystems. *Proceedings of the first ASWAS Conference*, December 8-15,1990,Al Ain, United Arab Emirates:163-166.

Alongi, D.M. (1994). The role of bacteria in nutrient recycling in tropical mangrove and other coastal benthic ecosystems. *Hydrobiologia.*, 285: 19-32.

Alongi, D. M., Pfitzner, J., Trott, L. A., Tirendi, F., Dixon, P., Klumpp, D. W. (2005). Rapid sediment accumulation and microbial mineralization in forests of the mangrove Kandelia candel in the Jiulongjiang estuary, China. *Estuar. Coast. Shelf Sci.* 63(4): 605-618.

Ananda, K., Sridhar, K.R. (2002). Diversity of endophytic fungi in the roots of mangrove species on the west coast of India. *Can. J. Microbiol.,* 48 (10): 871-878.

Ananda, K., Sridhar K. R. (2004). Diversity of filamentous fungi on decomposing leaf and woody litter of mangrove forests in the southwest coast of India. *Curr. Sci.,* 87(10): 1431-1437.

Ando, Y., Mitsugi, N., Yano, K., Hasebe, Y. (2001). Karube Initial fermentation of sea sludge using aerobic and thermophilic microorganisms in a mangrove soil. *Bioresour. Technol.,* 80(1):83-85.

Baker, T.A., Jones, E.B.G. MossLeano, S.T. (2001). Ultrastructure of ascus and ascospore appendages of the mangrove fungus Halosarpheia ratnagiriensis (Halosphaeriales, Ascomycota).*Can. J. Bot., - Revue Canadienne de Botanique,* 79(11): 1307-1317. .

Bashan, Y., Moreno, M., Troyo, E. Holugin, G. (2001). Growth promotion of the seawater-irrigated oilseed halophyte *Salicornia bigelovii* inoculated with mangrove rhizosphere bacteria and halotolerant *Azospirillum* spp. *Biol. Fertil. Soils,* 32(4): 265-272.

Bashan, Y., Holguin, G. (2002). Plant growth promoting bacteria: a potential tool for arid mangrove reforestation Trees. *Struct. Function,* 16(2-3):159-166.

Benka-Coker, M.O. and Olumagin, A. (1996). Effects of waste drilling fluid on bacterial isolates from a mangrove swamp oilfield location in the Niger delta of Nigeria. *Bioresour. Technol.,* 55:175-179.

Chaudhuri, S. and Sengupta, A. (1990). Mycorrhiyal relations of successional stages of mangrove vegetation at the Ganges river estuary in India. In : *Proc. Current Trends in Mycorrhizal Research* (Eds. B.L. Jalali and H. Chand), H.A.U., Haryana : 7.

Chinnaraj S. (1992). Higher marine fungi of Lakshdweep islands and a note on Q. kignalitis. Cryptog. *Mycol.* 13 : 313-139.

Chinnaraj S. (1993) Maglicolous fungi of atolls of Maldeives Indian Ocean. *Ind. J. Mar.* Sci., 22 : 141-142.

Das, S., Lyla, P.S. and Khan, A. (2006). Marine microbial diversity and ecology importance and future prospects. *Curr.Sci.,* 90:1325.

Dhanasekaran, D., Sivamani, P., Aruna girinathan, N., Panneerselvam, A. and Thajuddin, N. (2005). Screening and identification of antibiotic producing strains of marine streptomyces. *J. Microbial. World,* 7: 62-66.

Dhevendaran, K. and Annie, K. (1999). Antibiotic and L-asparaginase activity of *Streptomyces* isolated from fish, selfish and sediment of veli estuarine lake along Kerala coast. *Ind.J.Mar.Sci.,* 28:335-337.

Fisher, P. and Spadling, M.D. (1993). *Protected areas with mangrove habitate*. Draft report. World conservation Centre, Cambridge, U.K. : 60.

Griebler, C., Mindl, B. and Slezak, D. (2001). Combining DAPI and SYBR Green II for the enumeration of total bacterial numbers in aquatic sediments. *International Revi. of Hydrobiol.*, 86(4-5): 453-465.

Guan, S.H., Sattler, I., Lin, W.H., Guo, D.A. and Grabley, S. (2005). p-aminoacetophenonic acids produced by a mangrove endophyte: *Streptomyces griseus J. Natural Products*, 68(8):1198-1200.

Gupta, N., Mishra, S. and Basak, U.C. (2007a). Occurrence of *Streptomyces auranticus* in mangroves of Bhitarkanika. *Malaysian J Microbiol.*, 3(2):7-14.

Gupta, N., Sahoo, P. and Das, S. J (2006a). Selection and evaluation of antifungal metabolite from mangrove bacteria active against *Verticillium albo-atrum. Annals Pl. Protect. Sci.*, 14 (1): 267-69.

Gupta, N., Das, S. and Basak, U.C. (2009a). L asparaginase from fungi of Bhitarkanika Mangrove ecosystem. *Asia Pacific J. Molecul. Biol. Biotechn.*, 17(1): 27-30.

Gupta, N., Sabat, J., Basak, U. C. and Das, P. (2001). Rhizosphere microbial population in some tree mangroves of Bhitarkanika , Orissa. *J. Phytol. Res.* 14 (1):35-37.

Gupta, N., Das, S. and Basak, U.C. (2007b). Useful extracellular activity of bacteria isolated from Bhitarkanika mangrove ecosystem of Orissa coast. *Malaysian J. Microbiol.*, 3(2):15-18.

Gupta, N., Mishra, S. and Basak, U. C. (2009b). Occurrence of different morphotypes of *Streptomyces exfoliates* in mangrove ecosystem of Bhitarkanika, Orissa. India. *J. Culture Collections*, 6:21-27.

Gupta, N. and Das, S. (2008). Phosphate solubillizing fungi from mangroves of Bhitarkanika Orissa. *Hyati J. Biosci.*, 15 (2): 90-92.

Gupta, N., Routray, S., Basak, U. C. and Das, P. (2002a). Occurrence of arbuscular mycorrhizal association in mangrove forest of Bhitarkanika , Orissa , India. *Ind. J. Microbial.*, 42 (3) :247-248.

Gupta, N., Sahoo, D. and Basak, U.C. (2009c). Screening of *Streptomyces* for L asparagianse production. *Academic J. Cancer Res.*, 2 (2) : 92-93. IDOSI publications.

Gupta, N., Sabat, J. and Basak, U. C. (2005). Fungi associated with different plants of Bhitarkanika mangroves in Orissa. *Seshaiyana* , 13 (2) : 5-6.

Gupta, N., Sahoo, N. and Basak, U. C. (2006b). Distribution of Yellow and organge pigmented bacteria in brackish water of Bhitarkanika mangroves from Orissa. *Seshiyana*, 14(1):13.

Gupta, N., Sahoo, N. and Basak,. U.C. (2009d). Agarolytic bacteria from brackish water of Bhitarkanika mangroves, Orissa. *Seshiyana*, 17 (2):8-9.

Gupta, N., Shukla, R. and Basak, U. C. (2005). Antifungal activity of bacteria associated with mangrove trees. *Seshaiyana* , 13 (2):3-4.

Gupta, N., Das, S., Sabat, J. and Basak, U.C. (2007c). Isolation, Identification and growth of *Stachybotrys* obtained from mangrove ecosystem of Bhitarkanika , Orissa. *Int. J. Plant Sci.*, 2(1):64-68.

Gupta, N., Das, S., Sabat, J. and Basak, U.C. (2007d). Studies on *Pestalotiopsis* obtained from mangroves of Bhitarkanika , Orissa. *Int. J. Plant Sci.*, 2(1):90-93.

Gupta, N., Das, S. and Basak, U.C. (2010). TCP and rock phosphate solubilzation by mangrove fungi grown under different pH and temperature in liquid culture. *Int. J. Agricul. Technol.*, 6(3):6(3)421-428.

Gupta, N., Mishra, S. and Basak, U.C. (2009). Diversity of *Streptomyces* in mangrove ecosystem of Bhitarkanika . *Iranian J. Microbiology*, 1 (3) : 37-42.

Gupta, N., Mishra, S. and Basak, U.C. (2009). Microbial population in phyllosphere of mangroves grown in different salinity zones of Bhitarkanika. *Acta Bot. Malicitana*, 34:33-37.

Gupta, N., Basak, U.C. and Das, P. (2002b). Arbuscular mycorrhizal association of mangroves in saline and non saline soils. *Mycorrhiza News*,.13 (4):14-19.

Gupta, N. and Sabat, J. (2008). Isolation, characterization and evaluation of growth of salt tolerant mangrove microbes grown under NaCl stress. *Asian J. Bio Sci.*, 3(1): 84-86.

Gupta, N., Sahoo D. and Basak U.C. (2010). Evaluation of *in vitro* solubilization potential of phosphate solubilising *Streptomyces* isolated from phyllosphere of Heritiera fomes (mangrove), *Afr. J. Microbiol. Res.*, 4(3): 136-142.

Hartung, F., Worner, R., Hoque, M.I., Alam, S. S., Khan, S., Paul, A.R. and Muhlbach, H.P. (1998). Association of Phytopathogenic bacteria with top dying disease of Sundri tree (Heritiera fomes) in Bangladesh. *Angewandte-Botanic*, 72:48-55.

Hatano, K. (1997). Actinomycete population in mangrove rhizospheres. *IFO Res Commun.* 18: 26-31.

Holguin, G., Guzman, A. and Bashan, Y. (1992). Two nitrogen fixing bacteria from the rhizosphere of mangrove trees;their isolation,identification and invitro interaction with rhizosphere staphylococcus sp. *FEMS Microbiol.Ecol.*, 101:207-216.

Holguin, G. and Bashan, Y. (1996). Nitrogen fixation by *Azospirillum brasilense.* Cd is promoted when co-cultured with a mangrove rhizosphere bacterium (*Staphylococcus* sp.). Soil *Biol. Biochem.*, 28:1651-1660.

Imada, A., Igarasi, S., Nakahama, K. and Isono M. (1973). Asparaginase and Glutaminase activities of microorganism. *J. Gen. Microbiol.*, 76 : 85-89.

Inderbitzin, P. and Desjardin, D.E. (1999). A new halotolerant species of *Physalacria* from Hong Kong. *Mycologia*, 91(4): 666-668.

Jones, E. B. G., Abdel, M. A. and Wahab, S.A. (2005). Marine fungi from the Bahamas Islands. *Bot. Mar.*, 48(5-6): 356-364.

Jones, E. B. G., Abdel, M.A., Wahab, S. A., Alias and Hsieh, S. Y. (1999). *Dactylospora mangrovei* sp. nov. (Discomycetes, Ascomycota) from mangrove wood. *Mycoscience*, 40: 317-320.

Kathiresan, K . (2000). A review of studies on Pichavaram mangrove, southeast India. *Hydrobiologia*, 430 (1-3): 185-205.

Kathiresan, K. and Selvam, M. M. (2006) Evaluation of beneficial bacteria from mangrove soil. *Bot. Marina*, 49(1) 86-88.

Klekowski, E.J., Correder, J.E., Lowenfold, R., Klekowski, E.H. and Morell, J.M.(1994). Using mangroves to screen for mutagens in tropical marine environments. *Mar.Pollut.Bull.*, 28:346-350.

Kothamsi, D., Kothamasi. Shalini, Bhattacharya, A., Khuhad, R.C. and Babu, C. R. (2006). Arbuscular mycorrhiza and phosphate solubilising bacteria of the rhizosphere of the mangrove ecosystem of great nicobar islands India. *Biol. Fertil. Soils*, 42: 358-361.

Kristensen, E., Jensen, M.H., Banta, C.T., Hansen, K., Holmer, M. and King, G.M.(1998). Tranformation and transport of inorganic nitrogen in the sediments of a south-east asian mangrove forest. *Aqua.Microb.Ecol.*, 15:165-175.

Kumaresan, V. and Suryanarayanan, T.S. (2002) Endophyte assemblages in young, mature and senescent leaves of *Rhizophora apiculata*: evidence for the role of endophytes in mangrove litter degradation. *Fungal Diversity*, 9: 81-91.

Kumaresan, V. and Suryanaryan, T. S. (2001) Occurrence and distribution of endophytic fungi in a mangrove community. *Mycol. Res.*, 105(11): 1388-1391.

Lescic I., Zehl M., Muller R., Vukelic B., Abramic M., Pigac J., Allmaier G. Kojicprodic B. (2004). Structural characterization of extra cellular lipase from *Streptomyces rimosus* : assigment of disulphide bridge pattern by mass spectrometry. *Bio. Chem.*, 385 : 1147-1156.

Li, C.Y., Cheng, C.Y. and Chen, T.L. (2001). Production of *Acinetobacter radioresistens* lipase using Tween-80 as the carbon source. *Enz. Microb.Technol.*, 29:258-263.

Mann, E.D. and Steinke, T.D. (1993). Biological nitrogen fixation (acetylene reduction) associated with blue green algal (cyanobacterial community) in the Beachwood Mangrove Nature Reserve -II. Seasonal variation in acetylene reduction activity. *South Afr. J.Bot.*, 59:1-8.

Mishra, S., Basak, U.C. and Gupta, N. (2009). Characterization of *Streptomyces nogalator* isolated from some tree mangroves. *Inter. J. Applied Environ. Sci.*, 4(3):277-280.

Mishra, S. (2009). *Morphological, Biochemical and Physiological Characterization of Streptomyces Isolates from Mangrove Environment of Orissa*. Ph.D. thesis, Utkal University, Bhubaneswar, Orissa.

Mohanraju, R., Rajagopal, B.S., Daniels, L. and Natarajan, R. (1997). Isolation and characterization of a methanogenic bacterium from mangrove sediments. *J. Mar. Biotechnol.*, 5:2-3.

Namsaraev, B.B., Mitskevich, I.N., Miravet, M.E., Lukhioio, M. and Beiota, M. (1992). *Microbiological Processes in Cuban Mangroves*. Microbiology-New York., 61:92-98; translated from Mikrobiologiya., 61:123-130.

Obojska, A.B. and Leiczak, M. Kubrak (1999). Degradation of phosphonates by Streptomycete isolates. *Appl. Microbiol. Biotechnol.* 51:872-876.

Panchanadikar, V.V. (1993). Microbial activity in mangrove ecosystems of Goa and Konkan. *Int.J.Env.Stud.*, 43:297-305.

Patil, S. D. and Borse, B. D. (1985). Marine fungi from Indian mangroves. *The Mangroves : Proc. Nat. symp. Biol. Util. Mangroves.* (Ed.) L.J. Bhosle. : 151-152.

Poonyth, A.D., Hyde, K.D. and Peerally, A. (1999). A Intertidal fungi in Mauritian mangroves *Bot. Marina*, 42 (3): 243-252.

Pradhan, G. S. and Gupta, N. (2008). Studies on L asparaginase from *Streptomyces* sp. isolated from Bhitarkanika mangroves. *Proc. Symposium on Wetland and Mangrove Biodiversity in Orissa Coast,* RPRC, Bhubaneswar, (Eds.) N. Gupta and A. K. Mohapatra: 64-66.

Prasannarai, K. and Sridhar, K.R. (2001).Diversity and abundance of higher marine fungi on woody substrates along the west coast of India. *Curr. Sci.*, 81 (3): 304-311.

Purkayastha, R. P. and Pal, A. K. (1996). New foliicolous fungi from Indian mangroves (Sunderbans). *Ind. Phytopath.* 49(1):9-21.

Rao C.S.V.R. and Krishnamurthy K. (1994). Chemoheterotrophy in the mangrove environment. *Curr. Sci.*, 66:382-385.

Rapp, P. and Backhaus, S. (1992). Formation of extracellular lipases by filamentous fungi, yeasts and bacteria. *Enz. Microb. Technol.*, 14: 938-943.

Ravikumar, S., Ramanathan, G., Suba, N., Jeyaseeli, L. and Sukumaran, M. (2002). Quantification of halophilic *Azospirillum* from mangroves. *Ind. J. Mar. Sci.* 31(2): 157-160.

Ravishankar, T., Navamunlyammal, M.L., Gnanappazham, S., Nayak, G.C., Mahapatra and Selvam, V. (2004). *Atlas of mangrove wet lands of India Part -3 Orissa.* Publisher M.S. Swaminathan Research Foundation, Chennai.

Rojas, A., Holguin, G., Glick, B.R., Bashan, Y. (2001). Synergism between *Phyllobacterium* sp (N_2-fixer) and *Bacillus licheniformis* (P-solubilizer), both from a semiarid mangrove rhizosphere FEMS *Microbiol. Ecol.*, 35(2): 181-187.

Sabat, J. and Gupta, N. (2010). Nutritional Factors affecting the antifungal activity of *Penicillium steckii* of mangrove origin. *Afr. J. Microbiol. Res.*, 4(3) : 126-135.

Sabat, J. and Gupta, N. (2006). Antifungal activity of mangrove fungi against phyllosphere pathogen of rose. *Annals Plant Protec. Sci.*, 14(2):490- 491.

Sabat, J. and Gupta, N. (2009a). Iron ore solubilzation by *Penicillium restrictum* . Effect of carbon and incubation periods. *American-Eurasian J. Agron.*, 2 (1): 43-44.

Sabat, J. and Gupta, N. (2007). Antifungal activity of *Penicillium steckii* under different nutritional conditions. *J. Plant Dis. Sci.*, 2(1): 111-112.

Sabat, J. and Gupta, N. (2009b). Development of modified medium for the enhancement in antifungal activity of *Penicillilum steckii* (MF1 mangrove fungi) against *Verticillium* wilt pathogenic fungi of rose. *Braz. Arch. Biol. Technol.*, 52(4): 808-816.

Sabat, J., Basak, U.C. and Gupta, N. (2008). Difference in mineral solubilising activity of two strains of *Penicillium steckii* isolated from *Avicennia marina* and *Tamarix troupii* of mangrove ecosystem of Bhitrakanika . *Proc. symposium on wetland and mangrove biodiversity in Orissa Coast,* RPRC, Bhubaneswar, (Eds.) N.Gupta and A. K. Mohapatra: 49-51.

Sahu, M. K., Poorani, E., Sivakumar, K., Thangaradjou, T. and Sivakumar, K., (2007) Partial purification and anti-leukemic activity of L-asparaginase enzyme of the actinomycetes strain LA-29 isolated from the estuarine fish, Mugil cephalus (Linn.) *J. Environ. Biol.,* 28(3):645-650.

Sardessai, Y.N. and Bhosle, S. (2002). Organic solvent-tolerant bacteria in mangrove ecosystem. *Curr. Sci.,* 82, (6): 622-623.

Sarma, V. V., Hyde, K. D. and Vittal, B. P. R. (2001). Frequency of occurance od mangrove fungi from the east coast of India. *Hydrobiologia,* 455:41-53.

Sengupta, A. and Chaudhuri, S. (2002). Arbuscular mycorrhizal relations of mangrove plant community at the Ganges river estuary in India. *Mycorrhiza,* 12 (4):169-174.

Sengupta, A. and Chaudhuri, S. (1991). Ecology of heterotrophic dinitrogen fixation in the rhizosphere of mangrove plant community at the Ganges river estuary in India. *Occologia.,* 87:560-564.

Seshadri, S., Ignacimuthu, S. and Lakshminarsimhan, C. (2002). Variation of heterotrophic and phosphate solubilizing bacteria from Chennai Southeast coast of India. *Ind. J. Mar. Sci.,* 31: 69-72.

Sharma, J., Kumar, R. and Singh, A. (2004). Production of alkaline lipase by *Aspergillus fumigatus* using solid-state fermentation of products obtained from grain processing. *Ind. J. Microbiol.,* 44: 141-143.

Shome, R. and Shome, B.R. (2001). Microbial L-asparaginase from mangroves of Andaman Islands. *Ind.J. Mar. Sci.,* 30:183-184.

Sivakumar K., Sahu, M.K. and Kathiresan, K. (2005). Isolation and characterization of Streptomyces producing antibiotic from a mangrove environment. *Asian J. Microbiol. Biotechnol. Environ. Sci.,* 7: 457-464.

Sztajer, H., Maliszewska, I. and Wieczorek, J. (1988). Production of exogenous lipases by fungi, bacteria and actinomyces. *Enz. Microb. Technol.,* 10: 492-497.

Takeuchi, M. and Hatano, K. (1999). Phylogenetic analysis of actinobacteria in the mangrove rhizosphere. *IFO Res Commun.,* 19: 47-62.

Takeuchi, M. and Hatano, K. (1998). Gordonoa rhizosphere sp. nov. isolated from the mangrove rhizosphere. *Inter. J. Syst. Bacteriol.,* 48: 907-912.

Toledo, G., Bashan, Y. and Soeldner, A. (1995). Cyanobacteria and black mangrovesin northwestern Mexico: colonization and diurnal and seasonal nitrogen fixation on aerial roots. *Can.J.Microbiol.,* 41:999-1011.

Vazquez, P., Holguin, G., Puente, M. E., LopezCortes, A. and Basan, Y. (2000). Phosphate solublizing microorganisms associated with the rhizosphere of mangroves in a semiarid coastal lagoon. *Biol. Fertil. Soil.* 30(5-6): 460-468.

Vethanayagam, R.R.(1991). Purple photosynthetic bacteria from a tropical mangrove environment. *Mar.Biol.*, 110:161-163.

Vishwakiran, Y., Thakur, N. L ., Raghukumar, S., Yennawar, P. L . and Anil, A. C. (2001). Spatial and temporal distribution of fungi and wood-borers in the coastal tropical waters of Goa, India . *Bot. Marina*, 44(1): 47-56.

Walsh, G. E. (1974). *Mangrove: A review ecology of halophytes* . (Eds.) R. J. Reinhlod and W. H. Queen, New York. Academic Press: 51-174.

Wang Wu-Rong. (1993). Studies on the microbial flora of the Tansui river Mangrove forest. Institute of Botany. *Academia- Sinica- Monogra.*

Widiastuti, H., Darmono, T.W. and Hadioctomo, R.S. (1996). Characteristics of selected Bacillus thuringiensis isolates from Indonesia and their toxicity to Hyposidra talaca. *Monara -Porkobunan.*, 64: 65-78.

Woitchik, A.F., Ohowa, B., Kajungu, J.M., Rao, R.G., Goeyens, L. and Dehairs, F. (1997). Nitrogen enrichment during decomposition of magrove leaf litter in an east African coastal lagoon (Kenya): relative importance of biological nitrogen fixation. *Biogeochemistry*, 39:15-35.

□□□

Microbial Diversity and Functions, 2012
© D.J. Bagyaraj, K.V.B.R. Tilak, H.K. Kehri (eds.), pp. 291-307
New India Publishing Agency, New Delhi (India)
E-mail : info@nipabooks.com; Website : www.nipabooks.com

Chapter 12

Fungi of Indoor Environment - A Qualitative Study

Deepmala Pandey and Nisha Misra

ABSTRACT

Many airborne fungal spores are important as allergens, human and plant pathogens, in the bio-deterioration of stored materials like paper, leather, textile, paintings, historical monuments and spoilage of food stuffs. Surveys have been conducted to elucidate the magnitude of the problem. Indoor environment also plays an important role in the precipitation of allergic symptoms caused by fungal spores. The present review briefly highlights some of the qualitative investigations carried out on indoor environments like bakery, building, cold storage, cow shed, dormitory, educational institutions, garbage disposal plant, ware house, hospital, library, food grain godown, leather godown, house dust, poultry house, chick hatcheries, plastic industries and textile industries.

Keywords : Mycoflora, indoor environment

Introduction

The air near the earth's surface contains a rich population of spores of micro-organisms and pollen. This constitutes the air spore flora or 'air spora'. Hippocrates, the father of medical sciences who was aware of the fact that men were affected by epidemic fever when they inhaled air with such pollutants as are hostile to the human race. Probably it was the first instance which gave an idea that, atmosphere is a corridor and is not home for micro-organisms.

Modern aerobiology has its roots in the experiments of Spallanzini (1765) and Pasteur (1861). Their experiments disapproved the theory of spontaneous generation and gave foundation of germ theory of diseases.

The existing knowledge on the composition of the airspora started from 1872 when Ehrenburg first published information on the micro-organisms present in the dust of the atmosphere.

The term aerobiology came into use during 1930 as a collective term for studies in air spora, like airspora fungus spores, pollen grains and other micro-organisms.

Jacob (1951) elaborated the term to include dispersion of insect populations, fungal spores, pollen, bacteria and viruses. Edmonds (1973) defined aerobiology as a scientific and multidisciplinary approach focused on the transport of organisms and biologically significant materials.

With the inception of International Biological Programme (IBP) in 1964, the term has been further extended to include investigations of airborne materials of biological significance. The term thus embraces not only the study of microbes (Protozoa, Fungi, Algae, and minute Insects such as aphids) but also pollution by gases that exert specific biological effects.

Among the important particulate in aerobiology much consideration is given to smoke, dust, radionucleids and pesticides.

Modern aerobiological approach also takes consideration to man made pollutants such as gases which affect adversely the *via*bility of micro-organisms in the air and on the soil. Therefore pollutant gases and particles are also included in the aerobiological studies. The most important gases which affect organisms adversely are hydrogen sulphide (H2S), sulphur dioxide (SO2), carbon mono-oxide (CO), chlorine (Cl2), freon, hydrogen fluoride, ozone (O_3) and other photochemical pollutant such as PAN.

Aerobiology is concerned with the SRDDIBI *i.e.* Source, Release, Dispersion, Deposition, Identification, Behaviour and Impaction of organisms or materials and their effect on animals, human and plant systems. The environmental factors affect each of these stages *i.e.* SRDDIBI.

Aerobiology has been studied in another ways *i.e.* when pollutant gases and particles are included in aerobiological studies there is interaction of spores with these pollutant gases and particles then aerobiology is considered as atmospheric ecology or aeroecology. Likewise when the pollutant gases and micro-organisms of aerospora are studied with plant and human beings then this is considered as aeroplant pathology etc.

Aerobiological investigations has been broadly distinguished as-

(I) Outdoor or Extramural aerobiological investigations and

(II) Indoor or Intramural aerobiological investigations

The study of contamination in a closed system like buildings, hospitals, glass houses or industrial environment with reference to airborne microbial contaminants comes under indoor or intramural aerobiology as against the outdoor aerobiology which is concerned with the survey of biological material in open space like fields and forests.

The above said two types of aerobiological investigations have been carried out in different ways. These are as follows:

Botanical aerobiology deals with dispersal of micro-organisms causing plant diseases, dispersion of pollen causing allergenic reactions on animals or human beings (Allergology), and dispersion of seeds in an open or outdoor environment. On the other hand dispersion and impact of microbes in green houses, caves, glass houses, classrooms of educational institutions, grain storage godowns, temple, library etc. deals with closed or indoor environment. Thus Botanical aerobiology is either outer or indoor Botanical aerobiology.

Medical aerobiology deals with the influence of pollen, fungal spores, mites and dust on human beings and animals (allergology) and transmission of bacteria and viruses both in closed (indoor) as well as open (indoor) environments.

Technical or Industrial aerobiology is concerned with the influence of air pollutants on the environments *i.e.* in case of indoor environment with the influence of dust on human beings.

Experimental aerobiology deals with fundamental concepts like mathematical formations, development of methods, instrumentation and the methods of modelling of aerobiological systems. Most of the experimental aerobiology is carried out in indoors or in wind tunnels.

The fungal fragments and spores of air particularly airborne conidia and ejected ascospores and basidiospores adapted for aerial dispersal constitute the fungal component of indoor aerospora.

The number and types of fungal spores in indoor air depend on air exchange with the outside and the presence of indoor sources. Without such sources, the air spora will consists of the same species as outdoors in the same proportion, but their numbers are usually smaller. Spores may come from many sources within buildings. They may come from fungi growing in condensation of walls, food, mouldy stored products or spores may accumulate in house dust. They may become dispersed by human activity within the buildings.

In indoor with natural ventilation to the ambient atmosphere. The seasonally occurring pollen and spores from the air outside are present, but in very much lower concentrations (Davies, 1960). Addition of microbes coming in through the window, indoor air commonly has other derived materials from domestic sources. For example, it may have contaminated droplet nuclei, the minute residues left by evaporation of droplets expelled during sneezing or coughing (Gregory, 1973). Thus the microbial flora, of the indoor air depends on the number and kinds of organisms present and the mechanical movement within the enclosed space (Madeline and Linton 1974). The air inside the building is continuously contaminated with micro-organisms from human or animal occupants (Baruah, 1961; Sreeramulu, 1961).

Studies on mycoflora of indoor air are relatively few in India as those on outdoor air.

The present review briefly highlights some of the investigations carried out on indoor or intramural mycoflora. The first record on this aspect is the report of Cunningham (1873), who analysed micro-organisms of air of presidency jail at Calcutta, India. In the middle of 20th century Gregory published a review article on dispersion of air borne spores in 1945; Durham, 1942; Hyde and Williams, 1946 and Nilsby, 1949 on Allergy.

From the literature available it may be noted that there are several kinds of indoor environment eg. wards of hospitals, libraries, ware houses, classrooms, poultry houses, industries, glass houses, caves, cinema hall, archives, leather godown, cowshed, dormitory, cold storage, bakery, temples, factories, godown etc. The following table-1 summarises fungi reported from different indoor air environments by Singh, 1993; Pugalmaran and Vittal, 1997, 1999; Vittal and Rasool, 1995; Raghu and Vittal, 1992; Mani *et al.*, 1998; Vittal and Chinnaraj 1990; Vittal and Leela Glory, 1984; Shukla, 2002; Davies, 1960; Baruah, 1960; Sreeramulu, 1961; Frey *et. al.*, 1962; Noble and Clayton, 1963; Chute and Barden, 1964; Gravesen, 1972 and 1978; Tilak and Vishwe, 1975, 1976; Lumpkins and Corbit, 1976; Alteras and Lehrer, 1977; Marjut, 1977; Pohjola *et al.*, 1977; Tilak and Chakre, 1977; Burge *et al.*, 1978; Jayaprakash *et al.*, 1978; Levetin and Hurewitz, 1978; Rati and Ramalingam, 1979; Rati *et al.*, 1980; Tilak *et al.* 1980; Narania and Reddy, 1981; Singh, 1981; Calvo *et al.*, 1982; Vittal *et al.*, 1984; Banerjee *et al.*, 1987; Tripathi, 1987; Tilak Pillai, 1988; Singh *et al.*, 1989; Li and Kendrick, 1995; Nanda *et al.*, 1997; Arundhati *et al.*, 1998; Cheong *et al.*, 2004; Wu *et al.*, 2005 and Sandra *et al.*, 2009.

Table 1 : Fungi isolated from indoor air of different environments

Indoor Environment	Fungi isolated
Bakery	*Mucor racemosus, Rhizopus nigricans, Aspergillus flavus, A. fumigatus, A nidulans, A. niger A. terreus, A. versicolor, Drechslera australiensis, Gliocladium roseum, Penicillium citrinum, P. chrysogenum, Trichoderma viride.*
Building	*Cladosporium sp., Penicillium sp.*
Cold storage	*Mucor varians, Aspergillus sp.*
Cowshed	*Aspergillus flavus, A. fumigatus, Alternaria, Cladosporium sp., Curvularia sp.*
Dormitory	*Mucor recemosus, Rhizopus nigricans.*
Educational institutions	*Absidia sp., Acremonium rutilum, Alternaria alternata, A. brassicicola, A. carthami, A. cheiranthi, A. dianthi, A. dianthicola, A. fasciculata, A. geophila, A. humicola, A. longipes, A. radicina, A. sonchi, A. tenuissima, Alternaria sp., Arthroderma uncinatum, Aspergillus candidus, A. clavatus, A. flevipes, A. flavus, A. humicola, A. janus, A. koningi, A. nidulans, A. niger, A. ochraceous, A. phoenicis, A. sachari, A. sydowi, A. terreus, A. terricola, A. ustus, A. versicolor, Chaetomium cristatum, C. globosum, C. homophilatum, Chaetomium sp., Chrysosporium merdarium, Cladosporium herbarum, C. oxysporum, Curvularia geniculata, C. lunata, C. Oryzae, C. ovoidea, C. pallescens, C. trifoli, Curvularia sp., Drechslera australiensis, D. biseptata, Epicoccum nigrum Fusarium moniliforme, F. oxysporum, F. roseum, F. semitectum, Helminthosporium velutinum, Humicola fuscoatra, H.grisea, Monilia sp., Paecilomyces varioti, Penicillium chrysogenum, P.citrinum, P.islandicum, .P.italicum, P. ochraceum, P.oxalicum, Periconia sp., Rhizopus nigricans, Scopulariopsis, sp., Spegazzinia sp., Stemphylium sp., Syncephalastrum racemosum, S. Verruculosum, Trichothecium roseum, Ulocladium* sp., Sterile mycelium.
Garbage disposal plant	*Penicillium sp., Aspergillus sp., Cladosporium sp.*
Ware house	*Penicillium canescens Rhizopus stolonifer, Alternaria alternata, Curvularia, lunata (Cochliobolus lunatus), Cladosporium oxysporum, Aspergillus flavus, A. niger, Penicillium, Helminthosporium sp.*
Hospital	*Absidia glauca, Mucor racemosus, Rhizopus nigricans, Syncephalastrum racemosum, Alternaria alternata. Aspergillus flavus, A. fumigatus, A. japonicus, A. nidulans, A. niger, A. terreus, A. versicolor, Cladosporium cladosporioides, Curvularia lunata,*

	C. pallescens, Drechslera australiensis, D. hawaiiensis, Gliocladium roseum, Penicillium citrinum, P. chrysogenum, Pithomyces atroolivaceous, Stachybotrys parvispora.
Library	*Absidia glauca, Cunninghamella echinulata, Circinella simplex, mucor, hiemalis, Rhizopus nigricans, Syncephalastrum racemosum, Corynascus sepedonium, Alternaria tenuissima, A. padwicki, Aspergillus awamori, A. flavus, A. fumigatus, A. heteromorphus, A. japonicus, A. nidulans, A. niger, A. oryzae, A. sulphureus, A. terreus, Aureobasidium pullulans, Cladosporium cladosporioids, Corynespora cassiicola, Curvularia lunata, C. tuberculata, C. brachyspora, Drechslera halodes. D. hawaiiensis, D. australiensis, Fusarium oxysporum, F. semitectum, F. lateritium, Histoplasma capsulatum, Monodictys levis, Myrothecium verrucaria, Microsporum gypseum, Nigrospora sphaerica, Pestalotia sp., Penicillium purpurogenum, P. citrinum, P. rugulosum, P. implicatum, P. herquei, Trichoderma viride, Mucor racemosus, Alternaria alternata, Aspergillus versicolor, Chrysosporium sp., Cladosporium herbarum, Curvularia pallescens, Gliocladium roseum, Penicillium chrysogenum, Trichoderma harzianum, Periconia, Helminthosporsum, Bispora, Torula, Chaetonium, Stemphylium, Alternaria tenuis.*
Food grain Godown	*Physarum, Albugo, Choanophora, Mucor, Physoderma,* sp. of *Rhizopus, Sclerospora, Bitrimonospora, Botryosphaeria, Calospora, Chaetomium, Claviceps, Didymosphaeria, Erysiphe, Hypoxylon, Hysterium, Lepeosphaeria, Lophiostoma, Massaria, Meliola, Neurospora, Nodulosphaeria, Otthia, Parodiella, Passerinella, Pleospora, Podospora, Pringsheimia, Rosellina, Sordaria, Sporormia, Thielavia, Ganoderma, Smut spores, Teleutospores, Urediniospores, Alternaria, Aschersonia, Ascochyta, Aspergilli, Asperisporium, Asteromyces, Belanium, Beltrania, Beltraniella, Bipolaris, Bispora, Botryodiplodia, Botryosporium, Botrytis, Botrytrichum, Brachysporium, Camposporium, Ceratosporium, Cercospora, Cercosporella, Cercosporidium, Chaetomella, Cladosporiella, Cladosporium, Cordana, Corynespora, Curvularia. Deightoniella, Dendrographum, Dendryphion, Dendryphiopsis, Dictyoarthrinium, Dictyosporium, Diplocladiella, Diplodia, Diplodina, Drechslera, Endocalyx, Epicoccum, Excipularia, Exosporium, Fusarium, Fusicladium, Fusoma, Gilmaniella, Hadrotrichum, Haplographium, Haplosporella, Helicoma, Helicosporium, Helminthosporium, Hendersonula, Hendersonia, Heterosporium, Hirudinaria, Humicola, Lacellina, Lacellinopsis, Mammaria, Melancomium, Memnoniella, Microsporum, Nigrospora, Oidium, Papularia, Penicilli, Periconia, Periconiella, Pestalotia, Phoma, Phragmotrichum, Pithomyces,*

Pleurophragmium, Pleurothecium, Podosporium, Pseudobotrytis, Pseudotorula, Pyricularia, Sadasivania, Scolecosporium, Sclerotinia, Sclerotiopsis, Sclerotium, Sirodesmium, Sirosporium, Spegazzina, Sphaeropsis, Spicaria, Spondylocladiella, Sporidesmium, Steganosporium, Stemphylium, Stigmella, Tetrasporium, Tetracoccos porium, Tetraploa, Trichoderma, Trichophyton, Trichothecium, Torula, Tricocladium, Ulocladium, Ustiloginoidea, Zygosporium, Apiorynchostoma, Emericella, Coprinus, Monodictys, Thermomyces, Trichoconis, Absidia, Corymbifera, Syncephalastrum, racemosum, Alternaria alternata, Aspergillus candidus, A. clavatus, A. flavus, A. fumigatus, A. glaucus, A. japonicus, A. nidulans. A. niger, A. ochraceus, A. terreus, A. tamari, A. versicolor, Cladosporium, cladosporioides, Curvularia lunata, Curvularia pallescens, Drechslera, hawaiiensis, Fusarium oxysporum, Monilia sitophila, Nigrospora sphalrica, Paecilomyces variotii, Penicillium chrysogenum, P. citrinum, P. frequentans, P. oxalicum, P. purpurogenum, Trichoderma viride.

Leather godown

Absidia corymbifera, Mucor racemosus, Rhizopus stolonifer, Alternaria alternata, Aspergillus candidus, A. cervinus, A. clavatus, A. flevipes, A. flavus, A. fumigatus, A. glaucus, A. japonicus, A. nidulans, A. niger, A. ochraceus, A. oryzae, A. terreus, A. versicolor, Aureobasidium pullulans, Candida albicans, Cladosporium cladosporioides, Curvularia lunata, Drechslera hawaiiensis, Fusarium oxysporum, Monilia sp., Paecilomyces variotii, Penicillium citrinum, P. chrysogenum, P. expansum, P. funiculosum, P. janthinellum, P. oxalicum, Scopulariopsis brevicaulis, Trichoderma harzianum, Circinella simplex, Cunninghamella echinulata, Rhizopus nigricans, A. sydowi, Cladosporium oxysporum, Curvularia pallescens, Penicillium implicatum, P. javanicum.

House dust

Mucor mucedo, M. plumbeus, M. racemosus, Rhizopus stolonifer, Syncephalastrum racemosum, Chaetomium globosum, Acremonium sp., Alternaria alternata, Aspergillus cervinus, A. candidus, A. flavipes, A. flavus, A. fumigatus, A. japonicus, A. nidulans, A. niger, A. sydowi, A. terreus, A. versicolor, Cladosporium cladosporioides, C. herbarum, C. lunata, C. pallescens, C. tuberculata, Drechslera hawaiiensis, Fusarium lateritium, F. roseum, F. semitectum, Penicillium chrysogenum, P. citrinum, P. expansum, P. funiculosum, P. implicatum, Scopulariopsis brevicaulis, Stachybotrys parvispora, Sporothrix sp., Trichoderma viride, Phoma, Helminthosporium, Cephaliosporium, Nigrospora, Epicoccum, Leptosphaeria, Ganoderma, Coprinus, Epicoccum.

Poultry houseand chick hatcheries	*Absidia sp., Corymbifera sp., Mucor recemosus, Rhizopus stolonifer, R. nigricans, Acremonium sp., Alternaria alternata, Aspergillus candidus, A. carneus, A. cervinus, A. flevipes, A. flavus, A fumigatus, A. glaucus, A. japonicus, A nidulans, A. niger, A. ornatus, A. oryzae, A. terreus, A. versicolor, Aspergillus sp., Candida albicans, Cladosporium cladosporiodies, Curvularia lunata, Drechslera hawaiiensis, Fusarium oxysporum, Humicola sp., Monilia sp., Paecilomyces variotii, Penicillium chrysogenum, P. citrinum, P. expansum, P. funiculosum, Scopulariopsis brevicaulis, Trichothecium roseum, Cunninghamella echinulata, Mucor racemosus, Aspergillus wentii, Cephalosporium sp. , Cladosporium herbarum, Drechslera australiensis, Fusarium semitectum, Gliocladium roseum, Humicola fuscoatra, Pithomyces atroolivaceous.*
Plastic Industries	*Acremonium butyri, Alternaria alternata, A. brassicicola., A cheiranthi, A. dianthi, A. dianthicola, A. longipes, A. padwickii, A. radicina, A. raphani, A sonchi, A. tenuissima Alternaria sp., Aspergillus clavatus, A. flavipes, A. flavus, A. fresenii, A. fumigatus, A. funiculosus, A. humicola, A. janus, A. koningi, A. nidulans, A. niger, A. ochraceous,A. oryzae, A. parasiticus, A. phoenicis, A repens, A. sachari, A. sulphureus, A. sydowi, A. terreus, A terricola. A. unguis, A. variecolor, A. versicolor, A. violaceo-fuscus, A. wentii, Aspergillus sp. Beltrania sp., Botryotrichum piluliferum, Botryotrichum sp., Botrysporium sp(2). Botrytis cinerea, Brachysporium pendulisporum, Chaetomium eaprinum., C. funicola, C. globosum, C. homophilatum, Chrysosporium pannorum, Cladosporium cladosporoides, C. herbarum, C. lignicolum, C. oxysporum, C. resinae, Cordama pauciseptata, Cryptomela, Cunninghamella bertholletiae, C. echinulata, C. verticillata, Curvularia andropogonis, C. borreriae, C. clavata, C. inaequalis, C. lunata, C. oryzae, C. ovoidea, C. pallescens, C. prasadii, C. richardiae, C. senegalansis, C. stapeliae, C. trifolii. Drechslera, australiensis, D. biseptata, D. hawailensis, D. indica, D. iridis, Fusarium javanicum, F. moniliforme, F. oxysporum, F. roseum, F. semitectum, F. solani, Gonatobotryum apiculatum, Helminthosporium sp., Humicola grisea, Memnoniella echinata, Monilia brunnea, Monilia geophila, Penicillium atramentosum, P. chrysogenum, P. citrinum, P. notatum., ochraceum, P. oxalicum, Rhizopus nigricans, rhizopus sp., Spegazzinia intermedia, S. parkeri, S. tessarthra, Spegazzinia sp., Sporotrichum carnis, Sporotrichum sp., Stachybotrys lobulata, Stemphylium piriforme, Syncephalastrum racemosus, S. verruculosum, Thermomyces lanuguinosus, T. stellatus, Torula herbarum, Torula sp., Torulomyces sp., Trichothecium roseum, Ulocladium consortiale.*

Textile Industries	*Acremoniella atra, Acremonium butyri, Alternaria alternata, A. brassicicola, A. carthami, A. cheiranthi, a. dennisii, A. dianthi, A. longipes, A. radicina, A. raphani, A sonchi, A. tenuissima, Alternaria sp. Aspergillus amestelodami, Aspergillus clavatus, A. flavus, A. fresenii, A. fumigatus, A. funiculosus, A. humicola, A. janus, A. koningi, A. nidulans, A. niger, A. ochraceous, A. oryzae, A. parasiticus, A. phoenicis, A. repens, A. sachari, A. sulphureus, A sydowi, A. terreus, A. terricola, A ustus, A variecolor, A. versicolor, A violaceo-fuscus, A. wentii, Beltrania sp., Botryosporium sp.(2), Botrytis cineria, Cephalosporium asperum, Chaetomium funicola, C. globosum, Chaetomium sp., Chrysosporium merdarium, Cladosporium cladosporioides, C. herbarum, C. lignicolum, C. oxysporum, C. resinae, Cunninghamella bertholletiae, C. echinulata, C. verticillata, Curvularia andropogonis C. borreriae, C. clavata, C. geniculata, C. inaequalis, C. lunata, C. ovoidea, C. pallescens, C. penniseti, C. prasadii, C. protuberata, C. richardiae, C. trifolii, C. tuberculata, Drechslera australiensis, D. biseptata, D. dematioidea, D. indica, D. iridis, Fusarium bostrycoides, F. moniliforme, F. oxysporum, F. roseum, F. semitectum, Gonatobotryum apiculatum, Humicola nigriscens, Monilia brunnea, M. geophila, Mucor racemosus, M. sphaerosporus, Paecilomyces silvatica penicillium chrysogenum, P. citrinum, P. italicum, P. nigricans, P. notatum, P. ochraccum, P. oxalicum, Phoma sp., Pithomyces sp., Rhizopus nigricans, Rhizopus sp., Spegazzinia intermedia, S. parkeri, S. tessarthra, Spegazzinia sp., Syncephalastrum racemosum, S. verruculosum, Thermomyces lanuginosus, T. stellaltus, Torula gramininis, T. herbarun, Ulocladium Botrytis,* sterile mycelium.

Fungal flora occurring in air has been studied by various methods *e.g.* Petridish exposure method (gravity method) using Martin's peptone-dextrose agar, (Martin, 1950), nutrient plate, silicene, slide, culture plate technique, adhesive coated slides and Petridishes, Andersen sampler, Burkard Personal sampler, Rotorod sampler, Slit sampler (Pady and Kapica, 1956; Burge *et al.*, 1978; Raghu and Vittal, 1992; Pugalmaran and Vittal, 1997, 1999; Mani *et al.*, 1998; Shukla, 2002; Vittal *et al.*, 1990; Vittal and Rasool, 1995; Vittal and Glory, 1984; Noble and Clayton, 1963; Tilak Vishwe, 1975; Bhatia and Gaur, 1979; and Frey and Durie, 1962).

Petridish exposure method (gravity method) provides information on the relative abundance of numerous genera, particularly unspecialized saprophytes and of their numerical fluctuations (Turner, 1966) and on the number and nature of all *via*ble spores and hyphal fragments which will grow on a particular medium. A more precise account regarding identification of spores can be obtained by this method as compared to the direct examination of the spore catched on a sticky slide. If a daily samples is taken under the same conditions, gravity methods gives an excellent identification of seasonal changes (Dransfield, 1966). According to Gregory (1961), high accuracy is generally not required either in outdoor or in indoor work since the spore content of the atmosphere can vary enormously at different times of the year. This techniques was therefore, considered by him to be quite suitable for providing preliminary records from which long term changes in the atmospheric spore population may be inferred.

Biodeterioration which is an entirely new field of research of the application of aerobiology has emerged in relation to biodeterioration of materials in stores, equipments, paintings, library materials and frescoes. In this type of deterioration there is interaction between substrate and fungal organisms. This is menifested by staining or spoilage of materials mildewing or rotting, mechanical damage etc. During biodeterioration the fungal organisms cause decomposition of cellulose and binding materials thus causing spoilage of the substrate. Fungi bring about changes in the chemical composition and nutritive value of the materials which are stored there. These changes are significant to the extent that they decrease the market value of these products. There is probably very less record on this aspect of indoor aerobiology.

The fungi discharge their fragments and spores by various means eg. ascomycetous fungi discharge their spores to a height which carry them outside the laminar layer (ground and the surfaces of objects there on and which varies in thickness from less than a millimeter in day time to several meters in night). While dry spore forms are dispersed by wind.

The quality and quantity of indoor air of a place depend upon many factors *i.e.* speed of air, movement of individuals, hygienic conditions of individuals, height at which the place is situated, climate, atmosphere population and vegetation around the place. Therefore, it is necessary to keep in mind all the above mentioned factors while working/observing the infestation of indoor air by micro-organisms.

The fungal population of indoor air is derived mainly from vegetation, soil and from other substrate on which fungi grow. The fluctuation in the population of spores present in the indoor air depends on temperature, humidity, velocity by which wind enters inside, rainfall, local vegetation, surrounding of the place, number of ventilators, number and kinds of activities and the health of the individuals.

The presence of bioparticles in indoor is not only damaging problems to inanimates but also harmful to human or animal activities.

As early as (95-35 BC) Lucretius suggested that particles are carried by wind and concluded that influenza and cold spread due to inhalation. Since than the role of air causing diseases in human beings and animals gained importance. Many diseases of commercial and domestic animals eg. foot and mouth disease, ephemeral fever in cattle and fowl are caused by air borne spores. Spores of *Aspergillus fumigatus* cause putrient metritis in cows, Aspergillosis, mycotoxicosis, facial eczema and other fungal diseases are of great concern to veterinary scientists.

Many of such biotic elements attack the mucous membrane, causing seasonal rhinitis or conjunctivitis and is symptomized by intense sneezing, watering of eyes, nasal obstruction, itching of eyes and nose, redness of eyes and frequent coughing which may take place minutes after exposure to the offending allergens.

The following account enumerates fungi of indoor air which have been reported to cause allergic reactions, hypersensitivity or diseases in animals and human beings.

Table 2 : Fungi reported to cause allergic reactions, hypersansitivity or diseases in animals and human beings.

Author	Environment	Fungi
Vittal and Krishna moorthi (1988)	Madras City	*Nigrospora, Alternaria, Drechslera, Curvularia, Torula, Cladosporium, Periconia, Ganoderma, Coprinus and Leptosphaeria.*
Raghu and Vittal (1992)	House dust	*Mucor mucedo, M. plumbeus, M. racemosus, Rhizopus stolonifer, Syncephalastrum racemosum, Chaetonium globosum, Acremonium, sp., Alternaria alternata, Aspergillus cervinus, A. candidus, A. flavipes, A. flavus, A. fumigatus, A. japonicus, A nidulans, A. niger, A. sydowi, A. terreus, A. versicolor, Cladosporium, Cladosporioides, C. herbarum, Curvularia lunata, C. pallescens, C. tuberculata, Drechslera hawaiiensis, Fusarium lateritium, F. roseum, F. semitectum, Penicillium chrysogenum, P. citrinum, P. expansum, P. funiculosum, P. implicatum, Scopulariopsis brevicaulis, Stachybotrys parvispora, Sporothrix sp., Trichoderma viride.*
Baruah and Chetia (1960) and Agarwal *et al.* (1969)		Rust spores and Basidiospores.

Agarwal and Shivpuri (1974)	Clinical Study	*Rhizopus chinensis, Lacanidion, Aspergillus, Curvularia, Memnoniella, Helminthosporium alli, Spicaria pruinosum.*
Rati and Ramalingam (1979)	Poultry shed	*Aspergillus flavus.*
Singh *et al.* (1981)	Respiratory passage of allergic patients	*Aspergillus fumigatus, A niger.*
Singh and Singh (1999)	Susceptible individuals	*Aspergillus spp.*

This indicates that much less work has been reported on fungi of indoor air responsible for allergic reactions. Agarwal and Shivpuri (1974) and Singh *et al.* (1981) have made clinical studies. Others have reported only the allergic fungi from city, poultry shed and susceptible individuals.

Reports on disinfestations of indoor air of different environments are also few (Riley *et al.*, 1972; Aleksi, 1965; Upadhyay and Arora, 1975-76 and Shukla, 2002). These workers have applied physical means *i.e.* U.V. radiation, fumigation by leaves of neem and organic chemical compounds.

The aerobiological studies have generated information on-

1. Qualitative and quantitative composition of fungi of indoor air largely of food grains godowns, plastic factory, textile factory, educational institutions and libraries.
2. Methods of isolation/screening of fungi of indoor air.
3. Fungi responsible for allergic reaction.

More investigations are required on the following aspects:

1. Screening of mycoflora of less exposed indoor environments like temples, cinema halls, molls, hospitals, poultry house, cow shed, museum, industries like bakery, dairy, medicine, paint, leather, packing etc.
2. Changes caused by indoor fungi in the quality and quantity of the material stored there in.
3. Effect of environment (temperature, relative humidity) on the quality and quantity of indoor air of different places.
4. Qualitative and quantitative survey of indoor air borne fungi and their prevalence for allergenicity.
5. Disinfestation of indoor environment to eliminate the fungi:
   a. without harming human beings and animals living /dwelling there.
   b. without damaging the material kept there and indoor infrastructure.

Acknowledgement

The authors are grateful to the Head, Department of Botany, DDU Gorakhpur University, Gorakhpur, for providing research facilities and to DST, New Delhi, for proving financial support.

References

Agarwal, M.K., Mukherji, K.G. and Shivpuri, D.N. (1969). Studies on the allergenic fungal spores of Delhi, Indian metropolitan area, Botanical aspects. *J. Allergy*, 44: 193-203.

Agarwal, M.K. and Shivpuri, D.N. (1974). Fungal spores - Their role in respiratory allergy. *Adv. in Pollen Spore Res.*, 1: 78-128.

Aleksi Meskishvill, L.G. (1965). Mycoflora of an archives storeroom and results of testing some physical measure against it. *Soobshch Akhad. Nauk.Gruz. SSR*, 39:686-688.

Alteras, M.D. and Lehrer, N. (1977). Fungal flora in the air of a large hospital. *Castellania*, 11 : 217-219.

Arundhati, Baruah.(1998). A preliminary survey of fungal flora of Jorhat. *Adv.Pl.Sci.*, 11(1): 317-319.

Banerjee, U.C., Weber, P., Ruffin, J. and Banerjee, S. (1987). Airborne fungi of some residences in Durham, North Carolina, U.S.A. *Grana*, 26:103-108.

Baruah, H.K. and Chetia, M. (1960). Aerospora and allergic human diseases. A study of certain fungal spores and pollen grains of Gauhati. *J. Exp. Biol.*, 4: 236-238.

Bhatia, H.S. and Gaur, R.D. (1979). Studies on aerobiology atmospheric fungal spores. *New Phytol.*, 82: (2) 519-527.

Burge, H.P., Boise, J.R., Solomon, W.R. and Bandra, E. (1978). Fungi in libraries. An aerometric survey. *Mycopathologia*, 64: 67-72.

Calvo, M.A., Dronda, M.A. and Castello, R. (1982). Fungal spores in house dust. *Ann. Allergy*. 49: 213-219.

Cheong, C.D., Neumeister-Kemp, H.G., Dingle, P.W., Hardy, G. (2004). Intervention study of airborne fungal spores in homes with protable HEPA filteration units. *Moint J. Environ.*, 6:866-873.

Chute, H. L. and Barden. E. (1964). The fungus flora of chick hatcheries. *Avian Dis.*, 8: 13-19.

Cunningham. (1873). *Microscopic examination of Air*. Govt.Printers, Calcutta : 58.

Davies, R.R. (1960). *Via*ble moulds in house dust. *Trans. Brit. Mycol. Soc.*, 43(4): 617-630.

Di Meena, M.E. (1955). A quantitative study of air-borne fungus spores in Dunedin, New Zealand. *Trans. Brit. Mycol. Soc.*, 38: 119-129.

Dransfield, M. (1966). The fungal air-spora at Samaru,Northen Nigeria. *Trans. Br. Mycol. Soc.*, 49:121-132.

Durham, O.C. (1942). Air borne fungus spores as allergence. In: *Aerobiology*, Washington.

Edmonds, R. L. and Bennighoof, W.S. (1973). Aerobiology and its modern applications. *Report No.3 I B P Aerobiology Programme*. 17.

Foarde, K. and Berry, M. (2004). Comparison of biocontaminate levels associated with hard vs. carpet floors in non-problem schools. *J.Exposure Analysis Environ. Epedemiol.*, 14:541-548.

Frey, D., Cross, D.O. and Durie, E.B. (1962). Investigation of a series of samples of house dust for the presence of fungi: correlation with previous investigations of airborne fungi and sensitivity tests on patients. *Mycopath. Mycol. Appl.* 19: 83-86.

Gravesen, S. (1972). Identification and qualification of indoor airborne microfungi during 12th months from 44 Danish homes. *Acta. Allergol.*, 27: 337-354.

Gravesen, S. (1978). Identification and prevalence of culturable mesophilic microfungi in house dust from 100 Danish homes. *Allergy*, 33: 268-272.

Gregory, P. H. (1945).The dispersion of airborne spores. *Trans. Brit. Mycol. Soc.*, 28: 26-72.

Gregory, P.H. (1961). *The Microbiology of the Atmosphere*. Leonard Hill, London .251p.

Gregory, P.H. (1973). *Microbiology of the Atmosphere*. Leonard Hill Publishers, INC, London 2nd Edn.

Hyde, H. A. and William, D. A. (1946). A daily census of Alternaria spores caught from the atmosphere at Cardiff in 1942 and 1943.*Trans. Brit. Mycol. Soc.*, 29:78-85.

Jacob, W.C. (1951). Aerobiology in compendium of Meteorology. *Am. Meteorol. Soc. Botson* :1103-1111.

Jayaprakash, K.B., Rati, E. and Ramalingam, A. (1978). *Aspergillus flavus* in the air of Working environments. *Curr. Sci.*, 47: 920-921.

Levetin, E. and Hurewitz, D. (1978). A one year survey of the airborne molds of Tulsa. Oklahoma. II. Indoor survey. *Ann. Allergy*, 41: 25-27.

Li, D.W. and Kendrick, W.B. (1995). A year round comparision of fungal spores in indoor and outdoor air. *Mycologia*, 87 (2): 190-195.

Lucretius. In: *Microbiology*. By Pyatkin, K. and Krivoshein. Yu. M.I.R. Publisher Moscow.

Lumpkins, E.D.D. and Corbit, S.L. (1976). Air borne fungi survey. II. Culture plate survey of the home environments. *Ann. Allergy*, 36: 40-44.

Madaline, M.E. and Linton, A.H. (1974). In: *Microorganism function, form and environment*. Edward Arnold (Publication) Ltd., P. 492.

Mani, P., Vittal, B. P. R. and Pugalmaran, M. (1998). Aerometric study of *via*ble fungus spores in Poultry houses. In: *Environment and Aerobiology*. Research Periodicals and Book Publishing House Houston.U S A.

Marjut, K. (1977). Air borne spores in a mill and in a veneer factory. *Int. Aerobiol. News letter*, 6:8.

Martin, J. P. (1950). Use of acid, Rose Bengal and streptomycin in the plate method for estimating soil fungi. *Soil Sci.*, 69: 215-232.

Nanda, A., Nayak, B.K. and Behera, N. (1997). Airborne microfungi in the bakeries of Berhampur city, Orissa. *J.Mycopath. Res.*, 35(2):81-86.

Narania, K. and Reddy, S.M. (1981). Aeromycoflora of store houses of fruits in relation to post harvest diseases of Apples and Mangoes. *Proc. Nat. Conf. Env. Biol.*, 1: 169-172.

Nilsby, I. (1949). Allergy to moulds in Sweden. A Botanical and clinical study Acta. *Allergy*, 2: 57-90.

Noble, W.C. and Clayton, Y.M. (1963). Fungi in the air of hospital wards. *J. Gen. Microbiol.*, 32: 397-402.

O S H Answer: Indoor air quality- Moulds and Fungi. (2006). Canadian Centre for Occupational Health and Safety.

Pady, S.M. and Kapica, L. (1956). Fungi in air masses over Montreal during 1950 and 1951. *Can. J. Bot.*, 34: 1-15.

Pasteur, L. (1861). Memoire surles corpuscles organizes qui existent dans 1 atmosphere.Examen de la doctrine des generations spontaneous. *Ann. Sci. Nat.* (2001); 4c; Ser. 16: 5-98.

Pohjola, A.L., Auli, Rantio and Makiner, Y. (1977). Spore composition in a garbage disposal plants. Int. Aerobiol. *News lett.*, 6:10.

Pugalmaran, M. and Vittal, B.P.R. (1997). Aeromycological survey of indoor environment in leather godowns. 'Aerobiology' *Proceedings of 5th International Conference*, Bangalore. 1994. Oxford and IBA Publishing Co.1997.

Pugalmaran, M. and Vittal, B.P.R. (1999). Fungal diversity in the indoor and outdoor environments and dust of grain storage godowns. *Ind. J. Aerobiol.* Vol.12, Nos. 1 and 2, : 24-29.

Raghu, S. and Vittal, B. P. R. (1992). A survey of allergenic moulds in house dust and home environment. *Ind.J. Aerobiol.* Special Volume, : 161-165.

Rati, E.A. and Ramalingam, A. (1979). Toxic strains among airborne isolates of Aspergillus flavus link. *Ind. J.. Expt. Biol.*, 17: 97-98.

Rati, E., Jayaprakash, K.B. and Ramalingam, A. (1980). Airspora of a poultry shed at Mysore. *Ind. J. Microbiol.*, 20: 6-12.

Riley, R.L. (1972). The ecology of indoor atmosphere. Airborne infection in hospital. *J. Chronic Dis.* 25: 421-423.

Sandra V., McNeel, D.V.M., Richard A. and Kreutzer, M.D. (2009).Fungi and Indoor air quality. *J. Health and Environ. Digest*, 10:9-12.

Shukla, Deepmala (2002). *Studies on fungal flora of indoor air.* Ph. D. Thesis, University of Gorakhpur, Groakhpur, India.

Shukla Deepmala, Singh, Nisha, Singh Priyanka and Misra, Nisha (2003). Studies on the aspergilli in the indoor air of some educational institutions at Gorakhpur. *Proc. Nat. Acad. Sci. India,* 73:325-332.

Singh, B.P., Mukherjee, P.K. and Nath, P. (1981). Allergenic significance of airborne fungal spores of allergy patient's residences. *Proc. Nat. Acad. Sci., Part B. Biol. Sci.,* 47(1): 78-82.

Singh, A. and Singh, A.B. (1999). *Aspergillus* spp. As an important occupational risk factor among susceptible individuals. *Aerobiologia,* 15:233-240.

Singh, K. (1993). *Atmospheric concentration of microbes in indoor and outdoor environment* Ph.D. Thesis, University of Aurangabad, (M.S.)

Singh, Manju. Rai, Shashi and Rai, P.K. (1989). Studies on intramural aeromycology of warehouse at Bhopal. *Ind J. Appl. Pure Biol.,* 4 (2): 111-115.

Singh, N.I. (1981). Microbiology of the air inside the cinema halls. *Proc. Nat. Conf. Envi. Bio.* (Ed.) S.T. Tilak : 199-206.

Spallanzini. (1765). In: *Microbiology. Carpenter.* P. L.1972. W.B. Sander's Company Philadelphia.pp.28.

Sreeramulu, T. (1961). Concentration of fungal spores in the air inside the cattle shed. *Acta. Allergol.,* 16:337-346.

Tilak, S.T. and Chakre, O.J. (1977). Indoor microbial pollution of air and its relevance to storage diseases of food grains. *Proc. Symposium on Environmental Pollution and Toxicology,* Hissar, 45-50.

Tilak, S.T. and Pillai, S.G. 1(988). Fungi in library. An Aerobiological survey. *Ind.J. Aerobiol.,* 1: 92-94.

Tilak, S.T., Talib, S.H. and Babu, M. (1980). Allergenicity of certain air borne fungal spores of Hospital ward. *Environ. India* Vol. III: 27-30.

Tilak, S.T. and Vishwe, D.B. (1975). Microbial content of air inside Library. *Biovigyanam,* 1: 187-190.

Tilak, S.T. and Vishwe. D.B. (1976). Microbial content of air inside library. *Biovigyanam.,* 1 : 25-27.

Tripathi, R.N. (1987). Fungal air spora inside the Central Library of Gorakhpur University. *Water, Air, Soil Pollution,* 34: 125-134.

Turner, P. D. (1966). The fungal air spora of Hong Kong as determind by the agar plate method. *Trans. Brit. Mycol. Soc.,* 49: 225-267.

Upadhyay, R.K. and Arora, D.K. (1975-1976). Sporostatic nature of Neem smoke and its possible ecological influence on Air fungal flora of polluted site. *Jour. Sci. Res.,* 26: 125-129.

Vittal, B.P.R. and Chinnaraj, S. (1990). Mycological examination of the air straw store houses and airborne straw dust. *Ind. J. Aerobiol.* Vol.3 No.1 and 2.. pp. 11-14.

Vittal, B.P.R. and Glory, Leela, A. (1984). Airborne fungus spores of a library in India. *Grana*, 24: 129-133.

Vittal,B. P. R. and Krishnamoorthy, K. (1988). A census of airborne mold spores in the atmosphere of the city of Madras, India. *Ann. Allergy*, 60 : 99-101.

Vittal, B. P. R. and Rasool, S.K. (1995). Enumeration of airborne molds in some indoor environments of Madras city (India) by culturable and non-culturable volumetric samplers. *Aerobiologia*, 11: 201-204.

Wu, Li, Chiang, C.M., Huang, C.Y., Lee, C.C., Su, H.J. (2005). Changing microbial concentrations are associated with ventilation performance in Taiwan's air, conditioned office buildings. *J. Indoor Air*, 15:19-26.

□□□

Microbial Diversity and Functions, 2012
© D.J. Bagyaraj, K.V.B.R. Tilak, H.K. Kehri (eds.), pp. 309-326
New India Publishing Agency, New Delhi (India)
E-mail : info@nipabooks.com; Website : www.nipabooks.com

Chapter 13

Rhizosphere Biotechnology : Strategies for Biocontrol of Aflatoxin Production in Maize Crop

A.K. Roy, N.L. Mandal and S. Roy

ABSTRACT

Soil is the natural growth media for living plants and microbes, and rhizosphere is niche for a group of soil microorganisms inhabiting around the plant root zone. The warmed and humid agro-climatic conditions of the area aggravate A. flavus contamination and aflatoxin production in maize grains from field to storage. In this article aflatoxin contamination in maize, antagonistic activity of rhizosphere microbes against toxigenic A. flavus, through fungi, antagonistic activity of rhizosphere bacteria against toxigenic A. flavus, bacterization of stored maize seeds for aflatoxin control and inhibition of aflatoxin production in stored maize seed with rhizobacteria has been discussed in detail.

Keywords: Rhizosphere biotechnology, aflatoxin, biocontrol, maize

Soil is the natural growth media for living plants and microbes, and rhizosphere is niche for a group of soil microorganisms inhabiting around the plant root zone. The diversity in rhizosphere microbes though depends on various factors including plant species, soil texture but these are also known to influence the growth and development of plants by providing them growth promoting substances and disease controlling agents. Several species of *Aspergillus* are reported to constitute a major group of rhizosphere biota and some of them are aflatoxigenic to contaminate plant or plant parts with aflatoxin. The term aflatoxin is designated to a group of more than seventeen closely related secondary metabolites of *A. flavus* and *A. parasiticus* having furanocoumarine nature. Due to strong hepatocarcinogenic nature of aflatoxins in human and other domestic animals it attracts more attention than other mycotoxins. Since the discovery of Turky X disease in England, scientists of different discipline shared their ideas to control the elaboration of this hazardous mycotoxin with the help of different physico-chemical and biological methods.

Maize is an important food and is extensively cultivated throughout India particularly in North eastern parts including U.P., Bihar and Jharkhand states. Bihar state alone produces about 2 million ton of maize per year. Moreover both in Bihar and Jharkhand states maize is grown in all the seasons of the year because of its wide adaptability under diverse rainfall, temperature and soil types ranging from alluvial of the indo-Gangetic plains of (Bihar) to the black and red soil (Jharkhand). The warmed and humid agro-climatic conditions of the area aggravate *A. flavus* contamination and aflatoxin production in maize grains from field to storage. *A. flavus* inhabiting in rhizosphere zone of maize crop is also one of the sources of its contamination starting from seedling stage to grain setting. Several management strategies of aflatoxin control in corn have been reported but application of co-occurring rhizosphere micro flora for this purpose is a promising and eco- friendly approach, however, more investigations are still required for its sustenance. Therefore, an attempt has been made through present communication to achieve workable data for biological control of aflatoxin in maize.

Mycotoxin and mycotoxicoses

Mycotoxins comprising structurally diverse and chemically unrelated group of naturally occurring fungal metabolites have been strongly implicated as chemical precursors of toxicity in human and animals (Chauhan, 2004). The toxicity syndrome (mycotoxicoses) resulting from the intake of mycotoxin by man and animals is known due to its toxigenic, carcinogenic, mutagenic, teratogenic and estrogenic properties (Bennet, 1987). Though mycotoxicoses have been known for a long time, however, it got first recognition probably as ergotism in Europe under the name "Holy fire" which was resulted by the intake of contaminated grain with sclerotia of *Claviceps purpurea*. Despite the fact, mycotoxicoses were remained as "neglected diseases" until the out break of Turkey -X disease in Great Britain in the year 1960s which led to a multidisciplinary approach to investigate the causes of the problem in aflatoxin contaminated ground nut. At present more than 300 different mycotoxins produced by about 150 fungi are known, showing a large variety of chemical structure and their effect. Although many toxic metabolites have been isolated from laboratory cultures of moulds that occur on agricultural products so far only eight *viz.*, aflatoxins, zearalenone, ochratoxin, citrinin, trichothecene, patulin Fumonisins and Cyclopiazonic acid are found to have possible significant occurrence in naturally contaminated foods and feeds (Tables 1 and 2).

Table 1 : Natural contamination of mycotoxins and their effects

Mycotoxins	Mycotoxin producing fungi	Food commodities	Biological effects
Aflatoxins (B_1B_2,G_1,G_2)	*Aspergillus flavus* & *A. parasiticus*	Peanut, Corn, Wheat, Rice, Herbal druga and milk	Hepatotoxic,carc inogenic
Ochratoxin	*A. ochraceus* & *Penicillium viridicatum*	Cereal grains, Bean, peanuts, Wild seeds of medicinal use	Nephrotoxic
Citrinin	*P. citrinum* *P. viridicatum*	Wheat, barley, rice, Corn and herbals	Nephrotoxic
Patulin	*P . patulum*	Mouldy feed, Apple, Wheat straw residue	Edema, death of cattle, paralysis of motor nerve
Trichothecenes	*Fusarium tricinctum* & *F. solani*	Corn, Wheat, Cattle feed, mixed feed	Dermal necrosis, haemorrhages, digestive disorder
Zearalenone	*F. graminiarum*	Corn, mouldy hay, Commercial feed	Volvo vaginitis, estrogenic effect, atrophy of testis, ovaries, abortion
Fumonisins	*F. moniliforme*	Groundnut, Poultry feed	Leukoenuphalomalasa, abortion, carcinogenic
Cyclopiazonic acid	*P. cyclopium*	Corn, peanut, kodo millet	Human intoxication, muscle necrosis, haemorrhage and edema

Table 2 : Physical factors influencing mycotoxin production

Fungi	Mycotoxins	Nature of the substrate	Factors for optimum production	
			Moisture (%)	Temperature (^{0}C)
A. flavus	A flatoxins	Cereals	17-24	25-35
A. ochraceous	Ochratoxin-A	Maize ,Wheat	18.5	25
F. moniliforme	Zearalenone	Maize	20-30	24-27
F. roseum	Zearalcnonc	Oat,	17-24	24
		Barley	18-24	25
F. graminiarum	Zearalenone	Wheat	18-20	24

Aflatoxin and aflatoxicoses

Amongst all, **aflatoxins** are well known widely investigated mycotoxin and attracted worldwide attention since 1960 when more than one lakh Turkey birds and fourteen thousand ducklings were died in the poultry farm of England. Scientists from different discipline started investigation to isolate the pathogens, however, no bacteria, viruses, insecticides or pesticides were detected, therefore, this disease was named as "Turkey-X-disease"(X denotes unknown). Later on after examination of feeds sources, the imported Brazillian groundnut floors/meal was found heavily contaminated with *A. flavus* Link ex fries (Sargeant *et al.*, 1961) and when extracted, a highly toxic chemical compound was identified and named as **aflatoxins**. Presently three main genera of *Aspergillus viz., A. flavus, A. parasiticus* and *A. nomius* are reported having the aflatoxin production potentiality (Kurtzman *et al.*, 1987). Some other species like *A. oryzae* and *A. niger* are also reported to produce aflatoxins but authentic confirmation is still awaited.

Chemistry of aflatoxin

The aflatoxin is a generic term that refers to one or more (four) principle metabolites produced as B_1, B_2, G_1 and G_2. Chemically aflatoxin constitute a unique group of highly oxygenated coumarine derivative heterocyclic compounds which consist of more than 17 closely related compounds having their characteristic emission of blue green fluorescence under UV light. On the basis of blue green fluorescence and Rf (relative front) value, aflatoxin is categorized mainly into 2 major groups *i.e.* B_1 and B_2 for blue fluorescence and G_1 and G_2 for green fluorescence (1 and 2 stands for their Rf).The Rf value of B_1, B_2, G_1 and G_2 is in decreasing order from top to base on TLC plates. Another form of aflatoxin is M_1, a metabolic derivative of B_1 extracted in the milk of lactating animals. Among all categories of aflatoxins, B_1 is the most toxic followed by G_1, B_2 and G_2 in order of their decreasing potency which is known to be potent hepato-carcinogenic (Coker, 1999), teratogenic and immunosuppressive too (Abramson, 1998). Naturally occurring different combinations of aflatoxin have been classified as class I human carcinogenic (IARC, 1993).

Natural contamination of aflatoxin

Natural contamination of aflatoxins and other mycotoxins in agricultural commodities is a worldwide problem but greater in tropics and temperate zones. Reports (Bhat *et al.*, 1978; Mall *et al.*, 1983; Narita *et al.*, 1988; Sinha 1990) on mycotoxin contamination available so far clearly indicate that practically every edible material including fruits, vegetables, cereals, pulses, spices and plant drugs too is suitable substrate for mycotoxin elaboration.

India being a tropical country has great diversity in agro climatic condition due to unseasonable rains, high temperature and humidity, which adversely affect the both standing crops and stored food materials by favouring the growth of moulds. Extensive investigation on the natural occurrence of aflatoxin has also been made for more than any other mycotoxins because *A. flavus* is universally distributed in foods, feeds and other agricultural commodities. Aflatoxins are the most significant mycotoxin to be encountered as natural contaminants of maize grains (Mishra, 1977; Mall *et al.,* 1983 and Sinha, 1987), pulses (Sinha and Ranjan, 1989), oil seeds (Rao *et al;* 1965; Lalitha Kumari *et al.,* 1970; Sinha *et al.,* 1988; Sahay and Prasad, 1990), wheat (Mishra, 1977, Prasad *et al.,* 1982; Sinha, 1991), Sorghum grains (Tripathi, 1973; Bhadraiah and Rama Rao, 1982,83), pearl millet (Mishra and Daradhiyar, 1991), dry fruits and spices (Singh, 1983) under storage. Roy and his associates (1988, 89 and 90) have also reported the natural occurrence of aflatoxins in large number of crude herbal drugs *i.e.* seeds, fruits, barks, woods and leaves beyond the tolerance level as fixed by WHO *i.e.* 30 ppb in Indian context.

Aflatoxin contamination in maize

Among all food commodities, maize is one of the richest substrates for the growth of *A.flavus* and aflatoxin elaboration (Sinha 1980). Five mycotoxins *viz.* **Aflatoxin, Zearalenone, Ochratoxin** and two **trichothecenes: T2 toxin** and **deoxynivalenol** have been reported from maize grains. Though aflatoxin contamination in maize is worldwide (United States, Thailand, Uganda, Columbia, Brazil, Philippines), however, in India maize is grown in tropical as well as temperate regions of the country. The prevailing environmental conditions including flood, heavy rain and high temperature make the maize grains very prone to *A. flavus* contamination and aflatoxin production levels of which ranges from 25 to 12,500 mg/kg. Although aflatoxin production in maize was regarded primarily as a storage problems but it has been established that infection and aflatoxin formation may also occur in the standing maize crop (Resnik *et al.,* 1996; Setamou *et al.,* 1997). Mishra (1977) collected 135 stored maize samples and detected aflatoxin ranging from 8-1850 ppb in 35% samples. Bilgrami *et al.,*(1980, 82) after an exhaustive survey of almost all maize growing areas of Bihar recorded very high % incidence of aflatoxin in standing maize crop. Sinha (1983) reported natural occurrence of it in 25% samples of standing maize crops whereas 35% aflatoxin contamination was recorded in maize grains collected from storage conditions.

The aflatoxigenic fungi *i.e. A. flavus* and *A. parasiticus* are ubiquitous in nature and frequently present in air, soil and different agro ecosystem (Saito *et al.,* 1986; Nesci and Etcheverry, 2002). In rhizosphere a vast range of micro organisms such as bacteria, fungi and others are invariably present along with toxigenic and non toxigenic strains of *A. flavus* and *A. paraciticus*. Population of toxigenic *A. flavus* groups have been isolated from the rhizosphere zone of cotton, peanut and maize (Cotty, 1997; Horn and Dorner, 1998 and Wicklow *et al.*, 1998) and their presence in pre-harvest contamination has been expected. The rhizosphere strains may be one of the important sources of pre-harvest aflatoxin contamination in maize along with atmospheric strain. The rhizosphere strain of toxigenic *A. flavus* may also influence the inoculums load on standing crops or it may transit from rhizosphere to cob *via* vector such as insect, mites, bird etc. Aflatoxin production on standing maize crops is directly associated with landing of toxigenic inoculums of *A. flavus* on standing maize crops and their further germination and penetration. The constituent of maize kernels and suitable environmental condition such as high moisture content and temperature favours for mycelial growth and aflatoxin elaboration in standing crops. Maize grains in field condition even if aflatoxin is not developed in standing maize cobs; there is considerable scope for its production during harvest, transportation and storage whenever and wherever the conditions are condusively synchronized. Aflatoxin contamination in field samples may also be due to the physical and mechanical injuries caused by the bird, rodents, squirrels, insects or other agencies because the damaged maize crops becomes more receptive to fungal invasion and aflatoxin contamination in field (Bilgrami,1987).

Control strategies for aflatoxin contamination

Prevention of natural occurrence of aflatoxin in maize by aflatoxigenic fungi and their detoxification are cumbersome process, therefore, 100% safe food and feed supply is not effective. However, it may be possible either by preventing moulding of substrates or detoxifying the aflatoxin contaminated foods. In order to manage the aflatoxin problems different methods *viz.*, i) Cultural ii) Physical iii) Chemical iv) Botanicals and v) Biological control have been adopted in the recent past. Recognition of the need to control aflatoxin contamination of foods and feeds different researchers have applied their ideas to eliminate or to detoxify aflatoxin from susceptible crops *via* physical and chemical means (Goldflat, 1971; Basappa,1983; Bilgrami, 1990) however, the most promising and safest approach is through biological means where living micro organism such as bacteria, fungi, actinomycetes, yeasts and some botanicals are used to control these toxigenic strains either by the phenomenon of antibiosis, competition or parasitism. Prevention of aflatoxins contamination is thus prerequisite and their detoxification is secondary factor of its management. Decontamination and detoxification are two main strategies to control aflatoxin elaboration in foods, feeds and plant drugs. The contribution of earlier workers (Diener and Davis, 1967; Singh, 1983; Bilgrami, 1984 a and b) on this aspect cannot be ruled out. These methods undoubtedly show good result but are costlier affairs, time consuming and also alter the quality of the product. Pollution hazards by the use of chemicals/ pesticides also can not be ruled out. Biological control is ultimate mechanism against aflatoxin management which offers both potential strategies. The concept of Biological control based on the microbial antagonism to the plant pathogen/ toxigenic strains has opened a new vista under Control Program. Its application in the suppression or elimination of harmful microbes has been found significantly useful. In biological control strategy, living microbes (antagonist) are used to control harmful micro organism either by producing some toxic metabolites or causing disease in them without any ecological imbalance. The phenomenon of biological control is based on antibiosis (inhibition or destruction of one organism by the metabolic product of other), competition (injurious effect of one organism on other because of the utilization or removal of the same resource) or parasitism (where antagonist parasites on pathogen).

Therefore, attempts have been made in recent past to control aflatoxin problem in foods and feeds and herbal drugs by the application of bio agents (Mann 1977; Megaalla and Hafez 1982; Cotty, 1990; Roy and Chourasia 1990; Rosic *et al.*, 1991 ; Chourasia and Sinha 1994; Kumar and Roy 2001) have also reported earlier that microorganisms can inactivate or remove the aflatoxin. *Flavobacterium auratiacum* and selected acid producing moulds have been utilized for successful removal of aflatoxin form liquid culture media. Transformation of aflatoxin B_1 to aflatoxicol (less toxic then aflatoxin B_1) was observed in case of *Corynobacterium rubrum* (Mann and Rehm, 1976) and several sps of moulds belonging to the genera of *Aspergillus, mucor, Trichoderma* etc. (Detroy and Hesseltine, 1969; Robertson *et al.*, 1970; Mitsuo *et al.*, 1990).

Indian reports on aflatoxin management of agricultural commodities is meager; nevertheless Chourasia and Sinha (1994) have reported that non toxigenic strains of *A. flavus* could be the potential bio competitive agents to inhibit the growth and aflatoxin bio-synthesis by toxigenic strains in developing peanuts under green house condition. In addition to this, Roy and Chourasia (1990), Chourasia and Roy (1993), Chourasia (1995) have also found that several fungi and bacteria can break down aflatoxin in peanut and in aflatoxin containing liquid culture media.

Antagonistic activity of rhizosphere microbes against toxigenic A. flavus

Rhizosphere is a well characterized and specialized ecological niche for various microbes such as fungi, bacteria, actinomycetes, yeasts etc. Data on microbial association with different crops including maize has been investigated by several earlier workers (Harley 1948; Rovira, 1956, 1965; Verma, 1981; Khan and Prakash 1982). In the rhizosphere zone beneficial and deleterious microbes are continuously interacting with each other resulting their population and their varied rhizosphere effect. It also denotes interaction between soil and rhizosphere microbes and their ratio on the plant health (Manoharachary, 2007). Data on several earlier workers (Papa*viz*as 1983;Utkhede, 1992; Mehta *et al.*, 1995; Saxena *et al.*,2000) highlighted the importance of soil microorganisms as antagonists of several pathogenic and mycotoxigenic fungi, which directly or indirectly act as bio agents which may be exploited for Biological Control Program. The entire biological control program is based on the isolation of various microbes from different sources and their screening *in vitro* condition against various pathogenic microbes. During the screening of microbes as antagonists it is also imperative to study their mechanism because they interact in different manners such as parasitism, competition and antibiosis. It is also essential to notice that the degree of antagonistic potential of microbes is varied from species to species and also to strain. It is also important that some of the antagonists strain possesses more than one mode of action and also effective against many pathogens while other antagonist strains are specific to some pathogens. The crucial factor for the success of bio agents is their ability to colonize in the rhizosphere zone and they persist throughout the growing season. Sharma and Paul (1998) outlined some characteristics of an ideal biocontrol microbes are: i) Survival for prolonged periods in soil either in active or in passive forms. ii) Multiplication in laboratory conditions should be simple and in expensive. iii) be specific to the target organism if possible. iv) be active under the required environmental conditions. v) be efficient, economical and not be health hazards during handling and applications. In the last few years emphasis on the research and practice of biological control is steadily increasing as an important component of an integrated pest management in certain areas, however, a little work have been done for the biological control of aflatoxigenic *A. flavus* through rhizosphere microflora. This helps in reducing the dependency of fungicides and other chemicals for sustainable ecological damage. Thus, the careful screening is most successful steps for the success of Biocontrol program.

Through fungi

Fungi are extremely diverse group of organisms with about 75, 00,000 species were widely distributed in every ecosystem. Among all the sps of fungi, only limited number of sps is used as antagonists in biological control program. Biological control has played a major role in disease management since the beginning of agriculture. Farmers learn disease management practice through careful screening of antagonists by trial and error that takes advantage of natural biological control, even though they may not recognize its contributions to the well being of their crops. A critical analysis of the literature on biological control reveals a skewed research towards soil borne pathogens, as the response has been more positive in this area as compared to foliar pathogens. Among all microorganism, fungi are one of the potential candidates for biological control programme and several antagonist fungi *viz.*, *Trichoderma, Gliocladium, Penicillium, Aspergillus, Talaromyces, Trichothecium* sps inhabited in rhizosphere and other ecosystem have been exploited for the control of plant diseases. Several excellent examples of biological control of plant pathogens are available in India through fungi and are successfully implemented. Among the fungi, *Trichoderma* has been extensively used to manage a variety of plant pathogens. Several strains of *Trichoderma* have been commercially exploited and are available for use by the farmers. *T. harzianum* and *T. viridae* applied as seed dressing, effectively reduces mortality of ground nut seedlings due to stem rot and collar rot (Nagaraju and Urs 1998). Several sps of *Fusarium viz. F. semitectum, F. chlamydosporum* and non pathogenic *F. oxysporum* have been reported as bio control agents against various phyto pathogenic fungi (Gill and Chahal, 1988; Rao and Thakur, 1988). Hyperparasitic fungi that attack other fungi are known for several plant pathogens including some rusts, powdery and downy mildews. It is also well established fact that maize rhizosphere inhabits several fungi of different taxonomic group which may be one of the important sources for screening of antagonist fungi against toxigenic *A. flavus*. The antagonistic potential of fungi is achieved either by the production of chitinase or other cell wall degrading enzymes, production of antifungal chemical or competition for nutrients between pathogens and antagonist. Parasitism and hyper parasitism are very common in fungal isolates. Similarly mycotoxin producing fungi is greatly influenced by other co existing fungi competing for nutrients and space. Antagonistic effect of different fungi on aflatoxin production was also reported in different food commodities by earlier workers (Singh and Faull, 1982; Stack and Pettit, 1985; Singh *et al*., 1989; Mehan, 1992; Dubey and Patel, 2001). Therefore screening of more rhizosphere fungi is essentials for the inhibition of this hazardous micro organism.

In the present study all 30 fungi excluding *A. flavus* of different taxonomic group were isolated from different maize rhizosphere for screening their antagonistic activity against highly toxigenic *A. flavus i.e.* PAT-2. The antagonistic properties of each fungus and their potentials to inhibit the mycelial growth as well as aflatoxin production were recorded by dual culture technique in PDA Petri plates and SMKY medium respectively. The interaction behaviour of each fungi were determined by following the criteria laid down by Johnson and Curl (1972) and the % inhibition of radial mycelial growth, biomass of *A. flavus* and aflatoxin elaboration were recorded against each fungus (Table-3). Amongst all, 21 fungi were found to inhibit the radial growth of *A. flavus* on PDA petriplates which ranges from 15-57%. The maximum % inhibition (57%) in radial growth was recorded by the strain of *Cephalosporium roseum* and minimum (15%) in case of *Mucor mucedo.* The % inhibition of aflatoxin production and dry mycelial weight of *A. flavus* growth was also determined by co-culture technique in SMKY medium. The range of % inhibition of aflatoxin production was recorded between 30-90%. Out of all antagonists *A. niger* have the maximum (90%) potential to inhibit the aflatoxin production where as minimum (30%) was noticed in *Curvularia lunata* similarly % inhibition of dry bio mass of *A. flavus* was ranged from minimum (29%) by the sps of *Mucor mucedo* to maximum (86%) by the sps of *A. niger.*

Table 3 : Antagonistic activity of rhizosphere fungi against toxigenic *A. flavus*

Fungal isolates	Type of interaction	No. of fungi	% Inhibition of	
			aflatoxin production	dry mycelial wt.
A. nidulans, A. ochraceus, A. sydowi, Cladosporium herbarum, Mucor mucedo, Rhizoctonia solani and *Trichothecium roseum*	A	7	34-56	28-56
A. terreus, Alternaria alternata, Curvularia lunata, Fusarium chlamydosporum, Monilia brunnea and *Helminthosporium graminearum*	B	6	30-75	36-70
A. candidus, F. oxysporum and *F. solani*	C	3	40-51	46-52
A. niger, F. semitectum, and *P. rubrum*	D	3	43-90	45-46
Cephalosporium roseum and *P.chrysogenum*	E	2	71-82	72-81

Through Rhizobacteria

Over the past one century research has repeatedly demonstrated that out of diverse microorganism rhizobacteria can act as natural antagonists of various pathogenic microbes (Cook, 2000). Isolates of *Pseudomonas, Bacillus, Agrobacterium, Erwinia, Enterobacter* and *Serratia* have been isolated from different rhizosphere and were reported to have an important role in biological control of plant pathogens (Cook and Baker 1983). The biocontrol bacteria provide plant protection through induction of host plant defence mechanism, elimination of plant signals that trigger pathogen development or competition for nutrients or by production of antagonists compounds such as antibiotics, siderophore, cyanide and hydrolytic enzymes. The major success in the field of biological control through bacteria was the control of crown gall caused by *Agrobacterium tumefaciens* using *Azatobactor* strains 84 (Kerr, 1972). Since then several bacterial antagonists were proved for successful controlling of many post harvested diseases (Pathak, 1997). Among the bacterial antagonists sps of *Pseudomonas* are known to be highly potential. Fluorescent *Pseudomonas* has been reported to be antagonistic against many soil born and foliar pathogens. Podile and Prakash (1996) have studied in detail the mode of action of *Bacillus subtilis* AF strain, effective against *A. niger*. The bacterial cell adhered to the fungal mycelium, multiplied *in situ*, colonized and lysed mycelial surface. Extracellular chitinilytic enzymes of microorganism have a potential to suppress the activities of the pathogens by degrading the chitin in their cell walls and thus protected the plants from disease (Krishna *et al.*, 2003) Bacterial isolates *viz. Pseudomonas fluorescens*-2 and *Bacillus* isolates are found effective in reducing the population of *A. flavus*. These isolates also showed plant growth promoting efficacy (Weller 1988).

The present study is based on the screening of the effective RB which has ability to reduce the radial growth of *A. flavus* in dual culture on Nutrient Agar & Malt Extract Agar pour plate method or to reduce the mycelial growth and aflatoxin production in liquid Yeast Extract Sucrose medium. On the basis of dual culture method the bacteria which showed antagonistic efficacy against *A. flavus* were categorized into 3 categories such as poor antagonist (+), moderate (2^+) and good antagonist (3^+). After screening of all, 62 bacterial isolates showed antagonistic potential of which 13 showed higher, 18 moderate and 31 with poor antagonistic activities (Table-4). The RB antagonists were again screened for their effect on reduction in aflatoxin production and dry mycelial growth in liquid YES medium. The % inhibition of aflatoxin production was recorded maximum by 3^+ antagonists with a range of 70-86% followed by 2^+ with a range of 50-69% and the minimum by + antagonists between 12-49%. The % dry mycelial wt was also reduced proportionately (Table 10-14). Individual strain i.e, RB-116 showed maximum (86%) inhibition of aflatoxin production where as the minimum (16%) were recorded by RB strain 19. Similarly the maximum (84%) dry mycelial biomass reduction was recorded by RB-76 and minimum (18%) was recorded by RB strain 71.

Table 4 : Antagonistic activity of rhizosphere bacteria against toxigenic *A. flavus*

No. of antagonist RB	Antagonist Properties	% Inhibition of aflatoxin production	% inhibition of dry mycelial biomass
13	Highly(3+)	70-86	63-84
19	Moderate(2+)	50-69	48-70
30	Poor(+)	12-49	18-50

Bacterization of stored maize seeds for aflatoxin control

Of the all 13 highly antagonistic rhizobacteria *i.e.* 3^+ which showed potential of growth and aflatoxin reduction of toxigenic *A. flavus* under laboratory culture medium were further screened to study their effect for the inhibition of aflatoxin production in maize seeds, because all rhizobacteria were not capable to complete colonization on the seed surface after bacterization. Seed treatment with biocontrol agents is the easiest way of introducing sufficient quantity of these antagonists on the seed and protects against seed borne and soil borne diseases, *via* a variety of mechanisms. During the present study % inhibition of aflatoxin ranged from minimum (42%) by the RB -14 to maximum (73%) by the strain no 116 (Table -5). On the basis of above experiment, of the all 13 antagonist only 7 RB *viz.*, RB-1, RB-65, RB-76, RB-93, RB-116, RB-139, and RB-153 were considered as highly antagonist (above 50% inhibition) and were characterized for identification (Table -6).

Table 5 : Inhibition of aflatoxin production in stored maize seed with rhizobacteria

S. No	Strain No.	Aflatoxin B_1 in µg/kg	% inhibition
Control	560.0	-	
1	RB-1	201.6	64
2	RB- 14	324.8	42
3	RB -28	291.2	48
4	RB- 48	313.6	44
5	RB -65	246.4	56
6	RB- 76	173.6	69
7	RB- 81	302.4	46
8	RB- 93	207.2	57
9	RB -106	291.2	48
10	RB- 116	151.2	73
11	RB -129	308.0	45
12	RB -139	252.0	55
13	RB -153	263.2	53

Table 6 : Showing the list of identified potential antagonists Rhizobacteria

S.N	Strain designation	Identity
1	RB-1	*Psychrobacter pacificensis*
2	RB-65	*Bacillus subtilis*
3	RB-76	*Pseudomonas aeruginosa*
4	RB-93	*Pseudomonas cepacia*
5	RB-116	*Bacillus pumilus*
6	RB-139	*Pseudomonas putida*
7	RB-153	*Bacillus circulans*

Conclusion and Future Prospects

The survival of toxigenic strain of *A. flavus* and *A. parasiticus* in nature and their natural contamination in food commodities are very common. The discovery of antagonistic microbes such as bacteria, fungi and other have major role to manage the aflatoxin contamination but still there are some constrains which are to be resolved in future:-

1. Identification and selection of effective strain of natural antagonists.
2. To understand the biology, ecology, physiology, genetic behavior of bio control agents.
3. To identify most efficient host genotype-symbiont strains of antagonists.
4. To develop mass culture technique for field application of antagonists.
5. To demonstrate and assess the full benefits of bio control agents under field conditions.
6. To educate the public for effective utilization of bio control agent.
7. Seed bacterization and seed coating with antagonists micro organism to reduce the incidence of *A. flavus* and *A. parasiticus* in soil for check the aflatoxin production.
8. Identification and evaluation of aflatoxin inhibitor gene for reducing and checking the aflatoxin production
9. Making transgenic plants which have potential to resistance against *A. flavus and A. parasiticus.*
10. Extraction, purification and identification of chemical compounds from antagonists to achieve the control of *A. flavus* and *A. parasiticus* and handling as like pesticides.
11. Identification of detoxifying gene for aflatoxin control and making gene library.

References

Abramson, D. (1998). Mycotoxin formation and environmental factors. In: *Mycotoxins in Agriculture and food safety* (Eds.) K. K. Sinha and D. Bhatnagar, Marcel Dekker Inc., New York: 255-278.

Basappa, S.C. (1983). Physical methods of detoxification of aflatoxin contaminated materials. In: *Mycotoxins in food and feed* (Eds.) K.S.Bilgrami, T.Prasad and K.K.Sinha, Allied Press, Bhagalpur: 251-273.

Bennet, J.W. (1987). Mycotoxins, Mycotoxicoses and Mycotoxicology. *Mycopathologia,* 100: 3-5.

Bhadraiah, B. and Rama Rao, P. (1982). Isolates of *Aspergillus flavus* from *Sorghum* seeds and aflatoxin production. *Curr. Sci.,* 51:1116-1117.

Bhadraiah, B and Rama Rao, P. (1983). Aflatoxin production in pre-harvest and stored *Sorghum*. In: *Mycotoxins in food and feed* (Eds.) K. S. Bilgrami, T. Prasad and K. K. Sinha, Allied Press, Bhagalpur : 49-54.

Bhat, R.V., Nagrajan, V.and Tulpule, P.G. (1978. *Health hazards of Mycotoxins in India.* Indian Council of Medical Research, New Delhi: 58.

Bilgrami K.S., Prasad, T., Misra, R.S. and K.K. Sinha (1980). *Survey and study of mycotoxin producing fungi associated with grains in the standing maize crop*, reports of ICAR Project, Bhagalpur University, Bhagalpur, India.

Bilgrami, K. S., Mishra, R. S., Prasad, T. and Sinha, K. K. (1982). Screening of different varieties of maize for aflatoxin production by *Aspergillus parasiticus. Ind. Phytopath.*, 35: 376-378.

Bilgrami, K. S. and Sinha, K. K. (1984a). Mycotoxin contamination in food and its control. *Indian Rev. Life Sci.*, 4:19-36.

Bilgrami, K. S., Singh, Anjana and Ranjan, K.S. (1984b). Detoxification of aflatoxins in dry fruits and spices through light and heat treatments. *Nat.Acad. Sci. Letters*, 7: 273-274.

Bilgrami, K.S. and Sinha, K.K. (1987). Aflatoxin in India, In: *Aflatoxin in maize* (Eds.) M.S. Zuber, E.B. Lillehoj and B. L. Renfro, CIMMYT Mexico: 349-358.

Bilgrami, K.S. and Chaudhary, (1990). Incidence of *A. flavus* in the aerosphere of maize fields at Bhagalpur. *Ind. Phytopath.*, 43: 38-42.

Chauhan, R. K.S. (2004). 'Mycotoxins the hidden killers in foods and feeds' an over view and strategy for future. *J.Ind. Bot. Soc.*, 83 (4): 11-21

Chourasia, H.K. (1995). Kernel infection and aflatoxin production in peanut (*Arachis hypogaea* L.) by *Aspergillus flavus* in presence of Geocarposphere bacteria. *J. Food. Sci. Technol.*, 32(6): 459-464.

Chourasia, H.K. and A.K. Roy (1993). Growth and aflatoxin production by *Aspergillus parasiticus* with co-occurring fungi and bacteria. *J. Ind. Bot. Soc.*, 72:131-134.

Chourasia, H.K. and Sinha, R.K. (1994). Potential of the biological control of aflatoxin contamination in developing Peanut (*Arachis hypogaea* L.). *J. Food Sci. Technol.*, 31: 362.

Coker, R. (1999). Mycotoxins a global menace: A review. In: *Post harvest news letter* (Ed.) E. Highly, 51:12-13.

Cook, R. J. (2000). Advances in Plant health management in the twentieth century. *Annu. Rev. Phytopathol.*, 38: 95-116.

Cook, R. T. and Baker, H. F. (1983). *The nature and practice of Biological control of plant pathogens*. A.P.S. St. Paul, M.N. 539 pp.

Cotty, P. J. (1990). Effect of aflatoxigenic strains of *Aspergillus flavus* on aflatoxin contamination of developing cotton seeds. *Plant Dis.*, 74:233-235.

Cotty, P.J. (1997). Aflatoxin producing potential of communities of *Aspergillus* section Flavi from cotton producing areas in the United States. *Mycol. Res.*, 101, 698-704.

Detroy, R.W. and Hesseltine, C.W. (1969). Structure of a new transformation product of aflatoxin B_1. *Can. J. Biochem.*, 48: 830-832

Diener, U. L. and Davis, N.D. (1967). Biology of *A. parasiticus*. In: *Aflatoxin in Maize. Proceeding of the Workshop* (Eds.) M.S. Zuber, E.B. Lillehoj and B.L.Renfro: 33-40. Mexico City: CIMMYT.

Dubey, S.C. and Patel, B. (2001). Evaluation of fungal antagonists against *Thanatephorus cucumeris* causing web blight of Urd and Mung bean. *Ind. Phytopath.*, 54(2): 206-209.

Gill, K.S. and Chahal, S.S. (1988). Growth inhibition of *Claviceps fusiformis* with culture filtrate of *Fusarium chlamydosporum*. *Pl. Dis. Res.*, 3: 64-65.

Goldflat, L. A. (1971). Control and removal of aflatoxins. *J. Am. Oil Chem. Soc.*, 48: 605-610.

Harley, J. L. (1948). Biology of rhizosphere microorganism. In: *Recent advances in the biology of microorganism* (Eds.) K.S. Bilgrami and K. M. Vyas, Bishen Singh Mahendra Pal Singh, Dehradun, India,1980. *Biol. Rev.*, 23: 127-158.

Horn, B.W. and Dorner, J.W. (1998). Soil population of *Aspergillus* species from section Flavi along a transect through Peanuts growing region of the United States. *Mycologia*, 90: 767-776.

IARC (1993). International Agency for Research on Cancer. *Monograph on the Evaluation of Carcinogenic Risk to Human*, 56: 257-263.

Johnson, L.F. and Curl, E.A. (1972). *Methods of research on the ecology of soil borne plant pathogens*, Burgers Publicity Co Minneapolis MN PP 178.

Kerr, A. (1972). Biological control of crown gall: Seed inoculation. *J. Appl. Bacteriol.*, 35: 493-497.

Khan, M. A. A. and Prakash, D. (1982). Rhizosphere and rhizoplane mycoflora of Gram as affected by plant growth. *Ind. Phyopath.*, 35: 717-718.

Krishna Kishore, G., Pandey, S., Naryana, R.J. and Podile, A.R. (2003). Evaluation of chitinolytic strains of *Serratia marcescens* and *Bacillus circulans* for biological control of late leaf spot of groundnut. *ISOR National Seminar: Stress management in oilseeds.* Jan 28-30, 2003:19-20.

Kumar, S. and Roy, A.K. (2001). Inhibition of growth and aflatoxin production of *Aspergillus parasiticus* Speare by co-existing fungi *in vitro*. *J. Ind. Bot. Soc.*, 80: 149-151.

Kurtzman, C.P., Hon, B.W. and Hesseltine, C.W. (1987). *Aspergillus nomius*, a new aflatoxin producing species related to *A.flavus* and *A.tamari*. *Antonie van Leewenhock*, 53: 147-148.

Lalitha Kumari and Govindaswamy, C.V. (1970). Role of aflatoxin in groundnut seed spoilage. *Curr. Sci.*, 34: 306-309.

Mall, O.P., Ranawat, K.K. and Chauhan, S.K. (1983). Mould flora and aflatoxin contamination in maize kernels. In: *Mycotixin in food & foods* (Eds.) K.S. Bilgrami, T. Prasad and K.K. Sinha. Allied Press, Bhagalpur : 37-44.

Mann, R. (1977). Degradation of aflatoxin B_1 by Yeasts, Moulds and Bacteria. *Zleben Forsh.*, 163: 39-43.

Mann, R. and Rehm, H. (1976). Degradation products from aflatoxin B_1 by *Corynebacterium* rubrum, *Aspergillus niger, Trichoderma viridae* and *Mucor ambigeus. European. J. Appl. Microbiol.*, 2: 297-306.

Manoharachary, C. (2007). Rhizosphere-The forgotten ecological niche of microbes. *J. Ind. Phytopath.*, 60(3): pp.386.

Megalla, S.E. and Hafez, A.M. (1982). Detoxification of aflatoxin B_1 by acidogenous yoghurt. *Mycopathologia*, 77: 89-91.

Mehan, V.K. (1992). Management of aflatoxin contamination of ground nut. *J. of oil Seeds Res.*, 9(2): 276-285.

Mehta, R.D., Patel, K.A., Roy, K.K. and Mehta, M.H. (1995). Biological control of soil born plant pathogens with *Trichoderma harzianum. Ind. J. Mycol. Plant Pathol.*, 25: 1-2.

Mishra R.S. (1977). *Aflatoxin contamination of some agricultural commodities in IARI and biochemical effects of aflatoxin B1 on Maize seeds.* Ph.D. Thesis, G.B. Pant University of Agriculture & Technology, Pantnagar.

Mishra, N. K. and Daradhiyar, S. K. (1991). Mould flora and aflatoxin contamination of stored and cooked samples of Pearl millet in the paharia tribal belt of Santhal Pargana, Bihar, India. *Appl. Environ. Microbiol.*, 57: 1223-1226.

Mitsuo, N., Satoshi, M., Kazuo, S. Kengi, F. Taichiro, N. and Nabuhika, K. (1990). Interconversion of aflatoxin B_1 and aflatoxicol by several fungi. *Appl. Environ. Microbiol.*, 56: 1465-1470.

Nagaraju, P. and Urs, S.D. (1998). Comparative efficiency of fungicides and bioagents against *Aspergillus niger*, a causal agent of collar rot of groundnut. *Current Research*, University of Agricultural Science, Bangalore. 27: 137-139.

Narita, N., Suzuki, M., Udagawa, S., Sekita, S., Harada, M., Aoki, N., Tanaka, T., Hasegawa, A., Yamamoto, S., Toyzaka, N. and Matsuda, Y. (1988). Aflatoxin potential of *Aspergillus flavus* isolates from Indonesian Herbal drugs. *Proc. Jpn. Assoc. Mycotoxin Col.*, 27: 21-26.

Nesci, A. and Etcheverry, M. (2002). *Aspergillus* section Flavi populations from field maize in Argentina. *Lett. in Appl. Microbiol.*, 34: 343-348.

Papavizas, G.C. (1983). *Trichoderma* and *Gliocladium*: Biology, ecology and potential for biocontrol. *Annu. Rev. Phytopathol.*, 23: 23-54.

Pathak, V.N. (1997). Post-harvest fruit pathology- present status and future possibilities. *Ind. Phytopath.*, 50: 161-185.

Podile, A.R. and Prakash, A.P. (1996). Lysis and biological control of *Aspergillus niger* by *Bacillus subtilis* AFL. *Can. J. Microbiol.*, 42: 533-538.

Prasad T., Thakur, M.K., Prasad R.B. and Sahay, M. (1982). Aflatoxin contamination in wheat. *Biol. Bull. India,* 4(2): 107-108.

Rao, K. S., Madhawan, T. V. and Tulpule, P.G. (1965). Incidence of toxigenic strains of *Aspergillus flavus* affecting ground nut crop in certain coastal districts of India. *Ind. J. Med. Res.,* 53: 1169-1202.

Rao, V. P. and Thakur, R.P. (1988). *Fusarium semitectum* var. *majus*- a potential biocontrol agent of Ergot (*Claviceps fusiformis*) of Pearlmillet. *Ind. Phytopath.,* 41: 567-574.

Resnik, S., Neira, S., Pacin, A., Martinez, E., Apro, N. and Laterite, S. (1996). A survey of the natural occurrence of aflatoxins and zearalenone in Argentina field maize: 1983-94. *Food Addit. and Contamin.,* 13:115-120.

Robertson, J.A., Teunisson, D.J. and Bourdeaux, G.J. (1970). Isolation and structure of biologically reduced aflatoxin B_1. *J .Agric. Food Chem.,* 18: 1090-1091.

Rosic, J. L., Skrinjar, M. and Markor, S. (1991). Decrease of aflatoxin in yoghurt and acidified milks. *Mycopathologia,* 113: 117-119.

Rovira, A.D. (1956). A study of the development of the root surface microflora during the initial stage of plant growth. *J. Appl. Bacteriol.,* 19: 72-79.

Rovira, A.D. (1965). Interaction between plant roots and soil microorganism. *Ann. Rev. Microbiol.,* 19: 241-266.

Roy, A. K. (1989). Threat to medicinal plants and plant drugs by fungi. *J. Ind. Bot. Soc.,* 68: 149-153.

Roy, A. K. and Chourasia, H.K. (1990). Inhibition of aflatoxins production by microbial interaction. *J. Gen. Appl. Microbiol.,* 36: 59-62.

Roy. A. K., K. K. Sinha and H. K. Chourasia (1988). Aflatoxin contamination of some common Drug Plants. *Appl. Environ. Microbiol.,* 54(3): 842-843.

Sahay, S.S. and T. Prasad (1990). The occurrence of aflatoxin in mustard and mustard products. *Food Addit. Contamin.,* 7: 509-613.

Saito,M.,Tsuruta,O.,Siriacha,P.,Kawasugi,S.,Manabe,M. and Buangsuwon, D. (1986) Distribution and aflatoxin productivity of the atypical strains of *Aspergillus fluvus* isolates from soil in Thailand. *Proc. Japan Assoc. Mycotoxicol.,* 24: 41-46.

Sargeant, I, A. Sheridan, J.O. Kelly and R.B.A Carnatharn (1961). Toxicity associated with certain samples of groundnuts. *Nature,* 192: 1096 - 1097

Saxena, A. K. Pal, K.K. and Tilak, K.V.B.R. (2000). Bacterial biocontrol agents and their role in plant disease management, In: *Biocontrol Potential and its Exploitation in Sustainable Agriculture,*(Eds.) R.K.Upadhyay, K. G. Mukerji and B. P. Chamola, Kluwer Academic/Plenum Publisher, New York,pp.25-37.

Setamou, M., Cardwell, K.F., Schulthes, F. and Hell, K. (1997). *Aspergillus flavus* infection and aflatoxin contamination of preharvest maize in Benin. *Plant Dis.,* 81: 1323-1327.

Sharma, P.D. and Paul, P.K. (1998). Recent tactics of Biological Control. In: New Trends in Microbial Ecology, (Eds.) Bharat Rai and M. S. Dkhar, Dept. of Botany, NEHU, Shillong and ISCNR, Dept.of Botany,BHU,Varanasi,pp.272-286.

Singh, J. and Faull, J.L. (1982). Antagonism and biological control. In: *Biocontrol of Plant disease*, Vol.II (Eds.) K. G. Mukerji and K. L. Garg. CRC Press, Inc. 40: 167-176.

Singh, P. (1983). Mycotoxin contamination in dry fruits and spices. In: *Proc. Symp. Mycotoxin in Food and Feed*, Bhagalpur Univ., Bhagalpur. pp. 55-68.

Singh, Premlata, Ahmad, S.K. and Bhagat, Sita (1989). Microbial interaction and aflatoxin production. *J. Ind. Bot. Soc.*, 68: 169-171.

Sinha A.K. and Ranjan, K.S. (1989). Pre harvest aflatoxin problems in linseed (*Linum usitatissimum*) in Bihar. *J. Ind. Bot. Soc.*, 68:19-20

Sinha R.K., Bilgrami, K.S. and Prasad, T. (1988). Incidence of aflatoxins in mustard crop in Bihar. *Ind. Phytopath.*, 41:434-437

Sinha, A.K. (1991). *Studies on aflatoxin production in relation to varietal variations in wheat.* Ph.D.Thesis, T.M.Bhagalpur University, Bhagalpur.

Sinha, K.K. (1980). *Survey and study of aflatoxin producing isolates of A. flavus associated with maize grains in storage and standing crops.* Ph.D. Thesis, Bhagalpur University, Bhagalpur.

Sinha, K.K. (1983). Aflatoxin problem in storage and standing maize crop. In: *Proc. Symp. Mycotoxin in Food and Feed.* Bhagalpur:23-36.

Sinha, K.K. (1987). Aflatoxin contamination of maize in flooded areas of Bhagalpur, India, *Appl. Environ. Microbiol.*, 53:1391-1393.

Sinha, K.K. (1990). Incidence of mycotoxins in maize grains in Bihar State India, *Food Additi. Contamin.*, 7: 55-61.

Stack, J.P. and Pettit, R.E, (1985). Fungi affecting the germination of *Sclerotia* of *A. flavus* in soil *Proc. Am. Peanut Res. Educ. Soc.*, 16: 45.

Verma, R.N. (1981). Effect on plant age on rhizosphere microflora. *Biol. Bull. of India*, 3: 176-183.

Weller, D.M. (1988). Biological control of soil borne plant pathogens in the rhizosphere with bacteria. *Ann. Rev. Phytopathol.*, 26: 379-407.

Wicklow, D.T., McAlpin, C.E. and Platis, C.E. (1998). Characterization of the *Aspergillus flavus* population within an Illinois maize field. *Mycol. Res.*, 102: 263-268.

□□□

Microbial Diversity and Functions, 2012
© D.J. Bagyaraj, K.V.B.R. Tilak, H.K. Kehri (eds.), pp. 327-334
New India Publishing Agency, New Delhi (India)
E-mail : info@nipabooks.com; Website : www.nipabooks.com

Chapter 14

Algal Component in Lichen Formation : A Review

Navneet Kaur and M.P. Sharma

ABSTRACT

Different algal taxa involved in the formation of lichen/s are presented. To-date 43 algal genera, comprising members of Cyanophyta, Chlorophyta and one genus each of families Phaeophyceae (brown algae) and Xanthophyceae (yellow-green algae) are recorded, based on personal observations coupled with literature. The various algal taxa forming different lichens belonging to various families is presented. The taxa Trebouxia de Puymaly and Trentepohlia Born. amongst green algae; Scytonema Ag. and Nostoc Vauch. Among blue-green algae are dominant genera involved in the formation of lichens. The yellow and brown algae are only of rare occurrence.

Keywords: Blue-green, green, algal genera, lichen formation, rare taxa.

Introduction

Lichens are pioneers in rock disintegration leading to the formation of soil, comprises two components: algal partner (the phycobiont) and fungal partner (the mycobiont) living in symbiotic manner. The identification of algal partner is a difficult exercise; however, identification of fungal partner is quite helpful in lichen taxonomy. In this contribution, a review is presented based on the observations of large number of lichen specimens collected from different locations and substrata.

Algal Components

Forty three genera of algae have been recorded in the formation of different lichen taxa belonging to the different lichen families by several workers: Hale (1979); Tschermak (1988); Wischester (1988); Budel (1992); Gartner (1992); Ahmadjian (1993); Awasthi (2000 a, b); Sharma & Sharma (2000); Kirk *et al.* (2001); Prasher and Chander (2007) and Upreti & Nayaka (2008).

In all, 27 genera of Chlorophyceae (Green algae) have so far been recroded in different lichen families. These are either unicellular or multicellular (coenobial or filamentous) forms. Common filamentous algal taxa inhabiting lichens are: *Cephaleuros* Kunze. (Trentepohliaceae); *Cladophora* Kutz. (Cladophoraceae); *Trebouxia* de Puymaly (Chlorococcaceae) and *Trentepohlia* Born. (Trentepohliaceae). Among Cyanophyceae (Blue-green algae), 14 genera are involved. Common filamentous forms are *Scytonema* Ag. (Scytonemataceae) and *Nostoc* Vauch. (Nostocaceae), whereas, *Gloeocapsa* Kutz. is non-filamentous. It is pertinent to mention that Xanthophyceae (yellow-green algae) and Phaeophyceae (brown algae) are rarely involved in lichen formation. Each group is represented by only a single genus *Heterococcus* Chodat (Heteropediaceae) and *Pteroderma* Kuckuck (Lithodermataceae) respectively.

Various algal components in different lichen families are given in Table 1:

Table 1 : Phycobionts w.r.t. their occurrence in various lichen taxa/families:

Phycobiont Family	Phycobiont	Type	Lichen Taxa/Family
Chlorophyceae (Green phycobionts)			
Chaetophoraceae	*Coccobotrys* Chodat	Filamentous	*Lecidea* Ach., *Dermatocarpon miniatum* (L.) Mann., *Verrucaria nigrescens* Pers., Endolithic lichens
	Dilabifilum Tschermak-Woess (*Pseudopleurococcus* Snow)	Filamentous	*Arthopyrenia* Mass., *Verrucaria* Schrad.
	Diplosphaera Bial.	Unicellular or colonial	*Verrucaria* Schrad.
	Pseudopleurococcus Snow	Unicellular or colonial	*Verrucaria* Schrad.
Chlorococcaeae	*Asterochloris* Tschermak-Woess	Unicellular or colonial	*Varicellaria* Nyl.
	Chlorococcum Fr.	Unicellular or colonial	*Aspidothelium* Vain., *Bacidia* de Not.,

Contd...

			Bullatina Vezda & Poelt, *Conotrema* Tuck., *Gylectidium* Mull.Arg.
	Gloeocystis Naeg.	Unicellular or colonial	*Epigloea* Zuk., *Gyalecta* Ach., *Lecidea* Ach.
	Myrmecia Printz	Unicellular or colonial	*Bacidia* de Not., *Catapyrenium* Fw., *Catillaria* T. Fries, *Dermatocarpon hepaticum* (Ach.) T. Fries, *D. miniatum* (L.) Mann., *Lecidea* Ach., *Lobaria laetevirens* (Lightf.) Zahlbr., *L. palmonaria* (L.) Hoffm., *Phlyctis* Flot, *Psora* Hoffm., *Psoroma* Nyl., *Sarcogyne* Flot., *Verrucaria* Schrad.
	Trebouxia de Puymaly	Unicellular or colonial	*Alectoria nidulifera* Norrl., *Anaptychia palmatula* Vain., *Aspicilia* Massal, *Caloplaca* T.Fries, *Cetraria* Ach., *Cladonia* Hill, *Graphis* Adans., *Lecanora* Ach., *L. conizaeoides* Nyl., *Lecidea* Ach., *Lepraria chlorine* Nyl., *L. incana* (L.) Ach., *Parmelia furfuraceu* (L.) Ach., *P.saxatilis* Ach., *Ramalina* Ach., *Sphaerophorus globosus* (Huds.) Vain., *Stereocaulon* (Schreb.) Hoffm., *Umbilicaria* Hoffm., *U. pustulata* Hoffm., *Usnea* Dill. ex Adans., *Xanthoria* (Fr.) T. Fries., Physciaceae
Chlorosarcinaceae	*Chlorosarcinopsis* W. Herndon (Chlorosarcina Gern.)	Unicellular or colonial	*Lecidea* Ach.

Contd...

	Pseudoterbouxia Archib.	Unicellular or colonial	*Alectoria* Ach., *Buellia* de Not., *Caloplaca* T. Fries, *Cetraria* Ach., *Cladonia* Hill ex Browne, *Conotrema* Tuck.
Cladophoraceae	*Cladophora* Kutz.	Filamentous	*Blodgettia* E. Wright
Coccomyxaceae	*Coccomyxa* Schmidle	Unicellular or colonial	*Baeomyces roseus* Pers., *Icmadophila aeruginosa* (Scop.) Trevis., *Peltigera aphthosa* (L.) Willd., *Solorina* Ach.
	Nannochloris Nauman.	Unicellular or colonial	*Normandina* Nyl.
Mycoideaceae	*Phycopeltis* Millard.	Filamentous	*Arthonia* Ach., *Mazosia* Massal., *Opegrapha* Humb., *Porina lectissima* (Fries.) Zahlbr. *Trichothelium* Mull. Arg.
Oocystaceae	*Chlorella* Beij.	Unicellular or colonial	*Calicium* Pers., *Lecidea* Ach., *Lecidella* Korb., *Micarea* Fr., *Pseudocephellaria* Wain., *Trapelia* Choisy
	Dictyochloropsis Geitler	Unicellular or colonial	*Bacidia* de Not., *Brigantiaea* Trevis., *Catillaria* T. Fries, *Chaenotheca* T. Fries, *Megalospora* Meyen., Peltigeraceae
	Trochisia Kutz	Filamentous	*Polyblastia henscheliana* (Korb.) Lonnr.
Prasiolaceae	*Prasiola* Ag.	Filamentous	*Turgidosculum* Kohlm.
Protococcaceae	*Protococcus* Ag. *(Hyalococcus)*	Unicellular or colonial	*Dermatocarpon fluviatile* (Web.) T.Fries., *D. miniatum* (L.)Mann., *Endocarpon* Hedw., *Staurothele* Norm., *Thelidium* Mass., *Trapelia* Choisy
Trentepohliaceae	*Cephaleuros* Kunze.	Filamentous	*Raciborskiella* Hohnel, *Strigula* E. Fries.

Contd...

	Leptosira Borzi	Filamentous	*Thrombium* Wallr., *Vezdaea* Tschermak-Woess & Poelt
	Physolinum Printz	Filamentous	*Coenogonium* Ehrh.
	Trentepohlia Born.	Filamentous	*Chiodecton* Ach., *Coenogonium* Ehrh., *Dimerella* Trevis., *Glyphis* Ach., *Gyalecta cupularis* (Hedw.) Schaer., *Lecanactis stenhammeri* Arn., *Lecidea* Ach., *Roccella fusiformis* (L.) Lam. & DC., *R. montagnaei* Bel., *R. phycopsis* Ach., Arthoniaceae, Caliciaceae, Dirinaceae, Graphidaceae, Opegraphaceae, Porinaceae, Pyrenulaceae, Thelotremataceae, Many epiphyllous lichens
Ulotrichaceae	*Stichococcus* Naeg.	Filamentous	*Calicium* Pers., *Chaemotheca* T. Fries., *Coniocybe* Ach., *Endocarpon pusillum* Hedw., *Lepraria* Ach., *Staurothele* Norm.
Ulvaceae	*Blidingia* Kylin	Filamentous	*Turgidosculum* Kohlm.
Incertae Sedis	*Elliptochloris* Tschermak-Woess	Unicellular or colonial	*Baeomyces* Ehrh., *Catolechia* Korb., *Micarea* Fr., *Verrucaria* Schrad.
Cyanophyceae (Blue-green phycobionts)			
Capsosiraceae	*Hyphomorpha* Borzi	Filamentous with heterocysts	*Spilonema* Born.
Chroococcaceae	*Chroococcus* Naeg.	Unicellular or colonial	*Jenmania* Wacht., *Phylliscum* Nyl.
	Cyanosarcina Kovacik	Unicellular or colonial	*Psorotichia* Mass., *Synalissa* E. Fries., *Thyrea* Mass.

Contd...

	Gloeocapsa Kutz.	Unicellular or colonial	*Amygdalaria* Norm., *Anema* Nyl, *Gonohymenia* Stein, *Heppia* Naeg, *Huilia* A. Zahlbr., *Peccania* Mass., *Pyrenopsis* Nyl, *Stereocaulon* (Schreb.) Hoffm., *Thyrea* Mass.
Hyeliaceae	*Hyelia* Bornet & Flahault	Unicellular or colonial	*Arthopyrenia* Mass.
Microchetaceae	*Tolypothrix* Kutz.	Filamentous with heterocysts	*Hertella* Henssen
Microcystaceae	*Anacystis* Meneghini	Unicellular or colonial	*Peltula* Nyl.
Nostocaceae	*Nostoc* Vauch.	Filamentous with heterocysts	*Biatora* Korb., *Collema auriculatum* Hoffm., *Dendriscocaulon* Nyl., *Hydrothyria venosa* Russela., *Lempholemma* Korb., *Leptogium* Ach., *L. hildenbradis* (Garov.) Nyl., *Leightoniella* Henss., *Lobaria pulmonaria* (L.) Hoffm., *Peltigera canina* Willd., *P. horizontalis* (Huds.) Baumg. *P. polydactyla* Hoffm., *P. rufescens* (Weis.) Humb., *Sticta fuliginosa* (Dicks) Ach. Taxa of Heppiaceae, Pannariaceae and Nephromataceae
Rivulariaceae	*Calothrix* Ag.	Filamentous with heterocysts	*Coccotrema* Mull. Arg., *Lepolichen* Trevis., *Lichinia* C.A. Agardh., *Porocyphus* Korb.
	Dichothrix Zanardini	Filamentous with heterocysts	*Lichina* C.A. Agardh., *Placynthium* Ach.

Contd...

Scytonemataceae	*Scytonema* Ag.	Filamentous with heterocysts	*Coccocapia* Pers., *Cora pavonia* (Web.) E. Fries., *Degelia* Arvidsson & Galloway *Dendriscocaulon* Nyl., *Dictyonema* Agardh., *Erioderma* Fee, *Heppia* Naeg., *Hertella* Henssen, *Pannaria* Del., *Petractis* Fr., *Placynthium* Ach., *Stereocaulon* (Schreb.) Hoffm., *Thermutis* E.Fries., *Zahlbrucknerella* Herre
Stigonemataceae	*Stigonema* Ag.	Filamentous with heterocysts	*Amygdalaria* Norm., *Argopsis* T. Fries, *Ephebe* E. Fries, *Hullia* A. Zahlbr. *Pilophorus* T. Fries., *Spilonema* Born.
Xenococcaceae	*Chroococcidiopsis* Geitler	Unicellular or colonial	*Anema* Nyl., *Gonohymenia* Stein., *Peccania* Mass., Taxa of Lichinaceae
	Myxosarcina Printz	Unicellular or colonial	*Lichinella* Nyl., *Peccania* Mass., Peltula *Nyl.*
Phaeophyceae (Brown phycobionts)			
Lithodermataceae	*Petroderma* Kuckuck	-	*Verrucaria* Schrad.
Xanthophyceae (Yellow phycobionts)			
Heteropediaceae	*Heterococcus* Chodat	-	*Verrucaria* Schrad.

From the observations given in Table 1, it can be concluded that the members of Cyanophyceae are predominant in the formation of lichen thalli. Cyanophyceae is followed by Chlorophyceae, whereas, other groups are rarely involved in lichen formation. Phaeophyceae and Xanthophyceae are represented by one member each. *Trebouxia* de Puymaly and *Trentepohlia* Born. (both belonging to Chlorophyceae) are the most commonly occurring algal taxa in lichen thalli.

References

Ahmadjian, V. (1993). The Lichen Photobiont: What can it tell us about lichen systematics? *The Bryologist*, 96 (3): 310-313.

Awasthi, D. D. (2000a). *Lichenology in India Subcontinent.* Bishen Singh Mahendra Pal Singh Publishers, Dehradun : 125.

Awasthi, D. D. (2000b). *A Hand Book of Lichens.* Bishen Singh Mahendra Pal Singh Publishers, Dehradun : 157.

Budel, B. (1992.) Taxonomy of lichenized prokaryotic blue-green algae. In: *Algae and Symbioses* (Ed.). W. Reisser, *Bristol,*: 301-324

Gartner, G. (19920). Taxonomy of symbiotic eukaryotic algae. In: *Algae ans Symbiose* (Ed.) N. Resisser, Bristol, Biopress, : 325-339.

Hale, M. E. (1979). *How to know Lichens*, second edition. W. C. Brown Co., Dubugue Iowa : 246.

Kirk, P. M., Cannon, P. F., David, J. C. and Stalpers, J. A. (2001). *Dictionary of the Fungi* – 9th edn. (Ainsworth and Bisby). CAB International Bioscience, Egham : 655.

Prasher, I. B. and Hem Chander (2007). A Preliminary Report on Lichens and Macrofungi of Nanda Devi Biosphere Reserve (Uttaranchal). In: *Advances and Achievements in Mycology and Plant Pathology* (Eds.) I. B. Prasher and M. P. Sharma, Bishen Singh Mahendra Pal Singh Publishers, Dehradun : 293-302.

Sharma, M.P and Sharma, Anju (2000). Biodiversity in Lichens of the High Altitudes. In: *High altitudes of the Himalaya* Vol. II (Ed.) Y.P.S. Pangtey, Gyanodaya Prakashan, Nainital : 247-292.

Tschermak-Woess, E. (1988). The algal partner. In: Handbook of Lichenology (Ed.) M. Galun, *CRC Press, Inc. Boca, Raton Florida,* 1: 39-92.

Upreti, D. K. and Nayaka, S. (2008). Need for Creation of Lichen Gardens and Sanctuaries in India. *Curr. Sci.,* 10: 25-26.

Winchester, V. (1988). An assessment of Lichenometry as a method for dating recent stone measurements in two stone circles in Cumbria and Oxford Shire. *Bot. J. Linn. Soc.,* 96: 57-68.

□□□

Microbial Diversity and Functions, 2012
© D.J. Bagyaraj, K.V.B.R. Tilak, H.K. Kehri (eds.), pp. 335-352
New India Publishing Agency, New Delhi (India)
E-mail : info@nipabooks.com; Website : www.nipabooks.com

Chapter 15

Mycorrhizal Diversity in Different Ecosystems

R.S. Upadhyay, R. Raghuwanshi and Akhilesh Kumar

ABSTRACT

In nature the roots of most of the plants are infected by fungi to form mycorrhizal associations, which play a central role in nutrient capture from the soil. These fungi are found in almost every terrestrial ecosystem; even in pioneer zones or very saline environments, and are reported to represent the second largest biomass component of many terrestrial ecosystems. Mycorrhizal associations are found in a broad range of habitats ranging from aquatic to deserts and alpine altitudes. In natural ecosystems mixed population of mycorrhizal fungi co-exist, with certain fungi becoming dominant in a particular patch and subsequently being replaced as environmental conditions change. Once this equilibrium is disturbed, the population dynamics is disrupted and a biomass may develop towards a few or even for one dominant fungus. Different species of mycorrhizae respond differently in response to the environmental conditions. In the present paper, all major types of mycorrhizal associations in different ecosystems are discussed.

Keywords: Association, diversity, ecosystems, plants, mycorrhizae

Introduction

Mycorrhizae form close symbiosis between fungi and plant roots. There are two major categories of mycorrhizae namely, ectomycorrhizae and endomycorrhizae which are formed by mostly Basidiomycetous and Glomeromycetous fungi. The endomycorrhizae usually produce vesicles, arbuscles, inter and intra-cellular mycelium in the cortex of the host plants, and also produce extramatrical hyphae with spores and sporocarps. The endomycorrhizae are represented by *Acaulospora, Gigaspora, Glomus, Entrophospora, Scutellospora* and *Sclerocystis*. Daft and Nicolson (1974) placed the endomycorrhizae under Endogonaceae and later termed as vesicular-arbuscular mycorrhizae (VAM) or arbuscular mycorrhize (AM) (Smith, 1995)

AM fungi are an ubiquitous group of soil fungi which are known to colonize roots of plants belonging to more than ninety per cent of plant families (Trappe, 1987). They occur in a wide variety of environments such as arid, semiarid, aquatic, sand dunes, deserts, etc. This paper presents a brief summary of the distribution and diversity of mycorrhizal fungi in different ecosystems.

Mycorrhizae in Agro-ecosystem

AM fungi predominate in the roots and soils of agricultural crops and weed plants (Hayman, 1980; Trappe, 1981). In some natural ecosystems there is a positive correlation between plant cover and spore numbers (Anderson *et al.*, 1984). In natural soils dominated by perennial shrubs, spore numbers of AM fungi are generally smaller than in adjacent soils used for agriculture (Abbott and Robson, 1977). In contrast, in virgin native grasslands, the spore numbers were found much greater than those under adjacent wheat crops (Kucey and Paul, 1983). It has been suggested that the variation in AM species distribution could be related to host preference of the fungi, a concept referred to as "ecological specificity" by Mc Gonigle and Fitter (1990).

Cultural practices such as inter-cropping, rotation and tillage may have profound effect on the AM population e.g. cassava, which is highly dependent on mycorrhizal symbiosis for growth, developed higher levels of mycorrhizal infection when grown with legumes than when grown as a monoculture (Sieverding and Leihner, 1984). Disturbance of the top soil layers, which occurs during tillage, reduce the establishment and efficacy of the mycorrhizal symbiosis, and in some cases, can render the fungi completely ineffective in providing any plant growth benefit (Stahl *et al.*, 1988). Low tillage systems may, therefore, be beneficial for the cultivation of crops that exhibit large growth responses to inoculation with mycorrhizal fungi. The effect of soil disturbance can, however, differ according to the local circumstances. It has been suggested that where sustainable agriculture is desired it is necessary to maintain a diversity of mycorrhizal fungi as mixed crops will benefit more from a mixed population of the symbionts. Dodd *et al.* (1990) examined changes in spore numbers of individual AM fungi within the native population of savanna ecosystem in Columbia that resulted from different management practices. Twelve spore types were identified in the original soils, where it was noted that different spore populations developed rapidly under different crop regimes e.g. the spores of *Glomus occultum* and *Acaulospora myriocarpa* increased in sub-plots of shorghum, while those of *Entrophospora columbiana, Acaulospora mellea and Acaulospora morrowae* dominated in sub-plots where cowpea (*Vigna unguiculata*) was grown following a crop of Kudzu (*Pueraria phaseoloides*).

Other cultural practices also have effect on the AM fungal population. One of the most extensively studied of these cases is the effect of crop rotation. In semi-arid tropical soils, reduction in the numbers of mycorrhizal propagules by 40% was observed in soil, which was left fallow for a season and by 13% in soils where a non-mycorrhizal species such as mustard was cultivated (Harinikumar and Bagyaraj, 1988). Continuous cropping of winter wheat resulted in significantly reduced inoculum potential of AM fungi in the soil as compared to the soils which underwent a 4-year cereal crop rotation (Baltruschat and Dehne, 1988). It has not been adequately demonstrated whether such effects are due to the benefit of crop rotation or deleterious effects caused by continuous cropping. In an investigation of cropping systems with either continuous corn or continuous soybean, Johnson *et al.* (1991) observed that distinctly different mycorrhizal fungal community developed over time according to the crop species in cultivation. These results indicate that some selective preference or specificity of mycorrhizal fungal species with certain hosts occurs. Alterations in the structure of AM fungal communities by plant species indicate that understanding ecological interactions between both partners of this symbiosis such as selectivity and specificity of mycorrhizal fungi is fundamental to maintaining the most effective mycorrhizal fungal communities in agro-ecosystems.

Mycorrhizae in Aquatic Ecosystems

The first report of AM in shallow water aquatic plants was published by Sondergaard and Laegaard (1977) who observed these endophytes in five of the seven temperate aquatic plants growing in four oligotrophic soft water lakes in Denmark at the depth between 0.3-0.8 m. There are numerous reports on the incidence of occurrence of AM fungi in rice fields (Iqbal *et al.*, 1978; Mmbaga, 1979; Gangopadhyay and Das, 1982; 1984; 1988; Nopamornbodi *et al.*, 1985; 1987; Ilag *et al.*, 1987). Filer (1975) observed that flooding temporarily reduced the population of AM fungi and prevented the formation of new AM mycorrhizas and that the situation was restored after floodwater had receded. Sieverding (1979) found VA infection to be rapid at a soil water content less than the water holding capacity of the soil. Water logging seems to reduce, but not eliminate VA mycorrhizas in healthland vegetation (Malajczuk and Lamont, 1981). Read *et al.* (1976) found low levels of AM infection in marsh plants in an English Wetland. Thoen (1987) investigated the incidence of AM in the plants in Lake Retbe, Senegal and found no AM association in the hydrophytes in families Ruppiaceae and Lemnaceae or most of the hydrophytes of the families Cyperaceae, Lythraceae and Typhaceae). Although, presence of spores in aquatic sediments suggests that they probably enter the aquatic sediment through runoff from terrestrial ecosystems, information is accumulating that supports the hypothesis that there is significant genetic and physiological diversity in the population of AM fungi from dissimilar environments (Bethlenfalvay, 1992).

There are ectotypic variations among mycorrhizal fungi. Some mycorrhizal fungi are P-tolerant (Sylvia and Schenck, 1983), this may be due to genotypic adaptation (Hayman, 1982). Stahl and Christensen (1991) reported that populations of *Glomus mosscae* from different environments have genetically different ecotypes. These isolates were similar in morphology but different in their physiology.

Although, mycorrhizae have been reported to occur in water logged or flooded conditions, very few investigators have attempted to identify the species of the endophytes infecting water logged plants. Brown *et al.* (1988) identified 20 species belonging to four genera namely *Glomus, Sclerocystis, Gigaspora* and *Acaulospora,* which were associated with upland rice. Five AM fungal species were recovered from soil and sediment sample and from the rhizosphere of water logged *Casuarina cunninghamiana* (Khan, 1993c). Soil moisture seems to be important determinant of the species spectrum of fungi (Read and Boyd, 1986). Khan (1993a,c) found *Glomus mosseae* abundant in aerated soils, *G. fasciculatum* moderately abundant in swamps, *Sclerocystis rubiformis* rare in soil or absent in swamps and water. *Gigaspora margarita* was abundant in swamps and moderate in sediments. An unidentified *Scutellospora* sp. was moderately abundant in the sediments. Sward *et al.* (1978) reported spores of *G. margarita* from wetter areas of heath land soils in southeastern Australia. Saif *et al.* (1975) founded that *Gigaspora* sp. was predominant in sandy soils which remained at around 50-60% water holding capacity. Tests conducted by Anderson *et al.* (1984) on mycorrhizal associations also indicated that *G. margarita* was able to tolerate the wet sites and caused infection of plants growing on wet soils. Neither mycorrhizal associations in the roots nor mycorrhizal spores in the soil were found in 25 plants species from 12 families surveyed for AM in mangrove vegetation in India (Mohankumar and Mahadevan, 1986). These authors concluded that water logging substantially reduced the number of spores in mangrove soils and may abolish infection. Mycorrhizal colonization in four species of pioneer salt marsh plant species, namely *Arthrocnemum indicum, Portersia coarctata, Sesuvium portulacastrum* and *Suaeda maritima* (Sengupta and Chaudhuri 1990) and five common species of fungi namely *Glomus fasciculatum, G. macrocarpum, G. multicaulis, G. mosseae* and *G. margarita* were present in the rhizosphere soil samples.

Spore numbers and AM infection appear to be positively correlated to redox potential values (Khan, 1993b). Spores were found abundant in terrestrial soil, moderately abundant in swamps, rare or absent in water and moderately abundant to rare in sediments of aquatic trees (Khan, 1993a, and c).

Rivers, marshes, creeks and ponds are ecological habitats for plants adapted to withstand stress arising from water logging and high salinity. Mycorrhizae and nitrogen-fixing bacteria, in their root zones, enhance the ecological adaptations of these plants to such environments. As aquatic plants can have a profound impact on their environment (Carpenter and Lodge, 1986). Also, their metabolic activity can alter the physical and chemical surroundings (Gregg and Rose, 1982). Therefore, if aquatic macrophytes are managed properly they can be used in certain aquatic environments to provide oxygen to act as sinks for nutrients and possibly to deal with pollutants.

Cultivation of these aquatic plants for food, medicinal and miscellaneous commercial products may be exploited by using mycorrhizal endosymbionts. Lowland rice generally does not respond to P fertilization under flooded conditions (Alva *et al.*, 1980) perhaps because of mycorrhizal association. Therefore, mycorrhiza will be beneficial to rice under these conditions. These studies will provide information to our further understanding of the significance of mycorrhizae for management of aquacultural systems.

Mycorrhizae in Forest Ecosystems

Numerous species of mycorrhizal fungi are known to grow in forest soil. In beech wood soil, within square of 10 x 10 cm, population of up to 15 different species of ectomycorrhizas have been found in the organic layers (Brand, 1991). EM species of *Gautieria* and *Hysterangium*, that form distinctive hyphal or rhizomorph mats, have been observed in forest ranging from the subtropical (*Eucalyptus* in Australia) to boreal forests in Alaska (Castellano, 1988; Griffith *et al.*, 1991). The actual quantitative impact of these mat communities of the forest floor remains largely unknown, although Cromack *et al.*, (1979) reported up to 27% of temperate coniferous forest could be colonized by a single species *i.e.*, *H. setchelli*. Observation of high root densities in the top 5 cm of tropical wet forest soils (Janos, 1984) and the ubiquity of AM on rainforest plant species (Janos, 1980) suggest that AM fungi are always present in intact tropical forest.

A survey of literature shows that the members of Basidiomycetes constitute diverse population of ectomycorrhizal associates of conifers and other trees in the Himalayas (Lakhanpal, 1988). Many fungi such as species of *Boletus*, *Russula*, *Amantia*, *Cortinarius*, and *Tricholoma* are representative members of Basidiomycotina occurring in association with Himalayan conifers and other diverse hosts. They have been termed as broad-spectrum fungi. Other species such as *Scleroderma, Rhizopogon, Gauteria, Lycoperdon* and *Trappeinda* also form mycorrhizal association in tree species. The members of Gasteromycetes have been collected with specific trees such as *Cedrus deodara, Pinus roxburghii* and oak species. Such fungi are host specific fungi (Lakhanpal, 1988).

Mycorrhizae in Arid Environment

The literature on the occurrence of mycorrhizas in desert, arid and semi-arid regions, reveals that most of the plant taxa form associations with mycorrhizal fungi. Non-mycorrhizal taxa include the members of Cruciferae and Zygophyllaceae. Although Cactaceae, Chenopodiaceae, Cyperaceae, Amaranthaceae and Junaceae typically are thought to be non- mycorrhizal, most of the species were found to be infected under natural stressed rangeland conditions (Neeraj *et al.*, 1991). Root Systems of 108 plant species (29 families) were evaluated throughout the year for mycorrhization. Positive symbiosis was recorded in 86 plants in the Thar deserts of India (Neeraj *et al.*, 1991). Trappe (1981) reviewed the occurrence of mycorrhizas in rangeland plants, and noted that all mycorrhizal plants were infected with AM fungi.

Mycorrhizae in Stressed Ecosystems

There are several studies reporting the role of mycorrhiza in stressed habitats (Waaland and Allen, 1987). The mycorrhizae are very common in disturbed areas (Medve, 1984) which indicate their positive role in establishing and building the plant community in such habitats and that mycorrhizal associations are essential to the colonization of nutrient-deficient soil heaps left after mining. Although mycorrhizal potential was often ignored during establishment of vegetation on mine spoils (Danielson, 1985), their importance in this respect is now well recognized (Jasper, 1992 ; Kumar *et al.*, 1999). Daft and colleagues (Daft *et al.*, 1975; Daft and Hackskaylo, 1977) have shown that vesicular arbuscular infection of herbaceous plants is almost universal in the colonisation of coal mine spoils in Britain and USA and that some plants grew successfully on coal mine spoils only in a mycorrhizal state. There are evidences that AM determine the rate of succession in mined land (Doerr *et al.*, 1984). The proportion of mycorrhizal roots has been observed to increase along with plant cover during succession in many natural ecosystems (Lesica and Antibus, 1985). Many workers (Allen and Allen 1984; Miller, 1987) have shown that in tropical forest and arid shrub/grass community, the primary colonizers of disturbed sites are often non-mycorrhizal or facultative species, while obligatory mycorrhizal plants became dominant in later stages of succession. The presence or absence of AM fungi in secondary successional sites may determine the composition of the plant community that develops (Reeves *et al.*, 1979). Disturbed habitats lacking AM fungi have been found to be dominated in the early successional stages by facultative and non-mycotrophic plant species (Miller 1979; Reeves *et al.,* 1979).

The Hawaiian islands represent 30 million years of continuous volcanic activity and sand dune formations. Consequently, there has been continuous selection for adaptation that allows for the colonization of these primary successional sites. AM formation is unlikely after relatively large-scale succession-initiating disturbances such as volcanism (Gemma and Koske, 1990) and land slides. Over 75% of the 26 species colonizing volcanic substrates of various ages at Hawaii contained AM fungi. In the phosphate fixing Hawaiian soils (Foote *et al.,* 1972) mycotrophy appears to have been a valuable attribute among the earliest colonizers.

Mycorrhizal fungi have been reported on roots or in the root zone of cultivated and non-cultivated plants growing in disturbed and un-disturbed saline soils, including marshlands, river bank, road sides, and even on the edges of salt slick (Pond *et al.,* 1984; Siddhu and Behl, 1990; Janardhan *et al.* 1994; Hildebrandt *et al.,* 2001). These have been linked with increased plant biomass and development in saline soil (Poss *et al.,* 1985; Pfeiffer and Bloss, 1988; Ruizlozano and Azcon, 2000). Sporulation by AM fungi also does not appear to be affected by salinity. Population of chlamydospores, as high as 95/100 g soil, were recovered from saline soil in Pakistan (Khan, 1974). Thus in plants adapted to saline soil, salinity appears to have little effect on the formation of AM. However, salinity may dramatically affect mycorrhizal formation in plants unadapted to salt stress. One general concept about pH and AM fungi is that some AM fungi do not readily adapt to soil with a pH different from their soil of origin, and that pH change restricts AM establishment (Sylvia and Williams, 1992). Neutral to alkaline pH favours the germination of *Glomus mosseae* (Green *et al.,* 1976; Abbot and Robson, 1977; Porter *et al.,* 1987), while spores of *Gigaspora* germinated best between pH 5-6. Hepper (1984) determined the germination of *Acaulospora laevis* in soils having different pH and concluded that the optimum range for the germination was 4-5. In addition, a number of studies have shown that changing the soil pH affects the activity of indigenous as well as certain introduced AM fungi (Peuss, 1958; Hayman and Mosse, 1971; Mosse, 1972; Kruckelmann, 1973; Graw, 1979; Wang *et al.,* 1985). Although, there occurs a natural selection of AM species according to different soil pH conditions, there is no marked effect of soil pH on the extent of mycorrhizal infection in natural vegetation (Read *et al.,* 1976; Sparling and Tinker, 1978).

Mycorrhizae in Sand Dunes

The coastal sand dunes exhibit favorable condition for the association and development of AM fungi with plants since they are deficient in phosphorus (Koske and Halvorson, 1981; Ranwell 1972). The importance of AM fungi for the growth and succession of plant species in coastal sand dune was first recognized by Nicolson (1959). Aggregation of sand grains and colonization of AM fungi with dune plants significantly stabilizes the sand dunes (Koske and Polson 1984; Sutton and Sheppard, 1976). Several temperate locations e.g. Northeastern United States (Koske 1987; Koske and Halvorson, 1981), Italy (Giovannetti 1985; Giovannetti and Nicolson, 1983), Poland (Blaszkowski, 1997) and Scotland (Nicolson, 1960; Nicolson and Johnston 1979). Inventory in subtropical locations are quite recent: Southeastern United States (Sylvia, 1986), Gulf of Mexico (Corkidi and Rincon 1997 a, b), Baja, California (Siguenza *et al.*, 1996), Brazil (Sturmer and Bellei 1994), Japan (Abe *et al.*, 1994) and Australia (Koske 1975; Logan *et al.*, 1989). Hawaiian Island (Koske 1988; Koske and Gemma, 1996), India (Kulkarni *et al.*, 1997; Mohankumar *et al.*, 1988, Beena *et al.*, 2001) and Singapore (Louis, 1990) are the only three tropical locations surveyed so far for sand dune AM fungi. The members of the family Poaceae are the main sand dune stabilizing plants in temperate dune (Read, 1989), while in tropical locations plant species belonging to Asteraceae, Convolvulaceae, Fabaceae and Poaceae contribute towards the dune building processes alongwith AM fungi (Devall, 1992; Koske and Gemma, 1990; Kulkarni *et al.*, 1997; Moreno-Casasola and Espejel, 1986). AM fungi are a nearly constant component of the Hawaiian coastal strand vegetation. Over 70% of the 44 species of dune colonizing plants examined formed VA mycorrhizae (Gemma *et al.*, 1992). Knowledge of the plant diversity and their association with AM fungi on maritime sand dunes is of crucial importance for their efficient use in the conservation and management of strand ecosystem. Inventory of maritime sand dune AM fungi of Indian subcontinent are scanty (Kulkarni, *et al.*, 1997; Mohankumar, *et al.*, 1988). The high diversity of AM fungal flora has been recorded in the tropical (Kulkarni *et al.*, 1997; Mohankumar *et al.*, 1988) and subtropical sand dunes (Koske 1988; Koske and Gemma, 1996; Sturmer and Bellei, 1994). In a study (unpublished data) of Indian costal dunes the mean AM fungal species richness (number of AM fungi per plant species) for 28 plant species was 4.4 (range, 0-11) which is higher than the richness reported from Australia (1.5-2.4; Koske, 1975), Rhode Island (4.2; Koske, 1987) and Hawaiian Islands (2 and 2.4; Koske, 1988; Koske and Gemma, 1996). It was, however, lower than Cape Cod (5.3; Koske and Gemma, 1997), Virginia (6.3; Koske, 1987) and Brazil (5.9; Sturmer and Bellei, 1994). The frequently occurring AM fungi of west coast dunes of India were *Scutellospora crythropa*, *S. gregaria*, *Gigaspora margarita* and *Glomus albidum* (Kulkarni *et al.*, 1997). *Scutellospora gregaria* was found to be common on the dunes of Florida (Nicolson and Schenck, 1979), Hawaii (Koske and Gemma, 1996) and Brazil (Sturmer and Bellei, 1994). *Scutellospora gregaria* was found to be associated with *Ipomoea pes-caprae* in

Hawaiian dunes (Koske and Gemma, 1996). *Glomus intraradices* was isolated frequently in the dunes of Hawaii (Koske and Gemma, 1996). *Glomus aggregatum* is a common inhabitant of sand dunes in Hawaii (Koske, 1988), but is a minor species on the west coast dunes of India.

A wide variety of plant species have been established on the West Coast of India (sedges, scrubs, herbs, climbers and creepers) (Rao and Meher-Homji, 1985). Five plant species belonging to the so-called non-mycorrhizal family (Aizoaceae, Caryophyllaceae and Cyperaceae) were screened for AM fungal assocation. *Sesuvium portulacastrum* (Aizoaceae) a mat forming creeper helps in the stabilization of dunes (Rao and Meher-Homji, 1985). It was neither colonized by AM fungi, nor did possess spores in the rhizosphere. This plant species was also not colonized at the dunes of Hawaii (Koske and Gemma, 1990) and Singapore (Louis, 1990). However, on the dunes of Australia and Gulf of Mexico it was colonized up to 41 and 19%, respectively (Corkidi and Rincon 1997a; Logan *et al.*, 1989). *Polycarpaea corymbosa* (Caryophyllaceae) established mainly on the hind dunes. The rhizomatous plant species such as *Sporobolus virginicus* and *Jacquemontia sandwicensis* serve as good inoculum of AM fungi on Hawaiian dunes. The AM vegetative fragments and the spores in the roots of the former plant species retained viability and became potential inoculum in spite of sea water treatment, suggesting the oceanic dispersal of AM fungi (Koske and Gemma, 1990).

Ipomoea pes-caprae shows wide distribution in tropics and being a sand dune creeper it contributes substantially towards dune stabilization (Devall, 1992; Rao and Meher-Homji, 1985). Mycorrhizal association in *Ipomoea* has also been reported from India (Kulkarni *et al.*, 1997; Mohankumar *et al.*, 1988), Singapore (Louis, 1990), Gulf of Mexico (Corkidi and Rincon, 1997a) and Hawaiian Islands (Koske and Gemma, 1996). *Spinifex littoreus* (Poaceae) is dominant at Bengre St. 1 dunes in spite of heavy sand accretion. This plant species has been profusely colonized by AM fungi and distributed widely on the dunes under heavy accretion (Sridhar *et al.*, unpublished data).

Among the nitrogen fixing plant species on the dunes, the stoloniferous creepers *Canavalia rosea* and *Canavalia cathartica* show wide distribution on the west coast (Arun *et al.*, 1999). *Canavalia rosea* and *Canavalia cathartica* grow profusely across the dune along with *Ipomoea pes-caprae,* while *Canavalia cathartica* establishes independently on fore dunes during post monsoon season. Roots of *Canavalia rosea* and shown 35% and 46% colonization by AM fungi on the dunes of Australia (Logan *et al.*, 1989) and Gulf of Mexico, respectively (Corkidi and Rincon, 1997a). *Alysicarpus rugosus* found sparsely on the West Coast dunes in mixed vegetation (Arun *et al.*, 1999) which has possesses good sand binding ability owing to its fine fibrous roots.

Sand movement is an important factor affecting the distribution and composition of coastal plant communities (Martinez and Moreno-Casasola 1996; Moreno-Casasola, 1986). Stabilization of disturbed coastal ecosystem is dependent upon successful re-establishment of most effective plant communities (Louis 1990; Skujins and Allen, 1986). Based on the ecological amplitude, Rao (1977) suggested that mat-forming strand creepers might be used for the fixation of mobile dunes and moving sand. The literature reveals that these mat-forming strand communities (*Alysicarpus rugosus, Canavalia rosea, Canavalia cathartica, Ipomoea pes-caprae* and *Launaea sarmentosa)* are highly mycorrhizal and some of them are nitrogen fixers. Koske and Gemma (1997) and Miller (1985) suggested that for primary succession in sand dunes the restoration of above ground plant communities must include restoration of below ground AM fungal communities.

Conclusion

Mycorrhizal associations are the most widespread and common symbiosis, which have been well studied. Mycorrhizal associations are found in a broad range of habitats ranging from aquatic to deserts and alpine altitudes. These fungi are found in almost every ecosystem even in pioneer zones or very saline environments and are reported to represent the second largest biomass component of many terrestrial ecosystems. The presence and activity of mycorrhizae frequently determine the productivity, species composition and the diversity of natural ecosystems. In order to understand the functioning of the ecosystems it is thus necessary to study the diversity and role of these fungi in various ecosystems.

References

Abbott, L.K. and Robson, A.D. (1977). The distribution and abundance of vesicular-arbuscular endophytes in some Western Australian Soils. *Austr. J. Bot.*, 25: 515-522.

Abe, J.P., Masuhara, G.and Katsuya, K. (1994). Vesicular arbuscular mycorrhizal fungi in coastal dune plant communities I. Spore formation of *Glomus* sp. predominates under a patch of *Elymusmollis. Mycoscience,* 35: 233-238.

Allen, F.B. and Allen, M.F. (1984). Competition between plants of different successional stages: Mycorrhizae as regulators. *Can. J. Bot.*, 62: 2625-2629.

Alva, A.K., Larsen, S. and Bille, S. W. (1980). The influence of rhizosphere in rice crop on resin-extractable phosphate in flooded soils at various levels of phosphate application. *Plant Soil,* 56: 17-33.

Anderson R.C., Liberta, A.E. and Dickman, L.A. (1984). Interaction of vascular plants and vesicular-arbuscular mycorrhizal fungi across a soil moisture gradient. *Oecologia,* 64: 111-117.

Arun, A.B., Beena, N K., Raviraja, R..S. and Sridhar, K.R. (1999). Coastal sand dunes-A neglected ecosystem. *Curr. Sci.,* 77: 19-21.

Baltruschat, H. and Dehne, H.W. (1988). The occurrence of vesicular-arbuscular mycorrhiza in agrosystems: I. Influence of nitrogen fertilization and green manure in continuous monoculture and in crop rotation on the inoculum potential of winter wheat. *Plant Soil,* 107: 279-284.

Beena, K.R., Arun, A.B., Raviraja, N. S. and Sridhar, K. R. (2001). Association of arbuscular mycorrhizal fungi with plants of coastal sand dunes of west coast of India. *Trop. Ecol.,* 42: 213-222.

Bethlenfalvay, G. J. (1992). Mycorrhizae and crop production. In:, *Mycorrhizae in sustainable agriculture,* (Eds.) G.J. Bethlenfalvey, and R. G. Linderman *American Society of Agronomy Special Publication* 54, Madison: 1-27.

Blaszkowski, J. (1997). *Glomus gibbosum,* a new species from Poland. *Mycologia,* 89: 339-345.

Brand, F. (1991). Ektomykorrhizen an fagus sylvatica. Charakterisierung and identifizierung, Okologische Kennzeichnung and unsterile Kultivierung. Libri Botanici 2, IHW-Verlag, Eching.

Brown, M.B., Quimio, T.H. and Dc Castro, A.M. (1988). Vesicular arbuscular mycorrhizas associated with upland rice (*Oryza sativa* L.) *Philip. J. Agric.,* 71: 317-332.

Carpenter, S.R. and Lodge, D.M. (1986). Effects of submersed macrophytes on ecosystem processes. *Aquatic Bot.,* 26: 341-370.

Castellano, M.A. (1988). *The taxonomy of the genus Hysterangium (Basidiomycotina, Hysterangiaceae) with notes on its ecology.* Ph. D. Thesis. Oregon State University, Corvallis U. S. A.

Corkidi, L. and Rincon, E. (1997a). Arbuscular mycorrhizae in a tropical sand dune ecosystem on the Gulf of Mexico I. Mycorrhizal status and inoculum potential along a successional gradient. *Mycorrhiza,* 7: 9-15.

Corkidi, L. and Rincon, E. (1997b). Arbuscular mycorrhizae in a tropical sand dune ecosystem on the Gulf of Mexico II. Effects of arbuscular mycorrhizal fungi on the growth of species distributed in different early successional stages. *Mycorrhiza,* 7: 17-23.

Cromack, K., Sollions, P., Graustein, W.C., Speidel, K., Todd, A. R.G. W., Spycher, G., Li, C. Y. and Todd, R. L. (1979). Calcium oxalate accumulation and soil weathering in mats of the hypogeous fungi, *Hysterangium crassum. Soil Biol. Biochem.,* 11: 463.

Daft, M.J. and Hacskaylo, E. (1977). Growth of endomycorrhizal and nonmycorrhizal red maple seedlings in sand and anthracite spoil. *Forest. Sci.,* 23: 303-314.

Daft, M. J. and Nicolson, T. H. (1974). Arbuscular mycorrhizas in plants colonizing coal-wastes in Scotland. *New Phytol.,* 73: 1129.

Daft, M.J., Hacskaylo, E. and Nicolson, T.H. (1975). Arbuscular Mycorrhizas in plants colonizing mine spoils in Scotland and Pennsylvania. In: *Endomycorrhizas,* (Eds.) F.E. Sanders, B. Mosse and P.B. Tinker, Academic Press, London and New York: 561-580.

Danielson, R. M. (1985). Mycorrhizae and reclamation of stressed terrestrial environments. In: *Soil Reclamation Process-Microorganisms, Analysis and Applications.* (Eds.) R.L. Tate and D. A. Klien, Marcel Dekker, New York: 173-201.

Devall, M.S. (1992). The biological flora of coastal dunes and wetlands. 2. *Ipomoea pes-caprae* (L.) Roth. *J. Coastal Res.,* 8: 442-456.

Dodd, J.C., Arias, I., Koomen, I. and Hayman, D.S. (1990). The management of populations of vesicular-arbuscular mycorhizal fungi in acid-infertile soils of a savanna ecosystem II. The effects of pre cropping on the spore populations of native and introduced VAM fungi. *Plant Soil,* 122: 241-247.

Doerr, T.B., Redente, E.F. and Reeves, F.B. (1984). Effect of soil disturbance on plant succession and levels of mycorrhizal fungi in a sagebrush-grass land community. *J. Range Manag.,* 37: 135-139.

Filer Jr., T.H. (1975). Mycorrhizae and Soil microflora in a green-tree reservoir. *Flori. Sci.,* 21: 36-39.

Foote, D.E., Hill, E.L., Nakamura, S. and Stephens, F. (1972). *Soil Survey of Islands of Kauai, Oahu, Mani, Molokai and Lanai, State of Hawaii.* USDA Soil conservation service and University of Hawaii Agricultural Experiment station, Honolulu.

Gangopadhyay, S. and Das, K.M. (1984). Interaction between vesicular arbuscular mycorrhiza and rice roots. *Ind. Phytopathol.,* 37: 34-38.

Gangopadhyay, S. and Das, K.M. (1988). Control of soil borne diseases of rice through vesicular-arbuseular mycorrhiza. In: *Mycorrhiza round table,* (Eds.) A.K. Varma, A.K. Oka, K.G. Mukerji, K.V.B.R. Tilak and J. Raj, International Development Research Centre, Canada, Jawaharlal Nehru University, New Delhi: 560-580.

Gangopadhyay, S. and Das, K.M. (1982). Occurrence of vesicular-arbusecular mycorrhiza in rice in India. *Ind. Phytopath.,* 35: 83-85.

Gemma, J.N. and Koske, R.E. (1990). Mycorrhizae in recent volcanic substrates in Hawaii. *Am. J. Bot.,* 77: 1193-1200.

Gemma, J.N., Koske, R.E. and Flynn, T. (1992). Mycorrhizal in Hawaiian Pteridophytes: Occurrence and evolutionary significance. *Am. J. Bot.,* 79: 843-852.

Giovannetti, M. and Nicolson, T.H. (1983). Vesicular-arbuscular mycorrhizas in Italian sand dunes. *Trans. Brit. Mycol. Soc.,* 80: 552-557.

Giovannetti, M. (1985). Seasonal variations of vesicular-arbuscular mycorrhizas and endogonaceous spores in a maritime sand dune. *Trans. Brit. Mycol. Soc.,* 84: 679-689.

Graw, D. (1979). The influence of soil pH on the efficiency of vesicular-arbuscular mycorrhiza. *New Phytol.*, 82: 687-695.

Green, N.R., Graham, S.O. and Schenck, N.C. (1976). The influence of pH on the germination of VAM spores. *Mycologia*, 68: 929-934.

Gregg, W.W. and Rose, F.L. (1982). The effects of aquatic macrophytes on the stream micro-environment, *Aquatic Bot.*, 14: 309-324.

Griffiths, R.P., Castellano, M.A. and Coldwell, B.A. (1991). Ectomycorrhizal mats formed by *Grautieria monticola* and *Hysterangium setchellii* and their association with Douglasfir seedling, a case study. *Plant Soil*, 134: 255-259.

Harinikumar, K.M. and Bagyaraj, D.J. (1988). Effect of crop rotation on native vesicular arbuscular mycorrhizal propagules in soil. *Plant Soil*, 110: 77-80.

Hayman, D.S. (1982). Influence of Soils and fertility on activity and survival of vesicular-arbuscular mycorrhizal fungi. *Phytopath.*, 72: 1119-1125.

Hayman, D.S. (1980). Mycorrhiza and crop production. *Nature*, (London) 287: 487-488.

Hayman, D.S. and Mosse, B. (1971). Plant growth responses to vesicular-arbuscular mycorrhiza. I. Growth of Endogone-inoculated plants in phosphate deficient soils. *New Phytol.*, 70: 19-27.

Hepper, C.M. (1984). Regulation of spore germination of the vesicular-arbuscular mycorrhizal fungus *Acaulospora laevis* by pH. *Trans. Brit. Mycol. Soc.*, 83: 154-156.

Hildebrandt, U., Janetta, K., Quziad, F., Renne, B., Nawrath, K. and Bothe, H. (2001). Arbuscular mycorrhizal colonisation of halophytes in central European salt marshes. *Mycorrhiza*, 10: 4.

Ilag, L.L., Rosales, A.M., Elazegui, F.A. and Mew, T.W. (1987). Changes in the population of infective endomycorrhizal fungi in a rice-based cropping system. *Plant Soil*, 103: 67-73.

Iqbal, S.H., Tauquir, S., Aziz, Al., Ahmad, J.S. and. Iqbal, H.M (1978). A field survey of vesicular-arbuscular mycorrhizal associations in cereals. *Biol. Pakistan*, 24: 97-113.

Janardhan, K.K., Khaliq, A., Naushin, F. and Ramaswamy, K. (1994). VAM in an alkaline usar land ecosystem. *Curr. Sci.*, 67: 465-469.

Janos, D.P. (1984). Methods for vesicular-arbuscular mycorrhiza research in the lowland wet tropics. In: *Physiological Ecology of Plants of the Wet Tropics*. (Eds.) E. Medina, H.A. Mooney and C. Vasquez-Yares, De Junk, The Hague: 173-187.

Janos, D.P. (1980). Vesicular arbuscular mycorrhizae affect lowland tropical rain forest plant growth. *Ecology*, 61: 151-162.

Johnson, I.R., Melkonian, J.J., Thornley, J.M.H. and Riha, S.J. (1991). A model of water flow through plant incorporating shoot/root 'message' control of stomatal conductance. *Pl. Cell Environ.*, 14: 431-544.

Khan, A. G. (1993a). Occurrence and importance of mycorrhizae in aquatic trees of New South Wales, Australia. *Mycorrhiza*, 3: 31-38.

Khan, A.G. (1993b). The influence of redox potential on formation of mycorrhizae in trees from wetland and waterlogged areas of New South Wales, Australia, Abst, *9th North American Conference on Mycorrhizae*, Aug 8-12, 1993, University of Guelph, Guelph, Ontario: 18.

Khan, A.G. (1993c). Vesicular arbscular mycorrhizae (VAM) in aquatic trees of New South Wales, Australia, and their importance at land-water interface. In: *Wetlands and ecosystems*: (Eds.) B. Gopal, A. Hillbricht-Ilkowska and R.G. Wetzel, Studies on land water interactions. National Institute of Ecology, New Delhi: 173-180.

Khan, A.G. (1974). The occurrence of mycorrhiza in halophytes, hydrophytes and xerophytes and of *Endogone* spore in adjacent soils. *J. Gen. Microbiol.*, 81: 7-14.

Koske, R.E. and Gemma, J.N. (1990). VA mycorrhiza in vegetation of Hawaiian coastal strand: evidence for co-dispersal of fungi and plants. *Am. J. Bot.*, 77: 466-474.

Koske, R.E. and Halvorson, W.L. (1981). Ecological studies of vesicular-arbuscular mycorrhizae in a barrier sand dune. *Can. J. Bot.*, 59: 413-1422.

Koske, R.E. and Polson, W.R. (1984). Are VA mycorrhizae required for sand dune stabilization? *Bioscience*, 34: 420-424.

Koske, R.E. (1975). *Endogone* spores in Australian sand dunes. *Can. J. Bot.*, 53: 668-672.

Koske, R.E. (1987). Distribution of mycorrhizal fungi along the latitudinal temperature gradient. *Mycologia*, 79: 55-68.

Koske, R.E. (1988). Vesicular-arbuscular mycorrhizae of Hawaiian dune plants. *Pacific Sci.*, 42: 217-229.

Koske, R.E and Gemma, J.N. (1996). Arbuscular mycorrhizal fungi in Hawaiian sand dunes: Islands of Kauai. *Pacific Sci.*, 50: 36-45.

Koske, R.E and Gemma, J.N.1997. Mycorrhizae and succession in plantings of beach grass in sand dunes. *Am. J. Bot.*, 84: 118-130.

Kruckelmann, H.W. (1973). *Die vesikular-arbuskulare Mykorrhiza undihre Beeinflussung in landwirtschaftlichen Kultuven. Dissertation*, University of Braunschweig.

Kucey, R.M.N. and Paul, E.A. (1983). Vesicular-arbuscular mycorrhizal spore populations in various Saskatchewan soils and the effect of inoculation with *Glomus mosseae* on faba bean growth on greenhouse and field trials. *Can. J. Soil Sci.*, 63: 87-95.

Kulkarni, S.S., Raviraja, N.S.and Sridhar, K.R. (1997). Arbuscular mycorrhizal fungi of tropical sand dunes of West Coast of India. *J. Coastal Res.*, 13: 931-936.

Lakhanpal, T.N. (1988). Morphology and anatomy of mycorrhizal fungi from India. Presented at *National Workshop on mycorrhizae* of Jawaharlal Nehru University, New Delhi.

Lesica, P. and Antibus, R.K. (1985). Mycorrhizae of alpine fell-field communities on soils derived from crystalline and calcareous parent materials. *Can. J. Bot.*, 64: 1691-1697.

Logan, V.S., Clarke, P.J. and Allaway, W.G. (1989). Mycorrhizas and root attributes of plants of coastal sand dunes of New South Wales. *Austr. J. Plant Physiol.*, 16: 141-146.

Louis, I. (1990). A mycorrhizal survey of plant species colonizing coastal reclaimed land in Singapore. *Mycologia*, 82: 772-778.

Magurran, A.E. (1988). *Ecological Diversity and Its Measurement.* Croom Helm, London.

Malajczuk, N. and Lamont, B.B. (1981). Specialized roots of symbiotic origin in heathlands. In: *Heathlands and related shrublands of the world*: (Ed.) R.L. Specht, Analytical studies B. Elsevier, Amsterdam: 165-182.

Martinez, M.L. and Moreno-Casasola, P. (1996). Effects of burial by sand on seedling growth and survival in six tropical sand dune species from the Gulf of Mexico. *J. Coastal Res.*, 12: 406-419.

Mc. Gonigle, T.P and Fitter, A.H. (1990). Ecological specificity of vesicular-arbuscular mycorrhizal associations. *Mycol. Res.*, 94: 120-122.

Medve, R.J. (1984). The mycorrhizae of pioneer species in disturbed ecosystems in western pennsylvania. *Am. J. Bot.*, 71: 787-789.

Miller, F. M. (1979). Some occurrence of Vesicular-arbuscular mycorrhiza in natural and disturbed ecosystems of the Red Desert. *Can. J. Bot.*, 57: 619-623.

Miller, R. M. (1985). Mycorrhizae. *Restorat. Manage. Notes,* 3: 14-20.

Miller, R.M. (1987). The ecology of vesicular-arbuscular mycorrhizae in grass and shurb lands. In: *Ecophysiology of VA Mycorrhizal plants.* (Ed.) G. R. Safir, CRC press, Boca Raton, Florida: 135-170.

Mmbaga, M.T. (1979). A comparative study of VA mycorrhiza in paddy and upland rice in seven localities in Tanzania. In: *Abstr. 4th North American Conference on Mycorrhizae*, (Ed.) C.P.P. Reid, Fort Collins, Colorado: 45.

Mohankumar, V. and Mahadevan, A. (1986). Survey of vesicular arbuscular mycorrhizae in mangrove vegetation. *Curr. Sci.*, 55: 936.

Mohankumar, V., Ragupathy, S., Nirmala, C. B. and Mahadevan, A. (1988). Distribution of vesicular arbuscular mycorrhizae (VAM) in the sandy beach soils of Madras coast. *Curr. Sci.*, 57: 367-368.

Moreno-Casasola, P. and Espejel, I. (1986). Classification and ordination of coastal sanddure vegetation along the Gulf and Caribbean Sea of Mexico. *Vegetatio*, 66: 147-182.

Moreno-Casasola, P. (1986). Sand movement as a factor in the distribution of plant communities in a coastal dune system. *Vegetatio*, 65: 67-76.

Mosse, B. (1972). Effect of different *Endogne* strains on the growth of *Paspalum notatum*. *Nature*, 239: 221-223.

Neeraj, Shankar, A., Mathew, J. and Varma, A. K. (1991). Occurrence of VA mycorrhizal within Indian Semi-acid soils. *Biol. Fertil. Soils*, 11:140-144.

Nicolson, T.H. (1959). Mycorrhiza in the Gramineae. I Vesicular arbuscular endophytes with special reference to the external phase. *Trans. Brit. Mycol. Soc.,* 42: 132-145.

Nicolson, T.H. (1960). Mycorrhiza in the Gramineae. II. Development of different habitats, particularly sand duses. *Trans. Brit. Mycol. Soc.,* 43: 132-145.

Nicolson, T. H. and Johnston, C. (1979). Mycorrhizal in the Gramineae. III *Glomus fasciculatus* as the endophyte of pioneer grasses is a maritime sand dune. *Trans. Brit. Mycol. Soc.,* 72: 261-268.

Nicolson, T. H. and Schenck, N. C. (1979). Endogonaceous mycorrhizal endophyte in florida. *Mycologia,* 71: 178-198.

Nopamornbodi, O,, Schenck, N. C. and Vasuvat, Y. (1985). Incidence and survival of VA mycorrhizal fungi in paddy rice culture. In: *Proc 6th North American Conference on Mycorrhizae,* (Ed.) R. Molina, Forest Research Lab, Corvallis, Oregon: 333.

Nopamornbodi, O., Thumsurakul, S. and Vasuvat, Y. (1987). Survival of VA mycorrhizal fungi after paddy rice. In: *Mycorrhizae in the North American Conference on Mycorrhizae,* (Eds.) D.M. Sylvia, L.L. Hung and J.H. Graham, University of Florida, Gainesville, Florida: 53.

Peuss, H. (1958). Untersuchungen okologie and Bedeutung der Tabakmykorrhiza. archiv fur *Mikrobiologie,* 29: 112-142.

Pfeiffer, C.M. and Bloss, H.E. (1988). Growth and nutrition of guayule (*Parthenium argentatum*) in a saline soil as influenced by vesicular arbuscular mycorrhiza and phosphorus fertilization. *New Phytol.,* 108: 315-321.

Pond, E.C., Menge, J.A. and Jarrell, W.M. (1984). Improved growth of tomato in salinized soil by vesicular-arbuscular mycorrhizal fungi collected from saline soils. *Mycologia,* 76: 74-84.

Porter, W.M., Robson, A.D. and Abbott, L.K. (1987). Factors controlling the distribution of vesicular arbuscular mycorrhiza fungi in relation to soil pH. *J. Appl. Ecol..* 24: 663-672.

Poss, J.A., Pond, E., Meng, J.A. and Jarrel, W.M. (1985). Effect of salinity on mycorrhizal onion and tomato in soil with and without additional phosphate. *Plant Soil,* 88: 307-319.

Ranwell, D.S. (1972). *The Ecology of Salt Marshes and Sand Dunes.* Chapman and Hall, London.

Rao, T.A. (1977). Management of Coastal sandy biomass in India. *Seminar on Afforestation,* Institute of Public Health Engineers, Calcutta.

Rao, T.A. and Meher-Homji, V.M. (1985). Stand plant communities of the Indian Subcontinent. *Proc. Ind. Acad. Sci. (Plant Science),* 94: 505-523.

Read, D.J., Koucheki, H.K. and Hodgson, J. (1976). Vesicular-arbuscular mycorrhiza in natural vegetation systems. The occurrence of infection. *New Phytol.,* 77 : 641-653.

Read, D.J. and Boyd. R. (1986). Water relations of mycorrhizal fungi and their host plants. In: *Water, Fungi, and Plants*. (Eds.) P.G. Ayers and L Boddy, Cambridge University Press, Cambridge: 287-304.

Read, D.J. (1989). Mycorrhizal and nutrient cycling in sand dune ecosystem. *Proc. Royal Soc. Edinburgh*, 96: 89-100.

Reeves, F.B., Wagner, D., Moorman, T. and Kiel, J. (1979). The role of endomycorrhizae in revegetation practices in the semi-arid west. I. A comparison of incidence of mycorrhizae in severely disturbed vs. Natural environments. *Am. J. Bot.*, 66: 6-13.

Ruizlozano, J. M. and Azcon, R. (2000). Symbiotic efficiency and infectivity of an autochthonous arbuscular mycorrhizal *Glomus* sp. from saline soils and *Glomus deserticola* under salinity. *Mycorrhiza*, 10: (3)

Saif, S.R., Shakh, N.A. and Khan, A.G. (1975). Ecology of Endogone 1. Relationship of Endogone spore population with physical soil factors. *Islamabad J. Sci.*, 2: 1-5.

Sengupta, A. and Chaudhuri, S. (1990). Vesicular-arbuscular mycorrhiza (VAM) in pioneer salt marsh plants of the Ganges river delta in West Bengal (India). *Plant and Soil*, 122: 111-113

Siddhu, D.P. and Behl, H.M. (1990). Endomycorrhizal fungi from leguminous tree species for fuel wood plantation in alkaline soil sites. *Nitrogen Fixing Tree Res. Rep.*, 8: 34-36.

Sieverding, E. (1979). Einflu B der Bodenfeuchte auf die Effekivitat der VA-Mycorrhiza. (Influence of Soil moisture on effectiveness of VA-mycorrhizae). *Angew Bot.*, 53: 91-98.

Sieverding, E. and Leihner, D.E. (1984). Influence of crop and intercropping of Cassava with legumes on VA mycorrhizal symbiosis of Cassava. *Plant Soil*, 80: 143-146.

Siguenza, C. Espejel, I. and Allen. E.B. (1996). Seasonality of mycorrhizal in coastal sand dunes of Baja California. *Mycorrhiza*, 6: 151-157.

Skujins, J. and Allen, M.F. (1986). Use of mycorrhizal for land rehabilitation. *Mircen J.*, 2: 161-176.

Smith, S.E. (1995). Discoveries, discussion and directions in mycorrhizal research. In: *Mycorrhizae*, (Eds.) A. Verma and B. Hock, Springer, Verlage: 3-24.

Sondergaard, M. and Laegaard, S. (1977). Vesicular-arbuscular mycorrhiza in some aquatic plants. *Nature*, 268: 232-233.

Sparling, G. P. and Tinker, P. B. (1978). Mycorrhizal infection of Pennine grassland. I. Levels of infection in the field. *J. Appl. Ecol.*, 15: 943-950.

Stahl, P.D. and Christensen, M. (1991). Population variation in the mycorrhizal fungus *Glomus mosseae*-breadth of environmental tolerance. *Mycol. Res.*, 95: 300-307.

Stahl, P.D., Williams, S.E. and Christensen, M. (1988). Efficacy of native vesicular-arbuscular mycorrhizal fungi after severe soil disturbance. *New Phytol*, 110: 347-354.

Sturmer, S.L. and Bellei, M.M. (1994). Composition and seasonal variation of spore of arbuscular mycorrhizal fungi in dune soils on the island of Santa Catarina, Brazil. *Can. J. Bot.* 72: 359-363.

Sutton, J.C. and Sheppard, B.R. (1976). Aggregation of sand dune soil by endomycorrhizal fungi. *Can. J. Bot.* 54: 326-333.

Sward, R.J., Hallam, N.D. and Holl, A.A. (1978). Endogone spores in a heathland area of south-eastern Australia. *Austr. J. Bot.,* 26: 29-43.

Sylvia, D.M. (1986). Spatial and temporal distribution of vesicular arbuscular mycorrhizal fungi associated with *Vniola paniculata* in florida foredunes. *Mycologia,* 78: 728-734.

Sylvia, D.M. and Schenck, N.C. (1983). Application of superphosphate to mycorrhizal plants stimulates sporulation of phosphorus-tolerant vesicular-arbuscular mycorrhizal fungi. *New Phytol.,* 95: 655-661.

Sylvia, D.M. and William, S. E. (1992). Vesicular-arbuscular mycorrhizae in sustainable agriculture. In: *American Society of Agronomy,* (Eds.) G.J. Bethlenfalvag and R.G. Linderman, Madison: 101-124.

Thoen, D. (1987). First observations on the occurrence of vesicular arbuscular mycorrhizae (VAM) in hydrophytes, hygrophytes, halophytes and xerophytes in the region of Lake Retba (Cap-Vert, Senegal) during the dry season. *Memoirs Soc. Royal Bot. Belgium,* 9: 60-66.

Trappe, J. M. (1981). *Mycorrhizal and productivity of arid and semi-arid rangelands. Advances in food producing systems for arid and semi-arid levels.* Academic Press, New York: 581-703.

Trappe, J.M. (1987). Phylogenetic and ecologic aspects of mycotrophy in the angiosperms from an evolutionary standpoint. In: *Ecophysiology of VA Mycorrhizal plants,* (Ed.) G.R. Safir, CRC Press, BOCA Raton, FL. : 5-25.

Waaland, M.E. and Allen, E.B. (1887). Relationship between VA mycorrhizal fungi and plant cover following surface mining in wyoming. *J. Range Manage.,* 40: 271-276.

Wang, G.M., Stribley, D.P., Tinker, P.B. and Walker, C. (1985). Soil pH and vesicular-arbuscular mycorrhizas. In: *Ecological Interactions in Soil: Plants, Microbes and Animals,* (Ed.) A.H. Filter, Blackwell, Oxford: 219-224.

□□□

Microbial Diversity and Functions, 2012
© D.J. Bagyaraj, K.V.B.R. Tilak, H.K. Kehri (eds.), pp. 353-375
New India Publishing Agency, New Delhi (India)
E-mail : info@nipabooks.com; Website : www.nipabooks.com

Chapter **16**

Biodegradation of Green Manure by Soil Mycoflora

Asha Sinha and Manisha Srivastava

ABSTRACT

Various cropping systems and ecologies illustrate the positive use of chemicals both as fertilizers and pesticides and organic/biological sources within the frame work of Integrated Plant Nutrition and Plant Protection System (IPPS). While considering the organic/biological sources, green manures seem to be the best alternative that may largely suffice to the organic matter requirements of the soil under crop cultivation. Green manuring can be defined as a practice of ploughing or turning into the soil undecomposed green plant tissues for the purpose of improving physical structure as well as fertility of the soil. "Green manure" refers to fresh plant matter, which is added to the soil largely for supplying the nutrients contained in its biomass. The agricultural practices consisting of growing and ploughing under a green crop or addition of a green organic matter in the form of leaves and other succulent parts from outside to improve soil productivity is known as green manuring. In this review, advantages of green manuring, decomposition of green manure, process of decomposition, factors affecting decomposition and effects of green manuring have been discussed in detail.

Keywords: Green manures, biodegradation, soil mycoflora

Introduction

Various cropping systems and ecologies illustrate the positive use of chemicals both as fertilizers and pesticides and organic/biological sources with in the frame work of Integrated Plant Nutrition and Plant Protection

System (IPPS) (Roy, 1994). While considering the organic/biological sources, green manures seem to be the best alternative that may largely suffice to the organic matter requirements of the soil under crop cultivation.

Green manuring can be defined as a practice of ploughing or turning into the soil undecomposed green plant tissues for the purpose of improving physical structure as well as fertility of the soil. "Green manure" refers to fresh plant matter, which is added to the soil largely for supplying the nutrients contained in its biomass. The agricultural practices consisting of growing and ploughing under a green crop or addition of a green organic matter in the form of leaves and other succulent parts from outside to improve soil productivity is known as green manuring (Agarwal, 1967).

The practice of green manuring is adopted in various ways in different states of India. The green manuring can be differentiated into two ways:

a) *Green manuring in-situ*: In this system the green manure crops are grown and buried in the same field. The most common green manure crops grown under this system are sunnhemp (*Crotolaria juncea*), dhaincha (*Sesbania aculeata*), pillipesara (*Phaseolus trilobus*) and guar (*Cymopsis terragonoloba*).

b) *Green leaf manuring*: Green leaf manuring refers to turing into the soil green leaves and tender green twigs collected from shrubs and trees grown on bunds, waste lands and nearby forest areas. The common shrubs and trees used are glyricidia (*Glyricidia maculata*), *Sesbania speciosa*, karanj (*Pongamia pinnata*) etc.

The green manures are well known to improve the soil fertility and supplying a part of nutrients to the crop. Green manure must be plants of grain legume such as pigeon pea, green gram cowpea, soybean, groundnut, perennial woody multipurpose legumes, *viz.*, *Leucaena leucocephala* (Subabul), *Gliricidia sepium*, *Cassia siamea* or non-grain legumes like *Crotolaria*, *Sesbania*, *Centrosema*, *Stylosanthes*, *Desmodium* and forage crops like *Trifolium* (Berseem) and *Melilotus* (Senji) etc. (Yawalkar *et al.*, 1984, Palaniappan, 1994).

Higher availability of phosphorus from rock phosphate has been reported in rice due to green manuring (Ranjan and Kothandaraman, 1986). It is now a general recommendation to apply 30 kg P_2O_5/ ha to Dhaincha grown as the green manure for rice production. The green manure incorporation also resulted in 10-12% enhanced utilization of phosphorus and potassium (Lekha and Palaniappan, 1990). Enhanced availability of iron upon green manuring has also been reported (Palaniappan, 1994). Further studies are needed especially on the transformation and availability of other micronutrients.

Green manuring is now confined to certain areas of intensive agriculture. It was reported that the decomposition of green legume residues was much faster as compared to farmyard manure (FYM) where as cereal residues did not improve organic matter status of soil as compared to FYM and other residues. Gaur *et al.* (1970 a, b, 1971), Gaur and Mukherjee (1980) and Singh (1998) further indicated that legumes effect was more important than the increase in organic matter or nitrogen addition by green manuring. These experiments also showed that berseem, senji and pea left through their roots and stubble added 89.4, 53.1 and 20.7 kg nitrogen per hectare respectively. Agriculturists in India need to develop a suitable method to increase organic matter content of the soil. This can be achieved with the organic manuring including green manuring, composting etc. During the past few years the concept has evolved that the periodic addition of large quantities of either crop residue or green organic matter to the soil results in increased production of the plant material. Though, adequate inorganic fertilizers fulfill the requirement of nitrogen, nevertheless organic matter is still required to be maintained at high levels. The use of green residue or green manuring practices therefore needs to be developed for this purpose.

Table 1: List of common leguminous green manuring crops

Name	Botanical name	Growing season	Av. yield of green matter in quintals/ hectare	N percentage on green weight basis	N added in Kg. Per hectare
Sunnhemp	*Crotolaria juncea*	Kharif	152	0.43	84.0
Dhaincha	*Sesbania aculeata*	Kharif	144	0.42	77.1
Pillipesara	*Phaseolus trilobus*	Kharif	132	1.10	55.6
Mung	*Phaseolus aureus*	Kharif	57	0.53	38.6
Cowpea	*Vigna sinensis*	Kharif	108	0.49	56.3
Guar	*Cyamopsis tetragonoloba*	Kharif	144	0.34	62.3
Khesari	*Lathyrus sativus*	Rabi	88	0.54	61.4
Berseem	*Trifolium alexandrinum*	Rabi	111	0.43	60.7

T.J. Mirchandani and A.R. Khan, Green manuring, Indian Council of Agricultural Research, Rewiew series No.6, 1952.

Advantages of green manuring

The advantages generally attributed to green manuring are:

1. It adds organic matter to the soil. This stimulates the activity of soil microorganisms.
2. The green manure crops return to the upper top soil, plant nutrients taken up by the crop from deeper layers.
3. It improves the structure of the soil.
4. It facilitates the penetration of rainwater, thus decreasing run off and erosion
5. The green manure crops hold plant nutrients that would otherwise be lost by leaching.
6. When leguminous plants like sunnhemp and dhaincha are used as green manure crops they add nitrogen to the soil for the succeeding crop.
7. It increases the availability of certain plant nutrients like phosphorus (P_2O_5), calcium, potassium, magnesium and iron.

Green manuring increases crop yields

A large number of field experiments with green manuring have been conducted in various states of India. These experiments are conducted to study the effect of green manuring on the succeeding crops of paddy, wheat, sugarcane, cotton, peas, arhar, tea, coffee and various vegetables. The results obtained with paddy, wheat, sugarcane and cotton are given in Table 2. The beneficial effect of green manuring varies from crop to crop. Paddy is benefited most wheat is next while increase in cotton and sugarcane yield is relatively less.

Table 2 : Percentage increase in crop yield by adopting green manuring practice

Name of crop	Name of state	Name of green manure crop	Percentage increase in yield
Paddy	Tamil Nadu	Sunnhemp	24
	West Bengal	Sunnhemp	20
	Orissa	Dhaincha	21
	Uttar Pradesh	Dhaincha	24
	Bihar	Dhaincha	51
		Dhaincha	60
	Andhra Pradesh	Sunnhemp	63
		Sunnhemp	114
		Dhaincha	80
	Jammu and Kashmir	Lentil	54

Wheat	Punjab	Bhang	35
	Uttar Pradesh	Cowpea	21
		Sunnhemp	45
		Dhaincha	16
	Bihar	Sunnhemp	106
	Madhya Pradesh	Sunnhemp	13
	Delhi	Sunnhemp	35
Sugarcane	Uttar Pradesh	Sunnhemp	30
	Assam	Sunnhemp	9
		Dhaincha	8
Cotton	Gujarat	Dhaincha	21
	Maharashtra	Dhaincha	12
	Tamil Nadu	Sunnhemp	21

Complied from various sources

Decomposition of Green manure

Decomposition of plant materials depends on its chemical constituents as well as on the physical and biochemical conditions in the surrounding environments (Alexander, 1977: Subba Rao, 1988). Among the organic constituents of the plant material the water- soluble fraction is the least resistant component metabolized first, whereas cellulose and hemicellulose do not decompose as fast as the water-soluble substances. Lignin is the most resistant part and consequently, it forms most abundant in residual decaying organic matter. Thus, one fraction decomposes rapidly and the other decomposes slowly over a long period. Decomposition is the conversion of complex organic substances in to simpler ones; dead animals or plants are broken down into large particles and eventually into small molecules. The dead organic matter is thus gradually disintegrated until its structure can no longer be recognized and the complex organic molecules are broken down into carbon dioxide, water and mineral components. The term 'plant litter' is usually used in terrestrial systems and especially for materials derived from higher plants. Litter is the dead plant material not attached to the living plant. Decomposition of green manure serves three major functions: addition of mineral nutrients to the soil, thus increasing the productivity by making various plant nutrients available, providing energy for plant growth and supplying carbon for the formation of new cell materials especially those colonizing saprophytically on the decomposing green manure. The essential feature of the soil inhabitants is the acquisition of energy and carbon for new cell synthesis. Modification of soil environment by green manuring is in full accordance with the ecological principle. Soil contains five major group of organisms *viz.*, bacteria, actinomycetes, fungi, algae and protozoa or other members of animal kingdom such as nematodes, earthworms and arthropods. Among all organisms fungi play an important role in decomposition of many materials. Because of the vital role of fungi their activities have been considered as index of litter decomposition.

The organic substances play a direct role in the enhancement of soil fertility, as they are the sources of plant nutrients liberated in available form during the course of decomposition. Humus is the ultimate product, which can be considered as a storehouse of various plant nutrients essential for the plant growth. During the microbial decomposition of organic substance there is a release of nutrients with subsequent mineralization of carbon, nitrogen, sulfur, phosphorus and other elements.

Several studies have emphasized the importance of green manuring in plant nutrition maintenance of soil fertility as an essential component of IPNS and development towards sustainable agriculture (Panse *et al.*, 1947, Mukerji and Agarwal, 1950 and Palaniappan, 1994, Goyal, 1999; Jayapaul *et al.*, 2000; Sharma and Ghosh, 2000; Yadav *et al.*, 2000; Ed-Haun *et al.*, 2007; Singh *et al.*, 2007; Selvi and Kalpana, 2009 and Sinha *et al.*, 2009). Decomposition of either green or dry material is an integral part of soil microbiology. Several workers have given their views related to the decomposition of different green manure crops in their own ways but they all have visualized the role of microorganisms (Gaur and Mukherji, 1980, Subbarao 1988, Cosico, 1990, Aggarwal and Sekhon, 1991).

It was generally presumed that bacteria decomposed the organic materials and fungi can be found in the soil only as dormant spores deposited by chance from the air. However, Waksman's discovery (1938) of fungal mycelium in soil strongly suggests that fungi must have a role in the decomposition. He further postulated the existence of two ecological groups of fungi in soil 'soil inhabitants' and 'soil invaders'. Soil inhabitants are characteristically found in the soil where as the soil invaders are transient so journers with limited activity. The mainstream of litter decomposition study originates from the observation made by Muller (1974) on the Fagus wood, Quercus wood and Calluna heath of Denmark since then a number of studies have been undertaken by various workers on colonization, succession and decomposition of above ground parts of plant. The bacteria, which are involved to a significant extent in the decomposition of plant litter, can be classified in the orders Pseudomonadales, Eubacterials and Myxobacteriales. Genus *Pseudomonas* is the most important group in the order Pseudomonadales from the decomposition point of view. Two of the groups, *Cellivibrio* and *Alginomonas* are certainly involved in the degradation of complex organic molecules. The thermophilic bacilli produce highly thermoresistant spores, which may be an ecologically significant feature, particularly under natural conditions *viz.*, *Bacillus polymyxa, B. laterosporus, B. macerans, B. larvae, B. brevis, B. alvei, B. circulans* complex and some strains of *B. coagulans.* At the other end the ability of psychrophilic bacilli to maintain their metabolic activity at low temperatures points to their contribution to natural decomposition processes. The pectinolytic and cellulotytic species of *Clostridium* strictly anaerobic sporeformers group, deserve special attention in the plant litter decomposition studies. Apart from sporeformers group Enterobacteriaceae of order Eubacteriales is of interest in connection with plant litter decay. Among the aerobic bacilli, there are active decomposers of starch, pectins and xylans. The spore formers multiply during later stage of mineralization of plant litter and may be connected with the breakdown and recirculation of biomass. Cytophaga and Sporocytophaga are the important genera of fruiting myxobacteria that can be related with the litter breakdown process.

The actinomycetes are prokaryotic bacteria with elongated cells of filaments (0.5 - 2.0 μm diameter) usually showing some degree of true branching. They are usually considered to play a significant role in the breakdown of complex organic compounds along with other saprophytic organisms and in the cycling of nitrogen, sulphur and potassium. It is well known that addition of organic material to soil causes an increase in the number of actinomycetes (Singh, 1937) and there is some evidence that streptomycetes are common in soil rich in plant roots. These findings suggest that actinomycete are able to colonize litter in soil. Further, actinomycetes are most active in degrading lignin and starch and the decomposition of cellulose by individual actinomycetes is only exceeded by a mixed population of fungi. Theromophilic actinomycetes form a significant component of the population of thermophilic microorganisms. Such actinomycetes may be visible as chalk white deposits on composts or as clouds of fine dust released from mouldy hays. The ability to grow at temperatures above 40°C is found in several actinomycete genera especially in species of *Thermoactinomyces, Thermomonospora, Streptomyces, Pseudonocardia* and *Micropolyspora.* A useful working hypothesis is to consider that growth and sporulation of these strains occur at high temperatures in habitats such as composts and wet hay stacks and these spores, whose dormancy can be broken down by meeting high temperature conditions are widely dispersed.

The fungal species that play a large part in the plant litter decomposition are saprophytic in nature. On litter derived directly or indirectly from plants fungi are able to grow whenever they possess the necessary enzyme systems and when the environmental conditions allows within these conditions must be included in the biological component as the fungi must be good competitive saprophytes (Garrett, 1951) to be able to exploit the substrate fully. The ability of fungi to grow on a substrate will determine their role in the decomposition of the material. Fungi have successive phases of colonization, exploitation and exhaustion of organic substrates during decomposition. Initially the residue is colonized by weak parasites, which invade living tissue and pave the way for saprophytes. Later on the ground saprophytes colonize residue in successive and overlapping layers (Sadasivan, 1939 and Harper and Lynch, 1985). Each fungus that colonizes the residue after the microhabitat paves the way for the next (Nandi *et al.*, 1996). Litter quality variables such as total nitrogen, total phosphorus, total carbon, tissue nutrient concentrations explain most of the succession patterns and differences in the micro fungal assemblages of peat land plant litters (Thormann *et al.*, 2003). The fungal succession on the decomposing litter has been also studied by various workers (Hudson, 1968, Sharma, 1973, Jensen, 1974, Visser and Parkinson, 1975, Sinha, 1982, Sinha, 1992, Aneja, 1988, Sinha and Mahapatra, 1993, Sinha and Pathak, 1995, Dilly and Immuer, 1998 and Sinha *et al.*, 1998 a, b, Vibha and Sinha, 2009).

From time to time a number of reviews and articles have discussed the various aspects of microbiology of plant litter decomposition (Garrett, 1981, Sinha, 1993, Aneja, 1988, Steinke *et al.*, 1990 and Sarvanan *et al.*, 1995, Tian *et al.*, 1995, Kjoller and Struwe, 1997, Kuzyakov *et al.*, 1992, Pathak, and Sinha, 1995, Creus *et al.*, 1998, Sinha *et al.*, 1999, Goyal *et al.*, 1999, Lal *et al.*, 2000, Yadav *et al.*, 2002). Deuteromycotina mainly the dematiaceous hyphomycetes is by for the largest group of fungi involved in litter decomposition. Ascomycetes members are few while the Zygomycetes are rare in occurrence (Aneja, 1988). *Rhizophyctis* among Mastigomycotina is the genera that occurs on cellulosic substrates and has been shown to decompose cellulose in culture. *Pythium* and *Phytophthora* are two major Oomycetes that have a saprophytic phase in soil. *Absidia, Mucor, Mortierella, Zygorhynchus, Choanephora* and *Cunnighamella* are the soil fungi belonging to Mucorales. *Aspergillus* and *Penicillium* are the two ascomycetous genera that are widespread in soil, with *Aspergillus* more common in warmer soils and *Penicillium* more frequent in temperate and wetter soils. The third ascomycete genus is *Chaetomium* that is regularly isolated from soil, seeds and from almost any substrate that contains cellulose. Agaricales (toad stools) and Aphyllophorales or polyphorales (basket fungi) are the most commonly seen members of group Basidiomycotina that are associated with plant remains. They undoubtedly play a large part in litter breakdown in the area they prevail. Members of Deuteromycotina are very common and wide spread. *Macrophomina, Phoma, Septoria, Nigrospora, Pestalotia, Collectrotrichum, Gloeosporium* are major *Coelomycetes: Aspergillus, Penicillium, Torula, Trichoderma* and *Gliocladium* (Moniliaceae), *Alternaria, Curvularia, Cladosporium, Lacellina, Humicola, Memnoniella, Epicoceum, Bispora* and

Helminthosporium (Dematiaceae), *Fusaium, Myrothecium, Cephalosporium* (Tuberculariaceae) and Sterile mycelia are the principal hyphomycetes observed in crop litter decomposition.

Process of decomposition

Hudson (1968) and Garret (1970) have discussed the pattern of colonization of senescing and dead tissue (Table 3 and Table 4) respectively. Bell (1974) developed a model suggesting sequential colonization of senescent and dead tissue (Figure 1). This associated weak parasites with senescing tissues and a primary and secondary saprophytic flora with the utilization of simple carbohydrates and eventually of cellulose and lignin. The physical characteristics of plant material change as they decay and the changing physical conditions of a substrate probably have a very important influence

Table 3 : Hudson's Scheme of fungal succession on plant debris

Living	Senescent	Dead
PARASITES Ascomycetes and Fungi imperfecti; may be host specific or host restricted	RESTRICTED PRIMARY SAPROPHYTES Ascomycetes and Fungi Imperfecti *Cladosporium herbarum, Alternaria alternata, Epicoccum nigrum, Auriobasidium pullulans* and *Botrytis cinerea* (+ in tropics *Nigrospora* and *Curvularia sp.*)	SECONDARY SAPROPHYTES (I) Ascomycetes and Fungi imperfecti (II) Basidiomycetes (III) Soil inhabitants, eg. Mucorales, Penicillia, etc. (Hudson, 1968)

Table 4 : Garrett's trend of fungal succession

Senescent tissue	Dead tissue		
Stage la	Stage 1	Stage 2	Stage 3
Weak parasites	Primary saprophytic sugar fungi, living on sugars and carbon compounds simpler than cellulose	Cellulose decomposers and associated secondary saprophytic sugar fungi, sharing products of cellulose decomposition	Lignin decomposers and associated fungi (Garrett, 1970)

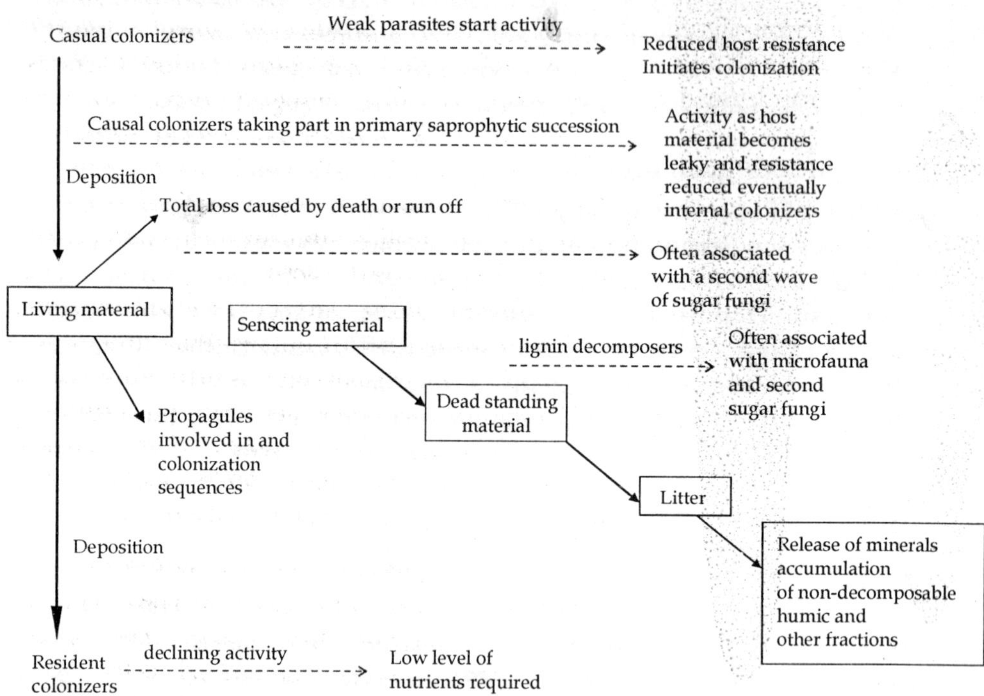

Fig. 1 : Sequences of development in colonization of living and dead tissue (Courtesy, Bell, 1974)

in fungal succession. This is particularly true with respect to their water relations. As cell wall polymers are hydrolyzed, the matric potential at any given time rises and this in turn increases the availability of water to microorganisms growing on the substrate. At the same time decayed materials retain more water. In the long run, such changes may make the substrate more susceptible to colonization by fungi and less tolerant to water stress (Aneja, 1988). The improvement in water availability through decay brought about by microbial pioneer colonizers may be one of the principal underlying forces, which gives direction to fungal succession on various substrates (Dix, 1985). The decomposition level of organic matter critically affects the composition of bacterial taxa as well as the populations and activities of biocontrol agents (Hoitink and Boehm, 1999).

Factors affecting decomposition

Addition of green manure increased the availability of phosphorus and potassium, the macronutrients required for balanced plant growth (Pervin *et al.*, 1995). Colonization of litter or the substrate by fungi is dependant on the environmental factors particularly the chemical constituents of the substrates and the competitive saprophytic ability of the invading fungi. The fungal population decreases during summer because of a rise in temperature causing a rapid loss of moisture from the litter. The maximum gain in the population of fungi may be observed during rainy season resulting into the rapid decomposition of manure or any organic material (Shukla *et al.*, 1978, Sinha, 1992). Pathak and Sinha (1995) reported that the improvement of moisture condition, moderation of temperature and higher relative humidity favour the maximum growth of fungi during rainy season. Moisture content of the decomposing material also plays an important role in colonization by microorganisms. (Webster and Dix, 1960, Hudson, 1962, Khanna, 1964 Sharma and Dwivedi, 1972, Sinha and Pandey, 1979, Dunn *et al.*, 1985, Magan *et al.*, 1989, Baker *et al., 1990*). Decomposition process is regulated by several biotic and abiotic factors *viz.*, carbon, nitrogen ratio, availability of oxygen, temperature, moisture, pH, percent of litter and availability of nutrient material which manipulate the population dynamics of microorganisms. (Partemuzoa, 1986, Steinke and Charles, 1986, Sinha *et al.*, 1999 and Sinha *et al.*, 2009).

Soil structure may affect the spread of microbial propagules, the growth and movement of organisms and of course, the growth of plant roots. Much of the influence of soil structure is operative through its effect on water relation, temperature and aeration, but the actual physical structure of the mineral horizons may affect the movement of soil animals. Crumb development in soil increases pore size, rainfall penetration and drainage and the interchange between the soil and aerial atmospheres.

Temperature affects the growth and activity of organisms directly and in soil it is also important because of the interactions with such factors as oxygen availability and relative humidity. Soil temperatures are affected by the slope and aspect of the land, the colour of the soil, the extent and nature of the vegetation cover and the water regime on the surface litter and the mineral soil (Dickinson, 1974). The most prominent effect of temperature is on the soil fauna.

The water regime, the concentration of salts and carbon dioxide in soil and the number of cations present affect soil pH. The rate of uptake of nutrients present in litter by decomposers is largely determined by the pH of the substrate. pH thus affects the growth and activity of the decomposers resulting ultimately in reduced assimilation of the substrate nutrients.

Radiation is important in terms of heating the soil but it may also have several more direct effects on the soil organisms. Many soil animals are negatively phototactic and they are therefore, absent from soil surface. Some fungi seem to need exposure to light before they sporulate. Most soprophilus fungi have a mechanism, which ensures that their spores are discharged towards a light source, eg. *Pilobolus* spp., *Sordaria* spp. and *Ascophanus* spp.

Spore germination is increased by light in most cases although the germ tube of *Botrytis cinerea* is negatively phototropic. In *Fusarium* spp. and *Trichoderma* spp. sporulation is dependent upon an exposure to light. The prevalence of dematiaceous fungi on the aerial plant parts speaks of their protection from ultraviolet radiation, which is absorbed by melanin present in the cell walls of such fungi.

Natural chemical constituents of the litter include simple inorganic salts, by - producers released during decomposition and exudates from living organisms. These substances may be toxic or growth supportive to the saprophytic activity. Addition of urea to the decomposing litter increases fungal colonization by balancing C: N ratio in favor of fungal growth. Nain (2000) reported that different components of cellulase enzyme complex have different levels of nitrogen requirement for their maximum activity. Ethylene has been shown to be produced in soil when oxygen pressure falls due to water logging. Beneficial compounds, such as hormones and vitamins are released during decomposition of many types of debris. A very long list of microbes has been shown to produce antibiotic like substances. Though, much of the studies are centered on the production of chemical substances by fungi and bacteria little is known about litter fauna concerning the subject.

Many toxic or synergistic interactions may be considered in terms of soil environment on a bulk, micro or molecular scale as the organisms affect one another through changes brought about within these habitats. Effects include the competition, parasitism, predation and antibiosis of fungi by other species of fungi. Such effects must be important in regulating the population of organisms in soil. Another type of interaction in the course of litter breakdown concerns the effect, which organisms have on litter in terms of its attraction for succeeding groups of saprophytes.

Pesticides are the chemicals that are meant to kill a pest applied against. Many such compounds persist in soil and frequently affect soil saprophytes. Alexander (1969) has reviewed the persistence of fungicides in detail. Persistent insecticides outlast fungicides and herbicides. DDT was such an example. Further, the cyclodienes including aldrin, dieldrin, heptachlor and endrin pose the problem of persistence in soil. The oganophosphorus compounds (eg. Parathion, malathion, diazinon) and carbamates (eg. Carbaryl) are relatively readily degraded in soil. The persistent insecticides may usually have little effect on the soil microbial population unless relatively high concentrations accumulate in soil *via* rainwash through the treated crop debris. Wilkinson and Lucas (1969) observed alteration of competitive success of various soil fungi by herbicides in terms of colonization of organic debris.

Field studies have provided evidences indicating that saprophytic activity in soil is governed by nutrient availability, particularly carbon and nitrogen availability as expressed in the C: N ratio. Other factors affecting the process of decomposition, its components and rate include the nature of the substrate, ratio of pentosans to lignin, placement of litter and others.

Effects of green manuring

Bhardwaj (1982) studied the effect of age on the decomposition period of dhaincha, while Elers and Hartmann (1988) observed that most of the nitrogen present in green manure crops became readily available following mineralization. The topsoil and green manure derived nitrogen provided the following crop with an average of 60 kg. N per hectare, this being adequate for production of spinach and lettuce without further addition of fertilizer. Subbarao (1988) studied the microbiological aspects of green manure. Marstorp and Kirchmann (1991) studied carbon and nitrogen mineralization and crop uptake of nitrogen from six green manure legumes decomposing in soil and found that around 30-35% of the total nitrogen in white clover (*Trifolium repens),* black medic (*Medicago lupulina*) and Subterraneum clover (*Trifolium subterraneum*) was mineralized. Khind *et al.* (1991) experimented for the effects of incorporation of *Sesbania* green manure and the period of decomposition on urea hydrolysis. Williams *et al.* (1995) investigated the effect of past soil management in the release of nitrogen from the green manure. Mugendi *et al.* (1994) reported that approximately 70-90 percent of the copings of *Cassia siamea* used as green manure decomposed exponentially within sixty days. Brar and Sidhu (1995) tested the effect of soil on pattern of release of NH_4 - N and NO_3 - N during decomposition of added green manure. Arunin *et al.* (1995) tested the effects of nitrogen fertilizer and amended *Sesbania* spp. on the N - balance and profitability of rice in saline and non - saline soil. Chuprova (1995) experimented for the effects of green manure crops on intensity of production destruction processes in agro ecosystems in Central Siberia. Beiderback *et al.* (1998) studied nitrogen benefits from four green manure legumes in dryland system. Mulleriyawa and Wetta Sinha (1997) reported increased fertility of rice soil through application of *Sesbania rostrata* green manure. Turkhede *et al.* (1998) studied the effects of green manuring and fertility level on yield and physico - chemical properties of the soil in low land paddy. Daimon (1999) reviewed the legumes used as green manure with special reference to *Crotolaria* and *Sesbania* spp. Sharma *et al.* (2000) found that the green manuring with *Crotolaria juncea* at 133.8 q dry weight/ha plus 40 Kg nitrogen incorporated in soil gave rice field equivalent to application of 120 Kg nitrogen. Based on this experiment he suggested that the green manuring with *Crotolaria Juncea* decreases the requirement of nitrogen in the rice field. Sharda *et al.* (1999) reported that Sunnhemp was effective in consuming soil moisture and reduction in green weed biomass. Jayapaul *et al.* (2000) reported that dhaincha together with 75% of the recommended nitrogen produced highest yield of Sugarcane. Gana *et al.* (2000) found that *Sesbania rostrata* incorporation enhanced crop vigour, cane height, single stalk weight and cane yield. Sharma *et al.* (2000) reported that the green manuring with sunnhemp significantly increased the organic carbon, total and available nitrogen, phosphorus and potassium and water holding capacity of the soil. They also reported greatest rice yield with older Sunnhemp combined with nitrogen or rhizobial inoculation.

Green manuring being a rapid decomposing and nutritional material have the advantage of regulating soil-borne plant pathogens. Hoitink (1986) in his article stated, "The activity of antagonists involved in biological control is affected by nutrients present in the substrate". Composts prepared from the tree leaves or rice hulls were effective in controlling pathogens. Vaughn *et al.* (1954) published the first report on the control of *Phytophthora* disease with tree bark on strawberry. Further it was shown that the addition of oilcake and green manure influenced the fungal population in various ways (Davely and Papa viz as, 1963). Lakshman and Chandrashekharan (1984) have reported that Sclerotia of *Rhizoctonia solani* could not germinate significantly in amended soils.

The induction of soil fungistasis and suppressiveness to various plant pathogens by soil amendments with organic manures has been reported by several investigators. Elad *et al.* (1980) have given the detailed account of effectiveness of *Trichoderma harzianum* against *Sclerotium rolfsii* and *Rhizoctonia solani* as a biological agent with the result of organic amendments. Bell *et al.* (1982) have also reported *in vitro* antagonism and induction of suppressiveness of *Trichoderma* spp. against six fungal soil - borne plant pathogens. *Trichoderma* spp. as biocontrol agents are widely studied by various workers (Roy 1977, Upadhyaya and Mukhopadhyay, 1986, Chet 1987, Kumar and Marimuthu 1994, Benhamous and Chet 1996, Stefanova *et al.* 1999, Mathivanan *et al.* 2000). Mukherjee *et al.* (1995) have shown antagonist properties of *Gliocladium virens* and *Trichoderma harzianum* on *Rhizoctonia solani* and *Sclerotium rolfsii.* Cook (1981) and Baker (1987) have reviewed the progress towards biological control of soil - borne plant pathogens. In some recent studies effect of organic amendment of soil on plant pathogens have also been observed (Papa viz as and Davey, 1960, Baker and Cook, 1974, Kelman and Cook, 1977, Chet and Baker, 1980, Elad *et al.*, 1980, 1983, Baker *et al.*, 1984, Hoitink, 1986, Dasgupta and Gupta, 1989, Kohl and Schlosser, 1989, Attalah and Lopez, 1991, Boparai and Singh, 1992, Manzali *et al,.* 1993 and Desai and Schlosser, 1999).

It is well established that the practice of green manuring not only provides a number of nutrients and organic matter to the plant at a very cheap cost but also helps in reduction of inoculum potential and disease incidence produced by several soil-borne plant pathogens. It also improves soil structure and weed and erosion control (Abdallahi and Dayegamiye, 2000, Blackshaw *et al.*, 2001). The several hazards of human and animal health as well as the environmental pollution problem resulting from use of chemical pesticides have now made us to look for the possible alternative with least or no hazards and have a longer effectiveness. This can be achieved either by using cultural, biological, mechanical and of course also chemical up to some extent or by harmonizing all the practices with the proper utilization of various resources as an integrated pest management scheme - a step ahead towards "Sustainable agriculture".

References

Abdallahi, M. M. and N. Dayegamiye, A. (2000). Effects of green manures on soil physical and biological properties and on wheat yields and N uptake. *Can. J. Soil Sci.*, 80: 81-89.

Agarwal, R.R. (1967). *Soil fertility in India.* Asia Publishing House, New Delhi.

Aggarwal, G.C. and Sekhon, N.K. (1991). Changes induced by cowpea green manure and farmyard manure in the timing of phonological events in maize. *J. Agric., Sci.*, 117.

Alexander, M. (1969). Microbial degradation and biological effects of pesticides in soil. In: *Soil Biology Reviews of Research.* Natural Resources Research, UNESCO 9: 209-240.

Alexander, M. (1977). *Introduction to Soil Microbiology.* (2nd Ed.) John Wiley, New York, 467 pp.

Aneja, K. R. (1988).Biology of litter decomposition fungi as decomposers of plant litter.In: *Prospective in Mycology and Plant Pathology,* (Eds.) V. P. Agnihotri, A. K. Sarbhoy and D. Kumar, Malhotra Publishing House, New Delhi:389-419.

Arunin, S., Annuluxtipun, Y. and Pongwichian, P. (1995). The effect of nitrogen fertilizer and amended Sesbania spp. on the N- balance and profitability of rice in saline and non - saline soil. In: *Soil Management in Sustainable Agriculture.* Proc. 3rd International Conf. On sustainable agriculture, Wye College University of London, U.K. 31 Aug to 14 Sep. 1993. Ashford, U.K. Wye College Press:300-301.

Attalah, T. and Lopez, J.M. (1991). Potential of green manure species in recycling of N, P, and K. *Biol. Agric. Hortic.*, 8 (1): 53-66.

Baker, K.F. and Cook, R. J. (1974). *Biological control of plant pathogens.* Freeman and Co. San Francisco.

Baker, R. (1987). Evolving concepts of biocontrol of plant pathogens. *Annu. Rev. Phytopath.*, 25: 67- 85.

Baker, R., Elad, Y. and Chet, I. (1984). The control experiment in the scientific method with emphasis on biological control. *Phytopath.*, 74: 1019- 1021.

Baker, T. J., Dyek, M. J., Barton, P. G., Oliver, G. R. and Nicholson, J. (1990). Effect of irrigation with sewage effluent on decomposition of litter in *Pinus radiata* forest. *Forest Ecol. Manage.*, 31(11):205-214.

Beiderback, V.O., Bouman, O. T., Campbell, C.A. Bailey, L.D. and Winkleman, G.E. (1998). Nitrogen benefited from four green manure legumes in dryland system. *Can. J. Pl. Sci.*, 76 (2): 307-315.

Bell, D.K., Wells, H. D. and Markham, C.R. (1982). In vitro antagonism of *Trichoderma* spp. against six fungal pathogens. *Phytopath.*, 72: 379-382.

Bell, M.K. (1974). Decomposition of herbaceous litter. In: *Biology of Plant Litter Decomposition*, (Eds.) C.H. Dickinson and G.J.F. Pugh, Academic Press, London and NewYork: 37-67.

Benhamous, N. and Chet, I. (1996). Parasitism of Sclerotia of *Sclerotium rolfsii* by *Trichoderma harzianum*: Ultrasturucture and cytochemical aspects of the interaction. *Phytopath.*, 86: 405-416.

Bhardwaj, K. K. R. (1982). Effect of the age and decomposition period of dhaincha on the yield of rice. *Ind. J. Agron.*, 27:284–285.

Blackshaw, R.E. Moyer, J.R. Doram, R.C. and Boswell, A.L. (2001). Yellow sweet clover, green manure and its residues effectively suppress weeds during fallow. *Weed Sci.*, 49: 406- 413.

Boparai, B.S. and Singh, Y. (1992). Effect of green manuring with Sesbania aculeate on physical properties of soil. *Arid Soil Res. Rehabil.*, 6 (2): 135-143.

Brar, D. S. and Sidhu, A. S. (1995). Effect of soil water on patterns of nitrogen release during decomposition of added green manure residue, *J. Ind. Soc. Soil Sci.*, 43(1):14-17.

Chet , I. (1987). *Trichoderma*: Application made of action and potential as a biocontrol agent of soil borne plant pathogenic fungi. In: *Innovative Approaches to Plant Disease Control.* John Wiley and Sons, New York. 137 – 160.

Chet, I. and Baker, R. (1980). Induction of suppressiveness to *Rhizoctonia solani.* Phytopathology, 70: 994- 998.

Chuprova, V.V. (1995). Effect of green manure crops on intensity of production – destruction process in agroecosystems of Central Siberia. *Agrikkimiya*, 11: 31-41.

Cook, R.J. (1981). Biological control of plant pathogens: Overview. In: *Biological Control in Crop Production*, (Ed.) G.C. Papavizas, Grenada, London: 23-44.

Cosico, W.C. (1990). Studies on green manuring in the Philippines. *Exten. Bullt.* No. 314 FFTC, Taiwan: 14.

Creus, C.J., Studdert, G.A., Echeverria, H.E. and Sanchez, S.R. (1998). Maize harvested residue decomposition and nitrogen dynamics in soil. *Cienciadel suelo*, 16 (2): 51-57.

Daimon, H. (1999). Advances in the study of tropical legumes used as green manure and their contribution to yield in various cropping systems. *Japanese J. Crop Sci.*, 68 (3): 337-347.

Dasgupta, A. and Gupta, P.K.S. (1989). Effect of different sol amendments on wilt of pigeonpea caused by *Fusarium udum*. *Beitragezur Tropischen Landwirtschaft und veterinarmedizir*, 27 (3): 341-345.

Davey, C.B. and Papavizas, G.C. (1963). Saprophytic activities of *Rhizoctonia* as affected by carbon nitrogen balance of certain organic amendments. *Soil Sci. Soc. Am. Proc.*, 27: 164-167.

Desai, S. and Schlosser, E. (1999). Parasitism of *Sclerotium rolfsii* by *Trichoderma*. *Ind. Phytopath.*, 52(1): 47-50.

Dickinson, C.H. (1974). Decomposition of litter in soil. In: *Biology of Plant Litter Decomposition*, (Eds.) C.H. Dickinson and G.J.F. Pugh, Academic Press, London and NewYork: 633-658.

Dilly, Q. and Immuer, U. (1998). Suppression in the food web during the decomposition of leaf litter in a black alder (Alnus glutinosa (Gaertn) L.) *Forest Pedobiol.*, 42 (2): 109-123.

Dix, N. J. and Webster, J. (1985). *Fungal Ecology*, Chapman & Hall, London.

Dunn, P. H., Barro, S. C. and Poth, M. (1985). Soil moisture affects survival of Microorganisms in heated chaparral soil. *Soil Biol. Biochem.*, 17(2): 143-148.

Ed-Haun, C., Ren-Shih, C. and Tsai, Y. H. (2007). Effect of different application rates of organic fertilizer on soil enzyme activity and microbial population. *Soil Sci. Pl. Nutri.*, 53: 132-140.

Elad, Y., Chet, I. and Henis, Y. (1983). Parasitism species on Rhizoctonia solani and Sclerotium rolfsii scanning electron microscopy. *Phytopath.*, 73: 85-88.

Elad, Y., Chet, I. and Katan, J. (1980). *Trichoderam harzianum*: a biocontrol agent effective against Sclerotium rolfsii and Rhizoctonia solanii. *Phytopath.*, 70: 119-121.

Elers, B. and Hartmann, H.D. (1988). Model studies on the mineralization of the green manure plants. *Landwirtschaftliche forschung*, 41(3-4): 246-252.

Gana, A.K., Busari, L.D. and Ibrahim, P.A. (2000). Assessment of the contribution of legumes to the nitrogen nutrition of sugarcane. *Acta Agron. Hungerica*, 48(1): 103-106.

Garrett, S.D. (1981). *Soil Fungi and Soil Fertility*. The Macmillan Company, New York: 66-77.

Garrett, S.D. (1970). *Pathogenic Root Infecting Fungi*. Cambridge Univ. Press, London.

Garrett, S.D. (1951). Ecological groups of soil fungi – A survey of substrate relationships. *New Phytol.*, 50: 149-166.

Gaur A.C. and Mukherjee, D. (1980). Recycling of organic matter through mulch in relation to chemical and microbiological properties of soil and crop yields. *Plant Soil*, 56: 273.

Gaur, A.C., Mathur, R.S. and Varshney, T.N. (1970 a). Decomposition of different types added organic matter in soil. *Agrochimia*, 14: 524.

Gaur, A.C., Sadasivan, K.V., Vimal, O.P. and Mathur, R.S. (1971). A study on the decomposition of organic matter in an alluvial soil; CO2 evolution, microbial and chemical transformations. *Plant Soil*, 35: 70.

Gaur, A.C., Sadasivan, K.V., Vimal, O.P. Mathur, R.S. and Kavinandan, S.K. (1970 b). Studies on the humification of the organic matter in a red sakar soil. *Zbl. Bakt. II Bd.* 128: 149.

Goyal, S., Chander, K., Mundra, M. C. and Kapoor, K. K. (1999). Influence of inorganic fertilizer and organic amendments on soil organic matter and microbial properties under tropical conditions. *Biol. Fertil. Soil*, 27(2):196-200.

Harper, S.T.H. and Lynch, J.M. (1985). Colonization and decomposition of wheat straw leaves, internodes and nodes. *J. Sci. Food Agricul.*, 32: 1057-1062.

Hoitink, H.A.J. (1986). Basis for the control of soilborne plant pathogens with composts. *Annu. Rev. Phytopath.*, 24: 93-114.

Hoitink, H.A.J. and Boehm, M.J. (1999). Biocontrol within the context of soil microbial communities: A substrate- dependent phenomenon. *Annu. Rev. Phytopathol.*, 37: 427-446.

Hudson, H.J. (1962). Succession of microfungi on aging leaves of *Saccharum officinarum*. *Trans. Br. Mycol. Soc.*, 41: 395-423.

Hudson, H.J. (1968). The ecology of fungi on plant remains above the soil. *New Phytol.*, 67:837-874.

Jayapaul, G. G. P., Duraisingh, R. R., Senthivel, T. and Joseph, M. (2000). Influence of population and stage of incorporation of intercropped green manure (Dhaincha) and nitrogen levels on yield and quality of sugarcane. *Ind. Sugar*, 49(12):989-991.

Jensen, V. (1974). Decomposition of angiosperm tree leaf litter. In: *Biology of Plant Litter Decomposition*, (Eds.) C.H. Dickinson and G.J.F. Pugh, Vol. II, 69-110. Academic Press, London.

Kelman, A. and Cook, R.J. (1977). Plant Pathology in the Peoples Republic of China. *Annu. Rev. Phytopath.*, 17: 409-429.

Khanna, P. K. (1964). *The Succession of fungi on some decaying grasses*, Ph.D. Thesis, Banaras Hindu University, Varanasi, India.

Khind, C.S., Garg, A. and Bajwa, M.S. (1991). Effect of sesbania, rice straw and pre-incubation on hydrolysis in wetland soil. *Int. Rice Res. Newslett.*, 16 (2): 18.

Kjoller, A. and Struwe, S. (1997). *Microbial diversity and its relationship to decomposition process.* Ecosystem Research Report- European commission, 24: 73-85.

Kohl, J. and Schlosser, E. (1989). Decay of sclerotia of Botrytis cinerea by Trichoderma spp. at low temperature. *J. Phytopath.*, 125: 320-326.

Kumar, A. and Marimuthu, T. (1994). Biological control of damping - off of *Eucalyptus kamaldulensis* caused by *Rhizoctonia* solani with *Trichoderma viride* and decomposed coconut coir pith. *Pl. Dis. Res.*, 9: 116-121.

Kuzyakov, Y., Ruhlmann, J. and Geyev, B. (1992). Kinetics and parameters of decomposition of vegetable crop residue during their incubation in soil. *Gartenbauwissenschogt*, 64 (4): 151-157.

Lakshman, P. and Chandrashekharan (1984). Effect of soil amendment on the viability of sclerotia of *Rhizoctonia solani* in soil. *Madras Agric. J.*, 71: 526-529.

Lal, J.K., Mishra, B. and Sarkar, A.K. (2000). Effect of plant residue incorporation on specific microbial groups and availability of some plant nutriens in soil. *J. Ind. Soc. Soil Sci.*, 48: 67-71.

Lekha, S. and Palaniappan, S.P. (1990). Radiotracer studies on P use efficiency in rice based cropping system. *J. Agron. Crop. Sci.*, 165: 14-18.

Lockwood, J.L. (1977). Fungistasis in soil. *Biol. Rev.*, 52: 1-43.

Magan, N.P., Hand, P.L. Kirkwood, I.A and Lynch J.M. (1989). Establishment of microbial inocula on decomposing wheat straw in soil of different water contents. *Soil Biol. Biochem.*, 21 (1): 15-22.

Manzali, D. Nipoli, P., Pisi, A. Filippini, G. And Ercole, D. (1993). Scanning electron microscopy study of *in vitro* antagonism of *Trichoderma* spp. strains against *Rhizoctonia solani* Kuhn. *Phytopath. Medit.*, 32: 1-6.

Marstorp, H. and Kirchmann, H. (1991). Carbon and nitrogen mineralixation and crop uptake of nitrogen from six green manure legumes decomposing in soil. *Acta Agricul. Scandinavica*, 41 (3): 243-252.

Mathivanan, N. Srinivasan, K. and Chelliah, S. (2000). Biological control of soilborne diseases of cotton, eggplant , okra and sunflower by *Trichoderma viride*. *Zeitschrift-fur-Pflanzenkrankheiten und pflanzeschutz*, 107 (3): 235-244.

Mugendi, D.N., Mochoge, B.O., Coulson, C.L. Stigter, C.J. and Sang, F.K. (1994). Decomposition of *Cassia siamea* loppings in semiarid Machakos, Kenya. *Arid Soil Res. Rehabilit.*, 8 (4): 363-372.

Mukerji, B.K. and Agarwal, R.R. (1950). Review of green manuring practices in India. *ICAR Misc. Bull.* No. 68.

Mukherjee, P.K., Mukhopadhyay, A.N., Sarmah, D.K. and Shreshtha, S.M. (1995). Comparative antagonistic properties of *Gliocladium virens* and *Trichoderma harzianum* on *Sclerotium rolfsii* and *Rhizoctonia solani* in relevance to understanding of mechanism of biocontrol. *J. Phytopath.*, 143: 275-279.

Muller, P.E. (1974). Tidsskr. Skovbr., 3: 1-124.

Mulleriyawa, R. and Wettasinha, C. (1997). Soil fertility management in irrigated rice field. *ILEIA Newslett.*, 13 (3): 18-19.

Nain, L.R. (2000). Effect of nitrogen supplementation on cellulose enzyme production from sorghum straw. *Ann. Agricul. Res.*, 21: 301-302.

Nandi, N., Hazra, J.N. and Sinha, N.B. (1996). Microbial synthesis of humus from rice straw following two-step composting process. *J. Ind. Soc. Soil Sci.*, 44(3): 413-416.

Palaniappan, S. P. (1994). Green Manuring: Nutrient potential and management. In: *Fertilizers, Organic manure, Recyclable waste and Biofertilizers.* Fertilizer development and Consultation Organization, New Delhi.

Panse, V.G., Dandavate, M.D. and Bokil, S.D. (1947). *Soil fertility investigations in India with special reference to manuring*, Army Press. Delhi.

Papavizas, G.C. and Davey, C.B. (1960). Rhizoctonia disease of bean as affected by decomposing green plant materials and associated microfloras. *Phytopath.*, 50: 404-410.

Partemuzoa, L.A. (1986). Abiotic and biotic factors during decomposition of organic residues. *Izv Sev Nauchn Tsentra vyssh Shk Estestv Nauki*, 3: 120-124.

Pathak, S. K. and Sinha, A. (1995). Leaf litter decomposing mycoflora safflower (*Carthamus tinctorius* L.) in relation to different climatic factors. *Crop Res.*, 9: 135-141.

Pervin, S. Hoque, M.S. , Jahiruddin, M. And Mian, M.H. (1995). The use of Sesbania as an alternative source of urea- N for BR-11 rice. *Pakistan J. Scientific Industr. Res.*, 38 (2): 85-87.

Ranjan, A.S. and Kothandaraman, G.V. (1986). Influence of organic matter on rock phosphate and its effect on yield and uptake of nutrients by rice. *Proc. Symp. Rockphosphate in Agriculture TNAU*: 161-163.

Kumar, R., Sinha, A., Srivastava, Seweta and Srivastava, Manisha (2010). Variation in soil mycobiota assested with decomposition of *Sesbania aculeata* (L.). *Asian J. Plant Pathol.*, Accepted.

Kumar, R., Srivastava, Seweta, Srivastava, Manisha and Sinha, A. (2010). Effect of organic amendments on soil Mycoflora. *Asian J. Plant Pathol.*, 4 (2): 73-81.

Roy, R.N. (1977). Parasitic activity of *Trichoderma viride* on the sheath blight fungus of rice (*Corticium sasakii*). *J. Pl. Dis. Prot.*, 84: 675-683.

Roy, R.N. (1994). Integrated plant nutrition system: An overview. In: *Fertilizers, organic manures, recyclable wastes and biofertilizers*, (Ed.) H.L.S. Tandon, Fertilizer development and Consultation Organization, New Delhi.

Sadasivan, T.S. (1939). Succession of fungi on decomposition of wheat straw in different soils with special reference to *Fusarium culmorum*. *Ann. Appl. Biol.*, 26: 497-508.

Sarvanan, A.Kaleeswari, R.K., Nambisan, K.M.P. and Sankaralingam, P. (1995). Leaf litter accumulation and mineralization pattern of hilly soils. *Madras Agric. J.*, 85 (3): 184-187.

Selvi, R. V. and Kalpana, R. (2009). Potentials of green manure in integrated nutrient management for rice - A review. *Agricul. Rev.*, 30(1):40-47.

Sharda, V.N., Sharma, N.K., Mohan, S.C. and Khybri, M.L. (1999). Green manuring for conservation and production in western Himalayas: 2. Effect on moisture conservation, weed control and crop yields. *Ind. J. Soil Conservation*, 27 (1): 31-35.

Sharma, A. R. and Ghosh, A. (2000). Effect of green manuring with Sesbania aculeata and nitrogen fertilization on the performance of direct-seeded flood-prone lowland rice. *Nutri. Cycl. Agroecosys.*, 57 (2): 141- 153.

Sharma, P. D. and Dwivedi, R. S. (1972).Succession of fungi on decaying Setaria glauca. *Trop. Ecol.*, 13: 183-201.

Sharma, P.D. (1973). Succession of fungi on decaying *Setaria glauca* Beauv. A qualitative analysis of the mycoflora. *Ann. Bot.* (Lond.), 37: 203-208.

Sharma, R., Dev, S. P. and Rameshwar (2000). Effect of green manuring on sunnhemp (*Crotalaria juncea* L.) on rice yield, nitrogen turnover and soil properties. *Crop Res.*, 19 (3): 418-423.

Shukla, A.N. Tandon, R.N. and Gupta, R.C. (1978). Phyllosphere mycoflora colonizing the leaf litter of sal (*Shorea robusta* Gaerth.) in relation to some of the environmental factors. *Trop. Ecol.*, 90: 1-6.

Singh, J. (1937). Soil fungi and actinomycetes in relation to manorial treatmen, season and crop. *Ann. Appl. Biol.*, 24: 154-158.

Singh, J.K. (1998). *Effect of green manure and nitrogen levels of crop yield and physico-chemical properties of soil under rice wheat cropping system.* Ph.D. Thesis, B.H.U., Varanasi, India.

Singh, S., Ghoshal, N. and Singh, K. P. (2007). Variation in soil microbial biomass crop roots due to deferring resource quality input in a tropical dry land agro ecosystem. *Soil Biol. Biochem.*, 39(1): 76-81.

Sinha, A. (1982). Litter decomposing mycoflora of savana in relation to different climatic factors. *Proc. Ind. Nat. Sci. Acad.*, 8: 138-146.

Sinha, A. (1992). Phyllosphere mycoflora colonizing the leaf litter of *Saccharumm officinarum* in relation to some climatic factors. *Crop Sci.*, 5 (3): 545-554.

Sinha, A. (1993). Release of nitrogen, phosphorus and potassium from decomposing litter of savanna. *J. Potassium Res.*, 9(1): 64-69.

Sinha, A. and Mahapatra, T. K. (1993). Studies on mycoflora of decomposing leaf litter of *Saccharum officinarum* L. in relation to different climatic factors. *Proc. Nat. Acad. Sci.*, 63:373-379.

Sinha, A. and Pandey, V. N. (1979). Studies on mycoflora of decomposing leaf litter of Diospyros melaxylon in relation to different climatic factors. *J. Sci. Res.*, 24: 1-8.

Sinha, A. and Pathak, SK. (1995). Relationship between microbial population and carbon dioxide evolution from decomposing leaf litter of teak (*Tectona grandis* L.). *New Agriculturist*, 6: 99-104.

Sinha, A., Kumar, R., Kamil, D. and Kapur, P. (2009). Release of nitrogen, phosphorus and potassium from decomposing *Crotalaria juncea* L. in relation to different climatic factors. *Environ. Ecol.*, 27(4B):2077-2081.

Sinha, A., Pathak, S.K. and Yadav, A.S. (1998 a). Studies on mycoflora of decomposing root litter of safflower (*Carthamus tinctorius* L.) in relation to different climatic conditions. *Crop Res.*, 16: 265-270.

Sinha, A., Singh A.K. and Yadav, A.S. (1998 b). Fungal decomposition of parthenium leaf in soil. *Ind. J. Weed Sci.*, 30: 76-78.

Sinha, A., Singh, A.K., Yadav, A.S. and Qaiser, J. (1999). Studies of mycoflora on decomposing leaf of parthenium in relation to different climatic factors. *Der Tropenlandwirt*, 100: 229-235.

Stefanova, M., Leiva, S. Larrinaga, L. and Koronado, M.F. (1999). Metabolic activity of *Trichoderma* spp. isolates for a control of soilborne phytopathogenic fungi. *Revista de la Facultad de Agronomia, Universaidad de Zulia.*, 16: 509-516.

Steinke T.D., Barnabas and Somaru, R. (1990). Structural changes and associated microbial activities accompanying decomposition of mangrove leaf in Engeni estuary (South Africa). *S. Afr. J. Bot.*, 56 (1): 39-48.

Steinke, T. D. and Charles, L. M. (1986). In vitro rate of decomposition of the mangrove (*Bugiera gymnorrhiza*) as affected by temperature and salinity. *S. Afr. J. Bot.*, 52(1) 39-42.

SubbaRao, N.S. (1988). Microbial aspects of green manure in lowland rice soils. *Proc. Symp. on Sustainable Agriculture*: The Role of green manure crops in rice farming system. IRRI, Philippines:131-149.

Thormann, M. N., Currah, R. S. and Bayley, S. E. (2003). Succession of microfungal assemblages in decomposing peat land plants.*Plant Soil*, 250:323–333.

Turkhede, A.B., Choudhary, B.T., Chore, C.N, Kalpande, H.V., Jiotode, D.J. and Kalpande, V.V. (1998). Effect of green manuring with glyricidia leaves and fertility level on yield and physico- chemical properties of the soil in lowland paddy. *Crop Res.*, 16 (3): 300-303.

Upadhyaya, J.P.and Mukhopadhyay, A.N. (1986). Biological control of *Sclerotium rolfsii* by *Trichoderma harzianum* in sugarbeet. *Trop. Pest Manag.*, 32: 215-220.

Vaughn, E.K. Roberts, A.N. and Melenthin, W.N. (1954). The influence of Douglas fir sawdust and certain fertilizer elements on the incidence of red stele disease of strawberry. *Phytopath.*, 44: 601-603.

Vibha and Sinha, A. (2009). Succession of fungi on decomposing rice stubbles in rice – wheat-cropping system. *Oryza*, 46 (2): 140-144.

Visser, S. and Parkinson, D. (1975). Fungal succession on aspen poplar leaf litter. *Can. J. Bot.*, 53: 1640-1651.

Webster, J. and Dix, N.J. (1960). Succession of fungi on decaying cocks foot culms III. Acomparison of the sporulation and growth of some primary saprophytes on stem, leaf-blade and leaf - sheath. *Trans. Br. Mycol. Soc.*, 43: 55-99.

Wilkinson, V. and Lucas, R.L. (1969). Effect of herbicides on the growth of soil fungi. *New Phytol.*, 68: 709-719.

Williams, S. Atkinson, D. and Sinclair, A. (1995). Effect of past soil management on the release of nitrogen from green manure. In: *Soil Management in Sustainable Agriculture, Proc. 3rd International Conf. on Sustainable Agriculture*, Wye College Press, University of London, U.K.

Yadav, A.S., Vibha and Sinha, A. (2002). Studies on decomposing mycoflora of *Crotolaria juncea* L. *Modern J. Life Sci.*, 1: 12-17.

Yadav, R. L., Dwivedi, B. S. and Pandey, P. S., (2000). Rice-wheat cropping system: assessment of sustainability under green manuring and chemical fertilizer inputs. *Field Crop Res.*, 65(1): 15-30.

Yawalkar, K.S., Agrawal, J.P. and Bokado, S. (1984). *Manures and fertilizers*. Agric. Horticultural Publishing House, Nagpur.

□□□

Microbial Diversity and Functions, 2012
© D.J. Bagyaraj, K.V.B.R. Tilak, H.K. Kehri (eds.), pp. 377-387
New India Publishing Agency, New Delhi (India)
E-mail : info@nipabooks.com; Website : www.nipabooks.com

Chapter **17**

Relationship of Soil Nutrients with Plant Pathogens and Diseases Caused by Them

M.N. Khare

ABSTRACT

Production technology has been worked out for crops and several inputs have been recommended which are to be applied in proper quantity at most suitable time to get the maximum return. Nutrition in the form of fertilizers is a must. However, injudicious use of agro-chemicals has resulted in diseases called 'iatrogenic plant diseases'. Under the circumstances it is important to know the relationship between soil nutrients and plant pathogens which are causing the diseases of economic importance. Author has given adequate emphasis to micronutrients as at a very minor quantity they reduce the disease to an acceptable level. He has also discussed in detail about all the major and minor nutrients required by the plants and how their deficiency or over doses affect the plant pathogens and in turn the diseases.

Keywords: Soil nutrients, plant pathogens, diseases, relationship

Introduction

Efforts are being made to increase agricultural production per unit area and per unit expenditure to feed the population in the country which has crossed one billion. The total area under crop production in the country is about 180 mha and the production has crossed 200 mt. It is not possible to further increase the area, hence vertical approach is essential to raise the production by increasing the productivity. Production technology has been worked out for crops and several inputs have been recommended which are to be applied in proper quantity at most suitable time to get the maximum return. Nutrition in the form of fertilizers is a must. The relationship of a crop and the pathogen is affected by the nutrients both required at macro and micro level.

The crops face both biotic and abiotic stresses. The reaction given by the plant is exhibited by development of certain abnormalities which are termed symptoms. Non-availability or excess of water, lack of proper aeration, high and low temperatures, light problem, lightning, wind, hail, deficiency or excess of essential elements, high or low pH are all abiotic factors which influence the crop growth, yield and quality of the produce. Injudicious use of agro-chemicals results in diseases called 'iatrogenic plant diseases'.

Agriculture is the most important source of renewable wealth and development of an integrated plant nutrient supply system involving an appropriate mix of organics, biological nitrogen fixation, phosphorus solubilizing microbes and need based chemical fertilizers would be crucial for the sustainability of production and soil as the resource base for it. The nutrients are depleted due to continuous cropping, erosion, leaching etc. Intensive cropping systems can remove annually 500-900 kg N + P_2O_5 + K_2O per ha per year along with substantial quantities of micronutrients. A paddy-potato-wheat rotation has been shown to remove Fe 4640 g, Mn 1243 g, Zn 615 g, Cu 325 g, B 305 g and Mo 17.5 g per hectare (Singh, 1996). High yielding varieties remove the essential nutrients very fast (Tisdale *et al.*, 1990).

Crops are attacked by various pathogens and insect pests which deteriorate the quality and quantity of the produce. Attempts have been made and continuous efforts are there to evolve varieties of crops resistant to such harmful agencies. The resistance of a crop variety is determined by its ability to limit the penetration, development and/or reproduction of invading pathogens. The resistance varies with genotype of the host and the pathogen, host age and environment. Race specific resistance is controlled by one or more major genes and nonspecific resistance is usually polygenic. The nonspecific resistance is influenced by several factors like environment and is degraded by deficiency of specific micronutrient in the host. In some cases host exhibited tolerance to certain level. The tolerance to pathogens is not a response specific to the pathogen but is a normal homeostatic response of the host that has been evolved to counter a wide range of environmental strains, mostly abiotic. Thus a genetically controlled ability to tolerate nutritional stress would also confer tolerance to a root rotting pathogen in a host growing in such soils, since the pathogen has the effect of reducing the nutrient absorbing surface (Graham and Webb, 1991).

Resistance and tolerance are relatively independent factors in a host pathogen relationship. The application of essential elements contributes to disease tolerance wherever the soil is deficient in that element, by increasing the vigour of the host plant. Last (1962) reported increased grain yield in cereal crop infected with powdery mildew grown in N deficient soil when N was supplied to the soil. The resistance of the host was lowered under N deficiency which improved on removal of the deficiency (Gour *et al.*, 2002).

The essential elements are classified into macronutrients like nitrogen, phosphorus, potassium, calcium, manganese, zinc, copper, molybdenum and chlorine. The nutrients are derived by the crop plants from the medium on which they grow through the root system. The availability of nutrients depends upon such factors like minerological composition of soil, pH and several other physico-chemical factors and the physiological activities in plants. Nutrients are essential for plant growth and required functions. Carbon, hydrogen and oxygen are the constituents of most of the organic compounds and are obtained by plants through air and water (Colhoun, 1973).

Macronutrients: NPK are considered as primary, and S, Ca, Mg secondary macronutrients.

Nitrogen: It is the component of proteins, amino-acids, chlorophyll, purines, pyrimidines and coenzymes. It increases plant growth, vigour and number of roots. Nitrogen decreases or increases disease intensity depending on other factors. It is mobile. Nitrogen deficiency leads to *chlorosis*, stunted plant growth, delayed flowering, sterility, no fruit formation and elongation of roots. Yellow berry of wheat is due to nitrogen deficiency. Gummosis and die back in citrus and stone fruits and firing or burning of leaves occurs due to excess nitrogen.

The forms of nitrogen influence the disease severity differently. The ammonium nitrogen increases cortical and root diseases due to the species of *Fusarium, Rhizoctonia, Aphanomyces, Cercosporella* and *Ophiobolus*, cereal rusts, mildews etc., however nitrate nitrogen has reversed the effect (Huber and Watson, 1974; Filho and Dhingra, 1980; Basuchaudhary and Gupta, 1988).

At higher doses of nitrogen host tissues develop larger thin walled cells which remain loose with larger intercellular spaces, the stomata remain wide open. Due to these factors susceptibility in host plants gets increased to many diseases like wilts, root rots, Alternaria leaf spots, bacterial blights, rusts, powdery mildew and paddy blast (Palti, 1981). In certain cases high dose of nitrogen reduces the disease incidence caused by *Pseudomonas syringae* and *Drechslera turcicum* in maize and *Fusarium moniliforme* in rice (Sunder and Satyavir, 1997). Heavy application of urea reduced tobacco wilt (*Pseudomonas solanacearum*).

Phosphorus: It is constituent of nucleoproteins, phospholipids - NADP, ADP, ATP etc. Phosphorus plays a vital role in energy transfer, respiration, photosynthesis, phosphorylation, starch hydrolysis, transamination and absorption of other nutrients. It is not mobile.

Phosphorus deficiency causes accumulation of soluble N compounds as in tomato, soybean etc. It results in purple pigmentation, stunted growth, less lateral shoots, retarded flowering and fruiting, premature death of buds, imperfect pollination.

Single foliar spray of phosphorus induced a systemic protection against powdery mildew in cucumber (*Sphaerotheca fuliginea*) and rust in maize (*Puccinia sorghi*) (Reuveni and Reuveni, 1998). Increased dose of P at 30, 60 kg ha^{-1} level decreased soybean rust (*Phakopsora pachyrhizi*) (Sharma *et al*., 1996), collar rot of lentil (*S. rolfsii*) (Khare, 1980), root rot of chickpea due to *Fusarium solani* (Gaur and Vaidya, 1982). However, higher dose of phosphorus increased root rot of cotton (*S. rolfsii*) (Jordan and Nelson, 1939) and wilt of sweet potato (Ngeve and Roncadori, 1984).

Potassium: The role of potassium in plants is to activate enzymes like pyruvic acid kinase. It affects mechanism of stomata, photosynthesis, starch synthesis, adenine synthesis, metabolic activities of chloroplasts and mitochondria, transpiration, water translocation and indirectly affects cell wall thickness. It is mobile, K makes the plant resistant to pathogens.

Under K deficiency loss of cell turgor occurs which facilitates penetration by fungi. Glutamine is high under K deficiency which stimulates spore germination as in *Pyricularia oryzae* on rice leaves (Suryanarayan, 1958). Leaf scorch of apple, potash hunger in potato, browning of leaves with necrosis giving rust like appearance in cotton. Seed often fails to mature and remains small.

K in split application, one at the basal stage and the other at maximum tillering in NPK combination of 40:20:20 kg ha^{-1} significantly reduced the incidence of sheath blight of rice (*R. solani*) and increased yield (Baruah, 1995). Application of K decreased black spot disease in rape (Sharma and Kolte, 1994). Ascochyta blight of chickpea (Singh and Chand, 1996), rice blast (Fillipi and Prabhu, 1998), *Rhizoctonia* disease of potato (Das and Western, 1959), root rots of sunnhemp caused by *R. solani* and *S. rolfsii* (Pal and Basuchaudhary, 1980), root and collar rot of lentil due to *R. solani* and *S. rolfsii* (Prasad and Basuchaudhary, 1984), cotton wilt (Miles, 1936). Powdery mildew of wheat was less when PK were applied (Yurina *et al*., 1997).

Magnesium: Chlorophyll contains Mg and it plays an important part in enzymatic activity in carbohydrate metabolism. Mg ions are essential for oxidative phosphorylation. It is mobile.

Mg deficiency results in interveinal chlorosis, anthocyanin pigment develops resulting in spots. Sand drown symptoms develop in tobacco leaf. Soybean is also affected. Magnesium deficiency usually occurs in acid, sandy, highly leached soils with low cation exchange capacities. The magnesium requirements vary in crops and their varieties. Cool, wet and cloudy weather are components for magnesium deficiency.

Calcium: Calcium plays an important role in the integrity of membranes and cell walls as calcium pectate is a constituent of middle lamella. Calcium promotes meristematic growth and also auxin action. It is immobile. It also plays a role in lipid synthesis.

The deficiency of calcium results in killing of growing points, roots are poorly developed. In potato, tuber and shoot formation is severely affected. Blossom end rot occurs in tomato. Die back symptoms appear in apple. In broad bean pods are deformed, wilted and blackened and seeds fail to develop. Application of gypsum, super-phosphate to *Ruscus* plants effectively reduced susceptibility of out branches to infection by *Botrytis cinerea* (Elad and Kirshner, 1992), pod rot due to *R. solani* and *Pythium myriotylum* in groundnut (Hallock, and Garren, 1968). Calcium deficiency occurs most commonly in strongly acid soils. Proper calcium quantity is required in alfalfa, cabbage, potato, sugarbeet, tomato, groundnut etc. Calcium uptake by plants is influenced by the ratios between calcium and other cations in the soil solution.

Sulphur: It is a constituent of S-containing amino-acids, tripeptides, proteins, vitamins, glutathione, lipoic acid, coenzyme A, glycosides and thiols. A portion of total S in plants is mobile and the movement is through phloem.

The deficiency of sulphur affects N metabolism, leaves become small, chlorotic, brittle and rolled. In soil, sulphur is available in different forms, organic and inorganic.

Micronutrients: The micronutrients are Mn, Cu, Zn, B, Fe, Mo, Cl, Ni and Si which play a definite role in plant metabolism. The micronutrients are required for growth of plants in lower quantity. They are constituents of prosthetic groups in metallo proteins and as activators of enzyme reactions. Some catalyse redox processes by electron transfer (Fe, Mn, Cu, Mo). Some form enzyme substrate complexes by coupling enzyme and substrate (Fe, Zn) or enhance enzyme reactions by influencing the molecular configuration of an enzyme or substrate (Zn). The non metals B and Cl are present in enzymes and essential organic compounds. The quantum of micronutrients of the host plant at the time of attack by a pathogen decide the defense capabilities. The effect is generally in deficiency range.

Manganese: Mn deficiency affects photosynthesis and O_2 evolution. It is an important contributor in the development of resistance in plants to both root and foliar diseases caused by fungi. Host plants require more quantity of Mn than fungi and bacteria. The effect of Mn is both in the deficiency range as well as in the sufficiency range. The Mn requirement of the host plant for disease resistance is higher than for yield and it somehow lowers the inoculum potential in case of soil borne pathogens. As the Mn concentration in host tissues decreases incidence of diseases increase. The root pathogens reduce the absorptive surface of the root which results in lesser absorption of Mn. The fungus also mobilizes Mn in the rhizosphere by oxidation, thereby decreasing the availability of Mn in soil. Mn gets concentrated around infection sites. Certain soil factors also affect the availability of soil Mn. Higher pH, addition of lime, nitrate forms of N, cold wet soils decrease whereas, ammonical forms of N, Cl fertilizers, green manures and irrigation increase the soil Mn availability. Several pathogenic diseases have been successfully controlled by Mn fertilizers like powdery mildew of cereals, downy mildew of sorghum, take all disease of wheat, common scab of potato. Mn deficiency develops marsh spot in pea.

As regards the mechanism of action of Mn in disease resistance, it plays a role in lignin synthesis which is a basis of resistance to powdery mildew and take all disease in wheat. The lignin response in roots has been reported greater than in shoots. Phenols are responsible for resistance to diseases. Mn is involved in biosynthesis pathway to phenol, its deficiency leads to a decrease in soluble phenols influencing resistance. Amino-peptidase is a host enzyme activated by a pathogen to supply essential amino acids for fungal growth. The induction of this enzyme is inhibited by Mn. Mn also inhibits pectinmethylesterase which is a fungal coenzyme for degrading host cell walls. The deficiency of Mn severely inhibits photosynthesis which influences root exudation and ultimately rhizoflora is affected influencing the pathogen. The root exudations increase Mn in soil making it toxic to the pathogen. Cotton cultivars tolerant to Mn toxicity are preferred as the high soil Mn concentrations controlled the inoculum density of *Verticillium alboatrum* by inhibiting secretial production which is necessary to saprophytic survival of the pathogen. The Mn deficiency results in greenish grey spots, flecks and strips in leaves of cereals and interveinal chlorosis of the younger and middle aged leaves in dicots. The tissues become necrotic. Alfalfa, citrus, oat, onion, potato, soybean, wheat, sugarbeet are sensitive to low levels of available manganese in soil.

Copper : Cu deficiency depresses reproductive growth, seed and fruit formation more than vegetative growth. It causes chlorosis leaf distation, and die back. Copper is used for the control of foliar pathogens by topical application in the form of salts. It is essential for plants. The fungi and bacteria need it in minute quantity and is toxic at higher concentrations which are tolerated by host plants. The foliar and root diseases are controlled by providing more Cu 10 to 100 times greater than those normally needed (0.1-0.2 kg ha^{-1}) to cure Cu deficiency. When Cu is applied to soil, the disease control is observed both in roots and leaves.

Copper fertilizer controlled powdery mildew of wheat (Graham, 1980). Wheat plant takes about six weeks to exhibit deficiency of Cu, consequently it affects disease resistance of more mature plants, when wheat normally acquires greater resistance called adult plant resistance. Several foliar diseases have been controlled with soil application of Cu as wheat powdery mildew, *Alternaria* of sunflower, brown rust of wheat, ergot of rye and barley, paddy blast and *Septoria* on wheat. Cu supplementation has suppressed many soil borne diseases like *Verticillium* wilt of tomato and cotton, potato scab, take all of wheat (Graham, 1983).

The mode of action of Cu in developing host resistance and reducing disease severity is based on direct toxicity. It denatures proteins, indirectly it helps as Cu plays a vital role in lignin synthesis. Lignin acts as a partial barrier in penetration of many pathogens. Cu is also involved in the synthesis of soluble phenols and in their oxidation to more toxic quinones which not only kill the invading microorganisms but also the surrounding cells of the host and give rise to a local lesion, an important auto-amino response which checks infections by obligate parasites like rust and powdery mildew.

Zinc : It functions as a divalent cation by coupling enzymes with corresponding substrates and forming tetrahedral chelates with different organic compounds including peptides. It is essential in protein synthesis. Plants generally contain 20-100 mg Zn per kg. Its deficiency results in less protein, auxins, stunted plants, low yields and prolonged duration of crops. Maize, rice, wheat, citrus, grapes, beans, apple etc. are sensitive to zinc deficiency.

Zinc increases the resistance to pathogens in host plants. It directly affects the pathogen as well through plant's metabolism. In *in vitro* tests Zn has checked the growth of the microorganisms. Zoosporangium and zoospore formation was inhibited in *Phytophthora megasperma* and growth of several other fungi. In high concentrations Zn checks several diseases. Pre-treatment of soil with $ZnSO_4$ decreased *Fusarium* wilt of cotton. Dipping of potato tubers in 0.05% $ZnSO_4$ plus 1% acetic acid for 15 minutes before planting checked black surf due to *R. solani*. Application of $ZnSO_4$ in soil also checked *Fusarium* root rot of chickpea, *R. bataticola* rot of groundnut Zn also plays a role in controlling fungal growth within the plant.

Zn is important in the stability of biological membranes and it helps in preventing root membranes from leaking. The Zn deficient roots leaked more carbohydrates and amino acids attracting more zoospores by chemotaxis which allow more successful invasion. In soybean collar rot caused by *S. rolfsii* high rates of $ZnSO_4$ (0.5%) decreased infection, it also increased the population of potential antagonistic microorganisms. The Zn efficiency is the factor responsible for tolerance to diseases. Zn checked blast in rice (Fillipi and Prabhu, 1998) and downy mildew of bajra.

Boron : The deficiency of boron (B) results in rosette like growth of foliage, cracking of internodes, withering or drying of growing points and reduced bud flower and seed production.

It has beneficial effects in reducing diseases like club root of crucifers, *R. solani* in green gram, pea, cowpea, *R. bataticola* in groundnut, *F. solani* in bean etc. B is involved in many physiological and biochemical functions.

Boron is involved in phenolic and lignin metabolism affecting membrane, cell wall development and pollen tube growth. It has been shown to decrease pathogens which do not involve vascular tissue like potato wart caused by *Synchytrium endobioticum* and club root of crucifers. In both the cases tumor or gall is formed. Pathogens invading vascular tissues like species of *Fusarium* and *Verticillium* enter the xylem vessels through invasion just behind the root cap where lignification is the least. B deficiency weakens the lignification of other parts of the root system where from invasion occurs later. B deficiency leads to hollow heart in groundnut resulting in discolouration and rotting of seeds. Premature fruit/seed drop also occurs.

Iron : The deficiency causes chlorosis of young leaves and cell division is inhibited. In soybean iron deficiency is common on high pH soils as solubility of iron is minimum between pH 7.4 - 8.5. Iron chlorosis is often associated with cool, rainy weather when soil moisture is high and soil aeration is poor. Applications of organic matter improves soil aeration and ultimately iron availability. Iron is transported to soybean roots by diffusion. Pineapple develops iron chlorosis when grown in soils high in manganese. At high concentrations Fe checks rust infection on wheat leaves, wheat smut and banana anthracnose. Foliar sprays of Fe increased resistance in apple and pear to *Sphaeropsis malorum* and tolerance of cabbage to *Olpidium brassicae*.

Fe plays key role in oxidative phosphorylation and is involved in plant synthesis directly or indirectly. High Fe requirement is there for phytoalexin synthesis. Production of antibiotics by several soil bacteria is highly promoted by Fe. Plant genotypes differ in iron uptake efficiency.

Molybdenum : It exists in plants as anion primarily in its highest oxidized form. Molybdenum containing enzymes are multicenter electron transfer proteins. The deficiency of molybdenum increases the concentration of soluble nitrogenous compounds like amides, whereas the protein concentration and alanine eminotransferese activity is decreased. Biological fixation is catalysed by molybdenum containing enzyme nitrogenase. The molybdenum required for root nodules in legumes and non legumes is high.

Deficiency caused poor and delayed flowering, reduced pollen viability. Application of Mo to tomato roots reduced *Verticillium* wilt. Mo had a direct effect by reducing the production of Roridin E., a toxin produced by *Myrothecium roridum*. Soil application of Mo decreased nematode populations. Interveinal mottle, scald, interveinal and marginal necrosis of leaves occurs in phaseolus bean.

Chlorine : It is involved in O_2 production during photosynthesis. Cl exists in soil plant system as the stable Cl anion. The chlorine anion is bound very slightly by most soils in the light acid to neutral pH and negligible at pH above 7. Chloride fertilizers as micronutrient have partially controlled several diseases like stalk rot of corn, stripe rust of wheat, take all of wheat, downy mildew of pearl millet, *Septoria* in wheat. Cl deficiency causes chlorosis of younger leaves in overall wilting due to possible effect of transpiration. Under chlorine excess, plants develop leaf burn symptoms as in fruit trees, vine crops and ornamental shrubs. Tobacco and tomato leaves get thickened and start to roll.

Nickel : Ni $(NO_3)_2$ at 70 mg/l as spray decreased leaf rust on wheat 77 to 0.1 pustule per leaf and entirely prevented stem rust infection. Ni has eradicative and protective action within a few minutes of its application. The efficacy of conventional organic fungicides has been improved by combining them with Ni salts.

The relatively low concentrations, short contact time and the protective effect of Ni suggest physiological mechanisms in the host leading to resistance. When Ni was supplied *via* roots in low concentrations the rust pustules were highly reduced in cowpea which suggests no direct toxic effect to the pathogens, it is involved in some biochemical pathway that leads to resistance.

Ni is an essential ultra micronutrient as it plays a vital role in the N metabolism of plants. The resistance pathway may have a higher requirement for Ni than growth itself.

Silicon : It is essential to biochemical pathways leading to resistance to certain pathogens. Silicates provided protection against rice blast fungus and brown spot. In certain soils low in silica, the application of silicate can suppress these pathogens, decreasing the disease and increasing the yield. High water content of soil encourages silicon uptake in crops like rice.

Other trace elements : Among other trace elements, Lithium and Cadmium through their marked suppressive effects on powdery mildews, are the most noteworthy. Li is supposed to catalyze a metabolic pathway which functions normally in defense.

Heavy metals Cd, Pb, Hg, Cr are increasing in soil due to increased industrialization, urban run off, sewage treatment plants, agricultural run off, domestic garbage dumps and mining operations. These elements have no known function and are toxic to plants at higher concentrations (Srivastava *et al.*, 2004).

It is evident from the above account that the addition of nutrients decreases the incidence of diseases in crop plants. The host plant tolerance and resistance to pathogens is most important. The response is more over the deficiency range for the element concerned. In any integrated crop protection programme adequate emphasis should be given to micronutrients as at a very minor quantity they reduce the disease to an acceptable level or at least to a level at which further control by other cultural practices or biocides becomes possible. It may be further controlled with least expenditure successfully. While recommending inorganic fertilizers, micronutrients may also be recommended after proper analysis. The objective has to be to get maximum yields from a crop.

References

Basuchaudhary, K.C. and Gupta, D.K. (1988).Role of fertilizers on plant disease management. In: *Perspectives in Mycology and Plant Pathology*, (Eds.) V.P. Agnihotri, A.K. Sarbhoy and Dinesh Kumar, Malhotra Publishing House, New Delhi: 84-91.

Colhoun, J. (1973). Effects of environmental factors on plant disease. *Ann. Rev. Phytopath.*, 11: 343-364.

Das, A.C. and Western, J.H. (1959).The effect of inorganic manures, moisture and inoculum on the incidence of root disease caused by *Rhizoctonia solani* Kunh in cultivated soil. *Ann. Appl. Biol.*, 47: 37-48.

Elad, Y. and Kirshner, B. (1992).Calcium reduces *Botrytis cinerea* damages to plants of *Ruscus hypoglossum*. *Phytoparasitica*, 20 : 285-291.

Filho, E.S. and Dhingra, O.D. (1980).Survival of *Macrophomina phaseolina* sclerotia in nitrogen amended soil. *Phytopath. Z.*, 97: 136-143.

Fillippi, M. C. and Prabhu, A.S. (1998).Relationship between panicle blast severity and mineral nutrient content to plant tissue in upland rice. *J. Plant Nutrit.*, 21: 1577-1587.

Graham, R.D. (1980). Susceptibility of powdery mildew of wheat plants deficient in copper. *Plant Soil*, 56: 181-185.

Graham, R.D. (1983). Effects of nutrient stress on susceptibility of pants to disease with particular reference to the trace elements. *Adv. Bot. Res.*, 10: 221-276.

Graham, R.D. and Webb, M.J. (1991). In: *Micronutrients in Agriculture*. II Edition. S.S.S.A. Book Series No. 4, pp. 329-370.

Hallock, D.L. and Garren, K.H. (1968). Pod breakdown,yield and grade of Virginia type of peanuts as affected by Ca, Mg and K sulphates. *Agron. J.*, 60: 253-257.

Huber, D.M. and Watson, R.D. (1974).Nitrogen form and plant diseases. *Ann. Rev. Phytopath.*, 12: 131-165.

Last, F.T. (1962).Effects of nutrition on the incidences of powdery mildew. *Plant Pathol.*, 11 : 133-135.

Ngeve, J.M. and Rocadori, R.W. (1984). In: Tropical *Root Crops : Production and uses in Africa*. Inter. Dev. Res. Centre, Ottawa, Canada: 197-202.

Palti, J. (1981). *Cultural practices and infectious crop diseases*, Springer Verlag Berlin Heidelberg, New York: 243.

Reuveni, R. and Reuveni, M. (1998).Foliar fertilizer therapy-a concept in integrated pest managenent. *Crop Protect.*, 17: 111-118.

Sharma, S.K., Verma, R.N. and Chauhan, S. (1996).Effect of NPK doses on yield and severity of three major diseases of soyabean at medium altitude of East Khashi Hills (Meghalaya). *J. Hill Res.*, 9: 279-295.

Sharma, S.R. and Kolte, S.J. (1994).Effect of soil-applied NPK fertilizers on severity of black spot disease (*Alternaria brassica*) and yield oilseed rape. *Plant Soil*, 167: 313-320.

Srivastava, A.K., Bhargava, P., Shukla, B. and Rai, L.C. (2004). Heavy metal induced oxidative stress and its protection in Cyanobacteria. In: *Vistas In Applied Botany* (Eds.) H.S. Prakash, S.R. Niranjana and K.R. Kini, Department of Applied Botany and Biotechnology, University of Mysore, Mysore: 272 – 296.

Sunder, S. and Satyavir (1997).Survival Of Fusarium moniliformae in soil enriched with different nutrient and their combinations. *Ind. Phytopath.*, 50 : 474-481.

Suryanarayana, S. (1958).Role of nitrogen in host susceptibility to *Pyricularia oryzae cav. Curr. Sci.*, 27: 447-448.

Tisdale, S.L., Nelson, W.L. and Beaton, J.D. (1990). *Soil Fertility and Fertilizers*, Macmillan Publishing Company, New York. pp. 112-413.

□□□

Microbial Diversity and Functions, 2012
© D.J. Bagyaraj, K.V.B.R. Tilak, H.K. Kehri (eds.), pp. 389-402
New India Publishing Agency, New Delhi (India)
E-mail : info@nipabooks.com; Website : www.nipabooks.com

Chapter 18

Mycorrhizal Symbiosis : A New Dimension for Agriculture and Environmental Development to Improve Production in Sustainable Manner

Kamal Prasad and A.K. Pandey

ABSTRACT

Much attention has been paid to mycorrhza as a tool for improving the growth and health of plants. The use of mycorrhiza in agricultureal, horticultural and forestry has been described. The significance of AM fungi in sustainable agriculture has been a subject of growing interest for the past several decades. In sustainable agriculture it is imperative to maximize benefits with low input costs. The fact remains that stable and lower human population is an integral component of sustainable agriculture and more so in an Indian context. Therefore, it is imperative to collect further information on the different aspects of AM fungal symbiosis as discussed in this article so as to utilize this plant microbe symbiotic system for the increased production and productivity in a sustainable manner. This may become possible when the integrated approach is made to study AM fungi right from the isolation of AM fungal spores to the high quality inoculum production and its application in the field. It is also imperative to stimulate new mutualism between mycorrhizal scientists and ecologists. It this regard, recent advancement made in the molecular techniques should be encouraged in order to obtain more authentic results. Adequate field testing of mycorrhizal inoculation and commercial exploitation of the potential benefits of mycorrhiza still rests on the development of suitable technology for mycorrhizal inoculum preparation.

Keywords: Mycorrhizal symbiosis, sustainable production, inoculum production

Introduction

Sustainable agriculture is vital in today's world as it offers the potential to meet our agricultural needs. An ideal agricultural system is sustainable, maintains and improves human health, benefits environment and produces enough food for increasing world population. Mycorrhizal fungal application technique is environment friendly and ensures safe and healthy agricultural products. Microbial populations are instrumental to fundamental processes that drive stability and productivity of agro-ecosystems. Several investigations addressed at improving understanding of the diversity, dynamics and importance of soil microbial communities and their beneficial and co-operative roles in agricultural productivity. The utilization of mycorrhizae as bio-fertilizer can decrease the use of nutrients, prevent the depletion of soil organic matter and reduce environmental pollution (Brazanti *et al.*, 1999; Prasad and Deploey, 1999; Guatam and Prasad, 2001,). Sustainable farming systems strive to minimize the use of synthetic pesticides and to optimize the use of alternative management strategies to control soil-borne pathogens. Mycorrhizal fungi are ubiquitous in nature and constitute an integral component of terrestrial ecosystems, forming symbiotic associations with plant root systems of over 90% of all terrestrial plant species, including many agronomically important species. Mycorrhizal fungi are particularly important in organic and/ or sustainable farming systems that rely on biological processes rather than agrochemicals to control plant. Mycorrhiza is a symbiotic, non-pathogenic association between soil-borne fungi with the roots of higher plants (Sieverding, 1991). They play an important role in the uptake of diffusion limited nutrients and water and protect plant health (Barea *et al.*, 2002; Ferrol *et al.*, 2002). Mycorrhizae have a wide range of application in sustainable low input agricultural systems (Singh and Adholeya, 2003). The use of mycorrhiza fungi may contribute to reducing chemical fertilizer inputs and sustaining plant productivity in agriculture and forestry (Mc Gonigle, 1998). They extend the effective root area many hundreds of times so plants grow faster, larger and stronger with less fertilizers and water. Mycorrhizae help host plants to grow better and healthier.

Mycorrhizae literally means 'fungus roots' resulting in a mutually beneficial symbiosis. Association of fungi and plant roots, is a universal phenomenon throughout the plant kingdom, and is beneficial and even indispensable for the life and healthy growth of the host plants. Since almost all the plant species of natural vegetation and the agricultural crop plants of the tropics live in mycorrhizal association with fungi, it is possible to increase productivity through manipulation of mycorrhiza. It is now a well established fact that mycorrhizal fungi function mainly in regulating phosphate absorption, translocation and transfer to host plant. The facilitation of plant nutrient uptake by fungi in symbiotic association with plant roots is emerging as one of the exciting frontiers for the enhancement of crop production (Addy *et al.*, 1997). Under normal conditions, concentration of available phosphate in the soil is very low. Besides this, it is also believed that mycorrhizae stimulate uptake of Zn, Cu, K, Ca as well as water. Prasad *et al.* (2006) demonstrated that mycorrhizal fungi enhanced biological nitrogen fixation. Mycorrhizal fungi have been found to be very effective under such nutrient deficient conditions because of their ability to mobilize even extremely small quantities of phosphorus and give it to the plants. That is the reason why AM fungi are being regarded as important biofertilizer and used for almost all the agricultural crops of tropical countries. An experiment performed by Plenchette and Morel (1996) showed that *Glycine max* colonized by AM fungus *Glomus intraradices* produced 80% of maximum growth at a soil solution P concentration of 0.110 mg ml^{-1}. Non-mycorrhizal plants required 0.148 mg ml^{-1} for the same rate of growth. This calculated to a saving of 222 kg P_2O_5 ha^{-1} to the farmer. Also, mycorrhizal fungi have given excellent results with woody ornamentals like liquid amber, *Ampelopsis* and Yew (Gianinazzi *et al.*, 1990) and vegetable crops like onion, asparagus and leek (Gianinazzi and Pearson, 1986). It has also been observed that dual inoculation of *Glomus intraradices* and *Bradyrhizobium japonicum* can enhance the soybean production considerably. Much has been discussed in various articles about this mycorrhizal symbiosis and immense informations have been gathered by the scientists till date. However, in this paper, it will not be our aim to review the literature but to highlight such important concerns of ectomycorrhizae/ AM fungal symbiosis, which we believe to be considered as the major thrust areas of research in the present scenario. These are: scientific classification of mycorrhizal fungi, ecological and physiological aspects, and systematic agricultural management practices.

In nature, the same effect is achieved through years of growth and decay and natural, cyclic conditions of weather and soil movement. The result is the creation of the pathways and receptors for nature's organic and inorganic elements to develop. Within this ecosystem are also certain types of fungi that serve to assist the natural processes of life and decay. These fungi, called mycorrhizae, form working partnerships with plants and transform base materials into food. Mycorrhiza is fungus that lives among and upon the roots of plants and trees. It exists to assist with the break down of complex nutrients, similar to enzymes and has the ability to create optimal conditions for delivery of food directly to plants' roots. In some rare cases, mycorrhizae acts as a pathogen, this exists only to feed itself at the expense of the plant. Mycorrhizae are grouped into several scientific categories owing to the types of plants and trees that they develop the strongest bonds and in the differences in growth and operation. Microscopic organisms develop into a self-sustaining ecosystem, which in turn transforms into an acceptable food source for plants.

Sustainable production using AM fungi : In natural ecosystems or low-tillage agriculture young seedlings can germinate and effectively "plug" into an already established "motorway" of hyphae of AM fungi which permeates the soil and links different plant species. The lack of host specificity is the secret to the success of AM fungi in mixed plant communities. The benefit to plants in natural plant communities is perhaps that less carbon from the plant photosynthate is needed by AM fungi colonization since it is plugged into a pre established mycelium. In contrast, crops in agricultural systems are frequently sown into tilled soil where this mycelium has been completely disrupted. Agriculture would therefore allow only those AM fungi with aggressive colonization strategies. There is a conflict of interest in the idea of maximizing plant production against an aim of maintaining a high biodiversity of AM fungi in soils. The latter maybe a necessity in natural ecosystems or restoration of degraded natural habitats but selection for efficient populations of AM fungi compatible with the aim of maximizing yields of certain crops may require a different management approach in agro-systems. Modern agricultural practices, such as high levels of fertiliser and pesticide inputs and long-term monocultures, have proven adverse effects on the diversity of soil microbiota which are at odds with the use of AM fungi. It is becoming clear that sustainable production practices, e.g. crop rotations with legumes, would benefit the survival of inoculum of AM fungi from season to season and hence increase or maintain inoculum of AM fungi for subsequent mycorrhizal crops. One potential weakness is that both systems are using varieties (genotypes) of crops bred for high inputs. This is a selection process driven by conventional plant production, and the varieties may not be suitable for optimal production under organic or other sustainable systems. There are examples of where modern lines of plants in commercial production appear to be less susceptible to colonization by AM fungi as a result of breeding programmes e.g. wheat 26. Other work has shown that the inbred lines of *Zea mays* L. with resistance to fungal pathogens were less able to form mycorrhizas compared with disease susceptible lines (Toth *et al.*, 1990). The relationship, however, between reduced colonization and nutrient uptake ability of AM fungi is uncertain and maybe uncoupled genetically. However, there is evidence for increased root fibrosity to compensate for the reduced role of AM fungi. These traits will operate fine under high input agricultural production but can the same varieties produce high yields under reduced input systems.

Past and future of AM fungi in plant production : Research on AM fungi in the 1970s and 1990s was dominated by the search for "super strains" capable of increasing plant biomass under any environmental and soil conditions. The desire to exploit AM fungi as a natural biofertilizers for the agricultural biotechnology industry was understandable, but it became clear that more knowledge was needed of the fungi themselves to allow commercial exploitation. Many inoculant companies have tried to commercialize the use of AM fungi with limited success. This has masked the importance of the symbiosis for normal plant growth and development in natural ecosystems where mycorrhizal plants dominate. Many inoculant companies use the same fungal consortia for all environments. The benefits of the symbiosis for nutrient uptake by plants in agro-systems is important but a more complete understanding of how to manage arbuscular mycorrhizas for optimum plant growth and development and general health is needed urgently, as high-input plant production practices are challenged by more sustainable approaches.

Mycorrhizal fungi and soil fertility : A key factor which affects the potential for mycorrhizas to benefit plants in particular sites is the supply of phosphate and nitrogen in soil (George *et al.*, 1994; George *et al.*, 1996; Chandreshekara *et al.*, 1995; Megavanshi *et al.*, 2010; Prasad, 2010b). Phosphorus is generally considered to be the most important plant-growth limiting factor which can be supplied by mycorrhizal associations, because of the many abiotic and biotic factors which can restrict its mobility in soils (Harley and Smith, 1983; Bolan, 1991; Marschner and Dell, 1994). Reductions in the benefit provided by mycorrhizal associations to plants are caused by increasing soil phosphorus levels (Birch, 1988; Jones *et al.*, 1990). High rates of P and N fertilizers suppress ectomycorrhiza development in the field (Menge *et al.*, 1977; Newton and Pigott, 1991). High concentrations of soil N can influence the relative abundance of different ECM types (Bowen, 1973; Brazanti *et al.*, 1999; Bougher, 1995).

Mycorrhizal fungi and adverse soil conditions: Land degradation due to salinity, water logging, erosion, etc. is serious and growing problems in all over the world. Excessive NaCl levels in soil inhibit mycorrhizal formation and restrict the activity of most mycorrhizal fungi, but some can tolerate these conditions (Malajczuk *et al.*, 1987, Read and Boyd, 1986; Juniper and Abbott, 1993). Observations in natural ecosystems have shown that plants with mycorrhizal associations are often less common than non-mycorrhizal species in soils which are waterlogged or saline, but that some mycorrhizal plants are normally present in even the worst soils (Brundrett, 1991). ECM fungi can be highly sensitive to water logging of soils, while AM fungi may be less sensitive (Prasad *et al.*, 2005; Bowen, 1993; Prasad, 2010a).

Mycorrhizal fungi and their roles in ecosystems : The ecology of mycorrhizal fungi is not well documented (Abbott and Gazey, 1994; Trappe, 1987; Francis and Read, 1995). Hence, in the discussion that follows, conclusions are mostly drawn from short-term studies with a small range of partnerships, often under experimental conditions. In nature, the situation is far more complex as a single tree may have fungal partners which can vary in time and space. The fungal/plant interface provides a conduit for the movement of carbon from the plant to the fungus, and for movement between plants linked by mycelia (Francis and Read, 1984; Read *et al.*, 1989; Van der Heijder, 1998; Simard *et al.*, 1997; Wu *et al.*, 2001). The nature of the interface and its mode of regulation are still being elucidated (Hall and Williams, 2000). It is generally believed that mycorrhizal plants direct more of their photosynthates into the soil than non mycorrhizal plants. This extra carbon accumulates in patches and at the edge of hyphal mats (Finlay and Read, 1986), and boosts the energy supply to the detrital food web, benefiting saprophytic microbes and other soil organisms (Barea, 2000). Because the chemical (Dieffenbach and Matzner, 2000) and physical environment around mycorrhizas (the mycorrhizosphere) differs from non mycorrhizas, presumably it provides microhabitats for soil biota that are not present in the rhizospere of nonmycorrhizal roots. Mycorrhizal fungi are estimated to consume from 15 to 50% of net primary production (Fogel and Hunt, 1979; Vogt *et al.*, 1982; Baltruschat and Dehne, 1988).

Nutrient cycling and nutrient conservation and mycorrhizal fungi : Fungi are crucial components of ecosystems as they transport, store, and release the nutrients cycle. A good example of the potential of mycorrhizal fungi to capture and deliver nutrients to their host comes from studies using inoculated eucalypts in field trials in sub-tropical China. Generally, trees in this region grow well below their potential. The main constraint to productivity appears to be low soil fertility (Dell and Malajczuk, 1994; Xu and Dell, 1998). Most of the land available for plantation forestry have been degraded over recent centuries with extensive loss of the A horizon caused by population pressure, inadequate management and over-harvesting. Topsoil crusting is common, contributing to enhanced erosion, reduced soil water storage, compaction and poor root development (Xu *et al.*, 2000; Bago, 2000). Low soil organic matter (SOM) content (<2%) also restrains productivity. As most soils for plantation eucalypts in southern China have lost their Ao layer, we need urgently to consider how to recover microbial biodiversity as there is no doubt that this will be important for improving long-term soil fertility. The capacity of some ECM fungi to promote both early growth and survival of eucalypts is very important for commercial plantations on these disturbed and difficult sites. Significant effects of ECM fungal inoculation on growth of plantation eucalypts were obtained at two sites in southern China (Xu *et al.*, 2001). Effects were isolate dependent with some isolates stimulating tree growth and some isolates depressing tree growth. Similar results were obtained in a trial in the Philippines where two isolates increased survival while one isolate decreased survival of *Eucalyptus urophylla* (Aggangan *et al.*, 1999). The improvement in growth could be attributed to the acquisition of P as other essential mineral nutrients were supplied at establishment. Generally, inoculation only increased stand volume under P-limiting soil conditions.

In forests, litter is an important nutrient reservoir. ECM fungi can mobilize P, N and other nutrients from litter to tree roots (Attiwill and Adams, 1993; Perez-Moreno and Read, 2000). Fogel and Hunt (1979) estimated that ECMs account for 43% of the annual turnover of N in a *Pseudotsuga menziesii* forest in Oregon. Litter type can affect the diversity and function of ECMs (Conn and Dighton, 2000). Buscot *et al.* (2000) propose that the high diversity of fungal partners that a tree may have allows optimal foraging and mobilization of various N and P forms from organic soil layers.

Mycorrhizal fungi and soil structure : It is obvious from the examination of ECM mycelial mats, that mycorrhizal fungi have a big impact on soil structure. Yet, there is scant information in the literature regarding soils in tropical ecosystems. In agricultural soils, AM fungi increase the formation of soil aggregates (Bethlenfalvay *et al.*, 1999).

Mycorrhizal fungi and value for human : In many upland forest regions of South East Asia, sporocarps of fungi, mostly basidiomycetes, have traditionally been collected for local consumption and trade (Dell *et al.*, 2000). Many of these fungi, especially members of the Amanitaceae, Boletaceae, Russulaceae, and Tricholomataceae, form ectomycorrhizal associations with trees in the families Dipterocarpaceae, Fagaceae and Pinaceae and are important for maintaining ecosystem function. The highest diversity of edible fungi is collected from mixed forests in China and the lowest diversity from areas of tropical pine and dipterocarps. In general, traded fresh sporocarps are 2 to 20 times more valuable, by weight, than local seasonal fruits and vegetables. International trade in a small number of species is having a major impact on the quality and sustainability of the mushroom harvest from some collecting sites. Forest fungi are also valued for medicine, for their aesthetics, as bio-indicators of environmental quality and for bio-remediation.

Mycorrhizal fungi and land management : All types of land management that involves tillage, timber harvesting, vegetation clearing or other forms of disturbance can affect mycorrhizal populations. Severe soil disturbance, such as fallowing agricultural soils (Thomson, 1987), crop rotation with non-host species (Gravito and Miller, 1998) or topsoil stripping and storage during mining (Jasper *et al.*, 1987; Gardner and Malajczuk, 1988; Bowen, 1973), markedly reduces populations of mycorrhizal fungi. Unlike AM fungi, ECM fungi may be able to quickly invade disturbed soils (Jasper, 1994). This is often the case for what have been termed "early colonising" genera such as *Laccaria, Pisolithus, Rhizopogon, Scleroderma* and *Thelephora*. Recolonisation mostly results from spore dispersal by wind and animal vectors from sporocarps in adjacent vegetation.

The mycelial network appears to be an important component of the inoculum potential of an undisturbed soil (Evans and Miller, 1990). Even minor soil disturbance can impact on the function of mycorrhizas. Severing mycelium reduces the extent of inter-root and inter-plant connections, thus reducing access to existing and new food bases (Read and Birch, 1988). The mycelial mats of ECM fungi, mentioned earlier, are vulnerable because they are easily disrupted by raking litter during the collection of leaves for fuel or the collection of edible fungi. Studies in a number of ecosystems (Reeves *et al.*, 1979; Allen *et al.*, 1987) show that in climax communities, normally dominated by species heavily colonised by mycorrhizal fungi, disturbance leads to a successional sequence in which re-colonisation is initiated by plant species which are non-mycorrhizal or little infected (Read and Birch, 1988). Gap-preferring species thus may have lower rates of mycorrhizal infection than species preferring undisturbed microsites (Onipchenko and Zobel, 2000). However, Brundrett *et al.*, (1995) found that for parts of tropical Australia, the activity of mycorrhizal fungi was higher in patches of early-successional vegetation than in undisturbed habitats. In a study in deciduous tropical forest in Mexico, Allen (1998) concluded that regrowth of vegetation in small gaps was not limited by mycorrhizal fungi, since they were still abundant after tree falls. However, recovery in pastures could be affected by low fungal diversity and dominance of grasses. Jasper *et al.*, (1991) found that disturbance of forest and heath land soils decreased colonization of test plants (clover) compared to disturbance of clover soil. They proposed that a larger number of propagules in the pasture soil may have allowed the pasture soil to maintain infectivity after disturbance. There is a suggestion that ecosystems with a high proportion of grasses and high numbers of AM spores may also be more tolerant of disturbance (Visser *et al.*, 1984). The abandonment of agricultural land in the Italian Alps resulted in succession from non mycorrhizal ruderal annuals to AM-colonized perennials and an increase in floristic richness (Barbi and Siniscalo, 2000). Over time, ECM hosts will increasingly dominate if old-field succession is allowed to continue. Use of fertilizers can affect mycorrhizal fungal populations (Baum and Makeschin, 2000; Pampolina *et al.*, 2001). Increasing soil fertility, especially P and N, can suppress mycorrhiza formation and/or mycorrhizal diversity but the effects are often host and fungal dependent. For example, the number of epigeous basidiocarps in an 11-year-old *Pinus taeda* stand in north California was reduced to 17 % following the addition of 25 kg P ha^{-1} (Menge *et al.*, 1977). In *Betula pendula*, the addition of 20 kg P ha^{-1} at 3 and 9 weeks after out planting reduced ectomycorrhizal root colonization (Newton and Pigott, 1991). By contrast, Fransson *et al.*, (2000) applied repeated balanced additions of nutrients to 36-year-old *Picea abies* and could find no measurable effect on morphotype richness (>60 ECM morphotypes) or total number of root tips. However, *Cenococcum* was more common in fertilized plots than in the controls. Soil P content is often negatively correlated with % root colonization (Maldonado *et al.*, 2000).

References

Abbott, L. K. and Gazey, C. (1994). An ecological view of the formation of VA mycorrhizas. *Plant Soil,* 159: 69-78.

Addy, H.D., Miller, M.H. and Peterson, R.L. (1997). Infectivity of the propagules associated with extraradical mycelia of two AM fungi following winter freezing. *New Phytol.,* 135:745-753.

Aggangan, N.S., Dell, B., Malajczuk, N. and dela Cruz, R.E. (1999). Comparative effects of ectomycorrhizal inoculation and fertilization on *Eucalyptus urophylla* S.T. Blake five years after outplanting in a marginal soil in the Philippines. In: *Challenges for Biotechnology in the Next Millenium.* Proceedings of the 7th International Workshop of BIO-REFOR, Manila, Philippines, November 3-5, 1998: 77-82.

Allen, E.B. (1998). Disturbance and seasonal dynamics of mycorrhizae in a tropical deciduous forest in Mexico. *Biotropica,* 30: 261-274.

Allen, E.B., Chambers, J.C., Connor, K.E., Allen, M.F. and Brown, R.W. (1987). Natural reestablishment of mycorrhizae in disturbed alpine ecosystems. *Arctic Alpine Res.,* 19: 11-22.

Attiwill, P.M., and. Adams, M.A. (1993). Nutrient cycling in forests. *New Phytologist,* 124: 561-582.

Baltruschat, H. and Dehne, H.W. (1988). The occurrence of vesicular-arbuscular mycorrhiza in agro-ecosystem, manure in continuous monoculture and crop rotation on the inoculum potential of winter wheat. *Plant Soil,* 107: 279-284.

Barea, J.M., Azcon, R, and Azcon-Aquilar, C. (2002). Mycorrhizosphere interaction to improve plant fitness and soil quality. *Anton. van Leeuwenhoek,* 81: 343-351.

Barbi, E., and Siniscalco, C. (2000). Vegetation dynamics and arbuscular mycorrhiza in oldfield successions of the western Italian Alps. *Mycorrhiza,* 10: 63-72.

Barea, J.M. (2000). Rhizosphere and mycorrhiza of field crops. *Biol. Res. Manage.* : 81-92.

Bago, B. (2000). Putative sites for nutrient uptake in arbuscular mycorrhizal fungi. *Plant Soil,* 226: 263-274.

Baum, C., and Makeschin, F. (2000). Effects of nitrogen and phosphorus fertilization on mycorrhizal formation of two poplar clones (*Populus trichocarpa* and *P. tremula* x *tremuloides*). *J. Pl. Nutrit. Soil Sci.,* 163: 491-497.

Bethlenfalvay, G.J., Cantrell, I.C., Mihara, K.L. and Schreiner, R.P. (1999). Relationships between soil aggregation and mycorrhizae as influenced by soil biota and nitrogen nutrition. *Biol. Fertil. Soils,* 28: 356-363.

Birch, C.P.D. (1988). The effects and implications of disturbance of mycorrhizal mycelial systems. *Proc. Royal Soc. Edinburgh,* 94B: 13-24.

Bolan, N.S. (1991). A critical review of the role of mycorrhizal fungi in the uptake of phosphorus by plants. *Plant Soil,* 134: 189–207.

Brazanti, M.B., Rocca, E. and Pisi, E. (1999). Effect of ectomycorrhizal fungi on chestnut ink disease. *Mycorrhiza,* 9: 103-109.

Bougher, N. L. (1995). Diversity of ectomycorrhizal fungi associated with eucalypts inAustralia. In: *Mycorrhizal Research for Forestry in Asia.* (Eds.) M. Brundrett, B. Dell, N. Malajczuk and M. Gong. ACIAR Proceedings No. 62, Canberra: 8-15.

Bowen, G.D. (1973). In: *Ectomycorrhizae – Their Ecology and Physiology,* (Eds.) G.C. Marks and T.T. Kozlowski, Academic Press, London: 151-205.

Brazanti, M.B., Rocca, E. and. Pisi, E. (1999). Effect of ectomycorrhizal fungi on chestnut ink disease. *Mycorrhiza,* 9: 103-109.

Brundrett, M., Bougher, N., Dell, B., Grove, T. and Malajczuk, N. (1995). *Working with Mycorrhizas in Forestry and Agriculture. ACIAR Monograph* 32, Australian Centre for International Agricultural Research, Canberra: 374-395.

Brundrett, M.C. (1991). Mycorrhizas in natural ecosystems. In: *Advances in Ecological Research,* (Eds.) A. MacFayden, M. Begon and A.H. Fitter, Academic Press, London: 171-133.

Buscot, F., Munch, J.C., Charcosset, J.Y., Gardes, M., Nehls, U. and. Hampp, R. (2000). Recent advances in exploring physiology and biodiversity of ectomycorrhizas highlight the functioning of these symbioses in ecosystems. *FEMS Microbiol. Rev.,* 24: 601-614.

Chandreshekara, C.P., Patil, V.C. and Sreenivasa, M.N. (1995). VA-mycorrhiza mediated P effect on growth and yield of sunflower (*Helianthus annus* L.) at different P levels. *Plant Soil,* 176:325-328.

Conn, C., and Dighton, J. (2000). Litter quality influences on decomposition, ectomycorrhizal community structure and mycorrhizal root surface acid phosphatase activity. *Soil Biol. and Biochemi.,* 32: 489-496.

Dell, B. and Malajczuk, N. (1994). Boron deficiency in eucalypt plantations in China. *Can. J. Forest Res.* 24: 2409-2416.

Dell, B., Malajczuk, N., Dunstan, W., Gong, M.Q., Chen, Y.L., Lumyong, S., P. Lumyong, S., Supriyanto and Ekwey, L. (2000). Edible forest fungi in SE Asia – Current practices and future management. *Proceedings of International Workshop BIOREFOR, Nepal,* 1999: 123-130.

Dieffenbach, A. and Matzner, E. (2000). In situ soil solution chemistry in the rhizosphere of mature Norway spruce (*Picea abies* [L.] Karst.) trees. *Plant Soil,* 222: 149-161.

Evans, D.G., and Miller, M.H. (1990). The role of the external mycelial network in the effect of soil disturbance upon vesicular-arbuscular mycorrhizal colonization of maize. *New Phytol.,* 114: 65-71

Ferrol, N., Barea, J.M., and Azcon-Aquilar, C. (2002). Mechanisms of nutrient transport across interfaces in arbuscular mycorrhizas. *Plant Soil,* 244:231-237.

Finlay, R.D., and. Read, D.J (1986). Hyphal mats. *New Phytol.*, 40: 276-279.

Fogel, R. and Hunt, G. (1979). Fungal and arboreal biomass in a western Oregon Douglas fir ecosystem: Distribution patterns and turnover. *Can. J. Forest Res.*, 9: 265-256.

Francis, R., and Read, D.J. (1984). Direct transfer of carbon between plants connected by vesicular-arbuscular mycorrhizal mycelium. *Nature* (London), 307: 53-56.

Francis, R. and Read D.J. (1995). Mutualism and antagonism in the mycorrhizal symbiosis, with special reference to impacts on plant community structure. *Can. J. Bot.–revue Canadienne de batanique*, 73: 1301-1309.

Fransson, P.M.A.,. Taylor, A.F.S and Finlay, R.D. (2000). Effects of continuous optimal fertilization on belowground ectomycorrhizal community structure in a Norway spruce forest. *Tree Physiol.*, 20: 599-606.

Gardner, J.H., and N. Malajczuk. (1988). Recolonisation of rehabilitated bauxite mine sites in Western Australia by mycorrhizal fungi. *Forest Ecol. Manage.*, 24: 27-42.

George, E., Romheld, V. and Marschner, H. (1994). Contribution of mycorrhizal fungi to micronutrient uptake by plants. In: *Biochemistry of Metal Micronutrient in the Rhizosphere.* (Eds.) J.A. Monthey, D.E. Crowely and D.G. Luster, Boca Raton FL, CRC Press, pp93-109.

George, E., Gorgus, E., Schmeisser, A. and Marschner, H. (1996). A method to measure nutrient uptake from soil by mycorrhizal hyphae. In: *Mycorrhizas in Integrated System from Genes to plant Development,* (Eds.) Aguilar Azcon and J.M. Beare, Luxembourg, European Community.

Gautam, S. P. and Prasad, K. (2001). VA mycorrhiza - Importance and biotechnological application. In: *Innovative Approaches in Microbiology,* (Eds.) D.K. Maheshwari, and R.C. Dube, Bishon Singh Mahendra Pal Singh, Dehradun: 83-114.

Gavito, M.E. and Miller, M.H. (1998). Changes in mycorrhiza development in maize induced by crop management practices. *Plant Soil*, 198: 185-192.

Gianinazzi, S., Gianinazzi - Pearson, V. and Trouvelot, A. (1990). In: *Biotechnology for fungi for Improving plant growth,* (Eds.). J.M. Whipps and B. Lumsden, Cambridge University Press, Cambridge, U.K.: 41-54.

Gianinazzi, S. and Gianinazzi- Pearson, V. (1986). *Symbiosis,* 2: 139-149.

Hall, J.L, and Williams, L.E. (2000). Assimilate transport and partitioning in fungal biotrophic interactions. *Austr. J. Plant Physiol.*, 27: 549-560.

Harley, J.L. and Smith, S.E. (1983). *Mycorrhizal Symbiosis.* Academic Press: London.

Jasper, D.A., Abbott, L.K and Robson, A.D. (1991). The effect of soil disturbance on vesicular-arbuscular mycorrhizal fungi in soils from different vegetation types. *New Phytol.*, 118: 471-476.

Jasper, D.A., Robson, A.D. and Abbott, L.K. (1987). The effect of surface mining on the infectivity of vesicular arbuscular mycorrhizal fungi. *Austr. J. Bot.*, 35: 642-652.

Jasper, D.A. (1994). Bioremediation of agricultural and forestry soils with symbiotic micro-organisms. *Austr. J. Soil Res.*, 32: 345-348.

Jones. M.D., Durall, D.M. and Tinker. P.B. (1990). Phosphorus relationships and production of extramatrical hyphae by two types of willow ectomycorrhizas at different soil phosphorus levels. *New Phytol.*, 115: 259-267.

Juniper, S. and Abbott, L. (1993). Vesicular-arbuscular mycorrhizas and soil salinity. *Mycorrhiza*, 4: 45-57.

Malajczuk, N., Trappe, J.M. and Molina, R. (1987). Interrelationships among some ectomycorrhizal trees, hypogeous fungi and small mammals: Western Australian and northwestern American parallels. *Austr. J. Ecol.*, 12: 53-55.

Maldonado, J.D., Tainter, F.H., Skipper, H.D. and Lacher., T. E. (2000). Arbuscular mycorrhiza inoculum potentia in natural and managed tropical montane soils in Costa Rica. *Trop. Agricul.*, 77: 27-32.

Marschner, H. and Dell, B. (1994). Nutrient uptake in mycorrhizal symbiosis. *Plant Soil*, 159: 89-102.

Mc Gonigle, T.P. (1998). A numerical analysis of published field trials with vesicular arbuscular mycorrhizal fungi. *Functional Ecol.*, 2:473-478.

Menge, J.A., Grand, L.F. and Haines, L.W. (1977). The effect of fertilization on growth and mycorrhiza numbers in 11-year-old loblolly pine plantations. *Forest Sci.*, 23: 37-44.

Meghvansi, M. K., Prasad, K., and Mahna, S.K. (2010). Symbiotic potential, competitiveness and compatibility of indigenous *Bradyrhizobium japonicum* isolates to three soybean genotypes of two distinct agro-climatic regions of Rajasthan, India. *Saudi J. Biol. Sci.*, 17: 303–310.

Newton, A.C. and Pigott, C.D. (1991). Mineral nutrition and mycorrhizal infection of seedling oak and birch. II. Effect of fertilizers on growth, nutrient uptake and ectomycorrhizal infection. *New Phytol.*, 117: 45-52.

Onipchenko, V.G., and Zobel, M. (2000). Mycorrhiza, vegetative mobility and responses to disturbance of alpine plants in the Northwestern Caucasus. *Folia Geobot. Phytotaxon.*, 35: 1-11.

Pampolina, N.M., Dell, B and. Malajczuk, N. (2001). Dynamics of ectomycorrhizal fungi in an *Eucalyptus globulus* plantation: effect of phosphorus fertilization. *Forest Ecol. Manage.*, 40: 45-49.

Perez-Moreno, J., and Read, D.J. (2000). Mobilization and transfer of nutrients from litter to tree seedlings *via* the vegetative mycelium of ectomycorrhizal plants. *New Phytol.*, 145: 301-309.

Plenchette, C. and Morel, C. (1996). External phosphorus requirement of mycorrhizal and nonmycorrhizal barley and soybean plants. *Biol. Fert. Soils*, 21:303-308.

Prasad, K. and Deploey, J.J. (1999). Incidence of arbuscular mycorrhizae and their effect on certain species of trees. *J. PA. Acad. Sci.*, 73(3): 117-122.

Prasad, K. and Meghvanshi, M.K. (2005). Interaction between indigenous *Glomus fasciculatum* (AM fungus) and Rhizobium and their stimulatory effect on growth, nutrient uptake and nodulation in *Acacia nilotica* (L.) Del. *Flora Fauna,* 11(1): 51-56.

Prasad, K. (2010a). Ectomycorrhiza Symbiosis: Possibilities and Prospects In: *Progress in Mycology,* (Eds.) M.K. Rao and G. Kovices, Scientific Publisher, Jodhpur: 290-308.

Prasad, K. (2010b). Responses of dual inoculation of arbuscular mycorrhizal fungi on the biomass production, phosphate, roots and shoots phenol concentrations of *Terminali arjuna* under field conditions. *Mycorrhiza News,* 22(2):13-17.

Read, D.J., and Boyd, R. (1986). Water relations of mycorrhizal fungi and their host plants.p.287-303. In: *Water, fungi and Plants,* (Eds.) P. Ayres and L. Boddy, Cambridge University Press, Cambridge.

Read, D. J. and Birch, C.P.D. (1988). The effects and implications of disturbance of mycorrhizal mycelial systems. *Proc. Royal Soc. Edinburgh,* 94B: 13-24.

Read. D.L., Leake, J.R. and Langdale, A.R. (1989). The nitrogen nutrition of mycorrhizal fungi and their host plants. In: *Nitrogen, Phosphorus and Sulphur Utilization by Fungi.* (Eds.) L. Boddy, R. Marchant and D.J. Read, Cambridge University Press, Cambridge: 181-204.

Reeves, B.F., Wagner, D., Moorman, T. and Keil, J. (1979). The role of endo-mycorrhizae in revegetation practices in the semi-arid west. I. A comparison of the incidence of mycorrhizae in severely disturbed vs natural environments. *Am. J. Bot.,* 66: 6-13.

Simard, S.W., Perry, D.A.,. Jones, M.D., Myrold, D.D,. Durall, D.M. and Molina, R. (1997). Net transfer of carbon between ectomycorrhizal tree species in the field. *Nature* (London), 388: 579-582.

Sieverding, E. (1991). Vesicular arbuscular mycorrhiza management in tropical agro systems. *Technical Cooperation, Federal Republic of Germany Eschbom.* ISBN 0-652840-652843.

Singh, R. and Adholeya, A. (2003). Effect of crop rotation on diversity and functionality of arbuscular mycorrhizal fungi (AMF). *Mycorrhiza News,* 15: 25–26.

Thompson, J.P. (1987). Decline of vesicular-arbuscular mycorrhizae in long fallow disorder of field crops and its expression in phosphorus deficiency of sunflower. *Austr. J. Agricul. Res.,* 38: 847-867.

Trappe, J.M. (1987). Phylogenetic and ecology aspects of mycotrophy in the angiosperms from an evolutionary standpoint. In: *Ecophysiology of VA Mycorrhizal plants.* (Ed.) G.R. Safir, CRC Press, Boca Roton: 2-25.

Toth, R., Toth, D. Stark, D. and Smith, D.R. (1990). Vesicular-arbuscular mycorrhizal colonisation in *Zea mays* affected by breeding for resistance to fungal pathogens. *Can. J. Bot.,* 68: 1039-1044.

Van der Heijden, M.G.A. (1998). Mycorrhizal fungal diversity determines plant biodiversity, ecosystem variability and productivity. *Nature,* 396 : 69-72.

Visser, S., Griffiths, C.L. and Parkinson, D. (1984). Topsoil storage effects on primary production and rates of vesicular-arbuscular mycorrhizal development in *Agropyron trachycaulum*. *Plant Soil,* 82: 51-60.

Vogt, K.A., C.C. Grier, C.E. Meir, and R.L. Edmonds. (1982). Mycorrhizal role in net production and nutrient cycling in Abies amabilis ecosystems in western Washington. *Ecology,* 63: 370-380.

Wu, B., Nara, K. and Hogetsu. T. (2001). Can 14C-labelled photosynthetic products move between *Pinus densiflora* seedlings linked by ectomycorrhizal mycelia? *New Phytol.,* 149: 137-146.

Xu, D., and Dell, B. (1998). Importance of micronutrients for productivity of plantation eucalypts in east Asia. In: *Overcoming Impediments to Reforestation. Proceedings of the 6 the International Workshop of BIO-REFOR, Brisbane, Australia,* December 2-5, 1997: 133-138.

Xu, D., Bai, J. and Dell, B. (2000). Overcoming the constraints to productivity of plantation eucalypts in southern China. In: *Mycorrhizal Fungal Biodiversity and Applications of Inoculation Technology. Proceedings of Guangzhou ACIAR International Workshop, 1998.* China Forestry Publishing House, Beijing: 93-100.

Xu, D., Dell, B. Malajczuk, N and Gong, M. (2001). Effects of P fertilisation and ectomycorrhizal fungal inoculation on early growth of eucalypt plantations in southern China. *Plant Soil,* 233: 35-44.

□□□

Microbial Diversity and Functions, 2012
© D.J. Bagyaraj, K.V.B.R. Tilak, H.K. Kehri (eds.), pp. 403-418
New India Publishing Agency, New Delhi (India)
E-mail : info@nipabooks.com; Website : www.nipabooks.com

Chapter 19

Botanical Pesticides An Eco-Chemical Natural Alternatives

N.K. Dubey, Ravindra Shukla, Priyanka Singh, Archana Singh and Bhanu Prakash

ABSTRACT

Higher plants can be exploited for the discovery of new bioactive products that could serve as lead compounds in pesticide development because of their novel modes-of-action. Different secondary metabolites of higher plants have been reported to exhibit efficacy as fungitoxicants as well as mycotoxin inhibitors. Pest controls using the botanicals are safer to the user and the environment because they break down into harmless compounds within hours or days in the presence of sunlight. The article presents different plant products found efficacious as antimicrobial and inhibitory to mycotoxin secretions by fungi. Their prospect as safe alternatives to synthetic antimicrobials in plant protection has been discussed.

Keywords: Antifungal; antimycotoxigenic; botanical pesticides; plant products

Plant-derived fungicides/preservatives and their possible application in protection of crops and their produce is being intensified as these are having enormous potential to inspire and influence modern agro-chemical research. There is a good reason to suppose that the secondary metabolites of plants have evolved to protect them from attack by microbial pathogens (Benner, 1993). In the past few years, due to concerns regarding the safety from synthetic antimicrobial agents there has been an increase in naturally developed substances which has resulted in a huge increase in the use of naturally derived compounds such as essential oils and plant extracts as potential antifungal agents. In recent years there has been considerable pressure by consumers to reduce or eliminate chemical fungicides in foods. Also, the plants have long been recognized to provide a potential source of chemical compounds or more commonly products, known as phytochemicals including essential oils and plant extracts (Negi *et al.*, 2005). There has been a growing interest on the research of the possible use of the essential oils and plant extracts, which can be relatively less damaging for pest and disease control (Costa *et al.*, 2000).

Higher plants produce some bioactive secondary metabolites having potential to protect the plants from different pests. A perusal of literature shows that several plants have been found to possess pronounced antimycotic activity against different moulds. Different higher plant essential oils exhibited active principles and have been reported from time to time to demonstrate pronounced antimicrobial activity. The presence of antimicrobial substances in higher plants is well known since ancient times. Such plant-derived chemicals may be exploited for their different biological properties. Many of these are thought to defend the plants producing them against herbivores and pathogens (Isman and Akhtar, 2007). There is growing evidence that most of these compounds are involved in the interaction of plants with other species-primarily the defense of the plant from plant pests (Tripathi *et al.*, 2004). Plants have been synthesizing chemicals for millions of years to protect them from predation by insects and infection from disease. However, it is only in recent years that detailed investigations have been made on higher plants to discover bioactive substances against different pests.

The present account deals with the glimpse of some earlier works carried out to explore some plant powders, essential oils, solvent extracts, their isolated compounds like anthraquinones, terpenoids, xanthones, flavonoides like gucosides and glycosides, phenolics, peptides, amides, saponins etc. as antifungal and antimycotoxigenic agents.

Antifungal activity of plant products and essential oils

A perusal of literature shows that several higher plants have been reported to possess pronounced fungitoxicity against mycelial growth or spore germination of different fungi. Mostly essential oils of different plants have been assayed by different techniques for antifungal screening.

Sato *et al.* (2000) investigated the hot water extract and the methanol extract of 29 samples for their antifungal activity against *Arthrinium sacchari* and *Chaetomium funicola* strains. Five samples, *Acer nikoense, Glycyrrhiza glabra, Lagerstroemia speciosa, Psidium guajava* and *Thea sinensis* showed high activity. $CHCl_3$-soluble fractions from *G. glabra* showed antifungal activity with minimum inhibitory concentrations (MICs) between 62.5 and 125 microg/ml against the above-mentioned two fungi.

Crude extracts of different parts of *Agapanthus africanus* were screened *in vitro* and *in vivo* against eight economically important plant pathogenic fungi like *Botrytis cinerea, Fusarium oxysporum, Sclerotium rolfsii, Rhizoctonia solani, Botryosphaeria dothidea, Pythium ultimum, Alternaria alternata* and *Mycosphaerella pinodes*. Radial mycelial growth was inhibited in five test organisms, while *Pythium ultimum*, and to a lesser extent *Fusarium oxysporum* and *Alternaria alternata*, showed a degree of tolerance. Neither of the extracts showed any phytotoxic reaction on the leaves, even at the highest concentration applied (Tegegne *et al.*, 2008).

The inhibitory effect of leaf powder of *Withania somnifera, Hyptis suaveolens, Eucalyptus citriodora*, peel powder of *Citrus sinensis, Citrus medica* and *Punica granatum*, neem cake and pongamia cake on the growth of *Aspergillus flavus* in soybean seeds during storage was investigated (Krishnamurthy *et al.*, 2008). All the tested plant materials showed remarkable antimicrobial activity.

Carvone derived from caraway seed showed antifungal activity against various fungal diseases during *in vitro* experiments (Hartmans *et al.*, 1995). In *in situ* experiments, promising antifungal activity was obtained against the potato storage diseases caused by *Fusarium sulphureum, Phoma exigua* var. *foveata* and *Helminthosporium solani*.

Quinoa (*Chenopodium quinoa* Willd) extracts due to its high content of triterpenoid saponins (20–30%) was investigated for its antifungal activity against *Botrytis cinerea*. The effect of alkaline treatment on its activity was also observed. Fungal membrane integrity experiments showed that alkali treated saponins caused membrane disruption, while non-treated saponins had no effects. The higher antifungal activity of alkaline treated saponins was probably due to the formation of more hydrophobic saponin derivatives that may have a higher affinity with the sterols present in cell membranes (Stuardo and Martin, 2008).

Antifungal but non-phytotoxic activities of *Origanum acutidens* oil and its aromatic monoterpene constituents *viz.*, carvacrol, p-cymene and thymol were determined by Kordali *et al.* (2008). The antifungal assays showed that oil, carvacrol and thymol completely inhibited mycelial growth of 17 phytopathogenic fungi and their antifungal effects were higher than commercial fungicide, benomyl. However, p-cymene possessed lower antifungal activity. The oil, carvacrol and thymol completely inhibited the seed germination and seedling growth of *Amaranthus retroflexus, Chenopodium album* and *Rumex crispus* and also showed a potent phytotoxic effect on these plants. However, p-cymene did not show any phytotoxic effect.

Three saponins, minutoside A, minutoside B, minutoside C, and two sapogenins, alliogenin and neoagigenin, were isolated from the bulbs of *Allium minutiflorum* and evaluated for their antimicrobial activity (Barile *et al.*, 2007). All the novel saponins showed a significant antifungal activity against soil-borne pathogens (*Fusarium oxysporum, F. oxysporum* f. sp. *lycopersici, F. solani, P. ultimum* and *Rhizoctonia solani*), air-borne pathogens (*Botrytis cinerea, Alternaria alternata* and *A. porri*) and the biocontrol fungus *Tricoderma harzianum* depending on their concentrations.

A flavonoid glycoside, kaempferol 3-O-ß-d-glucopyranosyl (1 → 2)-O-ß-d-glucopyranosyl (1→2)-O-[a-l-rhamnopyranosyl-(1→6)]-ß-d-glucopyranoside, along with two known C- and O-flavonoid glycosides, were isolated from *Dianthus caryophyllus* and exhibited antifungal activity against different *Fusarium oxysporum* f.sp. *dianthi* pathotypes (Galeotti *et al.*, 2008).

Navickiene *et al.* (2000) and de Silva *et al.* (2002) isolated several amides from Piperaceae plants like *Piper hispidum* and *Piper tuberculatum* bearing isobutyl, pyrrolidine, dihydropyridone and piperidine moieties and found their remarkable antifungal activity against *Cladosporium sphaerospermum* and *C. cladosporioides*.

The ethanolic extract of heartwood of *Calocedrus macrolepis* var. *formosana* was screened for antifungal compounds by agar dilution assay. Two compounds, γ-thujaplicin and γ-thujaplicin, responsible for the antifungal property of *C. macrolepis* were isolated and evaluated against total 15 fungi, including wood decay fungi, tree pathogenic fungi and molds. The hexane soluble fraction showed the strongest antifungal activity among all fractions. ß-Thujaplicin and γ-thujaplicin exhibited not only very strong antifungal activity, but also broad antifungal spectrum (Yen *et al.*, 2008).

A peptide designated cicerarin, isolated from seeds of the green chickpea (*Cicer arietinum*) exerted antifungal activity against *Botrytis cinerea, Mycosphaerella arachidicola*, and *Physalospora piricola* (Chu *et al.*, 2003).

An antifungal peptide isolated from dry seeds of the red lentil (*Lens culinaris*) inhibited mycelial growth in *Mycosphaerella arachidicola* with an IC50 of 36 μM. It also exhibited antifungal activity against *Fusarium oxysporum* (Wang and Ng, 2007).

Antifungal activity of lavender, rosemary, peppermint, sweet basil, rose, ginger, and thyme extracts alone at different concentrations was investigated against *Botrytis cinerea* (Rattanapitigorn *et al.*, 2006). The combination of thyme, lavender, and peppermint extracts with vanillin exhibited a marked antifungal activity against *B. cinerea*. Two naturally occurring lignans, methyl-nordihydroguaiaretic acid and nordihydroguaiaretic acid were extracted from

Larrea tridentate by assay-guided chromatography and their antifungal properties were evaluated against *Aspergillus flavus* and *Aspergillus parasiticus* by radial growth inhibition assay (Vargas-Arispuro et al., 2005). Methyl-NDGA was very effective in inhibiting mycelial growth of both fungi at 300 μg/ml, whereas 500 μg/ml of NDGA was necessary to completely inhibit the growth of the fungi. These compounds may have a biopesticidal potential as control agents for the aflatoxin secreting fungi.

The antifungal activity of acetone, hexane, dichloromethane and methanol leaf extracts of six *Terminalia* species (*Terminalia prunioides, Terminalia brachystemma, Terminalia sericea, Terminalia gazensis, Terminalia mollis* and *Terminalia sambesiaca*) were tested against five fungi (*Candida albicans, Cryptococcus neoformans, Aspergillus fumigatus, Microsporum canis* and *Sporothrix schenkii*). Most of the antifungal extracts had MIC values of 0.08 mg/ml, some were with MIC values as low as 0.02 mg/ml. Microsporum canis was the most susceptible microorganism and *Terminalia sericea* extracts were the most active against all microorganisms tested (Masoko *et al.,* 2005).

The leaf pulp of *Aloe vera,* and the bitter, yellow liquid fraction were evaluated for their inhibitory effect on the mycelial growth of three phytopathogenic fungi (*Rhizoctonia solani, Fusarium oxysporum* and *Colletotrichum coccodes*) and to determine the extract concentrations that can inhibit mycelial development. This was the first report of any Aloe liquid fraction activity against plant pathogenic fungi (Rodríguez *et al.,* 2005).

Abbassy *et al.* (2007) isolated two glucosides (simmondsin and simmondsin 2'-ferulate) from the seeds of jojoba plant, (*Simmondsia chinensis*) by bioassay-driven fractionations of the chloroform extract. The antifungal activity was studied against *Pythium debarianum, Fusarium oxysporum, Rhizocotonia solani* and *Botrytis fabae*. They showed moderate to high antifungal activity against four plant pathogenic fungi.

Bioactivity-guided fractionation of the ethyl acetate extract from leaves of *Piper crassinervium* yielded three prenylated hydroquinones together with two known flavanones *viz.* naringenin and sakuranetin. The antifungal activity was determined by direct bioautography against *Cladosporium cladosporioides* and *C. sphaerospermum.* (Danelutte *et al.,* 2003).

Acetone extracts from different parts of seven common invasive plant species occurring in South Africa *viz. Cestrum laevigatum* (flowers and leaves), *Nicotiana glauca* (flowers, leaves and seeds), *Solanum mauritianum* (fruits and leaves), *Lantana camara* (fruits, flowers and leaves), *Datura stramonium* (seeds), *Ricinus communis* (leaves) and *Campuloclinium macrocephalum* (leaves and flowers) were studied as potential sources of antifungal agents against phyto-pathogenic fungi (*Penicillium janthinellum, Penicillium expansum, Aspergillus niger, Aspergillus parasiticus, Colletotrichum gloeosporioides, Fusarium oxysporum, Trichoderma harzianum, Phytophthora nicotiana, Pythium ultimum* and *Rhizoctonia solani*. All extracts exhibited moderate to good activities on all tested fungi with minimum inhibitory concentrations (MICs) ranging from 0.08 mg/ml to 2.5 mg/ml (Mdee *et al.*, 2009).

Extracts of *Mitracarpus villosus* leaves and inflorescences were investigated for *in vitro* antifungal activities by agar-diffusion and tube-dilution techniques (Irobi and Daramola, 1993). Ethanolic extracts produced definite antifungal activities against *Trichophyton rubrum, Microsporum gypseum, Candida albicans, Aspergillus niger* and *Fusarium solani*. The minimum inhibitory concentration of the extracts ranged from 0.50 to 4.0 mg/ml while their minimum fungicidal concentration values ranged from 1 to 8 mg/ml. These results indicated that the extracts were fungistatic at lower concentrations and fungicidal at higher concentrations.

Natural tetranortriterpenoids such as cedrelone from *Toona ciliata*, azadiradione from *Azadirachta indica*, limonin, limonol and nomilinic acid from *Citrus medica*, along with some cedrelone derivatives were tested for their antifungal activity against *Puccinia arachidis*, a groundnut rust pathogen. Cedrelone was the most effective in reducing rust pustule emergence (Govindachari *et al.*, 2000).

Euclea natalensis (Ebenaceae), widely used for curing bronchitis, chronic asthma and urinary tract infections by the Zulus, in South Africa. Four triterpenes namely lupeol, betulin, ß-sitosterol, 20(29)-lupene-3ß-isoferulate and two naphthoquinones shinanolone and octahydroeuclein isolated from the ethanolic extract of *E. natalensis* root bark were investigated for their antifungal activity against *Aspergillus flavus, Aspergillus niger, Cladosporium cladosporioides* and *Phytophthora* sp. (Lall *et al.*, 2006).

Three biphenyls and four xanthones have been isolated from the aerial parts of *Monnina obtusifolia*. Their antifungal activity was determined by Pinto *et al.* (1994).The fungistatic activity of six aqueous extracts of chamomile (*Anthemis nobilis*), cinnamon (*Cinnamomum verum*), French lavender (*Lavandula stoechas*), garlic (*Allium sativum*), malva (*Malva sylvestris*) and peppermint (Mentha piperita) were tested against Aspergillus candidus, *A. niger, Penicillium* sp. and *Fusarium culmorum*. Highly concentrated extracts of chamomile and malva inhibited totally the growth of the tested fungi with malva being the most effective one (Magro *et al.*, 2006).

Antifungal activity of the leaf essential oil from *Calocedrus macrolepis* var. *formosana* Florin and its constituents were evaluated *in vitro* against six plant pathogenic fungi (Chang *et al.*, 2008). Sesquiterpenoids (T-muurolol and a-cadinol) were found more effective than monoterpenoid components of the oil and strongly inhibited the growth of *Rhizoctonia solani* and *Fusarium oxysporum*, with the IC50 values < 50 μg ml-1. These compounds also efficiently inhibited the mycelial growth of *Colletotrichum gloeosporioides, P. funerea, Ganoderma australe* and *F. solani*.

The essential oils from the aerial parts of *Achillea gypsicola, Achillea biebersteinii* and n-hexane extracts of their flowers were tested against 12 phytopathogenic fungi (Kordali *et al.*, 2009). The oils were found to be more fungitoxic as compared with hexane extracts of the plant samples. The antifungal activity of the oils were attributed to their relatively high content of oxygenated monoterpenes.

An in vitro initial screening of a range of 37 essential oils on inhibition of mycelial growth of *Fusarium verticillioides, F. proliferatum* and *F. graminearum* was done in 3% maize meal extract agar basic medium (Velluti et al., 2004). Cinnamon leaf, clove, lemongrass, oregano and palmarosa oils were the products tested suitable for being used as novel preservatives in the control of the three *Fusarium* species studied.

Essential oils of seven Moroccan Labiatae (*Calamintha officinalis, Lavandula dentata; Mentha pulegium, Origanum compactum, Rosmarinus officinalis, Salvia aegyptica, Thymus glandulosus*) were evaluated for their in vitro antifungal activity against *Botrytis cinerea*. Among them, *Origanum compactum* and *Thymus glandulosus* greatly inhibited the growth of the mycelium. *Mentha pulegium* exhibited moderate activity at 250 ppm. Thymol and carvacrol, the two main constituents of *Thymus glandulosus* and *Origanum compactum* exhibited the strongest antifungal activity with 100% of inhibition at 100 ppm (Bouchra *et al.*, 2003).

Essential oils from 25 species of medicinal plants were tested as mycelial growth inhibitors by agar dilution method against six fungal species namely *Fusarium oxysporum, Fusarium verticillioides, Penicillium expansum, Penicillium brevicompactum, Aspergillus flavus* and *Aspergillus fumigatus*. All essential oils affected growth of these fungi. The superior antifungal activity was finally proved in the case of *Pimenta dioica* on the basis of MIC values (Zabka *et al.*, 2009).

The inhibitory effects of 10 selected Turkish spices, oregano essential oil, thymol and carvacrol towards growth of nine food borne fungi were. The antifungal effects of sodium choride, sorbic acid and sodium benzoate and the combined use of oregano with sodium chloride were also tested under the same conditions for comparison. Oregano essential oil, thymol or carvacrol at concentrations of 0.025% and 0.05% completely inhibited the growth of all fungi, showing greater inhibition than sorbic acid at the same concentrations (Akgul and Kivanc, 1998).

Sahin *et al.* (2004) evaluated the antimicrobial activities of essential oils and methanol extracts of *Origanum vulgare* against 15 fungi and yeast species The essential oil *O. vulgare* possessed compounds with antimicrobial properties and, therefore, can be used as a natural preservative ingredient in food and/ or pharmaceutical industry.

Lemongrass (Cympopogon citratus L.) oil was tested for antifungal activity against *Colletotrichum coccodes, Botrytis cinerea, Cladosporium herbarum, Rhizopus stolonifer* and *Aspergillus niger in vitro*. Fungal spore production inhibited up to 70% at 25 ppm of lemongrass oil. At the highest oil concentration (500 ppm), fungal sporulation was completely retarded. Lemongrass oil reduced spore germination and germ tube length of al the fungi with the effects dependent on oil concentration (Nikos *et al.,* 2007).

Satureja subspicata essential oil, collected had a potential antifungal activity against 9 fungal strains, and therefore suggested as a potential source of antimicrobial ingredients for the food and pharmaceutical industry (Skocibusic, *et al.,* 2006).

Seven essential oils (ajowan, dill weed, Egyptian geranium, lemongrass, rosemary, tea tree, and thyme) were evaluated for their ability to inhibit growth of *Aspergillus niger, Trichoderma viride* and *Penicillium chrysogenum* (Yang and Clausen, 2007). Thyme and Egyptian geranium oil inhibited growth of all test fungi for 20 weeks. Likewise, dill weed oil vapors inhibited all test fungi for at least 20 weeks.

Hence, a perusal of literature on fungitoxicity of plant products shows that most of the phytochemicals have been found to exhibit strong antifungal activity with broad range of antifungal spectrum without phytotoxicity. Some of the products have been found to have better efficacy as fungitoxicant in comparision to synthetic fungicides tested. Keeping in view the demand of such products, there is need to record the efficacy of essential oils of traditionally used ethnomedicinal plants as fumigants in storage containers and their application in field conditions should be properly documented.

Antimycotoxigenic activity of plant extracts and essential oils

Although, a large number of plants and their products have been tested as growth inhibitory of fungi, only a few have been studied for their antiaflatoxigenic properties. Essential oil of *Cymbopogon flexuosus* and its major component, eugenol was found efficient in checking fungal growth and aflatoxin production. *C. flexuosus* essential oil absolutely inhibited the growth of *A. flavus* and aflatoxin B1 production at 1.3 µl ml-1 and 1.0 µl ml-1 respectively. Eugenol was more efficacious than the *Cymbopogon* oil as such which emphasizes masking of their efficacy when combined together with other constituents (Kumar *et al.,* 2009).

The effect of essential oils, ethanolic and aqueous extract of 41 vegetable species of Argentina on *Aspergillus* section *Flavi* growth was evaluated (Bluma *et al.,* 2008). Essential oils were found to be the most effective, controlling aflatoxigenic strains. Clove, mountain thyme and poleo essential oils showed the most antifungal effect under all growth parameters (percent germination, germ-tube elongation rate) as well as aflatoxin B1 accumulation.

Various individual and combined plant extracts were evaluated by Sidhu *et al.* (2009) for their efficacy against growth of *Aspergillus flavus* and aflatoxin production *in vitro*. Combinations of botanicals were found to be more effective in controlling fungal growth and aflatoxin production than individual extracts. Results suggested that synergistic effect of plant extracts can be used for control of fungal growth and aflatoxin production.

Essential oil extracted from the leaves of *Chenopodium ambrosioides.* was tested against the aflatoxigenic strain of test fungus *Aspergillus flavus* (Kumar *et al.,* 2007). The oil completely inhibited the mycelial growth at 100 µg/ml and exhibited broad fungitoxic spectrum against *Aspergillus niger, Aspergillus fumigatus, Botryodiplodia theobromae, Fusarium oxysporum, Sclerotium rolfsii, Macrophomina phaseolina, Cladosporium cladosporioides, Helminthosporium oryzae* and *Pythium debaryanum* at similar concentration. The oil also showed significant efficacy in inhibiting the aflatoxin B1 production by the aflatoxigenic strain of *A. flavus.*

The essential oils extracted from *Cymbopogon citratus, Monodora myristica, Ocimum gratissimum, Thymus vulgaris* and *Zingiber officinale* were investigated for their inhibitory effect by agar dilution technique against three food spoilage and mycotoxin producing fungi, *Fusarium moniliforme, Aspergillus flavus* and *Aspergillus fumigatus* (Nguefack *et al.,* 2004). The EO from *O. gratissimum, T. vulgaris* and *C. citratus* were the most effective and prevented conidial germination and the growth of all three fungi.

The antifungal activity of *Pimpinella anisum* (anise), Pëumus boldus (boldus), *Hedeoma multiflora* (mountain thyme), *Syzygium aromaticum* (clove), and *Lippia turbinate* (poleo) essential oils against *Aspergillus* section Flavi was evaluated in sterile maize grain under different water activity. Important reduction of AFB1 accumulation was observed in the majority of EO treatments at 11 days of incubation. Boldus, poleo, and mountain thyme EO completely inhibited AFB1 at 2000 and 3000 µg g-1 (Bluma and Etcheverry, 2008).

Essential oils of 12 medicinal plants were tested for inhibitory activity against *Aspergillus fluvus, A. parasiticus, A. ochraceus* and *Fusarium moniliforme.* The oils of thyme, cinnamon, marigold, spearmint, basil and caraway completely inhibited all the test fungi. However, chamomile and hazanbul at all concentrations were partially effective against the test toxigenic fungi (Soliman and Badeaa, 2002).

Essential oils of sweet basil (*Ocimum basilicum*), cassia (*Cinnamomum cassia*), coriander (*Coriandrum sativum*) and bay leaf (*Laurus nobilis*) in palm kernel broth inoculated with spore suspension (106/ml) of *Aspergillus parasiticus* were evaluated for their potential in the control of aflatoxigenic fungus *A. parasiticus* and aflatoxin production (Atanda, *et al.*, 2007). Sweet basil oil at 5% (v/v) was fungistatic and reduced aflatoxin production while oils of cassia and bay leaf stimulated the mycelia growth of the fungus in vitro but reduced the aflatoxin concentration 97.92% and 55.21% respectively,

Lippia rugosa essential oil was tested for its efficacy against *Aspergillus flavus* on artificial growth media. The effect of essential oil on aflatoxin B1 synthesis was evaluated in SMKY broth. Results showed that aflatoxin B1 synthesis was inhibited by 1000 mg l-1 L (Tatsadjieu *et al.*, 2009).

The effect of cinnamon, clove, oregano, palmarosa and lemongrass oils on zearalenone (ZEA), Fuminosin (FB1) and deoxynivalenol (DON) accumulation by isolates of *Fusarium graminearum* and *F. proliferatum* in non-sterilized naturally contaminated maize grain was evaluated at a 500 mg kg-1 level. Cinnamon, oregano and palmarose oils had significant inhibitory effect on FB1 production, while clove and lemongrass oils had only significant inhibitory effect (Velluti *et al.*, 2003., Marin *et al.*, 2004).

Antifungal activities of the *Rosmarinus officinalis* and *Trachyspermum copticum* oils were studied with special reference to the inhibition of *Aspergillus parasiticus* growth and aflatoxin production (Rasooli *et al.*, 2008). *T. copticum* L. oil showed a stronger inhibitory effect than *R. officinalis* on the growth of *A. parasiticus*. Aflatoxin production was inhibited at 450 ppm of both oils with that of *R. officinalis* being stronger inhibitor.

The effect of *Zataria multiflora* essential oil against growth, spore production and aflatoxin formation by *Aspergillus flavus* ATCC 15546 was investigated in synthetic media as well as Iranian ultra-filtered white cheese in brine. Oil effectively inhibited radial growth and spore production on potato dextrose agar (PDA) in a dose-dependent manner. At 200 ppm, the radial growth and sporulation was reduced by 79.4% and 92.5%, respectively. The growth was completely prevented at EO $\geq$ 400 ppm on PDA, and minimum fungicidal concentration (MFC) of the oil was estimated at 1000 ppm. The oil also significantly suppressed aflatoxin synthesis in broth medium at all concentrations tested (Gandomi *et al.*, 2009).

Moulding and mycotoxin production was studied during storage of rice following treatment with cinnamon and clove oils (Patkar *et al.*, 1994).

Aflatoxin B_1 and ochratoxin A were determined by monoclonal antibody-based enzyme-linked immunosorbent assay. Both moulding and mycotoxin were inhibited by 9 μl cinnamon oil. Clove oil was less effective than cinnamon oil in preventing moulding.

Antifungal activity of essential oils from *Thymus eriocalyx* and *Thymus xporlock* were studied with special reference to the inhibition of *Aspergillus parasiticus* growth and aflatoxin production. Static effects of the above oils against *A. parasiticus* were at 250 ppm and lethal effects of *T. eriocalyx* and *T. X-porlock* were 500 and 1000 ppm of the oils, respectively. Aflatoxin production was inhibited at 250 ppm of both oils with that of *T. eriocalyx* being stronger inhibitor (Rasooli and Owlia, 2005).

In order to find out plants useful in controlling aflatoxins, the essential oils from 12 medicinal plants were studied with special reference to the inhibition of *Aspergillus parasiticus* growth and aflatoxin production. Among plants tested, *Thymus vulgari* and *Citrus aurantifolia* were found to inhibit both *A. parasiticus* and aflatoxin production. The EOs from *Mentha spicata, Foeniculum miller, Azadirachta indica, Conium maculatum* and *Artemisia dracunculus* were only inhibited fungal growth, while *Carum carvi* effectively inhibited AF production without any obvious effect on fungal growth. The other plants including *Ferula gummosa, Citrus sinensis, Mentha longifolia* and *Eucalyptus camaldulensis* had no effect on *A. parasiticus* growth and AF production at all concentrations studied (Razzaghi-Abyaneh, 2009).

Thus the perusal of literature shows that most of the essential oils showed mycotoxin inhibitory activity at a concentration lower than the fungal growth inhibitory concentration .Hence their mode of action for mycotoxin inhibitory would be different from fungal growth inhibition. The oils showing inhibition of fungal growth and mycotoxin may be recommended for complete protection against biodeterioration of stored food grains.

Conclusion

Higher plants contain a wide spectrum of secondary metabolites such as phenolics, flavonoids, quinones, tannins, essential oils, alkaloids, saponins and sterols. Tens of thousands of secondary products of plants have been identified and there are estimates that hundreds of thousands of these compounds exist. These secondary compounds represent a large reservoir of chemical structures with biological activity. Therefore, higher plants can be exploited for the discovery of new bioactive products that could serve as lead compounds in pesticide development because of their novel modes-of-action (Regnault-Roger, 2005). Many of these are thought to serve an ecological function for the plants producing them, serving to defend the plants from herbivores and pathogens. Natural pest controls using the botanicals are safer to the user and the environment because they break down into harmless compounds within hours or days in the presence of sunlight. The plant pests and disease control either directly or indirectly using natural plant products/ botanicals, including essential oils, holds a good promise (Regnault-Roger, 1997; Isman, 2000; Isman, 2006; Bakkali *et al.*, 2008) It has been reported that natural plant products may successfully replace chemical fungicides and provide an alternative method to protect cereals, pulses and other agricultural commodities from aflatoxin B_1 production by *A. flavus* (Krishnamurthy *et al.*, 2008). Many tropical medicinal plants and spices have been used as pest control agents (Lale, 1992).

References

Abbassy, M.A., Abdelgaleil, S.A.M., Belal, A.S.H. and Abdel Rasoul M.A.A. (2007). Insecticidal, antifeedant and antifungal activities of two glucosides isolated from the seeds of *Simmondsia chinensis*. *Ind. Crop Prod.*, 26: 345-350.

Akgul, A. and Kivanç, M. (1998). Inhibitory effects of selected Turkish spices and oregano components on some foodborne fungi. *Int. J. Food Microbiol.*, 6: 263-268.

Bakkali, F., Averbeck, S., Averbeck, D. and Idaomar, M. (2008). Biological effects of essential oils – A review. *Food Chem. Toxicol.*, 46: 446–475.

Barile, E., Bonanomi, G., Antignani, V., Zolfaghari, B., Sajjadi, S.E., Scala, F. and Lanzotti, V. (2007). Saponins from *Allium minutiflorum* with antifungal activity. *Phytochem.*, 68: 596-603.

Benner, J.P. (1993). Pesticidal compounds from higher plants. *Pest Sci.*, 39: 95-102.

Bluma, R., Amaiden, M.R. and Etcheverry, M. (2008). Screening of Argentine plant extracts: Impact on growth parameters and aflatoxin B_1 accumulation by *Aspergillus* section *Flavi*. *Int. J. Food Microbiol.*, 122: 114-125.

Bluma, R.V. and Etcheverry, M.G. (2008). Application of essential oils in maize grain: Impact on *Aspergillus* section *Flavi* growth parameters and aflatoxin accumulation. *Food Microbiol.*, 25: 324-334.

Bouchra, C., Achouri, M., Hassani, L.M.I. and Hmamouchi, M. (2003). Chemical composition and antifungal activity of essential oils of seven Moroccan Labiatae against *Botrytis cinerea* Pers. Fr. *J. Ethnopharmacol*, 89: 165-169.

Chang, H-T., Cheng, Y-H., Wu, C-L., Chang, S-T., Chang, T-T. and Su, Y-C. (2008). Antifungal activity of essential oil and its constituents from *Calocedrus macrolepis* var. *formosana* Florin leaf against plant pathogenic fungi. *Bioresour. Technol.*, 99: 6266-6270.

Chu, K.T., Liu, K.H. and Ng, T.B. (2003). Cicerarin, a novel antifungal peptide from the green chickpea. *Peptides*, 24: 659-663.

Costa, T.R., Fernandes, F.L.F., Santos, S.C., Oliveria, C.M.A., Liao, L.M., Ferri, P.H., Paulo, J.R, Ferreira, H.D., Sales, B.H.N. and Silva, M.R.R. (2000). Antifungal activity of volatile constituents of *Eugenia dysenterica* leaf oil. *J. Ethnopharmcol.*, 72: 111-117.

Danelutte, A.P., Lago, J.H.G., Young, M.C.M. and Kato, M.J. (2003). Antifungal flavanones and prenylated hydroquinones from *Piper crassinervium* Kunth. *Phytochem.*, 64: 555-559.

de Silva, R.V., Navickiene, H.M.D., Kato, M.J., Bolzani, V.da.S., Meda, C.I., Young, M. C.M. and Furlan, M. (2002). Antifungal amides from *Piper arboreum* and *Piper tuberculatum*. *Phytochem.*, 59: 521-527.

Galeotti, F., Barile, E., Curir, P., Dolci, M. and Lanzotti, V. (2008). Flavonoids from carnation (*Dianthus caryophyllus*) and their antifungal activity. *Phytochem. Lett.*, 1: 44-48.

Gandomi, H., Misaghi, A., Basti, A.A., Bokaei, S., Khosravi, A., Abbasifar, A. and Javan, A.J. (2009). Effect of *Zataria multiflora* Boiss. essential oil on growth and aflatoxin formation by *Aspergillus flavus* in culture media and cheese. *Food Chem. Toxicol.*, 47: 2397-2400.

Govindachari, T.R., Suresh, G., Gopalakrishnan, G., Masilamani, S. and Banumathi B. (2000). Antifungal activity of some tetranortriterpenoids. *Fitoterapia*, 71: 317-320.

Hartmans, K.J., Diepenhorst, P., Bakker, W. and Gorris, L.G.M. (1995). The use of carvone in agriculture: sprout suppression of potatoes and antifungal activity against potato tuber and other plant diseases. *Ind. Crop Prod.*, 4: 3-13.

Irobi, O.N. and Daramola, S.O. (1993). Antifungal activities of crude extracts of *Mitracarpus villosus* (Rubiaceae). *J. Ethnopharmacol.*, 40: 137–140.

Isman, M.B. (2000). Plant essential oils for pest and disease management. *Crop Prot.*, 19: 603–608.

Isman, M.B. (2006). Botanical insecticides, deterrents, and repellents in modern agriculture and an increasingly regulated world. *Ann. Rev. Entomol.*, 51: 45–66.

Isman, M.B. and Akhtar, Y. (2007). Plant natural products as a source for developing environmentally acceptable insecticides. In: *Insecticides design using advanced technologies* (Eds.) I. Ishaaya, R. Nauen and A.R. Horowitz, Springer, Berlin, Heidelberg. pp. 235–248.

Kordali, S., Cakir, A., Akcin, T.A., Mete, E., Akcin, A., Aydin, T. and Kilic, H. (2009). Antifungal and herbicidal properties of essential oils and *n*-hexane extracts of *Achillea gypsicola* Hub-Mor. and *Achillea biebersteinii* Afan. (Asteraceae). *Ind. Crop Prod.*, 29: 562-570.

Kordali, S., Cakir, A., Ozer, H., Cakmakci, R., Kesdek, M. and Mete, E. (2008). Antifungal, phytotoxic and insecticidal properties of essential oil isolated from Turkish *Origanum acutidens* and its three components, carvacrol, thymol and *p*-cymene. *Bioresour. Technol.*, 99: 8788-8795.

Krishnamurthy, Y.L., Shashikala, J. and Naik, B.S. (2008). Antifungal potential of some natural products against *Aspergillus flavus* in soybean seeds during storage. *J. Stored Prod. Res.*, 44: 305-309.

Kumar, A., Shukla, R., Singh, P. and Dubey, N.K. (2009). Biodeterioration of some herbal raw materials by storage fungi and aflatoxin and assessment of *Cymbopogon flexuosus* essential oil and its components as antifungal. *Int. Biodeter. Biodegr.*, 63: 712-716.

Kumar, R., Mishra, A.K., Dubey N.K. and Tripathi, Y.B. (2007). Evaluation of *Chenopodium ambrosioides* as a potential source of antifungal, antiaflatoxigenic and antioxidant activity. *Int. J. Food Microbiol.*, 115: 159–164.

Lale, N. E. S. (1992). A laboratory study of the comparative toxicity of products from three spices to the maize weevil. *Post Harvest Biol. Tec.*, 2: 612-664.

Lall, N., Weiganand, O., Hussein, A.A. and Meyer, J.J.M. (2006). Antifungal activity of naphthoquinones and triterpenes isolated from the root bark of *Euclea natalensis. S. Afr. J. Bot.*, 72: 579-583.

Magro, A., Carolino, M., Bastos, M. and Mexia, A. (2006). Efficacy of plant extracts against stored products fungi. *Rev. Iberoam. Micol.*, 23: 176-178.

Marin, S., Velluti, A., Ramos, A.J. and Sanchis, V. (2004). Effect of essential oils on zearalenone and deoxynivalenol production by *Fusarium graminearum* in non-sterilized maize grain. *Food Microbiol.*, 21: 313-318.

Masoko, P., Picard, J. and Eloff J.N. (2005). Antifungal activities of six South African *Terminalia* species (Combretaceae). *J. Ethnopharmacol.*, 99: 301-308.

Mdee, L.K., Masoko, P. and Eloff, J.N. (2009). The activity of extracts of seven common invasive plant species on fungal phytopathogens. *S. Afr. J. Bot.*, 75: 375-379.

Navickiene, H.M.D., Alécio, A.C., Kato, M.J. Bolzani, V.S., Young, M.C.M., Cavalheiro, A.J. and Furlan, M. (2000). Antifungal amides from *Piper hispidum* and *Piper tuberculatum. Phytochem.*, 55: 621-626.

Negi, P.S., Chauhan, A.S., Sadia, G.A., Rohinishree, Y.S. and Ramteke, R.S. (2005). Antioxidant and antibacterial activities of various seabuckthorn (*Hippophae rhamnoides* L.) seed extracts. *Food Chem.*, 92: 119-124.

Nguefack, J., Leth, V., Zollo, P.H.A. and Mathur S.B. (2004). Evaluation of five essential oils from aromatic plants of Cameroon for controlling food spoilage and mycotoxin producing fungi. *Int. J. Food Microbiol.*, 94: 329-334.

Nikos, G. Tzortzakis and Costas, D. (2007). Antifungal activity of lemongrass (*Cympopogon citratus* L.) essential oil against key postharvest pathogens. *Innovat. Food Sci. Emerg Tech.*, 2: 253-258.

Patkar, K.L., Usha, C.M., Shetty, H.S., Paster, N. and Lacey, J. (1994). Effects of spice oil treatment of rice on moulding and mycotoxin contamination. *Crop Prot.*, 13: 519-524.

Pinto, D.C.G., Fuzzati, N., Pazmino, X.C. and Hostettmann, Kurt. (1994). Xanthone and antifungal constituents from *Monnina obtusifolia. Phytochem.*, 37: 875-878.

Rasooli, I. and Owlia, P. (2005). Chemoprevention by thyme oils of *Aspergillus parasiticus* growth and aflatoxin production. *Phytochem.*, 66: 2851-2856.

Rasooli, I., Fakoor, M.H., Yadegarinia, D., Gachkar, L., Allameh, A. and Rezaei, M.B. (2008) Antimycotoxigenic characteristics of *Rosmarinus officinalis* and *Trachyspermum copticum* L. essential oils. *Int. J. Food Microbiol.*, 122: 135-139.

Rattanapitigorn, P., Arakawa, M. and Tsuro, M. (2006). Vanillin enhances the antifungal effect of plant essential oils against *Botrytis cinerea. Int. J. Aromather.*, 16: 193-198.

Razzaghi-Abyaneh, M., Shams-Ghahfarokhi, M., Rezaee, M-B., Jaimand, K., Alinezhad, S., Saberi, R. and Yoshinari, T. (2009). Chemical composition and antiaflatoxigenic activity of *Carum carvi* L., *Thymus vulgaris* and *Citrus aurantifolia* essential oils. *Food Control*, 20: 1018-1024.

Regnault-Roger, C. (1997). The potential of botanical essential oils for insect pest control. *Integrated Pest Manag. Rev.*, 2: 25–34.

Regnault-Roger, C. (2005). Molecules allelochimiques et extraits vegetaux dans la protection des plantes: nature, role et bilan de leur utilisation au XX*e* siecle. In: *Enjeux Phytosanitaires pour l'Agriculture et l'Environnement* (Ed.) C.Regnault-Roger, Paris, Lavoisier : 625–650.

Rahin, F., Güllüce, M., Daferera, D., Sökmen, A., Sökmen, M., Polissiou, M., Agar, G. and Özer, H. (2004). Biological activities of the essential oils and methanol extract of *Origanum vulgare* ssp. *vulgare* in the Eastern Anatolia region of Turkey. *Food Control*, 15: 549-557.

Rodriguez, D.J., Hernandez-Castillo, D., Rodriguez-Garcia, R. and Angulo-Sanchez, J. L. (2005). Antifungal activity *in vitro* of *Aloe vera* pulp and liquid fraction against plant pathogenic fungi. *Ind. Crop Prod.*, 21: 81-87.

Sato, J., Goto, K., Nanjo, F., Kawai, S. and Murata, K. (2000). Antifungal activity of plant extracts against *Arthrinium sacchari* and *Chaetomium funicola. J. Biosci. Bioeng.*, 90: 442-446.

Sidhu, O.P., Chandra, H. and Behl, H.M. (2009). Occurrence of aflatoxins in mahua (*Madhuca indica* Gmel.) seeds: Synergistic effect of plant extracts on inhibition of *Aspergillus flavus* growth and aflatoxin production. *Food Chem. Toxicol.*, 47: 774–777.

Skocibusic, M., Bezic, N. and Dunkic, V. (2006). Phytochemical composition and antimicrobial activities of the essential oils from *Satureja subspicata* Vis. growing in Croatia. *Food Chem.*, 96: 20-28.

Soliman, K.M. and Badeaa, R.I. (2002). Effect of oil extracted from some medicinal plants on different mycotoxigenic fungi. *Food Chem. Toxicol.*, 40: 1669-1675.

Stuardo, M. and Martín, R.S. (2008). Antifungal properties of quinoa (*Chenopodium quinoa* Willd) alkali treated saponins against *Botrytis cinerea. Ind. Crop Prod.*, 27: 296-302.

Tatsadjieu, N.L., Dongmo, P.M.J., Ngassoum, M.B., Etoa, F.X. and Mbofung, C.M.F. (2009). Investigations on the essential oil of *Lippia rugosa* from Cameroon for its potential use as antifungal agent against *Aspergillus flavus* Link ex. Fries. *Food Control*, 20: 161–166.

Tegegne, G., Pretorius, J.C. and Swart, W.J. (2008). Antifungal properties of *Agapanthus africanus* L. extracts against plant pathogens. *Crop Prot.*, 27: 1052-1060.

Tripathi, P., Dubey, N.K., Banergi, R., and Chansuria, J.P.N. (2004). Evaluation of some essential oils as botanical fungitoxicants in management of post harvest rotting of citrus fruits. *World J. Microbiol. Biotechnol.*, 20: 317-321.

Vargas-Arispuro, I., Reyes-Báez, R., Rivera-Castañeda, G., Martínez-Téllez, M.A. and Rivero-Espejel, I. (2005). Antifungal lignans from the creosotebush (*Larrea tridentate*). *Ind. Crop Prod.*, 22: 101-107.

Velluti, A., Marín, S., Gonzalez, P., Ramos, A.J. and Sanchis, V. (2004). Initial screening for inhibitory activity of essential oils on growth of *Fusarium verticillioides, F. proliferatum* and *F. graminearum* on maize-based agar media. *Food Microbiol.*, 21: 649-656.

Velluti, A., Sanchis, V., Ramos, A.J., Egido, J. and Marín, S. (2003). Inhibitory effect of cinnamon, clove, lemongrass, oregano and palmarose essential oils on growth and fumonisin B_1 production by *Fusarium proliferatum* in maize grain. *Int. J. Food Microbiol.*, 89: 145-154.

Wang, H.X. and Ng, T.B. (2007). An antifungal peptide from red lentil seeds. *Peptides*, 28: 547-552.

Yang, V.W. and Clausen, C.A. (2007). Antifungal effect of essential oils on southern yellow pine. *Int. Biodeter. Biodegr.*, 59: 302-306.

Yen, T-B., Chang, H-T., Hsieh, C-C. and Chang, S-T. (2008). Antifungal properties of ethanolic extract and its active compounds from *Calocedrus macrolepis* var. *formosana* (Florin) heartwood. *Bioresour. Technol.*, 99: 4871-4877.

Zabka, M., Pavela, R. and Slezakova, L. (2009). Antifungal effect of *Pimenta dioica* essential oil against dangerous pathogenic and toxinogenic fungi. *Ind. Crop Prod.*, 30: 250–253.

□□□

Microbial Diversity and Functions, 2012

New India Publishing Agency, New Delhi (India)
E-mail : info@nipabooks.com; Website : www.nipabooks.com

Chapter 20

Chemistry and Fluorescent Behaviour of Some Lichens from Garhwal Himalaya

M.P. Sharma and Amit Jakhal

ABSTRACT

Eight species including 1 variety, 2 macro and 6 micro-lichens from Garhwal Himalaya, c 1500-2700 m, have been found to glow and emit fluorescent light, when exposed to long wave-length (254 nm) ultra-violet radiataions. Taxonomy and chemistry of all these species have been critically evaluated. The recorded taxa belong to families Physciaceae (4 species); Pertusariaceae, Rhizocarpaceae (1 species each) and Lichen imperfecti (2 species). The reported taxa are: Buellia hemispherica Singh and Awasthi, B. sub-glaziouana Singh and Awasthi; Pertusaria pertusa (Weig.) Tuck., Pyxine berteriana (Fee) Imsh. var. himalayensis Awasthi; P. cocoes (Swartz) Nyl., Rhizocarpon geographicum (L.) DC.; Lepraria taxonomic species I and II.

Keywords: Chemistry, UV-radiations effect, Fluorescence, Garhwal Himalayan lichens.

Introduction

During the course of investigations on lichens from Central Himalaya, particularly in temperate species from Garhwal area, it was observed that there are certain taxa, which glow and emit fluorescent light in presence of long wave-length (254 nm) ultra-violet radiations. In this contribution, 8 taxa including one variety, 2 species of macro and 6 species of micro-lichens showing fluorescent behaviour are reported. The reported taxa are: *Buellia hemispherica* Singh and Awasthi, *B. sub-glaziouana* Singh and Awasthi; *Pertusaria pertusa* (Weig.) Tuck., *Pyxine berteriana* (Fee) Imsh. var. *himalayensis*

Awasthi; *P. cocoes* (Swartz) Nyl., *Rhizocarpon geographicum* (L.) DC.; *Lepraria* taxonomic species I and II. One species *viz. Pyxine berteriana* (Fee) Imsh. var. *himalayensis* Awasthi is endemic to Himalaya; whereas two indeterminate *Lepraria* species may be new to science (pending observations and comparison with allied species and types). Chemical investigations on Indian lichens have been carried out by several workers *viz.* Aghoramurthy and Seshadri (1953); Aghoramurthy *et al.* (1954, 1961); Dhar *et al.* (1959); Grover and Seshadri (1959); Mittal and Seshadri (1954); Neelakantan (1952); Neelakantan *et al.* (1954); Sarin and Atal (1976); Seshadri (1949, 1953); Seshadri and Subramanian (1949); Subrahmanian and Ramakrishnan (1964). Awasthi (1988, 1991) and Upreti (1987, 1998) have used chemotaxonomy efficiently in the formulation of keys for the identification of lichen taxa. Occurrence of various acids and polysaccharides in different lichen taxa have been evaluated by Culberson (1966, 1970, 1972); Culberson and Kristinsson (1970); Elix (1998); Lamb (1951); Nishikawa *et al.* (1974); Orange *et al.* (2001); Takahashi *et al.* (1974) and Walker and James (1980, 1985).

Methods of Study

In each case the chemical tests (spot tests) were performed besides morphological and anatomical features in establishing the validity of the taxon. In the chemical tests, various reagents were applied on thallus cortex and medulla with the help of syringes. Various chemical substances (Phenolic metabolites) contained by lichen thallus give different reactions to the chemical tests, which are evident in the form of colour changes in cortex and medulla. These distinct colourations were compared with the colour standards*. The reagents prepared are as per Culberson (1972), Hale (1983), Awasthi (2000), Walker and James (1980) and Ovstedal and Lewis Smith (2001). The tests performed are as given:

K - Test : This test was performed by using 10-25% aqueous solution of potassium hydoxide (KOH).

C - Test : It was performed by using saturated aqueous solution of calcium hypochlorite ($Ca(ClO)_2$). A fresh solution having pungent smell was used each time because it loses its effect after some time.

KC - Test : This test was performed by using K–solution immediately followed by C–solution. Both K and C–solutions were prepared as given above.

P - Test : This test was performed by using 5% solution of paraphenylenediamine in ethyl alcohol. This solution was made freshly. A stable solution, Steiner's P was prepared by dissolving

1 g of paraphenylenediamine and 10 g of sodium sulphite in 100 ml of distilled water with 1 ml of liquid detergent.

*Colour standards are as per Munsell colour system as exemplified in the Munsell Book of Colour (Cabinet edition, 1963, Munsell Color Company, Baltimore, Md. U.S.A.)

I - Test : In this test, a solution of 0.5 g Iodine in 1.5 g potassium iodide with 200 ml distilled water was used.

For stained preparations, the stains were often combined with clearing agents such as lactic acid, lactophenol or glycerol to make temporary mounts for investigations. These were prepared as per the formulae given in "Dictionary of the Fungi" (Kirk *et al.*, 2001, 2008).

Thin layer chromatography was done as described by Culberson (1970, 1972); Culberson and Kristinsson (1970) and Walker and James (1980) for the identification of phenolic metabolites present in the lichen thallus. All the specimens were subjected to exposure of UV light. Thallus fluorescence is produced by the pigments present in the cortex or medulla. The pigments in the cortex may be xanthones, which fluoresce various shades of yellow, red or orange. The medulla contains depsides and depsidones, which fluoresce blue to white after removing cortex.

Taxonomic Account

Micro-Lichens

Buellia hemispherica Singh and Awasthi, *Biol. Mem.* 6(2): 186, 1981.

Plate 1, Figure A

Chemical tests: Thallus K+ yellow, C-, KC-, P+ deep-yellow becoming red; medulla I+ blue.

Thin layer chromatography: Contains lichenoxanthone and norstictic acid.

Fluorescent behaviour: Cortex UV+ yellow.

Collection examined: PAN 35125, on calcareous rocks, Khandolia, c 1800 m, Pauri, July 13, 2004.

Remarks: *Buellia quartziana* S. Singh and Awasthi and *B. nilgiriensis* S. Singh and Awasthi are closely related to *B. hemispherica* because all these show fluorescent behaviour in presence of long wave-length (254 nm) ultra-violet radiations. However, *B. hemispherica* shows UV+ yellow, whereas, the other two show UV+ red. Besides, medulla is I+ blue in *B. hemispherica*, it is I- in *B. quartziana*. On the other hand, *B. nilgiriensis* is differentiated from

B. hemispherica through its chemistry; the former contains salazinic acid besides norstictic acid and lichenoxanthone, whereas, the latter contains norstictic acid and lichenoxanthone only.

Species is reported only from the calcareous rocks from different localities of the study area.

Buellia sub-glaziouana Singh and Awasthi, *Biol. Mem.* 6 (2): 169-196, 1981.

Plate 1, Figure B

Chemical tests: Thallus K+ yellow turning reddish-brown, C-, KC-, P+ deep-yellow; medulla I+ blue.

Thin Layer Chromatography: Contains norstictic acid, baeomycesic acid and dichloro- lichenoxanthone.

Fluorescent behaviour: UV+ sordid-red.

Collection examined: PAN 35184, on calcarious rock, Town of Chamoli, *c* 980 m, October 3, 2005.

Distribution in Garhwal: Chamoli.

Remarks: Presence of dichloro-lichenoxanthone in the thallus of *B. sub-glaziouana* is the characteristic feature of the species, which is responsible for its fluorescent behaviour.

Buellia nilgiriensis S. Singh and Awasthi is closely related to *B. sub-glaziouana* in having medulla I+ blue, however, it differs from the latter in having brownish-black disc. The disc is dark-brown in *B. subglaziouana*. Both the species contain norstictic acid but *B. nilgiriensis* contains salazinic acid and *B. subglaziouana* contains baeomycesic acid and dichloro-lichenoxanthone besides norstictic acid.

Earlier record of the species is from district Hoshangabad, Madhya Pradesh (alt. 1050 m). It is a new record for the study area and Himalaya.

Pertusaria pertusa (Weig.) Tuck., Enum. *Nor. Amer. Lich.*: 56, 1895.
=*Lichen pertusus* L., Mantessa 1: 131, 1767.

Plate 1, Figure C

Chemical tests: Thallus K+ yellow-orange, C-, KC+ yellow, P+ orange-red.

Thin Layer Chromatography: Contains stictic, norstictic and constictic acids.

UV behaviour: UV+ pale-orange.

Collections examined: PAN 35048, 35049, on fallen twigs of *Pinus roxburghii* Sarg., Mussoorie road, Village Chamba, *c* 1600 m, Tehri, October 17, 2003;

PAN 35122, on bark of *Cedrela toona* Roxb., Khandolia, *c* 1700 m, Pauri, July 13, 2004; PAN 35123, on bark of *Pinus roxburghii* Sarg., Khirsu, *c* 1800 m, Pauri, July 15, 2004; PAN 35235, on bark of *Quercus semecarpifolia* Sm., Village Reni, *c* 1900 m, Nanda Devi Biosphere Reserve, Chamoli, October 4, 2005; PAN 35298, on bark of *Pinus roxburghii* Sarg., Gangnani, *c* 1850 m, Uttarkashi, October 14, 2005.

Remarks : The species is characterized by the presence of 2-many perithecioid ascocarps per verruca. These open by ostioles, which are sunken in depressed ostiolar region. Positive response of thallus to UV-light at 350 mm giving pale-orange fluorescence is a distinctive character of the species and is used as important identifying criterion.

Pertusaria pertusella Mull. Arg. Closely resembles to *P. pertusa* in having 2-spored asci and chemistry. Both contain stictic and constictic acids, whereas, verrucae are distinctly constricted at base in *P. pertusa*, these are not constricted at base in *P. pertusella.*

Species is widely distributed in different kinds of habitat from Eastern Himalayas to North-Western Himalaya and Madhya Pradesh.

Rhizocarpon geographicum (L.) DC., Flora Franc. Ed. 3 (2): 365, 1805.
=*Lichen geographicus* L., Spec. Pl.: 1140, 1753.

Plate 1, Figure D

Chemical tests : Thallus K-, C+ red, KC-, P+ orange-yellow.

Thin layer chromatography : Contains rhizocarpic, barbatic and gyrophoric acids.

Fluorescent behaviour : Cortex UV+ yellow.

Collections examined : PAN 35455, on quartzite rock, New Tehri, *c* 1700 m, Tehri, October 13, 2003; PAN 35456, on granite rock, Dunda, *c* 1200 m, Uttarkashi, October 15, 2005; PAN 35457, on phyllite rock, Chakrata, *c* 1800 m, Dehra Dun, November 15, 2006.

Remarks : The species is commonly known as "Map-lichen" due to its growing habit. Saxicolous habits, areolate nature of the thallus and muriform ascospores characterize the species. Moreover, the species exhibits yellow fluorescence, which is shown by cortex when exposed to long wave-length (254 nm) ultra-violet radiations.

Rhizocarpon viridiastrum (Wulfen) Korber closely resembles to *R. geographicum* in having 6-10-loculed muriform ascospores; however, the former have round or angular areoles, whereas, the latter have strongly angular areoles.

It is a very common species, which is widely distributed in various regions of Himalayas (tropical to temperate).

Lepraria sp. 1

Plate 1, Figure E

Chemistry : Thallus K+ yellow, C-, KC-, P+ yellow.

Thin layer chromatography : Contains atranorin and triterpenes.

UV behaviour : Thallus UV+ blue

Collections examined : PAN 35094, on bark of *Quercus semecarpifolia* Sm., Kholachauri, *c* 1800 m, Pauri, July 14, 2004; PAN 35095, on soil, Khirsu, *c* 1800 m, Pauri, July 16, 2004; PAN 35425, on Granite rock, Dhanolti, *c* 2250 m, Tehri, November 12, 2006.

Remarks : The present taxon is marked by its fluorescent behaviour, when exposed to long wave-length (254 nm) ultra-violet radiations. It is entirely different from known species of genus *Lepraria* in having indistinct prothallus.

The taxon has been collected from a number of localities from the study area occurring on variety of substrata such as bark of angiosperms and gymnosperms, rocks and soil.

Lepraria sp. 2

Plate 1, Figure F

Chemistry: Thallus K+ yellow, C-, KC-, P+ yellow.

Thin layer chromatography: Contains atranorin zeorin and triterpenes.

UV behaviour: Thallus UV+ pale-blue.

Collections examined: PAN 35013, on bark of *Pinus roxburghii* Sarg., New Tehri, *c* 1700 m, Tehri, October 14, 2003; PAN 35096, on bark of *Thuja orientalis* L., Khirsu, *c* 1800 m, Pauri, July 16, 2004; PAN 35360, on bark of *Pinus roxburghii* Sarg., Jabbar Khet (Mussoorie), *c* 1380 m, Dehra Dun, November 11, 2006.

Remarks: Yellowish-green to greenish-grey, powdery and sterile thallus are marking features of the present taxon. The characteristic Pale-blue fluorescent behaviour, when exposed to long wave-length (254 nm) ultra-violet radiations also characterize and differentiate the taxon from other known species of *Lepraria*.

The taxon is common in the study area in shady and moist locations.

Macro-Lichens

Pyxine berteriana (Fee) Imsh. var. *himalayensis* Awas., *Phytomorphology* 30: 366, 1980.

Plate 1, Figure G

Chemical tests: Thallus cortex K+ yellow, C-, KC-, P-; medulla K+ Yellow, C-, KC-, P- Thin layer chromatography: Contains atranorin and lichenoxanthone.

UV behaviour: Thallus UV+ yellow.

Collections examined : PAN 35350, on twigs of *Grevillea robusta* A. Cunn., Dhanolti, *c* 2250 m, Tehri, November 12, 2006; PAN 35150, on quartzite rock, Town of Pauri, *c* 1700 m, July 12, 2004; PAN 35151, on bark of *Cedrela toona* Roxb., Khandolia, *c* 1700 m, Pauri, July 13, 2004; PAN 35245, on twigs of *Pinus roxburghii* Sarg., Village Lata, *c* 2100 m, Nanda Devi Biosphere Reserve, Chamoli, October 5, 2005; PAN 35320, on bark of *Terminalia arjuna* (Roxb. ex DC.) Wt. and Arn., Town of Uttarkashi, *c* 1180 m, October 12, 2005; PAN 35321, on twigs of *Quercus leucotrichophora* A. Camus, Bhatwari, *c* 1200 m, Uttarkashi, October 13, 2005; PAN 35322, on bark of *Rhododendron arboreum* Sm., Gangnani, *c* 1850 m, Uttarkashi, October 14, 2005; PAN 35324, on calcarious rock, Barkot, *c* 2100 m, Uttarkashi, October 16, 2005; PAN 35396, on bark of *Ficus benghalensis* L., Chakrata, *c* 1800 m, Dehra Dun, November 15, 2006.

Remarks : It is one of the selected species, which shows fluorescent behaviour, due to the presence of lichen substance "lichenoxanthone", when exposed to long wave-length (254 nm) ultra-violet radiations.

Pyxine cognata Stirton is an allied species, which can be easily differentiated from *P. berteriana* by colour of medulla. Medulla is orange to rust coloured in the former, while, it is yellow in the latter.

The species is endemic to India. In N. W. Himalayas, it has been frequently collected on variety of hosts. From the study area, it has been collected on both soft, hard woods and rocks.

Pyxine cocoes (Swartz) Nyl., Mem. Soc. Imp. Sci. Nat. Cherbourg 5: 108, 1857.

=*Lichen cocoes* Swartz, Nov. Gen. Sp. Pl.: 146, 1788.

Plate 1, Figure H

Chemical tests : Thallus cortex K+ yellow, C-, KC-, P-; medulla K+ Yellow, C-, KC-, P-.

Thin layer chromatography : Contains atranorin, lichenoxanthone and triterpenes.

Fluorescent behaviour : Cortex UV+ yellow.

Collections examined : PAN 35066, on bark of *Quercus semecarpifolia* Sm., Village Chamba, *c* 1600 m, Tehri, October 10, 2003; PAN 35152, on Quartzite rock, Khandolia, *c* 1700 m, Pauri, July 13, 2004; PAN 35153, on bark of *Quercus semecarpifolia* Sm., Kholachauri, *c* 1800 m, Pauri, July 14, 2004; PAN 35154, on bark of *Thuja orientalis* L., Kholachauri, *c* 1800 m, Pauri, July 14, 2004.

Remarks : *Pyxine subcinerea* Stirton is closely allied species. It differs from *P. cocoes* in apothecia. Apothecia are lecanorine in *P. cocoes*, whereas, these are lecideine in *P. subcinerea.*

This *Pyxine* species has been collected from both sub-tropical to temperate areas in the study area. Earlier reports are from Central and Southern India (Awasthi, 2007), indicating it to be first report from Garhwal Himalaya.

Table : List of taxa showing fluorescence in presence of long wave-length (254 nm) ultra-violet radiations.

S. No.	Name of the species	Family	Nature of fluorescence in presence of UV radiations (254 nm)
MICRO-LICHENS			
1.	***Buellia hemispherica*** Singh and Awasthi	Physciaceae	Thallus UV+ yellow
2.	***Buellia subglaziouana*** Singh and Awasthi	Physciaceae	Thallus UV+ sordid-red
3.	***Pertusaria pertusa*** (Weig.) Tuck.	Pertusariaceae	Thallus UV+ pale-orange
4.	***Rhizocarpon geographicum*** (L.) DC.	Rhizocarpaceae	Thallus UV+ yellow
5.	***Lepraria*** sp. 1	Lichen impefecti	Thallus UV+ blue
6.	***Lepraria*** sp. 2	Lichen impefecti	Thallus UV+ pale-blue
MACRO-LICHENS			
7.	***Pyxine berteriana*** (Fee) Imsh. var. ***himalayensis*** Awas.	Physciaceae	Thallus UV+ yellow
8.	***Pyxine cocoes*** (Sw.) Nyl.	Physciaceae	Thallus UV+ yellow

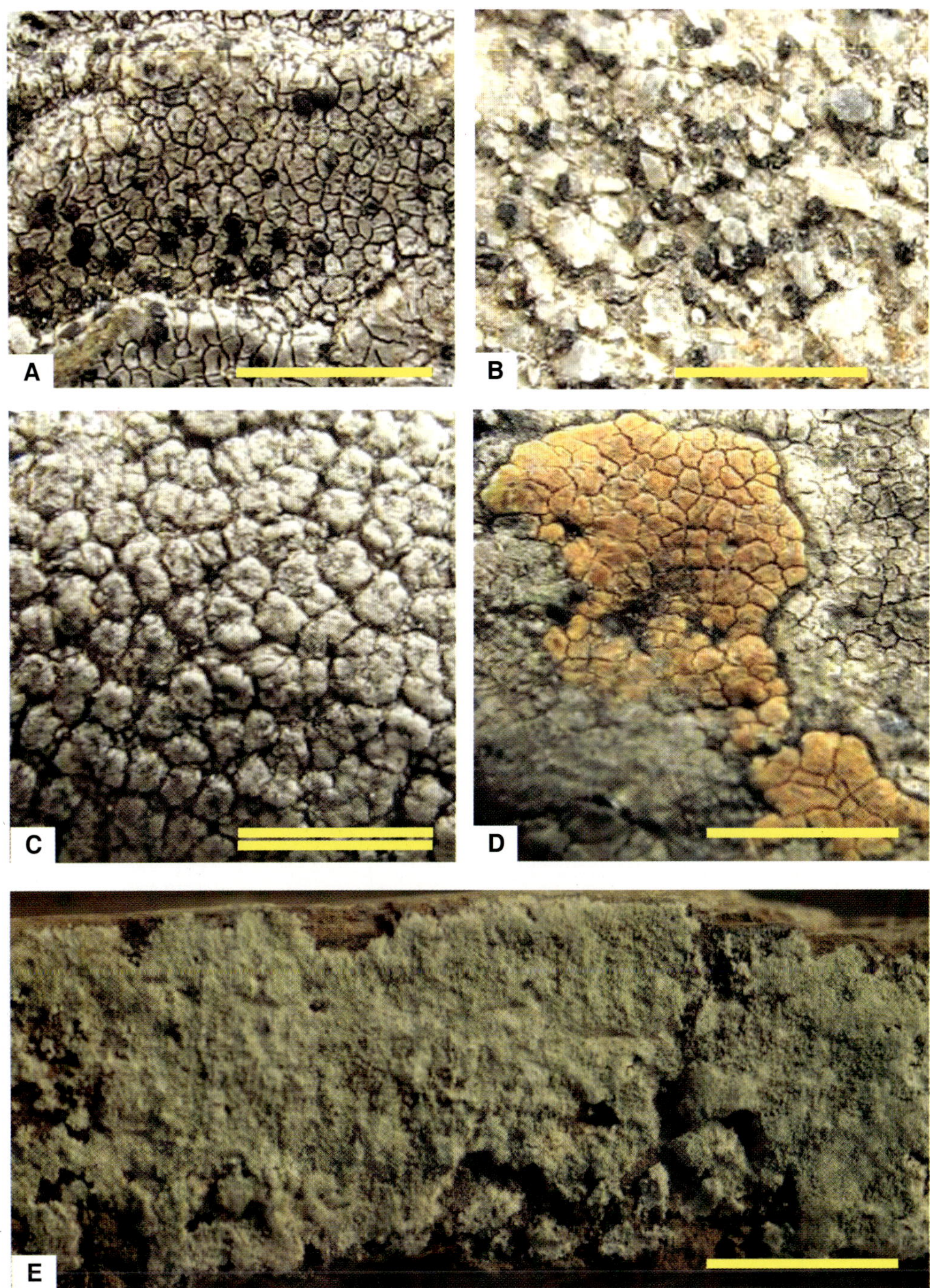

Fig. A – H: Thalli of A. *Buellia hemispherica*, B. *B. sub-glaziouana*, C. *Pertusaria pertusa*, D. *Rhizocarpon geographicum*, E. *Lepraria* sp. I

Fig. F. *Lepraria* sp. II, G. *Pyxine berteriana* var. *himalayensis*, H. *P. cocoes*; Scale: 5 mm.

Acknowledgements

The authors wish to thank Dr. D.K. Upreti, Deputy Director and Group Leader, Lichenology Laboratory, Plant Diversity and Conservation Biology Division, National Botanical Research Institute, Lucknow - 226001, for allowing the examination of certain specimens for comparison, besides helping in chromatographic observations. We are thankful to Prof. (Dr.) Ravinder Kumar, former Chairperson, Centre for Advance Studies in Geology, Department of Geology, Panjab University, Chandigarh - 160014 for his kind help in the identification of rock samples. Thanks are due to Mr. S. K. Rana, Chief Executive, Bio-Age Equipment and Services, Phase IX, Industrial Area, Mohali - 160062 (Punjab) for providing laboratory facilities.

References

Aghoramurthy, K. and Seshadri, T. R. (1953). Chemical investigation of Indian lichens - XV: A species of *Parmelia* containing lichenoxanthone. *J. Sci. Indus. Res.*, 12 B: 73-76.

Aghoramurthy, K., Sarma, K. C. and Seshadri, T. R. (1961). Chemical investigation of Indian Lichens: Part XXV - Chemical components of some rare Himalayan lichens. *J. Sci. Industr. Res.*, 20 B: 166-168.

Aghoramurthy, K., Neelakantan, S. and Seshadri, T. R. (1954). Chemical investigation of Indian lichens - XVII: Chemical components of some *Parmelia* lichens. *J. Sci. Industr. Res.*, 13 B: 326-328.

Awasthi, D. D. (1988). A key to the macrolichens of India and Nepal. *J. Hattori Bot. Lab.*, 65: 207-303.

Awasthi, D. D. (1991). A key to the microlichens of India, Nepal and Sri Lanka. *Biblioth. Lichenol.*, 40: 1-337 + Addendum.

Awasthi, D. D. (2000). *A Hand Book of Lichens*, Bishen Singh Mahendra Pal Singh Publ., Dehradun : 157.

Culberson, C. F. (1970). Supplement to "Chemical and botanical guide to lichen products". *The Bryologist*, 73: 177-377.

Culberson, C. F. (1972). Improved condition and new data for the identification of lichen products by a thin layer chromatographic method. *J. Chromatogr.*, 72: 113-125.

Culberson, C. F. and Kristinsson, H. (1970). A standardized method for the identification of lichen products. *J. Chromatgr.*, 46: 85-93.

Culberson, W. L. (1966). Chemistry and taxonomy of the lichen genera *Heterodermia* and *Anaptychia* in the Carolinas. *The Bryologist*, 69: 472-487.

Dhar, M. L., Neelkantan, S. Ramanujam, S. and Seshadri, T. R. (1959). Chemical investigation of Indian lichens - XXII. *J. Sci. Industr. Res.*, 18 B: 111-113.

Elix, J. A. (1998). Clarification of the synonymy and Chemistry of *Parmotrema zollingeri* and related species. *Austral. Lich.*, 42: 22-27.

Grover, P. K. and Seshadri, T. R. (1959). Chemical investigation of Indian Lichens – XXIII: Imperfect lichens. *J. Sci. Indus. Res.*, 18 B: 238-240.

Hale, M. E. (1983). *The Biology of Lichens*, third editon. Edward Arnold, London, 190 p.

Kirk, P. M., Cannon, P. F. David, J. C. and Stalpers, J. A. (2001). *Dictionary of the Fungi* – 9th edn. (Ainsworth and Bisby's). C.A.B. International Bioscience, Egham, 655 p.

Kirk, P. M., Cannon, P. F. Minter, D. W. and Stalpers, J. A. (2008). *Dictionary of the Fungi* – 10th edn. (Ainsworth and Bisby's). C.A.B. International Bioscience, Egham : 771.

Lamb, I. M. (1951). Biochemistry in the taxonomy of lichens. *Nature*, 168: 1-38.

Mittal, O. P. and Seshadri, T. R. (1954). Chemistry of lichenin and isolichenin. *J. Sci. Indus. Res.*, 13 A: 174-177.

Neelakantan, S. (1952). Chemical Investigation of Indian Lichens. *J. Sci. Industr. Res.*, 11 B: 338-340.

Neelakantan, S., Rao, V. S. and Rangaswami, S. (1954). Chemical components of some *Anaptychia* lichen of India. *Indian J. Pharm.*, 16: 173-175.

Nishikawa, Y., Ohki, K. Takahashi, K., Kurono, G., Fukuoka, F. and Emori, M. (1974). Studies on water soluble constituents of lichens II. Antitumor polysaccharides of *Lasallia, Usnea* and *Cladonia* species. *Chem. Pharm. Bull.*, 22 (11): 2691-2702.

Orange, A., James, P. W. and White, F. J. (2001). *Microchemical methods for identification of lichens*. British Lichen Society London.

Ovstedal, D. O. and Lewis Smith, R. I. (2001). *Lichens of Antarctica and South Georgia: A guide to their identification and ecology*. Cambridge Univ. Press : 424.

Sarin, Y. K. and Atal, C. K.(1976). Utilization of Indian Lichens for production of aromatic resinoids. *PAFAI, Souvenir*, Calcutta, 3: 35-40.

Seshadri, T. R. (1949). Chemical investigation of Indian Lichens – X: Chemical components of *Teloschistes flavicans*. *J. Sci. Indus. Res.*, 30 A: 67-73.

Seshadri, T. R. (1953). Chemical investigation of Indian lichens. *Indian J. Pharm.*, 15: 286-287.

Seshadri, T. R. and Subramanian, S. S. (1949). Chemical investigation of Indian lichens – VIII: Some lichens growing on Sandal trees (*Ramalina zayloriana* and *Roccella montagnei*). *J. Sci. Indus. Res.*, 30 A: 15-22.

Subrahmanian, S. S. and Ramakrishnan, S. (1964). Amino acids of *Peltigera cannina*. *Curr. Sci.*, 33: 522.

Takahashi, K., Takeda, T., Shibata, S., Inomata, M. and Fukuoka, F. (1974). Polysaccharides of lichens and fungi – VI. Antitumour active Polysaccharides of lichens of Stictaceae. *Chem. Pharm. Bull.*, 22 (2): 404-408.

Upreti, D. K. (1987). Key to the species of the lichen genus *Cladonia* from India and Nepal. *Feddes Repertorium*, 98: 469-473.

Upreti, D. K. (1998). A key to the lichen genus *Pyrenula* from India, with nomenclatural notes. *Nova Hedwigia,* 66: 557-576.

Walker, F. J. and James, P. W. (1980). A revised guide to microchemical techniques for the identification of lichen substances. *Bull. Brit. Lichen Soc.,* 46 (Suppl.): 13-29.

Walker, F. J. and James, P. W. (1985). A new quide to the microchemical technique for the identification of lichen substances. *Bull. Brit. Lich. Soc.,* 57 (Suppl): 1-41.

□□□

Microbial Diversity and Functions, 2012
© *D.J. Bagyaraj, K.V.B.R. Tilak, H.K. Kehri (eds.), pp. 433-445*
New India Publishing Agency, New Delhi (India)
E-mail : info@nipabooks.com; Website : www.nipabooks.com

Chapter 21

Candida albicans - Diversity, Identification, Genome and its Life Cycle

Kamal Rai Aneja, Chetan Sharma, Ashish Aneja and Parveen Surain

ABSTRACT

C. albicans is a diploid, dimorphic yeast producing three morphologic forms: yeast cells, pseudohyphae and true hyphae. It is an opportunistic pathogen causing various types of candidiasis which are on the increase around the globe. It is a polymorphic yeast with its genome organized in eight diploid chromosome with its genome size ranging between 14.3 and 16 Mb having 5,733 to 6, 318 genes. It has two mating-type like (MTL) alleles that control the mating behavior. In its genetic makeup it resembles with Saccharomyces cerevisiae hence classified in the Saccharomycetales. The chromosome number from tetraploid to diploid is brought about by the parasexual mechanism. Keeping in view the increase of incidence of candidiasis due to C. albicans around the globe, it is important to identify the causative organisms to the species level correctly. Morphological, biochemical, germ tube test and growth on different media are often used for identifying isolated yeasts which sometimes takes days to week to develop in culture hence, rapid identification of Candida species in clinical laboratory is becoming increasingly important. Molecular techniques utilizing amplification of target DNA or PCR amplification of conserved regions of the genome and sequencing of the resulting PCR product and restriction fragment length polymorphism (RFLP) analysis of the ribosomal DNA, provides quick and precise methods for diagnostics or, identification of Candida species.

Keywords : Candida albicans, diversity, identification, genome, life cycle.

Introduction

The genus *Candida* was erected by Berkhout in 1923 for those yeast-like fungi which were erroneously incorporated in the genus *Monilia* that was created in 1851. It is a large genus comprised of 355 species (Kirk *et al.*, 2008). Infections due to *Candida* and other fungi have increased dramatically in recent years and are of particular importance because of rising number of immuno-compromised patients (Nadeem *et al.*, 2010).*Candida albicans* was first demonstrated as the etiologic agent of the thrush in 1839 (Fenn, 2007). Hazen (1995) reported 17 species of *Candida* as the causative agent of disease in humans. Four *Candida* species, *C .albicans, C. glabrata, C. tropicalis and C. parapsilosis,* together account for about 95% of identifiable *Candida* infections (Pfaller and Diekema, 2007).Of these *C. albicans,* the most prevalent human fungal pathogen can cause life-threatening systematic infections, in addition to superficial mucosal conditions such as thrush and vaginitis.

C. albicans is seldom isolated outside the bodies of animals and is known from 58 species including wild and domesticated animals and birds. Many of us harbor this pathogen as part of our normal intestinal and urogenital mycota without symptoms of candidiasis. Compromised immune system and hormone changes due to HIV infection, leukemia, diabetes, drug therapy and pregnancy all may cause the yeast to infect us, hence referred to as an opportunistic pathogen. The infections often caused are superficial causing skin, mouth, throat and genital lesions. *C. albicans* is the 4th most common hospital acquired infection in the United States that show different levels of resistance to antifungal agents (Mousavi *et al.*, 2007). In India, *C. albicans* is considered to be the commonest and most virulent pathogenic species of the genus *Candida* (Basu *et al.*, 2003; Sengupta and Ohri, 1999; Prasad *et al.*, 1999).

Virulence is possible in *C. albicans* strains that have the ability to grow with the full repertoire of vegetative morphologic forms: yeast cells, pseudohyphal and true hyphal cells (Berman, 2006; Saville *et al.*, 2003; Gow *et al.*, 2002).Being a polymorphic yeast, *C. albicans* genome displays a very high degree of plasticity (Salmecki *et al.*, 2010).

Keeping in view the increase of incidence of candidiasis due to *C. albicans* around the globe, it is important to identify the causative organisms to the species level correctly. Morphological, biochemical, germ tube test and growth on different media are often used for identifying isolated yeasts which sometimes takes days to week to develop in culture hence, rapid identification of *Candida* species in clinical laboratory is becoming increasingly important. Molecular techniques utilizing amplification of target DNA or PCR amplification of conserved regions of the genome and sequencing of the resulting PCR product and restriction fragment length polymorphism (RFLP) analysis of the ribosomal DNA, provides quick and precise methods for diagnostics or, identification of *Candida* species (Borman *et al.*, 2008; Mousavi *et al.*, 2007; Cirak *et al.*, 2003) and candidiasis and hence leading to proper treatment. This review deals with the taxonomy, morphology, identification, genome and life cycle of *C. albicans* for the proper treatment of candidiasis in humans.

Systematic position

C. albicans is anamorphic yeast classified in the order *Saccharomycetales* along with the *Saccharomyces cerevisiae* because of its molecular closeness (Kirk *et al.*, 2008). In the classification system (Ainsworth, 1973), it has been classified in the class *Blastomycetes* (Table 1).

Table 1 : Classification of *C.albicans*

Domain	Kirk *et al.*, 2008	Ainsworth, 1973
	Eukarya	*Eukarya*
Kingdom	*Fungi*	*Fungi*
Phylum	*Ascomycota*	*Deuteromycota*
Class	*Saccharomycetes*	*Blastomycetes*
Sub class	*Saccharomycetidae*	-
Order	*Saccharomycetales*	*Cryptococcales*
Genus	*Candida*	*Candida*
Species	*albicans*	*albicans*

Morphological forms of *C. albicans*

Morphologically *C. albicans* is a dimorphic fungus mainly present in three forms- yeast cells, pseudohyphae and true hyphal cell (Fig. 1). Yeast cells are round to ovoid in shape and separated readily from each other. Pseudohyphae resemble elongated, ellipsoid yeast cells that remain attached to one another at the constricted septation site and usually growing in a branching pattern. True hyphal cells are long and highly polarized, with parallel sides and no constriction between cells.

An interesting feature of *C. albicans* is its ability to show the phenomenon of dimorphism i.e. to grow in two different ways; growth by budding and forming an ellipsoid bud (Fig. 1a) and in hyphal form, which can periodically fragment and give rise to new mycelia, or yeast-like forms (Fig.1b,c). Transitions between the two phenotypes can be induced *in vitro* in response to several environmental factors such as pH, temperature, or different compounds such as N acetylglucosamine or proline. However, perhaps the most critical criterion for pathogenicity is the induction of the mycelial form by serum or macrophages. In addition to the intrinsic biological interest of this dimorphism, its ability to switch between the yeast and the hyphal mode of growth has been implicated in its pathogenicity (Molero, 1998; Leberer, 1997; Cutler, 1991).

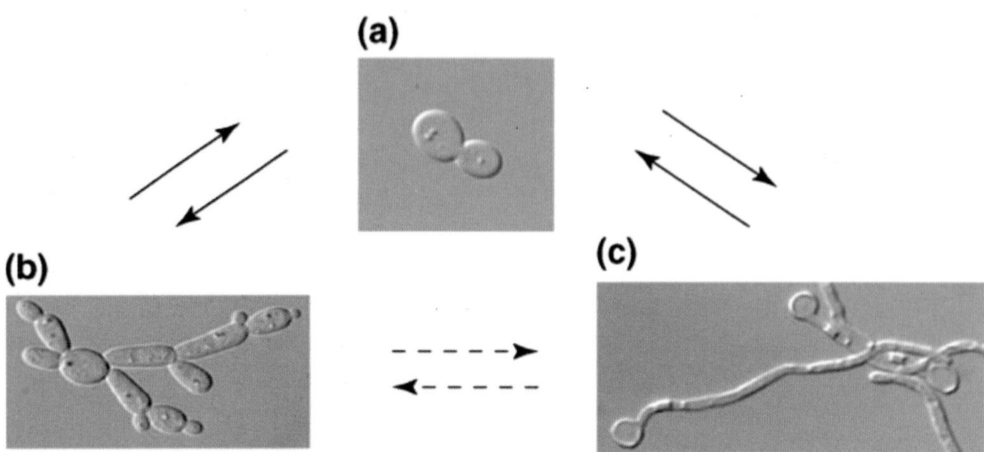

Fig. 1 : Dimorphism in *Candida albicans* and the main factors governing the dimorphism. In the presence of some environmental factors cylindrical outgrowth is initiated on the surface of a blastospore forming a germ tube. Germ tubes grow and septa are laid down behind the extending apical tip to form a hypha. Hyphal branches and/or secondary branches are produced just behind newly laid-down septa, constituting a mycelium. Secondary blastospores become separated from the filament (Berman, 2006; Molero *et al.*, 1998).

Identification of *C. albicans*

Candida albicans is the most often isolated yeast species from clinical specimens and is of clinical importance, so its rapid and true identification is the most important point for clinical laboratories and mycologists. Rapid presumptive identification of *C. albicans* is required for early diagnosis and therapy.

Priminarily identification of *C. albicans* in a clinical specimen begins with direct microscopic examination of stained and unstained samples of the specimen, which also includes their macroscopic features (morphology, colour, size, and texture) and other conventional methods such as germ tube formation test, growth on corn meal agar, culture media containing fluorogenic or chromogenic substrates specific for *C. albicans*, sugar fermentation and assimilation tests. Emphasis is being laid on the molecular techniques for its rapid identification (Babic and Hukic, 2010; Fenn, 2007).

I. Direct microscopy

Direct microscopy provides tentative diagnosis prior to growth in culture, and it may give enough information for the clinician to begin correct immediate patient management. Microscopic examination of Gram stained cells and KOH preparations are mainly used for the identification of yeasts (Bhavan *et al.*, 2010; Dinakar and Samyukta Reddy, 2010; Fenn, 2007).

(A) *Gram stain*: *C. albicans* cells stain purple and appear as round or oval, often with buds. Under oil immersion yeasts appear as giant Gram-positive cocci.

(B) *Potassium hydroxide (KOH)* : a wet mount: A 10% or 15% solution of KOH is used, which acts as a clearing agent of the tissues and cellular debris but does not damage the fungal cells. Specifically, the KOH digests proteinous debris, bleaches pigment, and dissolves the "cement" that holds keratinized cells together and make pseudohyphae, hyphae and spores more apparent.

(C) *Calcofluor white + KOH wet mount* : Calcofluor white is a fluorescent dye that binds to the cellulose and chitin in the cell walls of yeasts (and molds). When viewed under a fluorescent microscope, it is easy to see the intense yellow-green fluorescent yeast and mold elements. This is still a wet-mount preparation, and the KOH is necessary for its clearing ability. The calcofluor white greatly enhances the ability to detect fungi.

(D) *Potassium hydroxide + Lactophenol Cotton Blue (LPCB) wet mount* : This is used for the same purpose as the other KOH preparations. The LPCB stains the fungi, the lactic acid is a clearing agent, and phenol kills the fungi.

II. Colonial characteristics on different media

Five media commonly used for culturing and identification of *C. albicans* are Sheep blood agar, Sabouraud's dextrose agar (SDA),CHROM agar, methyl blue-Sabouraud agar and Candi *Select* 4. Of these, Sabouraud's dextrose agar is the most frequently used medium for primary isolation of *Candida* spp. On CHROMagar, *C. albicans* strains produce â-N-acetylgalactosaminidase, which interacts with the chromophore (chromonogenic hexosaminidase substrate) incorporated into the agar, and with incubation produces green colonies. This culture medium is also used to differentiate different species of *Candida.* In methyl blue-Sabouraud agar, the exact reaction between the methyl blue-Sabouraud dye and *C. albicans* is not known, there is a possible reaction with specific cell wall polysaccharides which produces the fluorescent metabolite (Fig 2). However, in case of Candi *Select* 4, which contains chromogenic substrates react with enzymes secreted by *C.albicans* that leads to specific enzymatic activity resulting in pink to purple colored colonies. On Sheep blood agar many isolates of *C.albicans* produce colonies with feet or extensions (Adam *et al.*, 2010; Nadeem *et al.*, 2010; Fenn, 2006; Yucesoy, 2001).

III. Production of chlamydospores on Corn meal Tween 80 agar

Corn meal agar with Tween 80 is used for demonstration of clamydospore production as it stimulates sporulation of *C. albicans.* The suspected *Candida* cultures are inoculated on corn meal Tween 80 agar medium (pH, 7) containing 4% corn meal powder, 1 % Tween 80 and 1.5 % agar in double distilled water. The plates are incubated at 37° C for 48 - 72 hr and observed for the production of chlamydospores, blastospores, branched pseudohyphae and true hyphae (Fenn, 2007; Slifkin, 2000).

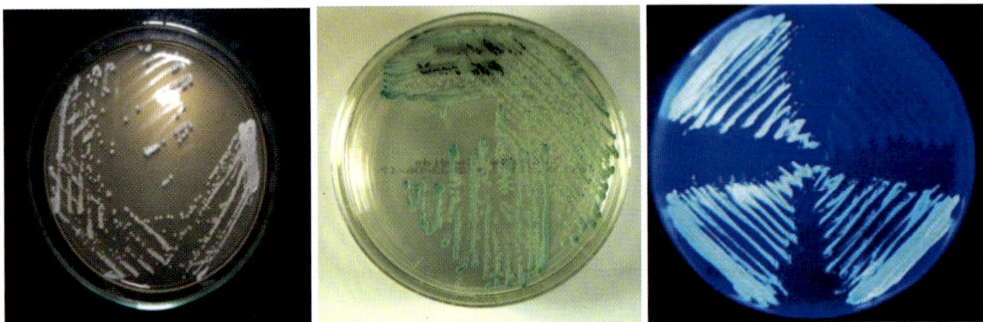

Fig. 2 : Colonial characteristics of *Candida albicans* on three agar media: (A) Growth on SDA; (B)Apple green colour colonies on CHROMagar; (C) Appearance of three *C.albicans* strains on methyl blue agar.

IV. Germ Tube Test

Germ tube test is a rapid screening procedure for differentiating *C. albicans* from other species (Fig 3). It provides a simple, reliable and economical procedure for the presumptive identification of *C. albicans*. A germ tube represents the initiation of a hypha directly from the yeast cell. It is a filamentous, cylindrical outgrowth from the yeast cell with no constriction present at the base. The suspected *Candida* cultures are grown on Sabouraud dextrose agar and inoculated into 0.5 ml of human serum in a small tube and incubated at 37° C for 2 hr. After desired period of incubation, a loop-full of

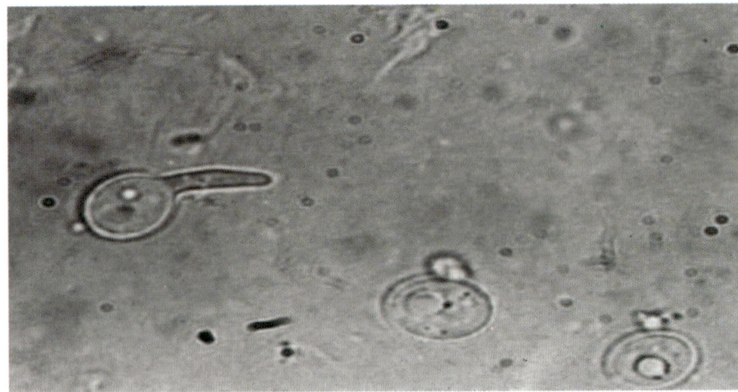

Fig. 3 : Germ tube production from a yeast cell of *C. albicans* in human serum.

culture is placed on a glass slide and overlaid with a cover-slip. The preparation is examined for the production of germ tubes (one/yeast cell) from the germinating blastospores (Sandven, 1990).

V. Physiological Tests

(A) Carbohydrate assimilation tests

The carbohydrate assimilation test measures the ability of yeast to utilize a specific carbohydrate as the sole source of carbon in the presence of oxygen. Carbohydrate assimilation profiles for *Candida* species can be obtained by examining zones of *Candida* growth around discs or wells impregnated with various sugars on basal agars. Sugars used for assimilation tests include dextrose, maltose, sucrose, lactose, galactose, melibiose, cellobiose, inositol (a form of sugar, carbocyclic polyol, cyclohexanehexol), xylose, raffinose, trehalose and dulcitol (or galactitol, a sugar alcohol, the reduction product of galactose). In this test, the *Candida* cultures are suspended in saline (NaCl) to which 1.5 ml of basal medium containing 67.8 % yeast nitrogen base is added. This suspension is then transferred to 13.5 ml of molten, cooled agar containing 2 % agar powder, mixed well, poured into a Petri dish and allowed to solidify. After the solidification of medium, 20 % sugar-soaked filter paper discs are placed on the medium. The plates are incubated at 25^0 C for 10 to 24 hr and observed for the zone of growth of *Candida* around the wells. Growth of the yeast indicates the assimilation of a sugar. Sugars utilized by *C.albicans* are given in table 2 (Bhavan *et al.*, 2010; Dinakar and Samyukta Reddy, 2010; Slifkin, 2000).

(B) Carbohydrate fermentation tests

Fermentation tests are generally performed in broth and to detect whether or not the yeast can produce acid or gas from the carbohydrate under anaerobic conditions (Table 2). Fermentative yeasts recovered from clinical specimens produce carbon dioxide and alcohol. Production of gas rather than a pH shift is indicative of fermentation. Dextrose, maltose, sucrose, lactose, galactose and trehalose are used in the test. The 5 ml of carbohydrate (pH, 7.4) containing 1 % peptone, 1 % sugar, 0.3 % beef extract and 0.5 % NaCl, 0.2 % Bromothymol blue in distilled water is dispensed in sterilized Durham tube and 0.2 ml of saline suspension of the test organism is added and incubated at 37^0 C for 10 days and observed for the production of gas. Fermentation of various sugars brought about by *C.albicans* is given in table 2 (Bhavan *et al.*, 2010; Dinakar and Samyukta Reddy, 2010).

Table 2 : Carbohydrate assimilation and fermentation test for *C.albicans* in different types of sugars

Sugars	Assimilation of	Fermentation of
Glucose	+	+
Maltose	+	+
Sucrose	+	-
Lactose	-	-
Galactose	+	+
Melibiose	-	Nt
Cellobiose	-	-
Inositol	-	Nt
Xylose	+	Nt
Raffinose	-	Nt
Trehalose	+	+
Dulcitol	-	Nt

+ Positive; - Negative; Nt- Not tested (Bhavan *et al.*, 2010; Lennette *et al.*, 1985)

VI. Molecular methods

Molecular techniques are considered to be more stable approach ES as of identification. It includes restriction fragment length polymorphisms (RFLPs) using gel electrophoresis or DNA-DNA hybridisation and polymerase chain reaction. These are mainly used for the identification of *Candida* strains but have been used less frequently for differentiation of species (Mirhendi *et al.*, 2006; Bautiste-Munoz *et al.*, 2003).

(A) Restriction fragment length polymorphisms (RFLPs) using gel electrophoresis

Restriction fragment length polymorphism is based on the digestion of DNA. Every organism possesses unique nucleotide sequences that distinguish it from every other organism on the basis of the number and size of the fragments. In this method, DNA is extracted from isolates and cleaved into fragments by restriction endonucleases, the fragments are separated by gel electrophoresis. This method has been successfully applied for the exact identification of *Candida* species. The differences in the restriction patterns for the rDNA regions of the various *Candida* species serve as a rapid means of differentiating among these organisms. A HaeIII digest is definitive for distinguishing *C. albicans* species from other non-*C.albicans* species. DdeI digestion also seems to be efficient to identify *C.albicans* species (Dinakar and Samyukta Reddy, 2010; Dembry *et al.*, 1994; Taylor *et al.*, 1999).

(B) Polymerase chain reaction

It has been used to identify *Candida* based on detection of candidal genes encoding for chitin synthase, actin and cytochrome P450LIAI. Ribosomal DNA (rDNA; genes encoding for ribosomal RNA) is also a frequent target in PCR systems since rDNA sequences exist in multiple copies within the genome, thereby offering greater sensitivity.

A relatively small divergence of the small subunit rDNA (16S-like) sequences has occurred during evolution and therefore conserved regions exist between distantly related species. However, spacer regions between rDNA conserved sequences evolve faster and variations in the primary structure of these spacers can exist between genera and species. Primers that target conserved rDNA sequence adjacent to the spacer regions permit the incorporation of the spacers into the PCR product. RFLP studies have shown that rDNA sequences differ between *Candida* species and PCR amplification of 18S rDNA, enables discrimination of *C. albicans, C. glabrata, C. tropicalis, C. krusei* and *C. dubliniensis.* (Dinakar and Samyukta Reddy, 2010; Cirak *et al.*, 2003; Wahyuningsih *et al.*, 2000).

Despite the advantages of PCR- based identification over phenotypic features that may be subject to variable expression and as such may lead to incorrect identification, the application of such molecular techniques is still relatively limited due to certain limitations such as lesser technical staff training and non availability of the facilities in the laboratories, especially in the developing countries of the world.

Genome in *C. albicans*

Genome is defined as the total inheritable genetic material of an organism, and a haploid set of chromosomes in eukaryote (Kirk *et al.*, 2008). *C. albicans* is polymorphic yeast with its genome organized in eight diploid chromosomes. In this yeast the genome is 14.3 to 16 Megabases (Mb) coding for 5,733 to 6318 genes (Table 3). The size of diploid species of *C. albicans* shows difference of nearly 50% with haploid species of *Candida* (e.g. *C. guilliermondii* and *C. lusitaniae*) having small genomes that ranges from 10.6 to 12.1 Mb (Selmecki *et al.*, 2010; Butler *et al.*, 2009).

Table 3 : Genome features of important *Candida* species (Butler *et al.*, 2009)

Species	Genome size (Mb)	No. of genes	Ploidy
C. albicans WO-1	14.4	6,159	diploid
C. albicans SC 5314	14.3	6,107	diploid
C. tropicalis	14.5	6,258	diploid
C. parapsilosis	13.1	5,733	diploid
C. guilliermondii	10.6	5,920	haploid
C. lusitaniae	12.1	5,941	haploid

C. albicans has two mating-type-like (MTL) alleles, MTLa and $MTL_{á}$. The MTL locus is on the left arm of chromosome 5 (Chr5), approximately 80 kbp from the centromere. Most *C. albicans* isolates are heterozygous for the MTL locus, but approximately 3 to 10% of clinical isolates are naturally homozygous at MTL. Mating can occur between strains carrying the opposite MTL locus, and most strains that were found to be naturally MTL homozygous are mating competent. MTL-homozygous strains have also been constructed from MTL-heterozygous strains by deletion of either the MTLa or $MTL_{á}$ locus or by selection for Chr5 loss on sorbose. Mating between these diploid strains of opposite mating type can occur both *in vitro* and *in vivo*. The products are tetraploid and do not undergo a conventional meiotic reduction in ploidy. Rather, they undergo random loss of multiple chromosomes, a process termed "concerted chromosome loss," until they reach a near-diploid genome content (Legrand *et al.*, 2004; Lachke *et al.*, 2003; Magee and Magee, 2003; Hull and Johnson, 1990).

Mating and Meiosis

C. albicans, a diploid and dimorphic yeast, reproduces asexually by budding (Fig. 4) and not known to reproduce sexually. But recent genomic studies have provided convincing evidence for the potential for a sexual cycle. A region of genome has been detected that encoded genes similar to those found at the mating locus of *Saccharomyces cerevisiae* hence classified in the *Saccharomycetales* by Kirk *et al.* (2008).

In *C. albicans* a well defined mating system has been identified that allows the conjugation of mating-type locus homozygous diploid cells (Fig 4), however, there is currently no evidence for a functional meiotic pathway that allows reductional division and a return to the diploid state from the tetraploid (Borman, 2008). *C.albicans* has a parasexual cycle (mating of diploid cells followed by mitosis and chromosome loss instead of meiosis) (Noble and Johnson, 2007). In this yeast, efficient mating occurs between α and β forms, yet the completion of the sexual cycle occurs by a parasexual mechanism of random chromosome lose rather than conventional meiosis (Bennett, 2009).In *C. albicans* mating occurs due to the presence of necessary mating genes where MTL_{a} locus expresses MTL_{a2}, which directs mating function, and the MTL_{α} expresses $MTL_{\alpha 1}$ controlling α mating functions. The resulting tetraploid products of mating breakdown to diploids through spontaneous chromosome loss, not meiosis. Diploids form from tetraploids through chromosome loss, indicating that mitotic crossing-over gave rise to progeny thus showing the operation of parasexual cycle (Whiteway and Bachewich, 2005).On the basis of genomic evidence, *C.albicans* either has a cryptic meiotic program that has yet to be identified, or conserved meiosis- specific genes have been programmed to function in the parasexual cycle (Butler *et al.*, 2009).

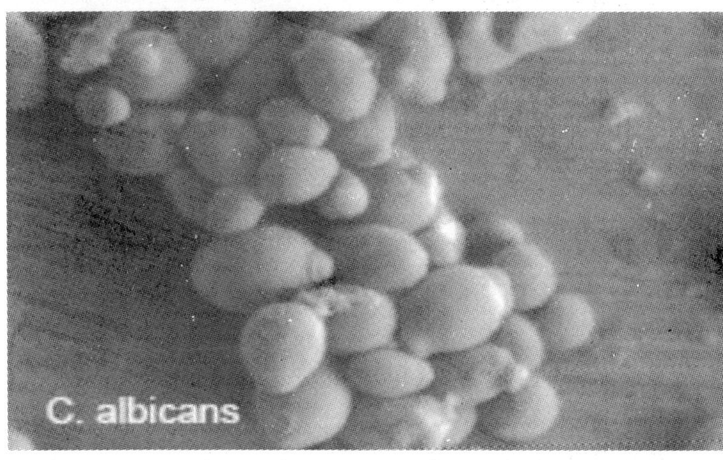

Fig 4 : Scanning electron microscopic image of *C.albicans* showing budding (Meurman *et al.*, 2007)

References

Adam, H.J., Richardson, S.E., Roscoe, M., Boroumandi, S., Gris, M. and Yau, Y.C.W. (2010). An implementation strategy for the use of chromogenic media in the rapid, presumptive identification of *Candida* Species. *The Open Mycol. J.*, 4: 33-38.

Ainsworth, G.C., Sparrow, F.K. and Sussman, A.S. (1973). *The Fungi- An Advanced Treatise.* Vol. IV A. Academic Press, New York.

Babic, M. and Hukic, M. (2010). *Candida albicans* and non-albicans species as etiological agent of vaginitis in pregnant and non- pregnant women. *Bosnian J. Basic. Med. Sci.,* 10: 89-97.

Basu, S., Gugnani, H.C., Joshi, S. and Gupta, N. (2003). Distribution of *Candida* species in different clinical sources in Delhi, India, and proteinase and phospholipase activity of *Candida albicans* isolates. *Res. Iberoam Micol.*, 20: 137-140.

Bautiste-Munoz, C., Boldo, X.M. and Tanaka, L.V. (2003). Identification of *Candida* spp. by ramdomly amplified polymorphic DNA analysis and differentiation between *Candida albicans* and *Candida dubliniensis* by direct PCR methods. *J. Clin. Microbiol.,* 41: 414-420.

Bennett, R.J. (2009). A *Candida* based view of fungal sex and pathogenesis. *Genome Bio.,* 10: 1-4.

Berman, J. (2006). Morphogenesis and cell cycle progression in *Candida albicans. Curr. Opinion Microbiol.*, 9:595-601.

Bhavan, P.S., Rajkumar, R., Radhakrishan, Seenivasan, C. and Kannan, C. (2010). Culture and identification of *Candida albicans* from vaginal ulcer and separation of enolase on SDS-PAGE. *Int. J. Biol.*, .2:84-93.

Borman, A.M., Linton, C.J., Miles, S.J., Johnson, E.M. (2008). Molecular identification of pathogenic fungi. *J. Antimicrobial Chem.*, S1: i7-i12.

Butler, G., Rasmussen, M.D., Lin, M.F., Santos, M.A.S., Sakthikumar, S. *et al.* (2009). Evolution of pathogenicity and sexual reproduction in eight *Candida* genomes. *Nature*. 459: 657-662.

Cirak, M.Y., Kalkanci, A. and Kustimur, S. (2003).Use of molecular methods in identification of *Candida* species and evolution of fluconazole resistance. *Mem. Inst. Oswaldo Cruz.,* 98: 1027-1032.

Cutler, J.E. (1991). Putative virulence factors of *Candida albicans*. *Annu. Rev. Microbiol.,* 45:187–218.

Dembry, L.M., Vazquez, J.A., Zervos, M.J. (1994). DNA analysis in the study of epidemiology of nosocomial candidiasis. *Infect Control Hosp. Epidemiol.,* 15: 48-53.

Dinakar, J. and Samyukta Reddy, B.V. (2010). Laboratory diagnosis of oral candidiasis. *J. Orofac. Sci.,* 2:70-74.

Fenn, J.P. (2007). Update of medically important yeasts and a practical approach to their identification. *Labmedicine,* 38: 178-183.

Gow, N., Brown, A. and Odds, F. (2002). Fungal morphogenesis and host invasion. *Curr. Opin. Microbiol.,* 5: 366-371.

Hazen K C. (1995). New and emerging yeast pathogens. *Clin. Microbiol. Rev. JAMA.,* 8:462–478.

Hull, C.M. and Johnson, A.D. 1999. Identification of a mating type-like locus in the asexual pathogenic yeast *Candida albicans*. *Science,* 285:1271–1275.

Kirk, P.M., Cannon, P.F., Minter, D.W. and Stalpers, J.A. (2008). *Ainsworth and Bisby's Dictionary of the Fungi*. 10th ed. Wallingford: CABI.

Lachke, S.A., Lockhart, S.R., Daniels, K.J. and Soll, D.R. (2003). Skin facilitates *Candida albicans* mating. *Infect Immun.,* 71:4970–4976.

Leberer, E., Ziegelbauer, K., Schmidt, A., Harcus, D., Dignard, D., Ash, J., Johnson, L. and Thomas, D.Y. (1997). Virulence and hyphal formation of *Candida albicans* require the Ste20p-like protein kinase CaCla4p. *Cur. Biol.,* 7:539–546.

Legrand, M., Lephart, P., Forche, A., Mueller, F.M., Walsh, T., Magee, P.T. and Magee, B.B. (2004). Homozygosity at the *MTL* locus in clinical strains of *Candida albicans*: karyotypic rearrangements and tetraploid formation. *Mol. Microbiol.,* 52:1451–1462.

Lennette, E.H., Balows, A., Hausler, W.J. and Shadomy, H.J. (1985). *Manual of Clinical Microbiology*.4th Ed. American society for Microbiology.Washington, D.C.

Magee, B.B. and Magee, P.T. (2000). Induction of mating in *Candida albicans* by construction of *MTLa* and *MTL*alpha strains. *Science,* 289:310–313.

Meurman, J.H., Siikala, E., Richardson, M. and Rautemaa. (2007). Non-*Candida albicans Candida* yeast of the oral cavity. Communicating *Cur. Res. Edu. Topics Trends App. Microbiol.,* 719-730.

Mirhendi, S.H., Makimura, K., Khoramizadeh and Yamaguchi. (2006). A one enzyme PCR-RFLP assay for identification of six medically important *Candida* species. *Jap. J. Med. Mycol.,* 47: 225-229.

Molero, G., Orejas, R.D., García, F.N., Monteoliva, L., Pla, J., Gil, C., Pérez, M.S. and Nombela, C. (1998). *Candida albicans*: genetics, dimorphism and pathogenicity. *Inter. Microbiol.*, 1:95–106.

Mousavi, A., Khalesi, E., Bonjar, G.H.S., Aghighi, S., Sharifi, F. and Aram, F. (2007). Rapid molecular diagnosis for *Candida* species using PCR-RFLP. *Biotechnol.*, 6: 583-587.

Nadeem, S.G., Hakim, S.T. and Kazmi, S.U. (2010). Use of CHROMagar Candida for the presumptive identification of *Candida* species directly from clinical specimens in resource-limited settings. *Libyan J. Med.*, 5: 1-6.

Noble, S.M. and Johnson, A.D. (2007). Genetics of *Candida albicans*, a diploid human fungal pathogen. *Annu. Rev. Genet.*, 41: 193-211.

Pfaller, M.A. and Diekema, D.J. (2007). Epidemiology of invasive candidiasis: a persistent public health problem. *Cli. Microbiol. Rev.*, 20: 133-163.

Prasad, K.N., Agarwal, J., Dixit, A.K., Tiwari, D.P., Dhole, T.N., Ayyagri, A. (1999). Role of yeasts as nosocomial pathogens and their susceptibility to fluconazole and amphotericin B. *Ind. J. Med. Res.*, 110: 11-17.

Sandven, P. (1990).Laboratory identification and sensitivity testing of yeast isolates. *Acta Odontol. Scand.*, 48: 27–36.

Saville, S.P., Lazzell, A.L., Monteagudo, C. and Lopez-Ribot, J.L. (2003). Engineered control of cell morphology *in vivo* reveals distinct roles of yeast and filamentous form of *Candida albicans* during infection. *Eukaryotic Cell*, 2: 1053-1060.

Selmecki, A., Forche, A. and Berman, J. (2010).Genome plasticity of the human fungal pathogen *Candida albicans*. *Eukaryotic cell*, 9:991-1008.

Sengupta, P. and Ohri, V.C. (1999). Study of yeast species isolated from clinical specimens. *Med. J. Armed Forces Ind.*, 55: 319-321.

Slifkin, M. (2000). Tween 80 opacity test responses of various *Candida* species. *J. Clinical Microbial.*, 38: 4626–4628.

Taylor, J.W., Geiser, D.M., Burt, A. and Koufopanou, V. (1999). The evolutionary biology and population genetics underlying fungal strain typing. *Clin. Microbiol. Rev.*, 12: 126-146.

Wahyuningsih R, Freisleben HJ, Sonntag HG, Schnitzler P. (2000). Simple and rapid detection of *Candida albicans* DNA in serum by PCR for diagnosis of invasive Candidiasis. *J. Clinical Microbiol.*, 38: 3016–3021

Whiteway M, Bachewich C. (2005). Fungal genetics. In: *Fungi: Biology and Applications,* (Ed.) K. Kavanagh, John. Wiley & Sons Ltd.: 35-64.

Yucesoy M, Esen N, Yulug N. (2001).Use of chromogenic tube and methyl blue-Sabouraud agar for the identification of *Candida albicans* strains. *Kobe. J. Med. Sci.*, 47: 161-167.

□□□

Microbial Diversity and Functions, 2012
© D.J. Bagyaraj, K.V.B.R. Tilak, H.K. Kehri (eds.), pp. 447-472
New India Publishing Agency, New Delhi (India)
E-mail : info@nipabooks.com; Website : www.nipabooks.com

Chapter **22**

Arbuscular Mycorrhizal Fungi in Agroforestry Systems

Ashok Shukla, Deepak Vyas and Anuradha Jha

ABSTRACT

Agroforestry is an intensive land use management system that integrates trees, shrubs, crops, etc. on a landscape level to achieve optimum benefits. Agroforestry is not only concerned with beneficial effects of one component on another, but it also involves the manipulations of negative effects to minimize their influence on the productivity of the overall system. In an agroforestry, trees and crops compete inevitably for light, nutrients, moisture and other resources. The interactions between tree and crop roots can also have an effect on soil organisms, which play an essential role in the functioning and productivity of agroforesty systems. Arbuscular mycorrhizal fungi (AMF) are one of the most important components of the soil biota in monoculture and agroforestry systems. AM fungi can rehabilitate degraded lands subjected to agroforestry systems and common mycorrhizal network may further enhance the benefits of agroforestry through vertical niche expansion of AMF. However, management practices of agroforestry affect mycorrhization, AMF community composition and diversity of AM fungi. This review focuses on AMF diversity, which plays a significant role in agroforestry systems.

Keywords : AM fungi, agroforestry sytems.

Introduction

Over the past century, conventional agriculture has successfully improved the productivity of annual crop production (Kumar and Shukla 2009). Agriculture impacts on biological, chemical and physical properties of soils, degradation of soil, erosion, soil coverage, biodiversity losses, changes in element and water balance of ecosystems, and contamination of groundwater

(Tilman *et al.*, 2002; Foley *et al.*, 2005). Intensive agricultural practices exploiting the land and other natural resources increased manifolds during the last 50 years and simultaneously the use of chemical fertilizers are also increased (de Carvalho *et al.*, 2009). Injudicious and untimely application of fertilizer in agriculture fields generated several environmental (Sharif *et al.*, 2010) and soil problems (Mahajan *et al.*, 2007; Yadav and Kumar 2009; Kumar and Singh 2010). In addition, there are some other side effects due to the practices of agricultural simplification (Tilman 1998; Pimentel *et al.*, 2005) where ecosystem services provided by the soil are increasingly bypassed (de Carvalho *et al.*, 2009; Wrage *et al.*, 2010). The perceived need to seek alternatives to current agricultural practices has resulted in an enhanced interest in agroforestry systems (Ingleby *et al.*, 2007; Kumar *et al.*, 2007a), which can conserve resources, improve environmental quality, rehabilitate degraded lands and provide multiple outputs to meet the daily demands of the rural population (Pande and Tarafdar 2004; Mridha and Dhar 2007; Muleta *et al.*, 2008). The need to increase food, fiber and fuel wood production to keep pace with the fast growing population is crucial (Wrage *et al.*, 2010). Under agroforestry, the needs for ecological sustainability can be reconciled with the needs for sustainable food production (Young 1997).

Agroforestry System

Agroforestry system is a combination of trees with crops, simultaneously or sequentially in the same area, and have as their major aim the optimization of beneficial ecological interactions among ecosystem components (Farrell and Altieri, 2002). Agroforestry systems put much emphasis on species diversity and on the interactions between these different species (de Carvalho, 2009). According to Thevathasan and Gordon (2004), agroforestry is the more sustainable land use system, which reduces the harmful effects of conventional agriculture. It provides number of ecological advantages such as increased carbon sequestration by improving soil physico-chemical quality, increased soil nutrient availability (Garg *et al.*, 2006), microbial activity in soil (Cardoso *et al* 2003a) and conserve biodiversity (Nair 1993). Cardoso *et al.*, (2003b) revealed that agroforestry systems are effective in soil conservation compared with conventional monocultural systems. In addition to maintaining above benefits, agroforestry systems show larger resilience and resistance to attacks by pest species and diseases (Young 1997; Ewel 1999; Van Noordwijk and Ong 1999; Souza 2006).

However, agroforestry is not only concerned with beneficial effect of one component on another, but also involves the manipulations of negative effects to minimize their influence on the productivity of the overall system. In an agroforestry system, belowground interactions occur due to the high density of tree roots found within the same region as the crop roots (Jose *et al.*, 2000 and 2004). The deep roots of trees can act as a 'safety net' by capturing nutrients that leach below the rooting zone of the alley crops and recycle them back into the system (Jose *et al.*, 2004; Thevathasan and Gordon 2004). As a result, inorganic fertilizer rates can be reduced in tree-based intercropping *i.e.* agroforestry systems, which in addition to the reduced nutrient leaching can decrease the amount of nitrous oxide emissions compared to conventional agricultural systems (Thevathasan and Gordon, 2004). At the tree-crop interface of agroforestry system, trees and crops compete inevitably for light (Shukla *et al.*, 2009), nutrients (Chirwa *et al.*, 2007), moisture (Everson *et al.*, 2009) and other available resources. The interactions between tree and crop roots can also have an effect on soil organisms, which play an essential role in the functioning and productivity of agroforesty systems. These include arbuscular mycorrhizal (AM) fungi which associate with the roots of most tree and crop species in agroforestry systems.

Arbuscular Mycorrhizal Fungi

AM fungi are characterized by fungal structures (arbuscules, vesicles and hyphae) that penetrate the cells of the host root and extraradical mycelium present in the soil (Smith and Read 2008). AMF are the important component of natural (Vyas *et al.*, 2005; Vyas *et al.*, 2006), agriculture (Gianinazzi *et al.*, 1989) and agroforestry systems (Leake *et al.*, 2004; Kumar and Shukla 2009), forming symbiotic associations with the majority of plants (de Carvalho *et al.*, 2009) and have been shown the multifunctional role in agro-ecosystems (Newsham *et al.*, 1995). Recently, Vyas *et al.*, (2007 and 2008a,b) reported mycorrhization in bryophytes and early land plants. Most obvious role of AMF is to increase the soil volume exploited by the host plant (Gianinazzi *et al.*, 2010) leading to increased water and nutrient uptake (Soni *et al.*, 2003; Vyas 2008), which in turn may enhance acquisition of nutrients *i.e.* phosphorus (Khaliq and Sanders 2000; Oehl *et al.*, 2002; van der Heijden *et al.*, 2006), nitrogen (N) through biological N fixation (Bolan 1991; Vance 2001; Garg *et al.*, 2006) and potassium (K; Dwivedi *et al.*, 2004). Other roles of AMF concern protection of the root system against pathogens (Garmendia *et al.*, 2004; Pozo and Azcon-Aguilar 2007; Elsen *et al.*, 2008; Garrido 2009; Singh and Vyas 2009; Singh *et al.*, 2010a and b; Vyas *et al.*, 2010), salinity (Tavares, 2007), phytotoxic elements (Rufyikiri *et al.*, 2000), or heavy metals (Andrade., 2003). AMF are also involved in the formation and maintenance of soil structure (Rillig and Mummey, 2006) and increase carbon input to soils (Rillig *et al.*, 2001; Zhu and Miller 2003), both of these effects contributing to reduce erosion (Rillig *et al.*, 2002; Nasim 2005). According to Mutuo *et al.* (2005) and Cardoso and Kuyper (2006), AM fungi can rehabilitate degraded lands subjected to agroforestry systems and common mycorrhizal network may further enhance the benefits of agroforestry through vertical niche expansion of AMF (Simard and Durall 2004; Cavagnaro *et al.,* 2005; Wilson *et al.,* 2009) and the low biomass production of agroforestry tree species in degraded areas can, therefore, be circumvented by the use of AM fungi (Shukla *et al.,* 2009). The role of perennial trees in maintaining AMF inoculum and in sustaining mycelial networks for short-lived crops may be an unintended benefit of agroforestry systems and provide an alternative approach to the use of cover crops to build up soil inoculum (Kumar *et al.,* 2007b). The importance of maintaining active populations of AM fungi in agroforestry soils in order to sustain crop productivity has also been demonstrated (Sieverding and Leihner 1984; Dodd *et al.,* 1990).

In contrast, monoculture may reduce the spectrum of AMF present in the soil after several years of continuous cultivation (Sieverding 1991). Greater management intensities of conventional agriculture such as tillage, fallow and fertilization strongly influence the abundance, species richness and AMF composition (Jansa *et al.,* 2002; Oehl *et al.,* 2003; Troeh and Loynachan 2003; Jumpponen *et al.,* 2005; Hijri *et al.,* 2006) and this could be due to number of factors including disturbance of AM fungal hyphal networks, changes in soil nutrient content (Mishra *et al.,* 2008), altered microbial activity, or changes in weed populations (Jansa *et al.,* 2003; Vyas *et al.,* 2008). Management strategies that require lower inputs such as organic farming tend to have a greater AM fungal diversity (Vyas *et al.,* 2002; Oehl *et al.,* 2004). Overall, conventional agricultural practices appear to decrease the abundance of AM fungi and cause a shift in species composition toward species that are inferior mutualists (Johnson 1993; Douds and Millner 1999). The efficiency of nutrient recycling in agroforestry systems and better understanding of the underlying biological processes is needed (Sieverding 1991). In this regard, studies on the distribution of AM fungi are vital because there is growing evidence that agroforestry practices may be important in maintaining the mycorrhizal inoculum potential in soils (Cardoso *et al.,* 2003a,b ; Ingleby *et al.,* 2007; de Carvalho *et al.,* 2009). On the other hand, knowledge on how to introduce and manage agroforestry systems is lagging behind, due to the specificity of each ecosystem and the great diversity and complexity of the interactions involved.

There are two strategies for managing AM populations in the agroforestry system, 1) to develop inoculation techniques with efficient AM fungi, adapted to the plant and the environment, and 2) to manage the indigenous AM fungi by agricultural practices in such a way that efficient AM fungi are enhanced (Sieverding 1991). AM fungi associated with agroforestry tree species may serve as an additional role by maintaining active AM propagules in the soil, which could then rapidly colonize roots of emerging crop seedlings (Haselwandter and Bowen 1996).

AM Fungi and Agroforestry Systems

A search on the Google Scholar with the words "*Agroforestry*" and "*Mycorrhiza*" yielded around 50 articles. However, fewer than 20 articles went beyond the evaluation of mycorrhizal responses of individual tree species. As a consequence of this meager database, many mycorrhizal aspects of agroforestry remain under investigated. Several reports are available on the influence of agroforestry systems on the AMF diversity and abundance. These investigations have found that agroforestry can have a positive effect on the AMF community, but there are cases where no significant effect or even a negative effect was observed. This may be due to the variety of locations and climates where these studies took place and the diverse combinations of tree and crop species used within these systems. However, these studies provide insight into the effect that tree-based intercropping systems can have on AM fungi. Shukla (2009) conducted an extensive study on the AMF diversity in agroforestry, effect of trees on AM colonization of intercrops and *vice versa*, cross-infectivity of AMF in agroforestry crops/plants, effect of management practices *viz.* effect of pruning of multipurpose tree species (MPTs) on their colonization, effect of P application, shade (i.e. sun light) and soil moisture etc.

Effect of Intercropping on Mycorrhization

In temperate region, effect of agroforestry systems on AM fungi is not well understood as only two studies to date have investigated AM fungi in these systems. Lacombe *et al.*, (2009) found a higher abundance of AM fungi in two different tree-based intercropping sites compared to adjacent conventional monocropping sites. In another study, Chifflot *et al.* (2009) recorded higher AMF diversity and spatial distribution of spores was significantly different in an agroforestry system compared to a monoculture forest plantation system. However, due to the higher adoption rate of agroforestry in tropical regions, there is evidence that tropical agroforestry systems support a diverse and abundant community of AM fungi. For example, few studies conducted in India found a positive influence of trees on spore abundance and AM fungal colonization of roots collected from the rhizosphere under tree canopies compared to samples collected out side tree canopies (Pande and Tarafdar 2004; Prasad and Mertia 2005). A survey of AM fungi in neem (*Azadirachta indica* L.) based agroforestry systems revealed a higher density of spores in samples collected from under tree canopies compared to samples collected 25 m from trees (Pande and Tarafdar 2004). Neem trees, which are highly AM dependent, had a higher level of root colonization compared to the crop and was likely responsible for the increased spore abundance in the rhizosphere under the tree canopies (Pande and Tarafdar 2004). A similar pattern was observed by Prasad and Mertia (2005) who found higher spore abundance and root colonization under the canopies of three agroforestry tree species that associate with AM fungi (*A. indica, Prosopis cineraria* (L.) Druce, and *Tecomella undulata* (Smith)) and three tree species that associate with AM and ECM fungi (*Acacia tortilis* (Forsk.) Hayne, *A. nilotica* (L.) Willd. Ex Del., and *Eucalyptus camaldulensis* (Dehnh.) compared to region beyond the tree canopies (12–15 m from trees). In addition, there was significant variability in the number of spores and percent root length colonization in the rhizosphere associated with each of the six tree species, but this did not appear to be correlated with the mycorrhizal status of the agroforestry tree species (Prasad and Mertia 2005).

Kumar *et al.* (2007a) recorded the root colonization of inter crops under and outside tree canopy in *kharif* (24 tree-crop combinations) and *rabi* (19 tree-crop combinations) seasons. They reported that AM colonization of crops was significantly higher under the tree canopy for all the 24 tree-crop combinations during the rainy season (*kharif*). It seems that the roots belonging to trees acted as source of inoculum for AM colonization of inter-crops as AM fungi are known to be non-specific in their selection of hosts. Some similar reports are available in literature for example Rajshekhara *et al.* (1989) have reported that colonization and sporulation of AMF in red gram and sunflower increased with decreasing distance from the trees *viz., Dalbergia sissoo, Dendrocalamus strictus, Tectona grandis, Leucaena leucocephala* and *Cassia equisitifolia.* Moreover, *kharif* crops were taken after a fallow of hot summer months, which has been reported to reduce the AM inoculum in field soils by many workers (Barbara and Hetrick 1984). Thus, the observed differences in colonization index of inter-crops during *kharif,* under and outside tree canopy can be explained on basis of availability of AM inoculum. During *rabi,* root colonization of inter-crops under and outside tree canopy showed a wide range of variation in agroforestry systems. Among tested tree-crop combinations, it was significantly higher only in five combinations than outside tree canopy. In general, development of AMF in roots of inter-crops was more under tree canopy as compared to open during initial stages of crop growth. During subsequent period, rate of root colonization of the crop plants was faster in sunny conditions (outside tree canopy) than under shade of trees. This is probably due to differences in light intensity under two conditions. Low light intensities have been reported to reduce AM colonization (Whitbeck 2001; Shukla *et al.,* 2009).

He *et al.* (2003) and Muleta *et al.* (2008) reported significantly higher spore density at the stem base of a tree compared with that in the canopy radius or away from trees radius. Asfaw (2003) also reported that proportion of colonized roots decreases with increasing distance laterally from the tree trunk. AMF survive better when they are close to the host crop on which they develop (Kabir 2005). This is probably because of their obligate biotrophic nature of existence, as close to tree trunks, there are more roots that can colonize by potential propagules and more spores are expected. According to Kumar *et al.* (2007a), in an agroforestry systems tree rhizosphere act as inoculum reservoir. Similarly, Mutabaruka *et al.* (2002) found that tree species (*Melia volkensii* Gurke, *Senna spectabilis* (DC.) Irwin and Barneby, *Gliricidia sepium* (Jacq.) Kunthex Walp, *L. leucocephala* (Lam.) de Wit) in agroforestry systems in Kenya had a variable effect on the abundance of AM fungal spores, with *M. volkensii*, *G. sepium*, and *L. leucocephala* having higher spore populations than monoculture plots. All four tree species were colonized by AM fungi and a significant correlation between levels of tree root colonization and spore abundance was observed (Mutabaruka *et al.*, 2002). In addition, an evaluation of seven different land use systems found that agroforestry sites and forest sites had the lowest AM fungal spore abundance, considerably lower than samples collected from pastures and cropped land (Leal *et al.*, 2009). Overall, incorporating trees in to agricultural systems can have a range of effects on the AM fungal community (Bainard *et al.*, 2010; unpublished). In temperate region, AMF communities are more abundant and diverse in agroforestry systems compared to monocultures. Similarly, in tropical regions, agroforestry systems generally had a positive effect on AM fungi, however, in some cases there was no effect or a negative effect of tree-based inter cropping systems on AM fungi compared to monoculture systems. The variable effect of agroforestry on AM fungi may be a function of the different cultivation techniques, climatic variation, or diverse tree-crop combinations used with in these agroforestry settings. However, it is unclear how this in turn influences the myorrhizal community that associates with the alley crops and whether this change is functionally beneficial to the crops.

Effect of Management Practices on Mycorrhization

Crop rotation: Crop rotation is one of the important management practice being implemented in agriculture/ agro-ecosystem which can affect the mycorrhizal status in the crop field (Vyas 2008). An *et al.* (1990a) studied the most probable number (MPN) of AM propagules in fescue and soybaen crop rotations. They reported six times more AMF spores from fescue plots as compared to continuous soybean cropping. In addition, the MPN/ trap culture method (An *et al.*, 1990b) yielded 13 species in the continuous soybean plots versus 16 species in the fescue. Work with other rotations showed *Glomus* species prevailing in rotation but *Gigaspora* species more numerous in continuous soybeans (An *et al.*, 1993a). The continuous soybean plots had lower species richness and diversity but higher dominance as compared to fescue. Change in crop host can cause some AMF to decline to levels undetectable by trap culture methods, or cause others to increase from previously undetectable levels (Guo *et al.*, 1993). Tall fescue (*Festuca arundinacea*) in rotation decreased populations of *G. macrocarpum*, whereas sorghum–sudangrass increased populations of *G. macrocarpum* (An *et al.*, 1993b; Hendrix *et al.*, 1995). Johnson *et al.* (1991) studied the effect of crop host and local edaphic factors upon the relative abundance of AM spores in fields planted to continuous maize or soybean for 5 years. The AMF that become numerous with continuous monocultures may contribute to the yield declines over time noted for such crops (Johnson *et al.*, 1992). Further, Douds *et al.* (1993 and 1995), observed greater levels of AMF spores and colonization of plants in soil managed with low-input methods versus conventional chemical inputs. Recently, Kumar *et al.* (2007b) studied the effect of crop rotations on inoculum potential in *Albizia procera* based intercropping system. For comparing agroforestry system to sole tree plantation, soil samples were collected from under and outside tree canopies of *A. procera* from three crop rotation treatments *viz.*, black gram + wheat (intercropped with *A. procera*), green gram + mustard (intercropped with *A. procera*) and control (sole *A. procera* plantation). Results indicated that intercropping increased the inoculum potential of AM fungi, under and outside tree canopy. Among tested treatments, maximum inoculum potential was recorded in crop rotation of black gram + wheat, followed by green gram + mustard and pure plantation of *A. procera*. Shukla (2009) studied the effect of intercropping on colonization of trees (*A. procera* and *E. tereticornis*) by AMF in two agroforestry settings. Intercropping significantly increased colonization index and spore counts in rhizosphere of both tree species with inter-crops (black-gram in *kharif* and wheat in *rabi*) as compared to pure plantations of *A. procera* and *E. tereticornis*.

Soil disturbance/ tillage: Since the extra-radical network of AMF mycelium functions as both the nutrient absorbing organ of the mycorrhiza and as inoculum for the colonization of new roots, it follows that soil disturbance from tillage may affect both colonization of plants sown into this soil and mycorrhizae mediated nutrient uptake. Miller *et al.* (1995) examined this phenomenon in field and growth chamber studies. A series of growth chamber experiments showed that maize plants grown in disturbed soil were less colonized by AMF and had lower shoot P and Zn concentrations than plants grown in undisturbed, field collected soil (Evans and Miller 1988; Fairchild and Miller 1988, Fairchild and Miller 1990). No effect of soil disturbance was seen for spinach (*Spinacea oleracea*) and rape (*Brassica napus*), two species which are not colonized by AMF (Evans and Miller 1988). Application of the fungicide benomyl to disturbed and undisturbed plots decreased the effect of disturbance by lessening the P uptake by mycorrhizae in undisturbed soils (Kunishi and Bandel 1991). These experiments suggest that a new seedling is benefited by a previously established AMF hyphal network in the soil (Evans and Miller 1990). In addition, if the availability of P in the soil were high enough to preclude any benefit from mycorrhizae, there should be no effect of soil disturbance on the P concentration of plants. This is what was found (Fairchild and Miller 1990), even though mycorrhiza formation was greater on plants in undisturbed soil. Zinc (Zn) absorption by plants was inhibited by disturbance in this experiment independent of P nutrition, as expected because of inhibition of formation of mycorrhizae from soil disturbance. Results of the growth chamber studies were verified in the field. Maize grown in no-till or ridge tillage management exhibited greater early season P absorption and mycorrhizal colonization than plants grown in moldboard plowed soils (McGonigle *et al.*, 1990; Mc-Gonigle and Miller 1993 and 1996). Although this shows that less P is needed in reduced tillage farm management, something as yet undetermined causes no-tilled maize to lose its early growth advantage and yield no better than plants in tilled soils (Miller *et al.*, 1995). Further, Galvez *et al.* (1995) observed that no tilled soils were more colonized by AMF than those in tilled soil. Tillage affected the distribution of spores and inoculum and affected the whole soil profile populations of spores of different spore type groups (Douds *et al.*, 1995).

Canopy management: Canopy management/ shoot pruning of tree component, a common practice of agroforestry systems which is employed to reduce ill effects of tree shade on growth of under storey crops (Nair 1993). According to Namirembe (1999) canopy management successfully reduced tree-crop competition. Mulching of crops with pruned shoots of leguminous trees is a good substitute for fertilizer application (Salami and Osonubi 2003) which enriches the soil. In India, where majority of the farmers cannot afford the use of inorganic fertilizers due to huge cost or logistic problems of distribution, mycorrhizal inoculation in agroforestry may provide a useful alternative. Tree shoot pruning led to increased fresh root yield and indicate that proper residue or tree pruning management with mycorrhizal inoculation management can ensure stable and high crop yields in the tropical soils even if at different sites. Shukla (2009) reported that pruning significantly reduced mycorrhizal colonization in five important agroforestry tree species (*Anogeissus latifolia, A. pendula, Albizia procera, Dalbergia sissoo* and *Tectona grandis*). Decrease in mycorrhizal colonization due to pruning in *Nardus stricta* L. and *Calluna vulgaris* (L.) Hull has been reported (Hartley and Amos 1999). This can be due to lower glucose content in roots of pruned trees than in the unpruned trees (Peter and Lehmann 2000). Similar results were also obtained by Whitcomb and Stutz (2001), who suggests that shoot pruning reduces tree root biomass and levels of AM colonization. However, it is not known whether tree shoot pruning, and the concomitant reduction of carbon (C) supplied to roots, would slow the spread of AM hyphae in the soil (Ingleby *et al.*, 2007). But, recently de Carvalho (2009) postulated that shoot pruning, did not restrict the mycorrhizal colonization of, or the spread of the mycelium network to, the annual crops. This effect may be of particular importance given that the mycelial network is responsible for the fast colonization of new roots, and thus for a growth stimulation and improved P absorption in young plants (Brundrett and Abbott 1994). However, the rate of spread of the AMF mycelium (1-3 mm day^{-1}) suggests that benefits from the network will develop slowly. But, how tree pruning would affect the C contribution of individual plants to the common mycorrhizal network, and how individual plants subsequently benefit from that network, remains to be determined.

Effect of fertilizers: Application of fertilizers is another important practice of agroforestry which generated several environmental (Sharif *et al.*, 2010) and soil problems (Tilman *et al.*, 2002; Foley *et al.*, 2005). Fertilizer application is also responsible for disturbing below ground microbial community. In an agroforestry system, fertilizers are being applied to the inter-crops not to tree, but tree component must receive these at sub optimal level, which might have improved the AM flora (Kumar *et al.*, 2007b). Sieverding (1991) has reported that small phosphate applications generally may improve AM fungal activities on infertile soils and their population benefits from N application probably due to longer maintenance of photosynthetic leaf area. The key function of AM fungi is the exploration of the soil beyond the range of roots for better plant growth and nutrition (Oehl *et al.*, 2002; van der Heijden *et al.*, 2006; Kapoor *et al.*, 2007). Eight years of fertilization of low nutrient soil caused populations of four species, including *Gigaspora gigantea*, to decline and *Glomus intraradices* to increase (Johnson 1993). This confirmed other work which showed *G. gigantea* more associated with natural or low-input systems than in conventional agriculture (Miller and Jastrow 1992; Douds *et al.*, 1993) and that *G. intraradices* is very tolerant of high nutrient situations (Sylvia and Schenck 1983). Communities from fertilized or unfertilized plots were collected and inoculated onto big bluestem grass (*Andropogon gerardii*) grown in the greenhouse. After 1 month of growth, plants inoculated with the 'unfertilized community' were larger than those inoculated with the 'fertilized community' (Johnson 1993). Microscopic observation of the mycorrhizas suggested AMF from the fertilized community may have been a greater C drain on their host plants relative to AMF from the unfertilized community. The former produced the same proportion of root length with vesicles as those from the unfertilized community, but a lower proportion of root length which had arbuscules, the site of nutrient transfer to the host. Interestingly, other isolates of *G. intraradices* are beneficial to the growth of plants (Graham and Timmer 1985; Hamel *et al.*, 1992). In a bioassay experiment, Kahiluoto *et al.*, (2000) observed that by increasing P supply there was a decrease in the colonization and the effectiveness of mycorrhizal colonization. According to Mohammad *et al.*, (2004), root colonization is inversely related to the P supply. The results are in agreement with many reports which suggest that addition of phosphate fertilizers above threshold levels results in a delay in infection and reduced chlamydospore production by AM fungi (Amijee *et al.*, 1989; Koide and Li 1990; Koide 1991; Thingstrup *et al.*, 1998; Li 2005).

Shukla *et al.* (2011a; unpublished) conducted a green house experiment in sand to investigate the effect of P applications (0, 5, 10, 20, 50 and 100 µg g^{-1}) on growth and AM colonization of two crops (*Phaseolus mungo* Roxb. and *Triticum aestivum*) and seedlings of two multipurpose tree species (*E. tereticornis* and *A. procera*). Significant increase in growth and P uptake was recorded after inoculations with *Acaulospora scrobiculata, Glomus cerebriforme* and *G. intraradices.* Results on effect of P application on mycorrhizal dependency (MD) of studied crop and tree species showed that decrease in MD with increase in P concentrations in non nitrogen fixing species (*T. aestivum* and *E. tereticornis*) was more than nitrogen fixing species (*P. mungo* and *A. procera*). Threshold P concentrations for maximum benefits from the AM symbiosis in above mentioned plant species varied from 5 to 20 µg g^{-1} and corresponding peaks of arbuscules, vesicles, sporocarp formation, colonization index and spore count per 100 g sand were noticed. Thus, the results showed that the recorded plant growth peaks were due to AM colonization of crops and tree rhizosphere. Inoculations with AMF were more important than P application (explaining 14 to 78% variation in plant growth) for *P. mungo, T. aestivum* and *A. procera* (forward selection method), whereas P application was more important for growth in *E. tereticornis.* Therefore, inoculating plants with a suitable AM inoculant could result in a benefit comparable to high P input and lead to a significant saving of inorganic P fertilizer. However, extrapolation of the results to the real conditions of agroforestry systems should be done with precaution because of differences in the substrate used in the present study.

Effect of irrigation: Suppressed crop yields and strong below ground competition for water are regular features of semi-arid agroforestry systems (Prasad and Mertia 2005). The role of the mycorrhizal network in hydraulic redistribution could be of particular importance in agroforestry systems. In literature, several reports are available on improved crop yields under drought conditions due to mycorrhizal colonization (Auge 2001; Al-Karaki *et al.*, 2004; Mendoza *et al.*, 2005; Wu *et al.*, 2006; Kungu *et al.*, 2008) and common mycorrhizal network may further enhance the benefits of agroforestry through vertical niche expansion of AMF (Cavagnaro *et al.*, 2005; Wilson *et al.*, 2009).

These fungi can cause changes in plant water relations and improve drought resistance (Boomsma and Vyn 2008). Under water-logged conditions, AM fungi improve plant growth through nutrient transport (Mendoz *et al.*, 2005), augmentation of root surface and phytohormone production (Bano 1987; Duan *et al.*, 1996). After deeply rooted plants have taken up water from profound soil layers, the activity of neighboring shallow rooted plants could be sustained by nocturnal water efflux coupled to water uptake and transfer by mycorrhizal fungi in superficial soil layers. This process of mycorrhiza mediated hydraulic redistribution has been demonstrated for ectomycorrhizal and arbuscular mycorrhizal systems (Allen 2007; Egerton-Warburton *et al.*, 2007), but its importance for agroforestry still needs experimental proof.

Water is the principal limiting factor for the growth of understory vegetation in agroforestry systems, where trees reduce the availability of water to the intercrops (Odhiambo *et al.*, 2001; Broadhead *et al.*, 2003; Livesley *et al.*, 2004; Everson *et al.*, 2009). To substantiate the agroforestry systems and role of water in growth and mycorrhization in an agroforestry setting, Shukla *et al.* (2011b; unpublished) carried out a green house trial to investigate the effect of soil moisture levels (field capacity [FC], half field capacity [HFC] and double field capacity [DFC]) on growth and AM colonization of *P. mungo*, *T. aestivum*, *E. tereticornis* and *A. procera*. Plant growth and P uptake increased significantly after AM inoculations (*A. scrobiculata*, *G. cerebriforme* and *G. intraradices*) in studied crop/tree seedlings. Efficiency of AM fungi depends upon soil moisture levels and plant species. These were less effective at HFC as compared to FC and DFC. AMF have been reported to improve plant growth over a wide range of soil water contents (Al-Karaki and Al-Raddad 1997; Al-Karaki *et al.*, 1998; Al-Karaki and Clark 1998; Karasawa *et al.*, 2000). Per cent increase in P uptake over control suggested that different AM fungi were effective at different soil moisture levels. They recorded maximum mycorrhization at FC in studied plants. Reduced but significant mycorrhizal activity was recorded at HFC and DFC, which were more or less similar. Reduced AM activity has been reported by several workers under conditions which are either too wet or too dry (Reid and Bowen 1979; Redhead 1975; Bohrer *et al.*, 2004; Escudero and Mendoza 2004; Fougnies *et al.*, 2007). While ranking the importance of two factors studied *viz.*, soil moisture levels and inoculation with AM fungi for their effect on the growth, AM inoculations entered at the first place (explaining 33 to 97% variation in plant growth) for *P. mungo*, *T. aestivum* and *A. procera* (forward selection method), whereas soil moisture levels dominantly explained the growth variation in *E. tereticornis*, except few cases. The second parameter also substantially increased the R^2 values. The results obtained in present study suggest that inoculation of plants with a suitable AM consortium could be more beneficial compared to individual

AM fungi, depending upon the soil moisture content of the fields and plant species.

Effect of light: Agroforestry is the most effective way for the restoration of disturbed land and development of local economy (Peng *et al.*, 2008). At tree crop interface of an agroforestry system, trees and crops compete for sun light (Nair 1993). There are many reports on level of light interception by trees in agroforestry systems under Jhansi conditions (Handa and Rai 2001; Newaj *et al.*,. 2004). Shading in agroforestry coffee (*Coffea arabica* L.) systems appear to have an effect on the AMF spore populations compared to monocultural coffee systems. Muleta *et al.* (2008) investigated the influence of 10 shade tree species on the density of AMF spores in agroforestry coffee systems compared to monocultural coffee systems. They found significantly higher spore densities in agroforestry compared to monocultural systems, and the highest spore populations were found in the top soil layer (0-30cm). In addition, there were variations in the spore abundance among the different shade trees with leguminous tree species (*Albizia* and *Acacia* species) maintaining a higher abundance of spores compared to *Ficus* species, which were not significantly different from monocultural systems. This effect may be attributed to the mycorrhizal associations of the tree species as *Ficus* species are typically less mycorrhizal dependent compared to leguminous tree species, which in turn likely impacts the AMF populations in the rhizosphere (Plenchette *et al.*, 2005). This effect was observed by Muleta *et al.* (2007) as they found higher AMF spores in the rhizosphere of coffee plants under leguminous shade trees compared to non-leguminous shade trees. Cardoso *et al.* (2003a,b) also found that the upper soil layers contained the highest spore densities in agroforestry and monocultural coffee systems. However, monocultural coffee systems had higher densities of spores in the upper soil layers and lower spore densities in the deeper soil layers compared to agroforestry coffee systems. This appeared to be due to the higher abundance of roots in the deeper soil layers in the agroforestry coffee system, which may enable these systems to maintain a higher abundance of AM fungal spores in the deeper soil layers. In contrast to these previous studies, there is evidence that some agroforestry systems may have no effect or in some cases a negative effect on the AM fungal community compared to monoculture systems despite utilizing tree species that associate with AM fungi. For example, Boddington and Dodd (2000) found no significant difference between AM fungal spore density and species richness in a *Peltophorum dasyrachis* Kurz ex Baker and maize (*Zea mays* L.) agroforestry system compared to a maize monoculture. However, they did find a higher concentration of extra-radical mycelium in the agroforestry system. A comparison of two agroforestry systems (*Sesbania macrantha* Welw. Ex E. Phillips and Hutch and *S. sesban* (L.) Merr.) intercropped with maize and a maize monoculture found that AM fungal species diversity did not differ significantly between the two agroforestry systems, but the *S. macrantha* agroforestry system had a lower species diversity compared to the maize monoculture (Jefwa *et al.*, 2006).

Shukla *et al.* (2009) carried out an experiment to investigate the effects of different light intensities (25, 50, 67 and 100% of full sun) on AM colonization and growth of *P. mungo, T. aestivum, E. tereticornis* and *A. procera* of Central India. Their results suggest that plant growth and P uptake were adversely affected by low light intensity. AM inoculations (*Acaulospora scrobiculata, Glomus intraradices* and an unidentified *Glomus* species) increased the plant growth under tested light conditions. Increase in growth of many plant species after AMF inoculation under different light intensities has been reported by several workers (Furlan and Fortin 1977; Paece and Grubb 1982; Son *et al.*, 1988; Harshi *et al.*, 2004). Mycorrhizal efficiency of different AMF varied in narrow range and mycorrization was delayed by lower light intensity. Root colonization and AM spore counts increased with increase in light intensity during successive months after AM inoculation. The stimulatory effect of light on the development of AMF has been shown by Furlan and Fortin (1977). AM fungi obtain their carbon source from the host plant and thus rely on both the photosynthetic ability of the plant and the translocation of photosynthates to the root. Hence, light can strongly affect the development of mycorrhizae. While ranking the importance of two factors studied i.e. light and AM fungi for their effect on the growth and P uptake, AM inoculations came in the first place (explained 50–82% variation) and light substantially increased the values of R^2 in stepwise regression analysis (forward selection). The results suggest that AM inoculation may enhance the growth and P uptake of intercrops under tree shade and the tree canopy management is likely to increase the efficiency of AM inoculants in agroforestry systems. Use of excessive shading (25% of full sun or more) in nurseries may be avoided and photosynthetically active radiation (PAR) lamps may be used to increase growth and colonization index of tree seedlings.

Conclusively, it can be stated that agroforestry systems can be a viable strategy for the preservation of natural resources while ensuring sustainable food production. Agroforestry systems are productive, in addition to containing a greater diversity of species than do simplified monoculture agro-ecosystems. AMF can rehabilitate degraded lands subjected to agroforestry systems and potentially improve physical, chemical and biological soil quality. Several of these functions are linked to the formation of a common mycorrhizal network, which may mediate the transfer of water and nutrients between different plants. The interactions between plants and soil as mediated by AM fungi are of prime importance due to the wide range of functions that these fungi perform. Agroforestry systems potentially maximize the benefits procured by

AMF, which in turn could mitigate negative interactions between trees and annual crops. This positive mycorrhiza-agroforestry feedback, and the common mycorrhizal network which produces it, deserve closer attention.

Acknowledgements

Authors are thankful to Head Department of Botany, Dr HS Gour Central University, Sagar, India. Ashok Shukla gratefully acknowledges University Grants Commission's Dr DS Kothari Post Doctoral Fellowship Scheme for providing financial support.

References

Al-Karaki, G.N. and Al-Raddad, A. (1997). Effects of arbuscular mycorrhizal fungi and drought stress on growth and nutrient uptake of two wheat genotypes differing in drought resistance. *Mycorrhiza,* 7:83–88.

Al-Karaki, G.N., Al-Raddad, A. and Clark, R.B. (1998). Water stress and mycorrhizal isolate effects on growth and nutrient acquisition of wheat. *J. Pl. Nutri.* 21: 891–902.

Al-Karaki, G.N. and Clark, R.B. (1998). Growth, mineral acquisition, and water use by mycorrhizal wheat grown under water stress. *J. Pl. Nutri.* 21:263–276.

Al-Karaki, G.N., McMichael, B. and Zak, J. (2004). Field response of wheat to arbuscular mycorrhizal fungi and drought stress. *Mycorrhiza,* 14:263-269

Allen, M.F. (2007). Mycorrhizal fungi: highways for water and nutrients in arid soils. *Vadose Zone J.,* 6:291–297

Amijee, F., Tinker, P.B. and Stribley, D.P. (1989). The development of endomycorrhizal root systems. VII. A detailed study of effects of soil phosphorus on colonization. *New Phytol.,* 111:435-446.

An, Z.Q., Grove, J.H., Hendrix, J.W., Hershman, D.E. and Henson, G.T. (1990a). Vertical distribution of endogonaceous mycorrhizal fungi associated with soybean as affected by soil fumigation. *Soil Biol. Biochem.,* 22:715-719.

An, Z.Q., Guo, B.Z. and Hendrix, J.W. (1993b). Mycorrhizal pathogen of tobacco: cropping history and current crop effects on the mycorrhizal fungal community. *Crop Protection,* 12:527-531.

An, Z.Q., Hendrix, J.W., Hershman, D.E., Ferriss, R.S. and Henson, G.T. (1993a). The influence of crop rotation and soil fumigation on a mycorrhizal fungal community associated with soybean. *Mycorrhiza,* 3:171-182.

An, Z.Q., Hendrix, J.W., Hershman, D.E. and Henson, G.T. (1990b). Evaluation of the most probable number (MPN) and wet sieving methods for determining soil borne populations of Endogonaceous mycorrhizal fungi. *Mycologia,* 82:576-581.

Andrade, S.A.L., Abreu, C.A., Abreu, M.F. and Silveira,. AP.D. (2003). Interacao de chumbo, da saturac¸ao por bases do solo e de micorriza arbuscular no crescimento e nutricao mineral da soja. *R. Bras. Ci. Solo.,* 27:945-954.

Asfaw, Z. (2003). *Tree species diversity, topsoil conditions and arbuscular mycorrhizal association in the Sidama traditional agroforestry land use, southern Ethiopia.* Doctoral dissertation, SLU, Acta Universitatis Agriculturae Sueciae, *Silvestria* vol 263.

Auge, R.M. (2001). Water relations, drought and vesicular arbuscular mycorrhizal symbiosis. *Mycorrhiza,* 11:3-42.

Bainard, L.D., Klironomos, J.N. and Gordon, A.M. (2010). Arbuscular mycorrhizal fungi in tree-based intercropping systems: A review of their abundance and diversity. *Pedobio-Int. J. Soil Biol., doi*:10.1016/j.pedobi.2010.11.001 (In press).

Bano, A. (1987). Role of vesicular arbuscular mycorrhiza in hydrophytic condition in relation to phytohormone. In: *Mycorrhiza Round table: proceeding of a workshop on mycorrhizae,* New Delhi, India, 13-15 March 1987.

Barbara, A. and Hetrick, D. (1984). Ecology of VA Mycorrhizal Fungi. In: *VA Mycorrhiza,* CRC Press, Inc. Boca Raton, Florida, pp 35-55.

Boddington, C.L. and Dodd, J.C. (2000). The effect of agricultural practices on the development of indigenous arbuscular mycorrhizal fungi. I. Field studies in an Indonesian ultisol. *Plant Soil,* 218:137-144.

Bohrer, K.E., Friese, C.F. and Amon, J.P. (2004). Seasonal dynamics of arbuscular mycorrhizal fungi in differing wetland habitats. *Mycorrhiza,* 14:329-337.

Bolan, N.S. (1991). A critical review on the role of mycorrhizal fungi in the uptake of phosphorus by plants. *Plant Soil,* 134:189–207.

Boomsma, C.R. and Vyn, T.J. (2008). Maize drought tolerance: Potential improvements through arbuscular mycorrhizal symbiosis? *Field Crops Res.,* 108:14-31.

Broadhead, J.S., Ong, C.K. and Black, C.R. (2003). Tree phenology and water availability in semi-arid agroforestry systems. *For. Ecol. Manage.,* 180:61-73.

Brundrett, M.C. and Abbott, L.K. (1994). Mycorrhizal fungus propagules in the jarrah forest. *New Phytol.,* 127:539-546.

Cardoso, I.M., Boddington, C., Janssen, B.H., Oenema, O. and Kuyper, T.W. (2003a). Distribution of mycorrhizal fungal spores in soils under agroforestry and monocultural coffee systems in Brazil. *Agroforest Syst.,* 58:33–43.

Cardoso, I.M., Janssen, B.H., Oenema, O. and Kuyper, T.W. (2003b). Phosphorus pools in Oxisols under shaded and unshaded coffee systems on farmers' fields in Brazil. *Agroforest Syst.,* 58:55-64.

Cardoso, I.M. and Kuyper, T.W. (2006). Mycorrhizas and tropical soil fertility. *Agri. Ecosyst. Environ.,* 116 (2006) 72–84.

Cavagnaro, T.R., Smith, F.A., Smith, S.E. and Jakobsen, I. (2005). Functional diversity in arbuscular mycorrhizas: exploitation of soil patches with different phosphate enrichment differs among fungal species. *Plant Cell Environ.,* 28:642–650.

Chifflot, V., Rivest, D., Olivier, A., Cogliastro, A. and Khasa, D. (2009). Molecular analysis of arbuscular mycorrhizal community structure and spores distribution in tree-based intercropping and forest systems. *Agric. Ecosyst. Environ.,* 131:32-39.

Chirwa, P.W., Ong, C.K., Maghembe, J. and Black, C. R. (2007). Soil water dynamics in cropping systems containing *Gliricidia sepium*, pigeonpea and maize in southern Malawi. *Agroforest Syst.*, 69:29-43.

de Carvalho, A.M.X., de Castro Tavares, R., Cardoso, I.M. and Kuyper, T.W. (2009). Mycorrhizal Associations in Agroforestry Systems. In: *Soil Biology and Agriculture in the Tropics*, (Ed.) P. Dion, Soil Biology 21, Springer-Verlag, Berlin Heidelberg: 185-208.

Dodd, J.C., Arias, I., Koomen, I. and Hayman, D.S. (1990). The management of populations of vesicular-mycorrhizal fungi in acid-infertile soils of a savanna ecosystem. I. The effect of pre-cropping and inoculation with VAM-fungi on plant growth and nutrition in the field. *Plant Soil*, 122:229–240.

Douds, D.D., Galvez, L., Jank, R.R. and Wagoner, P. (1995). Effect of tillage and farming system upon populations and distribution of vesicular-arbuscular mycorrhizal fungi. *Agric. Ecosys. Environ.*, 52:111-118.

Douds, D.D., Janke, R.R. and Peters, S.E. (1993). VAM fungus spore populations and colonization of roots of maize and soybean under conventional and low-input sustainable agriculture. *Agric. Ecosys. Environ.*, 43:325-335.

Douds, D.D. and Millner, P.D. (1999). Biodiversity of arbuscular mycorrhizal fungi in agroecosystems. *Agric. Ecosyst. Environ.*, 74:77-93.

Duan, X., Neuman, D.S. and Reiber, J.M. (1996). Mycorrhizal influence on hydraulic and hormonal factors implicated in the control of stomatal conductance during drought. *J. Exp. Bot.*, 47:1541-1550,

Dwivedi, O.P., Yadav, R.K., Vyas, D. and Vyas, K.M. (2004). Role of potassium in the occurrence of vesicular arbuscular mycorrhizal spores in the rhizosphere soils of *Lantana* species. In: *Microbiology and Biotechnology for Sustainable Development*, (Ed.) P.C. Jain, BS Publishers, New Delhi, : 248-253.

Egerton-Warburton, L.M., Querejeta, J.I. and Allen, M.F. (2007). Common mycorrhizal networks provide a potential pathway for the transfer of hydraulically lifted water between plants. *J. Exp. Bot.*, 58:1473–1483.

Elsen, A., Gervacio, D., Swennen, R. and De Waele, D. (2008). AMF-induced biocontrol against plant parasitic nematodes in Musa sp.: a systemic effect. *Mycorrhiza*, 18: 251–256.

Escudero, V. and Mendoza, R. (2004). Seasonal variation of arbuscular mycorrhizal fungi in temperate grasslands along a wide hydrologic gradient. *Mycorrhiza*, 15:291-299.

Evans, D.G. and Miller, M.H. (1988). Vesicular-arbuscular mycorrhizas and the soil-disturbance-induced reduction of nutrient absorption in maize I. Causal relations. *New Phytol.*, 110:67-74.

Evans, D.G. and Miller, M.H. (1990). The role of the external mycelial network in the effect of soil disturbance upon vesicular-arbuscular mycorrhizal colonization of maize. *New Phytol.*, 114:65-72.

Everson, C.S., Everson, T.M. and van Niekerk, W. (2009). Soil water competition in a temperate hedgerow agroforestry system in South Africa. *Agroforest Syst.*, 75:211-221.

Ewel, J.J. (1999). Natural systems as a model for the design of sustainable systems of land use. *Agroforest Syst.*, 45:1–21.

Fairchild, G.L. and Miller, M.H. (1988). Vesicular-arbuscular mycorrhizas and the soil-disturbance-induced reduction of nutrient absorption in maize II. Development of the effect. *New Phytol.*, 110:75-84.

Fairchild, G.L. and Miller, M.H. (1990). Vesicular-arbuscular mycorrhizas and the soil-disturbance-induced reduction of nutrient absorption in maize III. Influence of phosphorus amendments to soil. *New Phytol.*, 114:641-650.

Farrell, J.G. and Altieri, M.A. (2002). Sistemas agroflorestais. In: *Agroecologia: bases cientificas para uma agricultura sustentavel*, (Ed.) M.A. Altieri, Agropecuaria, Guaýba:413–440

Foley, J.A., DeFries, R., Asner, G.P., Barford, C., Bonan, G., Carpenter, S.R., Chapin, F.S., Coe, M.T., Daily, G.C., Gibbs, H.K., Helkowski, J.H., Holloway, T., Howard, E.A.,

Kucharik, C.J., Monfreda, C., Patz, J.A., Prentice, I.C., Ramankutty, N. and Snyder, P.K. (2005). Global consequences of land use. *Science*, 309:570–574.

Fougnies, L., Renciot, S., Muller, F., Plenchette, C., Prin, Y., de Faria, S.M., Bouvet, J.M., Sylla, S.N.D., Dreyfus, B. and Ba, A.M. (2007). Arbuscular mycorrhizal colonization and nodulation improve flooding tolerance in *Pterocarpus officinalis* Jacq. Seedlings. *Mycorrhiza*, 17:159–166.

Frossard, E. (2002). Phosphorus budget and phosphorus availability in soils under organic and conventional farming. *Nutr. Cycl. Agroecosyst.*, 62:25–35.

Furlan, V. and Fortin, J.A. (1977). Effects of light intensity on the formation of vesicular-arbuscular endomycorrhizas on *Allium cepa* by *Gigaspora calospora*. *New Phytol.*, 79:333-340.

Galvez, L., Douds, D.D., Wagoner, P., Longnecker, L.R., Drinkwater, L. and Janke, R.R. (1995). An overwintering cover crop increases inoculum of VAM fungi in agricultural soil. *Am. J. Alternative Agric.*, 10:152-156.

Garg, N., Geetanjali, K. and Amandeep, K. (2006). Arbuscular mycorrhiza: nutritional aspects. *Arch. Agron. Soil Sci.*, 52:593–606.

Garmendia, I., Goicoechea, N. and Aguirreolea, J. (2004). Effectiveness of three *Glomus* species in protecting pepper (*capsicum annum* L.) against *verticilium* wilt. *Biol. Control*, 31:296-305.

Garrido, J.M.G. (2009). Arbuscular mycorrhizae as defense against pathogens. In: *Defensive Mutualism in Microbial Symbiosis*. (Eds.) J.F. Jr., White and M.S. Torres, CRC Press, Boca Raton, FL, 183–198.

Gianinazzi, S., Gollotte, A., Binet, M.N., van Tuinen, D., Redecker, D. and Wipf, D. (2010). Agroecology: the key role of arbuscular mycorrhizas in ecosystem services. *Mycorrhiza*, 20:519-530.

Gianinazzi, S., Trouvelot, A. and Gianinazzi-Pearson, V. (1989). Conceptual approaches for the rational use of VA endomycorrhizae in agriculture: possibilities and limitations. *Agric. Ecosyst. Environ.*, 29:153–161.

Graham, J.H. and Timmer, L.W. (1985). Rock phosphate as a source of phosphorus for vesicular-arbuscular mycorrhizal development and growth of citrus in a soilless medium. *J. Am. Soc. Hortic. Sci.*, 110:489-492.

Guo, B.Z., An, Z.Q., Hendrix, J.W. and Dougherty, C.T. (1993). Influence of a change from tall fescue to pearl millet or crabgrass on the mycorrhizal fungal community. *Soil Sci.*, 155:393-405,

Hamel, C., Furlan, V. and Smith, D.L. (1992). Mycorrhizal effects on inter specific plant competition and nitrogen transfer in legume–grass mixtures. *Crop Sci.*, 32:991-996.

Handa, A.K. and Rai, P. (2001). *Annual report of National Research Centre for Agroforestry*. Jhansi, India : 12–14.

Harshi, K.G., Singhakumara, B.M.P. and Ashton, M.S. (2004). Effects of light and fertilization on arbuscular mycorrhizal colonization and growth of tropical rain-forest *Syzygium* tree seedlings. *J. Trop. Ecol.*, 20:525-534.

Hartley, S.E. and Amos, L. (1999). Competitive interactions between *Nardus stricta* L. and *Calluna vulgaris* (L.) Hull: the effect of fertilizer and defoliation on above- and below-ground performance. *J. Ecol.*, 87:330-340.

Haselwandter, K. and Bowen, G.D. (1996). Mycorrhizal relations in trees for agroforestry and land rehabilitation. *For. Ecol. Manage.*, 81:1–17.

He, X.H., Critchley, C. and Bledsoe, C. (2003). Nitrogen transfer within and between plants through common mycorrhizal networks (CMNs). *Crit. Rev. Plant Sci.*, 22: 531–567.

Hendrix, J.W., Guo, B.Z. and An, Z.Q. (1995). Divergence of mycorrhizal communities in crop production systems. *Plant Soil*, 170:131-140.

Hijri, I., Sykorova, Z., Oehl, F., Ineichen, K., Mader, P., Wiemken, A. and Redecker, D. (2006). Communities of arbuscular mycorrhizal fungi in arable soils are not necessarily low in diversity. *Mol. Ecol.*, 15:2277-2289.

Ingleby, K., Wilson, E.J., Munro, E.R.C. and Cavers, S. (2007). Mycorrhizas in agroforestry: spread and sharing of arbuscular mycorrhizal fungi between trees and crops: complementary use of molecular and microscopic approaches. *Plant Soil*, 294:125–136.

Jansa, J., Mozafar, A., Anken, T., Ruh, R., Sanders, I.R. and Frossard, E. (2002). Diversity and structure of AMF communities as affected by tillage in a temperate soil. *Mycorrhiza*, 12:225–234.

Jansa, J., Mozafar, A., Kuhn, G., Anken, T., Ruh, R., Sanders, I.R. and Frossard, E. (2003). Soil tillage affects the community structure of mycorrhizal fungi in maize roots. *Ecol. Appl.*, 13:1164-1176.

Jefwa, J.M., Sinclair, R. and Maghembe, J.A. (2006). Diversity of glomale mycorrhizal fungi in maize/sesbania intercrops and maize monocrop systems in southern Malawi. *Agroforest Syst.*, 67:107-114.

Johnson, N.C. (1993). Can fertilization of soil select less mutualistic mycorrhizae. *Ecol. Appl.*, 3:749-757.

Johnson, N.C., Copeland, P.J., Crookston, R.K. and Pfleger, F.L. (1992). Mycorrhizae: possible explanation for yield decline with continuous corn and soybean. *Agron. J.*, 84: 387-390.

Johnson, N.C., Pfleger, F.L., Crookston, R.K. and Simmons, S.R. (1991). Vesicular-arbuscular mycorrhizas respond to corn and soybean cropping history. *New Phytol.*, 117:657–664.

Jose, S., Gillespie, A.R. and Pallardy, S.G. (2004). Interspecific interactions in temperate agroforestry. *Agroforest Syst.*, 61:237–255.

Jose, S., Gillespie, A.R., Seifert, J.R. and Biehle, D.J. (2000). Defining competition vectors in a temperate alley cropping system in the Midwestern USA. 2. Competition for water. *Agroforest Syst.*, 48:41–59.

Jumpponen, A., Trowbridge, J., Mandyam, K. and Johnson, L. (2005). Nitrogen enrichment causes minimal changes in arbuscular mycorrhizal colonization but shifts community composition evidence from rDNA data. *Biol. Fertil. Soils*, 41:217–224.

Kabir, Z. (2005). Tillage or no-tillage: impact on mycorrhizae. *Can. J. Plant Sci.*, 85:23–29.

Kahiluoto, H., Ketoja, E. and Vestnerg, M. (2000). Promotion of utilization of arbuscular mycorrhizal through reduced P fertilization 1. Bioassays in a growth chamber. *Plant Soil*, 227:191-206.

Kapoor, R., Chaudhary, V. and Bhatnagar, A.K. (2007). Effects of arbuscular mycorrhiza and phosphorus application on artemisinin concentration in *Artemisia annua* L. *Mycorrhiza*, 17:581–587.

Karasawa, T., Arihara, J. and Kasahara, Y. (2000). Effects of previous crops on arbuscular mycorrhizal formation and growth of maize under various soil moisture conditions. *Soil Sci. Plant Nutri.*, 46:53–60.

Khaliq, A. and Sanders, F.E. (2000). Effect of vesicular arbuscular mycorrhizal inoculation on the yield and phosphorus uptake of field-grown barley. *Soil Biol. Biochem.*, 32:1691-1696.

Koide, R.T. (1991). Nutrient supply, nutrient demand and plant response to mycorrhizal infection. *New Phytol.*, 117:365-386.

Koide, R.T. and Li, M. (1990). On host regulation of the vesicular-arbuscular mycorrhizal symbiosis. *New Phytol.*, 114:59-65.

Kumar, A., Shukla, A., Hashmi, S. and Tewari, R.K. (2007a). Effect of trees on colonization of intercrops by vesicular arbuscular mycorrhizae in agroforestry systems. *Ind. J. Agric. Sci.*, 77(5):291-298.

Kumar, A., Hashmi, S., Shukla, A. and Chaturvedi, O.P. (2007b). Effect of intercropping on inoculum potential of vesicular arbuscular mycorrhizal fungi in agroforestry systems. *Ann. Arid Zone*, 46:31-36.

Kumar, A. and Shukla, A. (2009). Role of vesicular arbuscular mycorrhizal fungi in establishment of Agroforestry systems on marginal lands. In: *Agroforestry: Natural Resource Sustainability*, (Eds.) O.P. Chaturvedi and A.Venkatesh, Livelihood and Climate Moderation. Satish Serial Publishing House, New Delhi, India: 165-183.

Kumar, V. and Singh, G. (2010). Efficacy of fly ash based biofertlizers vs perfected chemical fertilizers in wheat (*Triticum aestivum*). *Mid. East J. Sci. Res.*, 6(2):185-188.

Kungu, J.B., Lasco, R.D., Cruz, L.U.D., Cruz, R.E.D. and Husain, T. (2008). Effect of vesicular arbuscular mycorrhiza (VAM) fungi inoculation on coppicing ability and drought resistance of *Senna spectabilis*. *Pak. J. Bot.*, 40:2217-2224.

Kunishi, H.M. and Bandel, V.A. (1991). Microbe enhanced P uptake by corn under no-till and conventional till. In: *The Rhizosphere and Plant Growth*, (Eds.) D.L. Keister and P.B. Cregan, Kluwer Academic Publishers, The Netherlands: 368.

Lacombe, S., Bradley, R.L., Hamel, C. and Beaulieu, C. (2009). Do tree-based intercropping systems increase the diversity and stability of soil microbial communities? *Agric. Ecosyst. Environ.*, 131:25-31.

Leake, J., Johnson, D., Donnelly, D., Muche, G., Boddy, L. and Read, D. (2004). Networks of power and influence: the role of mycorrhizal mycelium in controlling plant communities and agroecosystem functioning. *Can. J. Bot.*, 82:1016–1045.

Leal, P.L., Sturmer, S.L. and Siqueira, J.O. (2009). Occurrence and diversity of arbuscular mycorrhizal fungi in trap cultures from soils under different land use systems in the Amazon, Brazil. *Braz. J. Microbiol.*, 40:111-121.

Li, H. (2005). *Roles of Mycorrhizal Symbiosis in Growth and Phosphorus Nutrition of Wheat in a Highly Calcareous Soil*. Ph D Thesis, University of Adelaide, Australia.

Livesley, S.J., Gregory, P.J. and Buresh, R.J. (2004). Competition in tree row agroforestry systems. 3. Soil water distribution and dynamics. *Plant Soil*, 264:129-139.

Mahajan, S., Kanwar, S.S. and Sharma, P. (2007). Long-term effect of mineral fertilizers and amendments on microbial dynamics in an alfisol of Western Himalayas. *Ind. J. Micro.*, 47:86-89.

McGonigle, T.P., Evans, D.G. and Miller, M.H. (1990). Effect of degree of soil disturbance on mycorrhizal colonization and phosphorus absorption by maize in growth chamber and field experiments. *New Phytol.*, 116:629-636.

McGonigle, T.P. and Miller, M.H. (1993). Does the more effective mycorrhizal symbiosis found with reduced tillage increase corn yield? *1993 Annual Report*, Land Resource Science. University of Guelph, Guelph, Ont, Canada: 69-70

McGonigle, T.P. and Miller, M.H. (1996). Mycorrhizae, phosphorus absorption and yield of maize in response to tillage. *Soil Sci. Soc. Am. J.*, 60:1856-1861.

Mendoza, R., Escudero, V. and Garcia, I. (2005). Plant growth, nutrient acquisition and mycorrhizal symbioses of a waterlogging tolerant legume (*Lotus glaber* Mill.) in a saline-sodic soil. *Plant Soil*, 275:305-315.

Miller, R.M. and Jastrow, J.D. (1992). The role of mycorrhizal fungi in soil conservation. In: *Mycorrhizae in Sustainable Agriculture*, (Eds.) G.J. Bethlenfalvay and R.G. Linderman, Agron Soc Am Special Publication No 54, Madison, WI: 24-44.

Miller, M.H., McGonigle, T.P. and Addy, H.D. (1995). Functional ecology of vesicular-arbuscular mycorrhizas as influenced by phosphate fertilization and tillage in an agricultural ecosystem. *Crit. Rev. Biotechnol.*, 15:241-255.

Mishra, M.K., Dubey, A., Singh, P.K. and Vyas, D. (2008). Seasonal distribution of VAM fungi in Vindhyan soil of Madhya Pradesh, India. *Ind. Phytopathol.*, 61(6):360-362.

Mohammad, M.J., Mitra, B. and Khan, A.G. (2004). Effects of sheared root inoculum of *Glomus intraradices* on wheat grown at different phosphorus levels in the field. *Agric. Ecosyst. Environ.*, 103:245-249.

Mridha, M.A.U. and Dhar, P.P. (2007). Biodiversity of arbuscular mycorrhizal colonization and spore population in different agroforestry trees and crop species growing in Dinajpur, Bangladesh. *J. Forestry Res.*, 18: 91-96.

Muleta, D., Assefa, F., Nemomissa, S. and Granhall, U. (2007). Composition of coffee shade tree species and density of indigenous arbuscular mycorrhizal fungi (AMF) spores in Bonga natural coffee forest, southwestern Ethiopia. *Forest. Ecol. Manage.*, 241:145-154.

Muleta, D., Assefa, F., Nemomissa, S. and Granhall, U. (2008). Distribution of arbuscular mycorrhizal fungi spores in soils of smallholder agroforestry and monocultural coffee systems in southwestern Ethiopia. *Biol. Fertil. Soils*, 44:653–659.

Mutabaruka, R., Mutabaruka, C. and Fernandez, I. (2002). Research note: diversity of arbuscular mycorrhizal fungi associated to tree species in semiarid areas of Machakos, Kenya. *Arid Land Res. Manage.*, 16:385–390.

Mutuo, P.K., Cadisch, G., Albrecht, A., Palm, C.A. and Verchot, L. (2005). Potential of agroforestry for carbon sequestration and mitigation of greenhouse gas emissions from soils in the tropics. *Nutr. Cycl. Agroecosyst.*, 71:43–54.

Nair, P.K.R. (1993). *An Introduction to Agroforestry*, Kluwer Academic Publishers, Dordrecht.

Namirembe, S. (1999). *Tree shoot pruning to control competition for below-ground resources in agroforestry*. Ph.D. Thesis, University of Wales, Bangor, UK.

Nasim, G. (2005). The role of symbiotic soil fungi in controlling roadside erosion and the establishment of plant communities. *Caderno de Pesquisa Serie Biologia*, 17:119-136.

Newaj, R., Shanker, A.K. and Handa, A.K. (2004). *Annual report of National Research Centre for Agroforestry*, Jhansi, India, pp 5–7

Newsham, K.K., Fitter, A.H. and Watkinson, A.R. (1995). Arbuscular mycorrhiza protect an annual grass from root pathogenic fungi in the field. *J. Ecol.*, 83:991-1000.

Odhiambo, H.O., Ong, C.K., Deans, J.D., Wilson, J., Khan, A.A.H. and Sprent, J.I. (2001). Roots, soil water and crop yield: tree crop interactions in a semi-arid agroforestry system in Kenya. *Plant Soil*, 235:221-233.

Oehl, F., Oberson, A., Tagmann, H.U., Besson, J.M., Dubois, D., Mader, P., Roth, H.R. and

Oehl, F., Sieverding ,E., Ineichen, K., Mader, P., Boller, T. and Wiemken, A. (2003). Impact of land use intensity on the species diversity of arbuscular mycorrhizal fungi in agroecosystems of central Europe. *Appl. Environ. Microbiol.*, 69:2816–2824.

Oehl, F., Sieverding, E., Mader, P., Dubois, D., Ineichen, K., Boller, T. and Wiemken, A. (2004). Impact of long-term conventional and organic farming on the diversity of arbuscular mycorrhizal fungi. *Oecologia*, 138:574-583.

Paece, W.J.H. and Grubb, P.J. (1982). Interaction of light and mineral nutrient supply in the growth of *Impatiens parviflora*. *New Phytol.*, 90:361-362.

Pande, M. and Tarafdar, J.C. (2004). Arbuscular mycorrhizal fungal diversity in neem based agroforestry systems in Rajasthan. *App. Soil Eco.*, 26:233-241.

Peng, X.B., Cai, J., Jiang, Z.M., Zhang, Y.Y. and Zhang, S.X. (2008). Light competition and productivity of agroforestry system in loess area of Weibei in Shaanxi. *Ying Yong Sheng Tai Xue Bao*, 19:2414-9.

Peter, I. and Lehmann, J. (2000). Pruning effects on root distribution and nutrient dynamics in an acacia hedgerow planting in northern Kenya. *Agro. Sys.*, 50:59-75.

Pimentel, D., Hepperly, P., Hanson, J., Douds, D. and Seidel, R. (2005). Environmental, energetic, and economic comparisons of organic and conventional farming systems. *Bioscience*, 55:573–582.

Plenchette, C., Clermont-Dauphin, C., Meynard, J.M. and Fortin, J.A. (2005). Managing arbuscular mycorrhizal fungi in cropping systems. *Can. J. Plant Sci.*, 85:31-40.

Pozo, M.J. and Azcon-Aguilar, C. (2007). Unraveling mycorrhiza-induced resistance. *Curr. Opin .Plant Biol.*, 10:393–398.

Prasad, R. and Mertia, R.S. (2005). Dehydrogenase activity and VAM fungi in tree-rhizosphere of agroforestry systems in Indian arid zone. *Agroforestry Sys.*, 63: 219–223.

Rajshekhara, E., Sreenivasa, M.N. and Bhat, R.S. (1989). Effect of cropping system on vesicular- arbuscular mycorrhizal development in red gram and sunflower. *Kar. J. Agric. Sci.*, 2:231-3.

Redhead, J.F. (1975). Endotrophic mycorrhizas in Nigeria: Some aspects of the ecology of endotrophic mycorrhizal association of *Khaya grandiflora* C.D.D. In: *Endomycorrhizas*, (Eds.) F.E. Sanders, B. Mosse and P.B. Tinker, Academic Press, London: 447-459.

Reid, C.P.P. and Bowen, C.D. (1979). Effects of soil moisture on VA mycorrhiza formation and root development in *Medicago*. In: *The soil root interface*, (Eds.) J.L. Harley and R.S. Russell, Academic Press, London: 211.

Rillig, M.C. and Mummey, D.L. (2006). Mycorrhizas and soil structure. *New Phytol.*, 171:41–53.

Rillig, M.C., Wright, S.F. and Eviner, V.T. (2002). The role of arbuscular mycorrhizal fungi and glomalin in soil aggregation: comparing effects of five plant species. *Plant Soil*, 238:325–333.

Rillig, M.C., Wright, S.F., Nichols, K.A., Schmidt, W.F. and Torn, M.S. (2001). Large contribution of arbuscular mycorrhizal fungi to soil carbon pools in tropical forest soils. *Plant Soil*, 233:167–177.

Rufyikiri, G.S., Declerck, S., Dufey, J.E. and Delvaux, B. (2000). Arbuscular mycorrhizal fungi might alleviate aluminium toxicity in banana plants. *New Phytol.*, 148: 343–352.

Salami, A.O. and Osonubi, O. (2003). Influence of mycorrhizal inoculation and different pruning regimes on fresh root yield of alley and sole cropped cassava (*Manihot esculenta* Crantz) in Nigeria. *Arch. Agro. Soil Sci.*, 49:317-323.

Sharif, N.M., Rubina, K. and Burni, T. (2010) Occurrence and distribution of arbuscular mycorrhizal fungi in wheat and maize crops of malakand division of north west frontier province. *Pak. J. Bot.*, 42:1301-1312.

Shukla, A. (2009). *Effect of Tree Introduction in Agricultural Fields on Vesicular Arbuscular Mycorrhizal Status of Component Crops of Agroforestry Systems.* Ph. D. Thesis, Bundelkhand University, Jhansi, India

Shukla, A., Kumar, A., Jha, A., Ajit and Rao, D.V.K.N. (2011a). The effects of phosphorus application on growth and arbuscular mycorrhizal colonization of crops and tree seedlings. *Biol. Fertil. Soil*, (In Press)

Shukla, A., Kumar, A., Jha A. and Tripathi, V.D. (2011b). The effect of soil moisture on growth and arbuscular mycorrhizal colonization of crops and tree seedlings in alfisol. *Ind. Phytopathol.*, (In Press)

Shukla, A., Kumar, A., Jha, A., Chaturvedi, O.P., Prasad, R. and Ajit (2009). Effects of shade on arbuscular mycorrhizal colonization and growth of crops and tree seedlings in Central India. *Agroforest. Sys.*, 76:95-109.

Sieverding, E. (1991). *Vesicular Arbuscular Mycorrhiza Management in Tropical Agrosystems.* GTZ , Rossdorf, Germany.

Sieverding, E. and Leihner, D.E. (1984). Influence of crop rotation and intercropping of cassava with legumes on VA mycorrhizal symbiosis of cassava. *Plant Soil*, 80: 143–146.

Simard, S.W. and Durall, D.M. (2004). Mycorrhizal networks: a review of their extent, function, and importance. *Can. J. Bot.*, 82:1140–1165.

Singh, P.K., Mishra, M. and Vyas, D. (2010a). Effect of root exudates of mycorrhizal tomato plants on microconidia germination of *Fusarium oxysporum f. sp. lycopersici. Arch. Phytopathol. Plant Prot.*, 43:1495-1503.

Singh, P.K., Mishra, M. and Vyas, D. (2010b). Interaction of vesicular arbuscular mycorrhizal fungi with *Fusarium* wilt and growth of the tomato *Lycopersicon esculentum* (Mill.). *Ind Phytopathol.*, 63:30-34.

Singh, P.K. and Vyas, D. (2009). Biocontrol of plant disease and sustainable agriculture. *Proc. Nat .Acad. Sci., India, Sec B*, 79:110-128.

Smith, S.E. and Read, D.J. (2008). *Mycorrhizal symbiosis*. Academic Press Inc, San Diego.

Son, C.L., Smith, F.A. and Smith, S.E. (1988). Effect of light intensity on root growth mycorrhizal infection and phosphorus uptake in onions (*Allium cepa* L.). *Plant Soil*, 111:183-186.

Soni, A., Vyas, D. and Vyas, K.M. (2003). Effect of VAM fungi on growth and productivity of soybean (*Glycine max*). *J. Basic Appl. Mycol.*, 2:31-36.

Souza, H.N. (2006). *Sistematizacao da experiencia participativa com sistemas agroflorestais: rumoa sustentabilidade da agricultura familiar na Zona da Mata mineira*. Magister Scientiae dissertation, Universidade Federal de Vicosa, Vicosa

Sylvia, D.M. and Schenck, N.C. (1983). Application of superphosphate to mycorrhizal plants stimulates sporulation of phosphorus-tolerant vesicular-arbuscular mycorrhizal fungi. *New Phytol.*, 95:655-661.

Tavares, R.C. (2007). *Efeito da inoculacao com fungo micorrýzico arbuscular e da adubacao organica no desenvolvimento de mudas de sabia (Mimosa caesalpiniaefolia Benth.)*, sob estresse salino. Magister Scientiae dissertation, Universidade Federal do Ceara, Fortaleza

Thevathasan, N.V. and Gordon, A.M. (2004). Ecology of tree intercropping systems in the North temperate region: experiences from southern Ontario, Canada. *Agroforest. Syst.*, 61:257–268.

Thingstrup, I., Rubaek, G., Sibbesen, E. and Jakobsen, I. (1998). Flax (*Linum usitatissimum* L.) depends on arbuscular mycorrhizal fungi for growth and P uptake at intermediate but not high soil P levels in the field. *Plant Soil*, 203:37-46.

Tilman, D. (1998). The greening of the green revolution. *Nature*, 396:211–212.

Tilman, D., Cassman, K.G., Matson, P.A., Naylor, R. and Polasky, S. (2002). Agricultural sustainability and intensive production practices. *Nature*, 418: 671–677.

Troeh, Z.I. and Loynachan, T.E. (2003). Endomycorrhizal fungal survival in continuous corn, soybean, and fallow. *Agron. J.*, 95:224-230.

van der Heijden, M.G.A., Streitwolf-Engel, R., Riedl, R., Siegrist, S., Neudecker, A., Ineichen, K., Boller, T., Wiemken, A. and Sanders, I.R. (2006). The mycorrhizal contribution to plant productivity, plant nutrition and soil structure in experimental grassland. *New Phytol.*, 172:739–752.

Van Noordwijk, M. and Ong, C.K. (1999). Can the ecosystem mimic hypotheses be applied to farms in African savannahs? *Agroforest. Syst.*, 45:131–158.

Vance, C.P. (2001). Symbiotic nitrogen fixation and phosphorus acquisition. Plant nutrition in a world of declining renewable resources. *Plant Physiol.*, 127:390–397.

Vyas, D. (2005). Diversity and distribution of VA mycorrhiza in thirteen wheat cultivars. In: *Emerging Trends in Mycology Plant Pathology and Microbial Biotechnology*, (Eds.) G. Bagyanarayana, B. Bhadraiah and I.K. Kumar, BS Publications, Hyderabad: 49-66.

Vyas, D. (2008). Vesicular arbuscular mycorrhizal status in wheat cultivar 306 under different phosphorus concentration. *Ind. J. Agrofor.*, 10:62-64.

Vyas, D., Dubey, A., Singh, P.K., Mishra, M.K., Soni, A. and Soni, P. (2006). VA mycorrhizal fungi in tropical monsoonic grassland. *J. Basic Appl. Mycol.*, 5:78-81.

Vyas, D., Dubey, A., Soni, A., Mishra, M.K. and Singh, P.K. (2007). Arbuscular mycorrhizal fungi in early land plants. *Mycorrhiza News*, 19:22-24.

Vyas, D., Dubey, A., Soni, A., Mishra, M.K., Singh, P.K. and Gupta, R.K. (2008a). Vesicular arbuscular mycorrhizal association in bryophytes isolated from eastern and western himalayas. *Mycorrhiza News*, 19(4):16-18.

Vyas, D., Dwivedi, O.P., Yadav, R.K. and Vyas, K.M. (2002). Diversity of VAM fungi. In: *Frontiers of Fungal Diversity in India*, (Eds.)G,P.Rao, C.Manoharachari, D.J. Bhat, R.C. Rajak and T.N. Lakhanpal, International Book Distributing Company, Lucknow: 873-889.

Vyas, D., Singh, P.K., Mishra, M. and Gupta, R.K. (2010). Phytoprotection by arbuscular mycorrhizae. In: *Bioinoculants for integrated plant growth*, (Ed.) H.C. Lakshman, MD Publication Pvt. Ltd, New Delhi: 374-420.

Vyas, D., Singh, P.K., Mishra, M.K. and Dubey, A. (2008b). VA mycorrhizal association in weeds of semi natural grassland of Sagar. *Ind. J. Agrofor.*, 10:91-97.

Whitbeck, J.L. (2001). Effects of light environment on vesicular-arbuscular mycorrhiza development in *Inga leiocalycina*, a tropical wet forest tree. *Biotropica*, 33:303-311.

Wilson, G.W.T., Rice, C.W., Rillig, M.C., Springer, A. and Hartnett, D.C. (2009). Soil aggregation and carbon sequestration are tightly correlated with the abundance of arbuscular mycorrhizal fungi: results from long-term field experiments. *Ecol. Lett.*, 12:452–461.

Whitcomb, S.A. and Stutz, J. (2001). Effects of pruning on root length density, root biomass, and arbuscular mycorrhizal colonization in two landscape shrubs. In: *Proceedings of the 3rd international conference on Mycorrhizas* (ICOM3), Adelaide. Abstract: 1-157.

Wrage, N., Lardy, L.C. and Isselstein, J. (2010). Phosphorus, plant biodiversity and climate change. In: *Sociology, Organic Farming, Climate Change and Soil Science. Sustainable Agriculture Reviews 3*, (Ed.) E. Lchtfouse: 147-169.

Wu, Q., Xia, R. and Hu, Z. (2006). Effect of arbuscular mycorrhiza on the drought tolerance of *Poncirus trifoliata* seedlings. *Front. Forest China*, 1:100-104.

Yadav, D.S. and Kumar, A. (2009). Long term effects of nutrient management on soil health and productivity in rice (*Oryza sativa*)-wheat (*Triticum aestivum*) cropping system. *Ind. J. Agronomy*, 53(1):32-36.

Young, A. (1997). *Agroforestry for soil management*, 2nd edn. ICRAF and CAB International, Wallingford.

Zhu, Y.G. and Miller, R.M. (2003). Carbon cycling by arbuscular mycorrhizal fungi in soil-plant systems. *Trends Pl. Sci.*, 8:407–409.

❑❑❑

Microbial Diversity and Functions, 2012
© D.J. Bagyaraj, K.V.B.R. Tilak, H.K. Kehri (eds.), pp. 473-483
New India Publishing Agency, New Delhi (India)
E-mail : info@nipabooks.com; Website : www.nipabooks.com

Chapter 23

Diversity of AM Fungi in Rhizosphere of Tree Species of Indian Thar Desert

Nishi Mathur, Mehboob Chouhan, Mohnish Vyas, Shilpa Yadav, Minal Tamboli and Anil Vyas

ABSTRACT

A field study of seven arid districts of Rajasthan was undertaken to evaluate the occurrence of selected tree species and arbuscular mycorrhizal fungal (AMF) associations with them. Five genera were identified in the rhizosphere of these selected tree species. A high diversity of AMF was observed which varied between different host tree species. Among the five genera, Glomus occurred most frequently, with four species, Acaulospora and Gigaspora were found with three species, while Entrophospora and Scutellospora was detected with two species. Gigaspora margarita, Gigaspora rosea, Glomus deserticola, Glomus constrictum, Glomus fasciculatum, Glomus mosseae and Scutellospora calospora were the most dominant species. The AMF spore density was not clearly affected by the host tree suggesting that biotic factors may be relatively less important than abiotic/edaphic factors for establishing population pattern. The spore density of AMF had a strong positive correlation with soil pH and organic carbon content and a negative correlation with Olsen's P content of the soil. The association with AMF of these tree species native to the harsh environmental conditions of the Indian Thar Desert may play a significant role in the re-establishment and conservation of these multipurpose desert tree species.

Keywords : AM fungi, diversity, Indian Thar desert, tree species.

Introduction

AMF associations are ubiquitous and play an important role in ecosystem diversity. They can modify the structure and function of plant communities (Giovannetti and Gianinazzi-Pearson, 1994; Douds and Millner, 1999) and may be useful as indicators of ecosystem change (McGonigle and Miller, 1996). Arbuscular mycorrhizal fungi are frequently distributed in different areas of Indian Thar Desert (Mathur and Vyas, 2000). Studies on the distribution and activity of AMF can help to elucidate the ecological significance of AMF associations. The population of AMF varies greatly and their distribution is affected by various biotic and abiotic factors (Mohammad *et al.*, 2003). Preliminary studies have indicated that AMF are very common in arid soils and form associations with most of the plants growing in Indian desert (Mathur *et al.* 2009) reported better establishment of vegetation in arid areas by using AMF as these fungi may/often enhance plant absorption of P and other elements, improve water uptake and its transport to plants and enable the plants to withstand high temperatures. Further, Panwar and Vyas (2002) indicated the significance of AMF in re-establishment and conservation of endangered plants in arid areas. Turnau and Haselwandter (2002) also considered AMF as a tool for re-establishment of endangered plant species. In general, knowledge of the mycorrhization of desert tree species is lacking. On the other hand, information of this kind must be considered as a pre-requisite for making tree re-establishment programmes successful.

Prior to exploiting the biofertilizers potential of AMF in relation to these desert tree species, it is necessary to examine the spatial distribution and colonization of these microbes in soil, since AMF species vary with ecosystems (McGonigle and Miller, 1996) and are affected by edaphic factors.

Keeping these facts in mind, an extensive field investigation was carried out to evaluate spatial distribution and colonization of AMF species present in the rhizosphere of eight selected desert tree species and to study effects of edaphic factors on AMF populations in the rhizosphere. (Table 1)

Site Description

The Indian Thar Desert comprises about 70% part of the Western Rajasthan. An intensive field survey of these 7 districts was undertaken in order to find out occurrence of selected desert tree species *i.e., Acacia senegal, Acacia nilotica, Prosopis cineraria, Tecomella undulata, Anogeissus pendula, Anogeissus latifolia, Salvadora oleoides, Salvadora persica* and AMF associations with them. Important climatological characteristics of surveyed districts are summarized in Table 2.

Soil Sampling

Rhizosphere soil samples (soil adhering to the roots) were collected at 30-90 cm depths along with root samples in five replicates from each plant. Before sampling, the soils from the upper layer were scrapped off to remove foreign particles and litter. The collected soil and root samples were placed in an insulated carrier for transport and immediately refrigerated at 4°C upon arrival. The roots were processed immediately. All the soil samples collected from the rhizosphere of a particular plant species of a district were homogenized replication wise before processing by sieving (< 2 mm mesh size) to remove stones, plant material and coarse roots. Subsample of each soil was air dried and used for estimation of various physico-chemical properties and to establish successive pot cultures (trap cultures).

Table 1 : Different Attributes of AM Fungal Colonization in the Rhizosphere Soil from the Indian Thar Desert

S. No.	Studied species	Family	Location	No. of vesicles per cm root segment	Spore population per 10g soil	No. of AM species	Degree of colonization (%)
1.	*Acacia senegal*	Fabaceae	Jodhpur	8.2	20	7	30
			Jaisalmer	12.3	35	9	40
			Nagaur	11.2	27	8	35
2.	*Acacia nilotica*	Fabaceae	Chokha	15.3	26	9	35
			Pokran	12.4	30	8	38
3.	*Prosopis cineraria*	Fabaceae	New Campus	20	30	8	26
			Pali	22	27	10	31
			Barmer	26	32	12	33
4.	*Tecomella undulata*	Bignoniaceae	Jodhpur	15	22	7	26
			Mandore	18	27	9	38
			Sirohi	20	31	9	32
5.	*Anogeissus pendula*	Combretaceae	Kailana	12	22	7	24
			Ranakpur	16	25	9	31
			Bikaner	18	27	11	37
6.	*Anogeissus latifolia*	Combretaceae	Jodhpur	21	20	7	32
7.	*Salvadora oleoides*	Salvadoraceae	Jodhpur	24	27	9	37
8.	*Salvadora persica*	Salvadoraceae	Mathania	12	37	12	36
			Bilara	18	39	10	38
			Osian	23	42	13	32

Trap Cultures

Successive pot cultures (trap cultures) have been shown to be a useful tool in inducing sporulation of AMF from field soils in arid ecosystems to facilitate the detection of AMF species that are present in the rhizosphere and roots but do not sporulate readily in the field at the time of sampling (Stutz and Morton, 1996). To establish successive pot cultures, 500 g dry wt. field soil was mixed with autoclaved sand (1:1, v/v) and planted with surface-sterilized seeds (by 0.1% w/w mercuric chloride solution for 2 min and then washed with distilled water) of *Cenchrus ciliaris* L. as host.

Table 2 : Important site characteristics of surveyed districts in Rajasthan (India).

District	Latitude (N)	Longitude (E)	Total geographic area (sq.km)	Rain fall[a] (mm)	Mean max. temp (°C)[b]	Mean min. temp (°C)[b]	Relative humidity (%)[b]
Barmer	24°4′-26°32′	70°5′-72°52′	28,387	286.7	45.9	3.04	54.7
Bikaner	27°11′-29°03′	71°54′-74°12′	27,244	250.9	46.5	2.45	49.6
Jaisalmer	26°5′-28°0′	69°3′-70°0′	38,401	217.4	46.3	2.05	50.2
Jodhpur	26°0′-27°37′	72°55′-73°52′	22,850	389.1	46.1	4.51	48.7
Nagaur	26°25′-27°40′	73°1′-75°15′	17,718	431.6	42.9	1.44	47.8
Pali	24°45′-26°75′	72°48′-74°20′	12,387	427.2	44.8	2.20	55.9
Sirohi	24°20′-25°17′	72°16′-73°10′	5,136	450.5	47.0	2.50	62.0

[a]Average of last 10 years. [b]Average of last 5 years

Root Colonization by AMF

To determine the percent root colonization, root samples collected from different sites were washed in tap water and staining was done by the method of Philips and Hayman (1970) for rapid assay of mycorrhizal association. The root samples were cut into pieces of 1 cm length and placed in 10% KOH solution, which was kept at boiling point for about 10 min (depending upon the hardness of the root sample). The root samples were captured on a fine sieve and rinsed with distilled water until the brown colour disappeared. Post-clearing bleaching was done with alkaline hydrogen peroxide (0.5% NH_4OH and 0.5% H_2O_2 v/v in distilled water). Roots were rinsed with distilled water, treated with 1% HCl and stained with 0.05% w/v trypan blue in lactic acid-glycerol. Assessment of colonization was conducted on each sample by the glass slide method, in which 100 randomly selected root segments of each replication were determined microscopically. A segment was counted as infected when hyphae, vesicles, or arbuscules were observed. The infection percentage was determined by the method given by Biermann and Lindermann (1981).

Spore Extraction

Spores of AMF were extracted from the field and successive pot culture soils by the wet sieving and decanting technique of Gerdemann and Nicolson (1963). Total spore numbers of mycorrhizal fungi in the soil samples were estimated by the method of Gaur and Adholeya (1994) and spore densities were expressed as the number of spores per 100 g of soil. The isolated spores were picked up with needle under a dissecting microscope and were mounted in polyvinyl lactoglycerol (PVLG). However, PVLG was mixed with Meltzer's reagent (1: 1, v/v) in case of *Scutellospora* species. All the spores (including broken ones) were examined using Medilux-20 TR compound microscope. Taxonomic identification of spores up to species level was based on spore size, spore colour, wall layers and hyphal attachments using the identification manual of Schenck and Perez (1990) and the description provided by the International collection of AMF.

Soil Parameters

Soil samples were analysed for pH and electrical conductivity on 1: 2.5, soil: water suspension. Organic carbon was estimated by the method of Walkley and Black (1934) using 1 N potassium dichromate and back titrated with 0.5 N ferrous ammonium sulphate solution. Available phosphorus in soil was determined by extraction with 0.5 M sodium bicarbonate for 30 min (Olsen *et al.*, 1954). Soil texture was estimated gravimetrically by hydrometer method (Jackson, 1967).

Results and Discussion

Physicochemical properties of the soils of each site are presented in Table 3. Soil texture varies from sandy gravel to clay loam. The soil had a pH range from 8.16 to 8.42, organic carbon between 0.20 and 0.45% and Olsen P level of 4.2-7.7 mg kg^{-1}. In general, soils are alkaline in reaction, low in organic matter content and available P status.

Fourteen species of AMF were identified in the rhizosphere soils collected from field and successive pot cultures. It scattered over five genera *viz.*, *Acaulospora*, *Gigaspora*, *Glomus*, *Entrophospora* and *Scutellospora* (Table 4). *Glomus* species were most dominant and made up for more than 50% of the total isolates followed by *Acaulospora* (3 species), *Scutellospora* (2 species), *Gigaspora* (3 species) and *Entrophospora* (2 species). It is evident that the occurrence of various species of AMF varied considerably with different tree species.

In almost all the sites Glomus species predominated the AM population and contributed to 25 to 50 percent of the total (figure 1). Other genera found were *Acaulospora*, *Gigaspora*, *Entrophosphora* and *Scutellospora*. It is reported that genus *Glomus* to be the most common AMF genus distributed globally and it is also known to dominate in the tropical areas (Chaurasia, 2000) as well as temperate regions (Vestburg 1995) of the World. Its dominance under various climatic conditions ranging from tropical (Chaurasia, 2000) to high arctic region (Dalpe and Aiken, 1998) has been reported earlier. Wide occurrence of genus *Glomus* in the present study as well as reports of several workers suggested that genus *Glomus* has a very wide ecological amplitude, that is responsible for its adaptability and survival in different habitats and vegetational composition.

Table 3 : Physicochemical characteristics of site soils.

District	pH	EC (dSm-1)	OC (%)	Olsen P (mgkg-1)	Texture
Barmer	8.32	0.26 ± 0.02	0.45 ± 0.01	5.1 ± 0.03	Sandy gravel
Bikaner	8.34	0.24 ± 0.01	0.38 ± 0.01	4.9 ± 0.01	Sandy loam
Jaisalmer	8.42	0.25 ± 0.04	0.38 ± 0.01	4.6 ± 0.01	Sandy
Jodhpur	8.35	0.28 ± 0.03	0.38 ± 0.01	4.2 ± 0.04	Loamy sand
Nagaur	8.16	0.29 ± 0.01	0.20 ± 0.01	5.6 ± 0.01	Loamy sand
Pali	8.18	0.22 ± 0.02	0.32 ± 0.01	7.7 ± 0.01	Coarse loam
Sirohi	8.19	0.23 ± 0.01	0.35 ± 0.01	4.8 ± 0.01	Sandy loam

± Standered error of mean

Table 4 : Distribution of AMF species associated with different tree spp.

AMF species	Plant species					
	AS	AN	PC	TN	AP	SP
Acaulospora laevis Gerdmann & Trappe	++	++	++	+	++	++
Acaulospora morrawae Spain & Schenck	++	+	+	++	+	++
Acaulospora sporocarpia Berch	-	-	++	-	-	+
Gigaspora margarita Becker & Hall	++	++	+++	++	+	+
Gigaspora gigantea Nicol & Gerd	+	++	++	++	++	+++
Gigaspora rosea Nicol & Schenck	+	+	-	+	+	-
Glomus aggregatum Schenck & Smith	+++	+++	++	++	+	++
Glomus constrictum Trappe	++	+++	+++	++	+++	++
Glomus deserticola Trappe Bloss & Menge	+++	++	+++	+	++	++
Glomus mosseae Gerd. & Trappe	++	+	++	+++	++	++
Scutellospora calospora Walker & sanders	-	+	-	+	+	+
Scutellospora nigra Walker & senders	+	-	+	+	++	-
Entrophospora columbiana Spain and Schenck	+	-	-	+	-	+
Entrophospora sp.	+	+	+	-	+	-

- = Absent, + = Low (< 20%), ++ = Moderate (20-50%), +++ = High (>50%), AS= *Acacia senegal*, AN= *Acacia nilotica*, PC= *Prosopis cineraria*, TN= *Tecomella undulata*, AP = *Anogeissus pendula*, SP = *Salvadora persica*.

This reveals a high specific consortium to each rhizosphere with a high degree of variance in species composition. Hence, a very high AMF diversity index in Thar Desert soils was apparent. *Gigaspora margarita, Gigaspora rosea, Glomus constrictum, Glomus deserticola, Glomus fasciculatum, Glomus mosseae* and *Scutellospora nigra* were the most dominant species (Table 4). *Glomus* was the most abundant of all AMF genera under arid environment (Tarafdar and Kumar, 1996), which may be due to its resistance to high soil temperature.

The density of viable AMF spores recovered from the rhizosphere soil samples collected from field and successive pot cultures ranged between 20 and 50 spores 10 g^{-1} soil for studied plants. The spore density is relatively low, which is common for arid and semi-arid lands (Requena *et al.*, 1996). These findings agree with that of Al-Raddad (1993), who attributes these differences to the length of the growing season and the type of root systems of trees, which make the rhizosphere more favourable to spore propagation and AMF colonization. It is clear from the results that the rhizosphere soils collected from field and successive pot cultures in Jodhpur have higher AMF spore densities compared to other sites. This may be because of poor soil fertility (in terms of available phosphorus) which results in higher AMF populations (Norani, 1996).

Natural AMF colonization of root samples varied between 38 and 68%. Cleared and stained roots showed the presence of globose to subglobose or ellipsoid bodies (vesicles or spores), dichotomously branched structures (arbuscules) and hyphae in all the sites. Extramatrical hyphae bearing resting spores were also seen associated with the roots of selected plants during study. Considerable variation in percent root colonization and number of different AMF spores associated with plant rhizosphere was observed but no definite correlation could be established between them (Kalita *et al.*, 2002). However, contradictory results were reported by Mutabaruka *et al.* (2002), as a significant positive correlation and by Louis and Lim (1987), as a negative correlation between percent root colonization and AMF spores.

Table 5 : reveals correlation analysis between AMF spore population and different edaphoclimatic factors. It is evident from the results that AMF spore populations were affected by soil pH, organic carbon and Olsen P content. A significant positive correlation with pH ($r = 0.85$, $p<0.01$) and organic carbon ($r = 0.68$, $p<0.05$) was recorded during present investigation.

Table 5 : Relationship between AMF spore population and different edaphic factors

Edaphic factores	AMF spore population
pH	0.88**
EC	0.53
OC	0.68*
Olsen P	-0.86**
Temp (max)	0.14
Temp (min)	-0.24
RH	0.08

*p<0.05; **p<0.01; n = 12

In contrast, a strong negative correlation was observed with soil Olsen P content ($r = -0.86$, $p<0.01$). Blaszkowski (1993) while investigating plant communities in Poland, observed a significant positive correlation between AMF spore density and soil pH. A positive correlation with organic carbon content in soil coincide with the findings of Mohammad *et al.* (2003), who reported the same while investigating under semi-arid environment of Jordan. Organic matter content in the soil increases the water-holding capacity of the soil (Brady and Weil, 1996) and, therefore, may facilitate a more favourable soil moisture condition for the AMF population. When plants have high nutrient availability (especially phosphorus), a negative response and low AMF spore population should be expected.

Our results pioneered to identify the status and occurrence of the selected tree species in Thar Desert environment as well as AMF diversity with them, indicating the mycorrhizal dependency of these plants. *Glomus* is considered to be the most common arbuscular mycorrhizal genus in this region. No host plant or geographic location specificity was observed, suggesting the population of AMF species was affected mainly by edaphic factors. Recovery of large AMF diversity with these plants reveals the rich wealth of AMF diversity in harsh environmental conditions in Thar Desert. These native AMF isolates with the capacity to survive under stress conditions may be instrumental in the re-establishment of these halophylic plant species. Thus the present study of mycorrhizal status may well prove crucial for any attempts of re-establishment of such endangered plants. Appropriate strategies can be drawn for the artificial inoculation of one or some of these indigenous AMF, which would make the re-establishment and regeneration attempts ecologically and economically viable in such constrained ecosystems. These approaches will increase our scope to manipulate the symbiosis in conservation schemes.

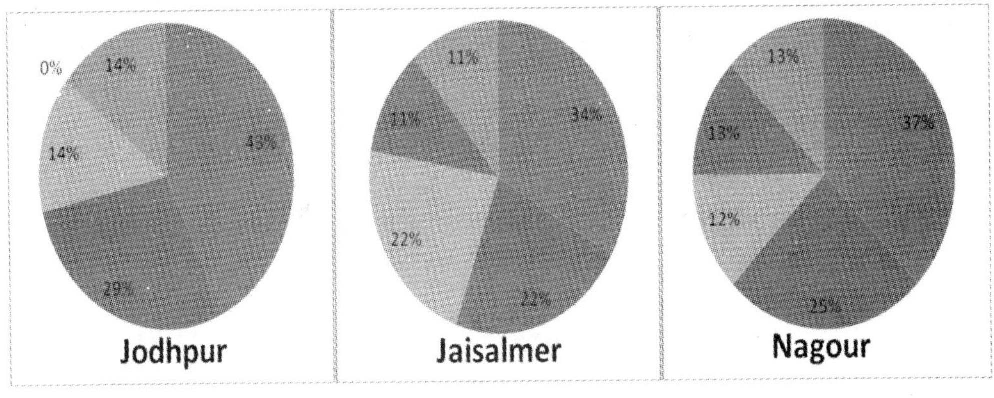

Acacia senegal

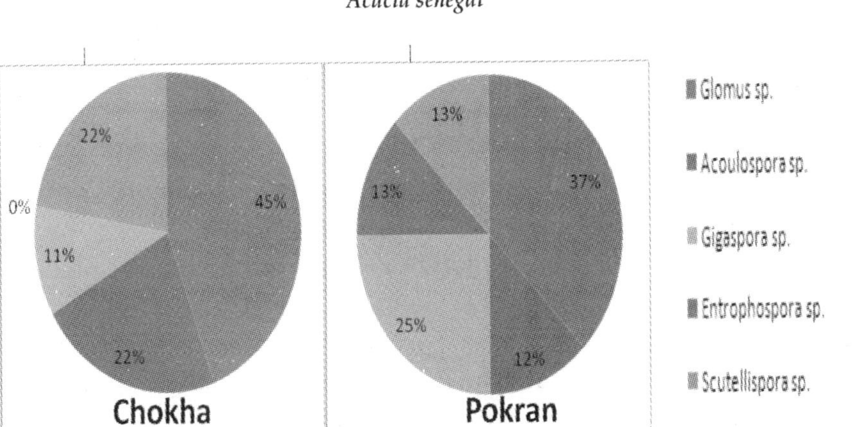

Acacia nilotica

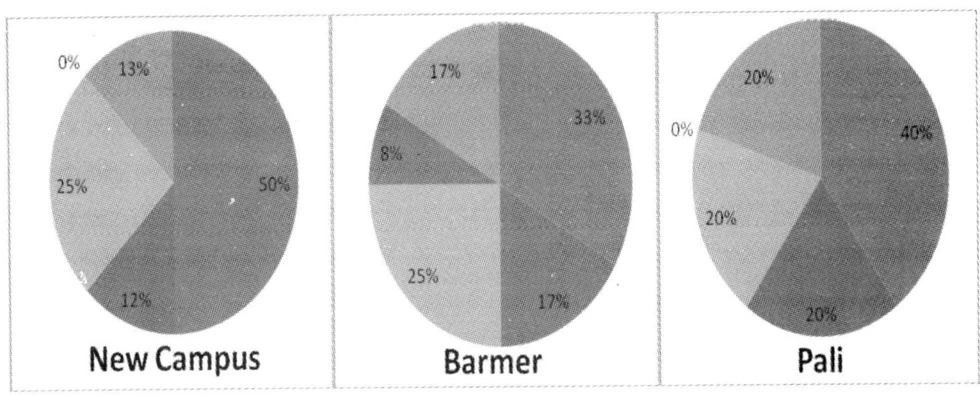

Prosopis cineraria

Fig. 1 : Arbuscular mycorrhizal fungal diversity in the rhizosphere soil of different tree species of Indian Thar desert.

References

Al-Raddad, A., (1993). Distribution of different *Glomus* species in rainfed areas in Jordan. *Dirasat*, 20: 165-182.

Biermann, B. and R.G. Lindermann, (1981). Quantifying vesicular arbuscular mycorrhizae: A proposed method towards standardization. *New Phytol.*, 87: 63-67.

Blaszkowski, J., (1993). The occurrence of arbuscular fungi and mycorrhizae *Glomales* in plant communities of maritime dunes and shores in Poland. *Bull. Polish Acad. Sci. Biol. Sci.*, 41: 377-392.

Brady, N.C. and R.R. Weil, (1996). The *Nature and Properties of Soils*. 11th Edn., Prentice Hall Incorporation, New Jersey, USA.

Chaurasia,B. (2000), *Ecological study of tropical forest trees with special reference to arbuscular mycorrhizal (AM) association*. Ph.D. Thesis, Dr. H.S.Gour Vishwavidyalaya, Sagar, M.P., India.

Dalpy, Y. and Aiken,S.G. (1998). Arbuscular mycorrhizal fungi associated with *Festuca* species in the Canadian High Arctic. *Can. J. Bot.*, 76: 1930-1938.

Douds, D.D. and P.D. Millner, (1999). Biodiversity of arbuscular mycorrhizal fungi in agroecosystems. *Agric. Ecol. Environ.*, 47: 77-93.

Gaur, A. and A. Adholeya, (1994). Estimation of VAMF spores in soil: A modified method. *Mycorrhiza News*, 62: 10-11.

Gerdemann, J.W. and T.H. Nicolson, (1963). Spores of mycorrhizal *Endogone* species extracted from soils by wet sieving and decanting. *Trans. Br. Mycol. Soc.*, 46: 235-244.

Giovannetti, M. and V. Gianinazzi-Pearson, (1994). Biodiversity in arbuscular mycorrhizal fungi. *Mycol. Res.*, 98: 705-715.

Jackson, M.L., (1967). *Soil Chemical Analysis*. 1st Edn., Prentice Hall of India Pvt. Ltd., New Delhi, India.

Kalita, R.K., D.P. Bora and D. Dutta, (2002). Vesicular arbuscular mycorrhizal associations with some native plants. *Indian J. For.*, 25: 143-146.

Louis, I. and G. Lim, (1987). Spore density and root colonization of vesicular arbuscular mycorrhizas in tropical soil. *Trans. Mycol. Soc.*, 88: 207-212.

Mathur, N. and Vyas, A. (2000). Mycorrhizal dependency of *Tamarix aphylla* in saline areas of Thar Desert. *Natutalia*, 25:105-110.

Mathur, N., Singh, J., Bohra, S., Bohra, A., Solanki, R., Vyas, A. (2009). Synergistic effect of inoculation of plant growth promotion rhizobacteria and arbuscular mycorrhizal fungus on productivity of sorghum. *J. .Mycol. Pl. Pathol.*, 39(1) : 59-65.

McGonigle, T.P. and M.H. Miller, (1996). Development of fungi below ground in association with plants growing in disturbed and undisturbed soils. *Soil Biol. Biochem.*, 28: 263-269.

Muhammad, M.J., S.R. Hamad and H.I. Malkawi, (2003). Population of arbuscular mycorrhizal fungi in semi-arid environment of Jordan as influenced by biotic and abiotic factors. *J. Arid., Environ.*, 53: 409-417.

Mutabaruka, R., C. Mutabaruka and I. Femandez, (2002). Diversity of arbuscular mycorrhizal fungi associated to tree species in semi-arid areas of Machakos, Kenya. *Arid Land Res. Manage.*, 16: 385-390.

Norani, A., (1996). An assessment and enumeration of vesicular arbuscular mycorrhizal propagules in some forest sites of Jengka. *J. Trop. For. Sci.*, 9: 137-146.

Olsen, S.R., C.V. Cole, F.S. Watanabe and L.A. Dean, (1954). *Estimation of available phosphorus in soils by extraction with sodium bicarbonate.* US Department of Agriculture Circular No. 939. US Government Printing Office, Washington, DC.

Panwar, J. and A. Vyas, (2002). AM fungi: A biological approach towards conservation of endangered plants in Thar Desert, India. *Curr. Sci.*, 82: 576-578.

Phillips, J.M. and D.S. Hayman, (1970). Improved procedures for clearing roots and staining parasitic and vesicular-arbuscular mycorrhizal fungi for rapid assessment of infection. *Trans. Br. Mycol. Soc.*, 55: 158-161.

Requena, N., P. Jeffries and J.M. Barera, (1996). Assessment of natural mycorrhizal potential in a desertified semiarid ecosystem. *Applied Environ. Microbiol.*, 62: 842-847.

Schenck, N.C. and Y. Perez, (1990). *Manual for the Identification of VA Mycorrhizal Fungi.* 3rd Edn., Synergistic Publications, Gainesville, Florida, USA., pp: 286.

Stutz, J.C. and J.B. Morton, (19960. Successive pot cultures reveal high species richness of arbuscular mycorrhizal fungi in arid ecosystems. *Can. J. Bot.*, 74: 1883-1889.

Tarafdar, J.C. and P. Kumar, (1996). The role of vesicular arbuscular mycorrhizal fungi on crop, tree and grasses grown in an arid environment. *J. Arid Environ.*, 34: 197-203.

Turnau, K. and K. Haselwandter, (2002). Arbuscular mycorrhizal fungi, an essential component of soil microflora in ecosystem restoration. In: *Mycorrhizal Technology in Agriculture,* (Eds.) S. Gianinazzi, H. Schiiepp, J.M. Barea and K. Haselwandter, Birkhiiuser Verlag, Switzerland: 137-149.

Vestburg, M. (1995). Occurrence of Some Glomales in Finland. *Mycorrhiza*, 5:329-336.

Walkley, A.J. and I.A. Black, (1934). Estimation of soil organic carbon by the chromic acid titration method. *Soil Sci.*, 37: 29-38.

□□□

Microbial Diversity and Functions, 2012
© D.J. Bagyaraj, K.V.B.R. Tilak, H.K. Kehri (eds.), pp. 485-505
New India Publishing Agency, New Delhi (India)
E-mail : info@nipabooks.com; Website : www.nipabooks.com

Chapter 24

Extremophiles : The Enigmatic Life

Shagun Sharma and Rohit Sharma

ABSTRACT

The ability of life to dwell in extreme environments on our mother planet earth right from a thirsty desert environment to highly water saturated oceans, gave the scientists a new area to work in during the 1990's. The discovery of microbial life which had the ability to survive in conditions which were highly inhospitable for the normal life gave rise to a revolution in the field of science and technology. Extremophiles are one of the most primitive forms of life on earth since they have accustomed themselves to the present day harsh conditions compared to the normal environment. The last few years have seen a massive increase in their discovery, since the areas which once were thought barren by the scientists are now found inhabited by the microbial life. The extremophiles endure conditions right from extreme temperatures, pressure, salt concentration to radiations, metal tolerance, pH etc. During the past few years the industrial applications associated with the extremophiles and their products have led to their intensive exploration. Keeping in view the present and the future scenario, extensive efforts are still required in order to exploit the extreme environments for extremophiles and their products.

Keywords : Archaea, applications, Extremophiles, Extremozymes and Environment.

Introduction

It is appropriately said that "Nature does not hurry, yet everything is accomplished". Our planet Earth supports wide diversity of life, right from organisms living in water to huge elephants on land. Life has been able to dwell in such extreme environments which once were unimaginable to the ancients of nineteenth century. Remarkably, the organisms not only endure their lot; in fact they execute best in their strenuous habitats. Romans, who were known for their excesses coined the term "extremus" which is a superlative of the word exter, which means:- being on the outside. The discovery of the organisms which had the capability to flourish in extreme environments was marvelous. It was during the 1980s and 1990s when scientists were able to discover the microbial life which was flexible enough to mold themselves and survive in conditions which were extremely inhospitable for the normal microbial life. "Extremophiles" - derived from Latin word *'extremus'* meaning 'extreme' and *'philos'* meaning 'love'. These extreme loving organisms were named so by Macerloy (1974).

The enhanced revolution in the field of science and discovery of extremophiles, shows the ability of the life forms to accustom to the different environments and there has been an upsurge in the prospects of these organisms to flourish elsewhere in the universe (Seckbach, 1997). Extremophiles are one of the most primitive forms of life on earth as they have adapted to harsh conditions as of today, compared to the present life forms. Most of the extremophiles generally include Archaea and Bacteria, since most of the higher organisms are less adaptive to the extreme variations in the environment from the normal conditions.

Classification of Extremophiles

It has been correctly said that extremophiles are the superheroes of tomorrow's world. These are uncanny microorganisms which dwell and flourish extremely well in the conditions which are considered harsh by the humans. The following few examples give us a reason to believe them as "superheroes". *Pyrodictium abyssi*, a microorganism which dwells near the volcanic vents found at the ocean beds. It is one of the members of thermophiles, since it can tolerate extremes of heat and can survive easily in boiling water. Another such example is *Desulforudis audaxviator*, which was found 1.7 miles below earth's surface in South Africa few years back. This microorganism survives in the absence of light, oxygen and is known to tolerate extremes of heat. It survives in such extreme conditions by combining other elements such as decaying uranium, which radiates from the surrounding rocks with water. These tiny little microorganisms are anticipated to soon affect the mankind in a gigantic way. In the last few years the hunt for these organisms has increased massively as the scientists have realized that the places which were once considered barren are abound with microorganisms. These organisms are also chased because of their massive potential which can be employed in number of industrial applications.

Table 1 : Table of classification of extremophiles along with respective examples (Brock 1986; Demirjian, 1999)

Environmental Factors	Category	Definition	Examples
Temperature	Psychrophile	<15 °C	*Psychrobacter*
	Mesophile	15-60 °C	*Serratia marcescens*
	Thermophile	60-80 °C	*Desulfotomaculum*
	Hyperthermophile	>80 °C	*Pyrolobus fumarii*
Radiations		up to 4 million rad of radiation	*Deinococcus radiodurans*
Pressure	Barophile Piezophile	Weight loving Pressure loving	unknown for microbes upto 130 MPa
Desiccation	Xerophiles	anhydrobiotic	*Artemia salina*, fungi, microbes
Salinity	Halophile	Salt loving microbes(2-5 M NaCl)	*Halobacteriacea, Dunaliella salina*
pH	Alkaliphile	pH >9	*Natronobacterium*
	Acidophile	lower pH	*Cyanidium caldarium*
Oxygen tension	Anaerobe Micro-aerophile Aerobe	cannot tolerate O_2 tolerates some O_2 requires O_2	*Methanococcus jannaschii* *Clostridium* *Micrococcus luteus*
Chemical extremes	gases metals	Able to tolerate high metal concentrations (metalotolerant)	*Cyanidium caldarium* (pure CO_2) *Ferroplasma acidarmanus*(Cu, As, Cd, Zn)

Extremophiles' characterization seem straightforward and clear but at the same point there are certain philosophical issues which require further exploration. First, what is 'extreme' exactly? The meaning of extreme depends on the beholder. Any condition that deviates from the normal environmental conditions is known as extreme environment, say for example- bacteria dwelling in soil in tropical areas, is growing under normal conditions but for a bacteria flourishing in Sahara deserts, is under extreme conditions, hence is an extremophile. All physical factors that change continuously and extremes in the conditions which make functioning of the organisms difficult are 'extreme'. For example, in an aqueous environment in order to maintain chemistry, cells require specific pH, temperatures and solutes, ability to repair damage and accurate control over the biomolecules. Specific conditions such as radiations, dessication and oxygen, destroy biomolecules. Reactive oxygen species are formed by oxygen which leads to oxidative damage of the nucleic acids, proteins and lipids (Tyrell, 1991; Newcomb *et al.*, 1998). The second issue rotates around the question that- is it important for an extremophile to love extreme environment or just tolerate it? Talking practically, latter seems a better option which can be determined experimentally, but, biologically, former serves as a better answer. In the last few years many extreme loving microbes have been found but at the mean time there are certain microorganisms that just endure harsh conditions instead of loving them. And the last issue is, whether it is important for an organism to be extremophile throughout its life cycle and under all conditions? It has been reported in *Deinococcus radiodurans*, which is radiation resistant, that the radiation resistance in *D. radiodurans* is strictly diminished during the stationary phase compared to the logarithmic phase growth (Minton, 1994), under Mn^{2+} concentrations more than normal (Chow and Tan, 1990), under limited nutrient conditions and with freezing or desiccation (Venkateswaran *et al.*, 2000). Vegetative forms are more susceptible to extreme environments compared to the spores (for example, *Bacillus subtilis*) which are far more resistant to environmental extremes.

Tardigrades ("water bears") – Are one of the most pliant organisms. They have the ability to go into a hibernation mode - known as the tun state, where by it can endure temperatures ranging from -253°C to 151°C, along with exposure to x-rays and vacuum conditions.

Extremophiles and Biotechnology

The massive industrial applications associated with the extremophiles and their products have led to their intensive exploration during the past few decades. With the isolation of the new strains, development of new pathways, identification of new products, molecular and biochemical characterization of the components of the cell, the expected potential of these microorganisms is increasing exponentially. Some of the extreme enzymes known as 'extremozymes' used commercially includes alkaline proteases which are used in detergents. Extremozymes possess massive potential in the area of pharmaceuticals and agriculture. They are stable at extreme incubation conditions; they increase catalytic activity and specificity in several biological processes (Chadha and Patel, 2008). Various thermophiles serve as a source of DNA Polymerases. These are generally derived from *Thermococcus littoralis, Thermus aquaticus, Thermotoga maritime, Pyrococcus woesii* and *P. furiosus* and these enzymes are then used in polymerase chain reaction, abbreviated as PCR. Thermostable enzyme known as alpha- amylases is used in the starch saccharification process, this enzyme does not require the presence of Ca^{2+} for their activity/stability. Other enzymes which are cost effective and are used in starch saccahrification process includes amylopullulanase, glucoamylase and glucose isomerase, they are active in low pH range (Sharma, 2000; Narang *et al.,* 2001; Rao *et al.,* 2003a, 2003b; Kumar and Satyanarayana, 2003; Satyanarayana *et al.,* 2004). Hyperthermophlic archaeon *Thermococcus kodakaraensis* KOD1 derived thermostable chaperonins CpkA and CpkB were immobilized and were found useful in stabilization of enzymes (Izumi *et al.,* 2001). Enzymes derived from thermophiles have also been used for the development of optical nanobiosensors, stable and non-consuming analyte. The basis of this device is the ability of the thermophilic enzymes to bind the substrate at room temperature without bringing about any change in the substrate (Staiano *et al.,* 2005). The binding of enzyme and substrate is monitored as enzyme fluorescence variation. A thermostable non-specific nucleases was isolated from a thermophilic bacteriophage GBSV1 in 2008 by Song's group. This enzyme was found to degrade nucleic acids of various types including single and double stranded DNA and RNA (Song *et al.,* 2008). Halophilic archaea product myc oncogene, a eukaryotic homologue, has been used for the screening of the cancer patients sera. A higher number of positive reactions were produced by archael homologue compared to that recombinant proteins produced in *E. coli.* Beta- carotene is manufactured by *Dunaliella bardawii,* a green alga. *Marinobacter hydrocarbonoclasticus,* a haloterant, have been found to degrade a variety of aliphatic and aromatic hydrocarbons (Gauthier *et al.,* 1992). A halophilic culture which was able to degrade benzene, toluene, ethylbenzene and xylene in 1–2 weeks was developed by Nicholson and Fathepure (2004). The applications associated with extremophiles and products derived from them are still limited and a massive increase is expected in the near future. A metalloprotease derived from *Bacillus stearothermophilus* was mutated in order to increase its thermal stability and it was found that the activity of the mutated protein was 340 times higher than normal wild type protein and was found functional in the presence of denaturing agents even at a temperature of 100°C, also it retained its activity at 37°C. The development of recombinant microbes is expected to give rise to further advances in the said area of extremophiles and their industrial use. *Deinococcus,* recombinant strain was capable of degrading organo-pollutants, found in radioactive mixed waste environment. The oxidation of toluene, chlorobenzene, 2, 3-dichloro-1- butene and indole were brought about in highly irradiating environment (6000 rad/h) by toluene dioxygenase, which was found in the recombinant strain. Also the recombinant strain was found more resistant to toluene and trichloroethylene at a concentration which was much higher than found in radioactive mixed waste. The above cited example shows the importance of extremophiles in bioremediation of the waste sites which are found to be contaminated with radioactive, heavy metals and organopollutants (Cavicchioli and Thomas, 2003).

One of the most common application is biomining which involves the use of microbial life for the extraction of metals such as gold, copper, cobalt etc. from their respective ores. (Rohwerder *et a*l., 2003; Olson *et al.*, 2003). The said area has recently received substantial interest and is propelled due to the enhanced development in the field of genomics which helps to study each microorganism and its community (Valenzuela *et al.*, 2006).

It has been indicated by research that besides possessing the ability to help us in developing myriad pharmaceuticals for the treatment of several diseases, these superheroes are also expected to possess the abilities which shall help us fighting the problems such as global warming and environment pollution.

Lechuguilla Cave in United States is the fifth longest cave in the world and is well thought-out as one of the most immaculate and inimitable area. Scientists collected the bacterial samples from the stagnant pools of water in the caves and after several preliminary tests found that these rock eating bacteria had the potential to cure breast cancer by killing the cancerous cells without harming healthy ones.

Another advanced use of extremophiles includes the use of *Deinococcus radiodurans* derived infrared light producing proteins, which are a replacement for GFPs from jelly fish as latter cannot survive for long in living cells. The infrared light produced by *D. radiodurans* can be used to light up the entire cell and track down the disease and perhaps develop new ways to cure the same.

It has been correctly stated that besides being a contributor in the field of medicines, extremophiles are not "one trick ponies" for the scientific world. They can provide service as far as our own environment is concerned. One of the extremophile isolated from Yellowstone National Park has shown to make industrial bleaching process environment friendly by using the hydrogen peroxide found in the waste water and thus neutralizing it and making the process clean and friendly. Another thermophile isolated from hot ocean depths has shown promise in utilizing waste products for the generation of electricity and removal of radioactive elements from the environment.

The other developments associated with extremophiles include the use of Noncoding RNAs (ncRNAs) especially from thermoacidophilic archaea, mainly from *Sulfolobus* species (Tang *et al.*, 2005; Omer *et al.*, 2006;

Ciammaruconi *et al.*, 2007; Lillestol *et al.*, 2006). These could be used for the biocatalytic steps transient control in bioprocess by giving a dose of small interfering RNA (siRNA).

Table 2 : The following table shows the potential biotechnological applications of extremophiles (Demirjian *et al.*, 2001; Cavicchioli and Thomas, 2003)

Source	Use
Thermophiles	
1. Alkaline phosphatase	Diagnostics
2.DNA polymerase	DNA amplification by PCR
3.Proteases and Lipases	Dairy products
4.Alcohol dehydrogenase	Chemical Synthesis
5. Antibiotics	Pharmaceuticals
6.Xylanases	paper bleaching
7. Oil degrading microorganisms	Surfactants for oil recovery
Psychrophiles	
1.Alkaline phosphatase	Molecular biology
2. Proteases	meat tenderizing and contact-lens cleaning solutions
3. Lipases and proteases	Cheese manufacture
4. Lipases, cellulases, and amylases	Detergents
5. Various enzymes	Modifying flavours
6.Methanogens	Methane production
7. Polyunsaturated fatty acids	Food additives, dietary supplements
Barophiles	
Oligotrophs/oligophiles	Microbially enhanced oil recovery process
	Drinking water assimilable organic carbon bioassay
Halophiles	
1.Eukaryotic homologues (e.g. myc oncogene product)	Antitumor drugs screening and Cancer detection
2.Rheological polymers	Oil recovery
3.Polyhydroxyalkanoates	Medical plastics
4. Lipids	Heating oil, Cosmetic packaging and Liposomes for drug delivery

Contd...

5.Membranes	Pharmaceuticals surfactants
Radiation-resistant microbes	Degradation of organopollutants in radioactive mixed-waste environments
Acidophiles	
1.Sulphur-oxidizing microorganisms	Metals recovery and coal desulfurication
2.Microorganisms	Solvents and Organic acids
Alkaliphiles	
1.Proteases, cellulases, xylanases, lipases and pullulanases	Detergents
2. Pectinases	Waste treatment, Fine papers and de-gumming
3.Proteases	Gelatin removal on X-ray film
4.Elastases, keritinases	Dehairing of hide
5.Proteases and Xylanases	Bleaching of pulp

Extreme Environments

Temperature- Temperature gives rise to a series of challenges, ranging from the structural devastation produced by the ice crystals at one extreme, to the biomolecules denaturation at the other. Water solubility of gasses is associated with temperature, creating troubles at elevated temperature for aquatic organisms which require O_2 or CO_2. Denaturation of proteins and nucleic acids takes place at temperatures approaching 100 °C; it also increases the membranes fluidity to toxic levels.

Yet, the thermal preferences in nature vary from hyperthermophilic (maximum growth >80 °C) to psychrophilic (maximum growth <15 °C) (Morita, 1975). Archaea are the most hyperthermophilic organisms, with *Pyrolobus fumarii* (Crenarchaeota), a chemolithoautotroph capable of reducing nitrate and growing at the temperatures as high as 113 °C (Blochl *et al.*, 1997). Microbes which are capable of growing at high temperatures include wide variety of prokaryotes and eukaryotes. Hence, Brock (1986) recommended a definition: 'a thermophile is an organism capable of living at temperatures at or near the maximum for the taxonomic group of which it is a part'. There are thermophiles amongst the phototrophic bacteria (cyanobacteria, green and purple bacteria), eubacteria (*Bacillus, Thermus, Clostridium, Desulfotomaculum, Thiobacillus,* lactic acid bacteria, spirochetes, actinomycetes and several other genera) and the archaea (*Pyrococcus, Sulfolobus, Thermococcus, Thermoplasma,* and the methanogens). In comparison, 60 °C is the upper limit for eukaryotes, a temperature appropriate for some protozoa, fungi and algae. The highest temperature is lower by another 10°C for mosses; it is about 48°C for vascular plants and is 40°C for fish, possibly due to the decreased solubility of oxygen at elevated temperatures.

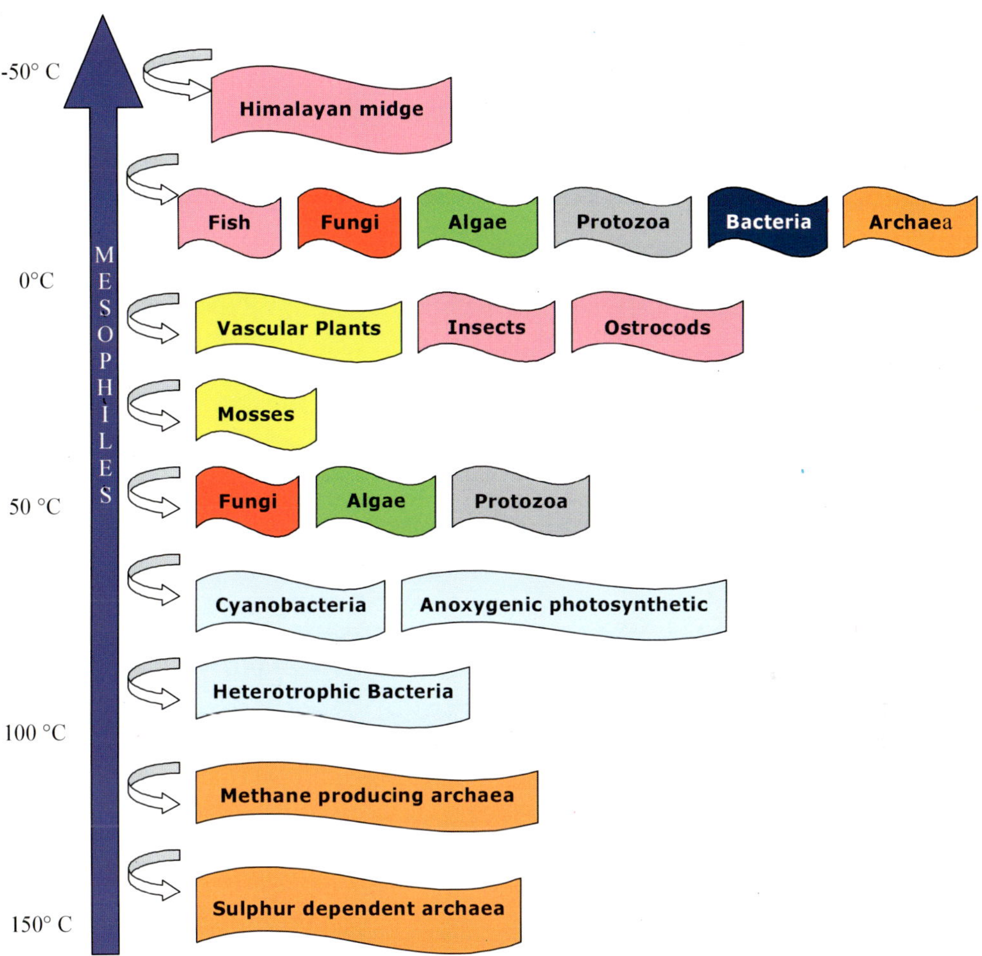

Fig. 1 : Shows the highest and lowest temperature for each major taxon. (Lynn and Rocco, 2001)

For any microbe, the nucleic acids, proteins and lipids are usually susceptible to high temperature and therefore, there is no particular factor that permits all thermophiles to grow at enhanced temperatures. The membrane lipids incase of thermophiles are more saturated and contain higher amount of straight chain fatty acids compared to the mesophiles. This allows thermophilic growth at elevated temperatures by providing the accurate degree of fluidity required for membrane to function. A paracrystalline surface layer (S-layer) with protein or glycoprotein is conatined by many archaeal species and this is expected to function as an external protective barrier. Certain histone-like proteins which bind to DNA have been recognized in hyperthermophiles, which are expected to protect DNA. Moreover, hyperthermophiles comprise of a reverse gyrase, a type1 DNA topoisomerase which is responsible for positive supercoiling and hence, stabilizing DNA. As with increasing temperature proteins begin to denature and becomes unstable, therefore, the stabilization and refolding of proteins is done by the heat shock proteins, chaperones. Certain protein properties such as an increased hydrogen bonds number,salt bridges, higher degree of structure in hydrophobic cores and elevated levels of thermophilic amino acids (e.g. proline residues with fewer degrees of freedom), are also known. A higher and lower amount of arginine and lysine respectively for thermostable proteins have been reported. Accumulation of solutes such as 2, 3-diphosphoglycerate, intracellular potassium and polyamines assist proteins to maintain their stability. At higher growth temperatures all the macromolecules such as RNA and DNA must be stable and functional.

As per the thermal adaptation hypothesis there exists a positive correlation between the content of G+C and growth temperature since G: C pair is more stable thermally than A: T pair. However, some authors could not make any correlation between temperature and GC composition. Later, it was claimed by Musto *et al.*, (2004) that there exists a positive correlation between genomic GC content and optimal growth temperature in most of the bacterial families studied by them; however various strong criticisms were put forward against their work (Marashi *et al.*, 2004; Basak *et al.*, 2005)

Psychrophilic microorganisms have been successful in colonizing almost all the permanently cold environments right from the deep sea to mountain and the polar regions. Depending on the optimal growth temperature some of these organisms are also known as psychrotolerant or psychrotroph (Morita, 1975). All major taxa representatives dwell in temperatures just below 0 °C. Till now, around 100 odd new species of Gram-positive and Gram-negative bacteria have been reported from a variety of habitats ranging from soil, fresh water, marine lakes, sea ice and oceans. Various species within the genera *Alcaligenes, Aquaspirillum, Brevibacterium, Arthobacter, Bacillus, Bacteroides, Methanosarcina, Moritella, Microbacterium, Micrococcus, Photobacterium, Shewanella, Polaribacter, Psychroserpens* and *Vibrio* have been reported as psychrophilic (Morita, 2001). Psychrophiles have effectively surmounted two main challenges: first, the low temperature, since any decline in temperature affects the rate of biochemical reactions exponentially; and second, the aqueous environment viscosity, which increases by a factor greater than two between

0 °C and 37 °C. With some organisms such as *Moritella profunda* remarkable adaptations have been observed. *Moritella profunda* is a psychropiezophile- a microorganism which is adapted to cold conditions and lives in the deep sea– showing a maximal growth rates at 2 °C and 12 °C as the maximum growth temperature (Xu *et al.*, 2003). This indicates that certain enzymes and supramolecular structures show a changed conformation at temperatures as low as 2 °C, that affects the metabolic flux negatively. Decrease in the temperatures possesses an adverse effect on functions of the membrane and the physical properties, in turn leading to reduced membrane fluidity, with the commencement of a gel-phase transition and, eventually, loss of function. The physical properties of the membranes are governed by the lipid composition and hence it is not astonishing that this concentration varies with the thermal territory of the microorganism. Generally, lower growth temperatures leads to the production of higher amount of unsaturated, polyunsaturated and methyl-branched fatty acids, and/or a shorter acyl- chain length, with studies reporting a high proportion of *cis*-unsaturated double-bonds and antesio-branched fatty acids (Chintalapati *et al.*, 2004; Russell, 1997). This altered composition is believed to have a major role in enhancing the membrane fluidity by the introduction of steric constraints which leads to change in the packing order or reduction in the number of membrane interactions. Other adaptations that have been recommended for increase in membrane fluidity take account of an enhanced content of large lipid head groups, non-polar carotenoid pigments and proteins (Chintalapati *et al.*, 2004). However, these adaptive strategies do not appear to be prevalent, and studies illustrate more compressed lipid head groups (Arthur and Watson, 1976) and decreased synthesis of non-polar carotenoid pigment (Fong *et al.*, 2001) in some psychrophiles.

Radiations : Radiation is a form of energy in transit, which is either as particles (for example, neutrons, protons, electrons, heavy ions or alpha particles) or as electromagnetic waves (such as, gamma rays, X-rays, infrared, ultraviolet (UV) radiation, microwaves, visible light, or radiowaves). Exceptional levels of radiation which are adequate to be eligible for 'extremophile' status – seldom occur naturally on the Earth, but intense UV levels and ionizing radiation are studied well because of their significance to medicine, warfare, energy production and space travel. The dangers arising because of UV and ionizing radiation range from diminished motility to photosynthesis inhibition, but the most severe damage is to nucleic acids. Direct DNA damage or reactive oxygen species induced indirect damage creates single- and double-strand breaks and modified bases.

Guinness Book of World Records has listed *Deinococcus radiodurans* as "the world's toughest bacterium." The microbe can endure drought conditions, nutrients limited conditions, and, most significantly, a thousand times more radiation than a person can. A proficient system for DNA repair makes the microbe so sturdy. High radiation doses destroy the genome of *D. radiodurans*, but the fragments are stitched back together by the organism at times in just a few hours. The repaired genome is as good as new. It can bear up ionizing radiation (up to 20 kGy of gamma radiation) and UV radiation (doses up to 1,000 J m^{-2}), but this amazing resistance is believed to be a consequence of extreme dessication resistance (Battista, 1997). Other organisms that are known for their ability to endure high radiation levels are the green alga *Dunaliella bardawil* (Ben-Amotz and Avron, 1990) and two species of *Rubrobacter* (Ferreira *et al.*, 1999). *Thermococcus littoralis* and *Pyrococcus furiosus*, the hyperthermophilic archaea are also known to endure high gamma- irradiation levels.

Desiccation : The unique properties of water make it the vital solvent for life. It possesses a high dielectric constant important for its solvent action along with high melting and boiling points with a broad range of temperature over which it remains liquid. Water forms hydrogen bonds and expands near its freezing point. No other compound is known to possess such traits. Thus, limitation of water is an extreme environment. Extreme dessication tolerant organisms enter anhydrobiosis, a state which is characterized by small amount of intracellular water and no metabolic activity. A variety of organisms including bacteria, fungi, yeast, insects, tardigrades , plants, mycophagous nematodes and the shrimp *Artemia salina* can become anhydrobiotic (Crowe, 1971; Wright, 1989; Glasheen and Hand, 1988; Potts, 1994).

Mechanisms of death as a result of anhydrobiosis comprise of irreversible phase changes to nucleic acids, lipids and proteins such as structural breakage through Maillard reactions denaturation and reactive oxygen species accumulation during drying, particularly under solar radiation (Cox, 1993; Dose *et al.*, 1995; Dose and Gill, 1994).

Metal tolerance : Some of the heavy metals are critical trace elements; but they can be toxic to all branches of life at high concentrations, including microbes, by the formation of complex compounds within the cell. Since heavy metals are found increasingly in the microbial habitats because of natural and industrial processes, therefore, the microbes have developed several mechanisms to bear the heavy metals presence (by complexation, efflux or metal ions reduction) or use metals in anaerobic respiration as the terminal electron. In order to have a toxic effect, heavy metal ions should first enter the cell. Since enzymatic functions and bacterial growth require the presence of heavy metals, certain mechanisms exist that permit the entrance and hence the uptake of metal ions into and by the cell. Two general uptake systems exists – the first is quick and unspecific, which is driven by development of chemiosmotic gradient across the cell membrane and thus requiring no ATP, and the second is slower and highly substrate-specific, driven by energy obtained from ATP hydrolysis. Though the first mechanism is highly energy efficient, but it results in an influx of heavy metals in huge amounts, and when present in high concentrations, they are expected to have lethal effects once inside the cell (Nies and Silver, 1995). In order to stay alive under metal-stressed conditions, bacteria have developed several types of mechanisms in order to tolerate the heavy metal ions uptake. These mechanisms comprise the accumulation and complexation of the metal ions inside the cell, efflux of metal ions outside the cell and heavy metal ions reduction to a less toxic state (Nies, 1999). Mergeay *et al.* (1985) tested several different metal ions minimal inhibitory concentrations (MICs) for *Escherichia coli* on agar medium, and found that mercury was the most toxic metal (with the lowest MIC), whereas manganese was the least toxic metal. *Staphylococcus aureus* was the first bacteria in which resistance to cadmium was identified.

Pressure : ZoBell and Johnson (1949) coined the term "Barophile" to describe bacteria which grow preferentially or exclusively at reasonably high hydrostatic pressures (Johnson *et al.*, 1954). Pressures of 200 to 600 atm adversely affect most terrestrial bacteria. The increase in hydrostatic pressure takes place at a rate of 10.5 kPa per metre depth, as compared to 22.6 kPa per metre for lithostatic pressure. Pressure decreases with altitude, such that with 10 km above sea level; the atmospheric pressure is approximately a quarter of that at sea level. There is an increase in the boiling point of water with pressure; therefore, at the bottom of the ocean water remains liquid at 400°C. At a temperature above approximately 100°C water does not exists as liquid, an increase in pressure increases the optimal temperature required for microbial growth, but usually by merely a few degrees. Challenge to life is enforced by pressure as it brings about a change in the volume. Pressure leads to decreased membrane fluidity due to compression in the packing of the lipids. If there is an increase in the volume as a result of a chemical reaction, as most do, an increase in pressure will inhibit the same. Although numerous organisms have accustomed themselves to very high pressures, an impulsive change can be fatal. Organisms which grow at high pressures, high and low temperatures, high and low organic nutrients are anticipated to flourish under deep-sea conditions. Bacteria (deep sea bacteria) have been isolated from the depth of 10,500m from the ocean and have been cultured at >100 Mpa at 2°C and 40 Mpa above 100 °C (Kato *et al.*, 1997). Most of the barotolerant and barophilic bacteria belong to γ-Proteobacteria (Raghukumar *et al.*, 1992). With increase in pressure, there is a relative increase in the amount of monounsaturation and polyunsaturation in the membrane. An increase in the unsaturated fatty acids in the cell membrane was observed in the barotolerant *Alteromonas* sp. (Nakasone *et al.*, 1998). The enhanced unsaturation leads to the production of more fluid membrane and hence counteracts the effects of the increased viscosity due to high pressure. Barophiles need dark or light-reduced environment, since they are tremendously sensitive to UV light and consequently dwell in the deep sea. Pressure may retard thermal denaturation and may stabilize proteins, as observed in DNA polymerases derived from hyperthermophiles *Pyrococcus* strain ES4, *P. furiosus* and *Thermus aquaticus*, the thermal inactivation of which is reduced by hydrostatic pressure.

Salinity : "Halophilism" is a remarkable existence of life in salty lakes, soils and various salted foods. Halophiles are found in environments with less free water that any human ever thought possible.Halophiles *Halobacterium* and *Halococcus* exist and reproduce within extremely salty waters of the Great Salt Lake and comparable sites and impart water a purple to red coloration. These halophiles can multiply and develop into intense populations of as many as 1,000,000 (1 million) to 100,000,000 (100 million) bacteria per milliliter. Paul *et al.* in 2008, from India compared the genomes (genes) and proteomes (protein patterns) of non-halophiles and halophiles. Halophilic species proteins have:

- Low hydrophobicity, i.e. more hydrophilicity than hydrophobic
- Number of acidic amino acids – especially Asp (aspartic acid); under-representation of Cys (cystine)
- Limited formation of helix and higher coiled structures occurrence

At the DNA level, of halophilic genomes the dinucleotide abundance profiles have some distinguishing and common characteristics, *i.e.*, specific DNA salt-adaptation signatures.

Thus, halophiles possess special internal and membrane proteins which enable them to dwell in the salty environments. Osmoprotectants have been discovered in moderate halophiles. In halophiles cytoplasm high concentrations of osmoprotectans such as glycine betaine, ectoine and hydroxyectoine have been found. These moderately simple, low-molecular-weight organic compounds surmount cellular osmotic stresses and keep up positive, cellular osmotic pressure. Enzymes can be protected from freeze- thaw damage by ectoine and hydroxyectoine- another very useful protectant effect. Moderate halophiles serve as a source for both osmotic and freeze-thaw stress protectants for industrial and research applications.

In all, halophiles possess special, osmoprotective biochemicals which enable them to endure the osmotic stress of high salt.

pH - pH is defined as $-\log_{10}[H^+]$. Biological processes generally occur towards the pH spectrum middle range and intracellular and environmental pH often fall in this range (for example, the pH of sea water is 8.2). However, in standard, pH can be high, as in case of soda lakes or drying ponds, or as low as 0 ([H^+] =1 M) and below. Proteins denature at exceptionally low pH.

Acidophiles are organisms that can endure and even flourish in acidic environments where the pH values range from 1 to 5. Acidophiles comprise of certain types of eukaryotes, bacteria and archaea which are found in a diverse acidic environments, including sulfuric pools and geysers, areas polluted by acid mine drainage, and even our own stomachs. Various types of acidophiles comprise of the green alga *Dunaliella acidophila* and red alga *Cyanidium caldarium*, which can exist below pH 1, as well as the fungi, *Acontium cylatium, Trichosporon cerebriae* and *Cephalosporium* sp., which grow near pH 0. The archean *Picrophilaceae* are the mainly acidophilic organisms which are known and are able to live at negative pH values. Generally, these low pH values can lead to coagulation and denaturation of enzymes and structural proteins. Denaturation of these proteins leads to the death of almost all higher plants and animals and bacterial cells follows. However, acidophiles resist proteins destruction by acid. Acidophiles have evolved numerous specialized mechanisms to maintain their internal cellular pH at a constant level (usually 7.2). These mechanisms involve "passive" regulation (not requiring the cell to expend energy) and "active" regulation (requiring energy). Passive pH regulation defenses depend primarily on enforcement of the cell membranes against unfavorable environment. At times some microorganisms secrete a biofilm which slows down the molecules diffusion into the cell, whereas others are able to bring about a change in their cell membrane in order to incorporate substances such as fatty acids that protect the cell. Another significant way by which microbes passively regulate their pH is by secretion of the buffer molecules that help raising the pH. Some have also evolved active pH regulation, which enables them to pump hydrogen ions at a constantly high rate out of their cells. This enables them to maintain their internal pH at around 6.5–7.0 (Gonzalez-Toril *et al.*, 2003).

Alkalophiles develop and multiply within the pH range of 10 to 12 or at times slightly higher pH. Such alkaline conditions are detrimental to most of the living organisms, yet alkalophiles resist and endure these conditions. Internal pH is maintained in alkaliphilic bacteria by both active and passive regulation mechanisms. The two modes of passive regulation include cytoplasmic pools of polyamines and low membrane permeability; whereas the active regulation is driven by sodium ion channels. In case of passive pH regulation, some cells have been found to contain cytoplasmic pools of polyamines. These cells are competent to buffer their cytoplasm in alkaline environments because they are rich in amino acids with positively charged side groups (lysine, arginine, and histidine). Another mode of passive regulation is low membrane permeability, which ensures that fewer protons move in and out of the cell. Usually, the rate at which the protons enter and leave a cell *via* diffusion is same; therefore it does not have much impact on homeostasis in an extreme pH environment. Some alkaliphilic bacteria, on the other hand, have developed sodium ion channels which actively drive the proton entry across the membrane, thus decreasing the overall pH of the cytoplasm. (Horikoshi, 1999)

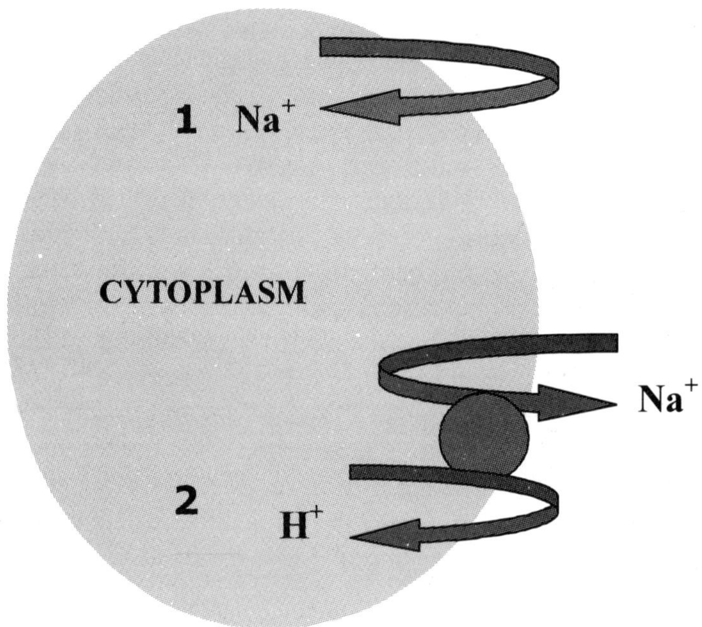

Fig. 2 : 1. A sodium-ion symporter regulates the mechanism for Na^+ entry, which is an essential antiport substrate 2. Na^+/H^+ antiport: with increase in concentration of Na^+ in the cytoplasm, it is expelled in exchange for protons (H^+)

The reversal of pH gradient is compensated by the alkaliphilic bacteria by possessing a high membrane potential or by coupling Na^+ expulsion to electron transport for pH homeostasis and energy transduction.

Oligotrophs/oligophiles : Organisms which are capable of growing in a medium containing 0.2–16.8 mg dissolved organic carbon per liter are known as Oligotrophs. Oligotrophs and Eutrophs (copiotrophs) coexist in natural ecosystems, and the ability of each to dominate a particular environment determines their proportion. An oligotrophic bacterium *Sphingomonas* sp. strain RB2256, was isolated from the Resurrection Bay, Alaska and irrespective of the nutrients concentration especially carbon and growth phase it was able to maintain its ultramicrosize. *Cycloclasticus oligotrophicus* RB1, another oligotroph was isolated from the Resurrection Bay, and it shared properties which were similar to *Sphingomonas* (e.g. relatively small genome size and single copy of the rRNA operon). Pramanik *et al.* (2003) while investigating the Leh soils for the diversity of these bacteria, reported abundance of oligophiles in nature. Various characteristics that are considered essential for oligotrophic microbes include a substrate uptake system that enables them to acquire nutrients from its surroundings. Therefore, oligotrophs ideally possess greater surface area to volume ratio, uptake systems with high-affinity and broad substrate specificities and inbuild mechanisms in order to resist environmental stresses (e.g. heat, hydrogen peroxide and ethanol). Number of microorganisms which are adapted to diminished nutrients conditions develop various structures which help them to increase their surfaces, e.g. *Caulobacter, Prosthecomicrobium, Hyphomicrobium, Ancalomicrobium.*

Future perspectives

World over extensive and intensive efforts in the understanding of the diversity of extremophilic microbes shows that we are still in the infancy and a lot more scientific exploration is still required. The isolation of extremophiles from the extreme environments as of now is the result of only sporadic attempts. Extensive funding, determined and collaborative efforts are required for understanding the microbes from extreme environments and hence exploiting them further.

References

Arthur, H. and Watson, K. (1976).Thermal adaptation in yeast: growth temperatures, membrane lipid, and cytochrome composition of psychrophilic, mesophilic, and thermophilic yeasts. *J. Bacteriol.*, 128: 56–68.

Basak, S., Mandal, S. and Ghosh, T. (2005).Correlations between genomic GC levels and optimal growth temperatures: some comments. *Biochem. Biophys. Res. Commun.*, 327: 969-970.

Battista, J. R. (1997). Against all odds: the survival strategies of *Deinococcus radiodurans. Annu. Rev. Microbiol.*, 51: 203-224.

Ben-Amotz, A. and Avron, M. (1990). *Dunaliella bardawil* can survive especially high irradiance levels by the accumulation of beta-carotene. *Trends Biotechnol.*, 8: 121-126.

Blöchl, E., Rachel, R., Burggraf, S., Hafenbradl, D., Jannasch, H.W., Stetter, K.O. (1997). *Pyrolobus fumarii*, gen. and sp. nov., represents a novel group of archaea, extending the upper temperature limit for life to 113 °C. *Extremophiles*, 1: 14-21.

Brock, T. D., (1986). Introduction: An overview of the thermophiles. In: *Thermophiles: General, Molecular and Applied Microbiology* (Ed.) T. D.Brock, John Wiley and Sons, New York: 1-16.

Cavicchioli, R. and Thomas, T. (2003). *Extremophiles*. The Desk Encyclopedia of Microbiology (Ed.) M. Schaechter, Elsevier, London: 436–453.

Chada, V. and Patel, B. K. C. (2008). Extremozymes belonging to glycosyl hydrolase family from *Halothermothrix orenii* (an extremophilic bacterium). *Proc. 7th International Symposium for Subsurface Microbiology*, Shizuoka:73.

Chintalapati, S., Kiran, M.D. and Shivaji, S. (2004). Role of membrane lipid fatty acids in cold adaptation. *Cell. Mol. Biol. (Noisy-le-grand)*, 50: 631–642.

Chow, F. I. and Tan, S. T. (1990). Manganese(II) induces cell division and increases in superoxide dismutase and catalase activities in an aging deinococcal culture. *J. Bacteriol.*, 172:2029-2035.

Ciammaruconi, A., Gorini, S. and Londei, P. (2007). A bifunctional archaeal protein that is a component of 30S ribosomal subunits and interacts with C/D box small RNAs. *Archaea*, 2:151-158

Cox, C. S. (1993). Roles of water molecules in bacteria and viruses. *Origins Life*, 23:29-36 .

Crowe, J. H. (1971). Anhydrobiosis: an unsolved problem. *Am. Nat.*, 105: 563-574.

Demirjian, D.C., Shah, P. and Morís-Varas, F. (1999). Screening for novel enzymes. *Top. Curr. Chemistry*, 200:1-29.

Demirjian, D.C., Morís-Varas, F. and Cassidy, C.S. (2001). Enzymes from extremophiles. *Curr. Opinion Chem. Biol.*, 5:144–151.

Dose, K., Dose, A.B., Dillmann, R., Gill, M., Kerz, O., Klein, A., Meinert, H., Nawroth, T., Risi, S. and Stridde, S. (1995). ERA-experiment: space biochemistry. *Adv. Space Res.*, 16(8), 119-129.

Dose, K. and Gill, M. (1994). DNA stability and survival of *Bacillus subtilis* spores in extreme dryness. *Origins Life*, 25: 277-293.

Ferreira, A.C., Nobre, M.F., Moore, E., Rainey, F.A., Battista, J.R. and da Costa, M.S. (1999). Characterization and radiation resistance of new isolates of *Rubrobacter radiotolerans* and *Rubrobacter xylanophilus*. *Extremophiles*, 3: 235-238.

Fong, N.J., Burgess, M.L., Barrow, K.D. and Glenn, D.R. (2001). Carotenoid accumulation in the psychrotrophic bacterium *Arthrobacter agilis* in response to thermal and salt stress. *Appl. Microbiol. Biotechnol.*, 56: 750–756

Gauthier, M. J., Lafay, B., Christen, R., Fernandez, L., Acquaviva, M., Bonin, P. and Betrand, J. (1992). *Marinobacter hydrocarbonoclasticus* gen. nov. sp. nov., a new extremely halotolerant, hydrocarbon–degrading marine bacterium. *Int. J. Syst. Bacteriol.*, 42: 568–576.

Glasheen, J. S. and Hand, S. C. (1988). Anhydrobiosis in embryos of the brine shrimp *Artemia*: characterization of metabolic arrest during reductions in cell-associated water. *J. Exp. Biol.*, 135:363-389.

Gonzalez-Toril, E., Llobet-Brossa, E., Casamayor, E.O., Amann, R. and Amils, R. (2003). Microbial ecology of an extreme acidic environment, the Tinto River. *Appl. Environ. Microbiol.*, 69: 4853-4865.

Horikoshi, K. (1999). Alkaliphiles: Some applications of their products for biotechnology. *Microbiol. Mol. Biol. Rev.*, 63(4):735-750.

Izumi, M., Fujiwara, S., Shikari, K., Takagi, M., Fukui, K. andImanaka, T. (2001). Utilization of immobilized Archaeal chaperonin for enzyme stabilization. *J. Biosci. Bioeng.*, 91: 316–318.

Johnson, F. H., Efyring, H. and Poiassar, Mi. J. (1954). The kinetic basis of molecular biology, John Wiley & Sons, Inc., New York. pp. 286-368.

Kato, C., Li, L., Tamaoka, J. and Horikoshi, K. (1997). The molecular biology of barophilic bacteria. *Extremophiles*, 1: 17–23.

Kumar, S. and Satyanarayana, T. (2003). Purification and kinetics of a raw starch-hydrolyzing, thermostable and neutral glucoamylase of the thermophilic mold *Thermomucor indicae-seudaticae*. *Biotechnol. Progr.*, 19: 936–944.

Lillestol, R.K., Redder, P., Garrett, R.A. and Brugger, K. (2006). A putative viral defence mechanism in archaeal cells. *Archaea*, 2:59-72.

Lynn, J., Rothschild and Mancinelli, R. L. (2001). Life in extreme environments. *Nature*, 409: 1092-1101.

Macelroy, R. D. (1974). Some comments on the evolution of extremophiles. *Biosystems*, 6:74-75.

Marashi, S.A. and Ghalanbor, Z. (2004). Correlations between genomic GC levels and optimal growth temperatures are not 'robust'. *Biochem. Biophys. Res. Commun.*, 325:381-383.

Meargeay, M., Nies, D., Schlegel, H.D., Gerits, J., Charles, P. and van Gijsegem, F. (1985). *Alcaligenes eutrophus* CH34 is a facultative chemolithtroph with plasmid-borne resistance to heavy metals. *J. Bacteriol.*, 162: 328-334.

Minton, K. W. (1994). DNA repair in the extremely radioresistant bacterium *Deinococcus radiodurans*. *Mol. Microbiol.*, 13: 9-15.

Morita, R. Y. and Moyer, C. L. (2001). Psychrophiles, origin of. In: *Encyclopedia of Biodiversity*, Academic Press, New York, vol. 4: 917–924.

Morita, R. Y. (1975). Psychrophilic bacteria. *Bacteriol. Rev.*, 39:144-167.

Musto, H., Naya, H., Zavala, A., Romero, H., Alvarez-Valín, F. and Bernardi, G. (2004). Correlations between genomic GC levels and optimal growth temperatures in prokaryotes. *FEBS Lett.*, :573:73

Nakasone, K., Akegami, A., Kato, C., Usami, R. and Horikoshi, K. (1998). Mechanisms of gene expression controlled by pressure in deep-sea microorganisms. *Extremophiles*, 2: 149–154.

Narang, S. and Satyanarayana, T. (2001). Thermostable *a*-amylase production by an extreme thermophile *Bacillus thermoleovorans*. *Lett. Appl. Microbiol.*, 32: 31–35.

Newcomb, T. G. and Loeb, L. A. (1998). DNA repair in prokaryotes and lower eukaryotes. *DNA Damage and Repair*, 1: 65-84.

Nicholson, C. A. and Fathepure, B. Z. (2004). Biodegradation of benzene by halophilic and halotolerant bacteria under aerobic conditions. *Appl. Env. Microbiol.*, 70:1222–1225.

Nies, D.H. (1999). Microbial heavy metal resistance. *Appl. Microbiol. Biotechnol.*, 51: 730-750.

Nies, D.H. and Silver, S. (1995). Ion efflux systems involved in bacterial metal resistances. *J. Indus. Microbiol.*, 14: 186-199.

Omer, A.D., Zago, M., Chang, A. and Dennis, P.P. (2006). Probing the structure and function of an archaeal C/D-box methylation guide sRNA. *RNA*, 12:1708-1720.

Olson, G.J., Brierley, J.A. and Brierley, C.L. (2003). Bioleaching review part B: progress in bioleaching: applications of microbial processes by the minerals industries. *Appl. Microbiol. Biotechnol.*, 63: 249-257.

Paul, S., Bag, S.K., Das, S., Harvill, E.T. and Dutta, C. (2008)."Molecular signature of hypersaline adaptation: insights from genome and proteome composition of halophilic prokaryotes." *Genome Biol.*, 9(4):R70.

Potts, M. (1994). Desiccation tolerance of prokaryotes. *Microbiol. Rev.*, 58, 755-805.

Pramanik, A., Gaur, R., Sehgal, M. and Johri, B. N. (2003). Oligotrophic bacterial diversity of Leh soils and its characterization employing ARDRA. *Curr. Sci.*, 84: 1550–1555.

Raghukumar, C., Raghukumar, S., Sharma, S. and Chandramohan, D. (1992). Endolithic fungi from deep-sea calcareous substrata; isolation and laboratory studies. *Oceanography of the Indian Ocean,* (Ed.) B.N. Desai: 3-9.

Rohwerder, T., Gehrke, T., Kinzler, K. and Sand, W. (2003). Bioleaching review part A: progress in bioleaching: fundamentals and mechanisms of bacterial metal sulfide oxidation. *Appl. Microbiol. Biotechnol.*, 63:239-248.

Russell, N.J . (1997). Psychrophilic bacteria – molecular adaptations of membrane lipids. *Comp. Biochem. Physiol. A. Physiol.*, 118: 489–493.

Satyanarayana, T., Noorwez, S. M., Kumar, S., Rao, J. L. U. M., Ezhilvannan, M. and Kaur, P. (2004). Development of an ideal starch saccharification process using amylolytic enzymes from thermophiles. *Biochem. Soc. Trans.*, 32 (2): 276–278.

Seckbach, J. (1997). "Search for life in the universe with terrestrial microbes which thrive under extreme conditions". In: *Astronomical and Biochemical Origins and the Search for Life in the Universe,* (Eds.) Cristiano Batalli Cosmovici, Stuart Bowyer and Dan Wertheimer, Milan: Editrice Compositori: 511.

Sharma, A. (2000). *Thermo-alkali stable celluase-free Xylanase of an Extreme Thermophile Bacillus (Geobacillus) thermoleovorans*, Ph D thesis, University of Delhi, New Delhi, pp. 177.

Song, Q. and Zhang, X.B. (2008). Characterization of a novel non-specific nuclease from thermophilic bacteriophage GBSV1. *BMC Biotechnol.*, 8: 43.

Staiano, M., Bazzicalupo, P., Rossi, M., D. and Auria, S. (2005). Glucose biosensors as models for the development of advanced protein-based biosensors. *Mol. Biosyst.*, 1: 354-362.

Tang, T.H., Polacek, N., Zywicki, M., Huber, H., Brugger, K., Garrett, R., Bachellerie, J.P. and Huttenhofer, A. (2005). Identification of novel noncoding RNAs as potential antisense regulators in the archaeon *Sulfolobus solfataricus*. *Mol. Microbiol.*, 55:469-481.

Tyrrell, R. M. (1991). UVA (320-380 nm) radiation as an oxidative stress. In: *Oxidative stress : oxidants and antioxidants*, (Ed.) H. Sies, London: Academic Press:57-83.

Uma Maheswar Rao, J. L. and Satyanarayana, T. (2003a). Enhanced secretion and low temperature stabilization of a hyperthermostable and Ca^{2+}-independent *a*-amylase of *Geobacillus thermoleovorans* by surfactants. *Lett. Appl. Microbiol.*, 36 : 191–196.

Uma Maheswar Rao, J. L. and Satyanarayana, T. (2003b). Statistical optimization of a high maltose-forming, hyperthermostable and Ca^{2+}independent *a*-amylase production by an extreme thermophile *Geobacillus thermoleovorans* using response surface methodology. *J. Appl. Microbiol.*, 95: 712–718.

Valenzuela, L., Chi, A., Beard, S., Orell, A., Guiliani, N., Shabanowitz, J., Hunt, D.F. and Jerez, C.A. (2006). Genomics, metagenomics and proteomics in biomining microorganisms. *Biotechnol. Adv.*, 24:197-211.

Venkateswaran, A., McFarlan, S. C., Ghosal, D., Minton, K. W., Vasilenko, A., Makarova, K., Wackett, L. P., Daly, M. J. (2000). Physiologic determinants of radiation resistance in *Deinococcus radiodurans*. *Appl. Environ. Microbiol.*, 66: 2620-2626.

Wright, J. C. (1989). Desiccation tolerance and water-retentive mechanisms in tardigrades. *J. Exp. Biol.*, 142: 267-292.

Xu, Y., Nogi, Y., Kato, C., Liang, Z., Ruger, H.J., De Kegel, D. and Glansdorff, N. (2003). *Moritella profunda* sp. nov. and *Moritella abyssi* sp. nov., two psychropiezophilic organisms isolated from deep Atlantic sediments. *Int. J. Syst. Evol. Microbiol.*, 53: 533–538.

ZoBell, C.E. and Johnson, F.H. (1949). The influence of hydrostatic pressure on the growth and viability of terrestrial and marine bacteria. *J. Bacteriol.*, 57: 179-189.

□□□

Microbial Diversity and Functions, 2012
© D.J. Bagyaraj, K.V.B.R. Tilak, H.K. Kehri (eds.), pp. 507-536
New India Publishing Agency, New Delhi (India)
E-mail : info@nipabooks.com; Website : www.nipabooks.com

Chapter 25

Innovative Package of Microbes and Fly Ash for the Reclamation of Salt Affected Soils

Rollie Varma and Avinash Pratap SIngh

ABSTRACT

Today more than 70% of energy is generated by coal-based thermal power plants. Indian coal contains about 40% ash, these power plants generate enormous amounts of fly ash, which is dumped in the nearby areas. This creates not only economy constraints but also imposes serious environmental problems, which are the potential hazards to the ecosystem. However, recent attempts in India and abroad are being made for biological association of fly ash e.g. in soil improvement, in wasteland management and also as a source of essential plant nutrients for nutrient deficient soil in terrestrial eco-system and agro-ecosystem. Fly ash has been found to have manifold advantages in agriculture especially in modification of soil texture and bulk density, in optimization of soil pH, in improvement of water holding capacity of soil, in the improvement of yield as a micronutrient supplement to soil and in creation of conducive condition for better plant growth. In this article utilization of fly ash along with VAM fungi and other microbes especially nitrogen fixers and phosphate solubilizers in removing the constraints of salt affected soils has been discussed.

Keywords: Fly ash, salt affected soils, reclamation, VAM fungi, N_2 fixers, phosphate solubilizers

Fly Ash Problem

Today more than 70% of energy is generated by coal-based thermal power plants. There are about forty major power plants in India and coal is the major commercial energy source. Generally powdered bituminous and sub-bituminous coal is combusted in electric generating plants. During this combustion huge amount of mineral residue is produced known as fly ash. It is also called as CCP (Coal Combustion Product). Since Indian coal contains about 40% ash, these power plants generate enormous amounts of fly ash, which is dumped in the nearby areas. A typical modern coal fired power station in India burns 0.7 T/MW/Hr of coal, consequently generating ash at the rate of 0.28 T/MW/hr. As per available estimates, the production of coal ash in India, including both fly ash and bottom ash, is likely to touch 140 million tones per year by 2020. Disposal of such huge amounts of fly ash not only requires more than 30,000 hectares of agricultural and forest lands, but also becomes a potential source of metal contamination in surface and ground water, threatening human health. Fly ash disposal poses problems in the form of land use, health hazards and environmental dangers. In India, the prevalent practice of fly ash disposal is to dump in lagoons, ponds, semi-agriculture lands and wastelands, which had lain to waste thousands of hectares all over the country. This creates not only economy constraints but also imposes serious environmental problems, which are the potential hazards to the ecosystem. To prevent fly ash from getting airborne, the dumping has to be constantly kept wet by sprinkling water over the area. The coal industry in USA spends millions of dollars on lining fly ash dumping grounds. But in India, these sites are not lined and it leads to seepage, contaminating surface and sub-surface water and soil. It lowers soil fertility and contaminates surface and ground water as it can leach into subsoil some toxic metals like cadmium, chromium, arsenic and lead etc., contained within the waste. When fly ash gets into the natural draining system, it results in siltation and clogs the system and reduces the pH and potability of water making it turbid. To minimize all these effects, large scale utilization of coal ash in different areas is gaining global attention. Since, international and national regulations are being formulated aiming at environmental protection and eco-system conservation, the disposal of waste has become more and more costly. Efforts are being made to reutilize fly ash for more useful and profitable exploitations and commercial purposes.

At present, large quantity of fly ash is being dumped in slurry form in large areas close to the power plants without being put to gainful use. Only a very small percentage (<3.5%) of fly ash generated in India is being used for gainful applications, whereas the corresponding figures of other countries may vary from 20 to 80%. Owing to its pozzolanic property, the main emphasis up to now in our country has been towards the utilization of fly ash for low and medium value applications like bricks, part replacement of cement and in mass concrete for dams, for paving roads and airport runways, embankments, sea-port fill etc. However, recent attempts in India and abroad are being made for biological association of fly ash e.g. in soil improvement, in wasteland management and also as a source of essential plant nutrients for nutrient deficient soil in terrestrial eco-system and agro-ecosystem. Fly ash has been found to have manifold advantages in agriculture especially in modification of soil texture and bulk density, in optimization of soil pH, in improvement of water holding capacity of soil, in the improvement of yield as a micronutrient supplement to soil and in creation of conducive condition for better plant growth.

Problems of salt affected soils

Land is a finite resource and due to increasing population and escalating demands associated with accelerated needs, there is a considerable pressure on land on account of the competing land uses. It is therefore, natural as well as necessary that our country with a very high man-to-land and animal-to-land ratios should turn its attention to the less fertile and problem soils which are at present lying fallow.

Large areas of land in the country called wastelands are degraded and lying unutilized due to various constraints. Out of India's total geographical area of 329 mha, 175 mha is either completely or partially unproductive. Part of it called culturable has the potential for the development of vegetation cover. In addition to these natural wastelands, there are large tracts of wastelands created due to human activities such as mining, deforestation etc. Such wastelands are also culturable and a vegetation cover can be developed on them after removing the constraints with some efforts. In view of the increasing shortage of plant resources due to population explosion, it has become imperative that all the wastelands are put to use by developing vegetation cover. The problem of wastelands in the country is of great concern to all of us. Absence of ground cover on such lands is causing not only soil erosion and silting of reservoirs but also posing a major threat to the agricultural production through deposition of sand over the fertile land adjoining them. The recovery of such lands in the country needs a high priority.

Most of the wastelands suffer from various kinds of stresses e.g., extremely poor nutritional status of the soil often associated with conditions of acidity or salinity, unfavourable climatic conditions and other agronomic constraints. Salt affected soils occupy approximately seven percent of the earth's land surface and agronomic and forest crop production on these sites is relatively low.

Salt-affected soils occupy extensive area in the world and in India as well, and present a serious impediment to crop production. The area of salt-affected soils in the world is about 952 mha as per map prepared by UNESCO/FAO. Out of which about 7 mha of salt-affected soils are found in India, out of which, 2.5 mha occurs in Indo-Gangetic plains covering the states of Uttar

Pradesh, Haryana, Punjab, Delhi and parts of Bihar. In Uttar Pradesh alone about 1.29 mha is salt affected and commonly known as '*usar*' or '*reh*' in local language.

These soils are completely barren and practically produce nothing. Plant growth and development in such soils is adversely affected either due to excessive amounts of neutral soluble salts or high exchangeable sodium or both. In India approximately 7 mha land is salt affected out of which, 2.5 mha occurs in Indo-Gangetic plains covering the states of Uttar Pradesh, Haryana, Punjab, Delhi and parts of Bihar. In Uttar Pradesh alone about 1.29 mha is salt affected and commonly known as '*usar*' or '*reh*' in local language.

Salt affected soils can be classified into three different categories, *viz.* saline, alkaline/sodic and saline-alkaline/saline-sodic. Saline soils are the soils for which the electrical conductivity of the saturation extract is more than 4.0 $m.mhos.cm^{-1}$, exchangeable sodium percentage is less than 15% and pH is less than 8.5. Whereas, in case of alkaline/sodic soils electrical conductivity of the saturation extract is less than 4.0 $m.mhos.cm^{-1}$, exchangeable sodium percentage is more than 15% and pH usually ranges between 8.5-10. Contrary to this, the term saline-alkaline is applied to the soils for which electrical conductivity of the saturation extract is greater than 4.0 $m.mhos.cm^{-1}$ and exchangeable sodium percentage is also greater than 15%, whereas pH is seldom higher than 8.5. These soils are usually formed as a result of combined processes of salinization and alkalization.

Extensive occurrence of alkaline soils has been reported from the Indo-Gangetic plains of Northern India. The agricultural history of the region suggests that these high alkaline and sodic lands have been left unproductive in this area for a long time. The area experiences intermittent dry and rainy seasons throughout the year and considerable rainfall (160–200 cm/yr) which is followed by fast evaporation leaves behind a white encrustation on the soil surface. The efflorescent crust consists predominantly sodium carbonate minerals ($Na_2CO_3.NaHCO_3.2H_20$) with minor thermonatrite ($Na_2CO_3.H_2O$). Being highly soluble, the presence of these minerals alone explains the high alkalinity of the soil (upto pH 10.5).

High alkalinity and high exchangeable sodium percentage (ESP) of the soil render it inhospitable for normal crop production and there is minimal bioproductivity in such soil. Accumulation of CO_3^- and HCO_3^- ions has a serious impact on the growth of vegetation due to their toxic effects. High pH affects nutrient solubility and high ESP decreases the availability of important nutrient elements, such as Ca, Mg and K. The organic matter content of such soil is also low and ranges between 0.1–0.2% only. Soil dispersion due to high

ESP deteriorates the physical condition of the soil and affects the air and water permeability. Due to low infiltration capacity, rain water stagnates on the soil easily and, in dry periods, irrigation is hardly possible. Agriculture in such soils is limited only to the crops tolerant to surface waterlogging. Soil compaction adversely affects root growth and development due to which seedlings often fail to germinate and establish. Thus overall plant growth is adversely affected under such extreme stress conditions.

Research efforts over the past two decades have resulted in evolving a number of practices for the reclamation and management of such soils for sustained crop production. Reclamation with raising moderately salt tolerant plants in combination with organic/chemical amendments is a common practice at governmental as well as non-governmental levels. However, most of the reclamation programs are not successful and ending into high rate of mortality. Development of all such land is urgently needed to relieve the pressure on the arable lands.

Development of plant cultivars that can produce economic yields under such soil conditions, as well as phytoremediation procedures that increase, by whatever mechanism, plant tolerance to salt stress are more permanent and complementary solutions. In this respect, biological processes such as mycorrhizal application to alleviate salt stress and use of moderately salt-tolerant species are better options.

Successful plantation on such lands requires a well-planned programme involving use of convenient and economical scientific inputs and strategies able to overcome the adversaries associated.

In this article the utilization of fly ash (as nutrient supplement) along with the microbial inoculants especially VAM fungi, nitrogen fixers and phosphate solubilizers in removing the constraints of salt affected soils has been discussed.

Fly Ash vs Plant Growth

Field and greenhouse studies both indicate that many chemical constituents of fly ash may benefit plant growth and can improve agronomic properties of soil. The effects on plants were primarily due to a shifted chemical equilibrium induced by the fly ash added to the soils. Kalra *et al.* (2003) reported that the silt dominant texture of fly ash improved loamy sand to sandy loam textures of the surface soils. The increased growth in yield of crops with fly ash incorporation was possibly due to modifications in soil moisture retention and transmission characteristics, bulk density, physico-chemical characters such as pH and EC and organic carbon content (Phung *et al.*, 1978). Adriano and Weber (2001) reported that fly ash amendment in soil at unusually higher rates substantially improved water holding capacity and plant available water without adversely affecting the growth. Lau and Wong (2001) reported that the addition of weathered coal fly ash at 5% resulted in higher seed germination rate and root length of lettuce (*Lactuca sativa*). Tripathi and Sahu (1997) reported that addition of increasing percentage of fly ash to soil resulted in increased germination, plant growth and yield of wheat. They reported that fly ash applied at the rate of 50% caused a considerable increase in pH, organic carbon, total nitrogen and available P of the soil. Thus, fly ash was suitable to be used as land fertilizer in lieu because of its essential micro elements, which can improve the quality of soil and crop production. Availability of greater amount of P, K, Ca, B, Mg, Mn, etc. (Wong and Wong, 1989; Khan and Khan, 1996) in ash amended soil are responsible for the improved growth and flowering of the plants. Khan (2002) proposed that the soil having 25-75% ash suited best either in maintaining a normal growth and flowering in lonia, lineria and gomphrina or inducing some enhancement. Khan and Khan (1996) recorded increase in both flowering and fruiting rates of tomato by the application of fly ash. A gradual increase in the concentration of fly ash from 40-80% induced an appreciable increase in the number of flowers/plant compared to the control (0% fly ash).

Higher yields of several agronomic crops were recorded on fly ash application (Page *et al.*, 1979; Plank and Martens, 1974; Elseewi *et al.*, 1978 b). These yield increases were attributed to increased availability of S to plants from fly ash and to an alleviation of B deficiency by field application of fly ash. Mahalingum (1973) obtained substantial increases in rice (*Oryza sativa*) yields by applying low rates of fly ash to a saline-alkali soil, thus, lowering the soil pH from 9.1 to 7.5. The S contents of lettuce, Swiss chard (*Beta vulgaris*), corn (*Zea mays*) and beans (*Phaseolus vulgaris*) were also increased by fly ash treatment irrespective of soil and ash type (Elseewi *et al.*, 1978 a). Kalra *et al.* (2003) reported that the shoot and root growth and yield of wheat (*Triticum aestivum* L.), mustard (*Brassica juncea* L.), lentil (*Lens esculenta* Moench.), rice (*Oryza sativa* L.) and maize (*Zea mays* L.) increased by the addition of varying levels of fly ash at the time of sowing/transplanting. Fly ash amendments have corrected plant nutritional deficiencies of B, Mg, Mo, S and Zn. This resulted in better growth and yield of plants.

Fly Ash vs Organic Matter

Organic amendments improve soil conditions by increasing cation-exchange capacity and organic matter content of the soil, thereby resulting in increased immobilization of toxic elements, higher fertility and enhanced microbial activity. Due to stress conditions and poor substrate availability on fly ash dump, it was necessary to give some supportive source of nutrient as organic amendment. Presence of organic matter has an additive effect as it reduces the concentration of toxic metals through sorption, lowers the C/N ratio and provides organic compounds, which promote microbial proliferation and diversity (Pitchel and Hayes, 1990). Certain inhibitory effects of soil microbes by toxic components of fly-ash may, furthermore, be attenuated by the application of organic materials. The strong correlation between pH correction and nutrient availability in the soil suggests that in fly ash, most elements are associated with the mineral phase. One can therefore, expect that interaction between the predominantly inorganic fly ash and organic matter may further enhance its beneficial effect on plant growth in problem soils (Page *et al.*, 1979).

Juwarkar and Jambhulkar (2007) reported that inoculation of biofertilizer and application of organic matter (FYM) helped in reducing the toxicity of heavy metals such as cadmium, copper, nickel and lead which were reduced by 25, 46, 48 and 47%, respectively, due to the increased organic matter content in the fly ash which complexes the heavy metals thereby decreasing of metals. Amendment of fly ash with FYM and biofertilizer helped in profuse root development showing 15 times higher growth in *Dendrocalamus strictus* plant as compared to the control. Addition of organic rich amendments to soils resulted in increased values of water-holding capacity, wilting point and available water capacity, surface stability, aeration and water penetration by alteration of the soil structure.

Vesicular Arbuscular Mycorrhizal Fungi

It has been established during the last few decades that vesicular arbuscular mycorrhizal (VAM) association is helpful in improving the overall performance of the plants. This association is gaining importance in the reclamation of wastelands and researches on VAM fungi have reached a stage of practical application in many countries for reclaiming the adverse sites. VA mycorrhizae have a number of assets to prove it a magic tool e.g. VA mycorrhizae increase the absorptive surface of roots manifold and improve the uptake of nutrients and water resulting in better performance of host plants. Mycorrhization ensures better supply of water and nutrients especially phosphorus, calcium and zinc to the plants, thereby improves growth and productivity. VAM fungi have been reported to increase the activity of nitrogen fixing organisms. Vesicular arbuscular mycorrhizae play a vital role in mineral cycling, energy flow and plant establishment in disturbed ecosystems, arid and semi arid zones as well as wastelands. They have been proved to be a potential tool for reclamation of alkali, calcarcous, saline and other types of problematic soils. Their ability to initiate plant succession in virgin lands, to increase tolerance to heavy metals and drought has been recognized. VAM fungi have been shown to decrease transplanting shocks and improve soil structure. Their association with plants has been reported to affect osmotic balance and hydraulic conductivity of roots and improve plant-water relations. It has also been reported to increase lignifications of root tissue, increase chitinolytic activity, higher concentration of phosphorus and amino acids or production of phytoalexins. Mycorrhizal association has been shown to affect the hormone levels, initiate rooting and help in the root development. VAM technology has proved to be effective even under unfavorable conditions related to nutrients or water. The technology has proved to be helpful not only in the establishment but also the growth and productivity of plants. There are conclusive evidences to show the importance of VAM fungi in reclamation of salt affected soils. Mycorrhizal advantage occurs mainly through the increased concentration of phosphate and decreased concentration of sodium in shoot tissues compared to non-inoculated plants, or due to diurnal changes in the leaf thickness or increased tolerance to temperature fluctuations and pH of the soil.

Saline-alkaline soil supports poor vegetation owing to high pH, high salt content and low nutrient level. VAM play a significant role in the establishment of plant communities in the disturbed and stressed ecosystems but vary considerably in their symbiotic effectiveness according to soil conditions and cultural stresses. The magnitude of the effect of mycorrhizal infection is generally correlated with the extent of colonization of the root system by VAM fungi.

Mycorrhizal association in plants has been reported to play a key pivotal role in coping with various stresses posed by physical and chemical environment of soil like nutrient deficiency. Selection of efficient VAM fungi is a new tool in dry land agriculture and agro-forestry. VAM enhance the soil water holding capacity and improve plants water status under drought condition besides increasing uptake of macro- and micronutrients. In infertile habitats like eroded and degraded soils, the mycorrhizae confer greater root longevity by absorbing nutrients beyond the depletion zone. In some soils, they play a direct role in soil aggregation. The role of VAM fungi in land rehabilitation is well established through previous reports. Salinity problem in arid and semi arid conditions is caused due to nutritional imbalances e.g., excess of Cl^- can interfere with NO_3^- and PO_4^{2-} uptake and that of Na^+ can inhibit Ca_2^+ and Mg_2^+. VAM fungi have been reported to increase plant tolerance to salinity. It has been suggested that the tolerance of salinity in mycorrhizal plants is due to improvement of 'P' uptake and increase of K^+ content in plants. Decrease in Cl^- concentration in mycorrhizal plants has also been associated with the tolerance of plants to the salinity; however, the mechanism is not clear. Improved salt tolerance of VAM plants has been attributed to enhanced mineral nutrition, improvement in physiological processes like photosynthesis or water use efficiency, and production of osmoregulators. Some studies show that native VAM can grow and function well in soils from which they are isolated e.g., polluted soils, semi arid degraded soils and agricultural systems. VAM ecotypes occurring in the saline-alkaline soils are hypothesized to be more adapted to high pH and salinity than VAM found in non-stressed areas. The isolation and propagation of these native VAM could improve success in the production of tolerant plants, which can be outplanted into disturbed polluted ecosystems. The role of VAM in saline and drought conditions is very important for increasing fertility. The potential of mycorrhizal inoculum can be utilized for the growth and development of plants in the stressed conditions through various ways, such as (a) enhancing establishment and survival of transplanted seedlings in adverse conditions, (b) improving plant growth rate by increasing nutrient uptake, (c) improving ability of host plant to compete root/soil borne plant pathogens, (d) boosting capacity of host plant against stresses, (e) increasing efficiency of nutrient recycling and (f) stabilization and aggregation of the soil. Mycorrhizal technology therefore, has assumed greater relevance in crop production in stressed conditions. They play significant role in the establishment and growth of plants/seedlings.

VA Mycorrhizal Fungi vs Salt-Affected Soils

Nearly seven percent of the land at the global level is saline due to the presence of excess soluble salts in the soil (Dudal and Purrell 1986). Salt stress has three fold effects: it reduces water potential, causes ion imbalance or disturbance in ion homeostasis and toxicity. Salt stress involves both osmotic and ionic stresses (Hayashi and Murata, 1998; Munns, 2002; Benlloch-Gonzales *et al.*, 2005), so growth suspension is directly related to total concentration of soluble salts or osmotic potential of soil water (Flowers, 2004). There are conclusive evidences to show the importance of VAM fungi in reclamation of saline soils (Pfieffer and Bloss 1988; Rozema, *et al.*, 1986). Mycorrhizal advantage accrues through increased concentration of phosphate and a decreased concentration of sodium in shoot tissues in *Parthenium ageratum* plants inoculated with *Glomus intraradices* compare to non-inoculated plants (Pfieffer and Bloss, 1988), due to diurnal changes in the leaf thickness as in *Aster tripolium* (Rozema, *et al.*, 1986) or increased tolerance to fluctuations in temperature and pH of the soil (Poss, *et al.*, 1985). The mycorrhizal plants under high salinity, however, have greater concentrations of potassium, sodium and phosphorus (Allen and Cunningham, 1983) where as VAM plants maintained a steady Na/K balance than the non-mycorrhizal plants. Thapar *et al.* (1991) made a survey of native VAM fungi in sodic soils of Haryana State and mentioned the presence of 13 *Glomus* spp., 1 *Scutellaspora* sp. and

Sclerocystis sp. with *Glomus* most abundant. Thaper and Uniyal (1996) studied the effect of VAM fungi and *Rhizobium* on growth of *Acacia nilotica* in sodic and New Forest soils and reported significant improvements in growth after inoculation with VAM alone. Mathur and Vyas (2000) studied the mycorrhizal dependency of *Tamarix aphylla* in saline areas of Thar Desert and reported that *G. margarita* was the most efficient VAM fungi for *T. aphylla* as it significantly increased the plants biomass and nutrient uptake by effectively conlonizing the roots. Giri and Mukherji (1999) suggested improved growth and productivity of *Sesbania grandiflora* (Pers) under salinity stress through mycorrhizal technology. Jindal *et al.* (1992) seen the effect of NaCl salinity on metabolism of vesicular arbuscular mycorrhizal inoculated moong plants and reported that mycorrhizal moong plants had higher concentrations of plant phosphorus, total sugars, K, Zn, Fe and leaf nitrate in comparison to non-mycorrhizal plants. Rinalolelli and Mancuso (1998) studied the short-term and long-term response of mycorrhizal and non-mycorrhizal olive plants *(Olea aeuropae* L.) to saline condition and seen that leaves of mycorrhizal plants showed a higher K/Na ratio than those of nonmycorrhizal plants, confirming the existence of a efficient root Na-exclusion mechanism in mycorrhizal roots. Asgharzadeh *et al.*(2001) worked on occurrence of arbuscular mycorrhizal fungi in saline soils of Tabriz Plain in relation to some physical and chemical properties of soil and concluded that glycophytes could benefit from VAM fungi even in saline conditions and halophytes have a positive effect on spore number and survival under salinity stresses. Cantrel and Linderman (2000) studied that preinoculation of lettuce or onion plants with mixtures of VAM fungi from saline or non-saline soils was an effective means of increasing plant tolerance to salt toxicity. Abo-Ghalia, (2001a) has seen synergetic interaction between the VA-mycorrhiza *Glomus intraradices* and *Saccharomyces cerevisiae* upon the growth and nutrition of *Sorghum bicolar* grown under saline conditions and reported that the mixture of two organisms was more effective in over-coming the salinity stress than each one alone. Abo- Ghalia (2001b) again worked on synergistic interaction between the VAmycorrhiza *Glomus intraradices* and *Saccharomyces cerevisiae* II on the metabolic activities of *Sorgham bicolar* grown under saline conditions and reported increase in accumulation of total soluble sugars, proline and total free amino acids especially, when used together. Such increases were related to plant tolerance to salinity. Yuan and Runjin (2002) studied the arbuscular mycorrhizal fungi in saline–alkaline soils of Yellow River Delta of China and isolated two species of *Archaespora,* twenty four species of *Glomus* and six species of *Acaulospora.* Landwehr *et al.* (2003) worked on the VAM fungus *Glomus geosporum* in European saline sodic and gypsum soils. Yano-Melo *et al.* (2003) have been seen the tolerance of mycorrhizal banana *(Musa* sp.cv. Pacovan) plantlets to saline stress and concluded that inoculation with specific VAMF constitutes an alternative method to reduce banana plant stress caused by soil salinization. Many studies have demonstrated that inoculation with VAM fungi improves growth of plants under a variety of salinity stress conditions (Diallo *et al.*, 2001; Burke *et al.*, 2003; Tain *et al.*, 2004). Recently Rabie, 2005 suggested that VAM fungi protected the host plants against the detrimental effects of salt. Mycorrhizae were involved in protection against salt stress, *via* better access to nutritional status (Zandavalli *et al.*, 2004) and modification of plant physiology *i.e.*, osmotic modifications (Rao and Tak, 2002) and photosynthesis (Merguihur *et al.*, 2002). To some extent, these VAM fungi have been considered as bio-ameliorators of saline soils (Yano- Melo *et al.*, 2003; Tain *et al.*, 2004).

VAM Fungi vs Fly Ash

Plants growing in fly ash showed a significant increase in the shoot and root dry mass in comparison to uninoculated plants (Garampalli *et al.*, 2005; Juwarkar and Jambhulkar, 2007; Ammaiyappan and Ayyamperumal, 2002; Kulshreshtha and Khan, 1999; Reddy and Garampalli, 2002). Mycorrhizal fungi, through their mycelia network, accumulate heavy metals from fly ash and retain them within their cells or carry them on their body surface when they form association with the plants. These mycelia threads, along with dense root biomass, assist in binding ash particles (Singh *et al.*, 2006).

Bi *et al.* (2003) showed that the root colonization by *Glomus mosseae* and *G. versiforme* gave higher maize yields, higher plant uptake of most of the nutrients studied and protected the plants from excessive accumulation of Na in the shoots when grown in soil overlying fly ash. The effect of mycorrhizal colonization on the distribution of Na between roots and shoots was very striking. This may have been due to interaction with the increased K absorption by mycorrhizal plants, but this interpretation would require further studies for verification. Although, the mycorrhizal plants had shoot P concentrations slightly below the critical level (0.25%), they had about twice the concentrations of non-mycorrhizal controls. Concentrations of other nutrients examined were always higher than the critical levels. *G. mosseae* generally enhanced plant growth more than *G. versiforme*, indicating that different VAM fungal spores have different mycorrhizal effectiveness. The use of relatively undisturbed top-soil will introduce a suite of indigenous VAM fungi, but more degraded materials will need to be inoculated with an effective fungal strain or a mixture of fungal strains. The roots had penetrated the fly ash in all treatments, but further work is required to differentiate between uptake from the soil and from the fly ash. Soil microorganisms are important components in re-vegetation of disturbed and potentially toxic environments because they can contribute to nutrient availability, immobilize heavy metals in the soil, and bind soil particles into stable aggregates that improve soil structure and reduce erosion potential. They further indicated that successful growth of maize is possible in soil overlying coal fly ash and can be improved by colonization of the plant roots by mycorrhizal fungi. Thus, remediation of areas in-filled with coal ash may be possible using either undisturbed soil containing *via*ble propagules of indigenous VAM fungi or disturbed soil inoculated with effective strains of VAM fungi. The fungi can assist plants in the exploitation of soil and possibly fly ash, nutrients and may help them to resist excessive salt (Na) accumulation in the shoots. VAM fungi may also contribute to the re-establishment of a general microflora, and of a sustainable agricultural system combined with the appropriate use of fertilizers. According to them the plants can absorb P directly from the soil, and the external hyphae of the mycorrhizal fungi were also reported to contribute to P uptake from the substrate, possibly including the fly ash. Li *et al.*, (1991) reported that VAM hyphae can lower the pH of rhizosphere soil, leading to an increase in plant uptake of soil. The mycorrhizal fungi led to an increase in K uptake from soil, and possibly from the fly ash.

Singh *et al.* (2006) reported that VAM inoculation in *Jatropha curcas* plants growing in fly ash showed a significant increase in the shoot and root dry mass in comparison to uninoculated plants. P concentration in shoots and roots of *J. curcas* was found to be influenced by AM inoculation, and was significantly higher in the mycorrhizal plants in comparison to the un-inoculated plants. Mycorrhizal fungi, through their mycelia network, accumulate heavy metals from fly ash and retain them within their cells or carry them on their body surface when they form association with the plants. These mycelia threads, along with dense root biomass, assist in binding ash particles.

The Centre for Mycorrhizal Research at TERI used the mycorrhizal technology at the fly ash dumping sites of two power stations in Badarpur (Delhi) and Korba (Chattisgarh). The plant species which form mycorrhizal association showed better plant growth, tolerance against biological and environmental stresses, and seedling survival. These micro-organisms provide nutrition to the plant by sequestering the nutrients from the soil and translocating them to the plant; in return, they get carbon from the plants. This makes the utilization of the nutrients highly efficient and reduces the dependence on external chemical inputs. They recorded a significant restoration in soil quality with respect to percentage nitrogen, available phosphorus, available potassium, organic carbon, dehydrogenase activity and total microbial population after two years of plantation. Porosity and water holding capacity also increased significantly over zero time. Organic matter (farmyard manure and compost) was added to help fly ash retain moisture and support growth. With small quantities of compost (no inorganic fertilizer was added) and benefits of mycorrhiza association, these plants can grow and establish themselves on fly ash overburdens. As they grow, their roots hold the fly ash together, making it less prone to being air-borne. In addition, the heavy metal toxicity in fly ash also decreased after two years of plantations. The mycorrhizae, through their mycelia network, accumulate heavy metal from fly ash and retain them within their cells. Besides, supplying nutrients, it produces acids that combine with some heavy metals to form compounds that are less mobile and less likely to pollute groundwater and surface runoff. The heavy metal contents in the leachates also decreased drastically as a result of reclamation activities. The height of various tree species *viz. Shorea robusta, Tectona grandis, Dalbergia sissoo, Albizzia procera, Casuarina equisetifolia, Melia azadirach, Acacia nilotica, Dendrocalamus strictus, Populus euphratica, Eucalyptus tereticornis, Bombex ceiba, Gmelina arborea, Mentha arvensis, Vetiver zizanoides, Polianthes tuberose, Helianthus* sp., *Tagetes erecta, Sesbania aculeata and Agave sisalana* at Korba showed a tremendous improvement during the growth period. Thus, mycorrhizal technology offers a biological means of assuring plant production at a low cost.

Garampalli *et al.* (2005) reported that the response in the pigeon pea (*Cajanus cajan* (L.) Millsp.) cv Maruti plants in terms of higher dry weight, under the influence of fly ash amendment in VAM infested soils was found to be considerably less (though not significant enough) when compared to the control plants (without fly ash) that have otherwise shown significant increase in growth over plants without *Glomus aggregatum* inoculation. However, fly ash amendment without VAM inoculation was also found to enhance the growth of plants as compared to control plants (without fly ash and VAM inoculum).

Reddy and Garampalli (2002) studied the effect of fly ash amendment at three concentrations (10, 20 and 30%) on the infectivity of *Glomus aggregatum* in low fertile soil using pigeon pea (*Cajanus cajan*) cv. Maruti and chickpea (*Cicer arietinum*) cv. Annigeri. They recorded a significant decrease in VAM colonization with increase in fly ash concentration. The effectiveness of *G. aggregatum* under the influence of fly ash was found significantly affected compared to the control, when judged by the growth response of pigeon pea. However, in chickpea, VAM association could slightly increase the growth over its control. Fly ash amendment alone also showed positive influence on the growth of both the crops over their controls (without VAM association). The influence of fly ash amendment together with the influence of VAM fungi, as bioremediation agents can be exploited suitably in the reclamation of wastelands and soils over burdened with fly ash.

Kulshreshtha and Khan (1999) studied the impact of fly ash obtained from a thermal power plant at Aligarh, on *Glomus caledonium* and *Rhizobium* sp. on the roots of *Vigna mungo*. They demonstrated that mycorrhizas and root nodulating bacterium protected the plants from some of the harmful effects caused by fly ash.

Ammaiyappan and Ayyamperumal (2002) surveyed the presence of arbuscular mycorrhizal (AM) species in an abandoned lignite fly ash pond, overburden dumps and reclaimed overburden dumps. From ash pond they isolated 15 VAM fungi (*Acaulospora gerdemannii, Gigaspora decipiens, Gigaspora gigantean, Gigaspora margarita, Glomus citricola, Glomus fasciculatum, Glomus formosanum, Glomus fulvum, Glomus maculosum, Glomus magnicaule, Glomus mosseae, Glomus tenebrosum, Sclerocystis pachycaulis, Scutellospora coralloidea, Scutellospora erythropa, Scutellospora persica and Sclerocystis sinuosa*). In all the sites, *Glomus mosseae* was the dominant VAM fungus.

Juwarkar and Jambhulkar (2007) recorded an increase in phosphate is due to addition VA mycorrhizae, which helps in phosphate immobilization and solubilization. The results observed are well supported by the findings of Kumar and Mishra (1991) where they reported the beneficial role of pressmud application to rice and maize crops but the optimum dose reported was 10 t/ha as against the50 t/ha FYM observed in the present study.

Phosphate Solubilizing Microorganisms

In case of wastelands generally the soils are poor in phosphorus, especially the available phosphorus, limiting the growth of transplanted seedlings. There are a number of soil microbes *viz., Pseudomonas striata, Bacillus subtilis, B. polymyxa, B. megaterium, B. pulvifaciens, B. circulans, Citrobacter* sp., *Aspergillus awamori, Penicillium digitatum, Aspergillus niger* etc. which solubilize insoluble Aluminium and Calcium phosphates and rock phosphate and increase the availability of phosphates to the plants. Increase in plant performance, yield and P uptake has been reported by several workers. Though most of the work has been done on agricultural crops, inoculation of the nursery seedlings with such microbes along with the rock phosphate may be helpful in removing the phosphorus deficiency and improving their performance under extremely stress conditions.

Application of fly ash at 40 t/ha in conjunction with phosphate solubilizer, *Pseudomonas striata* improved the bean yield and phosphorus uptake by grain and fly ash did not exert any detrimental effect on the population of *P. striata* in soil (Gaind and Gaur, 2002). Amendment of Class F, bituminous fly ash to soil at a rate of 505 Mg/ha did not show any detrimental effect on soil microbial communities. Analysis of community fatty acids indicated elevated populations of fungi, including arbuscular mycorrhizal fungi and Gram-negative bacteria. Fly ash and its different combinations with soil (w/w) were tested for use as a carrier for diazotrophs (*Azotobacter chrococcum, Azospirillum brasilense)* and phosphobacteria *(Bacillus circulans, P. striata*) (Gaind and Gaur, 2003 a, b) which showed their maximum *via*bility in fly ash alone or soil-fly ash (1:1) combination. The total occurrence of P in fly ash may be as high as the concentration observed in many organic manures, the P availability in fly as generally remains very low (Kumar *et al.,* 1998). While discussing P insolubility in fly ash (FA), Adriano *et al.* (1980) stated that most studies indicate that FA application caused no substantial changes and in some cases even lowered plant tissue P concentrations. Therefore, to increase fly ash acceptability as a source of plant nutrients, it is necessary to increase its phosphorus bioavailability. Microbial management may be an effective means for increasing such P bioavailability. However, the major problem with microbial management of fly ash is the sensitivity of different microorganisms to fly ash stated by Pitchel and Hayes (1990). To study the bioavailability of P in fly ash Bhattacharya and Chattopadhyaya (2002) conducted an experiment having five different combinations of fly ash (FA) and cow dung (CD). Incorporating organic matter with FA in different combinations resulted in increased P availability. Beneficial effects of organic matter on P transformation through various pathways have been discussed by Mandal (1964), the accelerating effects of vermicomposting may be due to rich microbial populations in the earthworm intestines (Edwards and Lofty, 1972), which might have played an important role in solubilizing P from unavailable forms. To establish this hypothesis, an attempt was made to study the occurrences of PSB in the two series of treatments after 50d of incubation. The inclusion of earthworms in the waste materials resulted in many-fold increases in the concentrations of PSB over the respective series without worms. These microorganisms tended to increase the P solubility in the waste materials. However, such increment was marginal in the FA-only series. The bacterial population probably could not get sufficient energy from FA, which was practically devoid of organic matter. Pitchel and Hayes (1990) have also reported restricted microorganism growth in FA. However, in the present study, when FA was treated with organic material in the form of cow dung at different concentrations, the PSB populations were active and they tended to solubilize P from fly ash to a considerable extent. The total P amount fixed in three major inorganic forms, two Ca and one Fe bounded, tended to decline in the vermicomposted series. This was attributed largely to increased effect of PSB on solubilizing different inorganic forms in this series. The contribution of fly ash to increased availability of P under different treatments was calculated by taking into account the available phosphorus status under treatment after 50 d of incubation. The contribution was lowest under the treatment with FA alone and increased with use of organic materials in different combinations. The FA+CD (1:1) vermicomposted treatment appeared to be the most efficient, probably due to the occurrence of the highest concentrations of PSB in this treatment due to sufficient availability of organic materials. Presence of a good amount of FA in this treatment might have also allowed PSB to extract higher amounts of P from FA into a solubilized form.

Rhizobium

Nitrogen fixing plant species have been emphasized amongst the pioneer communities for the wasteland development. Most of the wastelands are deficient in organic matter and nitrogen, causing a danger for the survival of the transplanted tree seedlings. The nitrogen fixing microbes play an important and decisive role in the restoration of such wastelands. Fixation by free-living bacteria is often negligible because of lack of suitable carbon substrates. However, nodulated plants can be of great use in building up soil fertility. A number of nitrogen fixing tree plants have already been suggested by the scientists for afforestaion of semi arid tracts with partial shifting sand dunes and other wastelands as pioneers.

It is estimated that 30% of legume nitrogen becomes available through decay of root nodules. Failure in effective nodulation results in stunted tree growth, low protein in foliage and consequently poor productivity. Therefore, heavy pelleting on seeds with lime, acid and alkali tolerant strains of rhizobia is recommended to form a normal practice in raising tree seedlings on wastelands to ensure adequate and healthy development of nodules in plants at an early age. The comparative efficiency of different legume species in building up nitrogen contents and increasing growth in a companion grass on mine spoils have been investigated by some researchers. Introduction of free-living nitrogen fixing bacteria in the rhizosphere of revegetated perennial grass species on strip-mined land has been advocated.

Fly ash has a vast potential for use in agronomy as an amendment, especially due to the physical condition and the presence of macro and micronutrients, which are conducive for plant growth; however, it is deficient in N and P. If these nutrient deficiencies are corrected, fly ash lagoons can be covered vegetatively. Many plants belonging to the family leguminosae were demonstrated to have a high tolerance and survival in arid, infertile and metal-contaminated areas (Musil, 1993; Vajpayee *et al.*, 2000; Cheung *et al.*, 2000). In addition, legume trees and symbiotic N2-fixing bacteria can improve N content of infertile soil and amended lagoon ash (Sanginga *et al.*, 1994; Cheung *et al.*, 2000). The use of *Rhizobium* has been reported to be a key factor for plant establishment under xeric and nutrient unbalanced conditions (Requena *et al.*, 1996; Barea *et al.*, 1996). For revegetating fly ash landfills, an integrated approach that envisages appropriate blending of fly ash with organic waste from domestic and industrial sources and the application of biotechnological means involving biological N_2 fixation and phosphate mobilization was considered important in plantation. Application of various nitrogen-fixing cynobacteria and leguminous trees inoculated with *Rhizobium* have been used to enhance N and P status and in reducing toxicity of fly ash landfills (Banerjee and Deb, 1993; Rai *et al.*, 2000, 2004).

Rai *et al.*, (2004) reported that the *Rhizobium* (PJ-1) inoculation (inoculums size 6.4 x 108 cells ml-1) increased the growth of *Prosopis juliflora* in 100% fly ash. A significant increase in the biomass on intact plant, root and shoot, photosynthetic area and nodulation frequency were observed after 45 days of transplantation under field conditions compared to the uninoculated plants grown in 100% fly ash. Similarly, the *Rhizobium*-inoculated plant exhibited more contents of photosynthetic pigments, protein and in vivo NR activity over uninoculated plants. The increased growth and metabolic activity might have facilitated more accumulation of metals in *Rhizobium*-inoculated *P. juliflora*. These results seem possible due to the increased tolerance of the inoculated plants to the stress caused by fly ash. The plants inoculated with *Rhizobium* seem to be more tolerant than the uninoculated plant, which may be due to increased N_2 fixation. Similar results have been reported in the case of *Cassia surattensis*, which proved to be successful candidate in vegetating fly ash lagoons (Vajpayee *et al.*, 2000). Growth promotion of nonleguminous canola and lettuce by application of *Rhizobium leguminosarum* under stress conditions was also given by Noel *et al.*, (1996). *Rhizobium* inoculation enhanced the growth of the plant in nitrogenlimited fly ash, thus it can be concluded that *Rhizobium* inoculation would be beneficial for revegetation as well as decontamination of fly ash contaminated soil and landfills using *P. juliflora*.

Juwarkar and Jambhulkar (2007) recorded an increase in the N content of the fly ash when amended with biofertilizers, which helps in biological nitrogen fixation and is a major source of N input. Biologically fixed nitrogen can thus, contribute to the needs of a growing plant, thus contributing its fertility in long run and in a sustainable manner.

Microbial Inoculants vs Salt affected soils/Fly Ash: A case study

Verma (2008) amended the salt-affected soils of Handia, Allahabad with different concentrations of fly ash (15, 30 and 45%) and organic matter (*Cynodon*/compost 2% w/w) and provided with consortium of VAM fungi, PSB (*Bacillus subtilis*) and N_2-fixer (*Rhizobium spp.*) alone as well as in combination and studied its effect on the microbial population and plant growth performance.

Salt affected site of Handia was divided into four different zones consisting of barren unproductive land (Zone I), grassy patches (Zone II), scanty vegetation (Zone III) and cultivated fields (Zone IV). The spore population in the I[st] zone ranged from 0-4 spores/100 g air dried soil. In the IInd zone population ranged from 2-10 spores/100 air dried soil while in the IIIrd zone it ranged from 10-12 spores/100g air dried soil. Maximum spore population was recorded in the IVth zone which is mainly of agricultural fields and the population ranged from 20- 25 spopre/100 g air dried soil. Among VAM fungi isolated from this ecosystem *Glomus* constituted the dominant genus. VAMF in different soil regions and their relations to soil properties and native plants have been investigated by several researchers (Kim and Weber, 1985; Rozema *et al.*, 1986; Cook *et al.*, 1993; Hoefangels *et al.*, 1993; Hilderbrandt *et al.*, 2001).

Populations of VAMF in saline-alkaline soils were variable and affected by many factors. For example, Hilderbrandt *et al.* (2001) reported mycorrhizal colonization as well as a high spore density in Central European salt marshes. Landwehr *et al.* (2003) found the number of spores in samples saline, sodic and gypsum soils to be rather variable but high on average, with *Glomus geosporum* dominant. Cooke *et al.* (1993) observed the vertical distribution of vesicular arbuscular mycorrhizae (VAM) in roots of salt marsh growing in saturated soils. Aliasgharzadeh *et al.* (2001) reported that the spore number of VAMF in saline soils of the Tabriz Plain was not correlated with soil salinity but suffered adverse effects of the accumulation of some anions and cations. Our results are in conformation with the earlier reports that VAMF populations are small in salt- affected soils as reported by some researchers (Hirrel, 1981; Barrow *et al.*, 1997) but others have reported relatively large populations (Sengupta and Chaudhary, 1990; Bhaskaran and Selvaraj, 1997). Spore density and root colonization showed significant correlation with plant species and soil properties (Aliasgharzadeh *et al.*, 2001), and the high soil salinity, poor plant diversity and low vegetation cover in salt-affected soils severely restricted colonization and diversity of VAMF.

VAM Inoculant in Improving the Performance of *Cicer arietinum* and *Phaseolus mungo* in Salt-Affected Soils

Performance of the *C. arietinum* and *P. mungo* plants raised in the salt-affected soils of Handia, Allahabad amended with different concentrations of fly ash (15, 30 and 45%) and organic matter (*Cynodon*/compost 2% w/w) and provided with consortium of indigenous VAM fungi, PSB (*Bacillus subtilis*) and N_2-fixer (*Rhizobium spp.*) alone as well as in combination was assessed. High rate of mortality was recorded at the time of germination. However, the magnitude of mortality varied with the treatments. 100% mortality was recorded in the control series, where the soil was neither amended with fly ash/organic matter nor inoculated with PSM/N_2-fixer/VAM. The salt-affected soils are characterized by high pH (11.4), free sodium ion and high electric conductivity (4.06m.mho.cm-1). Deposition of calcium clay makes it compact and impervious. Hence, its drainage and aeration are poor. It is harmful to the plants on accounts of its excessive concentration of soluble salts and exchangeable sodium, which retard their growth, by limiting the absorption of water and other nutrients. The accumulation of sodium in plant cells may cause toxic effect and injury by increasing osmotic pressure and plasmolysis. Alkalinity of soils results in unavailability of nutrients like nitrogen, phosphorus, potassium, iron, manganese, boron etc. to the plants. The presence of exchangeable sodium reduces the availability of calcium and prevents its intake by the plants.

A comparison of the characteristics of the soils of fertile agricultural fields with those of the Handia soils clearly shows that the soil type included in the study was not having a pH appropriate for the normal growth of plants. Same was true for the conductivity. Total soluble salts measured in terms of conductivity were 0.23m.mhos.cm-1 in the soils of fertile agricultural fields while they were 4.06m.mhos.cm-1 in case of soils of Handia showing that the soils are not supportive for growth of the plants.

Salinity also exhibits its adverse effects on plants through the decreased water potential of the root medium, through ion toxicity due to excessive Na+ or Cluptake, and through nutrient ion imbalance by the disturbance of essential intracellular ion concentrations (Greenway and Munns, 1980; Gorham *et al.*, 1985). These effects can perturb the physiological and biochemical functions of the cell and lead eventually to cell death. The mortality in all the treatments was reduced, minimum reduction being in the 15% series and maximum in the 45% series. In comparison to control all the microbial inoculants singly as well as in combination caused an appreciable reduction in the rate of mortality in all the three concentrations of fly ash (15, 30 and 45%).

A careful analysis of the findings shows that the microbial inoculants improved not only the mycorrhizal status of these two crops, but also their biomass and yield; however, the degree of improvement varied with the type of soil treatment and the microbial inoculants. The improvement in the performance of *C. arietinum* and *P. mungo* due to consortium of VAM fungi showed wide variations in the improvement in root/shoot biomass and yield.

The root and shoot biomass of the crops, *C. arietinum* and *P. mungo,* increased tremendously with increasing concentrations of fly ash. *C. arietinum* showed maximum improvement in root biomass in 45% triple inoculation

(VAM+PSB+*Rhizobium*) series amended with *Cynodon*, whereas in *P. mungo* maximum improvement in the root biomass was observed in 45% triple inoculation series supplemented with compost. But the highest improvement in shoot biomass of both the crops was recorded in 45% triple inoculation series amended with *Cynodon*. Maximum yield in both the crops and nodulation in *P. mungo* was recorded in the 45% triple inoculation series inoculated with VAM, PSB and N_2-fixer supplemented with *Cynodon*.

VAM fungi improve growth of plants under a variety of salinity stress conditions (Diallo *et al.*, 2001; Burke *et al.*, 2003; Tain *et al.*, 2004). VAM fungi protect the host plants against the detrimental effects of salt *via* better access to nutritional status (Zandavalli *et al.*, 2004; Rabie, 2005) and modification of plant physiology i.e. osmotic modifications (Roa ad Tak, 2002) and improved photosynthesis (Merguihur *et al.*, 2002). To some extent, these VAM fungi have been considered as bio-ameliorators of saline soils (Tain *et al.*, 2004).

The results of the study are in conformation with these reports, where mycorrhizal inoculation improved the mycorrhizal status, microbial population, root/shoot biomass, nodulation (*P. mungo*) and yield of the crops, *C. arietinum* and *P. mungo*. Vesicular arbuscular mycorrhizal fungi (VAMF) have been shown to decrease yield losses of plants in saline soils (Hirrel and Gerdemann 1980; Poss *et al.* 1985). This may be due to increased uptake of nutrients with low mobility, such as P, Zn and Cu (Al-Karaki and Al-Raddad 1997; Al-Karaki and Clark 1998) and improved water relations (Syl*via et al.*, 1993; Al-Karaki and Clark 1998). This may lead to increased growth and subsequent dilution of toxic ion effects (Juniper and Abbott 1993).

VAM inoculant have also been reported to overcome the constraints of salt affected soils (Cantrell and Linderman, 2000) by increasing their tolerance to high pH and temperature fluctuations (Poss *et al.*, 1985, Medeiros *et al.*, 1994), by bringing the pH to a favourable scale through the production of organic acids (Bolan *et al.*, 1984), by making the soil more porous (Logan *et al.*, 1989), by helping the plants in the uptake of calcium and by reducing the alkalinity of soil by increasing the production of alkaline phosphatase. Changes in the physicochemical characteristics of the soils of Handia because of different microbial inoculants and organic matter input may also be responsible for overcoming the constraints of the soils and thereby improving the performance of these crops.

VAMF also increases the resistance of plants to stresses such as extremes of pH and salinity due to higher accumulation of soluble sugars, proline and total free amino acids in roots (Rosendahl and Rosendahl, 1991; Abo-Ghalia, 2001 a, b; Feng *et al.*, 2000, 2002), increased concentration of phosphate and a decreased concentration of sodium in shoot tissues and a steady Na/K ratio

(Allen and Cunningham, 1983; Pfieffer and Bloss, 1988) and due to diurnal changes in the leaf thickness as in *Aster tripolium* (Rozema, *et al.*, 1986).

In the present study maximum microbial population was recorded in 45% triple inoculation series (PSB+VAM+*Rhizobium*) when amended with *Cynodon*. Increased microbial activity and population of fungi, including arbuscular mycorrhizal fungi and Gram-negative bacteria were reported for ash amended soils containing sewage sludge by Pitchel and Hayes, 1990. Moreover, the microorganisms invariably adapt to the stressed conditions and show a gradual increase in respiration after an initial lag. Gaind and Gaur (2002, 2003 a, b) reported that fly ash helped in enhancing the microbiological properties without having any detrimental effect on either C:N ratio or microbial population.Application of fly ash at 40 t/ha in conjunction with phosphate solubilizer, *Pseudomonas striata* improved the bean yield and phosphorus uptake. The supply of simple organic compounds and N are crucial for sustaining rich microbial population at these sites.

For revegetating fly ash landfills, application of biotechnological means involving biological N_2 fixation and phosphate mobilization was considered important in plantation. Application of various nitrogen-fixing cynobacteria and leguminous trees inoculated with *Rhizobium* have been used to enhance N and P status and in reducing toxicity of fly ash landfills (Banerjee and Deb, 1993; Rai *et al.*, 2000, 2004). Many plants belonging to the family leguminosae were demonstrated to have a high tolerance and survival in arid, infertile and metal-contaminated areas (Musil, 1993; Vajpayee *et al.*, 2000; Cheung *et al.*, 2000). In addition, legume trees and symbiotic N_2-fixing bacteria can improve N content of infertile soil and amended lagoon ash (Sanginga *et al.*, 1994; Cheung *et al.*, 2000). A significant increase in the biomass on intact plant, root and shoot, photosynthetic area and nodulation frequency were observed after 45 days of transplantation under field conditions in the plants inoculated with *Rhizobium* in comparison to the uninoculated plants grown in 100% fly ash (Rai *et al.*, 2004; Vajpayee *et al.*, 2000; Noel *et al.*, 1996).

In dual inoculation series, where the soils were inoculated either with VAM and PSB or with VAM and N_2-fixer, showed improvement in mycorrhization, microbial population, root/shoot biomass and yield in comparison to non-inoculated control and single inoculation. The results are in conformation with the reports of Delorenzini *et al.* (1979), Krone *et al.* (1987), Jalali and Thareja (1981), Jalali *et al.* (1990), Azcon- Aguilar *et al.* (1986), Lee and Bagyaraj (1986), Singh and Kapoor (1999), Amora- Lazecano *et al.* (1998), Johansson *et al.* (2004). Many bacteria produce organic acids and solubilize inorganic and organic forms of phosphorus that are unavailable to plants. VAM fungi, on the other hand, do not take up insoluble P, but more efficiently utilize labile forms. Barea *et al.*, (1975) suggested the possibility of a synergistic interaction between mycorrhizal fungi and phosphate solubilizing bacteria and observed an increase in dry weight and P uptake in maize grown in a low P soil. According to some scientists the interactions observed might be due to the production of plant hormones such as IAA, gibberellins and cytokinins (Barea *et al.*, 1976) or vitamins by PSB (Baya *et al.*, 1981), rather than P solubilization. VAM fungi are known to enhance nodulation and nitrogen fixation by legumes (Amora-Lazecano *et al.*, 1998; Johansson *et al.*, 2004). Moreover, VAM fungi and nitrogen-fixing bacteria often act synergistically on infection rate, mineral nutrition, plant growth and disease resistance (Jalali and Thareja, 1981, Jalali *et al.*, 1990, Rabie and Almadini, 2005).

In triple inoculation series, where the soils were inoculated with VAM, PSB and N_2-fixer, showed improvement in microbial population and all the growth parameters in comparison to non-inoculated control, single or dual inoculation. Juwarkar and Jambhulkar (2007) also recorded a 4.5 times increase in the nitrogen content due to addition of *Bradyrhizobium* and *Azotobacter* species, while phosphate content was increased by 10.0 times due to addition of VAM, which helps in phosphate immobilization. Due to biofertilizer inoculation different microbial groups such as *Rhizobium, Azotobacter* and VAM spores, which were practically absent in fly ash improved to 7.1 x 107, 9.2 x 107 CFU/g and 35 VAM spores/10 g of fly ash, respectively. Kumar *et al.* (2001) also reported better growth performance and nutrition in cowpea when inoculated with VAM, *Azotobacter* and *Rhizobium*.

Besides microbiological properties of the soils, physico-chemical characteristics also play an important role in improving the growth parameters of the plants. In the present study pH of the soil was reduced from 10.00 in the control (without any treatment) series to 8.00 in the 45% triple inoculation series supplemented with *Cynodon* (2% w/w) for *Cicer arietinum*. Similarly the pH reduced from 10.10 in control to 7.80 in the 45% triple inoculation series supplemented with *Cynodon* (2% w/w) in the case of *Phaseolus mungo*. Similarly maximum reduction in EC from 1.80 in control to 0.13m.mho.cm-1 in the 45% triple inoculation series (compost 2% w/w) was recorded in *C. arietinum*. In *P. mungo* EC reduced from 1.80 to 0.23m.mho.cm-1.

Better performance of the plants in salt-affected soils may be primarily due to shifting chemical equilibrium induced by fly ash. Kalra *et al.* (2003) reported increased growth and yield of the crops with fly ash incorporation possibly due to modifications in bulk density and physico-chemical characteristics of soil, such as pH, EC and organic carbon content.

The abovementioned reports clearly show that the problems of wasteland and fly ash may be managed with an appropriate blending of fly ash with organic matter and the application of VA Mycorrhizal fungi along with nitrogen fixer and phosphate solubilizers.

References

Abo-Ghalia, H.H. (2001a). Synergistic interaction between the VA-mycorrhiza *Glomus intraradices* and *Sachromyces ceravisae*. I upon the growth and nutrition of *Sorghum bicolar* grown under saline condition. *Afr. J. Mycol. Biotechnol.*, 9 (3): 27-41.

Abo-Ghalia, H.H. (2001b). Syergistic interaction between VA- mycorrhiza *Glomus intraradices* and *Saccharomyces cerevisiae* I. upon the growth and nutrition of *Sorghum bicolar* grown under saline conditions. *Afr. J. Mycol. Biotechnol*, 9 (3): 43-52.

Adriano, D. C. and Weber, J. T. (2001). Influence of fly ash on soil physical properties and turf grass establishment. *J. Environ. Qual.*, 30: 596-601.

Adriano, D. C., Page, A. L., Elseewi, A. A., Chang, A. C. and I. Straughan, I. (1980). Utilization and disposal of fly ash and other coal residues in terrestrial ecosystems – A review. *J. Environ. Qual.*, 9: 333-334.

Aliasgharzadeh, N., Rastin, N.S., Towfighi, H. and Alizadeh, A. (2001). Occurrence of arbuscular mycorrhizal fungi in saline soils of the Tabriz Plain of Iran in relation to some physical and chemical properties of soil. *Mycorrhiza*, 11: 119-122.

Al-Karaki, G.N. and Al-Raddad, A. (1997) Effects of arbuscular mycorrhizal fungi and drought stress on growth and nutrient uptake of two wheat genotypes differing in drought resistance. *Mycorrhiza* 7:83–88.

Al-Karaki, G.N. and Clark, R.B. (1998) Growth, mineral acquisition, and water use by mycorrhizal wheat grown under water stress. *J. Pl. Nutr.* 21:263–276.

Allen, F.B. and Cunningham, G.L. (1983). Effect of vesicular arbuscular mycorrhizae on *Distichlis spicata* under three salinity levels. *New Phytol.*, 93: 27-236.

Amora-Lazecano, E., Vazquez, M. M. and Azcon, R. (1998). Response of nitrogen-transforming micro organisms to arbuscular mycorrhizal fungi. *Biol.Fertil. Soils*, 27: 65-70.

Ammaiyanppan, S. and Ayyamperumal, M. (2002). Distribution of mycorrhizas in an abandoned fly ash pond and mined sites of Neyveli Lignite Corporation, Tamil Nadu, India. *Basic Appl.Ecol.*, 2(3): 277-284.

Azcon-Aguilar, C., Gianinazzi-Pearson, V., Fardeau, J.C.and Gianinazzi, S. (1986). Effect of vesicular-arbuscular mycorrhizal fungi and phosphate-solubilizing bacteria on growth and nutrition of soybean in a neutral-calcareous soil amended with ^{32}P-^{45}Ca-tricalcium phosphate. *Plant Soil*, 96:3-15.

Banerjee, M. and Deb, M. (1993). Fly-ash induced changes in growth and protein content of *Spirulina platensis*. *Chem. Environ. Res.*, 2: 151-154.

Barea, J.M., Azcon, R. and Hayman, D. (1975). Possible synergistic interactions between *Endogone* and phosphate-solublizing bacteria in low-phosphate soil. In : *Endomycorrhiza*. (Eds.) B. Mosse, F.E. Sanders and P.B. Tinker, Academic Press, New York: 409-417.

Barea, J.M., Navarro, E. and Montoya, E. (1976). Production of plant growth regulators by rhizosphere phosphate solubilizing bacteria. *J. Appl. Bacteriol.*, 40.

Barea, J. M., Tobar, R. M. and Azcon-Aguilar, C. (1996). Effect of genetically modified *Rhizobium melioloti* inoculants on the development of arbuscular mycorrhizas, root morphology, nutrient uptake and biomass accumulation in *Medicago sativa. New Phytol.*, 134, 361-369.

Barrow, J.R., Havstad, K. M. and McCaslin, B.D. (1997) .Fungal root endophytes in four-wing saltbush, *Altiplex canescens*, on arid rangeland of southwestern USA. *Arid Soil Res Rehabil.*, 11:177–185.

Baya, A.M., Boethling, R.S. and Ramos-Cormenzana (1981). Vitamin production in relation to phosphate solubilization by soil bacteria. *Soil Biol. Biochem.*. 13: 527-531.

Benlloch-Gonzaiez, M., Fournier, J. Ramos, J. and Benlloch, M. (2005). Strategies underlying salt tolerance in halophytes are present in *Cynara ardunculus. Plant Sci.*, 168(3): 653-659.

Bhaskaran, C. and Selvaraj, T. (1997). Seasonal incidence and distribution of VA-mycorrhizal fungi in native saline soils. *J. Environ. Biol.*, 18(3): 209-212.

Bhattacharya, S. S. and Chattopadhyay, G. N. (2002). Increasing bioavailability of phosphorus from fly ash through vermicomposting. *J. Environ. Qual.*, 31: 2116-2119.

Bi, Y. L., Li, X. L., Christie, P., Hu, Z. Q. and Wong, M. H. (2003). Growth and nutrient uptake of arbuscular mycorrhizal maize in different depths of soil overlying coal fly ash. *Chemosphere*, 50: 863-869.

Bolan, N.S.; Robson, A.D. and Barrow, N.W. (1984). Increasing phosphorus supply can increase the infection of plant roots by vesicular-arbuscular mycorrhizal fungi. *Soil Biol. Biochem.*, 16: 419-420.

Burke, D. J., Hamerlynck, E.P. and Hahn, D. (2003). Interactions between the salt marsh grass season. Spartina patens, arbuscular mycorrhizal fungi and sediment bacteria during the growing. *Soil Biol. Biochem.* 35: 501-511.

Cantrel, I.C. and Linderman, R.G. (2000). Preinoculation with VA-Mycorrhizal fungi increases plant tolerance to soil salinity. *In challenges facing irrigation* and *drainage in the new millennium. Proceedings U.S. committee on Irrigation and Drainage.* 27-30.

Cheung, K. C., Wong, J. P. K., Zhang. Z. Q., Wong, J. W. C. and Wong, M. H. (2000). Revegetation of lagoon ash using legume species *Acacia auriculiformis* and *Leucaena leucocephala. Environ. Pollut.*, 109, 75-82.

Cook, J.C., Butler, R.H. and Madol (1993). Some observations on the vertical distribution of VAM in roots of salt marsh grasses growing in saturated soils. *Mycologia* 85: 547–550.

Delorenzini, G., Barea, J.M. and Olivares, J. (1979). Fertilizacion biologica (micorrhiza + *Rhizobium* - phosphobacterias) de *Trifolium pratense* en diferentes condicions de cultivo. *Rev. Lat-amer, Microbiol.*, 21: 129-134.

Diallo, A., Samb, P. and Macauley, C. A. (2001). Water status and stomatal behaviour of cowpea, *Vigna unguiculata* (L) Walp, plants inoculated with two *Glomus* species at low soil moisture levels. *Eur. J. Soil Biol.*, 37: 187-196.

Dudal, R. and Purnell, M.F. (1986). *Reclamahon* and *Revegetahon Research* 5:169. *Ecodevelopement* of usar land at Banthra. (Ed.) T.N. Khoshoo. In: *Proceedings of the symposium on Inputs Sciences and Technology of theDevelopment Wastelands* (Eds.) M. Sharma, T.N. Khoshoo and U.S. Srivastava, *Proc. Nat. Acad. Sci.*, Allahabad, India: 17-32.

Edwards, C. A. and Lofty, J. R. (1972). *Biology of earthworms*. Chapman and Hall,

Elseewi, A. A., Bingham, F. T. and Page, A. L. (1978 a). Growth and mineral composition of lettuce and Swiss chard grown on fly ash amended soils. In: *Enviromental chemistry and cycling processes*. (Eds.).D. C. Adriano and I. L. Brisbin, CONF-760429. U.S. Dep. Commerce, Springfield. 568-581.

Elseewi, A. A., Bingham, F. T. and Page, A. L. (1978 b). Availability of sulphur in fly ash to plants. *J. Environ. Qual.* 7: 69-73.

Feng, G., Li, X.L., Zhang, F.S. and Li, S.X. (2000). Effect of phosphorus and AM fungus on response of maize plant to saline environment. *Pl. Resour. Environ.*, 9: 22-26.

Feng, G., Zhang, F., LI, X., Tian, C., Tang, C. and Rengel, Z. (2002). Improved tolerance of maize plants to salt stress by arbuscular mycorrhiza is related to higher accumulation of soluble sugars in roots. *Mycorrhiza*, 12: 185-190.

Flowers, T. J. (2004). Improving crop salt tolerance. *J. Exp. Bot.*, 55: 307-319.

Gaind, S. and Gaur, A. C. (2002). Impact of fly ash and phosphate solubilizing bacteria on soybean productivity. *Bioresource Technol.*, 85: 313-315.

Gaind, S. and Gaur, A. C. (2003 a). Evaluation of fly ash as a carrier for diazotrophs and phosphobacteria. *Bioresource Technol.*, 95: 187-190.

Gaind, S. and Gaur, A. C. (2003 b). Quality assessment of compost prepared from fly ash and crop residues. *Bioresource Technol.*, 87: 125-127.

Garampalli, R. H., Deene, S. and Reddy, C. N. (2005). Infectivity and efficacy of *Glomus aggregatum* and growth response of *Cajanus cajan* (L.) Millsp. in fly ash VAMended sterile soil. *J. Environ. Biol.*, 26(4): 705-708.

Giri, B. and Mukerji, K.G. (1999). Improved growth and productivity of *Sesbania egyptiaca* (Pers.) under salinity stress through mycorrhizal technology, *J. Phytological Res.*, 12(1/2): 35-38.

Gorham, J., Wyn, Jones, R. G., MacDonnell, E. (1985). Some mechanisms of salt tolerance in crop plants. *Plant Soil.* 89: 15-40.

Greenway, H. and Munns, R. (1980). Mechanisms of salt tolerance in non halophytes. *Ann. Rev. Pl. Physiol.* 31: 149-190.

Hayashi, H. and Murata, N. (1998). Genetically engineered enhancement of salt tolerance in higher plants. In: *Stress Response of photosynthetic Organisms: Molecular Mechanisms and Molecular regulation*. (Ed.) N. Sato Murata, Elsevier, Amsterdam: 133-148

Hildebrandt, U., Janetta, K., Ouziad, F., Renne, B., Nawrath, K. and Bothe, H. (2001). Arbuscular mycorrhizal colonization of halophytes in Central European salt marshes. *Mycorrhiza,* 10 :175–183.

Hirrel, M.C. (1981). The effect of sodium and chloride salts on the germination of *Gigaspora margarita. Mycologia,* 73 :610–617.

Hirrel, M.C. and Gerdemann, J.W. (1980). Improved growth of onion and bell pepper in saline soils by two vesicular-arbuscular mycorrhizal fungi. *Soil. Sci. Soc. Am. J.,* 44: 654–655.

Hoefnagels, M.H., Broome, S. and Shafer, S.R. (1993). Vesicular-arbuscular mycorrhizae in salt marshes in North Carolina. *Estuaries,* 16 :851–858.

Jalali, B. L. and Thareja, M. L. (1981). Suppression of *Fusarium* wilt of chick pea in vesicular arbuscular mycorrhizal inoculated soils. *Internat. Chickpea Newsletter,* 4: 21-22.

Jalali, B.L., Chhabra, M.L. and Singh, R.P. (1990). Interaction between vesicular-arbuscular mycorrhizal endophyte and *Macrophomina phaseolina* in mungbean. *Ind. Phytopath.,* 43: 527-530.

Jindal, V., Atwal, A., Sekhon, B.S. and Singh, R. (1992). Effect of NaCl salinity on metabolism of vesicular-arbuscular mycorrhizae inoculated moong plants. M*ycorrhiza News,* 4: 3.

Johansson, J. F., Paul, L. R. and Finlay, R. D. (2004). Microbial interactions in the mycorrhizosphere and their significance for sustainable agriculture. *FEMS Microbiol. Ecol.,* 48: 1-13.

Juniper, S. and Abbott, L. (1993). Vesicular arbuscular mycorrhizas and soil salinity. *Mycorrhiza,* 4: 45-57.

Juwarkar, A. A. and Jambhulkar, P. H. (2007). Restoration of fly ash dump through biological interventions. *Environ. Monit. Assess.,* DOI 10.1007/s 10661-007-9842-8.

Kalra, N., Jain, M. C., Joshi, H. C., Chaudhary, R., Kumar, S., Pathak, H., Sharma, S. K., Kumar, V., Kumar, R., Harit, R. C., Khan, S. A. and Hussain, M. Z. (2003). Soil properties and crop productivity as influenced by fly ash incorporation in soil. *Environ. Monit. Assess.,* 87(1): 93-109.

Khan, M. R. (2002). Effects of soil application of fly ash on the growth and flowering of seasonal ornamental plants. *Environ. Biol. and Conserv.,* 7: 81-84.

Khan, M. R. and Khan, M. W. (1996). The effect of fly ash on growth and yield of tomato. *Environ. Pollut.,* 92: 105-111.

Kim, C.K. and Weber, D.J. (1985). Distribution of VA mycorrhiza on halophytes on inland salt playas. *Plant Soil,* 83: 207-214.

Krone, W., Bichler, B., Viebrock, E. and Moawad, A.M. (1987). Interactions between VA mycorrhiza and phosphate solubilizing bacteria. In: *7th North American Conference on Mycorrhizae,* University of Florida, Gainesville: 328.

Kulshreshtha, Madhu and Khan, M. W. (1999). Impact of fly ash application in soil on root colonization by a VAM fungus and root nodulation by *Rhizobium*. *Ind. Phytopath.*, 52 (2): 185-187.

Kumar, A., Sarkar, A. K., Singh, R. P. and Sharma, V. N. (1998). Characterization of fly ash from steel plants of eastern India. *J. Ind. Soc. Soil Sci.*, 46: 459-461.

Kumar, V. and Mishra, B. (1991). Effect of two types of press mud cake on growth of rice-maize and soil properties. *J. Ind. Soc. Soil Sci.*, 39: 109-113.

Landwehr, M., Hildebrandt, U., Wilde, P., Nawrath, K., Toth, T., Biro, B. and Bothe, H. (2003). The arbuscular mycorrhizal fungus *Glomus geosporum* in European saline, sodic and gypsum soils. *Mycorrhiza*, 12:199–211

Lau, S. S. S. and Wong, J. W. C. (2001). Toxicity evaluation of weathered coal fly ash amended manure compost. *Water, Air Soil Pollut*. 128: 243-254.

Lee, A. and Bagyaraj, D.J. (1986). Effect of soil inoculation with vesicular arbuscular mycorrhizal fungi and other phosphate rock dissolving bacteria or thiobacilli on dry matter production and uptake of phosphorus by tomato plants. *New Zealand J. Agricul. Res.*. 29: 525-531.

Li, X.L.; George, E. and Marshher, H. (1991). Phosphorus depletion and pH decrease at the root-soil and hyphae-soil interfaces of VA mycorrhizal white clover, fertilized by VAMmonium. *New Phytol.*, 119: 397-404.

Logan, V.S., Clarke, P.J. and Allaway, W.G. (1989). Mycorrhizas and root attributes of plants of costal sand dunes of New South Wales. *Aust. J. Plant Physiol.*, 16:141-146.

Mahalingum, P .K. (1973). "Ameliorative properties of lignite fly ash in reclaiming saline and alkali soils." *Madras Agric. J*. 60: 469-487.

Mandal, L. N. (1964). Effect of time, starch and lime on the transformation of inorganic phosphorus in a water-logged rice soil. *Soil Sci.*, 97: 127-132.

Mathur, N. and Vyas, A. (2000). Mycorrhizal dependency of *Tamarix aphylla* in saline areas of Thar Desert. *Naturalia* (*Sao Paulo*), 25:105-110.

Medeiros, C.A.B., Clark, R.B. and Ellis, J.R. (1994). Growth and nutrient uptake of *sorghum* cultivated with vesicular arbuscular mycorrhizae isolates at varying pH. *Mycorrhiza*, 5: 185-191.

Merguihur, A. E., Burity, H. A., Tabosa, J. N. and Maia, F.L. (2002). Salt stress response and proline accumulation in *Brachiaria humidicola*. *Revista, Argentina de Microbiologia*, Argentina A.34 (2): 77-82.

Munns, R. (2002). Comparative physiology of salt and water stress. *Plant Cell Environ*. 20: 239-250.

Musil, C. F. (1993). Effect of invasive Australian acacias on the regeneration, growth and nutrient chemistry of South African lowland fynbos. *J. Appl. Ecol.*, 30, 361-372.

Noel, T. C., Sheng, C., Yost, C. K., Pharis, R. P. and Hynes, M.F. (1996). *Rhizobium leguminosarum* as a plant growth promotion of canola and lettuce. *Canad. Journ. Of Microbiol.,* 42: 279-283.

Page, A. L., Elseweei, A. A. and Straughan, I. (1979). Physical and chemical properties of fly ash from coal-fired power plants with reference to environmental impacts. *Residue Rev.* 71: 83-120.

Plank, C. O. and Martens, D. C. (1974). Boron availability as influenced by application of fly ash to soil. *Soil Sci. Soc. Am. Proc.* ,38: 974-977.

Pfeiffer, C.M. and Bloss, H.E. (1988). Growth and nutrition of guayule (*Parthenium argentatum*) in a saline soil as influenced by vesicular arbuscular mycorrhiza and phosphorus fertilization. *New Phytologist,* 108(3): 315-321.

Pitchel, J. R. and Hayes, J. M. (1990). Influence of fly ash on soil microbial activity and populations. *J. Environ. Qual.,* 19: 593-597.

Phung, H. T., Lund, L. J. and Page, A. L. (1978). Potential use of fly ash as a liming material. In D. C. Adriano & I. L. Brisbin (Eds.), *Environmental chemistry and cycling processes* (pp. 504-515). Springfield, VA: CONF-760429- U.S. Department of Commerce.

Poss, J.A., Pond, E., Menge, J.A. H. and Arrell, W.M. (1985). Effect of salinity on mycorrhizal onion and tomato in soil with and without additional phosphate. *Plant Soil,* 88:307–319.

Rabie, G. H. and Almadini, A. M. (2005). Role of bioinoculants in development of salt-tolerance of *Vicia faba* plants under salinity stress. *Afr. J. Biotechnol.,* 4 (3): 210-222.

Rai, U. N., Pandey, K., Sinha, S., Singh, A., Saxena, R. and Gupta, D. K. (2004). Revegetating fly ash landfills with *Prosopis juliflora* Linn: impact of different amendments and *Rhizobium* inoculation, *Environ. Int.,* 30 (3): 293-300.

Rai, U. N., Tripathi, R.D., Singh, N., Kumar, A., Ali, M. B., Pal, A. and Singh, S. N. (2000). Amelioration of fly ash by selected nitrogen fixing blue green algae. *Bull. Environ. Contam. Toxicol.,* 64: 294-301.

Rao, A. V. and Tak, R. (2002). Growth of different tree species and their nutrient uptake in limestone mine spoil as influenced by arbuscular mycorrhizal (AM) fungi in Indian arid zone. *J. Arid Environ.,* 51: 113-1119.

Reddy, C. N. and Garampalli, H. R. (2002). Effect of fly ash on VAM formation and growth response of pulse crops infected with *Glomus aggregatum* in sterile soil. *In: Frontiers in microbial biotechnology and plant pathology* (Eds.) C. Manoharachary, D. K. Purohit, S. R. Reddy, Singara, M. A. Charya and S. Girisham, Jodhpur, India: Scientific Publishers (India).

Requena, N., Jeffries, P. and Barea, J. M. (1996). Assessment of natural mycorrhizal potential in a desertified semiarid ecosystem. *Appl. Environ. Microbiol.,* 62, 842-847.

Rinalolelli, E. and Mancuso, S. (1998). Short-term and long-term response of mycorrhizal and non-mycorrhizal olive plants (*Olea eusopaea* L.) to saline conditions. *Olive,* 74: 45-49.

Rosendahl, C.N. and Rosendahl, S. (1991). Influence of vesicular-arbuscular mycorrhizal fungi (*Glomus* spp.) on the response of cucumber (*Cucumis sativis* L.) to salt stress. *Environ. Exp. Bot.*, 31: 313–318.

Rozema J., Arp, W., Diggelen, J.V., Esbroek, M.V., Broekman, R. and Punte, H. 1986). Occurrence and ecological significance of VAM in the salt marsh environment. *Acta Bot. Neerlandica*, 35(4): 457–467.

Sanginga, N., Danso, S. K. A., Mulongoy, K, Ojeifo, A. A. (1994). Persistence and recovery of introduced *Rhizobium* ten years after inoculation on *Leucaena leucocephala* grown on an Alfisol in southwestern Nigeria. *Plant Soil*, 159, 199-204.

Sengupta, A. and Chaudhuri, S. (1990). Vesicular arbuscular mycorrhiza (VAM) in pioneer salt marsh plants of the Ganges delta in West Bengal (India). *Plant Soil*. 122: 113-115.

Singh, S. and Kapoor, K.K. (1999). Inoculation with phosphate solubilizing microorganisms and a vesicular-arbuscular mycorrhizal fungus improves dry matter yield and nutrient uptake by wheat grown in a sandy soil. *Biol. Fertil. Soils*, 28(2): 139-144.

Singh, Reena, Sharma, M. P. and Adholeya, Al. (2006). Screening of *Jatropha curcas* germplasm from different provenances for cultivation in fly ash overburdens using arbuscular mycorrhiza fungi. *Mycorrhiza News*, 18 (3): 24-27.

Syl*via*, D. M.; Hammond, L. C.; Bennett, J. M.; Haas, J. H. and Linda, S. B. (1993). Field response of maize to a VAM fungus and water management. *Agron. J.* 85:193–198.

Tain, C. Y., Feng, G., Li, XL. And Zhang, F. S. (2004). Different effects of arbuscular mycorrhizal fungal isolates from saline or non-saline soil on salinity tolerance of plants. *Appl. Soil Ecol.* 26(2): 143-148.

Thapar, H. S. and Uniyal, K. (1996). Effect of VAM fungi and *Rhizobium* on growth of *Acacia nilotica* in sodic and New Forest soils. *Ind. Forester*, 122(11): 033-1039.

Thapar, H. S., Uniyal, K. and Verma, R.K. (1991). Survey of native VAM fungi of sodic soils of Haryana State. *Ind. Forester*, 117(12): 1059-1069.

Tripathi, Anuradha and Sahu, R. K. (1997). Effect of coal fly ash on growth and yield of wheat. *J. Environ.Biol.*, 18: 131-135.

Vajpayee, P., Rai, U. N., Choudhary, S, K., Tripathi, R. D. and Singh, S. N. (2000). Management of fly ash landfills with *Casia surattensis* Burm: A case study. *Bulletin Environ. Contamin. Toxicol.*, 65, 675-682.

Verma, Rollie (2008). *Management of flyash and wasteland soils through VAM technology*. D. Phil. Thesis, University of Allahabad, Allahabad, India.

Wong, M. H. and Wong, J. W. C. (1989). Germination and seedling growth of vegetable crops in fly ash amended soils. *Agricul. Ecosys. Environ.t*, 26, 23-35.

Yano-Melo, A.M., Saggin Junior, O. J. and Maia, L. C. (2003). Tolerance of mycorrhizal banana (*Musa* sp. cv. *Pacovan)* plantlets to saline stress. *Agricul. Ecosys. Environ.,* 95 (1): 343-348.

Yuan Wang Fa and Runjin Liu. (2002). Arbuscular mycorrhizal fungi in saline – alkaline soils of Yellow River Delta. *Mycosystema,* 21(2): 196-202.

Zandavalli, R. B., Dillenburg, L.R. and Paulo, V. D. (2004). Growth responses of *Araucaria angustifolia* (Araucariaceae) to inoculation with the mycorrhizal fungus *Glomus clarum. Appl. Soil Ecol.,* 25 (3): 245-255.

□□□

Microbial Diversity and Functions, 2012
© D.J. Bagyaraj, K.V.B.R. Tilak, H.K. Kehri (eds.), pp. 537-557
New India Publishing Agency, New Delhi (India)
E-mail : info@nipabooks.com; Website : www.nipabooks.com

Chapter **26**

Taxonomic Diversity of AM Fungi in Alkaline Soils of Upper Gangetic Plains of Allahabad

Varun Khare and Pooja Rai

ABSTRACT

Extensive occurrence of alkaline soils has been reported from the Indo-Gangetic plains of Northern India, which is commonly known as 'usar' or 'reh'. High alkalinity and high exchangeable sodium percentage (ESP) of such soil render it inhospitable for normal crop production. Research efforts over the past few decades have resulted in evolving a number of practices for the reclamation and management of such soils. However, application of biological inputs in combination with some moderately salt tolerant plant species seems to be a better option. AM fungi is well known to enhance the ability of plants to cope with environmental stresses generally prevalent in the degraded ecosystems and stress conditions by providing a number of nutritional and non-nutritional benefits. Since isolates of AM fungi differ in their tolerance to adverse physical and chemical conditions in soil and are better adapted to their native stressed microhabitat, selection of appropriate native AMF isolate is necessary for any large scale reclamation program. Taxonomic diversity of AM fungi in the alkaline soils of Upper Gangetic plains of district Allahabad and adjoining areas has been investigated and it was found that such soils have a detrimental effect on the AM spore population, distribution and diversity. Maximum spore population occurred in winter and minimum in summer. Glomus followed by Acaulospora was the most dominant genus. Acaulospora scrobiculata and Glomus fasciculatum were the most frequent species followed by Glomus mosseae and Glomus aggregatum. Glomus fasciculatum was the most abundant species followed by Glomus mosseae and Acaulospora longula.

Keywords: Alkaline Soil, Arbuscular Mycorrhizal Fungi (AMF), Taxonomic Diversity, Frequency, Relative Abundance, Seasonal Variation.

Introduction

India, which constitutes only 2% of the world's total land, provides shelter to the 16% of the world population. Because of this high man to land ratio, there is a considerable pressure on land on account of the competing land uses. It is therefore, natural as well as necessary that the focus should be turned towards the less fertile and problem soils, which are at present lying fallow. Recently, National Wasteland Development Board (NWDB) and National Remote Sensing Agency (NRSA) in a joint project (National Wastelands Inventory Project) have estimated that about 63.85 mha (20.17%) land in the country is wasteland under thirteen different categories (www.nrsa.gov.in). However, in Uttar Pradesh, out of 83 districts surveyed, 3.87 mha (13.17%) was recorded as wasteland.

One of the important categories of wasteland is salt affected wasteland, which occupies extensive area in the world and in India as well, presenting a serious impediment to crop production. These soils are completely barren and practically produce nothing. Plant growth and development in such soils is adversely affected either due to excessive amounts of neutral soluble salts or high exchangeable sodium or both. In India approximately 7 mha land is salt affected out of which, 2.5 mha occurs in Indo-Gangetic plains covering the states of Uttar Pradesh, Haryana, Punjab, Delhi and parts of Bihar. In Uttar Pradesh alone about 1.29 mha is salt affected and commonly known as '*usar*' or '*reh*' in local language (Janardhanan *et al.*, 1994; Chauhan and Singh, 1999).

Salt affected soils can be classified into three different categories, *viz.* saline, alkaline/sodic and saline-alkaline/saline-sodic. Extensive occurrence of alkaline soils has been reported from the Indo-Gangetic plains of Northern India (Shahid *et al.*, 1994; Sharma *et al.*, 2000). The agricultural history of the region suggests that these high alkaline and sodic lands have been left unproductive in this area for a long time. The area experiences intermittent dry and rainy seasons throughout the year and considerable rainfall (160–200 cm/yr) which is followed by fast evaporation leaves behind a white encrustation on the soil surface (Datta *et al.*, 2002). The efflorescent crust consists predominantly sodium carbonate minerals ($Na_2CO_3.NaHCO_3.2H_2O$) with minor thermonatrite ($Na_2CO_3.H_2O$). Being highly soluble, the presence of these minerals alone explains the high alkalinity of the soil (up to pH 10.5) (Datta *et al.*, 2002).

High alkalinity and high exchangeable sodium percentage (ESP) of the soil render it inhospitable for normal crop production and there is minimal bioproductivity in such soil (Chhabra, 1995). Accumulation of CO_3^- and HCO_3^- ions has a serious impact on the growth of vegetation due to their toxic effects.

High pH affects nutrient solubility and high ESP decreases the availability of important nutrient elements, such as Ca, Mg and K. The organic matter content of such soil is also low and ranges between 0.1–0.2% only. Soil dispersion due to high ESP deteriorates the physical condition of the soil and affects the air and water permeability. Due to low infiltration capacity, rain water stagnates on the soil easily and, in dry periods, irrigation is hardly possible. Agriculture in such soils is limited only to the crops tolerant to surface waterlogging. Soil compaction adversely affects root growth and development due to which seedlings often fail to germinate and establish. Thus overall plant growth is adversely affected under such extreme stress conditions (Chauhan and Singh, 1999).

Research efforts over the past two decades have resulted in evolving a number of practices for the reclamation and management of such soils for sustained crop production. Application of biological inputs in combination with some moderately salt tolerant plant species seems to be a better option for the reclamation and management of such soils (Cantrell and Linderman, 2001; Giri and Mukerji, 2004). In the past few decades, it has been well established that the AM fungi enhance the ability of plants to cope with environmental stresses generally prevalent in the degraded ecosystems (Rosendahl and Rosendahl, 1991; Ruiz-Lozano, 2003; Giri and Mukerji, 2004; Al-Karaki, 2006; Khare *et al.*, 2008). This association helps in alleviating stress conditions by providing a number of nutritional as well as non-nutritional benefits. Under stressed conditions, mycorrhizal association improves rooting and root hair production in host, increases the absorptive surface of roots manifold for the uptake of nutrients and water, thereby helps in the establishment and survival of the seedlings (Bieleski, 1973; Barea *et al.*, 1983; Harley and Smith, 1983; Bolan and Robson, 1987; Smith and Gianinazzi-Pearson, 1988). Advantage of AM fungi in reclamation of various salt-affected soils has been attributed to better transport of phosphorus than sodium ions to host plants (Pfeiffer and Bloss, 1988; Smith and Smith, 1990).

Recent investigations have shown that AM fungi alter the physiology of the host as well for their better survival under stressed conditions. Symbiosis often results in altered rates of water movement into, through and out of host plants, with consequent effects on tissue hydration, hormone production and plant physiology (Danneberg *et al.*, 1992; Ruiz-Lozano and Azcon 1995; Auge, 2001; Ruiz-Lozano *et al.*, 2001 a, b). Other mechanism may include osmotic adjustment, which assists in the maintenance of leaf turgor, and effects on physiological processes such as photosynthesis, transpiration, conductance and water use efficiency (Ruiz-Lozano *et al.*, 1996; Duan *et al.*, 1996).

In this respect AM fungi have emerged as a potential biofertilizer and an effective bioinoculant in the recent past and are important in the establishment and survival of plants in a wide range of habitats (Pfleger *et al*; 1994; Enkhtuya *et al.*, 2003). With different levels of compatibility between host plant and AMF (Klironomos, 2003), appropriate isolates of AMF must be selected before any reclamation program at large scale, especially when native or non-native isolates are being considered (Dodd and Thomson, 1994). Most of the studies show that especially in stressed conditions and degraded soils native AM fungi can grow and function better possibly as they are better adapted to their stressed microhabitat (Enkhtuya *et al.*, 2000; Caravaca *et al.*, 2003; Calvente *et al.*, 2004). In addition, species and isolates of AM fungi differ in their tolerance to adverse physical and chemical conditions in soil (Sengupta and Chaudhuri, 1990; Juniper and Abbott, 1993; Joshi and Singh, 1995; Aliasgharzadeh *et al.*, 2001).

Therefore, it is important to study the diversity of native AM fungal species for any reclamation program, as they are better adapted to survive under stressed conditions prevailing there than any exotic AM fungal species. Therefore, the taxonomic diversity of AM fungi in the alkaline soils of Upper Gangetic planes has been investigated especially with reference to the district Allahabad and adjoining areas. A brief description of the few dominant and frequent species of AM fungi occurring in alkaline soils of Upper Gangetic planes of Allahabad has been given below.

Taxonomic Diversity of AM Fungi in Alkaline Soils of Upper Gangetic Planes of Allahabad

A systematic survey of the alkaline soils of Allahabad and adjoining areas was undertaken to assess the population and diversity of arbuscular mycorrhizal fungi in such soils. AM spore population was determined in 50g air-dried soil by wet sieving and decanting method (Gerdemann and Nicolson, 1963). AM spores were mounted in PVLG and PVLG + Melzer's reagent (1:1 v/v) and identified to the species level using the synoptic keys of Trappe (1982), Schenck and Perez (1990) and INVAM species guide (http://invam.caf.wvu.edu).

Acaulospora bireticulata Rothwell & Trappe

Distinguishing Features : Sporocarps unknown. Azygospores formed singly in soil, sessile, borne laterally on a hyaline, thin walled hypha tapered from 2.5-7.5μm dia. at its base to 10-30μm dia. near its terminus. Globose to sub-globose vesicle 127-135μm dia. Spore bearing hypha with emergent, branched, flagellate hyphae and collapsing by maturity. Spores globose, 150-155μm dia., sub-hyaline when young becoming light brown by maturity. Spore surface ornamented with polygonal greyish green sides and a paler, depressed central stratum. Ridges occasionally branched towards the center of polygons or forming irregular, isolated projections at polygon centers. Polygons 6-18μm long, the enclosed spore surface beset with round-tipped four to six sided processes of 1.0x1.0μm to give the appearance of an inverted reticulum. Spore walls of three layers, each about 1.0μm thick, the outer layer dark greyish green to greyish brown, the inner layers hyaline.

Distribution : Occurred with 66.7% frequency of occurrence and 3.5% relative abundance. Population was recorded as 6 spores/50g soil in winter, 5 spores/50g soil in rainy season and 2 spores/50g soil in summer.

Acaulospora denticulata Sieverding & Toro

Distinguishing Features : Sporocarps unknown. Spores yellow brown to dark brown, globose to sub-globose, (112-) 130-170 (-175)μm dia. Spores formed laterally on the 4-28μm dia. tapering neck of a globose to sub-globose sporiferous saccule, 80-160μm dia. with 1-2μm thick wall; spores formed on neck 50-90μm from sporiferous saccule; at point of spore formation the saccules neck is 14-25μm dia. Sporiferous saccule and neck breaking off after spore formation. Composite spore wall consists of four walls (walls 1-4) in two separable groups (groups A & B). Wall group A is composed of one wall (wall1), yellow brown to red brown, 2.5-10μm thick; the wall consists of inseparable polygonal segments, four five or six sided, 3-6x5-10μm dia., (1-) 4-6 (-8)μm thick, each segment has a circular or ellipsoidal projection, 3-6x3-4μm wide and 1-2μm high with a depression in its centre, 1.2-2x2-3μm wide and 0.5-1.5μm deep; wall 1 readily separates from other walls when spore is broken open. Wall group B is composed of three hyaline membranous walls (walls 2,3 and 4), each 0.5-1.5μm thick; walls may all separate or all three may remain attached to wall 4. Spore content hyaline.

Distribution : Occurred with 66.7% frequency of occurrence and 5.2% relative abundance. Population was recorded as 11 spores/50g soil in winter, 6 spores/50g soil in rainy season and 2 spores/50g soil in summer.

Acaulospora laevis Gerdemann & Trappe

Distinguishing Features : Spores forming singly in soil, sessile, borne laterally on a wide thin walled hypha 30-40μ dia. that terminates into a nearby globose, thin walled vesicle. Vesicle, approximately the same size as the spore, develops to full size prior to spore formation, with dense, white content, becoming empty and shrunken at spore maturity and then usually lost in sieving. Spores smooth, 119-300 x 119-500μ, globose to sub-globose, ellipsoid to irregular sometimes, dull yellow when young becoming deep yellow-brown to red-brown at maturity. Spore wall continuous except for the occluded opening consisting of three layers: A rigid, yellow-brown to red-brown outer wall 2-4μ thick, and two hyaline inner membranes, the innermost sometimes minutely roughened. Wall in older spores at times become minutely perforated. Spore content globose to somewhat polygonal.

Distribution : Occurred with 58.3% frequency of occurrence and 4.1% relative abundance. Population was recorded as 8 spores/50g soil in winter, 5 spores/50g soil in rainy season and 2 spores/50g soil in summer.

Acaulospora longula Spain & Schenck

Distinguishing Features : Azygospore formed singly in soil, borne laterally on hyphae slightly tapering to a swollen, globose to sub-globose hyphal terminus (60-)70-90(-110)µm dia. with walls 0.5µm thick. Hyphae at the point of spore attachment 6-12µm dia. Distance of connecting hypha between spore and terminus 100-200µm. Hyphal terminus often collapsing after spore formation and usually becoming detached from the mature spore in soil. Spores are dull, sub-hyaline to pale yellow in soil but light yellow and shiny from loss of outer mucilaginous wall. Spores globose to sub-globose (55-)75-90(-100)µm dia. occasionally ellipsoidal to irregular. Composite spore wall 2.5-5.0µm thick of separable portions distinguishable in broken spores. Outer wall mucilaginous, ephemeral, 0.5-3.0µm thick. Wall two 2.0-3.0µm thick, inseparable from wall three, which is 0.5µm thick. Wall four hyaline, 0.5-1.0µm, usually attached to wall five, which is membranous and 0.5-1.0µm thick, turning light purple in Melzer's reagent. Walls are most obvious in broken and stained spores. Spore content hyaline to sub-hyaline.

Distribution : Occurred with 75.0% frequency of occurrence and 9.0% relative abundance. Population was recorded as 14 spores/50g soil in winter, 12 spores/50g soil in rainy season and 7 spores/50g soil in summer.

Acaulospora scrobiculata Trappe

Distinguishing Features : Sporocarps unknown. Azygospores forming singly in soil, sessile, borne laterally on a wide thin walled hyaline hypha that terminates nearby in a thin walled vesicle. Vesicle globose, 100-160µm in dia., becoming empty and collapsing by spore maturity. Spores globose to broadly ellipsoid, 100-240x100-220µm sub-hyaline when young becoming light brown at maturity. Spore surface evenly pitted with depressions 1.0-1.5x1.0-3.0µm, separated by ridges 2.0-4.0µm thick. The mouths of the depressions circular to elliptical or occasionally linear to Y-shaped. Spore wall continuous except at the circular, rimmed vesicle attachment, about 15µm dia., consisting of four layers. Layer one rigid, pitted, sub-hyaline to light greenish yellow, 3.0-6.0µm thick. Layer two adhering but separable, smooth, hyaline, 0.2-0.5µm thick. Layer three adhering but separable, smooth, hyaline, 0.5-1.0µm thick. Layer four separated, sometimes minutely roughened, hyaline 0.2-1.0µm thick.

In Melzer's reagent, outer three spore wall layers yellow, while innermost becomes deep red. Spore contents of small, relatively uniform guttules.

Distribution : Occurred with 92.5% frequency of occurrence and 8.8% relative abundance. Population was recorded as 12 spores/50g soil in winter, 5 spores/50g soil in rainy season and 5 spores/50g soil in summer.

Glomus aggregatum Schenck & Smith

Distinguishing Features : Chlamydospores formed in loose clusters or in sporocarps without a peridium. Sporocarps of variable size ranging from 660-1800x330-1400µm. Sporocarps hyaline to light yellow with a greenish tint in transmitted light, becoming yellow with age. Chlamydospores globose, sub-globose, obovate, cylindrical to irregular. Spore dia. average (50.4-) 72.5 (-91.2) µm when globose and (7.2-) 87 (-110.4) x (57.6-) 72 (-79.2) µm when subglobose. Spores hyaline to yellow. Spore walls yellow to yellow-brown, varying from 1.2-2.4µm thick consisting of an outer wall slightly thicker and lighter in color than the inner wall. Walls separable with slight pressure and most apparent in stained preparations. Hyphae at the point of spore attachment 4.8-12µm wide. Spore contents confluent with hyphal contents on young spores but separated from hyphae on older spores by inner spore wall. Pore not occluded by hyphal wall thickening. Hyphal attachment straight or recurved sharply at the spore base. Most hyphae 4.8-7.2µm dia.

Distribution : Occurred with 83.3% frequency of occurrence and 8.4% relative abundance. Population was recorded as 13 spores/50g soil in winter, 11 spores/50g soil in rainy season and 7 spores/50g soil in summer.

Glomus caledonium (Nicol. & Gerd.) Trappe & Gerdemann

Distinguishing Features : Chlamydospores formed singly in soil or in sporocarps. Sporocarps up to 6 mm dia., compact, sub-globose, with a pallid peridium of hyaline thin-walled hyphae 8-25µm dia. and a light brown gleba. Spores dull yellow to brown, generally globose to sub-globose but sometimes ellipsoid or irregular, 130-279x120-272µm. Spore wall 6-10(-16)µm thick, composed of a hyaline outer layer 1-4(-8)µm thick and a yellow to brown inner layer 4-8(-10)µm thick. Outer layer easily separable, often thickened at the hyphal attachment, and extending along the attached hyphae for some distance. Inner wall thickening extending into the attached hypha a short distance. Spore contents separated from attached hypha by a thin, yellow, curved wall formed at the hyphal attachment or occasionally as much as 15µm down the attached hypha from the point of attachment.

Distribution : Occurred with 25.0% frequency of occurrence and 1.2% relative abundance. Population was recorded as 3 spores/50g soil in winter and 1 spore/50g soil in rainy season. However, no population was recorded in summer.

Glomus canadens (Thaxter) Trappe & Gerdemann

Distinguishing Features : Sporocarp subspherical or irregularly lobed with soft but rather firmly coherent peridial layer and dark brown gleba. Spores distinguished by a septum, ovoid to ellipsoid, or somewhat asymmetrical, 70-80x54-58μm, very rarely 100x65μm. Wall hyaline or pale yellowish, 4μm thick. Hyphae 8-14μm of the usual type with occasional clearly defined septa. Sporophores characteristically slender, 5-6μm.

Distribution : Occurred with 41.7% frequency of occurrence and 4.0% relative abundance. Population was recorded as 5 spores/50g soil in winter and 5 spores/50g soil in rainy season. However, no population was recorded in summer.

Glomus claroideum Schenck & Smith

Distinguishing Features : Chlamydospores formed singly or in loose clusters in the soil and infrequently as single spores in the root. Chlamydospores globose, 70-180μm diameter, occasionally subglobose to irregular: 59-126x72-145μm. Spore wall (4.5-10.5) μm consisting of 1 or 2 walls with the outer wall laminate and usually thicker than the inner wall. Sopre wall hyaline to yellow, becoming yellow-brown with age. Outer spore wall smooth. Spore content globular, hyaline to yellow to light yellow, retained by the membrane on young spores and occluded by spore walls on older spores. Spore subtending hyphae 7.5-15μm diameter at the spore attachment.Walls of subtending hypha on young spores 3.0-5.5μm thick and 1.5-3.0μm thick on older spores at the point of attachment, usually abruptly tapering below the spores, considerable branching of subtending hyphae usually occurs 50-150μm below the spores.

Distribution : Occurred with 58.3% frequency of occurrence and 3.5% relative abundance. Population was recorded as 5 spores/50g soil in winter, 2 spores/50g soil in rainy season and 6 spores/50g soil in summer.

Glomus constrictum Trappe

Distinguishing Features : Chlamydospores naked, formed singly or in lose clusters in soil, sub-globose to globose, 150-330μm dia., dark brown to black, shiny-smooth. Spore walls 7-12(-15)μm thick, dark brown, one-layered or occasionally seeming two-layered; base straight or occasionally with a short funnel-shaped projection; attachment occluded by wall thickenings; contents of oil globules of widely varying sizes. Reaction to Melzer's reagent not distinctive. Attached hypha straight to recurved and with the following features appearing in sequence away from the spore. Point of attachment with dark brown walls 3-6μm thick. Just beyond the point of attachment the hypha is constricted to 10-17 (-22) μm dia. Just beyond the constriction the hypha is inflated to 15-30μm dia. with yellow to yellow-brown walls 2-3μm thick, from which often grow several hyaline to yellow, fragile, thin-walled hypha 5-7μm dia. Just beyond the inflated segment there is often a thick-walled septum and beyond the inflated segment the hypha is dichotomously forked.

Distribution : Occurred with 50.0% frequency of occurrence and 5.2% relative abundance. Population was recorded as 11 spores/50g soil in winter and 8 spores/50g soil in rainy season. However, no population was recorded in summer.

Glomus fasciculatum (Thaxter) Gerdemann & Trappe

Distinguishing Features : Chlamydospores borne free in soil, in loose aggregations, in small compact clusters and in sporocarps. Sporocarps up to 8x5x5mm, irregulary globose or flattened, tuberculate, greyish brown. Peridium absent. Chlamydospores 35-105μm dia. when globose, 75-150x35-100μm dia. when sub-globose to obovate, ellipsoid, sub-lenticular, cylindrical or irregular. Spore walls highly variable in thickness (3-17μm), hyaline to light yellow or yellow brown. The thicker walls often minutely perforate with thickened inward projections. Hyphal attachments 4-15μm dia., occluded at maturity. Walls of attached hyphae often thickened 1-4μm near the spore.

Distribution : Occurred with 92.5% frequency of occurrence and 16.1% relative abundance. Population was recorded as 32 spores/50g soil in winter, 20 spores/50g soil in rainy season and 7 spores/50g soil in summer.

Glomus invermaium Hall

Distinguishing Features : Spores hypogeous, globose, 50-75μm dia., light brown to brown formed in loose sporocarps up to 1mm across. Peridium lacking. Spore wall double, outer colourless, 1-1.5μm thick, inner light brown to brown, 3-6μm thick. Outer wall extending down the subtending hypha for up to 100μm. Walls inseparable. Subtending hyphae 6-13μm dia., colourless to brown, slightly pinched-in at the point of attachment. Pore 1-4μm wide, without septum.

Distribution : Occurred with 58.3% frequency of occurrence and 3.3% relative abundance. Population was recorded as 9 spores/50g soil in winter and 3 spores/50g soil in rainy season. However, no population was recorded in summer.

Glomus macrocarpum Tul. & Tul.

Distinguishing Features : Sporocarps are fragmentary, none of the species more than 5mm diameter. Spores are usually slightly longer than wide, subglobose or globose to regular (90-130μm). Spore wall is composed of two distinct layers, outer wall is thin (1-2μm) and hyaline, inner wall layer is yellow in section, 6-12μm thick, with a series of laminations occasionally visible or rarely appearing as two distinct layers, Spores taper to the point of the attachment of the single persistent hypha. The average diameter of the hypha at this point is 16μm. the inner wall at maturity thickens to occlude the poreof the attached hypha and the wall thickening continues into the subtending hypha for up to 90μm of spore. Infrequently the pore seems to be closed by a septum that is thinner than the normal occluding wall thickening. Spores characteristically bear a straight, long subtending hypha which may extend up to 100μm before branching or breaking.

Distribution : Occurred with 16.7% frequency of occurrence and 1.4% relative abundance. Population was recorded as 4 spores/50g soil in winter and 1 spore/50g soil in summer. However, no population was recorded in rainy season.

Glomus mosseae (Nicol. & Gerd.) Gerdemann & Trappe

Distinguishing Features : Sporocarps 1-10 spored, globose to ellipsoid, up to 1mm dia. Peridium of loosely interwoven, irregularly branched, hyaline, septate hyphae 2-12μm dia., the walls up to 0.5μm thick, frequently anastomosing to form a thin network, enclosing the chlamydospores entirely, incompletely or with some spores unenclosed. Chlamydospores yellow to brown, globose to ovoid, obovoid or somewhat irregular, 105-310x110-305μm, with one or occasionally two funnel-shaped bases 20-30(-50)μm dia., divided from subtending hyphae by a curved septum. Walls 2-7μm thick with a thin, often barely perceptible hyaline outer membrane and a thick, brownish yellow inner layer.

Distribution : Occurred with 83.3% frequency of occurrence and 13.9% relative abundance. Population was recorded as 23 spores/50g soil in winter, 18 spores/50g soil in rainy season and 10 spores/50g soil in summer.

Glomus occultum Walker

Distinguishing Features : Sporocarps unknown. Chlamydospores borne singly or in loose clusters or in compact clusters, often broader than long, ovoid to obovoid, subangular to irregular, less frequently globose to sub-globose, 15-100x20-120μm, hyaline to white. Subtending hypha funnel-shaped to simple, 5-50μm long, 3-10μm wide at spore base, tapering to 2-5μm; attached axially or eccentrically and recurved to straight, sometimes closed distally by a septum. Spore wall one to two layered with an additional rough outer deposit of granular material which sloughs with age. Outer deposit up to 2μm thick except at the spore base, where it may be greatly thickened. Outer wall, when present, less than 1μm thick, often indistinct. Inner wall (1-)1.5-2.5(-5) μm thick, usually of two sometimes indistinct laminations.

Distribution : Occurred with 25.0% frequency of occurrence and 1.4% relative abundance. Population was recorded as 4 spores/50g soil in winter and 1 spore/50g soil in rainy season. However, no population was recorded in summer.

Glomus reticulatum Bhattacharjee & Mukerji

Distinguishing Features : Chlamydospores borne freely and singly in the soil, not known to occur in sporocarps, dark brownish black, globose, 130-170μm in dia. Wall 10-15μm thick, clearly differentiated into an outer and inner wall. Outer wall 5-7μm thick, two layered and fissured, outermost layer 1-2μm thick and inner layer 4-5μm thick. Inner wall with regular geometric reticulate markings (5-10μm apart) on its outer surface. Subtending hypha funnel shaped, 8-10μm wide at the point of attachment, tapering up to 5μm after a distance of 50-80μm; pore open. Wall thickening extends down the subtending hypha to a distance of 40μm.

Distribution : Occurred with 41.7% frequency of occurrence and 4.0% relative abundance. Population was recorded as 7 spores/50g soil in winter and 3 spores/50g soil in rainy season. However, no population was recorded in summer.

Glomus tortuosum Schenck & Smith

Distinguishing Features : Sporocarps unknown. Chlamydospores borne singly in soil but occasionally gathering in pairs. Immature spores sub-hyaline without a mantle, thin-walled (0.4-5μm dia.) and attached to extramatrical hyphae up to 500μm long. Mature spores yellow to dull grey-brown with a mantle of sinuous hyphae closely oppressed to the spore and flattened, 4-10μm wide, forming layers of hyphae on the spore surface up to 25μm thick. Occasionally, mantle extends down the hyphal attachment. Mantle hyphae hyaline when young, acquiring a brownish segment with age, and originating from swellings on the hyphal attachment 10-20μm below the spore or arising from other hyphae adjacent to the spore. Mantle frequently with adhering debris and soil particles. Chlamydospores largely globose (120-) 160 (-210) μm dia., some sub-globose 94-180x112-230μm excluding the mantle. Spores with a single, laminate, thin wall, 0.5-2μm dia. Spore contents globular but usually obscured by the mantle. Width of hyphal attachment at the spore base 9-20 (-26) μm. Extramatrical hyphae 2-11μm dia., hyaline to light yellow.

Distribution : Occurred with 33.3% frequency of occurrence and 2.7% relative abundance. Population was recorded as 8 spores/50g soil in winter and 2 spores/50g soil in rainy season. However, no population was recorded in summer.

Sclerocystis dussii (Pat.) von Hohn

Distinguishing Features : Sporocarps 263-540µm broad, sub-globose to hemispheric, pulvinate or pyriform, single to fused together laterally and vertically to a depth of three or four, forming a tuberculate crust. Whitish when young, becoming tan by maturity. Upper surface of crust covered with thin walled vesicles upto 340-77µm, globose when young, becoming obovoid to ellipsoid to clavate or broadly clavate and rounded at the tip, at times constricted near middle, collapsing when dried. Peridium of individual sporocarps 20-60µm thick, composed of thick walled interwoven hyphae. Chlamydospores 50-80x32-54µm, obovate to clavate, ellipsoid or oblong-ellipsoid, cut off from subtending hyphae by septa just below the spore bases, tightly grouped in a single layer in a hemisphere around a central plexus of hyphae. Spores absent at the sporocarp base. Chlamydospore wall up to 3µm thick at base and 2µm at apex, brown in color.

Distribution : Occurred with 8.3% frequency of occurrence and 0.8% relative abundance. Population was recorded as 2 spores/50g soil in winter season. However, no population was recorded in rainy or summer seasons.

Discussion

Distribution of AM fungi in different ecological regions and their relations to soil properties and native plants have been investigated by several researchers (Kim and Weber, 1985; Rozema *et al.*, 1986; Cook *et al.*, 1993; Janardhanan *et al.*, 1994; Aliasgharzadeh *et al.*, 2001; Wang *et al.*, 2004; Shi *et al.*, 2007). It is established that variation in AM distribution, spore density and colonization with different host plant species is generated by a variety of mechanisms, including variation in host species and their phenology, mycorrhizal dependency, soil properties, host plant-mediated alteration of the soil microenvironment, or other unknown host plant traits (Hayman, 1982; Eom *et al.*, 2000; Wang *et al.*, 2004).

In various salt-affected soils, especially saline soils, very small or even zero spore population has been reported earlier by Kim and Weber (1985), Barrow *et al.* (1997), Carvalho *et al.* (2001) and Wang *et al.* (2004). Alkaline soils also have a detrimental effect on the AM spore population, distribution and diversity and similar case in the alkaline soils of Upper Gangetic plains. High pH, poor plant diversity and low vegetation in such soils may severely restrict colonization and diversity of arbuscular mycorrhizae (Wang *et al.*, 2003). AM spore population also shows changes with seasonal variation and host plant phenology (Bohrer *et al.*, 2004). Maximum spore population occurs in winter and minimum in summer. Number of spores increases after a period of maximum root growth. Studies by various authors have shown maximum spore count in winter (21-25^{0}C) and minimum in monsoon (28-34^{0}C) and summer (28-42^{0}C) months (Saif and Khan, 1975; Giovannetti, 1985; Singh and Varma, 1985; Syl*via*, 1986).

Glomus species is most dominant in the alkaline soils of Upper Gangetic plains followed by *Acaulospora* species. Predominance of *Glomus* species may be attributed to the fact that these are the most widespread AM fungi and occur in neutral and alkaline soils (Mosse, 1973). *Glomus* species has been reported to have a wide pH tolerance range. These are also resistant to the high soil temperatures (Al-Raddad, 1993; Shi *et al.*, 2007). Another reason for the predominance of *Glomus* species may be the ability of members of Glomaceae (unlike Gigasporaceae) to recolonize roots from mycelia fragments, coupled with the ability of *Glomus* species, unlike *Gigaspora* or *Scutellospora*, readily to establish anastomoses between separate mycelia. These species may therefore, be particularly well adapted to recolonize roots following disruption of the soil mycelium.

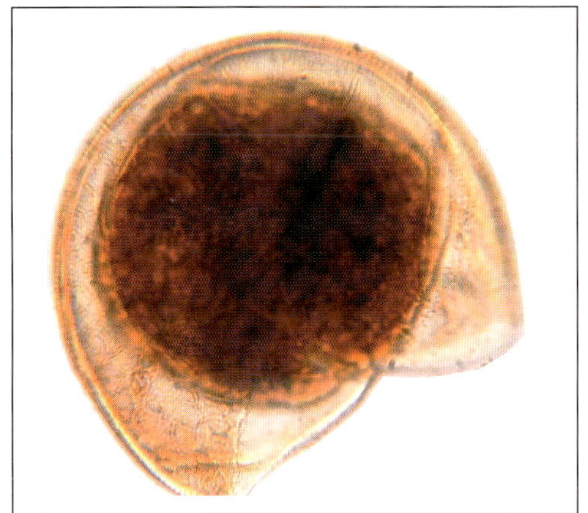

Acaulospora bireticulata

Acaulospora denticulata

Acaulospora laevis

Acaulospora longula

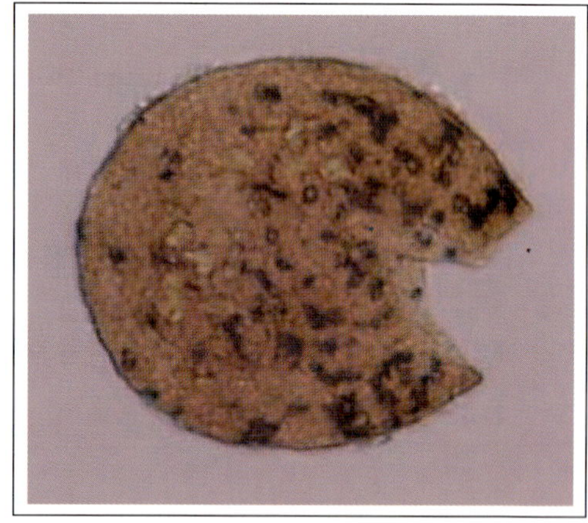

Acaulospora scrobiculata

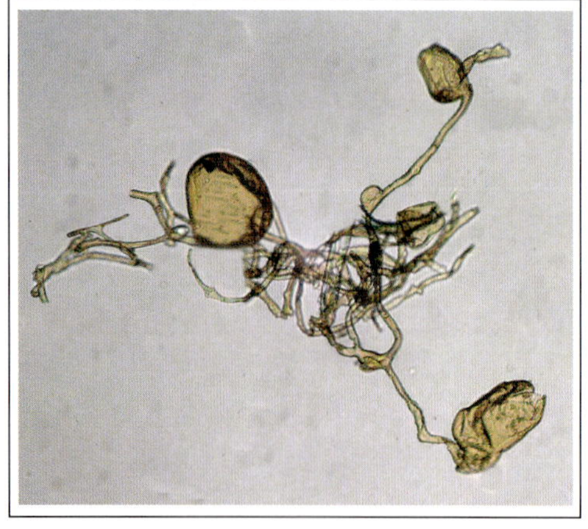

Glomus aggregatum

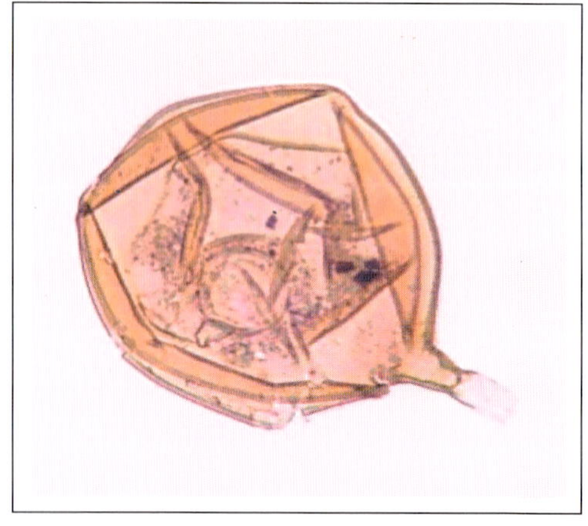

Glomus caledonium

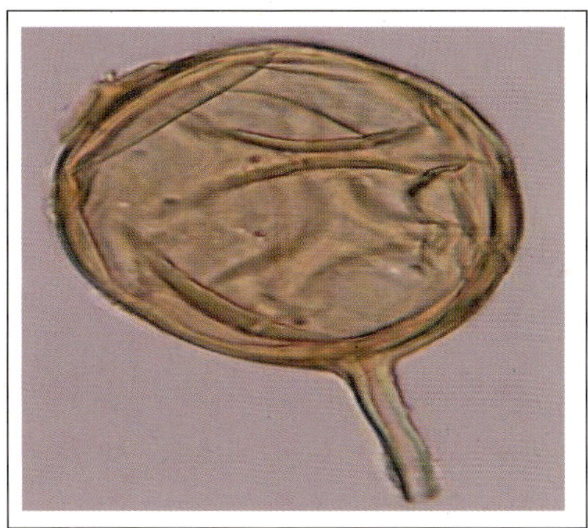

Glomus canadens

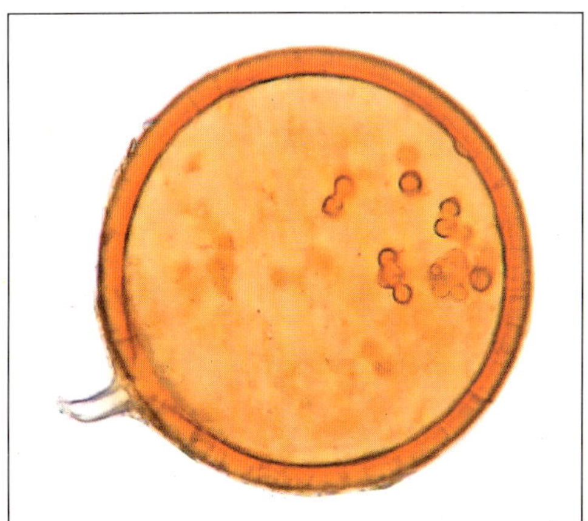

Glomus claroideum

Glomus constrictum

Glomus fasciculatum

Glomus invarmaium

Glomus macrocarpum

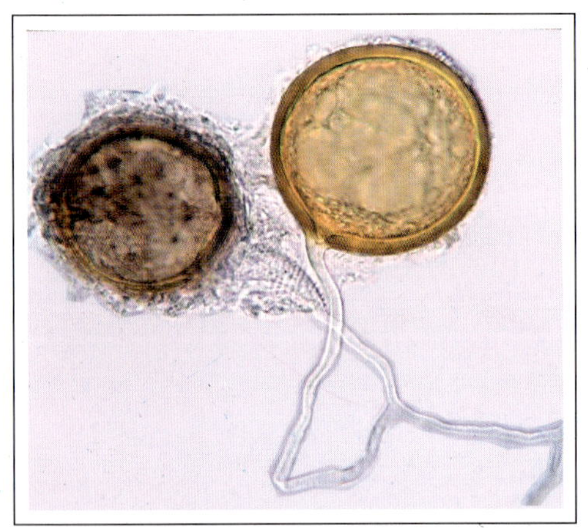

Glomus mosseae

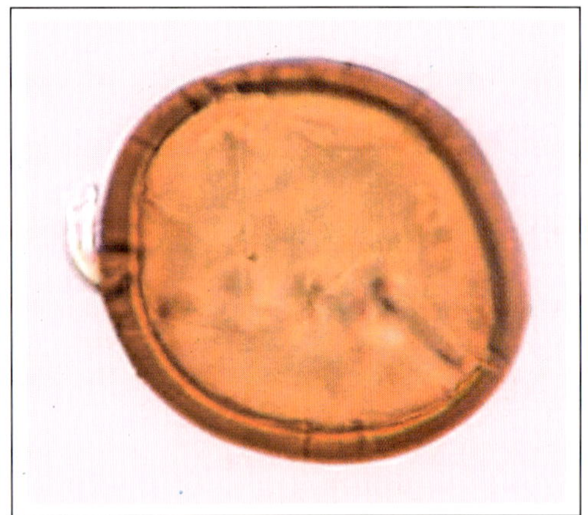

Glomus occultum

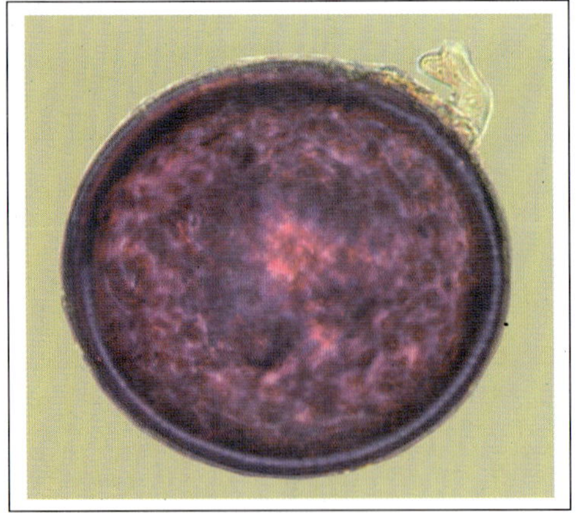

Glomus reticulatum

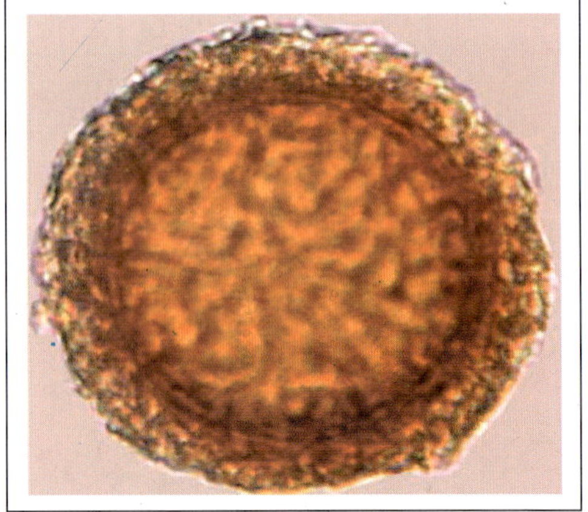

Glomus totuosum

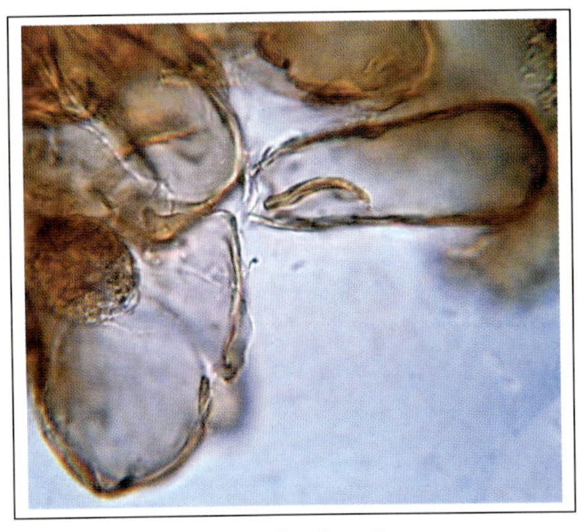

Sclerocystis dussi spore

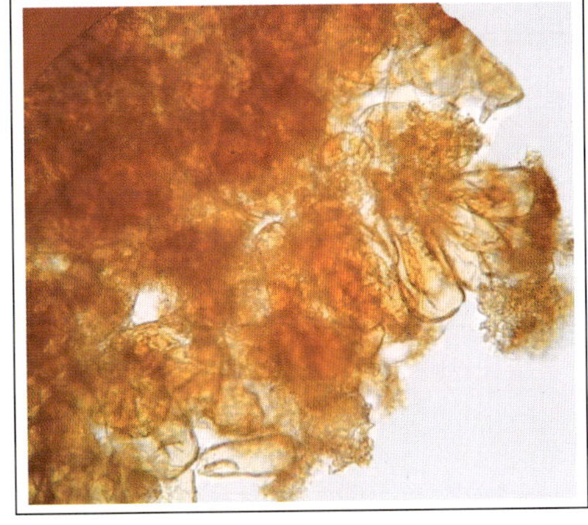

Sclerocystis dussi sporocarp

Most of the AM fungal species occurring in alkaline soils are small in size and have thick walls. These characters may provide them more adaptability under stress conditions. Similar results indicating a tendency in species composition towards mostly small spore fungi belonging to the genus *Glomus* have previously been reported in the stressed habitats (Janardhanan *et al.*, 1994; Stutz *et al.*, 2000; Wang *et al.*, 2004; Shi *et al.*, 2007), as these species may be more adaptable in adjusting patterns of sporulation to environmental stress conditions (Jacobson, 1997).

References

Al Karaki, G.N. (2006). Nursery inoculation of tomato with arbuscular mycorrhizal fungi and subsequent performance under irrigation with saline water. *Scientia Horticul.*, 109: 1-7.

Aliasgharzadeh, N., Rastin, N.S., Towfighi, H. and Alizadeh, A. (2001). Occurrence of arbuscular mycorrhizal fungi in saline soils of the Tabriz Plain of Iran in relation to some physical and chemical properties of soil. *Mycorrhiza*, 11: 119-122.

Al-Raddad, A.M. (1993). Distribution of different *Glomus* species in rainfed areas in Jordan. Dirasat-Ser. B: *Pure Appl. Sci.*, 20: 165-182.

Auge, R.M. (2001). Water relations, Drought and VA mycorrhizal symbiosis. *Mycorrhiza*, 11: 3-42.

Barea, J.M., Azcon, R. and Azcon-Gdeaquilar, C. (1983). Interactions between phosphate solubilizing bacteria and VA mycorrhiza to improve the utilization of rock phosphate by plants in non acidic soils. In: *3rd International Congress on Phosphorus Compounds. Brussels*: IMPHOS: 127-144.

Barrow, J.R., Havstad, K.M. and McCaslin, B.D. (1997). Fungal root endophytes in four-wing saltbush, *Altiplex canescens*, on arid rangeland of southwestern USA. *Arid Soil Res. Rehabil.*, 11: 177-185.

Bieleski, R.L. (1973). Phosphate pools, phosphate transport and phosphate availability. *Ann. Rev. Pl. Physiol.*, 24: 225-252.

Bohrer, K.E., Karl, F.F. and Amon, J.P. (2004). Seasonal dynamics of arbuscular mycorrhizal fungi in differing wetland habitats. *Mycorrhiza*, 14: 329-337.

Bolan, N.S. and Robson, A.D. (1987). Effects of vesicular arbuscular mycorrhiza on the availability of iron phosphates to plants. *Plant Soil*, 99: 401-410.

Calvente, R., Cano, C., Ferrol, N., Azcon-Aguilar, C. and Barea, J.M. (2004). Analyzing natural diversity of arbuscular mycorrhizal fungi in olive tree (*Olea europea* L.) plantations and assessment of the effectiveness of native fungal isolates as inoculants for commercial cultivars of olive plantlets. *Appl. Soil Ecol.*, 26: 11-19.

Cantrell, I.C. and Linderman, R.G. (2001). Pre-inoculation of lettuce and onion with AM fungi reduces deleterious effects of soil salinity. *Plant Soil*, 233: 269-281.

Caravaca, F., Barea, J.M., Palenzuela, J., Figueroa, D., Alguacil, M.M. and Roldan, A. (2003). Establishment of shrub species in a degraded semiarid site after inoculation with native or allochthonous arbuscular mycorrhizal fungi. *Appl. Soil Ecol.*, 22: 103-111.

Carvalho, L.M., Cacador, I. and Martins-Loucaoma (2001). Temporal and spatial variation of arbuscular mycorrhizas in salt marsh plants of the Tagus estuary (Portugal). *Mycorrhiza*, 11: 303-309.

Chauhan, R.P.S. and Singh, B.B. (1999). Management of salt-affected soils for sustainable crop production. In: *Modern Approaches and Innovation in Soil Management*. (Eds.) D.J. Bagyaraj, A. Varma, K.K. Khanna and H.K. Kehri, Rastogi Publications, Meerut: 205-223.

Chhabra, R. (1995). Nutrient requirement for sodic soil. *Fertil. News*, 40(10): 13-21.

Cook, J.C., Butler, R.H. and Madol (1993). Some observations on the vertical distribution of VAM in roots of salt marsh grasses growing in saturated soils. *Mycologia*, 85: 547–550.

Danneberg, G., Latus, C., Zimmer, W., Hundeshagen, B., Schneider-Poetsch, H.J. and Bothe, H. (1992). Influence of vesicular-arbuscular mycorrhiza on phytohormone balances in maize (*Zea mays* L.). *J. Plant Physiol.*, 141: 33–39.

Datta, S., Thibault, Y., Fyfe, W.S., Powell, M.A., Hari, B.R., Martin, R.R. and Tripthy, S. (2002). Occurrence of trona in alkaline soils of the indo-gangetic planes of Uttar Pradesh (UP) India. *Episodes*, 25(4): 236-240.

Dodd, J.C. and Thomson, B.D. (1994). The screening and selection of inoculant arbuscular-mycorrhizal and ectomycorrhizal fungi. *Plant Soil*, 159: 149-158.

Duan, X., Neuman, D.S., Reiber, J.M., Green, C.D., Saxton, A.M. and Auge, R.M. (1996). Mycorrhizal influence on hydraulic and hormonal factors implicated in the control of stomatal conductance during drought. *J. Exp. Bot.*, 47: 1541-1550.

Enkhtuya, B., Oskarsson, U., Dodd, J.C. and Vosatka, M. (2003). Inoculation of grass and tree seedlings used for reclaiming eroded areas in Iceland with mycorrhizal fungi. *Folia Geobot.*, 38: 209-222.

Enkhtuya, B., Rydlova, J. and Vosatka, M. (2000). Effectiveness of indigenous and non-indigenous isolates of arbuscular mycorrhizal fungi in soils from degraded ecosystems and man-made habitats. *Appl. Soil Ecol.*, 14: 201-211.

Eom, A.H., David, C., Hartnett, A., Gail, W.T. and Wilson, C. (2000). Host plant species affects on arbuscular mycorrhizal fungal communities in tall grass prairie. *Oecologia*, 122: 435-444.

Gerdemann, J.W., Nicolson, T.H. (1963). Spores of mycorrhizal *Endogone* species extracted from soil by wet sieving and decanting. *Trans. Br. Mycol. Soc.*, 46: 235-244.

Giovannetti, M. (1985). Seasonal variations of vesicular arbuscular mycorrhiza and endogonaceous spores in maritime sand dunes. *Trans. Br. Mycol. Soc.*, 84: 678-684.

Giri, B. and Mukerji, K.G. (2004). Mycorrhizal inoculant alle*via*tes salt stress in *Sesbania aegyptica* and *Sesbania grandiflora* under field conditions: evidence for reduced sodium and improved magnesium uptake. *Mycorrhiza*, 14: 307-312.

Harley, J.L. and Smith, S.E. (1983). *Mycorrhizal Symbiosis*. Academic Press.

Hayman, D.S. (1982). Influence of soils and fertility on activity and survival of VAM fungi. *Phytopath.*, 72: 1119-1125.

Jacobson, K.M. (1997). Moisture and substrate stability determine VA mycorrhizal fungal community distribution and structure in an arid grassland. *J. Arid Environ.*, 35: 59-75.

Janardhanan, K.K., Abdul-Khaliq, Naushin Fauzia, Ramaswamy, K. (1994). Vesicular-arbuscular mycorrhiza an alkaline usar land ecosystem. *Curr. Sci.*, 67(6): 465-469.

Joshi, K.C. and Singh, H.P. (1995). Interrelationships among VA mycorrhiza population, soil properties and root colonization capacity of soil. *J. Ind. Soc. Soil Sci.*, 43: 204-207.

Juniper, S. and Abbott, L. (1993). Vesicular arbuscular mycorrhizas and soil salinity. *Mycorrhiza*, 4: 45-57.

Khare, V., Singh, S.K., Singh, S. and Kehri, H.K. (2008). Efficacy of AM fungi in improving the growth performance of *Cyamopsis tetragonoloba* under saline soil conditions. *Mycorrhiza News*, 20(2): 17-20.

Kim, C.K. and Weber, D.J. (1985). Distribution of VA mycorrhiza on halophytes on inland salt playas. *Plant Soil*, 83: 207-214.

Klironomos, J.N. (2003). Variation in plant response to native and exotic arbuscular mycorrhizal fungi. *Ecology*, 84: 2292-2301.

Mosse, B. (1973). Advances in the study of vesicular-arbuscular mycorrhiza. *Ann. Rev. Phytopath.* 11: 171-196.

Pfeiffer, C.M. and Bloss, H.E. (1988). Growth and nutrition of guayule (*Parthenium argentatum*) in a saline soil as influenced by vesicular arbuscular mycorrhiza and phosphorus fertilization. *New Phytol.*, 108 (3): 315-321.

Pfleger, F.L., Steward, E.L. and Noyd, R.K. (1994). Role of VAM fungi in mine land revegetation. In: *Mycorrhizae and Plant Health* (Eds.) F.L. Pfleger and R.G. Linderman, APS, St. Paul.: 47-82.

Rosendahl, C.N. and Rosendahl, S. (1991). Influence of vesicular-arbuscular mycorrhizal fungi (*Glomus* spp.) on the response of cucumber (*Cucumis sativis* L.) to salt stress. *Environ. Exp. Bot.*, 31: 313–318.

Rozema, J., Arp, W., Diggelen, J.V., Esbroek, M.V., Broekman, R. and Punte, H. (1986). Occurrence and ecological significance of vesicular arbuscular mycorrhiza in the salt marsh environment. *Acta. Bot. Neerl.*, 35: 457-467.

Ruiz-Lozano, J.M. (2003). Arbuscular mycorrhizal symbiosis and alle*via*tion of osmotic stress. New perspectives for molecular studies. *Mycorrhiza*, 13: 309-317.

Ruiz-Lozano, J.M. and Azcon, R. (1995). Hyphal contribution to water uptake in mycorrhizal plants as affected by the fungal species and water status. *Physiol. Plant*, 95: 472-478.

Ruiz-Lozano, J.M., Azcon, R. and Gomez, M. (1996). Alle*via*tion of salt stress by arbuscular mycorhhizal *Glomus* sp. in *Lactuca sativa* plants. *Phisiologia Plantarum*, 98: 767-772.

Ruiz-Lozano, J.M., Collados, C., Barea, J.M., Azcon, R. (2001a). Cloning of cDNAs encoding SODs from lettuce plants which show differential regulation by arbuscular mycorrhizal symbiosis and by drought stress. *J. Experiment. Bot.*, 52: 2241–2242.

Ruiz-Lozano, J.M., Collados, C., Barea, J.M., Azcon, R. (2001b). Arbuscular mycorrhizal symbiosis can alle*via*te drought-induced nodule senescence in soybean plants. *New Phytol.*, 151: 493–502.

Saif, S.R. and Khan, A.G. (1975). The influence of season and stage of development of plant on Endogone mycorrhiza of field grown wheat. *Can. J. Microbiol.*, 21: 1021-1024.

Schenck, N.C. and Perez, Y. (1990). *Manual for the identification of vesicular arbuscular mycorrhizal fungi*. INVAM, University of Florida, Gainesville, Fla, USA.

Sengupta, A. and Chaudhuri, S. (1990). Vesicular arbuscular mycorrhiza in the primary salt marsh plants of Ganges river delta in West Bengal (India). *Plant Soil*, 122: 111-113.

Shahid, S.A. and Jenkins, D.A. (1994). Mineralogy and micromorphology of salt crusts from the Punjab, Pakistan. in soil micromorphology. *Develop. Soil Sci.*, 22: 799-810.

Sharma, D.P., Singh, K. and Rao, K.V.G.K. (2000). Subsurface drainage for rehabilitation of water logged saline lands: example of a soil in semi arid climate. *Arid Soil Res. Rehabil.*, 14: 373-384.

Shi, Z.Y., Zhang, L.Y., Li, X.L., Feng, G., Tian, C.Y. and Christie, P. (2007). Diversity of arbuscular mycorrhizal fungi associated with desert ephemerals in plant communities of Junggar Basin, northwest China. *Appl. Soil Ecol.*, 35: 10-20.

Singh, K. and Varma, A.K. (1985). Association of bacteria with endogonaceous spore *Glomus macrocarpus* var. *geosporus* extracted from xerophytic plants. *Trans. Jpn. Soc. Mycol.*, 21: 39-44.

Smith, S.E. and Gianinazzi-Pearson, V. (1988). Physiological interactions between symbionts in vesicular-arbuscular mycorrhizal plants. *Ann. Rev. Pl. Physiol. Molec. Biol.*, 39: 221-224.

Smith, S.E. and Smith, F.A. (1990). Structure and function of the interfaces in the biotrophic symbiosis as they relate to nutrient transport. *New Phytol.*, 114: 1-38.

Stutz, J.C., Copeman, R., Martin, C.A. and Morton, J.B. (2000). Patterns of species composition and distribution of arbuscular mycorrhizal fungi in arid regions of southwestern North America and Namibia. *Can. J. Bot.*, 78: 237-245.

Syl*via*, D.M. (1986). Spatial and temporal distribution of vesicular arbuscular mycorrhizal fungi associated with *Uniola paniculata* in Florida foredune. *Mycologia*, 78: 728-734.

Trappe, J.M. (1982). Synoptic keys to the genera and species of zygomycetous mycorrhizal fungi. *Phytopath.*, 72: 1102-1108.

Wang, F.Y., Liu, R.J., Lin, X.G. and Zhou, J.M. (2004). Arbuscular mycorrhizal status of wild plants in saline-alkaline soils of the Yellow River delta. *Mycorrhiza*, 14: 133-137.

Wang, S.Q., Jiang, J., Lei, J.Q., Zhang, W.M. and Qian, Y.B. (2003). The distribution of ephemeral vegetation on the longitudinal dune surface and its stabilization significance in the Gurbantunggut desert. *Acta Geogr. Sin.*, 58: 598-605.

□□□

Microbial Diversity and Functions, 2012

New India Publishing Agency, New Delhi (India)
E-mail : info@nipabooks.com; Website : www.nipabooks.com

Chapter **27**

Role of Plant Growth Promoting Rhizobacteria as Biocontrol Agents

C. Manoharachary and K.V.B.R. Tilak

ABSTRACT

Microorganisms that harbour in the root region are the most befitting candidates for use as biocontrol agents. The rhizosphere provides the frontline defense for root against attack by pathogens. Pathogens face a complicated phenomena of antagonism from roots. They also compete with each other for site of colonization on the root surface and plant nutrients. The ideal biocontrol agent introduces and /or promotes the antagonists only, whenever required and are most effective in minimizing the wasteful application of inoculum to non-targets. Many genera belonging to bacteria, fungi, actinomycetes and viruses are used as biocontrol agents to combat several important plant diseases. The present review focuses on the use of bacterial antagonists ,particularly rhizobacteria, as biocontrol agents of fungal diseases of plants , their mechanism of action, molecular and genetic basis of antagonistic property and possible genetic manipulations to bring about desirable changes.

Keywords: PGPR, Biocontrol, Plant diseases, mechanism

Introduction

The group of beneficial, root associative bacteria that stimulates the growth of plants are known as plant growth-promoting rhizobacteria (PGPR) or plant health promoting rhizobacteria (PHPR) (Kloepper *et al.*, 1989). Fluorescent pseudomonads and bacilli comprise major group among PGPR along with other bacteria like *Acetobacter, Actinoplanes, Agrobacterium, Alcaligenes, Arthrobacter, Azospirillum, Azotobacter, Bacillus, Cellulomonas, Clostridium,*

Enterobacter, Erwinia, Flavobacerium, Pasteuria, Serratia and *Xanthomonas.* The beneficial rhizosphere microorganisms also include rhizobia and bradyrhizobia, which establish symbiotic relationship with leguminous plants. These bacteria generally, improve the plant growth through direct effects on plant by producing plant growth-promoting substances, increasing the availability and uptake of nutrients and suppressing soil-borne plant pathogens (Dutta and Podile 2010, Tilak *et al.*, 2010, Wu *et al.*, 2009, Lugtenberg and Kamilova, 2009; Nautiyal and Tilak, 2009).

Over the last decade, understanding of rhizosphere biology has progressed with the discovery of PGPR that colonizes plant roots and promote plant growth. These PGPR could compete with other rhizosphere microorganisms most effectively, leading to increased plant growth. Application of plant growth promoting rhizobacteria has been shown to increase legume growth and development in terms of plant nodulation and nitrogen fixation under normal growth conditions. PGPR have also been shown to increase plant yields in non-legume crops. (Angaw *et al.*, 2011, Tilak *et al.*, 2010, Vogeti *et al.*, 2009, Podile and Kishore, 2006, Gupta *et al.*, 2003).

The suppression of growth of soil-borne and root-borne plant pathogens by the use of antagonistic microorganisms to reduce diseases is termed as biocontrol. National Academy of Sciences (USA) defined biocontrol as "the use of natural or modified organisms, genes or gene products to reduce the effects of undesirable organisms (pests) and to favour desirable organisms such as crops, trees, animals and beneficial insects and microorganisms". Wilson (1997) defined biological control as "the control of a plant disease with a natural biological process or with the product of a natural biological process." This definition allows the inclusion of biological chemicals produced by living organisms and extracted from them, host resistance (constitutive and induced) and antagonistic microorganisms.

The rhizosphere bacteria are the ideal biocontrol agents as they can provide the front line defence for plant roots against the attack by various plant pathogens. Disease suppression by biocontrol agents occurs due to interactions among the biocontrol agents and the members of the spermosphere, rhizosphere or phyllosphere community. The microbes used in biocontrol have various advantages namely: (1) these organisms are considered safer than the chemicals as they do not accumulate in the food chain, (2) self replication circumvents repeated applications, (3) unlike chemical agents, target organism seldom develop resistance (4) chemical control agents along with biocontrol agents are advocated in integrated plant disease control management and (5) properly developed biocontrol agents are not considered harmful to the population and functional dynamics of the soil microorganisms in the rhizosphere. The major disadvantages include variability of field product.

Also, the effectiveness of a given biocontrol agent may be restricted to a specific location, due to the effects of soil and climate. Moreover, biological control depends upon the establishment and maintenance of a threshold population of bacteria on planting material or in soil, and a drop in *via*bility below that level may eliminate the possibility of biological control (Weller, 1988). Many soil edaphic factors, including soil temperature, moisture, pH, clay content, interactions of biological disease control microorganisms with other rhizosphere bacteria and with pathogens also affect their *via*bility and tolerance to adverse conditions once applied. Concentration of O_2 and CO_2 in the soil is also one of the major factors that affects activity of biocontrol agent in the rhizosphere (Nautiyal, 1997; Goel *et al.*, 2001).

Several rhizosphere bacteria have been demonostrated to posses biocontrol potential. *Pseudomonas* spp. make up a dominant population in the rhizosphere and seems to be one of the most appealing for the biological control of plant diseases. The worldwide interest in the *Pseudomonas* spp. as bocontrol agents was started in 1970's with the studies conducted at the University of California, Berkeley, USA (Weller, 1988) and several companies now have developed biocontrol agents as commercial products. Fluorescent pseudomonads posses several properties that have made them biocontrol agents of choice. These include (a) efficient-colonization of roots, tubers, hypocotyl etc., (b) ability to utilize a large number of organic substrates commonly found in roots and root eudates, (c) relatively easily cultivated under lab conditions, (d) production of variety of secondary metabolites which are toxic to bacterial and fungal pathogens and (e) compatibility with commonly used pesticides and other biocontrol agents.

Despite the extensive research where biological agents have been used to control plant diseases, there have been limited commercial successors. An efficient biocontrol agent must meet the requirements of a good colonizer and critical competitor in the rhizosphere. Root colonization is a pre-requisite for a strain to act successful as a biocontrol agent. Colonization is an active process, which involves the proliferation of microorganisms in/on and around the growing roots (Johri *et al.*, 1997). Microorganisms compete with each other for carbon source, mineral nutrients and infection sites on the roots. Competition between the biocontrol agent and pathogen can result in displacement of the latter (Osburn *et al.*, 1989). This review encompasses the role of pseudomonads as biocontrol control agents against phyto-pathogens.

Rhizobacteria as biological control agents of plant pathogens

Fluorescent pseudomonads have revolutionized the field of biological control of soil-borne plant pathogenic fungi. Most of them fall either in *fluorescens* or *putida* group. During the last three decades, they have emerged as the largest potentially most promising group of plant growth promoting rhizobacteria involved in the biocontrol of plant diseases (Cook, 1993 ; Pierson and Weller, 1994; Barbosa *et al.*, 1995; Gomes *et al.*, 1996; Wei *et al.*, 1996; Compant *et al.*, 2005). Fluorescent pseudomonads have received the most attention for several compelling reasons. First, they readily colonize roots in nature, where they are frequently the most common among microorganisms (Weller, 1988) .The simple nutritional requirement and the ability to use many carbon sources that exude from roots and to compete with indigenous microflora, may explain their ability to colonize the rhizosphere (Mazzola and Cook, 1991). Additionally, pseudomonads are amenable to genetic manipulation. These characteristics make them useful vehicle for the delivery of antimicrobial and insecticidal compounds and plant hormones to the rhizosphere. O'Sullivan and O'Gara (1992) reviewed the traits of fluorescent pseudomonads such as production of antibiotics, hydrogen cyanide, siderophores which are involved in suppression of plant root pathogens.

There are numerous examples of biocontrol of several devastating fungal plant pathogens of important crops by fluorescent pseudomonads and has been reviewed from time to time (O'Sullivan and O'Gara, 1992; Kumar and Dube,1992; Weller and Thomashow, 1993; Krishnamurthy and Gnanamanickam, 1997; Pierson and Weller, 1994; Saxena *et al.*, 2000; Pal *et al.*, 2001; Duffy *et al.*, 2004; Compant *et al.*, 2005). Natural disease suppression involving pseudomonads have been reported by many workers. In disease suppressive soils, though the pathogen initially causes disease, on continued cropping it fails to do so. Naturally occurring soil pseudomonads are important elements in these soils suppressive to diseases.

A number *of Pseudomonas* strains have been used as biological control agents in green house and field conditions against an array of plant pathogens. In many cases they not only help in suppressing the pathogens, but also improved the plant yield by acting as plant growth promoters (O'Sullivan and O'Gara, 1992; Dowling and O'Gara, 1994). Even though there are problems like variable results obtained in different soil types and inadequate survival on seeds prior to planting, *Pseudomonas* spp. have great potential in biological control of plant pathogens (Pal *et al.*, 2001). Few examples of PGPR's as biocontrol agents against plant pathogens are enlisted in Table 1.

Table 1 : Biocontrol of pathogens by PGPR

Biocontrol organism	Suppressed pathogen	Crop
Pseudomonas fluorescens	*Erwinia* spp.	Potato
	Erwinia carotovora	Cassava
	Fusarium spp.	Raddish

Contd...

	Thievalovioposis basicola	Tobacco
	Rhizoctonia solani	Peanut
	F. oxysporum f. sp. *ciceri*	Chickpea
	Pythium ultimum	Pea
	Xanthomonas malvacearum	Cotton
	Botrytis cinerea	Petunia
	Macrophomina phaseolina	Chickpea
	G. graminis var. *tritici*	Wheat
	Sarocladium oryzae	Rice
Pseudomonas putida	*Fusarium* spp.	Raddish
	Ewinia carotovora	Potato
	F. oxysporum	Flax
	F. oxysporum f. sp. *lycopersici*	Tomato
	F. solani	Beans
	Xanthomonas campestris	Potato
P. aureofaciens	*G. graminis* var. *tritici*	Wheat
	Phytophthora megasperma	Asperagus
Pseudomonas (Burkholderia) cepacia	*Fusarium* spp.	Tomato
	F. graminearum	Wheat
	F. moniliforme	Maize
	Rhizoctonia solani	Cotton
	Botrytis cinerea	Apple
	Penicillum expansum	Apple
	Sclerotinia sclerotiorum	Sunflower
	Heterodera glycines	Soybean
	Meloidogyne incognita	Soybean
Pseudomonas spp.	*Fusarium oxysporum*	Carnation
	F. moniliforme	Maize
	Pythium ultimum	Sugarbeet
	Rhizoctonia solarni	Cowpea
	Agrobacterium tumefaciens	Grapievine
Bacillus subtilis	*Fusarium roseum*	Corn
Bacillus spp.	*G. graminis var. tritici*	Wheat
	Pythium spp.	Wheat
	Rhizoctonia spp.	Wheat
Rhizobium &	*Macrophomina phaseolina*	Soyabean
Bradyrhizobium spp.	*Rhizoctonia solani*	Mungbean
	Fusarium solani	Sunflower

Mechanisms involved in biological control

Soil pseudomonads involved in plant disease control or suppression produce a number of metabolites. It is important to know which metabolite is involved in disease suppression so as to use the strain efficiently and for further genetic modifications. Biocontrol activity is mediated by production of allelochemicals including siderophores, antibiotics, biocontrol volatiles, lytic enzymes and detoxification enzymes (Haas *et al.*, 2000 ; Glick, 2003 ; Bais *et al.*, 2006).

There are two approaches to know the exact mechanism of disease suppression by pseudomonads. One requires a purified compound of known structure. The disease suppressive ability of the compound is compared with that of the producer strain in a conducive soil. If the compound mimics the disease suppression by the disease controlling strain, it is assumed that the metabolite has a role to play in disease suppression (Kloepper *et al.*, 1980). The second approach involves generation of mutants from the *Pseudomonas* strain, either by random transposon mutagenesis or site directed mutagenesis, having no more production of an extracellular antifungal metabolite. The effectiveness of these mutants in suppressing the disease is compared with that of the wild type strain. The loss of ability to suppress a disease by a mutant defective in production of a particular metabolite indirectly suggests the role of that metabolite in disease control.

The possible mechanisms of disease suppression by pseudomonads as suggested by several workers (O'sullivan and O'Gara, 1992; Dowling and O'Gara, 1994) are as follows :

(i) Competition for Fe^{3+} by producing siderophores

(ii) Producing volatile or diffusible metabolites : antibiotics, HCN, ammonia

(iii) Induction of systemic resistance in plants

(iv) Successful root colonization

The various criteria have been suggested for ideal antagonists for the biological control of various plant diseases. These include (a) genetic stability (b) high consistent efficiency (c) ability to survive under adverse environmental conditions (d) effectiveness against a wide range of pathogens on a variety of fruits and vegetables. (e) amenability for growth on an inexpensive medium in fermentors (f) non-production of secondary metabolites that might be toxic to humans (g) resistance to standard fungicides (h) compatibility with other chemical and physical treatments of the commodity such as heating and waxing. Many biological agents do not perform better in the field due to the complexity and variability of physical, chemical, microbiological and environmental factors in the field and the sensitivity of biocontrol microorganisms to these factors.

Generally, microorganisms isolated from the rhizosphere of a specific crop are better adapted to that crop and may provide better control of disease than organisms, originally isolated from other species (Cook, 1993). The use of combination of multiple antagonistic organisms has been found to provide improved disease control over the use of single organisms (Duffy and Weller, 1995 ; Pierson and Weller, 1994 ; Larkin and Fravel, 1998). Multiple organisms may enhance the level and consistency of control by providing multiple mechanisms of action, a more stable rhizosphere community, and effectiveness over a wide range of environment conditions.

Siderophore production and disease suppression

Though iron is the fourth abundant element on the earth's crust, it is almost unavailable to living organisms due to its insoluble nature. In aerated and oxidized soils iron is present as Ferric (Fe^3) form which is less soluble. Its solubility in water is very less (10^{-18}M), which is too low to support microbial requirements. Since iron is a very essential component in the metabolic processes, its unavailability acts as a limiting factor for growth of all organisms. Rhizosphere is a region where there is a very high competition for nutrients among the heterogeneous microflora present. Under these conditions some microorganisms secrete siderophores, which are iron binding ligands and sequester iron in the ferric form and outgrow other microflora (Neilands and Leong, 1986; Briat, 1992). Using a very specific receptor on the outer membrane, these organisms take up the siderophore-iron complex which other organisms cannot use.

Fungal siderophores generally have a lower affinity than the siderophores of plant growth promoting rhizobacteria (PGPR) (Raaijmakers *et al.*, 1995; Whipps, 1997; Loper and Henkels, 1999), hence, they are better competitors for available iron.

Siderophores produced by fluorescent pseudomonads mediated iron regulated antagonism against a variety of phytopathogens in culture (Kloepper *et al.*, 1980; Duffy *et al.*, 2004) and in soil (Schipper *et al.*, 1987). The fluorescent siderophores termed pyoverdines, pyochelin and pseudobactin represent only one class of sideropores produced by fluorescent *Pseudomonas* spp. (Compant *et al.*, 2005). Pyoverdines produced by *Pseudomonas* spp. may deplete the medium of available iron, thereby inhibiting the growth of an indicator strain, while ferric complexes have no effect. Unlike microbial phytopathogens, plants are not adversely affected by localized depletion of iron caused by PGPR because plants can grow at much lower concentration than microorganisms

(O'Sullivan and O'Gara, 1992). Also certain plants use microbial Fe (III) siderophores such as ferrioxamine B (Crowley *et al.*, 1988), the ferrichromes and rhodotorulic acid (Crowley *et al.*, 1988). The mechanism is that the plants transport Fe from microbial Fe (III) siderophores by reduction to Fe (II) and ligand exchange with phytosiderophores respectively, or by the direct uptake of the Fe (III) siderophores (Wang *et al.*, 1989; Bar-Ness *et al.*, 1992; Visca *et al.*, 1993). Although the concentrations of pyoverdines produced in the rhizosphere by *Pseudomonas* spp. are adequate, a number of factors like host plant, the target phytopathogen, edaphic and other environmental parameters and the affinity of the specific siderophore for iron determine its effectivity as a biocontrol agent. Some examples of rhizobacteria producing siderophores are given in Table 2.

Table 2 : Siderophores produced by plant associated rhizosphere bacteria.

Siderophores	Producing rhizobacteria
Catechols	
i. Agrobactin	*Pseudomonas fluorescens*
ii. Chrysobactin	*Erwinia chrysanthami*
iii. Enterobactin	Entrobacteriaceae
iv. Pyochelin	*Pseudomonas aeruginosa*
v. 2,3 Dihydroxybenzoic acid	*Azotobacter vinelandii*
vi. Azotochein	*A. vinelandii*
vii. Aminohelin	*A. vinelandii*
Hydroxamates	
i. Aerobactin	*Erwinia carotovora*
	Enterobacter cloacae
ii. Ferroxamine E	*Pseudomonas cepacia*
iii. Ferrioxamine E	*Erwinia herbiocola*
iv. Hydroxamate K	*Rhizobium leguminosarum*
v. Nacardamine	*P. stutzeri*
vi. Frankobactin	*Frankia* spp.
vii. Schizokinien	*Bacillus megaterium*
Catechol and Hydroxamate type	*Azotobacter chroococcum*
	Azospirillum lipoferum D2
	Azospirillum lipoferum M
	Pseudomonas spp.
Pyoverdin	*Pseudomonas fluorescens*
	P. aeruginosa

Contd...

	P. putida
	P. syringae
Other types	
i. Rhizobactin	*Rhizobium meliloti*
ii. Citric acid	*Bradyrhizobium japonicum*
iii. Anthranilic acid	*Rhizobium leguminosarum*
iv. Azotobactin	*Azotobacter vinelandii*

Antibiotic mediated suppression

In vitro inhibition of the growth of pathogen by the biocontrol agent is the first evidence of role of antibiotic production in the disease suppression. The approaches to confirm the involvement of an antibiotic in the disease suppression include;

a) Application of the purified compound in the place of the biocontrol agent in *in vivo* conditions and comparing with the efficiency of biocontrol agent, provided the chemical structure of the compound is known.

b) Isolation of the inhibitory compound from the rhizosphere.

c) Isolation of mutants defective in the production of inhibitory compound and comparison of disease control ability of the mutant with that of wild type.

The polyketide antibiotic 2,4-diacetylphloroglucinol (DAPG) has received particular attention as it plays a key role in the ability of introduced *Pseudomonas fluorescens* strains to suppress a broad spectrum of crop diseases (Haas *et al.*, 2000; de Souza *et al.*, 2003; Duffy *et al.*, 2004). Pseudomonads produce a wide range of antibiotics such as phenazine carboxylic acid, pyoluteorin, pyrolnitrin, acetophloroglucinols, salicylic acid, oomycin etc. (Defago, 1993; Raaijmakers *et al.*, 2002; Pal *et al.*, 2000; de Souza *et al.*, 2003)

Table 2 enlists examples of certain antibiotics produced by pseudomonads and their effect on disease suppression. The antibiotics pyoluteorin (Plt), pyrrolnitrin (Prn), Phenazine-1-carboxylic acid (PCA), and 2-4-diacetyl phloroglucinol (Phl) are currently a major focus of research in biological control. The polyketide antibiotic 2,4- diacetylpholoroglucinol (DAPG) has received particular attention as it plays a key role in the ability of introduced *Pseudomonas fluorescens* strains to suppress a broad spectrum of crop diseases.

Conservation of *phl* genes for biosynthesis of DAPG among ecologically and geographically diverse antagonistic pseudomonads further supports the global importance of DAPG production in biocontrol (Keel *et al.*, 1996; Wang *et al.*, 2001). Keel *et al.*, (1996) reported that DAPG producers carry *hcn* genes for producing broad spectrum hydrogen cyanide (HCN) indicating a relationship between the two metabolites.

Genes for Antifungal Toxins

There are number of antifungal toxins produced by biological control agents including pseudomonads. These toxins include pyocyanine, pyoluteorin, pyolnitrin,oomycin A, phnazine1-carboxylic acid and its derivatives etc. Further, literature shows that *Pseudomonas glumae* strain EM 85 produce antifungal antibiotic gene which was found to be located on a 23 kb fragment of chromosomal DNA (Anith *et al.*, 1998).

Burkhead *et al.* (1994) reported that *Pseudomonas cepacia* B 37 W produced pyrrolnitrin antibiotic inhibitory to *F. sambucinum*. *P.fluorescens* F113 lac zy produces an antifungal phloroglucinols which protects sugarbeet against *Pythium* mediated damping-off. The work of Carruthers *et al.* (1995) conclusively implicated the role of antibiotic in disease suppression. *Pseudomonas aureofaciens* PA 147-2 produces an antibiotic (AF++) which inhibits the growth of phytopathogens *in vitro* (Carruthers *et al.*, 1995). They developed Tn5 antibiotic deficient mutant PA109 and tested along with the wild type for their ability to suppress root rot of *Asparagus officinalis*. Seedlings coinoculated with the pathogen and wild type strain showed a significantly reduced level of infection and disease severity compared to seedlings inoculated with the pathogen alone. However, all seedlings treated with Af – mutant PA 109 were diseased (Carruthers *et al.*, 1995).

Enhanced antibiotics pyoluteorin and 2,4-diacetylphloroglucinol production in mutants of *Pseudomonas fluorescens* strain CHAO resulted in corresponding improvement in suppression of *Pythium ultimum* induced diseases of cucumber plants. Introduction of cosmid pME3090 carrying 22kb of CHAO DNA into strain CHAO enhanced antibiotic pyoluteorin (Plt) and 2,4-diacetyl phloroglucinol (Phl) production by 3-5 fold *in vitro* and in an artificial wheat rhizosphere (Mauhofer *et al.*, 1995). The difference between the amounts of plt and phl produced by two strains was greater in presence than in absence of *P. ultimum*. Presence of phytopathogen induces the host plant to produce more nutrients in root exudates which in turn may stimulate *Pseudomonas fluorescens* CHAO/pME 3090 to overproduce the antibiotics Plt and Phl. Differential protection offered by mutants to different plants show that bacterial antibiotic production varies in the rhizosphere of different plants due to difference in root exudates (Maurhofer *et al.*, 1995). The correlation between six to seven fold over production of antibiotic phloroglucinol *in vitro* and in wheat rhizosphere environment with a corresponding enhancement in the biocontrol activity was observed (Bonsall *et al.*, 1997; Maurhofer *et al.*, 1995).

Complementation of specific mutants defective in antifungal property have been employed for identification of gene(s) responsible for antifungal toxin biosynthesis (Carruthers *et al.*, 1995; Bangera and Thomashow 1996; Hill *et al.*, 1997). Genetic analysis show that the production of 2-4 diacetylphloroglucinol by *Pseudomonas aureofaciens* Q2-87 is coded by 4.8 kb DNA fragment and in *P.fluorescens* Q2-87 by a 5 kb DNA segment (Bangera and Thomashow, 1996).

Anith *et al.* (1999) while studying the analysis of mutation affecting antifungal property of a fluorescent *Pseudomonas* sp. during cotton-*Rhizoctonia* interaction observed that the mutant AN 21 generated from the wild type strain by chemical mutagenesis showed no antifungal property. The mutant was unable to produce an antifungal antibiotic and failed to inhibit the fungal growth. *In vivo* experiment also established that antifungal antibiotic was responsible for the disease control ability of the strain.

A number of antifungal metabolites produced by pseudomonads appears to be regulated by a global regulator (gac A) (Gaffney *et al.*, 1993). This leads to difficulty in identification of structural genes for individual antifungal factors by mutagenesis, since regulatory mutants affected in the global regulation of all antifungal factors are commonly found (Hill *et al.*, 1997).

Another strategy to genetically engineered *Pseudomonas* spp. with improved biocontrol activity is through increasing the amount of antibiotic produced by a given strain or by transferring biosynthetic loci into a heterologous strain (Thomasshow and Weller, 1996). In another study, inactivation of pqq genes, responsible for biosynthesis of pyrroloquinoline quinine, a co-factor of different hydrogenases in *Pseudomonas fluorescens* CHAO stimulated the production of antibiotic pyoluteorin (Schnider *et al.*, 1995). It is proposed that the possible mechanism of stimulation of antibiotics production is through shunting metabolites from other metabolic pathways to pyoluteorin biosynthesis pathway.

HCN Production

In the late exponential phase of growth many pseudomonads produce HCN (Askeland and Morrison, 1983). Antagonistic effect of *P. flurescens* strain CHA0 on *Thielaviopsis basicola* in the rhizoplane of tobacco appears to be the result of HCN production (Stutz *et al.*, 1986; Voisard *et al.*, 1989). It has also been proposed that HCN may induce plant defense mechanism.

An HCN deficient mutant *P. fluorescens* derived from HCN producer strain CHAO, is less effective than the wild type in suppressing tobacco black root rot caused by *Thielaviopsis basicola* (Stutz *et al.*, 1986). This deficiency can be restored by a plasmid carrying a 5.0 kb genomic fragment from strain CHAO. The fragment contains three continuous open reading frames designated as ABC which encode cyanide synthase (Voisard *et al.*, 1994). Three genes lem A, gac A and anr have shown to be involved in the regulation of cyanogenesis in *Pseudomonas* spp. The lem and gac A genes globally control antibiotic regulator of arginine deaminase pathway and cyanogenesis (Voisard *et al.*, 1994; Zimmermann *et al.*, 1991). *Pseudomonas putida* strain BK 8661 which produces siderophore(s), antibiotics(s) and low levels of hydrogen cyanide (HCN) suppressed growth of *Septoria tritici* and *Puccinia recondita* f. sp. *tritici in vitro* and on wheat leaves. HCN over producing derivatives, the pleiotropic mutant of strain BK 8661 are deficient in siderophore and antibiotics production. HCN produced by the over producing bacterial strains resulted in a small but significant increase in the suppression of symptoms caused by *S. tritici* and *P. recondita* f. sp. *tritici* of wheat seedling leaves.

Lytic Enzymes

Production of lytic enzymes like chitinases and ß-1,3-glucanases by certain bacteria forms the basis of control of plant-pathogenic fungi in the rhizosphere. Koby *et al.*, (1994) reported the introduction of Tn7 based chiA gene into *P. fluorescens* and the construct could improve the biocontrol activity against *Rhizoctonia solani*.

Enzyme ß 1,3 glucanase produced by *Pseudomonas stutzeri* YPL-1 could suppress *Fusarium solani, Sclerotium rolfsii* and *Pythium ultimum* by 85, 48 and 71% respectively (Fridlender *et al.*, 1993). Production of hydrogen cyanide and ammonia has been reported as a mechanism of disease suppression by few bacteria (Stutz *et al.*, 1986).

Induction of systemic resistance

Plants develop systemic resistance to a variety of pathogens when inoculated previously with a pathogen or with inducing chemical compounds (Ward *et al.*, 1991; Liu *et al.*, 1995; Krishna Murthy and Gnanamanickam, 1997; Rama Moorthy *et al.*, 2001; Ryu *et al.*, 2004a). It is speculated that the aggressive colonization by the beneficial microorganisms, which itself is not a pathogen, may generate some signals that induce plant defence response.

Induced systemic resistance increases the activity of chitinases, β-1, 3-glucanases, protease, lipases, peroxidases and other pathogensis related proteins (PR). These PR proteins show antimicrobial activity *in vitro* . Some biocontrol PGPB strains have been found to produce chitinase, β-1,3-glucanase, protease and lipase that aids pathogen cell lysis (Chet and Inbar 1994). *Pseudomonas stutzeri* secreting extracellular chitinase and laminarinase could digest and lyse *Fusarium solani* mycelia resulting in suppression of root rot disease. Similarly suppression of diseases caused by *Rhizoctonia solani, Sclerotium rolfsii* and *Pythium ultimum* was mediated by β-1,3 glucanase producing strain of *Pseudomonas cepacia* (Fridlender *et al.*, 1993). Transfer of chitinase gene (Chi A) into a PGPR *P. fluorescens* resulted in gene expression, secretion of chitinase enzyme and biocontrol of *Rhizoctonia solani* (Koby *et al.*, 1994). PGPR induced ISR in *Arabidopsis* occurs without induction of transcripts of PRs (Pieterse *et al.*, 1996). Similarly in radish, *Pseudomonas fluorescens* strain WCS 417 mediated ISR is not associated with accumulation of PRs (Harrison *et al.*, 1993).

Ryu *et al.* (2004a) opined that bacterial volatiles induced systemic resistance in *Arabidopsis thaliana.* They further demonstrated that the plant growth promoting bacteria induced systemic acquired resistance (SAR) in *A. thaliana* against cucumber mosaic virus by a salicylic acid and NPRI-independent and jasmonic acid-dependent signaling pathway (Ryu *et al.*, 2004 b).

Role of Ethylene

One mechanism by which plants respond to fungal phytopathogen infection is by synthesizing "stress" ethylene. Ethylene can induce some of the PR proteins e.g. β-1-3-glucanase and chitinase (Abeles *et al.*, 1992). Structural reinforcements of cell wall such as lignification and accumulation of hydroxyproline rich cell wall proteins are also enhanced by ethylene (Boller, 1990). However, reports exist that ethylene is not the signal involved in induction of systematic resistance (SAR) because SAR gene expression in ethylene-insensitive mutant of *Arabidopsis* is similar to the wild type plants (Chang *et al.*, 1993). Majority of plant growth promoting bacteria appear to trigger induced systemic resistance *via* a SA-independent pathway involving jasmonate and ethylene signals (Pieterse *et al.*, 1996).

Root Colonization and Competition

Extensive colonization of root surface by pseudomanads is essential for any efficient biocontrol activity because unsuccessful biocontrol field experiments are often correlated with poor colonization of the root system by the biocontrol bacteria .A positive correlation was observed between population size and biocontrol activity of *Pseudomonas fluorescens* strain 2-79 against *Gaeumannomyces graminis* var. *tritici,* the take - all pathogen of wheat (Bull *et al.*, 1991). The application of an increased population of *Pseudomonas putida* on seed pieces of potato to control the pre-emergent decay caused by *Erwinia carotovora* sub sp. *atroseptica* had a positive effect as reported by Xu and Gross (1986). Similar observations were reported in case of *Pythium* damping off disease of cucumber controlled by *Pseudomonas cepacia* isolate 808 (Elad and Chet, 1987). Mutational studies to elucidate the molecular basis of rhizospheric colonization show that presence of flagella ,the presence of O-antigen of lipopolysaccharides (LPS) and ability to synthesize amino acids are important colonization traits.

The process of root colonization involves migration of the introduced bacterial strain towards the plant root, attachment, distribution along the elongating root and proliferation. Though *Pseudomonas* strains have been shown to migrate towards soybean seeds in soil, no correlation between chemotactic ability and root colonization has been reported (Scher *et al.*, 1988). Transfer of site-specific recombinase gene for a rhizosphere-competent *Pseudomonas fluorescens* into a rhizosphere-incompetent *Pseudomonas* strain enhanced the ability to colonize the root tips (Dekkers *et al.*, 1998).

Attachment of the bacterial cells to the root surface by agglutination has been suggested as an important aspect in the process of successful disease suppressive ability of the introduced strain.

Brand *et al.* (1991) reported competition with respect to nutrient utilization in the rhizosphere as a major attribute required by the biocontrol agent, which is brought about by the root colonization ability and establishment.

Microbial characteristics such as nutrient availability, chemotaxis toward seed or root exudates,growth rate,antibiotic production, siderophore production, cell surface properties like outer membrane proteins, exopolysaccharides, fimbriae and flagella. Tolerance to dry soil and low osmotic potential, tolerance to fungicides or other chemicals play an important role in rhizosphere colonization by PGPR's (Goel *et al.*, 2001). Besides, plant characteristics like plant species, varieties of same plant species and altered plant mutants and root morphology is also equally important in colonization and establishment of rhizobacteria (Dey *et al.*, 2011). Soil and environment factors *viz.*, soil type and texture, soil moisture, pH, temperature, fertility of soils and applied pesticides play a key role in root colonization by PGPRs (Dey *et al.*, 2011; Barriuso *et al.*, 2008).

Genetic studies on antagonistic PGPRs with respect to disease suppression

Disease suppression by pseudomonads being a complicated process, mutant analysis has been used to establish the relative role of each mechanism exhibited by a particular bio-control agent. Two important requirements as far as this approach is concerned are:

(i) The production of mutants lacking ability to produce the antagonistic metabolite which is supposed to be responsible for biocontrol activity, and

(ii) A system which measures the disease suppression quantitatively in terms of plant growth and yield.

Genetic analysis and localization of genes responsible for the production of various antifungal metabolites produced by *Pseudomonas* sp. has been done by many workers (Vincent *et al.*, 1991; Carruthers *et al.*, 1995; Bangera and Thomashow1996; Anith *et al.*, 1998).

The mutants generated should be impaired only in the antifungal metabolite production and otherwise isogenic to the wild type strain. Once the desired mutants are obtained either by chemical mutagenesis or by biological mutagenesis, the next step is to identify the DNA fragment which has been mutated. This could be achieved through a complementation approach, in which a cosmid or plasmid- based genomic library of the wild type strain is constructed and mobilized into the defective mutant and looked for complementation of the lost character.

A genomic library of an organism contains the whole genome of it, cloned to a specific vector molecule in an overlapping fashion of DNA fragments, and established and maintained in a bacterial host, usually *Escherichia coli* Overlapping genomic DNA fragments of the whole genome are generated by partial restriction of the genomic DNA with a specific restriction enzyme followed by size fractionation using ultra centrifugation. The fragments thus obtained are then cloned to a specific purpose vector linearized with the same restriction enzyme used for generation of the genomic fragments. The cloned molecules are then mobilized and stored in *Escherichia coli* and maintained (Old and Primrose, 1989).

Cosmids are usually used as cloning vectors for establishment of gene libraries. They combine the characters of plasmids and bacteriophage vectors. They have the genetic information necessary for *in vitro* packaging of DNA into 1 phage heads in *cis* configuration. They are capable of carrying large fragments of DNA, and can be propagated in *E. coli* and handled as plasmids.

The molecular basis of antifungal toxin production by fluorescent *Pseudomonas* strain EM 85 was studied by Anith *et al.* (1999). The bacterial strain showed *in vitro* inhibition of growth of *Rhizoctonia solani.* NTC mutagenesis of the wild type strain helped in the isolation of an antifungal-toxin-defective mutant AN 21. A genomic library of the wild type strain was constructed by Anith *et al.*, (1998) in the cosmid vector PIAFR 1 and maintained in an *E. coli* background. Complementation analysis with cosmid library resulted in the isolation of a cosmid clone which complemented the defective character in the mutant AN 21. The size of the complementing DNA fragment was found to be 23.5 kb.

Acknowledgement

One of the authors (KVBR Tilak) is grateful to The National Academy of Sciences, India for providing financial support as Senior Scientist. C.Manoharachary is thankful to UGC, New Delhi.

References

Abeles, F.B., Morgan, P.W. and Saltveit, M.E. Jr. (1992). Regulation of ethylene production by internal, environmental and stress factors. In: *Ethylene in Plant Biology*: II ed Academic Press II ed. :56-119.

Angaw, T., Tilak, K.V.B.R. and Saxena, A.K. (2011). Field response of legumes to inoculation with plant growth promoting rhizobacteria. *Biol.Fertil.Soils* (in press).

Anith, K.N., Tilak, K.V.B.R., Khanuja, S.P.S. and Saxena, A.K. (1998). Cloning of genes involved in the antifungal activity of a fluorescent *Pseudomonas* sp. *World J. Microbiol. Biotech.*, 14: 939-941.

Anith, K.N., Tilak, K.V.B.R., Khanuja, S.P.S. and Saxena, A.K. (1999).Molecular basis of antifungal toxin production by fluorescent *Pseudomonas* sp. strain EM 85- A biological control agent. *Curr. Sci.*, 77(5):671-677.

Askeland, R.A. and Morrison, S.M. (1983). Cyanide production by *Pseudomonas fluorescens* and *Pseudomonas aeruginosa*. *Appl. Environ. Microbiol.*, 45 : 1802-1807.

Bais, H.P., Weir, T.L., Perry, L.G., Gilroy, S. and Vivanco, M. (2006). The role of root exudates in rhizosphere interactions with plants and other organisms. *Ann. Rev. Pl. Biol.*, 57: 233-266.

Bangera, M.G. and Thomashow, L.S. (1996). Characterization of a genomic locus required for synthesis of the antibiotic 2,4-diacetylphloroglucinol by the biological agent *Pseudomonas flurescens* Q2-87. *Mol. Plant-Microbe Interact.*, 9 : 83-90.

Barbosa, M.A.G., Michereff, S.J., Mariano, R.L.R. and Maranthao, E. (1995). Biocontrol of *Rhizoctonia solani* in cowpea by seed treatment with fluorescent *Pseudomonas* spp. *Summa-Phytopath.*, 21 : 151-157.

Bar-Ness, E., Hadar, Y., Shanzer, A. and Libman, J. (1992). Iron uptake by plants from microbial siderophores, A study with 70-nitrobenz-2 oxa-1,3-diazole-des-ferrioxamine as fluorescent ferrioxamine B analog. *Pl. Physiol.*, 99 : 1329-1335.

Barriuso, J., Solano, B.R., Lucas, J.A., Lobo, A.P. ,Garcia-Villaraco, A. and Gutierrez Manero, F.J. (2008). Ecology, genetic diversity and screening strategies of plant growth promoting rhizobacteria (PGPR). In: *Plant-bacteria interaction, strategies and techniques to promote plant growth* (Eds.) J. Ahmad, J. Pichtel, and S. Hayat, Weinheim: Wiley. VCH GMBH &Co.:1-13.

Boller, T. (1990). Ethylene and plant-pathogen interactions. *Curr. Top. Pl. Physiol.*, 5: 138-145.

Bonsall, R.F., Weller, DM. and Thomasshow, L.S. (1997). Quantification of 2,4-Diacetylphloroglucinol produced by fluorescent *Pseudomonas* spp. *in vitro* and in the rhizosphere of wheat. *Appl. Environ. Microbiol.*, 63 : 951-955.

Brand, I., Luttenberg, B.J.J., Glandorf, D.C.M., Baker, P.A.M.H., Schippers, B. and de Weger, L.A. (1991). Isolation and characterization of a superior potato root colonizing *Pseudomonas* strain. In : *Plant growth promoting rhizobacteria – Progress and Prospects.* (Eds.) C. Keel *et al.*, IOBC/WPRS Bulletin: 350-354.

Briat, J.F. (1992). Iron assimilation and storage in Prokaryotes. *J. Gen. Microbiol.*, 38:2475-2483.

Bull, C.T., Weller, D.M. and Thomashow, L.S. (1991). Relationship between colonization and suppression of *Gaeumannomyces graminis* var. *tritci* by *Psedomonas fluorescens* strain 2-79. *Phytopath.*, 81:954-959.

Burkhead, K.D., Schisler, D.A. and Slininger, P.J. (1994). Pyrrolnitrin production by biological control agent *Pseudomonas cepacia* B 37w in culture and in colonized wounds of potatoes. *Appl. Environ. Microbiol.*,60 : 2031 - 2039.

Chang, C., Kwok, S.F., Blucker, A.B. and Meyerowitz, E.M. (1993). *Arabidopsis* ethylene - response gene ETR-1 similarity of product to two-component regulators. *Science*, 262 : 539-544.

Carruthers, F.L., Shum-Thomas, T., Conner, A.J. and Mahanty, H.K. (1995). The significance of antibiotic production by *Pseudomonas aureofaciens* PA 147-2 for biological control of *Phytophthora megasperma* root rot of *Asparagus*. *Plant Soil*, 170 : 339-344.

Chet, I. and Inbar, J. (1994). Biological control of fungal pathogens. *Appl. Biochem. Biotechnol.*, 48:37-43.

Compant, S., Duffy, B., Nowak, J., Clement, C. and Essaid Ait, B. (2005). Use of plant growth-promoting bacteria for biocontrol of plant diseases, mechanisms of action and future prospects. *Appl. Environ. Microbiol.*, 71(9):4951-4959.

Cook, R.J. 1993. Making greater use of introduced microorganisms for biological control of plant pathogens. *Ann. Rev. Phytopath.*, 31 : 53-80.

Crowley, D.E., Reid, C.P.P. and Szaniszlo, P.J. (1988), Utilization of microbial siderophores in iron acquisition by oat. *Plant Physiol.*, 87 : 680-695.

Defago, G. (1993). 2,4-diacetylphloroglucinol, a promising compound in biocontrol. *Pl. Path.*, 42: 311-312.

Dekkers, L.C., van der Bij, A.J., Mulders, I.H.M., Phoelich, C.C., Wentwood, R.A.R., Glandorf, D.C.M., Wijffelman, A. and Lugtenberg, B.J.J. (1998). Role of the O-antigen of lipopolysaccharide, and possible roles of growth rate and of NADH ubiquinone oxidoredutase (*nuo*) in competitive tomato root-tip colonization by *Pseudomonas fluorescens* WCS 365. *Mol. Plant-Microbe Interact.*, 11: 763-771.

De Souza, J.T., de Boer, M., de Waard, P., van Beek, T.A. and Raaijmakers, J.M. (2003). Biochemical, genetic and zoosporicidal properties of cyclic lipopeptide surfactants produced by *Pseudomonas fluorescens*. *Appl. Environ. Microbiol.*, 69: 7161-7172.

Dey, R., Pal, K.K., Manoharachary, C. and Tilak, K.V.B.R. (2011). Influence of soil and plant types on diversity of rhizobacteria. *Proc. Nat. Acad.Sci. India* (in press).

Dowling, D.N. and O'Gara, F. (1994). Metabolites of *Pseudomonas* involved in biocontrol of plant diseases. *TIBTECH,* 12 : 133-141.

Duffy, B., Keel, C. and Defago, G. (2004). Potential role of pathogen signaling in multitrophic plant-microbe interactions involved in disease protection. *Appl. Environ. Microbiol.,* 70:1836-1842.

Duffy, B.K. and Weller, D.M. (1996). Biological control of take-all of wheat in the pacific northwest of the USA using hypovirulent *Gaeumanomyces graminis* var.*tritici* and fluorescent pseudomonads. *J. Phytopath.,* 144: 585-590.

Dutta, S. and Podile, A.R. (2010). Plant growth promoting rhizobacteria (PGPR): the bugs to debug the root zone. *Crit. Rev. Microbiol.,* 36(3) : 232-244.

Elad, Y. and Chet, I. (1987). Possible role of competition for nutrition in biocontrol of *Pythium* damping off by bacteria. *Phytopath.,* 77:190-195.

Fridlender, M., Inbar, J. and Chet, I. (1993). Biological control of soil borne plant pathogens by a β-1,3-glucanase producing *Pseudomonas cepacia. Soil Boil. Biochem.,* 25 : 1211-1221.

Gaffney, T., Friedrich, L., Vermooij, B., Negrotto, B., Nye, G., Ukness, S., Ward, E., Kessmann, H. and Ryals, J. (1993). Requirement of salicyclic acid for the induction of systemic acquired resistance. *Science,* 261:754-756.

Glick, B.R. (2003). Plant growth promoting bacteria. In *Molecular Biotechnology –Principles and Applications of Tecombinant DNA* . (Eds.) B.R. Glick and J.J. Pasternak, ASM Press Washington DC USA :436-454.

Goel, A.K., Sindhu, S.S. and Dadarwal, K.R. (2001). Application of plant growth promoting rhizobacteria as inoculants of cereals and legumes. In: *Recent Advances in Biofertilizer Technology* (Eds.) A.K.Yadav, M.R. Motsara and S. Ray Chauduri, Society for Promotion and Utilization of Resources and Technology: 207-256.

Gomes, A.M.A., Peixoto, A.R., Mariano, R.L.R. and Michereff, S.J. (1996). Effect of bean seed treatment with fluorescent *Pseudomonas* spp. on *Rhizoctonia solani* control. *Aruivas-de-Biologiae-Technologia* , 39 : 537-545.

Gupta, A., Gopal, M., Saxena, A.K. and Tilak, K.V.B.R. (2003).Effects of coinoculation of plant growth promoting rhizobacteria and *Bradyrhizobium* sp. *(Vigna)* on growth and yield of greengram (*Vigna radiata* (L.)Wilczek). *Trop. Agric.*(Trinidad), 80(1): 28-35.

Haas, D., Blumer, C. and Keel, C. (2000). Biocontrol activity of fluorescent pseudomonads genetically dissected: Importance of positive feedback regulation. *Curr. Opinion Biotechnol.,* II: 290-297.

Harrison, L.A., Letendre, L., Kovacevih, Pierson, E.A. and Weller, D.M. (1993).Purification of an antibiotic effective against *Gaeumannomyces graminis* var. *tritici* produced by a biocontrol agent *Pseudomonas aureofaciens. Soil Biol. Biochem.* 25:215-221.

Hill, S.E., Hammer, P. and Ligon, J. (1997). The role of antifungal metabolites in biological control of plant disease, In: *Technology Transfer of Plant Biotechnology*. (Ed.) P.M. Greshoff, CRC Press Inc Boca Raton Florida.

Johri, B.N., Rao, C.V.S. and Goel, Reeta (1997). Fluorescent pseudomonads in plant disease management. In: *Biotechnological Approaches in Soil Microorganisms for Sustainable Crop Production*. (Ed.) K.R. Dadarwal, Scientific Publishers, Jodhpur: 193-221.

Keel, C., Weller, D.M., Natsch, A., Defago, G., Cook, R.J. and Thomashow, L.S. (1996). Conversion of the 2,4-diacetylphloroglucinol biosynthesis locus among fluorescent *Pseudomonas* strains from diverse geographic locations. *Appl. Environ .Microbiol.*, 65: 552-563.

Kloepper, J.W., Leong, J., Teintze, M. and Schroth, M.N. (1980). *Pseudomonas* siderophores : A mechanism explaining disease suppression in soils. *Curr. Microbiol.*, 4: 317-330.

Kloepper, J.W., Lifschitz, R. and Zablotowicz, R.M. (1989.). Free-living bacterial inocula for enhancing crop productivity. *Trends Biotechnol.*, 7 : 39-44.

Koby, S., Schickler, H., Chet, I. and Oppenheim, A.B. (1994). The chitinase encoding Tn7-based *chi* gene endows *Pseudomonas fluorescens* with the capacity to control plant pathogens in soil. *Gene*, 147 : 81-83.

Krishna Murthy, K. and Gnanamanickam, S.S. (1997). Biological control of sheath blight of rice : Induction of systemic resistance in rice plant associated *Pseudomonas* spp. *Curr. Sci.*, 72: 331-334.

Kumar, B.S.D. and Dube, H.C. (1992). Seed bacterization with a fluorescent *Pseudomonas* for enhanced plant growth ,yield and disease control. *Soil Biol. Biochem.*, 24: 539-542.

Larkin, R.P. and Fravel,, D.R. (1998).Efficacy of various fungal and bacterial biocontrol organisms for control of *Fusarium* wilt of tomato. *Plant Dis.*, 82: 1022-1028.

Liu, L., Kloepper, J.W. and Tuzan, S. (1995). Induction of systemic resistance in cucumber against *Fusarium* wilt by growth promoting rhizobacteria. *Phytopath.*, 85 : 695-698.

Loper, J.E. and Henkels, M.D. (1999). Utilization of heterologous siderophores enhances ;evels of iron available to *Pseudomonas putida* in the rhizosphere. *Appl. Environ. MIcrobiol.*, 65: 5357-5363.

Lugtenberg, B.J. and Kamilova, F. (2009). Plant growth promoting rhizobacteria. *Ann. Rev. Microbiol.* 63: 541-556.

Maurhofer, M., Keel, C., Haas, D. and Defago, G. (1995). Influence of plant species on disease suppression by *Pseudomonas fluorescens* strain CHAO with enhanced antibiotic production. *Pl. Pathol.*, 44:40-50.

Mazzola, M. and Cook, R.J. (1991). Effects of fungal root pathogens on the population dynamics of biocontrol strains of fluorescent pseudomonads in the wheat rhizosphere. *Appl. Environ. Microbiol.*, 57 : 2171-2178.

Nautiyal, C.S. (1997). Selection of chickpea-rhizosphere-competent *Pseudomonas fluorescens* NBRI 1303 antagonistic to *Fusarium oxysporum* f.sp. *Curr. Microbiol.*, 35 : 52-58.

Nautiyal, C.S., Tilak, K.V.B.R. (2009).Agriculturally important rhizobacteria as bioinoculants for enhancing plant growth and soil health In: *Agriculturally Important Microorganisms* Vol 11 (Eds.) G.G. Khachatourians, D.K. Arora, T.P. Rajendran and A.K. Srivastava. Academic World International, New Delhi, India: 77.

Neilands, J.B. and Leong, S.A. (1986). Siderophores in relation to plant growth and disease. *Ann. Rev. Pl. Physiol.*, 37:187-208.

Old, R.N. and Primorse, S.B. (1989). *Principles of gene manipulation.* Blackwell Scientific Publications, Oxford.

Osburn, R.M., Schroth, M.N., Hancock, J.G. and Hendson, M. (1989). Dynamics of sugar beet seed colonization by *Pythium ultimum* and *Pseudomonas* species : effects on seed root and damping-off. *Phytopath.*, 79 : 709-716.

O'Sulivan, D.J. and O'Gara, F. (1992). Traits of fluorescent *Pseudomonas* spp. involved in suppression of plant root pathogens. *Microbiol. Rev.*, 56 : 662-676.

Pal, K.K., Tilak, K.V.B.R., Saxena, A.K., Dey, R. and Singh, C.S. (2001). Suppression of maize root diseases caused by *Macrophomina phaseolina, Fusarium moniliforme* and *Fusarium graminearum* by plant growth promoting rhizobacteria. *Microbiol. Res.*, 156: 209-223.

Pierson, E.A. and Weller, D.M. (1994). Use of mixtures of fluorescent pseudomonads to suppress Take-all and improve the growth of wheat, *Phytopath.*, 84 : 940-947.

Pieterse, C.M.J., van Wees, M., Hoffland, E., Van Pelt, J.A. and Van Loon, L.C. (1996). Systemic resistance in *Arabidoposis* induced by biocontrol bacteria is independent of salicyclic acid accumulation and pathogensis related gene expression. *Plant Cell*, 84:940-947.

Podile, A.R. and Kishore, G.K. (2006). Plant growth-promoting rhizobacteria. .In: *Plant Associated Bacteria* (Ed.) S.S. Gnanamanickam, Netherlands, Springer: 195-230.

Raaijmakers, J.M., Vandersluis, I., Koster, M., Bakker, P.A.H.M., Weisbeek, P.A. and Schippers, B. (1995). Utilization of heterologous siderophores and rhizosphere competence of fluorescent *Pseudomonas* spp. *Can. J. Microbiol.*, 41:126-135.

Raaijmakers, J.M., Vlami, M. and de Souza, J.T . (2002). Antibiotic production by bacterial biocontrol agents. *Antonie Leeuwenhoek*, 81: 537-547.

Ramamoorthy, V., Viswanathan, R., Raghuchander, T., Prakasam, V. and Samaiyappan, R. (2001). Induction of systemic resistance by plant growth-promoting rhizobacteria in crop plants against pests and diseases. *Crop Prot.*, 20: 1-11.

Ryu, C.M., Farag, M.A., Hu, C.H., Reddy, M.S., Kloepper, J.W. and Pare, P.W. (2004 a.). Bacterial volatiles induce systemic resistance in *Arabidopsis. Pl. Physiol.*, 134:1017-1026.

Ryu, C.M., Murphy, J.F., Mysore, K.S. and Kloepper, J.W. (2004b). Plant growth promoting rhizobacteria systemically protect *Arabidopsis thaliana* against cucumber mosaic virus by a salicylic acid and NPRI-independent and jasmonic acid -dependent signaling pathway. *The Plant J.,* 39:381-392.

Saxena, A.K. , Pal, K.K. and Tilak, K.V.B.R.(2000). Bacterial biocontrol agents and their role in plant disease management. In: *Biocontrol Potential and its Exploitation in Sustainable Agriculture. Vol.1. Crop Diseases, Weeds and Nematodes.* (Eds.) R.S., Upadhyay, K.G. Mukerji and B.P. Chamola, Kluwer Academic/Plenum Publishers New York: 25-37.

Schinder, U., Keel, C., Voisard C., Defago, G. and Hass, D. (1995). Tn5-directed cloning of pqq genes from *Pseudomonas fluorescens* CHAO : Mutational inactivation of the genes results in overproduction of antibiotic pyluteorin. *Appl. Environ. Microbiol.,* 61:3856-3864.

Schippers, B., Bakker, A.W. and Bakker, P.A.H.M. (1987). Interactions of deleterious and beneficial rhizosphere microorganisms and the effect of cropping practices. *Ann. Rev Phytopathol.,* 25: 339-358.

Stutz, E.M., Defago, G., and Kerri, H. (1986). Naturally occurring fluorescent pseudomanads involved in suppression of black root rot of tobacco, *Phytopath.,* 76 : 181-185.

Thomashow, L.S. and Weller, D.M. (1996). Current concepts in the use of introduced bacteria for biological disease control : mechanisms and antifungal metabolites. In: *Plant-Microbe Interactions Vol.1* (Eds.) G. Stacey and N. Keen, Champman and Hall New York :187-235.

Tilak, K.V.B.R., Pal, K.K. and Dey, R. (2010). *Microbes for Sustainable Agriculture.* IK International Publ., New Delhi, India: 200.

Vogeti, S.,Brunda Devi, K., Tilak, K.V.B.R. and Bhadraiah, B. (2009)_. Associative effects of *Glomus fasciculatum* and *Pseudomonas aeruginosa* on nitrogen and protein cpntents of sweet potato (*Ipomoea batatas* L.). *Ind. Phytopath.,* 62(2):258-260.

Voisard, C., Bull, C.T., Keel, C., Laville, J., Maurhofer, M., Schinider, U., Defago, G. and Hass, D. (1994). Biocontrol of root diseases by *Pseudomonas fluorescens* CHAO current concepts and experimental approaches. In: *Molecular Ecology of Rhizosphere Microorganisms.* (Eds.) F.O.Gara, D.N. Dowling and B. Boesten, VCH Weinhaim, Germany: 67-89.

Voisard, C., Keel, C., Hass, D. and Defago, G. (1989). Cyanide production by *Pseudomonas fluorescens* helps suppress black rot of tobacco under gnotobiotic system. *EMBO J.,* 8 : 351-358.

Vincent, M.V., Harrison, L.A., Brackin, J.M., Kovacevic, P.A., Mukerji, P., Weller, D.M. and Pierson, E.A. (1991). Genetic analysis of the antifungal activity of a soilborne *Pseudomonas aeurofaciens* strain. *Appl. Environ. Microbiol.,* 57 : 2928-2934.

Visca, P., Ciervo, A., Sanfilippo, V. and Orsi, N. (1993). Iron regulated salicylate synthesis by *Pseudomonas* spp. *J. Gen. Microbiol.,* 139 : 1995-2001.

Wang, J., Budde, A.D. and Leong, S.A. (1989). Analysis of ferrichrome biosynthesis in the phytopathogenic fungus *Ustilago maydis* : cloning of an ornithine-N-Oxygenase gene. *J. Bacteriol.,* 171:2811-2818.

Wang, C.N., Ramette, A., Panjasamernwong, P., Zala, M., Nalseh, A., Moeme-Loecoz, Y. and Defago, G. (2001) . Cosmopolitian distribution of phlD-containing diacetylphloroglucinol crop associated pseudomonads of worldwide origin. *FEMS Microbiol. Ecol.,* 37:105-116.

Ward, E.R., Uknees, S.J., Williams, S.C., Dincher, S.S., Weiderhold, D.L., Alexander, D.C., Ahl-Goy, P., Metraux, J.P. and Ryals, J.A. (1991). Coordinate gene activity in response to agents that induce systemic acquired resistance. *Plant Cell,* 3: 1085-1094.

Wei, G., Kloepper, J.W. and Tuzun, S. (1996). Induced systemic resistance to cucumber diseases and increased plant growth by plant growth promoting rhizobacteria under field conditions. *Phytopath.,* 86 : 221-224.

Weller, D.M. (1988). Biological control of soil borne plant pathogens in the rhizosphere with bacteria. *Ann. Rev. Phytopath.,* 26 : 379-407.

Weller, DM and Thomashow, LS. (1993). Use of rhizobacteria for biocontrol. *Curr. Opinion Biotechnol.,* 4:306-311.

Whipps, J.M. (1997). Developments in the biological control of soil-borne plant pathogens. *Adv. Bot. Res.,* 26: 1-133.

Wilson, C.L. (1997). Biological control and plant disease – a new paradigm. *Ind. Microbiol. Biotechnol.,* 19 : : 158-159.

Wu, C.H., Bernard, S., Mersen, G.L. and Chen, W. (2009). Developing microbe-plant interactions for application in plant growth promotion and disesease control ,production of useful compounds, remediation and carbon sequestration. *Microbiol. Biotechnol.,* 2: 428-440.

Xu, G.W. and Gross, D.C. (1986). Field evaluation of interactions among fluorescent pseudomonads, *Erwinia carotovora* and potato yields. *Phytopath.,* 76 : 423-430.

□□□

Microbial Diversity and Functions, 2012
© D.J. Bagyaraj, K.V.B.R. Tilak, H.K. Kehri (eds.), pp. 581-601
New India Publishing Agency, New Delhi (India)
E-mail : info@nipabooks.com; Website : www.nipabooks.com

Chapter 28

Soil Microbial Diversity in Natural and Man Made Forage Ecosystems

Pradeep Saxena, Diwakar Bahukhandi and Sharmila Roy

ABSTRACT

The traditional forage production systems and grazing lands are under extreme pressure due to the population growth of both human and livestock, decreasing area and inputs. In the central India, especially the Bundelkhand region, the soil is highly weathered, degraded and have low inherent nutrient, which was further worsened due to use of chemical fertilizers and pesticides, for enhancing crop production to meet the escalating demand of the population.

The fertility of soil is innermost to the sustainability of both natural and managed ecosystems. The natural ecosystems maintain their production through well balanced above and below ground processes, which serve to conserve the nutrients, water and soil organic matter within the systems. But man made systems like fodder cultivation, the attention has been given to the production side of the problem with the focus on crop varieties, agronomical practices, fertilizer applications, and use of pest management strategies. Similarly, soil degradation is seldom attributed to the decline of soil biota and reduction in their activities. However, recent research results have shown that the process which eliminate the beneficial soil flora and fauna communities have resulted in low productivity and have degraded the soil fertility especially in low input systems. As the biological activities of soil are closely linked with the maintenance of soil structure and productivity, sound understanding of various aspects of soil flora and fauna in maintenance of soil fertility in fodder production systems is vital and is discussed in this article.

Keywords: Soil microbes, diversity, forage ecosystem

Introduction

Indian agriculture and livestock is an integral part of rural living. The country accounts for 15 per cent of the total world's livestock population with only 2 per cent of the world's geographical area, which plays an important role in country's endeavor and in meeting the demand for milk, meat, wool, hides and bone manures etc.

The animals not only provide livestock products for human consumption but also a major energy source for draught power in agricultural operations. But the productivity/strength of the livestock is quite low. The major constraint in livestock productivity is shortage in availability and poor quality of feeds and fodder. To compensate for the fall in productivity, the villagers alternatively maintain large herd of livestock. This has further complicated the situation.

Available resources in our country can meet only half of the present requirement. There is a shortage of 40% dry fodder, 25% green fodder and 47% of concentrate in country. However, the extent of deficit varies from state to state. The requirement of dry fodder, green fodder and concentrates are 650.7, 761.5 and 79.4 million tones. Thus livestock is under fed. This can be met only through increasing the land productivity and improved fodder production systems.

The traditional forage production systems and grazing lands are under extreme pressure due to the population growth of both human and livestock, decreasing area and inputs. The area under agriculture has increased by 18.6 per cent and the livestock population by 61.2 per cent. Only 6.9 million hectare (4.4% of the total cropped area) is under fodder cultivation and there is little scope for extension in area, further wastelands are put for fodder crops and forage grasses and legumes. This limited fodder resources of the country are unable to meet the requirement of ever increasing livestock population.

In the central India, especially the Bundelkhand region, the soil is highly weathered, degraded and have low inherent nutrient, which was further worsened due to use of chemical fertilizers and pesticides, for enhancing crop production to meet the escalating demand of the population. Their indiscriminate use has also resulted in a decline in crop yields and the deleterious effects of this improper use of the chemicals extended to water bodies, human bodies and the air we breathe.

The fertility of soil is innermost to the sustainability of both natural and managed ecosystems. The natural ecosystems maintain their production through well balanced above and below ground processes, which serve to conserve the nutrients, water and soil organic matter with in the systems. But man made systems like fodder cultivation, the attention has been given to the production side of the problem with the focus on crop varieties, agronomical practices, fertilizer applications, and use of pest management strategies. Similarly, soil degradation is seldom attributed to the decline of soil biota and reduction in their activities. However, recent research results have shown that the process which eliminate the beneficial soil flora and fauna communities have resulted in low productivity and have degraded the soil fertility especially in low input systems.

Some Terms and Concepts

Biodiversity, biological diversity and ecological diversity

We all are well aware that species are distributed unevenly across the earth's surface but magnitude of difference between the two systems or regions varies and can not compare on written accounts alone. Ecologists have always been intrigued by patterns of species abundance and diversity (Rosenzweig, 1995; Hawkins, 2001). Some questions raised by these patterns, such as the diversity of island assemblages, having latitudinal gradients of diversity, or distribution of commonness and rarity in ecological communities, continue to challenge investigators (Brown, 2001). It is often assumed that the tem "Biological diversity" was coined in the early 1980s. Izsak and Papp (2000) credit it to Lovejoy (1980a). Harper and Hawksworth (1995) note that the term is of older provenance but also date its renaissance to 1980 (Lovejoy, 1980a; 1980b; Norse and McManus, 1980). Gerbilskii and Petrunkevitch (1955) first mention biological diversity in context of interspecific variation in behavior and life history. Since than, the word is used more widely. Norse *et al.* (1986) were the first to describe biological diversity in to three components i.e. genetic diversity (within species diversity), species diversity (number of species) and ecological diversity (diversity of communities).

The tem "Biodiversity" is more recent origin. This contraction of "biological diversity" can be traced to a single event "Biological diversity" means the variability among living organisms from all sources including, inter alia, terrestrial, marine and other aquatic systems and the ecological complexes of which they are part; this includes diversity within species, between species and of ecosystems. Hubbell (2001) defined biodiversity as "synonymous with species richness and relative species abundance in space and time". . Biodiversity and Biological diversity can be used interchangeably. Magurran (2003) defined biological diversity as "the variety and abundance of species in a defined unit of study"

"Ecological diversity" is defined by Pelou (1975) as "the richness and variety... of natural ecological communities". Norse and McManus (1980) treated ecological diversity equivalent to species richness.

In general, Biodiversity is a comparative term to know whether a domain is more diverse than another or diversity has changed over time due to some processes, climatic conditions or geographical regions/boundaries. In addition to this, communities are further identified by the presence of ecological interactions among the constituent species. A community is the arena within which competition, predation, parasitism and mutualism are played out. Indeed, the relationship between resources, species interactions and species abundance is the key to explaining the characteristic patterns of diversity.

Fauth *et al.* (1996) describe the associations of organisms in the context of three overlapping sets delineated by phylogeny, geography and resource (Fig. 1). Set A (Phylogeny) encompasses species of common descent. Communities (set B) are defined as collections of species occurring at a particular place and time. To meet up this operational definition it is essential to identify geographical boundary of the community. This boundary may either be natural or arbitrary. According to Fauth *et al.* (1996), the crucial point is that communities are not delimited either by phylogeny or resource use. Guilds belong to the third set and define groups of organisms that exploit the same resources in a similar manner. The interactions of the sets offer clarification of other widely used terms and concepts. An assemblage consists of phylogenetically related members of the community. Local guild embraces species that share resources and belong to same community. Ensembles include interacting species that share ancestry as well as resource.

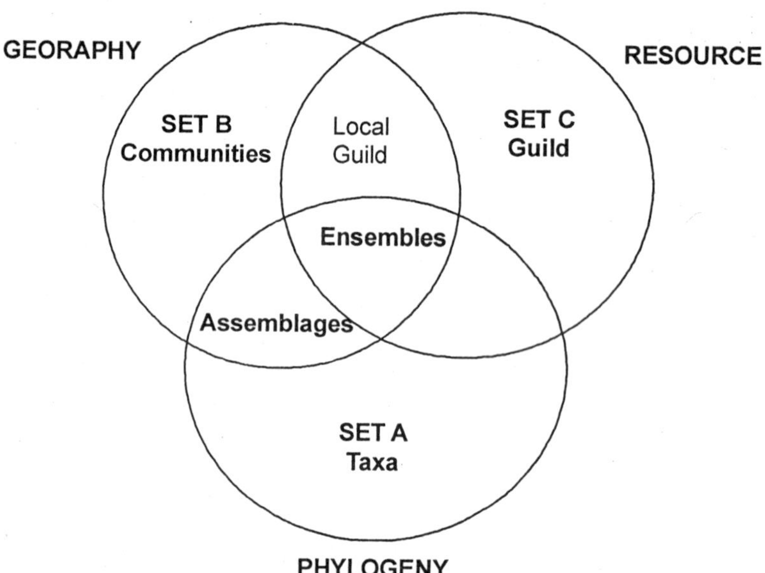

Fig. 1 : Venn diagram to assign groups of organisms to three ecological sets (After Fauth et al., 1996)

What are Ecosystems and Ecosystem Functions ?

The soil-habitat of soil biota

In an attempt to better conceptualize soil systems into biologically relevant regions on the basis of their spatial and temporal heterogeneity, the concept of "sphere of influence" has been proposed (Coleman *et al.*, 1994; Beare *et al.*, 1995). The soil profile may be divided into five major layers *viz.*, (i) Detritosphere - above soil surface, comprise of litter fermentation, humification, which have considerable root mycorrhizal and saprophytic activity and grazing by the soil fauna. (ii) The drilosphere - The soil layer influenced by the activities of earthworm and other insects; (iii) The porosphere - which is the region of water films occupied by bacteria, protozoa, nematodes and of channels between aggregates occupied by micro-arthropods and the aerial hyphae of fungi; (iv) the aggregatosphere - the region where the activity of microbes and fauna is concentrated in the voids between microaggregates and macroaggregates and (v) the rhizosphere - the zone of soil influenced by roots associated by mycorrhizal hyphae and other products. The spores are formed and maintained by biological influence that operates at different spatial and temporal scales. Moreover, each sphere has distinct properties that regulate interactions among organisms and biological properties that they mediate.

The variation in soil profile, resource availability, microclimate, chemical and physical structure influence the soil biota size, composition and distribution (Buscot and Verma, 2005). The profile of an undisturbed mature soil is divisible into a number of strata, which also includes surface organic layers (hemiedaphon) and the underlying mineral soils (enedaphon). The vegetation layer above the soil surface is termed as epigeon. The Hemiedaphon may be further divided into hydrophile, mesophile and xerophile based on the moisture content.

The temporary and permanent members of soil fauna can be distinguished based on their period /stages of life cycle spent in soil. The permanent inhabitants of soil are termed as 'geobionts'. The temporary category belongs to the insects that enter in soil as adults and escape during unfavorable conditions or insects that undergo part of their development, as eggs or larvae, in the soil, termed as 'geophiles'. For example, coleoptera, thysanoptera, heteroptera, diptera etc. The geophiles are further distinguished as inactive and active geophiles. Inactive geophiles include adult insects which seeks the shelter afforded by loose or decaying leaf litter/wood and in surface soil. The temporary geophiles have little or no contribution in soil structure. The active geophiles pass different stages of development in soil and are closely associated with soil. The inactive geophiles are also termed as transients and the active geophiles as periodic and temporary based on their period of presence in soil.

Diverse groups of soil organisms

The modern studies have shown the living organisms can be broadly classified into two radically different kinds; prokaryotes (less complex cell structure) and eukaryotes (organisms with true nucleus), which are further divided into many other forms. Over the times the living things have occupied different ecological niche. These niches are determined by competition between species determined by the availability of nutrients, water availability and soil environment such as temperature, pH, and salt concentration. The microorganisms employ different strategies to survive and prosper in different environments. They may be combative strategies (c-selected) which maximizes occupation and exploitation of resources under non stressed conditions; stress (s-selected) strategies which have involved the development of adoption which allow survival and endurance of continuous stress environment and ruderal (r-selected) strategies characterized by a short span with a high reproductive potential which often enables success in severely disturbed but favorable for a short period. These three strategies can merge to give secondary strategies (C-R, S-R, C-S and CSR) which form part of a continuum with transition zone between them.

The free-living components of soil biota are bacteria, fungi, algae and the fauna. The soil flora and fauna comprise thousands of species with a wide range of ecological strategies. They may be classified either on the basis of body width *viz.*, micro, meso and macro-organisms (Table 1) or on the basis of functional groups *viz.*, mycophagous/herbivores, omnivores and predators, period of soil inhabitance, habitat preference or biological activity.

Table 1 : Classification of soil biota based on body size.

Organism	Size (mm)	Examples
Microflora	< 1	Bacteria, algae, fungi, actinomycetes
Microfauna	< 2	Protozoa, nematode
Mesofauna	2-10	Collembola, acarina
Macrofauna	>10	Earthworms, termites, snails, arachnids

Feeding and locomotion are the other two main activities that divide the organisms in different groups. Based on locomotion the soil animals can be distinguished as burrowing ones from the others that move on the soil surface or through the pore spaces/channels/cavities in soils. In terms of feeding activity the soil animals can be classified into five major groups (Table 2).

Table 2 : Classification of soil biota based on their activity

Organism	Activity
Carnivores	Predator, Animal Parasites
Phytophagous	Above ground green plant material Root systems Wood material
Saprophagous	Caprophages Xylophages Necrophages Detrivores
Symbionts	VAM and Endophytes
Microphytic feeder	Fungal hyphae, spores, algae, lichens, bacteria feeder
Miscellaneous feeders	Omnivores (varies from site to site, available as food)

All living things can be classified into one of the five fundamental kingdoms of life namely, Monera, Protista, Fungi, Plantae, Animalia and are well represented in soil ecosystem.

Bacteria are the most abundant group of microorganism in soil, probably half of the total microbial biomass in the soil. Their population ranges from one hundred thousand to several hundred million per gram of soil, depending on the condition of the soil. Cocci, bacilli, spirilli are the common groups of bacteria. Among these bacilli are common whereas spirilli are very rare. The most common soil bacteria come under the genera *Pseudomonas, Arthrobacter, Clostridium, Achromobacter, Bacillus, Micrococcus, Flavobacterium, Chromobacterium, Sarcina, Aerobacter etc.* All autotrophs use CO_2 of the atmosphere as carbon source, while photoautotrophs derive their energy from the oxidation of simple inorganic matter. Examples of this kind are *Nitrosomonas, Nitobacter etc.* Majority of soil bacteria are heterotrophs. They utilize atmospheric nitrogen *e.g. Rhizobium, Azotobacter, Beijerinckia, Derxia, Azospirillum etc.*

Fungi are eucaryotes and referred as moulds, rusts, mildew, mushroom *etc.* They are organotrophs and are primarily responsible for the decomposition of organic residues, though in plate count bacteria outnumber them. Fungi form hyphae which may be septate or non septate and commonly multinucleate.

Actinomycetes are commonly regarded as an intermediate group between the bacteria and fungi. They produce hyphae and conidia like fungi but the width of the mycelium is around the width of bacterial cell. Actinomycetes do not have chitin and cellulose in the cell wall, a common feature of fungi. They grow slowly in agar media compared to bacteria and form powdery colonies. *Micromonospora, Nocardia, Streptomyces, Streptosporangium and Thermoactinomyces* are some of the important genera of soil actinomycetes.

Protozoa are unicellular microorganisms; vary greatly in morphology and feeding habits. Free-living protozoa in soil feed on dissolved organic substances and on other organisms. Many feed wholly by grazing and predation. The soil ciliates depend primarily on bacteria as food. Amoebae feed on bacteria, other protozoa, and fungal spores. Protozoa occurring in soil are presented by small flagellates (*Oikomonas, Bodo*), amoebae (*Vahlkampfia, Naegleria, Hartmannella, Nuclearia, Dictyoste*), Ciliates (*Colpoda, Leptopharynx, Chilodenella, Cyrtolophosis, Uroleptus, Keronopsis, Gonostomum, Vorticela striata, Spathidium, Bresslana, Tetrahymena rostrata, Podophrya*) and Testacea (*Cyclopyxix, Phryganella, Trinema, Corythion, Centropyrix, Nebela, Difflugia, Arceil, Microchlamys*). The microfauna can directly affect the structure of microbial communities.

Many small soil dwelling invertebrates, such as millipedes, nematodes, centipedes, mites, annelids, spiders, insects etc. have their effect on the physical and chemical properties of soil and detritus and on the structure of microbial communities. Nematodes are one of the oldest existing life forms on the basis of fossil specimens discovered in Scotland. Nematodes feed, move and reproduce like protozoa in water films around soil particles. Following the protozoa, nematodes are the second most dominant group of soil fauna in numbers and biomass. Most of the terrestrial nematodes are <2mm long and 0.05 mm wide.

The most abundant soil arthropods, in terms of number of individuals and species, are the acarina (mites) and collembolans (springtails), these are found in a great variety of habitats *viz.*, tropical zone, desert, temperate zone, arctic region etc. They are also located at high altitudes on the mountains and are early colonizers of embryo soils.The predominant macroarthropods populations were coleoptera and hymenoptera while the dominant micro arthropods were collembolan and acari. The density of gastropods, isopods diplopods and chilopods were relatively low in arable soil because of their sensitivity to agricultural practices. Earthworms, ants, and termites do fragmentation and transportation of organic matter in deeper soil layers. Millipedes, mites, isopods, and collembolans also participate in litter fragmentation. Invertebrates that channel through the soil, or mix litter with soil, affect water intake and percolation rates. Those that fragment litter greatly increase the litter surface area for the activity of soil micro-flora.

The soil biota and ecosystem function

Soil organisms are an integral part of agricultural and forestry ecosystems. They play a significant role in maintaining soil health, ecosystem functions and production. Each organism has a specific role in the complex web of life in the soil. The interacting functions of soil organisms and the effects of human activities in managing land for agriculture and forestry affect soil health and quality. The sustained use of the agriculture and water resources is dependent upon maintaining the health of the living biota that provides critical processes and ecosystem services.

The microorganisms and invertebrate animals play key role in the primary productivity of the ecosystem they inhabit. Their diverse role can be categories into (a) facilitating nutrient acquisition by the vegetation through the mycorrhiza and N- fixing organisms, (b) regulating the flow of nutrients through decomposition, mineralization and immobilization, (c) mediating and breakdown of organic matter, (d) modification of soil structure which influence water availability to the plants, (e) modifying the health of the plants by parasitism and pathogenesis.

Soil flora is the primary consumer in the detritus food web. They act on the organic wastes and convert them either into useful or into innocuous and less harmful substance. In nature fungi, bacteria, and invertebrates interact and influence decomposition and functioning of rhizosphere.

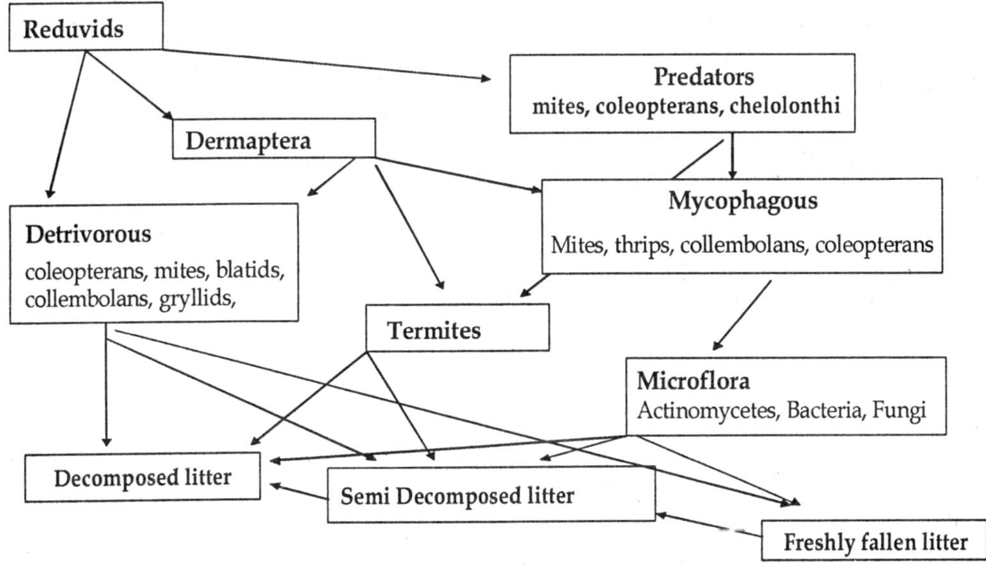

Fig. 2 : A Typical food Chain of forest soil ecosystem (Ananthakrishnan & Thomas, 1993)

Soil fauna governs the distribution, abundance and activity of soil fungi and bacteria. In saprophytic succession, six mechanisms of interaction are important to control fungal distribution and abundance (i) selective grazing of fungi, (ii) dispersal of fungal inoculum to stimulate microbial activity; (iii) direct supply of mineral nutrients in urine and faeces, (iv) stimulation of bacterial activity by faunal activity, (v) compensatory fungal growth due to periodic grazing, and (vi) release of fungi from competitive stasis. In rhizosphere, the mechanisms of interactions are dispersal and selective grazing of flora. Micro and meso fauna carry fungal propagules including root pathogens, to root surface. They also graze on root surface, and they selectively graze saprophytic fungi. It has been shown that dispersal of pathogens to the rhizosphere is less important than preferential grazing on pathogens.

Soil organisms can also be used to reduce or eliminate environmental hazards resulting from accumulations of toxic chemicals or other hazardous wastes. This action is known as bioremediation. A number of soil organisms can be detrimental to plant growth, for example, the build up of nematodes under certain cropping systems. However, they can also protect crops from pest and disease outbreaks through biological control and reduced susceptibility.

Table 3 : Fodder production systems

Fodder	**Cultivated fodder**		**Grassland/pasture**		**Tree based fodder**	
Production Systems	Intensive fodder production systems	Food Fodder System	Cultivated grasslands or Improved pastures	Natural grassland	Silvipasture systems	Horti-pasture Systems
Purpose	Round the year availability of green, nutritious fodder	Availability of green fodder between the major food crops	For quality grass as fodder	To improve the fodder quality, soil restoration	Fodder availability in lean period, conserve soil and moisture	Additional fodder incomes from orchards
Clients	Intensive Dairy Farmers	House hold livestock	Grazing based animal husbandry		Farmers having wastelands, forest departments	

Forage Production Systems

The animal husbandry in Central Indian region is primarily grazing based. The traditional forage production systems are low in productivity and under great strain, due to both land and input constrains. Extensive studies have been carried out for fodder and other biomass production from such systems and it has been concluded that depending on the site conditions increase in fodder and other biomass is several times. In view of this, such systems are finding place in a variety of government and other projects on development and rehabilitation of degraded lands.

The cultivated fodder especially sorghum in kharif and berseem in rabi are popular in this region. The other cultivated fodders are lucerne, cowpea, oat, maize, and bajra. Amongst the perennial cultivated forages napier-pearl millet hybrid, guinea grass and para grass are popular with the farmers. A number of fodder cropping sequences are developed and promoted for the lands exclusively used for fodder production. Augmentation of forage production from cultivated land depend on the high yielding pest and disease resistance varieties. Since the region is semiarid and wherever the prospects of fodder crop cultivation is limited, perennial species based systems like grassland and silvipasture are recommended for degraded forests and other wastelands. These systems are popular with the farmers and with the developmental agencies such as wasteland and forest departments.

Soil Microbial Biodiversity and the Forages

As the biological activities of soil are closely linked with the maintenance of soil structure and productivity, sound understanding of various aspects of soil flora and fauna in maintenance of soil fertility in fodder production systems is vital. In view of this, a number of studies have been made to understand various aspects of soil biota and their role in sustainable productivity.

Soil Biodiversity Characterization and Taxonomy

Soils contain enormous numbers of diverse living organisms assembled in complex and varied communities and are incomprehensible. However, the observed diversity and the mathematical theories guide in exploring the below ground diversity to some extent. Soil biodiversity reflects the variability among living organisms in the soil - ranging from the myriad of invisible microbes, bacteria and fungi to the more familiar macro-fauna such as earthworms and termites. Plant roots can also be considered as soil organisms in view of their symbiotic relationships and interactions with other soil components. These diverse organisms interact with one another and with the various plants and animals in the ecosystem forming a complex web of biological activity. Environmental factors, such as temperature, moisture and acidity, as well as anthropogenic actions, in particular, agricultural and forestry management practices, affect to different extent soil biological communities and their functions.

The works related to soil biodiversity in this region have revealed rich community diversity associated with different fodder production systems. Among the arthropods 148 species were encountered in different systems studied. They belonged to four classes *viz.*, Insecta (17 orders), Arachnida (3 orders), Crustacea (1 order) and Myriapoda (4 orders). Out of all encountered species, 81 per cent were identified. The species that more frequently occurred constituted 28 per cent of the total soil biodiversity while 18 per cent were in the rare category. Remaining 54 per cent were of medium order. The absolute species occurring in all the seasons and in all the systems, including bare land, were 4 *Formica,* 3 Isoptera and 2 cryptostigmata (Scheloribatidae, Epilohmanniidae) species.

Among the micro-flora ten actinomycetes genera (11 species), seven bacterial strains and thirty eight fungal genera (79 species) were isolated and identified from different fodder systems.

a. Effect of Land use/ forage production systems on diversity spectrum

The land management practices alter soil conditions and the soil community of micro, meso and macro-organisms. However, the relationship between specific practices and soil functions is less clear. In general, the structure of soil communities is largely determined by ecosystem characteristics and land use systems. The activity of different species depends on specific management practices as these affect the micro-environment conditions, including temperature, moisture, aeration, pH, pore size, and type of food sources e.g. in dry land situations where earthworms are few, but termites, ants and other invertebrates are abundant, perform similar functions.

Since soil biota community change in response to various land management practices, an attempt was made to study the soil biota associated with various forage production systems. The studies indicated that the perennial systems, especially natural grassland, supported higher species biodiversity compared to the intensive forage cultivation. The forage cropping system that characterized high dominance of only a few species matched with the degraded land situation (Fig.3).

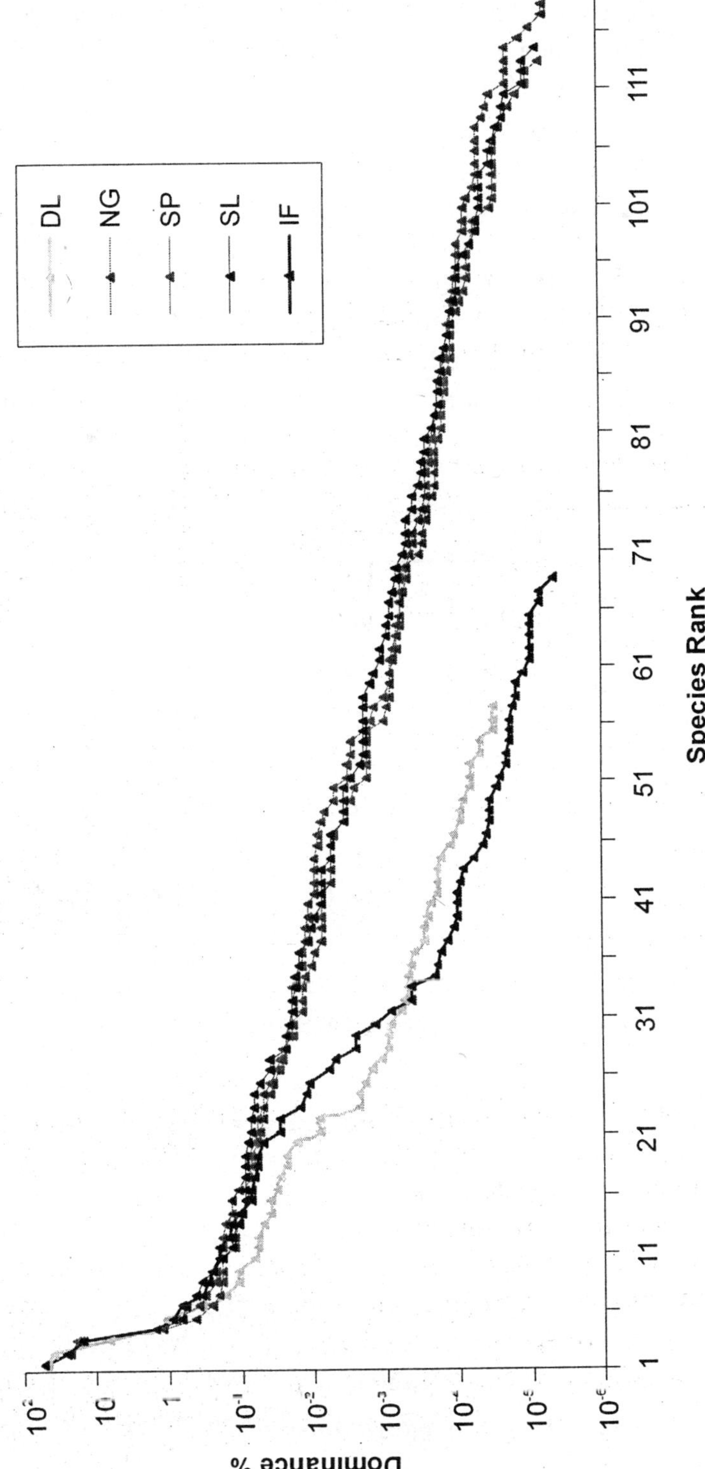

Fig. 3 : Species Diversity of soil biota in different land use systems

DL = Degraded Lands; SP = Sown Pasture; NG = Natural Grassland; SL = Silvipasture; IF = Intensive fodder crop cultivation

The superiority of perennial systems in terms of high species richness and medium abundance is a reflection of the high level of maturity attained, allowing other species to flourish and thus leading towards a sound ecosystem (Fig. 4).

The silvipasture systems being popular fodder production systems promoted for lean period and conservation of degraded lands. A study has been conducted to evaluate the four silvipasture systems (three were single tree species based while fourth was multiple tree species based) for the various soil fertility aspects. It has been found that besides the good fodder yield they support high biodiversity and density of the soil biota. It has been found that the higher above ground plant diversity higher the belowground diversity (Table 4).

Table 4 : Species Diversity of soil biota in different Silvipasture systems

Soil Biota	Silvipasture Systems			
	SL-1	SL-2	SL-3	SL-4
Collembola	10	9	7	13
Cryptostigmata	20	13	19	27
Prostigmata	5	5	3	7
Mesostigmata	5	9	9	12
Astigmata	1	1	-	3
Other arthropods	20	22	20	48
Micro-flora	20	19	20	22

(SL 1= *Acacia tortilis*; SL 2= *Albiziz amara*; SL 3= *Hardwikia binata*; SL 4= Multispecies Silvipasture)

Mycorrhizae are associations between higher plants and fungi and are usually beneficial. Species diversity of VAM .associated with 20-year old fodder trees in five silvopastoral systems *viz., Acacia tortilis* + *Cenchrus ciliaris, Leucaena leucocephala* + *Panicum maximum, Hardwickia binata* + *Sehima nervosum, Albizia amara* + *Cenchrusciliaris, Albizia lebbeck* + *Sehima nervosum* was studied at IGFRI (Table 5). Eight species, belonging to four genera of AM fungi were found to be associated with five species of fodder trees. *Glomus* appeared to be most dominant genus followed by *Gigaspora, Acaulospora* and *Sclerocystis* (Hasan *et al.,* 2004).

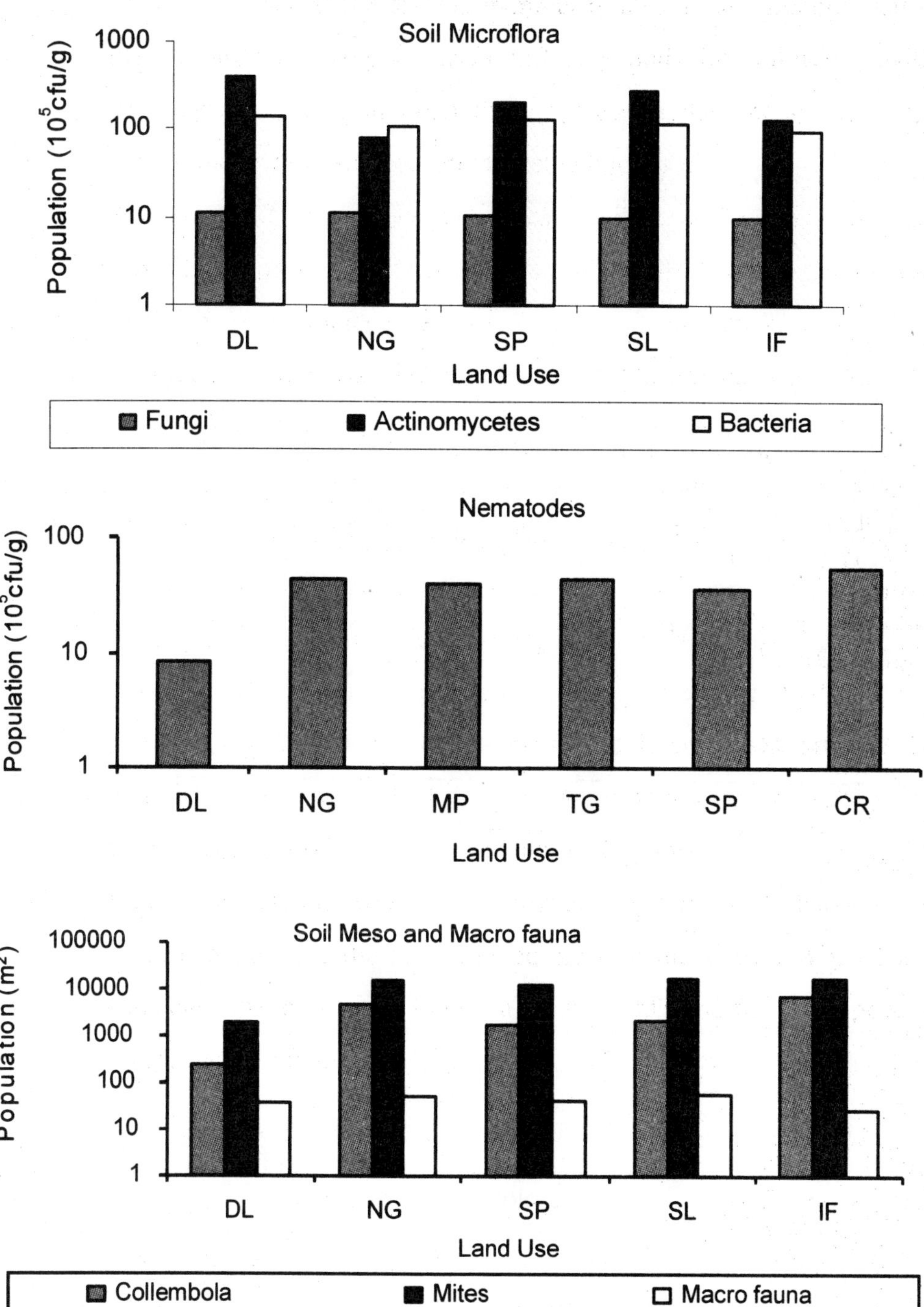

Fig. 4 : Population density of Soil Biota in different land use
DL = Degraded land, NG = Natural Grassland, SP = Sown Pasture, SL = Silvipasture system, IF= Intensive Forage Cultivation

Table 5 : Species diversity of AM fungi associated with fodder trees

Tree Species	Spore Density (/100g soil)	Myconhizal Colonization (%)	AM Species Density
Leucaena leucocephala	200	79	2.63
Albizia amara	170	40	1.05
A. lebbek	197	52	1.72
Hardwickia binata	118	35	1.44
Acacia tortilis	167	43	1.04

b. Forage production systems on abundance of soil biota

Different land use systems changes the soil environment and thus influence the habitat of soil biota and their dynamics (Kim *et al.*, 2000; Smith *et al.* 2003). The Whittaker's β diversity and Index of dissimilarity are the ways to quantify such impacts. The decreasing index value of Whittaker (β_w) represents the increasing similarity, while the dissimilarity value is directly reflecting the result (Table 6).

Table 6: Similarity Indeces (β_w) in various production systems

	DL	NG	SP	SL	IF
DL		0.59	0.60	0.56	0.29
NG	0.11		0.07	0.08	0.44
SP	0.05	0.06		0.06	0.47
SL	0.14	0.03	0.09		0.42
IF	0.11	0.12	0.18	0.15	

Values above the diagonal are for fauna and below the diagonal are for flora (where 0 = similar and 1= dissimilar)

The complex and mature habitats like natural grassland, sown pastures and silvipasture system are closely related than the simplified systems (crop cultivation and degraded lands plot).

c. Seasonal dynamics and effect of environmental factors and soil nutrients

The observation at specific time of year provide information on the population structure in an ecosystem which enables evaluation of the importance of particular functional group during the crop growth and hence provide an understanding of effects of management practices.

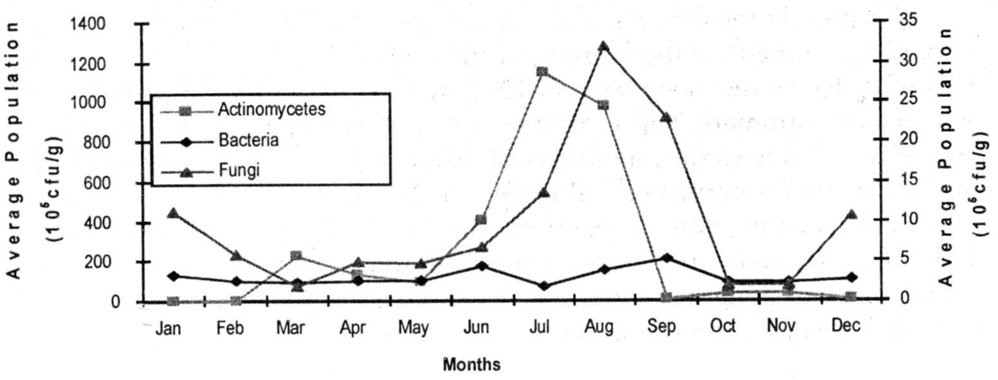

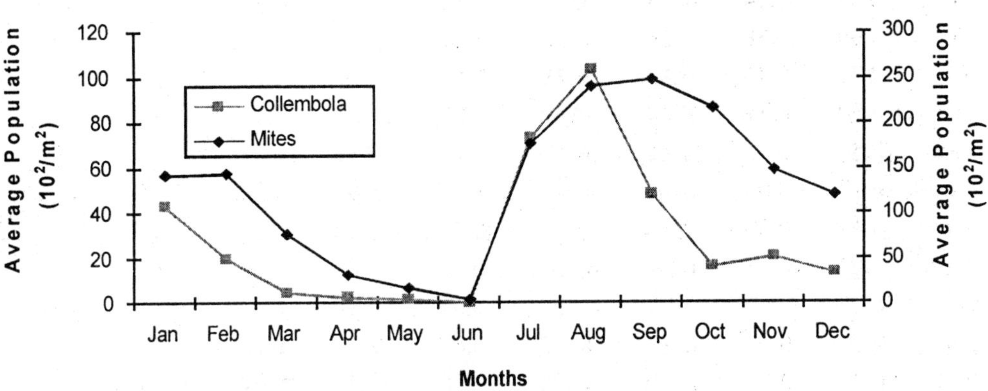

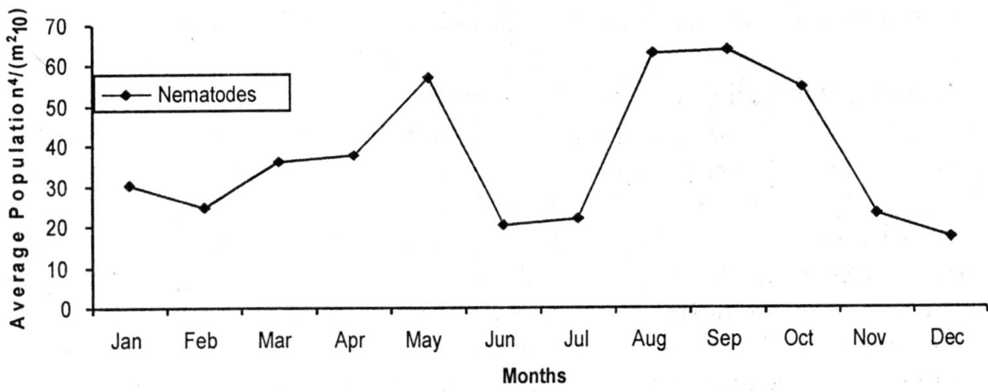

Fig. 6 : Seasonal abundance of various soil biota

The population density and activity patterns of soil meso-fauna were strongly governed by the seasons in this region. The maximum population was recorded in monsoon season. The low population was recorded in peak winters and summers. This indicates that the population build up of different soil biota is influenced by a number of factors climate/microclimate, physical and chemical properties of soil (Table 7). The most influencing parameters were soil moisture, soil temperature, rainfall. The different organisms have been found related with the soil nutrients parameters, but to understand their role in nutrient dynamics intensive research programmes are required for the red soil type of the central India.

Table 7 : Correlation of different climatic and soil nutrient parameters with the soil biota

	C	DM	PM	OM	TM	N	A	B	F
ST	-0.08	**-0.24**	**-0.26**	**-0.31**	**-0.32**	0.18	**0.29**	0.02	-0.02
SM	**0.60**	**0.58**	**0.58**	0.07	**0.54**	**0.42**	0.06	0.09	**0.39**
RH	**0.31**	**0.45**	**0.49**	**0.25**	**0.48**	0.20	-0.05	0.15	**0.46**
CO_2	**0.50**	**0.66**	**0.73**	**0.42**	**0.73**	**0.61**	0.18	0.17	**0.68**
Rain	**0.37**	**0.41**	**0.41**	**0.33**	**0.47**	**0.34**	0.31	**0.36**	**0.72**
Av N	0.07	0.04	0.08	-0.16	-0.01	**0.32**	0.07	0.12	0.08
Av P	0.18	**0.31**	**0.36**	0.01	**0.28**	**0.42**	-0.20	-0.20	-0.13
Av K	-0.17	**-0.27**	**-0.26**	**-0.39**	**-0.36**	0.03	**0.36**	0.11	-0.01
OC	-0.05	-0.01	0.11	-0.20	-0.06	0.13	0.12	0.16	0.01
pH	0.07	0.01	-0.02	**-0.27**	-0.07	0.11	0.08	0.05	0.19
EC	-0.05	0.12	-0.10	**-0.24**	0.00	-0.15	**0.29**	-0.10	0.11

R (5%) = 0.23; r (1%) = 0.30
(ST = Soil Temperature; SM = Soil Moisture; RH = Relative Humidity; Rain = Rain Fall; Av. N,P,K = Available Nitrogen, Phosphorus, Potash; OC = Organic Carbon; EC = Electrical Conductivity ; C = Collembola; DM = Fungivorous/Detrivorous Mites; PM = Predator Mites; OM = Other Mites; TM = Total Mites; N = Nematodes; A = Actinomycetes; B = bacteria; F = Fungi)

d. Litter decomposition and soil biodiversity

The decay/decomposition can be broadly defined as the decrease in the mass of litter. The reduction in mass results from (a) leaching of soluble minerals (b) catabolism or oxidation of organic matter (c) physical breakdown (i.e. comminution) of the substrate. As a result of comminution, leaching and oxidation rates are often greatly enhanced. The decay of plant litter, in nature occurs with the active involvement of soil microarthropods and microorganisms besides various abiotic environmental factors (Buscot and Verma, 2005). The impact and role of litter inhabiting arthropods in the process of litter decay is also well documented.

The litter mass loss of respective systems in different periods of year showed a positive trend with the population build up of soil biota, suggesting their potential role in maintaining soil nutrient dynamics (Fig. 7). In all the systems of study, increase in litter mass loss was accompanied by a significant increase in the number of collembola and acarines. The collembola were the early colonizers and their population was very high at the initial three months of decomposition in all the litter types. Among the microflora the fungi population was positively and significantly correlated with the litter mass loss.

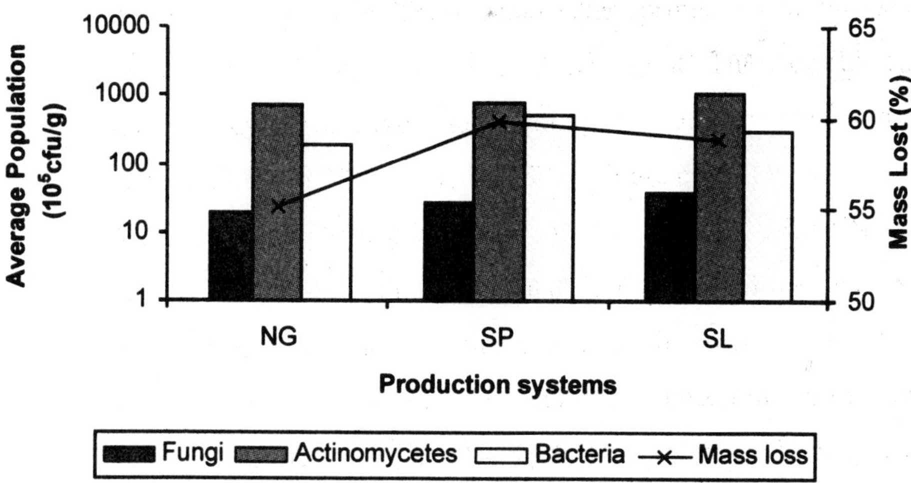

Fig. 7 : Litter mass loss and soil micro flora and fauna in different systems

Future research strategy

The management of soil fertility depends on the mechanistic understanding of the biological process regulating organic matter dynamics and soil physical structure modification. The management practices must be implemented at the ecosystem level by integrating soil biological process with the human decision-making relating to allocation of resources between all components of ecosystem. The future research and developmental programmes need to be oriented in a way to integrate process research and system research. The research approach should be directed towards (i) adoption of an integrated approach involving ecology and agriculture in land management and (ii) intensification of process level studies.

The first step would be to identify the species involved in biological process, wherever possible. Secondly, in order to achieve the imperative resolution necessary to describe decomposition and nutrient cycling, sufficient life history and trophic information must be obtained for these organisms. Thirdly, the interactions of the soil organisms must be explored.

Greater emphasis is required on long-term studies instead of short-term experimentation for validation of various hypotheses. Similarly, designing of agricultural systems/techniques that either prevent destruction of beneficial soil biota or provide favorable environmental condition to facilitate their activities should receive adequate emphasis. Finally, careful integration of local knowledge of soil organism in relation to fertility in designing agricultural practices with a view to improve its technical feasibility, economic profitability and adoptability by farmers is to developed as a package.

The soil biological research needs to be emphasized on the role of soil flora and fauna regulating nutrient release for particular SOM pool and improving soil physical structure. The synchronization of availability of nutrients through SOM with plant growth also needs to be study in depth. The amalgamation of traditional knowledge with scientific information may provide the basic framework for soil management practice.

References

Anantha Krishnan, T.N. and Thomas, S.K. (1993). Chemical ecology of two heterogenous forest litter ecosystems and related trophic interactions of insect communities. *Internat. J. Ecol. Environ. Sci.*, 19: 143–60.

Beare, M.H., Coleman, D.C., Crossley, D.A. Jr., Hendrix, P.F. and Odum, E.P. (1995). A hierarchieal approach to evaluating the significance of soil biodiversity to biochemical cycling In: *The significance and regulation of soil Biodiversity*. (Eds) H. P. Collins, G. P. Robertson and M. J. Klug, Kulwar Amesterdum: 5-22.

Buscot, F. and Verma, A. (2005). *Microorganisms in Soils: Roles in genesis and functions.* Springer-Verlag, Berlin Heidelberg, Germany. 419pp.

Brown, J.H. (2001). Towards the general theory of biodiversity. *Evolution*, 55: 2137-2138.

Coleman, D.C. and Crossley, D.A. (1996). *Fundamentals of soil ecology*. San Diego, Academic Press.

Fauth J.E., Bernardo, J., Camara, M., Resetarits, W.J., Van Buskirk, J and McCollim S.A. (1996). Simplifying the jargon of community ecology: a conceptual approach. *Am. Nat.*, 147: 282-286.

Gerbilskii, N.L. and Petrunkevitch, A. (1955). Intraspecific biological groups of acipenserines and their reproduction in low regions of rivers with biological flow. *Systematic Zool.*, 4: 86-92.

Harper J.L. and Hawksworth, D.L. (1995). Preface. In: *Biodiversity: Measurement and estimation*. (Ed.) D.L. Hawksworth): 5-12. London: Chapman & Hall.

Hasan, N., Suresh, G. and Khan, T.A. Mychorrhizal Diversity of fodder trees under silvipastoral systems. *Grassland Fodder News* 10(4): 6.

Hawkins B.A. (2001). Ecology's oldest pattern. *Trends Ecol. Evol.*, 16: 470.

Hubbell, S.P. (2001). The unified neutral theory of biodiversity and biogeography. In Princeton, NJ : Princeton University Press, Princeton NJ.

Izsak, J. and Papp, L. (2000). Measuring β diversity indices and measures of biodiversity. *Eco. Modelling*, 130: 151-156.

Lovejoy, T.E. (1980a). Changes in biological diversity. In: *The Global 2000. Report to the President* Vol. 2. The Technical Report (Ed. G. O. Barney): 327-332. Harmondsworth, UK: Penguin.

Lovejoy, T.E. (1980b). Forward. In: *Conservation Biology: an evolutionary-ecological perspective* (Eds.) M.E. Soule and B.A. Wilcox):v-ix. Sinauer, Sunderland, MA.

Magurran, A.E. (2003). *Measuring Biological Diversity*. Blackwell Science Ltd. Malden MA.

Norse E.A., Rosenberg, K.L. and Wilcove, D.S. (1986). *Conserving biological diversity in our national forests*. Washington, DC: The Wilderness Society.

Norse, E.A. and McManus, R.E. (1980). Ecology and living resources biological diversity. In: *Environmental quality 1980: The eleventh annual report of the council on environmental quality*. Washington, DC: Council of Environmenal Quality.

Pelou, E.C. (1975). *Ecological diversity*. Wiley Interscience, NewYork.

Rosenzweig, M.L. (1995). *Species Diversity in Space and Time*. Cambridge, UK, Cambridge University Press.

□□□

Microbial Diversity and Functions, 2012
© D.J. Bagyaraj, K.V.B.R. Tilak, H.K. Kehri (eds.), pp. 603-639
New India Publishing Agency, New Delhi (India)
E-mail : info@nipabooks.com; Website : www.nipabooks.com

Chapter 29

Phycotoxins

Richa Tandon, V.K. Dwivedi and G.L. Tiwari

ABSTRACT

Photosynthetic blooms of toxin producing marine algae are often referred to as "red tides". Proliferations of freshwater toxin-producing cyanobacteria are simply called "cyanobacterial blooms" or "toxic algal blooms." Biotoxins of cyanobacteria are water-soluble and heat stable and they are released upon aging or lysis of the cells. The primary types of cyanobacterial biotoxins involves three types of toxins: hepatotoxins (microcystin; cylindrospermopsin and nodularin), which are taken up by the liver and cause weakness and anorexia (loss of appetite); neurotoxins (usually anatoxin and saxitoxin), which effect the nervous system and dermatotoxins (aplysiatoxin and lyngbyatoxin, LPS), which cause skin and mucous irritations upon contact. Unlike freshwater toxic algal blooms, marine algal toxins become a problem primarily because they may concentrate in shellfish and fish that are subsequently eaten by humans, causing syndromes known as paralytic shellfish poisoning/PSP (paralysis due to saxitoxin and gonyaulatoxins), diarrhetic shellfish poisoning/DSP (diarrhea due to okadaic acid), amnesic shellfish poisoning /ASP(memory disturbances due to domoic acid), neurotoxic shellfish poisoning/NSP (paresthesia due to brevetoxin; also bronchial spasms). Coral reef fishes are the vectors of another algal poisoning, the ciguatera fish poisoning/CFP (paresthesia, warm-cold sense inversion etc., due to ciguatoxins and maitotoxin).

Among chlorophyta Caulerpa (caulerpicin and caulerpin), Chaetomorpha and Ulva spp. (hemolysins) are also known to represent as toxin producers. Prymnesium parvum (prymnesin) and Ochromonas spp. are the member of chrysophyceae which have been shown to contain extractable toxins. Chronic toxicity of carrageenans have been recorded from the red algae (Chondrus, Gigartina, Hypnea, Rhodymenia etc.).

Keywords: Biotoxins, hepatotoxins, neurotoxins, dermatotoxins, saxitoxin, brevetoxin and ciguatoxin.

Introduction

The present review deals with contemporary knowledge on phycotoxins (algal toxins) of fresh-water and marine habitats.The term algae mostly refers to microscopically small, unicellular or unbranched filaments organisms, some of which form colonies and thus reach sizes visible to the naked eye as minute green particles. These organisms are usually finely dispersed throughout the water bodies and may cause considerable turbidity if they attain high densities. In marine habitats, dinoflagellates are said to be the most important toxin producers in its devastation of marine waters. However, in the freshwater environment, toxins have been identified from certain cyanobacteria. Most importantly for human health, cyanotoxins, which are produced within algal cell walls and released after cell death, present grave immunological challenges.Toxins are lethal or debilitating, even in small amounts. Vulnerable populations, like children, the elderly, or those with lowered immune systems are particularly susceptible to diseases caused by algal toxins. The origin of marine algal toxins is mostly from unicellular algae. In response to favorable conditions they, may proliferate and/or aggregate to form dense concentrations of cells or "blooms." In many cases, toxic species are normally present in low concentrations, with no environmental or human health impacts; toxicity in general depends on their presence in high cell concentrations. Phytoplankton species that produce toxins, currently included under the broad term harmful algal blooms (HABs), previously they were called red tides. The effects of algal blooms vary widely. Some algae are toxic only at very high densities, while others can be toxic at very low densities (a few cells per liter). Some blooms discolor the water (thus the terms "red tide" and "brown tide"), while others are almost undetectable with casual observation (Shumway, 1990). World-wide, diatoms provide a major food resource for zooplankton and also produce atmospheric oxygen. Some marine diatoms can produce a toxin which can accumulate in shellfish and poison humans. However, freshwater diatoms do not produce such toxins.

Wide varieties of toxins are produced by five phyla of algae: Cyanophyta (cyanobacteria), Chlorophyta (green algae), Chrysophyta (diatoms, yellow-green and golden algae), Pyrrhophyta (dinoflagellates), and Rhodophyta (red algae). The types of molecules involved are diverse, going from simple ammonia to complicated polypeptides and polysaccharides. The physiological effects are also varied, ranging from the acute toxicity of paralytic shellfish poison of *Gonyaulax*, leading to death in a short period of time, to the chronic toxicity of carrageenans from red algae, which induce carcinogenic and ulcerative tissue changes over long periods of time. Perhaps the only link among the wide variety of toxins is that each is produced by some form of algae. Filter-feeding shellfish, zooplankton, and herbivorous fishes ingest these algae and act as vectors to humans either directly (e.g., shellfish) or through further food web transfer of sequestered toxin to higher trophic levels. Consumption of seafood contaminated with algal toxins *viz.*, saxitoxins, okadaic acids, domoic acids, brevetoxins etc., results in five seafood poisoning syndromes known as paralytic shellfish poisoning (PSP), diarrheic shellfish poisoning (DSP), amnesic shellfish poisoning (ASP), neurotoxic shellfish poisoning (NSP) and ciguatera fish poisoning (CFP). All are caused by toxins synthesized by dinoflagellates, except for amnesic shellfish poisoning (ASP), which is produced by diatoms of the genus *Pseudonitzschia.* Ciguatera fish poisoning (CFP) is caused by benthic dinoflagellate toxins in coral reef communities.

The following text is divided into two parts. The first part deals with cyanobacterial toxins which are mostly reported from fresh water habitats and a few from marine habitats. The second part deals with algal toxins of marine habitats and it includes mainly green algae, golden algae, diatoms, dinoflagaellates, red algae and shellfish poisioning.

Cyanobacterial toxins

Several cyanobacterial species commonly forming mass occurrences (blooms) in fresh, brackish and marine waters produce toxins. They have caused death of animals all over the world and may be a health hazard for humans. The toxins are classified, according to the target of their toxic action, as hepatotoxins, neurotoxins and dermatotoxins. In fresh and brackish waters, the most frequently found hepatotoxins are cyclic hepta- and pentapeptides, microcystins and nodularins, respectively. In tropical waters, an alkaloid toxin cylindrospermopsin commonly occurs. Four types of neurotoxins are known: anatoxin-a, homoanatoxin-a, anatoxin-a(s) and paralytic shellfish poisions (PSPs). In marine waters, neurotoxins and dermatotoxins have been reported. Most species producing these toxins are planktonic, but anatoxin-a is produced by benthic *Oscillatoria* and PSPs and dermatotoxins by *Lyngbya.*

The number of cyanobacterial toxins isolated and characterized over the last decade had increased rapidly, but many new compounds are probably still to be discovered. Cyanobacteria also serve as a rich source of novel bioactive metabolites, including many cytotoxic, antifungal and antiviral compounds (Patterson *et al.,* 1994) Table-1.

Neurotoxins : Those which interfere with the functioning of neuromuscular system of the animals and often result in paralysis and very fast death are 'neurotoxins'. Neurotoxins are alkaloids and they all block the neurotransmission: anatoxin-a by mimicking the effect of acetylcholine, anatoxin-a(s) as an anticholinesterase and PSPs by blocking the sodium channel.

Table 1 : Cyanobacterial Toxins

Taxa	Toxin	Structure	Symptoms
Anabaena flos-aquae, A. planktonica, Oscillatoria spp., *Aphanizomenon* spp., *Anabaena lemmermanni*	Neurotoxin Anatoxin-a	Alkaloids (secondary amines)	Primary target: Nerve synapse; Blocking of post-synaptic depolarization; mimicking the effect of acetylcholine
Oscillatoria formosa	Neurotoxin Homo-anatoxin-a	Secondary amines	Depolarizing neuromuscular blocking agent
Anabaena flos-aquae	Neurotoxin Anatoxin a(s) (unique organophosphate)	Phosphate esters of a cyclic N-hydroxyguanine	Primary target: Nerve synapse; Blocking of acetylcholinesterase
Anabaena circinalis Aphanizomenon flos-aquae, Lyngbya wollei	Neurotoxin Saxitoxin(SXT) Paralytic shellfish poisoning (PSP)	Carbamate alkaloids	Primary target: Nerve axons; Sodium channel blocking agent-acute poisioning result in death by paralysis and respiratory failure
Cylindrospermopsis raciborskii, Umezakia natans, Aphanizomenon ovalisporum, Anabaena bergii	Hepatotoxin Cylindrospermopsin	Alkaloids	Cytotoxic; Liver, kidney and other organ damage; genotoxic
Microcystis aeruginosa, M. viridis, Oscillatoria agardhii, Nostoc spp., *Planktothrix agardhii, P. rubescens, Anabaenopsis milleri, Hapalosiphon hibernicus*	Hepatotoxin Microcystins	Cycic heptapeptide	Liver damage; inhibition of protein phosphatase activity, Tumour promotion in animals
Nodularia spumigena	Hepatotoxin Nodularin	Cyclic pentapeptide	Hemorrhaging of the Liver; Tumour promotion in animals
Microcystis flos-aquae, M. aeruginosa, Anabaena circinalis, Cylindrospermopsis raciborskii, Phormidium spp.	Endotoxin Lipopolysaccharides (LPS)	Alkaloid	Gastro-intestinal disorders, skin, eye irritation, respiratory symptoms, affects any exposed tissue
Lyngbya majuscula (Marine)	Dermatotoxins Lyngbyatoxin-a, Debromoaplysiatoxin Aplysiatoxin	Alkyl phenols	Acute dermatitis, Protein Kinase C activators; inflammatory activity
Schizothrix calcicola (Marine), *Oscillatoria nigroviridis* (Marine)	Aplysiatoxin	Alkyl phenols	Inflammatory activity
Trichodesmium (Marine)	Neurotoxic factor		

Paralytic shellfish poisons (PSPs)/ Saxitoxins : Of the cyanobacterial neurotoxins, the most widely known are a group of carbamate toxins known as saxitoxins. These are known for paralytic shellfish poisions (PSP), and a group of 21 structurally-related saxitoxins variants is currently recognized. These toxins known to be produced by marine dinoflagellates (red tide organisms), are concentrated in mussels, clams and oysters, and to have caused human poisionings (Anderson, 1994). They were first found in *Aphanizomenon flos-aquae* blooms (Gentile & Maloney, 1969; Jackim & Gentile, 1968; Ikawa *et al.*, 1982; Mahmood & Carmichael, 1986). More recently, PSPs have been known to be produced by *Anabaena circinalis* (Humpage *et al.*, 1994). Their major mode of action regarding toxicity in vertebrates is *via* the blockage of voltage-gated sodium channels, resulting in paralysis and in acute cases, death. In North America the benthic freshwater cyanobacterium *Lyngbya wollei* produced PSPs (Carmichael *et al.*, 1995). In marine *Trichodesmium*, a neurotoxic factor was reported that was not anatoxin-a or anatoxin-a(s).There are also some indications that ciguatera-type toxins typically found in tropical marine fish may be of cyanobacterial origin (Hahn & Capra, 1992; Endean *et al.*, 1993).

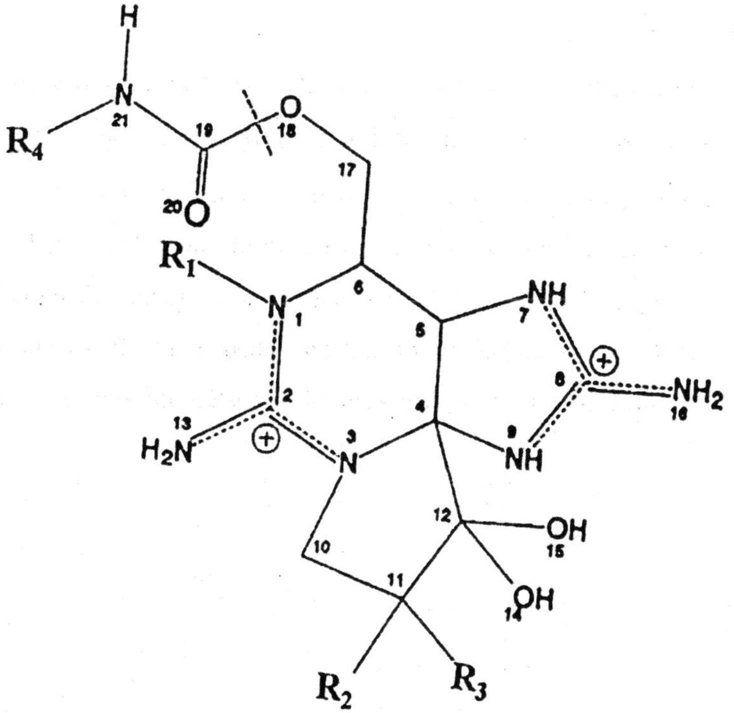

Structure of the PSPs found in dinoflagellates and cyanobacteria. Toxins detected in the cyanobacteria *Aphanizomenon flos-aquae* and *Anabaena circinalis* are indicated (+).

Anatoxins : The name of toxins is often based on source of algal genus known to produce the particular toxins. The name anatoxin is based on the genus *Anabaena*. Anatoxins are a group of low molecular weight alkaloids produced only by cyanobacteria. Anatoxin-a (MW 165) is first described in the fresh-water cyanobacteria *Anabaena flos-aquae* from Canada (Carmichael *et al.*, 1975; Devlin *et al.*, 1977) and later in Finnish strains of *Anabaena* spp. (*flos-aquae*/*lemmermannii*/*circinalis* group), *Oscillatoria* sp., *Aphanizomenon* sp. and *Cylindrospermum* sp. (Sivonen *et al.*, 1989) and in benthic *Oscillatoria* (Edwards *et al.*, 1992). This toxin is a secondary amine and its molecular mode of toxic activity is as a post-synaptic acetylcholine antagonist, and is thus able to bind to and open the ion channel. It cannot, however, be deactivated by acetylcholinesterase, and so the ion channel stays open. The muscle cell continues to contract until it fails from exhaustion, resulting in paralysis, asphysixiation and death.

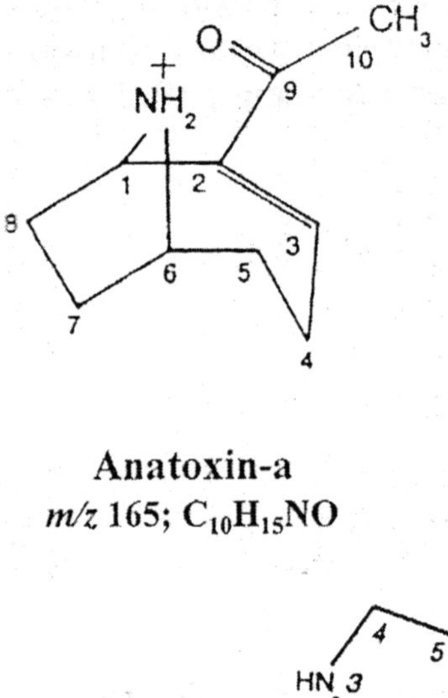

Anatoxin-a
m/z 165; $C_{10}H_{15}NO$

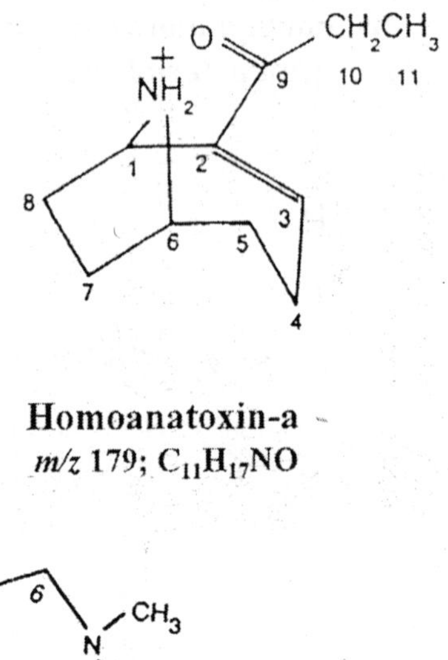

Homoanatoxin-a
m/z 179; $C_{11}H_{17}NO$

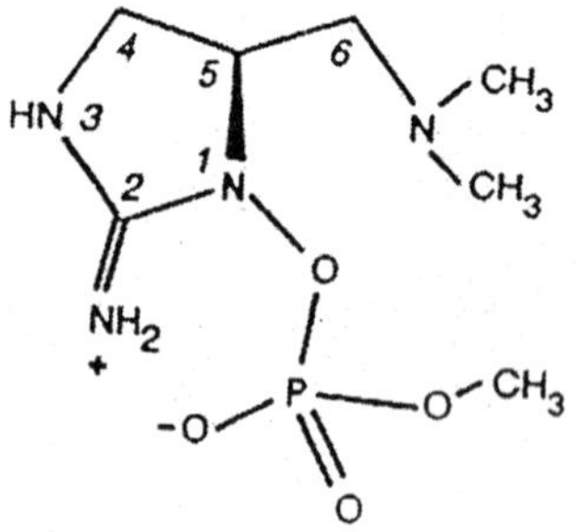

Anatoxin -a(S)
m/z 252; $C_7H_{17}N_4O_4P$

The chemical structure of anatoxin-a (2-acetyl-9-azabicyclo (4-2-1) non-2-ene), homoanatoxin-a and anatoxin-a(S), (N-hydroxyl-guanide methyl phosphate ester), cyanobacterial neurotoxins found in fresh waters.

Five naturally occurring structural variants, including homoanatoxin-a (MW 179) are known from a strain of *Oscillatoria formosa* Bory from Norway (Skulberg *et al.*, 1992), with some of them thought to be degradation products of the parent toxin. Anatoxin-a (s) is a unique phosphate ester of a cyclic N-hydroxyguanine (MW 252) structure produced by Canadian *Anabaena flos-aquae* strain NRC 525-17. This toxin is similar in its action to synthetic organophosphate pesticides such as parathion and malathion. Anatoxin-a(s) is the only natural organophosphate known, although similar in name to anatoxin-a, is quite different in terms of structure and toxicity. The suffix (s) stands for salivation and also indicates a comparison with synthetic organophosphorus pesticides. Its one of the symptoms of intoxications is hypersalivation. Since it is more water soluble than synthetic organophosphates, with consequently less tendency to bioaccumulate and stay in fat cells and cell membranes, it may yield a less nefarious pesticide than its synthetic brethren. It acts as an irreversible peripheral anticholinesterase (Mahmood *et al.*, 1988; Cook *et al.*, 1989). So far, no other variants of this neurotoxin have been discovered.

Hepatotoxins : The most commonly occurring groups of cyanobacterial poisions are the cyclic peptide hepatotoxins of the microcystin and nodularin family.They are named hepatotoxins as they damage the hepatocytes of the liver. Intrahepatic haemorrhage and hypovolemic shock leads to death. Cylindrospermopsin is a unique alkaloid hepatotoxin. It showed congestion in kidney and heart along with hepatic necrosis.

Microcystins : The cyanobacterial hepatotoxins appear to be the most widely distributed types of cyanobacterial toxins in aquatic environments. Of these, the most commonly encountered in freshwaters and extensively studied, are the heptapeptide microcystins (Carmichael *et al.*, 1988). The toxins have been found and characterized from *Microcystis aeruginosa* (Kutzing) Lemmermann, *M. viridis* Lemmermann, *Anabaena flos-aquae* Brebission, *Anabaena* sp., *Oscillatoria agardhii* Gomont and *Nostoc* sp. (A. Brown) Lemmermann, *Anabaenopsis milleri* Voronichin and also from a terrestrial cyanobacterium *Hapalosiphon hibernicus* W. et G.S. West. *Microcystis aeruginosa* has worldwide occurrence.The molecular weight of MCYSTs varies from 909 to 1067 depending upon the variable amino acids present. Seventy structural variants of this toxin are currently known. The most commonly found microcystin variant is microcystin-LR. The two letter suffix is derived from the variability of the molecule, as a result of aminoacid substitution at position 2 & 4 of the heptapeptide ring, with L-leucine and L-arginine occupying these positions in this particular microcystin variant, and it co-occurs with microcystin-RR.

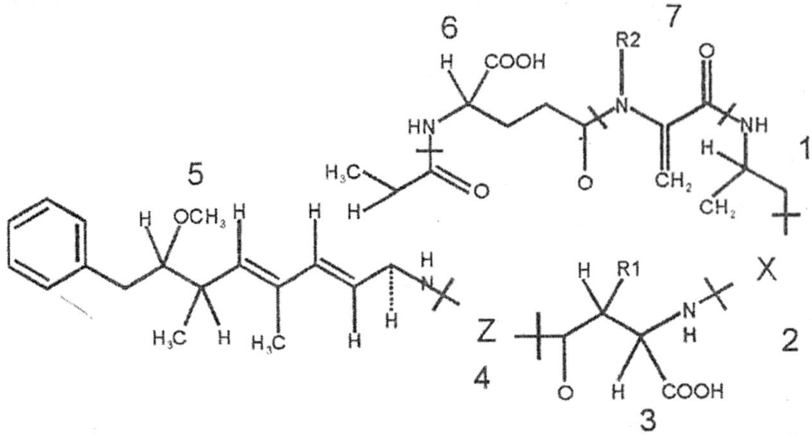

Structure of microcystin
1. D-Alanine 2. Variable L-amino acid 3. D-Methylaspartic acid 4. Variable L-amino acid 5. 3-amino-9-methoxy-2,6,8-trimethyl-10-phenyldeca-4,6-dienoic acid (Adda) 6. D-Glutamic acid 7. N-Methyldehydroalanine

Nodularins : Similar in structure to the microcystins are the nodularins. These are cyclic peptides containing only 5 aminoacids in total, as compared to 7 in microcystins. Although microcystins are produced by a number of freshwater cyanobacterial genera, nodularins have only been documented in *Nodularia spumigena* Martens, a bloom forming cyanobacterium that occurs

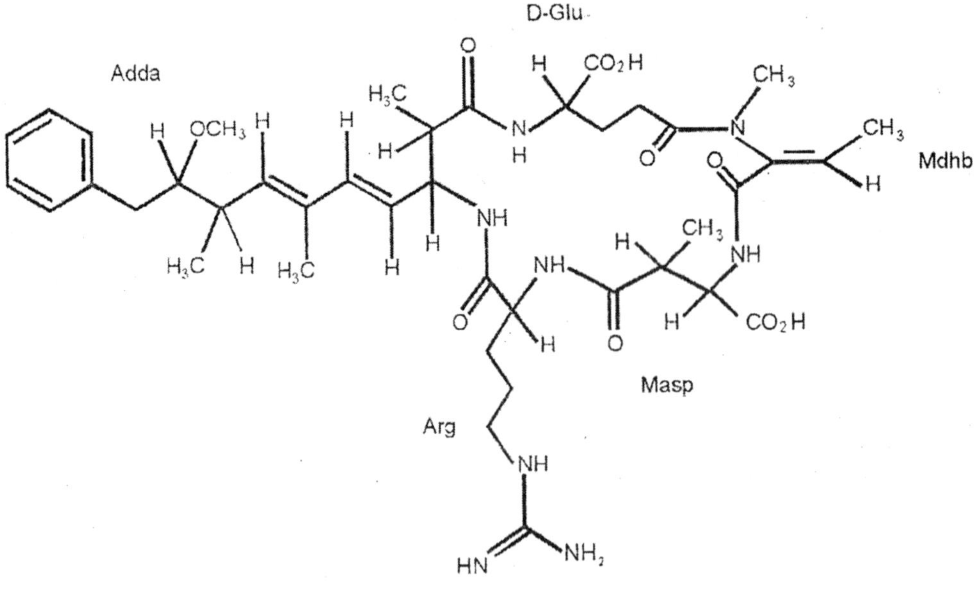

Structure of nodularin

generally in brackish and marine waters. The chemical structure of nodularin is cyclo- (D-MeAsp^{1} -L-arginine^{2}-Adda^{3}-D-glutamate^{4}-Mdhb^{5}) in which Mdhb is 2-(Methylamino)-2-dehydrobutyric acid.Toxicity of microcystins and nodularins vary from high to nondetectable. Bioactive microcystins and nodularins inhibit protein phosphatases in the liver (Eriksson *et al.*, 1990; Honkanen *et al.*, 1990, 1994; MacKintosh *et al.*, 1990; Matsushima *et al.*, 1990), leading to depolymerization of intermediate filaments and microfilaments (Falconer and Yeung, 1992; Eriksson and Goldman, 1993; Runnegar *et al.*, 1993). This causes shrinkage of the cells controlling circulation of blood in the liver, resulting in hemorrhages (Carmichael, 1994). Microcystins are potent tumour promoters (Falconer and Buckley, 1989; Nishiwaki-Matsushima *et al.*, 1992) and nodularin may also be a carcinogen (Ohta *et al.*, 1994).

Cylindrospermopsin : Cylindrospermopsin is one of the most recently discovered cyanobacterial hepatotoxins. It was identified after a human poisioning incident at an Australian drinking water reservoir in 1979. The toxicity was attributed to a bloom of the cyanobacterium *Cylindrospermopsis raciborskii* (Woloszynska) Seenaya et Subba Raju in the drinking water supply. It is an alkaloid cytotoxins with molecular weight 415 (Ohtani *et al.*, 1992) and it affects the liver, the kidneys, the thymus and the heart (Hawkins *et al.*, 1985; Terao *et al.*, 1994). Subsequently, cylindrospermopsin was found in the cyanobacteria, *Umezakia natans* Watanabe (Harada *et al.*, 1994) and *Aphanizomenon ovalisporum*. Cylindrospermopsin inhibits protein synthesis (Terao *et al.*, 1994) and also causes gene damage.

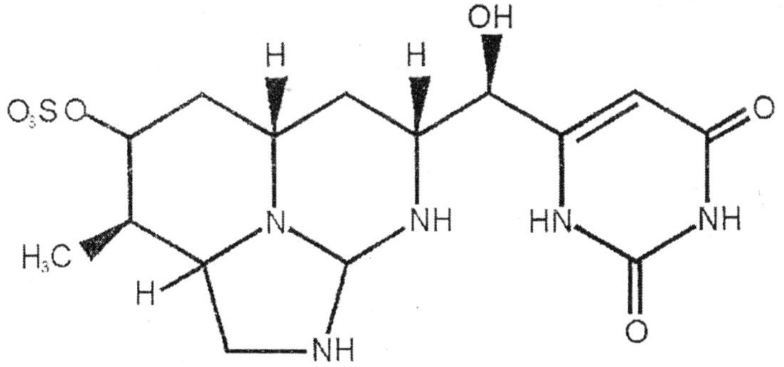

Structure of cylindrospermopsin

Dermatotoxins and gastrointestinal toxins : Marine cyanobacterium such as *Lyngbya majuscula* Harvey produces a toxin called aplysiatoxins which are commonly known for their dermatotoxic activity causing inflammation of the skin (Moikeha & Chu, 1971; Hashimoto *et al.*, 1976). They are also potent tumour promoters and protein kinase C activators (Fujiki *et al.*, 1990). Aplysiatoxins and debromoaplysiatoxin have been found associated with filamentous cyanobacterial species including *Schizothrix calcicola* (C. Agardh) Gomont (Entzeroth *et al.*, 1985) and *Oscillatoria nigroviridis*. The chemically different Lyngbyatoxin-a (Cardellina *et al.*, 1979), found in another strain of *Lyngbya majuscula*, has caused dermatitis and severe oral and gastrointestinal inflammation (Sims Zandee Van Rilland, 1981; Moore *et al.*, 1993).

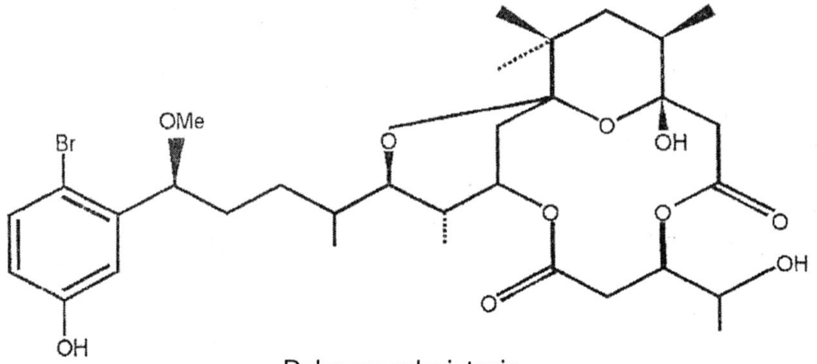

Debromoaplysiatoxin

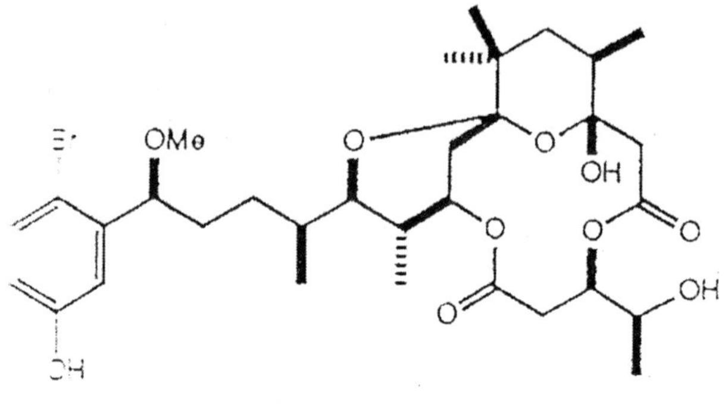

Aplysiatoxin

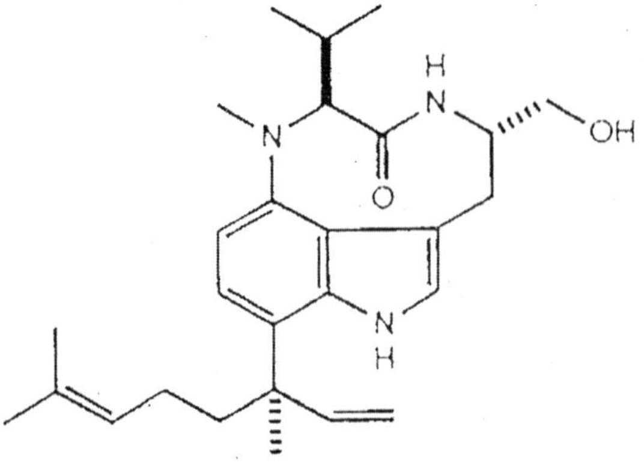

Lyngbyatoxin A

A standard feature of Gram-negative prokaryotes, including enteric bacteria, is the potential to produce lipopolysaccharides (LPS). Cyanobacteria are no exception. Lipopolysaccharides, found in the outer layer of the cell wall of all cyanobacteria where they form complexes with proteins and phospholipids, are pyrogenic and toxic. LPS is heat stable and toxic to mammals, with potency varying between different sources of bacteria and cyanobacteria. LPS molecules consist of three main parts: O antigen, Core polysaccharides and Lipid A moieties. The lipid A region is responsible for biological responses and symptoms of LPS-exposure including fever, diarrhea, vomiting and hypertension. Fevers are generally caused by the release of pyrogenic compound by the host body in response to LPS ingestion, with haemodialysis water and aerosolized LPS being important potential sources of exposure.

Phycotoxins from marine habitat

Chlorophyta: A few rare instances of toxin production have also been reported for the members of Chlorophyta.

Caulerpa spp.: The marine benthic green alga *Caulerpa* is responsible for the production of two toxic substances, namely, caulerpicin and caulerpin (Doty & Aguilar-Santos, 1966, 1970). Both of these compounds have demonstrated toxicity in mice. They were originally isolated from *C. racemosa* but were also identified in *C. sertulariodes, C. lentillifera, and C. lamourouxii*. It is interesting to note that *Caulerpa* is probably the most popular edible alga in the Philippines but becomes toxic during the rainy months. The toxicity is believed to derive from the agitation of the plant during the rainy season (Doig *et al.*, 1973). The infrared spectrum of caulerpicin indicated that this compound was a long-chain saturated hydroxy amide. Aguilar-Santos and Doty (1968) hypothesized from the spectral data that the structure was as follows: (n=23, 24, 25).

$$CH_3-(CH_2)_{13}-\underset{\displaystyle |}{\overset{\displaystyle CH_2OH}{CH}}-NH-CO-(CH_2)_n-CH_3 \quad \text{Caulerpicin}$$

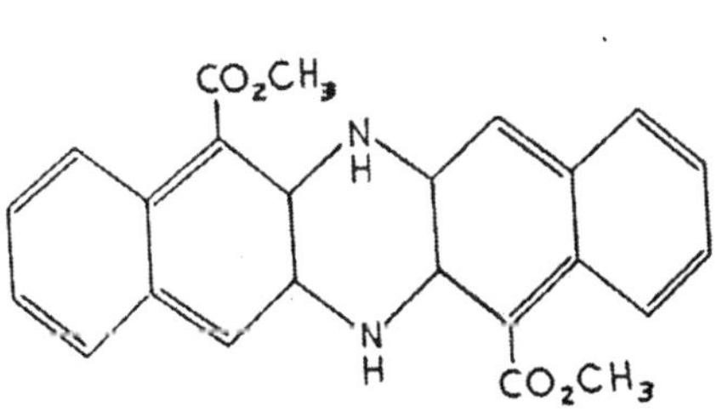

CO_2CH_3 ... NH ... NH ... CO_2CH_3

Caulerpin (from Scheuer, 1973)

The human physiological symptoms associated with caulerpicin ingestion include numbness and a cold sensation of the extremities, rapid and difficult breathing, slight depression, and eventually loss of balance. Depending on the dose, the effects are usually gone within a couple of hours to a day. Caulerpin was found to be a heterocyclic, red substance after it was crystallized from ether extracts of the alga. It is a pyrazine derivative. Both spectral analysis and degradation reactions were used to determine the hypothesized structure of dimethyl 6, 13-dihydrodibenzo [b, i] phenazine-5, 12-dicarboxylate (Scheuer, 1973) Table-2.

Cheatomorpha minima: The organic extract from the green alga *Cheatomorpha minima* has been experimentally shown to have hemolytic activity and fish toxicity or ichthyotoxicity (Fusetani *et al.*, 1976). The infrared spectrum of the organic extract was typical for fatty acids, and the predominant species detected by gas-liquid chromatography were palmitic (33%); palmitoleic (12%); oleic, elaidic, and/or vaccenic (14%); and linoleic (10%) acids. The active, purified organic extract was obtained as a colorless solid which killed killifish (*Oryzias latipes*) in 120 min at 5µg/ml and had a hemolytic activity of 1.99 saponin units (a quantitative measure of hemolytic activity for solid compounds) per mg.

Ulva sp.: Another genus of the green algae which has hemolytic activity is *Ulva* (activity found in three separate species in the nondialyzable fraction of a 70% ethanolic extract). After chemical separations, a total of three distinct hemolysins were isolated from *Ulva pertusa* (Fusetani & Hashimoto, 1976). Of the three hemolysins isolated, two were water soluble and the third was fat soluble. The fat-soluble hemolysin was identified as palmitic acid (a C_{16} saturated fatty acid), with a hemolytic activity of 0.24 saponin unit per mg. On the basis of infrared spectra and combustion data, one water-soluble substance is believed to be a galactolipid with a formula of $C_{31}H_{58}O_{14}$. This substance had a hemolytic activity of 1.44 saponin units per mg, as measured by the method of Hashimoto and Oshima (1972). The second substance is believed to be a sulfolipid with a formula of $C_{25}H_{47}O_{11}SK$ and a hemolytic activity of 2.01 saponin units per mg. Both of the water-soluble hemolysins were tested on sea urchin (*Hemicentrotus pulcherrimus*) (Ruggieri and Nigrelli, 1960).

Chrysophyta : The phylum Chrysophyta is composed of three different classes of algae. These three are: Xanthophyceae (Heterokontae or yellow-green algae); Chrysophyceae (golden algae); and Bacillariophyceae (diatoms). There is no documented case of a toxic member of the Xanthophyceae. Almost all of the published work on toxins from this phylum is related to the single species *Prymnesium parvum*, which is a member of the Chrysophyceae. Another genus of the Chrysophyceae, *Ochromonas*, has been shown to contain extractable toxins. There is only one documented case of health impairment due to an alga from the Bacillariophyceae (Wolke & Trainor, 1971).

Chrysophyceae (goden algae): Table-3a.

Prymnesium parvum: *P. parvum* is a marine alga of the family Chrysophyceae. *P. parvum* is a relatively small alga (10 µm in diameter) that has two flagella, a haptonema, and a scaly surface (Shilo, 1972). This alga has been associated with ichthyotoxicity (Otterstrom & Steemann-Nielsen, 1940; Reich and Aschner, 1947). There is evidence to show that several different toxic components are present in *P. parvum*. These toxins are capable of eliciting three separate physiological responses which are used to characterize the toxins; namely, ichthyotoxic activity, hemolytic activity, and antispasmodic effects on smooth muscle (Skilo, 1967; Ulitzur & Shilo, 1970). Paster (1968) isolated a glycolipid that was biologically alkaline labile (at pH 8 for 24 h) and was stable under acidic conditions (pH 5) and low temperatures (Paster, 1973). Ulitzur and Shilo (1970) isolated a lipoprotein-carbohydrate molecule which had stable or slightly activated ichthyotoxicity under alkaline conditions (0.5 N NaOH). Consequently, the environmental conditions also determine the extracellular stability of the toxins, in addition to affecting biosynthesis. As mentioned, the toxic compound (prymnesin) isolated by Paster (1968) was a glycolipid. Ultracentrifugation showed the molecular weight to be 23,000 ± 1,800, and this compound was the only one found in both ultracentrifugation and electrophoresis. Alternatively, the compound isolated by Ulitzur and Shilo (1970) was a lipoprotein carbohydrate molecule.This compound was named toxin B and consists of all six of the components that were separated *via* thin-layer chromatography. Of the six separate components, only three showed any protein content, whereas all six of the toxins contained phosphate. The entire six-component toxin was comprised of 22% protein, 0.47% phosphate, and 10 to 12% hexose sugars. In terms of chemical partitioning, toxin B was found to be similar to acidic polar lipids (Shilo, 1972); however in terms of many solubility characteristics, the molecule appeared to behave like a proteolipid compound. Proteolipids are differentiated from lipoproteins because they are soluble in organic solvents but not soluble in water. Toxin B was found to differ from proteolipids because it was soluble in dimethyl sulfoxide and methanol. Many effects of *Prymnesium* toxins, including ichthyotoxicity, cytotoxicity, and hemolytic activity, are predicated on a single mechanism. This mechanism is the change in the permeability of cell membranes (Imae & Inoue, 1974) which causes the membranes to become cation permselective. As a result of this change, certain cells are lysed, which results in many of the different biological activities of the toxin. A second important biological activity of the toxin, which may or may not be related to the change in membrane permeability, is the blockage of the neuromotor impulse at the postsynaptic membrane of the neuromuscular junction (Parnas and Abbott, 1965). *Prymnesium* toxins block the neuromuscular junction without causing depolarization.

Ochromonas spp.: Certain species of *Ochromonas*, have been found to be toxic (*O. malhamensis, O. danica*, and *O. minuta*) (Halevy and Avivi, 1968). Spiegelstein (1969) extracted two separate fractions from *Ochromonas*, which he labeled Ta and Tm. Both Ta and Tm show ichthyotoxic, hemolytic, and antispasmodic activities, like *P. parvum* toxin. However, in contrast to *Prymnesium* toxin, these toxins are not inactivated by either visible or ultraviolet light. Ta is heat, alkaline, and acid labile, whereas Tm is heat, alkaline, and acid stable. The fish lethality for these compounds occurs for both intraperitoneal (i.p.) administration and immersion. As with *Prymnesium* toxin, cation activation when the fish are immersed has been reported for both of these substances.

Bacillariophyceae: Table-3b.

Pseudonitzschia pungens, P. multiseries, P. australis and *P. pseudodelicatissima* produce domoic acid, a toxic amino acid. The role of this substance plays in the metabolism of the bacterium itself is still unclear. More details regarding the diatoms that prodoce toxins are further included in amnesic shellfish poisoning (see later).

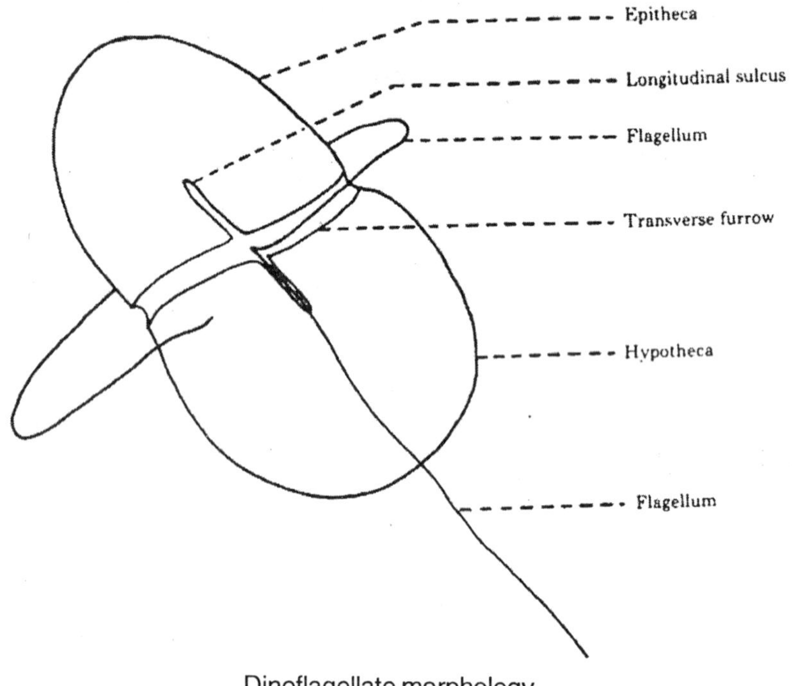

Dinoflagellate morphology

Table 2 : Chlorophyta (Green algae) Toxins

Taxa	Toxin	Structure	Symptoms
Caulerpa spp. *C. racemosa*, *C. sertulariodes*, *C. lentillifera*, *C. lamourouxii*	Caulerpicin Caulerpin	Long chain saturated hydroxyl amide Pyrazine derivative	Human physiological symptoms associated: Numbness, cold sensation, difficult breathing, slight depression , loss of balance
Chaetomorpha sp. *C. minima*	Fatty acid Hemolysin		Hemolytic activity/Icthyotoxic
Ulva sp. *U. pertusa*	Hemolysins of the 3 components, 2 are water soluble & the 3rd was fat soluble	Of the two water soluble, 1 is galactolipid ($C_{31}H_{58}O_{14}$); Other sulpholipid ($C_{25}H_{47}O_{11}SK$); Fat soluble identified as palmitic acid	Hemolytic activity
Heterosigma akashiwo Raphidophyte			Catastrophic losses of cultured and wild fish, particularly in the pacific north-west

Table 3a : Chrysophyta-Chrysophyceae/golden algae

Taxa	Toxin	Structure	Symptoms
Prymnesium parvum	Prymnesin and Toxin B	Glycolipid lipoprotein carbohydrate molecule	Ichthyotoxic activity, hemolytic activity, and antispasmodic effects on smooth muscle
Ochromonas malhamensis, *O. danica*, *O. minuta*	Ta and Tm		Ichthyotoxic, hemolytic, and antispasmodic activities

Table 3b : Chrysophyta- Bacillariophyceae/Diatoms

Taxa	Toxin	Structure	Symptoms
Pseudo-nitzschia multiseries, *P. seriata*, *P. australis*, *P.pseudodelicatissima*, *P. delicatissima*, *P. turgidula* and related diatoms	Domoic acid Amnesic shellfish poisionong (ASP)	Water soluble tricarboxylic amino acid	Gastrointestinal: diarrhea, vomiting, abdominal pain, nausea Neurological: dizziness, headache, seizures, disorientation, short-term memory loss, respiratory difficulty, coma
Chaetoceros sp.			Clog gill fish

Pyrrhophyta : The algal members of the Pyrrhophyta are also known as the dinoflagellates. This phylum has a unique morphology that differentiates its members from those of the other phyla. The cells are divided into two sections (the epitheca and the hypotheca) by a groove that circumscribes the algal cell (the transverse furrow). In the most cases, the cells have two flagella; one is directed longitudinally and provides the general directional locomotion, whereas the other circles around the transverse furrow and causes the cell to rotate. Toxic dinoflagellates are often marine organisms. Most of the documented causes of paralytic shellfish poisoning have involved *Gonyaulax* genus, but another dinoflagellate, *Pyrodinium phoneus,* is also known to produce this type of poisoning in the North Sea (Schantz, 1971). There is one more species of dinoflagellate, found in Japan, which has been shown to be hepatotoxic and nephrotoxic in animals, namely, *Exuviella mariae-lebouriae.*

Recently, there has been work to substantiate the hypothesis that ciguatera (a type of fish poisoning) is due to a toxic dinoflagellate (Moore, 1977). A single investigation has concluded that *Diplosalis* is very likely to be the cause of ciguatera (Yasumoto *et al.,* 1977). Table-4.

Peridinium polonicum: There has been evidence to incriminate the freshwater dinoflagellate *Peridinium polonicum* as an ichthyotoxic species (Hashimoto *et al.,* 1968, Matida *et al.,* 1967). It was originally isolated from Lake Sagani, Tokyo, Japan and identified as *Glenodinium gymnodinium.* As with all algal toxins, environmental conditions were critical in determining the extent of toxicity. The toxic compound was an alkaloid (Mosher *et al.,* 1964). About the same time, the mass spectrum of the purified hydrochloride was found to be very similar to the mass spectrum of 12-methoxyibogamine (Bltha *et al.,* 1972). No further chemical analysis has been published.

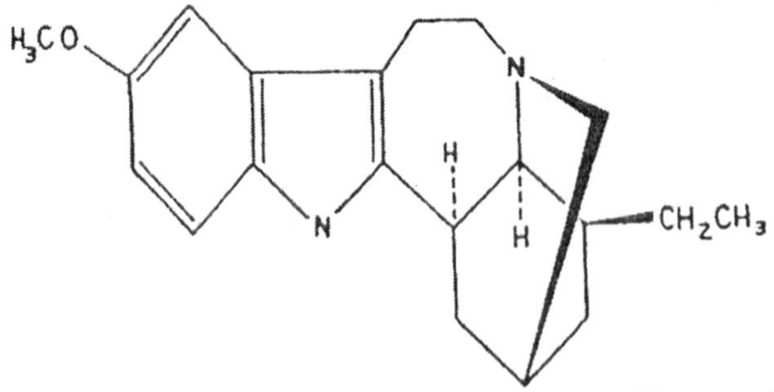

12-Methoxybogamine (from Bltha *et al.,* 1972)

Amphidinium spp.: Several species (*A. klebsii* and *A. rhynchocephalum, A. carteri*) within the dinoflagellate genus *Amphidinium* have been implicated as ichthyotoxic. Ikawa and Taylor (1973) hypothesized that the toxin was choline-like. By combining the use of assay procedures and chromatographic work, it was determined that *A. carteri* has three nonlipoidal choline derivatives. One of the substances has been identified as choline O-sulfate, and another behaves physiologically, chemically, and nutritionally like a choline ester. At least one of the unknown compounds exhibited acetylcholine-like activity on a number of heart systems (Ikawa & Taylor, 1973). The toxicological properties of these choline-like algal extracts were not tested.

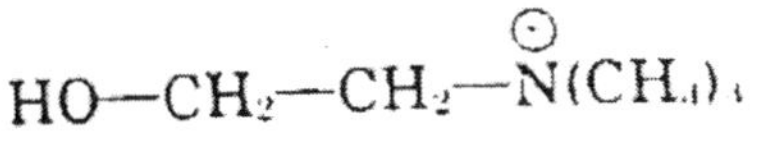

$$HO{-}CH_2{-}CH_2{-}\overset{\oplus}{N}(CH_3)_3$$

Choline

$$^{\ominus}O{-}\overset{\overset{O}{\|}}{\underset{\underset{O}{\|}}{S}}{-}O{-}CH_2{-}CH_2{-}\overset{\oplus}{N}(CH_3)_3$$

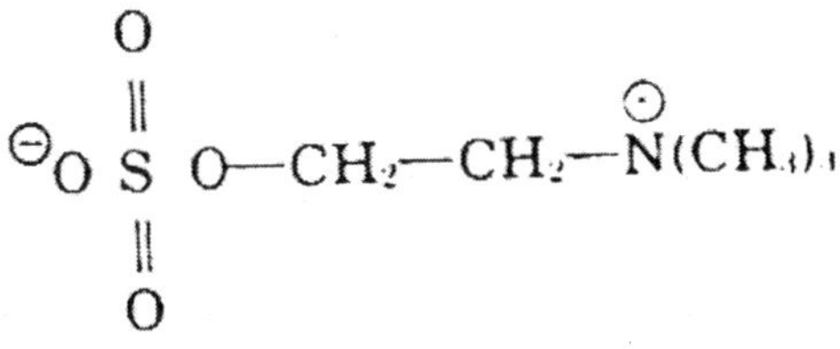

Choline O-sulfate

Noctiluca miliaris: The dinoflagellate *Noctiluca miliaris* is responsible for many red tides. According to Okaichi and Nishio (1976) the toxicity of *Noctiluca* is due to ammonia.

Gymnodinium spp.: The nonthecate marine dinoflagellate *Gymnodinium* has two different species which have exhibited toxic properties. *G. breve* is found predominantly in the Gulf of Mexico and is responsible for the red tides in this area, whereas with *G. veneficum* of the English Channel, which has been found to contain toxic substances. *G.breve* produces at least two brevetoxins. These are fat-soluble complex molecules (polyketides). The red tides due to *G. breve* are lethal to a large variety of aquatic and terrestrial biota. Also, vertebrates are more susceptible than invertebrates be*via*se the toxin acts on the nervous system, which is more specifically defined in vertebrates. Humans are susceptible to the toxic effects of *G. breve* red tides primarily through the food chain (especially shellfish/Neurotoxic shellfish poisioning/NSP). The toxic effects produced in humans are neither severe nor of long duration. No human deaths have been reported as a result of ingestion of seafood exposed to *G. breve* (Steidinger *et al.*, 1973). *Gymnodinium* is not the alga that is generally associated with paralytic shellfish poison (PSP). Another type of human health problem occurs when *G. breve* cells are lysed and become airborne due to sea spray. In this instance the cells are responsible for an odorless eye and respiratory irritant (Ingle, 1954; Woodcock, 1948). Several basic parameters differentiate *G. breve* toxin (NSP) from the paralytic shellfish poison (PSP) that is generally attributed to *Gonyaulax*. *Gymnodinium* toxin is an endotoxin, whereas *Gonyaulax* toxin is an exotoxin. *Gymnodinium* toxin is not water soluble, but *Gonyaulax* toxin is water soluble (Schantz, 1971). The physiological action of *G. breve* toxin is as much of an enigma as the chemical structure of this compound. The dominant hypothesis as to the mode of action of *G. breve* toxin is that it acts as a depolarizing agent, similar to the toxin from *G. veneficum* (Abbott and Ballantine, 1957). *Gymnodinium veneficum* toxin increases neurotransmission before blockage and causes depolarization. Saxitoxin, which has been isolated from both *Gonyaulax* and *Aphanizomenon*, produces its effects through a change in sodium conductance, whereas the toxin from *Prymnesium* has no effect on sodium conductance. Humans who are exposed to *G. breve* toxins *via* ingestion develop symptoms of central nervous system poisoning (hot and cold reversals, vertigo, and slowed pulse), as well as symptoms of peripheral nervous system toxicity (ataxia, dilated pupils, mild diarrhea, and tingling sensations). The rapid mechanism was of the neuromuscular type, which ended with respiratory failure, but the slow mechanism was characterized by a distended bladder and the dilation of capillaries. *G. catenatum* produces saxitoxin, a purine alkaloid which is associated with paralytic shellfish poisoning (PSP). Saxitoxin blocks sodium channels, which leads to paralysis.

Gonyaulax (*Alexandrium*) spp.: There are six different species of the marine dinoflagellate *Gonyaulax* which have been documented as toxic species. The three species *G. catenella, G. acatenella, and G. tamarense* have been shown to cause paralytic shellfish poisoning (PSP), whereas the species *G. monilatum* and *G. polygramma* have been related to ichthyotoxic incidents. *Gonyaulax* species associated with paralytic shellfish poisoning have received more experimental attention than have the other species. Paralytic shellfish poisoning occurs when an organism feeds on a shellfish that has filtered a toxic species of the dinoflagellates (most frequently *Gonyaulax*). The shellfish are not adversely affected by the alga. There are more than 220 human fatalities which have occurred as results of paralytic shellfish poisoning (Halstead, 1965). Most species of shellfish bind *Gonyaulax* toxins in the hepatopancreas, where it does not cause any harm to the shellfish. *G. tamarense* is considered a more severe practical problem, because the red tides of this alga occur more frequently and cover a larger area than the tides of *G. catenella* (Ghazarossian *et al.*, 1974). There have been seven separate toxins isolated from *G. tamarense*: saxitoxin, gonyautoxin 1, gonyautoxin 2, gonyautoxin 3, gonyautoxin 4, gonyautoxin 5, and neosaxitoxin. For only three of these toxins have the chemical structures been fully described (one of the three is the compound saxitoxin, which was previously described as being isolated from *Aphanizomenon flos-aquae*); however, the other four toxins are hypothesized to have similar structures. The chemical characteristics of all seven of these toxins are very similar. They are all very hygroscopic and highly water soluble. Saxitoxin is the most potent algal toxin on record.

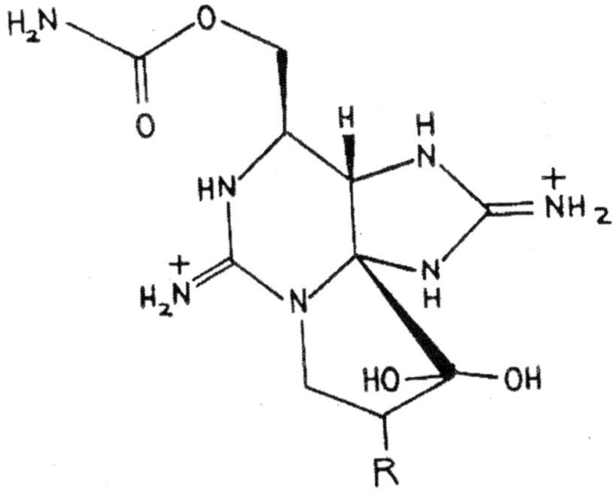

Gonyaulax toxins: saxitoxin (R = H), gonyautoxin 2 (R = a-OH), and gonyautoxin 3 (R =, - OH) (from Shimizu *et al.*, 1976)

Rhodophyta : The carrageenans are heterogenous mixtures of anionic polysaccharides which are extracted from certain algae of the Rhodophyta. The chemical compositions of these molecules include alternating derivatives of galactose, 3, 6-anhydrogalactose, and sulfated galactose. Molecules with the basic structure of carrageenan have been isolated from the following genera of the Rhodophyta: *Chondrus, Gigartina, Hypnea, Rhodymenia, Irideae, Gracilaria, Furcellaria, Polydes,* and *Eucheuma* (DiRosa, 1972). There are essentially two types of carrageenans, which are differentiated according to molecular weights. The native, undegraded, foodgrade carrageenan has a high molecular weight (100,000 to 800,000), includes both kappa and lambda chains, and is derived primarily from the alga *Chondrus crispus*. This type of carrageenan is used as an emulsifier, stabilizer, or thickener in many types of foods because of its ability to bind casein Table-5.

The carrageenan used in the therapy of peptic ulcers has a comparatively low molecular weight (5,000 to 30,000), consists of only iota chains, and is derived primarily from the alga *Eucheuma spinosum*. The iota, kappa, and lambda galactose units differ in the number of sulfate groups present and in the types of chemical linkages (Engster and Abraham, 1976). The reason for mentioning the carrageenans in this paper is that they represent an algal polysaccharide with the capability to induce serious physiological effects. They have a wide spectrum of biological actions which include the induction of acute (edema) and chronic (granuloma) inflammatory responses, inhibition of the Cl component of the complement sequence, activation of Hageman factor (coagulation factor XII is a plasma protein) selective toxicity to macrophages (Allison *et al.*, 1966; Catanzaro *et al.*, 1971) and strong

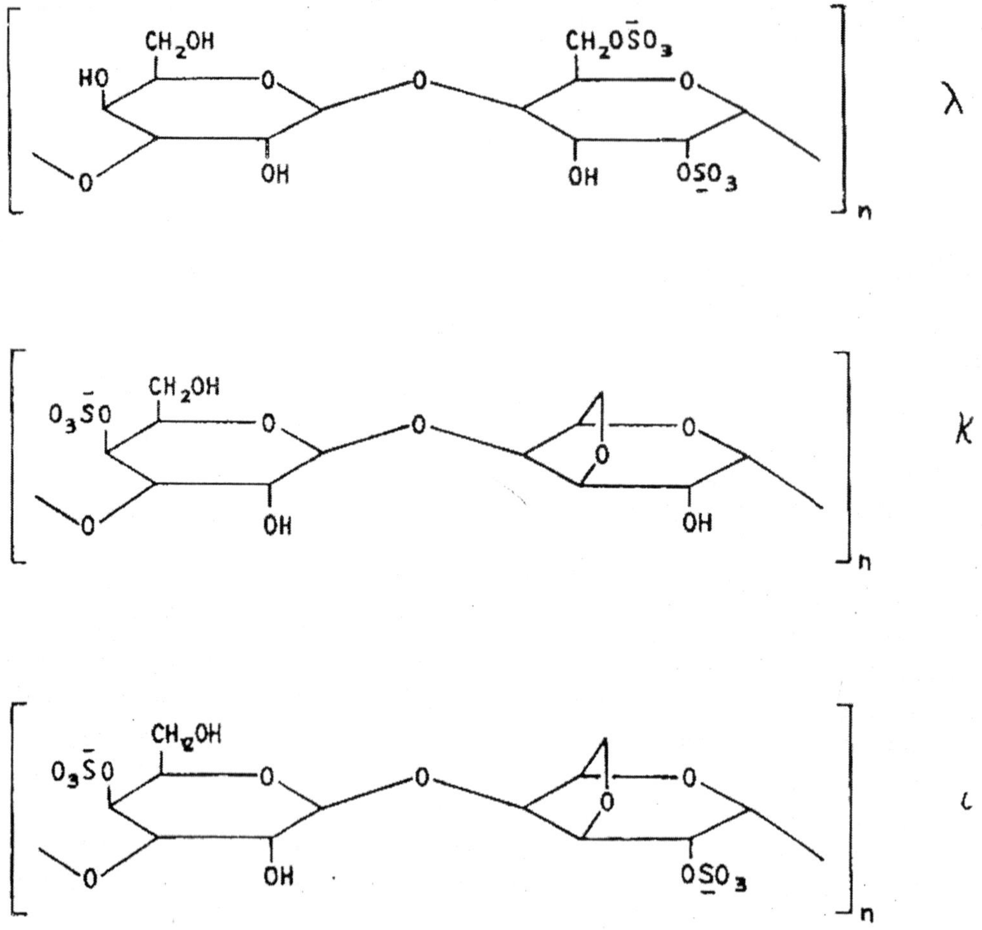

Idealized structures of the different carrageenans (from Mueller and Rees, 1968)

immunosuppression (Aschheim and Raffel, 1972; Bice *et al.*, 1971). Many physiological effects of carrageenan coupled with its widespread use as a food additive and as a drug have led to a series of experiments to determine the health implications of its use. Until recently, the only deaths produced by carrageenans were through the intravenous route of administration. At 15 mg/kg, the carrageenan fraction from *Gigartina acicularis* caused two out of three dogs to die within 24 h (Houck *et al.*, 1957). Similarly, in a rabbit study the minimum lethal dose, when administered intravenously, to cause death within 24 h was 1 to 5 mg/kg for lambda-carrageenan and 3 to 15 mg/kg for kappa-carra'geenau; these fractions were taken from *C. crispus*. Thomson and Horne (1976) also made several observations concerning the toxic mechanisms of carrageenans. Since carrageenans activate Hageman factor, they also activate kinins, which enhance thrombosis. They also found that the carrageenans were hepatotoxic. Several studies have shown that the high-molecular-weight species do not produce major tissue disruption (*i.e.*, ulceration) (Bonfils, 1970; Maillet *et al.*, 1970). Alternatively, the low-molecular-weight carrageenans have produced ulcerative colitis, mucosal erosions, and other tissue changes in laboratory animals (Marcus and Watt, 1969).

Marine phycotoxins

Three types of illness are associated with ingestion of seafood: allergic, toxic and infectious. In the following paragraphs some specific disorders caused by biotoxins are discussed. There are various micro-organisms in the phytoplankton which produce toxins.

Ciguatera Fish Poisoning (CFP)

- Toxins of a microscopic dinoflagellate: *Gambierdiscus toxicus.*
- Poison present in certain tropical sea fish, especially coral reef fish.
- Nausea, vomiting, paresthesia, warm-cold sense inversion, pruritus, headache, shock.
- Treatment with mannitol IV in the acute stage

Table-4.

Ciguatera fish poisoning (CFP) is a seafood intoxication due to consumption of a gastropod (*Livona pica*) which was known locally as "cigua" in Cuba. In this way the name ciguatera was introduced. Intoxication caused by ladderlike polyether toxins, primarily attributed to the dinoflagellate, *Gambierdiscus toxicus* (Yasumoto *et al.*, 1977) which grows as an epiphyte on filamentous macroalgae associated with coral reefs and reef lagoons. On this type of substrate intense competition between various algae takes place, including other toxic dinoflagellates (*Amphidinium, Ostreopsis, Prorocentrum, Coolea*). It results in a dynamic balance between the various species. *Gambierdiscus toxicus* is responsible for the production of two kinds of toxins: maitotoxins and ciguatoxins. The lipophilic precursors to ciguatoxin produced by alga enter the food web when these algae are grazed upon by herbivorous fishes and invertebrates. These precursors are biotransformed to ciguatoxins (Lewis and Holmes, 1993) and bioaccumulated in the highest trophic levels. Large carnivorous fishes associated with coral reefs are a frequent source of ciguatera.

Table 4 : Dinoflagellate/Pyrrhophyta

Taxa	Toxin	Structure	Symptoms
Peridinium polonicum (FW) (*Glenodinium gymnodinium*)	Glenodine	alkaloids	Icthyotoxic
Amphidinium spp. *A. klebsii, A. carteri, A. rhynchocephalum*	Choline esters Choline-O-sulphate		Fish mortality/ Icthyotoxic
Noctiluca miliaris (responsible for red tides)	Ammonia		Fish & shellfish kills due to high levels of ammonia after red tide
Gymnodinium breve (*Ptychodiscus brevis*)	Brevetoxin Aerosolized Neurotoxic shellfish poisioning (NSP)	Lipophilic polyether	To impact human health through bronchial spasms, Gastrointestinal: diarrhea, vomiting Neurological: Parasthesia (sensation of numbness or tingling on the skin)
Gymnodinium catenatum, Cochlodinium catenatum Gymnodinium veneficum	Saxitoxin (SXT) Paralytic shellfish poisioning (PSP)	Purine alkaloid	Neurological: respiratory paralysis, tingling, burning, numbness, drowsiness, incoherent speech, rash Increases neurotransmission before blockage and causes depolarization
Alexandrium (*Gonyaulax*) *catenella, A. acatenella, A. tamarense, A . minutum, A. ostenfeldii, A. monilatum*	Saxitoxin (SXT) Gonyautoxin (GTX 1-5) Neosaxitoxin (NSXT) Paralytic shellfish poisioning(PSP)		Neurological: respiratory paralysis, tingling, burning, numbness, drowsiness, incoherent speech, rash
Gonyaulax polygramma			Fish and shellfish kills due to anoxia after red tides
Pyrodinium bahamense, P. phoneus	Saxitoxin Gonyautoxin(GTX) Paralytic shellfish poisioning (PSP)		Paralytic shellfish poisioning in the north sea
Gambierdiscus toxicus	Ciguatoxin - Maitotoxin scaritoxin Gambiertoxin Ciguatera fish poisioning (CFP)		Gastrointestinal: diarrhea, vomiting, abdominal pain Neurological: headache, numbness reversal of hot & coldsensations,Cardiovascul ar: Irregular heart rhythm, low bp

Contd...

Goniodoma sp.	Goniodomin		Icthyotoxic
Dinophysis fortii, *D. acuminata*, *D. acuta*, *D. hastata*, *D. rotundata*, *D. norvegica*	Dinophysitoxin-1(DTX1) Okadaic acid (OA) Diarrhetic shellfish poisioning (DSP)	Acidic polyether	Gastrointestinal: diarrhea, vomiting, nausea, abdominal pain, chills, headache, fever
Prorocentrum lima, *P. elegans*, *P. hoffmannianum*, *P. concavum*	Okadaic acid (OA) Diarrhetic shellfish poisioning (DSP)		Diarrhea, vomiting, nausea, abdominal pain, chills, headache, fever
Gyrodinium aureolum	1-acyl-3-digalactosylglycerol, octadecapentaenoic acid		Icthyotoxic
Pfiesteria piscicida (Achlorophyllous)	Toxins as yet unidentified; May be volatilized Dermonecrotic neurologic toxins		To impact human health through respiratory route

CiguateraToxins

Maitotoxin : This poison was first isolated in 1971 by Yasumoto in Japan, from the intestine of the black surgeon fish (*Ctenochaetus striatus*), known as "Maito" in Tahiti. The poison is not found in other tissues of these animals. In Tahiti this fish is eaten after grilling but without being eviscerated, which means that clinical problems may follow due to ingestion of maitotoxin. There is respiratory and cardiac arrhythmia, areflexia and muscular atonia, followed by cyanosis and death without convulsions. Maitotoxin is a complex molecule in the form of a long chain with many cyclical ethers. The molecular weight is 3422 Dalton (C_{164}, H_{256}, O_{164}). The structure was elucidated in 1992. It is one of the most powerful non-protein toxins that has ever been discovered (50 times more powerful than tetrodotoxin, a neurotoxin present in fish order of Tetraodontiformes) and is only surpassed by palytoxin, a polyketide present in some sea anemones (*Palythoa* spp.) and certain crabs. Maitotoxin is a powerful activator of calcium channels.

Ciguatoxin : Ciguatoxin was originally isolated by Scheuer in Hawaii in 1967. The structural formula of ciguatoxin was discovered on the basis of 350 μg of poison originating from 830 kg of Javanese giant moray eels (*Gymnothorax javanicus*). The toxin is present in low concentrations, but is extremely powerful. The toxins form a family of very closely related structures with a molecular weight of 941-1117 Dalton. There are a number of variants, depending on whether certain chemical groups (-H, $-CH_3$, etc) are present or not. It is a heat-resistant, fat-soluble polyheterocyclic molecule structurally related to brevetoxin. The poison binds to voltage-dependent sodium channels in muscle and nerve cells, so that they remain open.

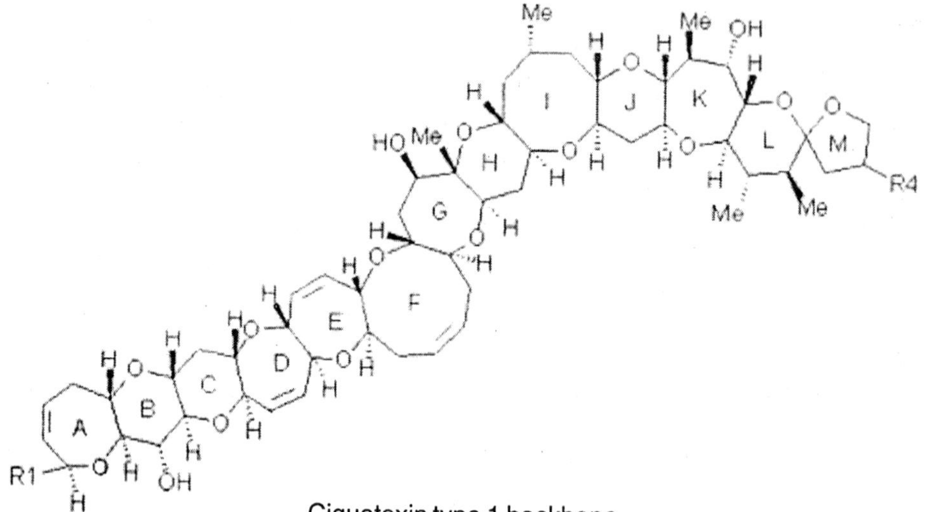

Ciguatoxin type 1 backbone

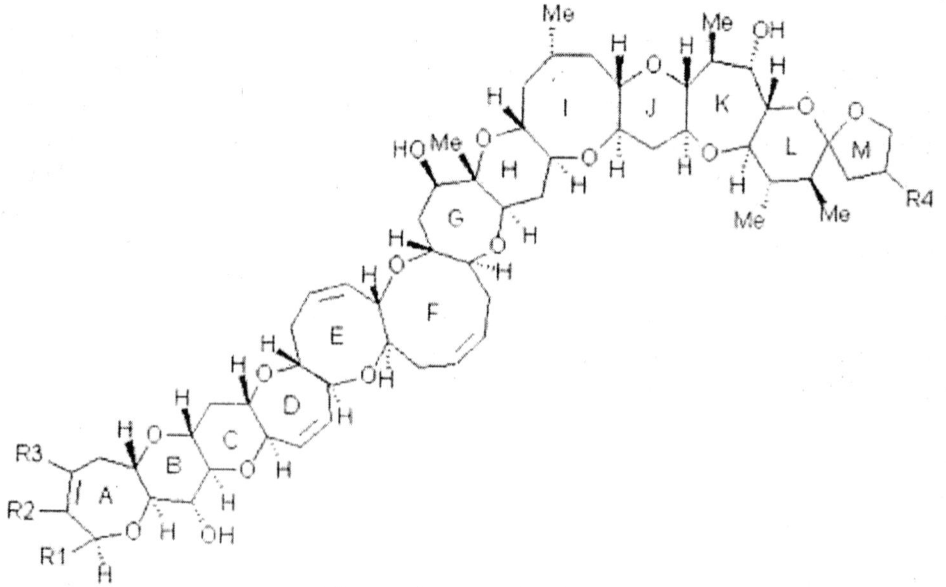

Ciguatoxin type 2 backbone

The symptoms are gastro-intestinal, cardiological and neurological in nature and are generally self-limiting. Ciguatera characterized by numbness and tingling around the mouth, hands, and feet; joint and muscle pains with weakness or cramps; vomiting, diarrhea, chills, itching, headache, sweating, and dizziness; and reversal of temperature sensation, where cold things feel hot and hot things feel cold. These symptoms all appear to arise because of ciguatoxin's ability to activate the sodium channels found in nerves and muscles.

Shellfish -associated biotoxins

Problems caused by phycotoxins in mussels, oysters and other edible seafood: Table-4

- PSP: paralysis due to saxitoxin and gonyaulatoxins
- NSP: paresthesia due to brevetoxin; also bronchial spasms
- DSP: diarrhea due to okadaic acid
- ASP: memory disturbances due to domoic acid

Separate problem

- *Pfiesteria*: skin ulcers and lesions of the central nervous system. Toxins unclear.

Many shellfish -certainly bivalves- often filter enormous amounts of seawater (a mussel filters 50-150 litres of seawater each day). The smallest particles of the plankton, including dinoflagellates, algae and diatoms, remain behind as food. In this manner shellfish concentrate toxins. Most shellfish are not themselves sensitive to these toxins. Toxins may be further modified chemically within the shellfish. The toxins are heat-stable and are not destroyed by boiling, although they may leak into the cooking water. They must not be confused with the toxins of some freshwater cyanobacteria, such as *Phormidium* spp. or the hepatotoxic microcystines of *Microcystis aeruginosa* which may be found in drinking water.

Paralytic shellfish poisoning (PSP) : PSP is a global problem and is caused by the consumption of molluscan shellfish contaminated with a suite of heterocyclic guanidines collectively called saxitoxins (STXs). Saxitoxin takes its name from the Alaskan butter clam *Saxidomus giganteus*. This animal can harbour very large amounts of toxin, and is responsible for a great deal of morbidity. In freshwater, blue-green algae, namely *Anabaena circinalis*, manufacture the toxins and may transfer the toxins to freshwater shellfish (for example *Alathyria condola*) although no reports exist of intoxication from freshwater sources of these toxins. In ocean, saxitoxin is produced by dinoflagellates, particularly *Alexandrium* (*Gonyaulax*) *tamarense*, *Alexandrium catenella*, *Pyrodinium bahamense*, *Gymnodinium catenatum* and *Cochlodinium catenatum*. Many derivatives of saxitoxin are known as gonyautoxins. The name refers to *Gonyaulax*, the former name of *Alexandrium* dinoflagellates. The basic chemical stucture of these gonyautoxins is identical, but they are distinguished by chemical side-chains such as: -H, $-OSO_3$, $-CONH_2$, $-CONHSO_3$). Saxitoxin ($C_{10}H_{17}N_7O_4$; MW = 299) blocks sodium channels, which leads to paralysis. Deaths resulting from saxitoxin are known. Sometimes the patient requires mechanical ventilation. The lethal dose for humans is 0.1 to 1 mg (Levin, 1992). Consequently the toxin is extremely powerful (as toxic as tetrodotoxin). It is even regulated under the Chemical Weapons Convention. The toxins are heat-stable and water-soluble. Differentiation between PSP, tetrodotoxin poisoning and ciguatera is not easy.

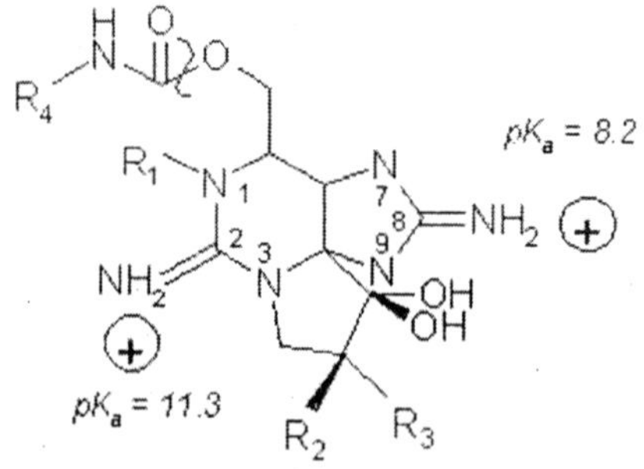

In the core of the saxitoxin family of toxins, variation may occur at the depicted R groups and the carbamoyl group may be replaced at the wavy line.

There is no antitoxin for PSP. Treatment is based on symptomatic care and the avoidance of complications. Inducing vomiting is dangerous due to the risk of aspiration due to loss of the gag reflex. In case of respiratory depression artificial respiration is necessary. Oxygen should be administered. Whether vitamin B injections are beneficial is still an open question.

Neurotoxic Shellfish Poisoning (NSP) : Gymnodinium breve (previously known as *Ptychodiscus brevis*) from which the toxin name is derived, is found in the Caribbean and the Gulf of Mexico. This dinoflagellate produces at least two brevetoxins (PbTx).The brevetoxins are lipophilic 10- and 11-ring polyether chemicals. They disturb neuromuscular transmission (Rein *et al.*, 1994). After being inhaled as aerosol they cause bronchial spasms. This may be manifested as an "asthma" crisis, rhinitis, sneezing, cough or burning eyes after walking on the beach while a strong breeze which splashes up water (with the toxin). This kind of aerosol is facilitated by the fact that *Gymnodinium* is a very fragile organism which easily breaks in the surf, releasing the endotoxins. The *Alexandrium* sp. in the Pacific or in the North Atlantic are much less fragile and do not cause irritation *via* aerosol. Brevetoxins may be present in molluscs (oysters, mussels) during an algal bloom, but are not present in fish, crabs or snails. If the toxins are absorbed in the intestine, nausea and vomiting, abdominal pain and diarrhea occur. There then follows paresthesia around the mouth, which extends further to the throat, trunk and limbs. Ataxia, mydriasis, vertigo, breathing difficulties, headache and bradycardia may follow. As yet no deaths due to NSP have been reported. The diagnosis is clinical. There is no antidote.

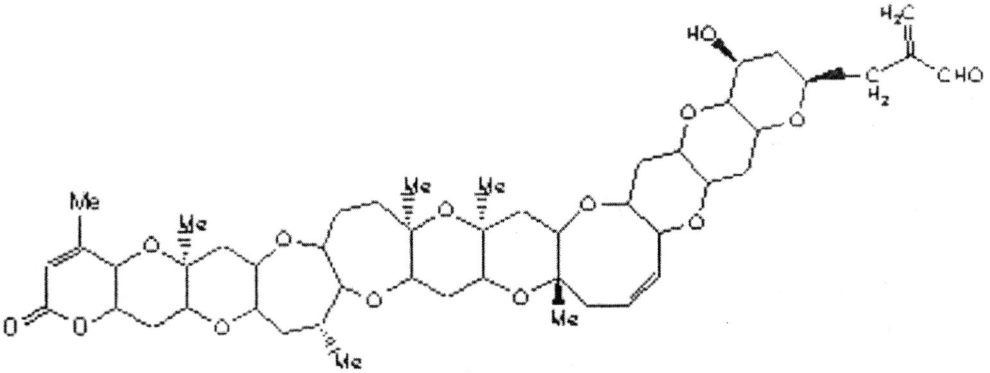

Brevetoxin-A, a type I brevetoxin

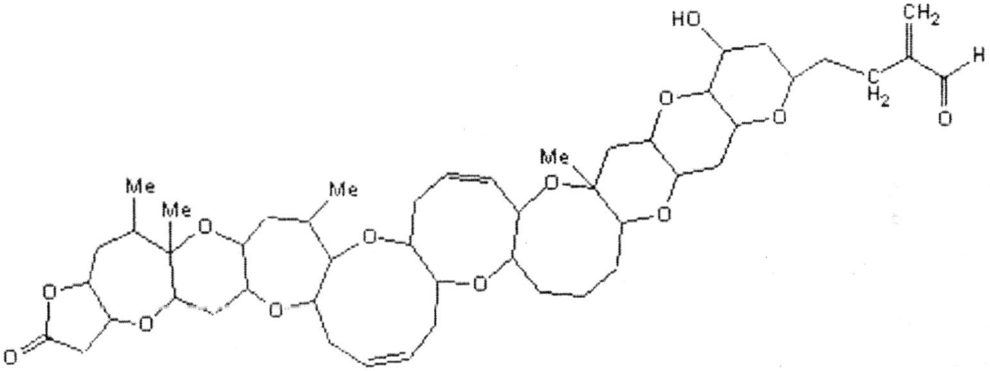

Brevetoxin-B, a type II brevetoxin

Diarrhetic Shellfish Poisoning (DSP) : Various *Dinophysis* spp. and *Prorocentrum* spp. (Lawrence *et al.*, 1998) produce okadaic acid and derivatives (polyketides). The substance takes its name from the marine sponge *Halichondria okadai,* from which it was first isolated. These marine sponges are cultivated in Japan and New Zealand and also contain halichondrine, an antitumoural substance (possibly active against melanoma). Okadaic acid has several derivatives. They are known as dinophysitoxins and pectenotoxins (Yasumoto *et al.*, 1979, 1980). The toxins are powerful inhibitors of protein phosphatases 1A and 2A. They are possibly carcinogenic (Fujiki *et al.*, 1988). Severe diarrhea results from acute intoxication. Not all diarrhea after eating seafood is the result of this toxin. Molluscs can also contain viruses (Norwalk agent) and bacteria (*Salmonella, Vibrio* sp.).

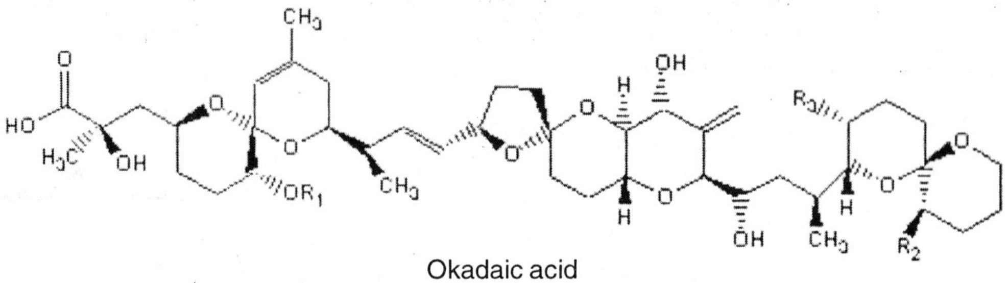

Okadaic acid

Okadaic acid

Amnesic Shellfish Poisoning (ASP) : ASP is caused by domoic acid (DA), a neurotoxic tricarboxylic amino acid which is structurally related to glutamic acid. It was chemically identified after its isolation in 1958 from the seaweed *Chondria armata,* found off the coast of Japan. In 1987, more than 100 people became ill and several people died following the consumption of blue mussels caught off Prince Edward Island, Canada. Canadian scientists found that domoic acid had entered the food chain when the mussels fed on a toxic algal bloom of the pennate diatom *Pseudonitzschia pungens* forma *multiseries* (Subba-Rao *et al.,* 1988; Bates *et al.,* 1989). This is therefore not a toxin of a dinoflagellate, but from diatoms. Domoic acid (MW=311; $C_{15}H_{21}NO_6$) is known to occur at low concentrations in various red algae (*Chondria armata, Alsidium corallinum* and *Digenea simplex*) (Murakami *et al.,* 1953). If the concentration of domoic acid is more than 20 ppm, the seafood is unsuitable for human consumption. After eating toxic mussels, people experience an initial feeling of nausea and diarrhea, together with hyperexcitation, followed by symptoms attributable to necrosis of certain parts of the brain such as the amygdala and parts of the hypothalamus. Disturbed behaviour and loss of memory, as well as involuntary facial grimaces, convulsions, coma and death may follow. Sometimes the chronic symptoms are similar to those of Alzheimer's disease.

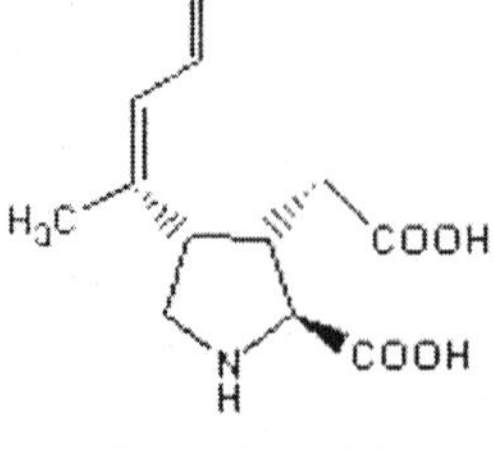

Domoic acid

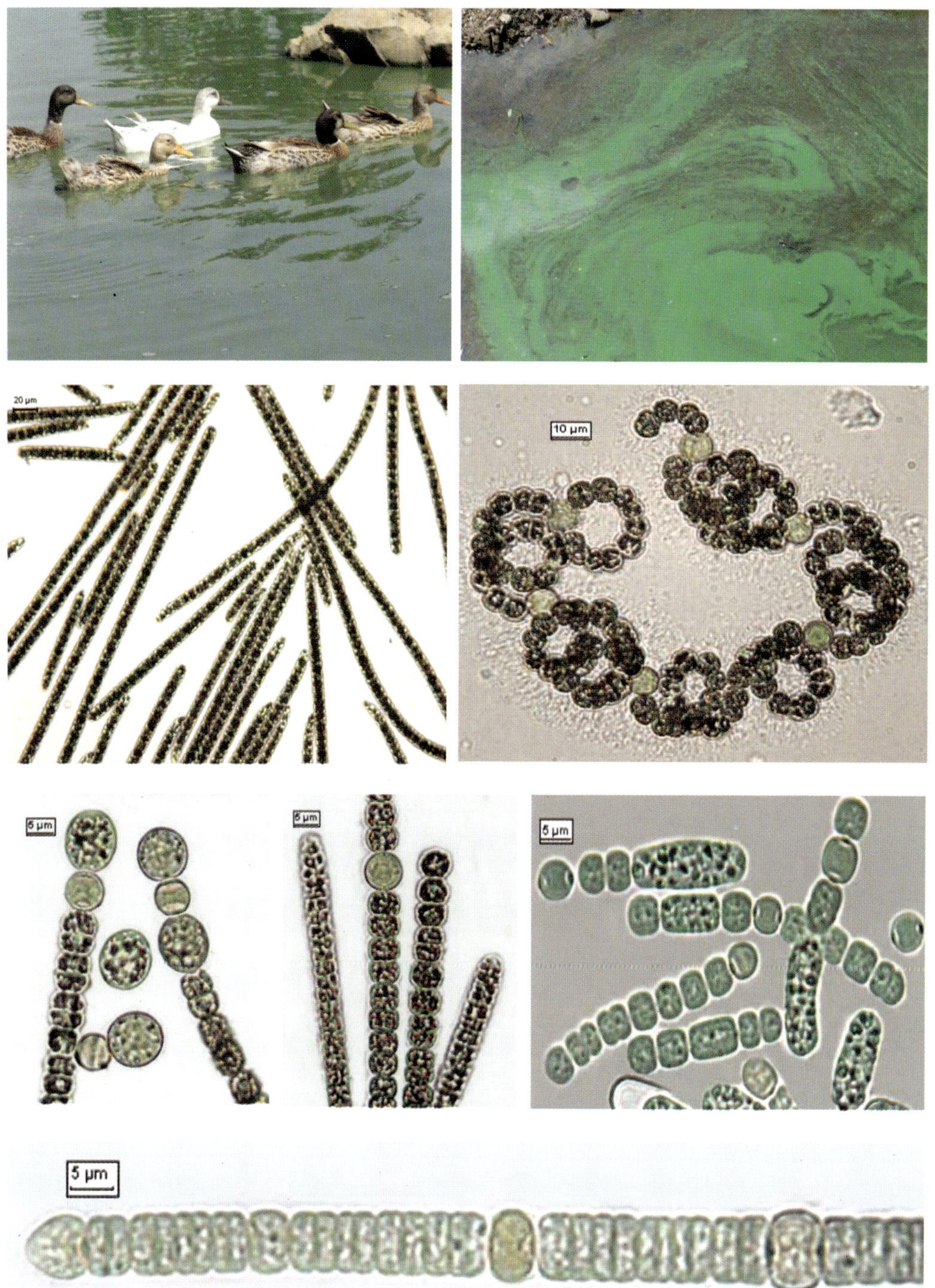

Plate : 1. Bloom of *Microcystis* 2. Bloom of *Planktothrix* 3. *Planktothrix* spp. 4-7. *Anabaena* spp. 8. *Nodularia* spp.

Table 5 : Rhodophyta (red algae)

Taxa	Toxin	Structure	Symptoms
Chondria armata *Alsidium corallinum* *Digenea simplex*	Domoic acid Amnesic shellfish poisoning (ASP)		Diarrhea, vomiting, abdominal pain, nausea
Chondrus, Gigartina, Hypnea, Rhodymenia, Iridea, Gracilaria, Furcellaria, Polydes, Eucheuma	Toxic properties of Carageenans	Heterogenous mixture of anionic polysaccharides	Induction of acute (edema) & chronic (granuloma), inflammatory responses, activation of Hageman factor, immunosuppression, toxicity to macrophages

Pfiesteria and Estuary-Associated Syndrome : Pfiesteria differs from the previously discussed dinoflagellates in that it is a nonphotosynthetic, heterotrophic dinoflagellate (Burkholder *et al.,* 1997). This toxin-producing unicellular alga has an improbably complex life cycle and many (24) morphological forms. In recent years it has caused massive fish death in rivers and estuaries in North Carolina and parts of Mexico. There are two kinds of toxins: water-soluble and fat-soluble. The fat-soluble one causes skin lesions in fish. The water-soluble toxin affects the central nervous system. It stuns the fish so that they do not swim away, and then the fat-soluble toxin can do its work. The fish skin is damaged, ulcers result and the internal salt balance is disturbed. This alga has not been known to date outside North American waters, but account must be taken of the possibility that in future it may appear on other coasts, for example *via* ballast water from freighters. Humans may have problems if they breathe in the toxins *via* aerosol or if they get it on their skin.

Acknowledgements

The authors are thankful to the Head, Department of Botany, University of Allahabad for providing laboratory facilities. One of us (Richa Tandon) thankfully acknowledges the financial support from DST, New Delhi.

References

Abbott, B. and Ballantine, D. (1957). The toxin from *Gymnodinium veneficum* Ballantine. *J. Mar. Biol. Assoc.,* U.K. 36:169-189.

Aguilar-Santos, G. and Doty, M. (1968). Chemical studies on three species of the marine algal genus *Caulerpa*, p. 173-176. In: *Drugs from the sea.* (Ed.) H. Freudenthal, Marine Technology Society, Washington, D.C.

Allison, A.C., Harrington, J.S. and Birbeck, M. (1966). An examination of the cytotoxic effects of silica on macrophages. *J.Exp. Med.*, 124 : 141-154.

Anderson, D.M. (1994). Red tides. *Scientific American August,*: 52-56.

Aschheim, L. and Raffel, S. (1972). The immunodepressant effect of carrageenan. *RES J. Reticuloendothel. Soc.*, 11:253-262.

Bates, S.S., Bird, C.J., Defrietas, A.S.W., Foxall, R., Gilgan, M., Hanic, L.A., Johnson, G.R., McCulloch, A.W., Odense, P. and Pocklington, R. (1989). Pennate diatom *Nitzschia pungens* as the primary source of domoic acid, a toxin in shellfish from eastern Prince Edward Island, Canada. *J. Fish Aquat. Sci.*, 46(7):1203-1215.

Bice, D., Schwartz, H., Lake, W. and Salvaggio, J. (1971). The effect of carrageenan on the establishment of delayed hypersensitivity. *Int. Arch. Allergy Appl. Immunol.*, 41:628-636.

Bltha, K., Koblicova, Z. and Trojanek, J. (1972). Absolute configuration of the iboga and voacanga alkaloids: *Chiroptical Approach Tetrahedron Lett.*, 27:2763-2766.

Bonfils, S. (1970). Carrageenan and the human gut. *Lancet* ii: 414.

Burkholder, J.M. and Glasgow, H.B. Jr. (1997). *Pfiesteria piscicida* and other *Pfiesteria*-like dinoflagellates: behavior, impacts, and environmental controls. *Limnol. Oceanogr.*, 42:1052-1075.

Cardellina, II J.H., Marner, F.-J. and Moore, R.E. (1979). Seaweed dermatitis: structure of lyngbyatoxin A. *Science*, 204: 193-195.

Carmichael, W.W. (1994). The toxins of cyanobacteria. *Scientific American (January)*: 65-72.

Carmichael, W.W., Beasley, V.R., Bunner, D.L, Eloff, J.N., Falconer, I., Gorham, P., Harada, K.-I., Krishnamurty, T., Yu M.-J., Moore, R.E., Rinehart, K., Runnegar, M., Skulberg, O.M. and Watanabe, M. (1988). Naming of cyclic heptapeptide toxins of cyanobacteria (blue green algae). *Toxicon*, 26: 971-973.

Carmichael, W.W., Biggs, D.F. and Gorham, P.R. (1975). Toxicology and pharmacological action of *Anabaena flos-aquae* toxin. *Science*, 197: 542-544.

Carmichael, W.W., Evans, W.R., Yin Q.Q., Bell, P. and Mocauklowski, E. (1995). Evidence for paralytic shellfish poisions in the mat forming cyanobacteria *Lyngbya wollei. VII International Conference on Toxic Phytoplankton*, 12-16 July 1995. Sendai, Japan.

Catanzaro, P., Schwartz, H. and Graham, R. (1971). Spectrum and possible mechanism of carrageenan cytotoxicity. *Am. J. Pathol.*, 64: 387-404.

Cook, W.O., Beasley, V.R., Lowell, R.A., Dahlem, A.M. and Carmichael, W.W. (1989). Consistent inhibition of peripheral cholinesterases by neurotoxins from the freshwater cyanobacterium *Anabaena flos-aquae*: studies of ducks, swine, mice and a steer. *Environ. Toxicol. Chem.*, 8: 915-922.

Devlin, J.P., Edwards, O.E., Gorham, P.R., Hunter, M.R., Pike, R.K. and Stavric, B. (1977). Anatoxin-a, a toxic alkaloid from *Anabaena flos-aquae* NCR-44h. *Can. J. Chem.*, 55: 1367-1371.

DiRosa, M. (1972). Biological properties of carrageenan. *J. Pharm. Pharmacol.*, 24:89-102.

Doig, M., Martin, D. and Padilla, G. (1973). Marine bioactive agents: chemical and cellular correlates, p. 1-37. In: *Marine Pharmacognosy.* (Eds.) D. Martin and G. Padilla Academic Press Inc., New York.

Doty, M. and Aguilar-Santos, G. (1966). Caulerpicin, a toxic constituent of *Caulerpa. Nature* (London), 211:990.

Doty, M. and Aguilar-Santos, G. (1970). Transfer of toxic algal substances in marine food chains. *Pac. Sci.*, 24:351-355.

Edwards, C., Beattie, K.A., Scrimgeour, C.M. and Codd, G.A. (1992). Identification of anatoxin-a in benthic cyanobacteria (blue-green algae) and in associated dog poisionings at Loch Insh, Scotland. *Toxicon,* 30:1165-1175.

Endean, R., Monks, S.A., Griffith, J.K. and Llewellyn, L.E. (1993). Apparent relationship between toxins elaborated by the cyanobacterium *Trichodesmium erythraeum* and those present in the flesh of the narrow-barred spanish mackerel *Scomberomorus commersoni. Toxicon,* 31: 1155-1165.

Engster, M. and Abraham, R. (1976). Cecal response to different molecular weights and types of carrageenan in the guinea pig. *Toxicol. Appl. Pharmacol.*, 38:265-282.

Entzeroth, M., Blackman, A.J., Mynderse, J.S. and Moore, R.E. (1985). Structures and Stereochemistries of oscillatoxin B, 31-noroscillatoxin B, oscillatoxin D, and 30-methyloscillatoxin D. *J. Organic Chem.*, 50: 1255-1259.

Eriksson, J.E. and Goldman, R.D. (1993). Protein phosphatase inhibitors alter cytoskeletal structure and cellular morphology. *Adv. Protein Phosphatases*, 7: 335-357.

Eriksson, J.E., Toivola, D., Meriluoto, J.A.O., Karaki. H., Han, Y.-G. and Hartshoorne, D. (1990). Hepatocyte deformation induced by cyanobacterial toxins reflects inhibition of protein phosphatases. *Biochem. Biophysic. Res. Commun.*, 173: 1347-1353.

Falconer, I.R. and Buckley, T.H. (1989). Tumour promotion by *Microcystis* sp., a blue-green alga occurring in water supplies. *Med. J. Austr.*, 150: 351.

Falconer, I.R. and Yeung, D.S.K. (1992). Cytoskeletal changes in hepatocytes induced by *Microcystis* toxins and their relation to hyperphosphorylation of cell proteins. *Chemic. Biol. Interactions,* 81: 181-196.

Fujiki, H., Suganuma, M., Suguri, H., Yoshizawa, S., Tagai, K., Uda, N., Wakamatsu, K., Yamada, K., Murata, M. and Yasumoto, T. (1988). Diarrhetic shellfish toxin, dinophysistoxin-1, is a potent tumor promotor on mouse skin. *Jpn. J. Cancer Res.*, 79:1089-1093.

Fujiki, H., Suganuma, M., Suguri, H., Yoshizawa, S., Takagi, k., Nakayasu, M., Ojika, M., Yamada, K., Yasumoto, T., Moore, R.E. and Sugimura, T. (1990). New tumor promoters from marine natural products. In: *Marine Toxins: Origin, Structure and Molecular Pharmacology*, Vol. 418 (Eds.) S. Hall and G. Strichartz, American Chemical Society, Washington, DC. 232-240.

Fusetani, N. and Hashimoto, Y. (1976). Hemolysins in a green alga *Ulva pertusa*, p. 325-332. In: *Animal, Plant and Microbial Toxins,* (Eds.) A. Ohsaka, K. Hayashi, and Y. Sawai, vol. 1. Plenum Publishing Corp., New York.

Fusetani, N., Ozawa, C. and Hashimoto, Y. (1976). Fatty acids as ichthyotoxic constituents of a green alga *Cheatomorpha minima. Bull. Jpn. Soc. Sci. Fisheries,* 42:941.

Gentile, J.H. and Maloney, T.E. (1969). Toxicity and environmental requirements of a strain of *Aphanizomenon flos-aquae* (L) Ralfs. *Can. J. Microbiol.,* 15: 165-173.

Ghazarossian, V., Schantz, E., Schnoes, H. and F. Strong. (1974). Identification of a poison in toxic scallops from a *Gonyaulax tamarense* red tide. *Biochem. Biophys. Res. Commvia,* 69:1219-1225.

Hahn, S.T. and Capra, M.F. (1992). The cyanobacterium *Oscillatoria erythraea*-a potential source of toxin in the ciguatera food-chain. *Food Addit. Contamin.,* 9:165-173.

Halevy, S. and Avivi, L. (1968). Isolation of hemolysins from *Ochromonas* spp., *Prymnesium parvum* and *Trypanosoma ranarum. J. Protozool.,* 15(Suppl.): 45.

Halstead, B. (1965). Poisonous and venomous marine animals, vol. 1. U.S. Government Printing Office, Washington, D.C. Ghazarossian, V., Schantz, E., Schnoes, H. and Strong, F. 1974. Identification of a poison in toxic scallops from a *Gonyaulax tamarense* red tide. *Biochem. Biophys. Res. Commun.,* 69:1219-1225.

Harada, K.-I., Ohtani, I., Iwamoto, K., Suzuki, M., Watanabe, M.F., Watanabe, M. and Terao, K. (1994). Isolation of Cylindrospermopsin from a cyanobacterium *Umezakia natans* and its screening method. *Toxicon,* 32: 73-84.

Hashimoto, Y. and Oshima, Y. (1972). Separation of gramiistins A, B and C from a soapfish *Pogonoperca punctata. Toxicon* 10: 279-284.

Hashimoto, Y., Okaichi, T., Dang, L. and Noguchi, T. (1968). Glenodinine, an ichthyotoxic substance produced by a dinoflagellate *Peridinium polonicum. Bull. Jpn. Soc. Sci. Fisheries,* 34:528-534.

Hashimoto,Y., Kamiya, H., Yamazato, K. and Nozawa, K. (1976). Occurrence of a toxic blue-green alga including skin dermatitis in Okinawa. In: *Animal, Plant, and Microbial Toxins,* Vol. 1 (Eds.) A. Ohsaka, K. Hayashi and Y. Sawai, Plenum Publishing, New York:333-338.

Hawkins, P.R., Runnegar, M.T.C., Jackson, A.R.B. and Falconer, I.R. (1985). Severe hepatotoxicity caused by tropical cyanobacterium (blue-green alga) *Cylindrospermopsis raciborskii* (Woloszynska) Seenaya and Subba Raju isolated from a domestic water supply reservoir. *Appl. Environ. Microbiol.,* 50: 1292-1295.

Honkanen, R.E, Codispoti, B.A., Tse, K. and Boynton, A.L. (1994). Characterization of natural toxins with inhibitory activity against serine/threonine protein phosphatase. *Toxicon,* 32: 339-350.

Honkanen, R.E, Zwillers, J., Moore, R.E., Daily, S.L., Khatra, B.S., Dukelow, M. and Boynton, A.L. (1990). Charcterization of microcystin-LR, a potent inhibitor of type 1 and type 2A protein phosphatases. *J. Biol. Chem.,* 265: 19401-19404.

Houck, J., Morris, R. and Lazaro, E. (1957). Anticoagulant, lipemia clearing and other effects of anionic polysaccharides extracted from seaweed. *Proc. Soc. Exp. Biol. Med.,* 96: 528-530.

Humpage, A.R., Rositano, J., Bretag, A.H., Brown, R., Baler, P.D., Nicholson, B.C. and Steffensen, D.A. (1994). Paralytic shellfish poisions from Australian cyanobacterial blooms. *Austr. J. Marine Freshwater Res.,* 45: 761-771.

Ikawa M., Wegener K., Foxall T.L. and Sasner J.J. JR. (1982). Comparizon of the toxins of the blue-green alga *Aphanizomenon flos-aquae* with the *Gonyaulax* toxins. *Toxicon,* 20: 747-752.

Ikawa, M. and Taylor, R. (1973). Choline and related substances in algae, p. 203-240. In: *Marine Pharmacognosy,*(Eds.) D. Martin and G. Padilla, Academic Press Inc., New York.

Imae, M. and Inoue, K. (1974). The mechanism of the action of *Prymnesium* toxin on membranes. *Biochem. Biophys. Acta,* 352:344-348.

Ingle, R. (1954). Irritant gases associated with red tide. *Univ. Miami Mar. Lab. Spec. Serv. Bull.,* 9:1-4.

Jackim. E. and Gentile, J. (1968). Toxins of a blue-green alga: similarity to saxitoxin. *Science,* 162:915-916.

Lawrence, J.E., Bauder, A.G., Quilliam, M.A. and Cembella, A.D. (1998). *Prorocentrum lima*: a putative link to diarrhetic shellfish poisoning in Nova Scotia, Canada. In: *Harmful Algae,* (Eds.) B.Reguera, J.Blanc, M.L. Fernandez, T. Wyatt and Santiago del Compostella: Xunta de Galacia and Intergovernmental Oceanographic Commission, 1998: 78-79.

Levin, R.E. (1992). Paralytic shellfish toxins: their origins, characteristics, and methods of detection: a review. *J Food Biochem.,* 15:405-417.

Lewis, R.J. and Holmes, M.J. (1993). Origin and transfer of toxins involved in ciguatera. *Comp. Biochem. Physiol.,* 106C:615-628.

MacKintosh, C., Beaattie, K.A., Klumpp, S., Cohen, P. and Codd, G.A. (1990). Cyanobacterial microcystin-LR is a potent and specific inhibitor of protein phosphatase 1 an 2A from both mammals and higher plants. *FEBS Letters,* 264: 187-192.

Mahmood, N.A. and Carmichael, W.W. (1986). Paralytic shellfish poisions produced by the freshwater cyanobacterium *Aphanizomenon flos-aquae* NCR-525-17. *Toxicon,* 25: 1221-1227.

Mahmood, N.A., Carmichael, W.W. and Pfahler, D. (1988). Anticholinesterase poisionings in dogs from a cyanobacterial (blue green algae) bloom dominated by *Anabaena flos-aquae. Am. J. Veterinary Res.,* 49: 500-503.

Maillet, M., Bonfils, S. and Lister, R. (1970). Carrageenan: effects in animals. *Lancet* ii: 414-415.

Marcus, R. and Watt, J. (1969). Seaweeds and ulcerative colitis in laboratory animals. *Lancet* ii: 489-490.

Matida, Y., Kimura, S., Yashimuta, C., Kumada, H. and Tokunaga, E. (1967). A toxic freshwater algae, *Glenodinium gymnodinium* Penard caused fish kills in artificially impounded Lake Sagami. *Bull. Freshwater Fish. Res. Lab.*, 17:73-77.

Matsushima, R., Yoshizawa, S., Watanabe, M.F, Harada, K.-I., Furusawa, M., Carmichael, W.W. and Fujiki, H. (1990). *In vitro* and *in vivo* effects of protein phosphatase inhibitors, microcystin and nodularin, on mouse skin and fibroblasts. *Biochem. Biophys. Res. Commun.*, 171: 867-874.

Moikeha, S.N. and Chu, G.W. (1971). Dermatitis-producing alga *Lyngbya majuscula* Gomont in Hawaii.II. Biological properties of the toxic factor. *J. Phycol.*, 7: 8-13.

Moore, R. (1977). Toxins from blue-green algae. *BioScience*, 27:797-802.

Moore, R.E., Ohtani, I., Moore, B.S., DE Koning, C.B., Yoshida, W.Y., Runnegar, M.T.C. and Carmichael, W.W. (1993). Cyanobacterial toxins. *Gazzetta Chimica Italiana*, 123: 329-336.

Mosher, H., Fuhrman, F., Buchwald, H. and Fisher, H. (1964). Tarichatoxin-tetrodotoxin: a potent neurotoxin. *Science*, 144:1100.

Murakami, S., Takemoto, T. and Shimizu, Y. (1953). Studies on the effective principles of *Diagenea simplex* Aq. I: Separation of the effective fraction by liquid chromatography. *J. Pharmacol. Soc. Jpn.*, 73:1026-1028.

Nishiwaki-Matsushima, R., Ohta, T., Nishiwaki, S., Suganuma, M., Kohyama, K., Ishikawa, T., Carmichael, W.W. and Fujiki, H. (1992). Liver tumour promotion by the cyanobacterial cyclic peptide toxin microcystin-LR. *J. Cancer Res. Clinical Onchol.*, 118: 420-424.

Ohta, T., Sueoka, E., Iida, N., Komori, A., Suganuma, M., Nishiwaki, R., Tatematsu, M., Kim, S.-J., Charmichael, W.W. and Fuziki, H. (1994). Nodularin, a potent inhibitor of protein phosphatases 1 and 2A, is a new environmental carcinogen in male F344 rat liver. *Cancer Res.*, 54: 6402-6406.

Ohtani, I., Moore, R.E. and Runnegar, M.T.C. (1992). Cylindrospermopsin: a potent hepatotoxin from the blue-green alga *Cylindrospermopsis raciborskii*. *J. Am. Chem. Soc.*, 114: 7941-7942.

Okaichi, T. and Nishio, S. (1976). Identification of ammonia as the toxic principle of red tide *Noctiluca miliaris*. *Bull. Plankton Soc. Jpn.*, 23:25-30.

Otterstrom, C. and Steemann-Nielsen, E. (1940). Two cases of extensive mortality in fishes caused by the flagellate *Prymnesium parvum Carter*. *Rep. Dan. Biol. Sta.*, 44:5. England.

Parnas, L. and Abbott, B. (1965). Physiological activity of the ichthyotoxin from *Prymnesium parvum*. *Toxicon*, 3:133-145.

Paster, Z. (1968). Prymnesin; the toxin of *Prymnesium parvum Carter*. *Rev. Int. Oceanogr. Med.*, 10:249-258.

Paster, Z. (1973). Pharmacognosy and mode of action of prymnesin, p. 241-263. In: *Marine pharmacognosy*, (Eds.) D. Martin and G. Padilla, Academic Press Inc., New York.

Patterson, G.M.L., Larsen, L.K. and Moore, R.E. (1994). Bioactive products from blue-green algae. *J. Appl. Phycol.*, 6: 151-157.

Reich, K. and Aschner, M. (1947). Mass development and control of the phytoflagellate *Prymnesium parvum* in fishponds in Palestine. Palest. *J. Bot. Jerusalem Ser.*, 4:14-23.

Rein, K., Lynn, B., Gawley, R. and Baden, D.G. (1994). Brevetoxin B: chemical modifications, synaptosome binding, toxicity, and an unexpected conformational effect. *J. Org. Chem.*, 59: 2107-2113.

Ruggieri, G. and Nigrelli, R. (1960). The effects of holothurin, a steroid saponin from the sea cucumber, on the development of the sea urchin. *Zoologica* (N.Y.), 45:1-16.

Runnegar, M.T., Kong, S. and Berndt, N. (1993). Protein phosphatase inhibition and in vivo hepatotoxity of microcystins. *Am. J. Physiol.*, 265: 224-230.

Schantz, E. (1971). The dinoflagellate poisons, p. 3-26. In: *Microbial toxins*, (Eds.) S. Kadis, A. Ciegler, and S. Ajl, vol. 7. Academic Press Inc., New York.

Scheuer, P. (1973). Chemistry of marine natural products. Academic Press Inc., New York. Wolke, R. and Trainor, F. 1971. Granulomatous enteritis in *Catostomas commersoni* associated with diatoms. J. *Wildl. Dis.*, 7:76-79.

Shilo, M. (1972). Toxigenic algae. *Prog. Incl. Microbiol.*, 11:235-265.

Shumway, S.E. (1990). A review of the effects of algal blooms on shellfish and aquaculture. *J. World Aquacul. Soc.*, 21:65-104.

Sims, J.K. and Zandee Van Rilland, R.D. (1981). Escharotic stomatitis caused by the "stinging seaweed" *Microcoleus lyngbyaceus* (formerly *Lyngbya majuscula*). Case report and literature review. *Hawaii Med. J.*, 40: 243-248.

Sivonen, K., Himberg, K., Luukkainen R., Niemela, S.I., Poon, G.K. and Codd, G.A. (1989). Preliminary characterization of neurotoxic blooms and strains from Finland. *Toxicity Assessment*, 4: 339-352.

Skilo, M. (1967). Formation and mode of action of algal toxins. *Bacteriol. Rev.*, 31:180-193.

Skulberg, O.M., Carmichael, W.W., Anderson, R.A., Matsunaga, S., Moore, R.E. and Skulberg R. (1992). Investigations of a neurotoxic oscillatorian strain (Cyanophyceae) and its toxin. Isolation and characterization of homoanatoxin-a. *Environ. Toxicol. Chem.*, 11: 321-329.

Spiegelstein, M., Reich, K. and Bergmann, F. (1969). The toxic principle of *Ochromonas* and related Chrysomonadina. (Translated from German.) *Verh. Intern. Ver. Limnol.*, 17:778.

Steidinger, K., Burklew, M. and Ingle, R. (1973). The effects of *Gymnodinium breve* toxin on estuarine animals, p. 179-202. In: *Marine Pharmacognosy*, (Eds.) D. Martin and G. Padilla, Academic Press Inc., New York.

Subba-Rao, D.V., Quilliam, M.A. and Pocklington, R. (1988). Domoic acid B a neurotoxic amino acid produced from the marine daitom *Nitzschia pungens* in culture. *Can. J. Fish Aquat. Sci.,* 45:2076-2079.

Terao, K., Ohmori, S., Igarashi, K., Ohtani, I., Watanabe, M.F., Harada, K.-I., Ito, E. and Watanabe, M. (1994). Electron microscopic studies on experimental poisioning in mice induced by cylindrospermopsin isolated from blue-green alga *Umezakia natans. Toxicon,* 32: 833-843.

Thomson, A. and Horne, C. (1976). Toxicity of various carrageenans in the mouse. *Br. J. Exp. Pathol.,* 57:455-459.

Ulitzur, S. and Shilo, M. (1970). Procedure for purification and separation of *Prymnesium parvum* toxins. *Biochem. Biophys. Acta,* 201:350-363.

Wolke, R., and Trainor, F. (1971). Granulomatous enteritis in *Catostomas commersoni* associated with diatoms. *J. Wildl. Dis.,* 7:76-79.

Woodcock, A. (1948). Note concerning human respiratory irritation associated with high concentrations of plankton and mass mortality of marine organisms. *J. Mar. Res.,* 7:56-62.

Yasumoto, T., Nakajima, I., Bagnis, R. and Adachi, R. (1977). Finding of a dinoflagellate as a likely culprit of ciguatera. *Bull. Jpn. Soc. Sci. Fish,* 43:1021-1026.

Yasumoto, T., Oshima, Y. and Yamaguchi, M. (1979). Occurrence of a new type of shellfish poisoning in Japan and chemical properties of the toxin. In: *Toxic Dinoflagellate Blooms,* (Eds.) D.Taylor and H.H. Seliger, Amsterdam: Elsevier, 1979: 495-502.

Yasumoto, T., Oshima, Y., Sugawara, W., Fukuyo, Y., Oguri, H., Igarishi, T. and Fujita, N. (1980). Identification of *Dinophysis fortii* as the causative organism of diarrhetic shellfish posioning. *Bull. Jpn. Soc. Sci. Fish,* 46:1405-141.

□□□

Microbial Diversity and Functions, 2012
© D.J. Bagyaraj, K.V.B.R. Tilak, H.K. Kehri (eds.), pp. 641-667
New India Publishing Agency, New Delhi (India)
E-mail : info@nipabooks.com; Website : www.nipabooks.com

Chapter **30**

AM Fungi : Importance, Nursery Inoculation and Performance after Out Planting

D.J. Bagyaraj and H.K. Kehri

ABSTRACT

The word "mycorrhiza" literally means "fungus root" to describe the mutualistic association between roots of higher plants and certain fungi. Arbuscular mycorrhizae (AM) are the most common and widely occurring of all the mycorrhizal associations and have great economic significance. They belong to the phylum Glomeromycota. Mycorrhizal associations are potential factors in determining the diversity in ecosystems. They can probably modify the structure and functioning of a plant community in a complex and unpredictable way. Currently there is considerable resistance against the use of chemical pesticides and fertilizers, because of their hazardous influence on the environment, and on soil, plant, animal, and human health. Hence, use of biofertilizers and biocontrol agents are recommended in practical agriculture/ horticulture/forestry.

Plants multiplied by seedlings such as vegetables, fruits, forest, and ornamental plants can be inoculated with AM fungi due to the possibility of using a small amount of inoculum and to the easy handling, disinfecting and adjustment of the fertility of the substrate. There are many reports on nutritional improvement after inoculation of AM fungi and on the increase in the growth of seedlings, accelerating their formation and reducing the time and cost of nursery. Seedlings thus raised will be colonized by the introduced fungus and then can be planted out in the field. In annuals, seedlings raised in nursery beds or containers supplied with selected AM fungi and transplanted to the field have produced economic growth responses. There are only a few reports on the post-transplant effect of AM fungi inoculated seedlings of perennial plants like plantation crops, fruit trees and trees important in forestry. The most promising field for use of seedlings inoculated with AM fungi is in those seedlings that are meant for reforestation of degraded areas

Recent studies have shown that inoculation with microbial consortia consisting of an efficient AM fungus together with a nitrogen fixer, P solubilizer and PGPR carefully screened and selected for a particular crop plant or forestry species is more beneficial than AM fungus alone. This simple nursery technology will not only help to produce healthy, vigorously growing seedlings in the nursery but will also reduce transplant shock and ensure better survival, growth and productivity when out planted in the field. Such studies in perennials are very few and should be pursued in future. This technology will not only improve plant growth and productivity but will also reduce the usage of fertilizers and pesticides thus minimizing environmental pollution.

Keywords: AM fungi, nursery inoculations, performance

Introduction

Frank (1985), the German botanist first coined the term "mycorrhiza" which literally means "fungus root" to describe the mutualistic association between roots of higher plants and certain fungi. These associations are grouped, based on morphological and anatomical characters, as ectomycorrhizae and endomycorrhizae. Ectomycorrhizae are common among temperate forest tree species and can be cultured on laboratory media. Endomycorrhizae, include arbutoid, monotropoid, ericoid, orichid and arbuscular mycorrhizal (AM) forms. Arbuscular mycorrhizae are the most common and widely occurring of all the mycorrhizal associations and have great economic significance. They cannot be cultured on laboratory media.

AM fungi are said to establish a mutualistic relationship with 80% of vascular plants (Gianinazzi and Gianinazzi pearson, 1986; Lakshmipathy *et al.*, 2007). Plants that rarely form AM fungal association include Caryophyllaceae, Brassicaceae, Chenopodiaceae and Cyperaceae (Hirrel *et al.*, 1978). In addition to their widespread distribution throughout the plant kingdom, arbuscular mycorrhizae are ubiquitous and occur in plants grown in arctic, temperate and tropical regions (Mosse, 1981). They have been reported to be associated with plants grown in sand dunes, coal mines (Khan, 1978) and aquatic environments. Blaszkowski (1994) observed variations in AM fungal diversity with the changes in plant species. Plants of a particular family are colonized by specific types of AM fungi and a few AM fungal genera were found only in the plants of a particular family.

AM fungi have the widest host range and distribution of all the mycorrhizal associations. It is estimated that about 90% of vascular plants normally establish mutualistic relationships with AM fungi. AM have been observed in 1000 genera of plants representing some 200 families. There are at least 300,000 receptive hosts in the world flora and there are about 220 species of AM fungi. If the hosts are divided up evenly among the fungi, with no overlap in host range, each fungus would have more than 1,360 potential partners. We know that the host range overlaps extensively, suggesting that some individual AM fungi may well have access to thousands of host.

Arbuscular mycorrhizal fungi: classification, importance and occurrence

Arbuscular mycorrhizal fungi are obligate symbionts and cannot be cultured on synthetic media. Their penetration takes place through root hairs or epidermal cells and then grows intercellularly or intracellularly in the root cortex, ultimately developing short haustoria like structures called arbuscules within the cortical cells. These arbuscules function as sites of nutrient exchange between the fungus and host roots. Vesicles are formed in the cortical cells, which are thin walled structures of various sizes and shapes and function as storage organs. The presence of vesicles and arbuscules is the criteria for identifying AM fungus in the roots.

AM fungi belong to the phylum Glomeromycota, which has a single class Glomeromycetes with 4 orders Glomerales, Diversisporales, Paraglomerales and Archaeosporales. There are 11 families, 17 genera and 228 species of AM fungi (Schüßler and Walker, 2010). The commonly occurring genera of AM fungi are *Glomus, Gigaspora, Scutellospora, Acaulospora* and *Entrophospora*.

Improved plant growth due to inoculation of soil with AM fungi has been demonstrated especially under P deficient conditions (Mosse, 1977). The growth improvement is mainly because of enhanced P uptake. AM fungi can also enhance tolerance or resistance to root pathogens (Borowicz, 2001) and abiotic stresses such as drought and metal toxicity (Meharg and Cairney, 2000). AM fungi play a role in the formation of stable soil aggregates, build up a macroporous structure of soil that allows penetration of water and air and prevents erosion (Miller and Jastrow, 1992). There is well documented evidence that AM fungi have important effects on plant P uptake. Greater soil exploration by mycorrhizal roots as a means of increasing phosphate uptake is well established. In phosphate deficient soils immobile phosphate ions develop a phosphate depletion zone around the roots. The hyphae spread beyond this zone and directly translocate nutrients from the soil to the root cortex (Hayman, 1983). Experiments with ^{32}P labeled phosphate indicate that AM fungal hyphae obtain their extra phosphate from the labile pool rather than by accessing insoluble phosphate by solubilizing it (Raj *et al.* 1981). Sparingly soluble rock phosphate is better utilized by the hyphae by closer physical contact with the ions dissociating at the particle surface.

The increased growth of plants inoculated with AM fungi is not only attributed to improved phosphate uptake but also to better availability of other elements like Zn, Cu, K, Al, Mn, Fe etc. AM fungi affect the levels of plant hormones. Allen *et al.* (1991) measured levels of plant hormones like cytokinins and gibberlin-like substances. AM fungi can tolerate a wide range of soil water regimes and also improve water relationships of many plants. It is still unclear whether the observed effects are directly due to the fungus itself or indirectly due to some alteration in host physiology as a result of improved P nutrition. Anatomical and other physiological studies have brought out that mycorrhizal plants have increased rates of respiration, photosynthesis and increased amounts of sugars, amino acids, RNA etc. and larger and/or more number of chloroplasts, mitochondria, xylem vessels, motor cells, etc. Changes in the root exudations and altered rhizosphere microorganisms (which also affect plant growth) may result because of colonization of roots by AM fungi (Caroline Machado and Bagyaraj, 1995).

Mycorrhizal colonization may also allow introduced populations of beneficial soil organisms like *Azotobacter. Azospirillum* and phosphate solubilizing bacteria to be maintained in high numbers than around non-mycorrhizal plants and to exert synergistic effects on plant growth. It is apparent from the investigations on AM fungi-plant pathogen interaction that AM fungi can usually (though not always) deter or reduce the severity of disease caused by soil-borne pathogens. All these studies bring out that AM fungi help the host plant in more than one way and that AM fungal inoculation helps plants growth.

AM fungi, in addition to their widespread distribution throughout the plant kingdom, are also geographically ubiquitous and occur in plants growing in arctic, temperate and tropical regions (Mosse, 1981). As an explanation for their remarkably widespread distribution, Trappe (1977) proposed that AM fungi were disseminated intercontinentally prior to the continental drift. The super continent Gondwanaland is thought to have begun to break apart and drift north about 125 million years ago. Fossil records of plants containing AM like structures have occurred as Trappe suggested. In general, AM fungal population is more in cultivated soil and their numbers decrease markedly below the top 15cm (Redhead, 1977). They are normally not found in depths beyond the normal root range of plants (Mosse, 1981). Although AM fungi are ubiquitous in soils, the patterns or distribution of individual species have not been fully understood. Studies on the distribution of species have either covered large geographical areas (Hall, 1977) or smaller regions (Abbott and Robson, 1977). The distribution of species of AM fungi varies with climatic and edaphic environment as well as with land use. For example, *Acaulospora laevis* is common in Western Australia (Abbott and Robson, 1977) but occurs less frequently in soils of eastern Australia. *Glomus* spp. appear to have the widest distribution. *Gigaspora* and *Sclerocystis* spp. are more common in tropical soils. *Acaulospora* seems to be better adapted to soils with pH < 5.0. Infact, certain AM fungi have been linked to particular kind of soil: *Glomus mosseae* with fine textured, fertile high pH soils; *Acaulospora laevis* with coarse textured, acid soils; and *Gigaspora* species with sand dune soils (Kendrick and Berch, 1985).

A fundamental problem in studies on the distribution of AM fungi lies in identifying the fungi. It relies mostly on spore morphology (Hall, 1977), which may change with spore age (Abbott and Robson, 1979). Description of many of the taxa has been based on single collections of uncertain age. This, coupled with the possibility of finding undescribed taxa, can make the identification of spores collected from field samples very difficult. Another difficulty in determining the distribution of AM fungi is that spores of all species are not equally easy to recover from soil. Soil often contains spores of more than one species of AM fungus (Abbott and Robson, 1981). Moreover, the same root may become colonized by more than one species of AM fungi (Mosse, 1977). Some AM fungi develop characteristic morphological features within plant roots (Abbott and Robson, 1979). This can allow quantitative estimates to be made on the total amounts of coarse or fine AM fungal hyphae within plant roots (Sparling and Tinker, 1978).

Ecosystems and AM fungi

Arbuscular mycorrhizal fungi form the main component of soil microbiota in most agro-ecosystems. Hence AM fungi have been shown to have a strong influence on plant species diversity. Since these fungi are obligate symbionts, their population and diversity are determined by the plant species present in the given ecosystem. Apart from plant species, human activities also affect these fungi. These modify the structure and functioning of plant communities in a complex and unpredictable way (Grime *et al.*, 1987). AM fungi, by forming an extended, intricate hyphal network, can absorb mineral nutrients from the soil and deliver them to their host plants in exchange for carbohydrates. Facilitated nutrient uptake, particularly with respect to immobile nutrients such as phosphorus, is believed to be the main benefit of this symbiosis for plants (George *et al.*, 1995). Apart from this, they can also enhance drought resistance, resistance against root pathogens and tolerance to heavy metal toxicity in plants (Borowcz, 2001). AM fungi also play a role in the formation of stable soil aggregates, building up of macroporous structures of soils that allow penetration of water and air, and prevent erosion (Miller and Jastrow, 1992). In a given ecosystem these fungi play an important role in carbon allocation, nutrient cycling and maintenance of diversified ecosystems (Doss and Bagyaraj, 2001). The presence of these fungi and their genetic and functional diversities are important for both plant community and ecosystem productivity (Bidartondo *et al.*, 2002).

The composition of the plant community may also affect mycorrhizal fungi causing differential reproduction and survival which will definitely act as a selective force on the composition of AM population in soil (Sanders and Fitter, 1992). It is important to develop AM fungal management strategies for sustainable low input but reasonably productive and ecologically sound agriculture (Muthukumar and Udaiyan, 2002). Hence AM fungi are very important for the functioning of terrestrial ecosystems, environmental conservation and sustainable agriculture.

Human activities like application of fertilizer, crop rotation and soil management may also alter the population and diversity of AM fungi (Giovannetti and Gianinazzi-Pearson, 1994). Modern intensive farming practices are evidently a threat to AM fungi, as indicated by studies on these fungi (Mader *et al.*, 2000). However, little is known about the effect of management practices on the species diversity of these fungi. Oehl *et al.* (2003) observed decrease in species richness of AM fungi with increased land use intensity in central Europe. In recent days, AM fungal population and diversity are declining because of agricultural/land use intensification.

Ecosystems and diversity of AM fungi

Glomalean fungi have following unique biological and ecological properties:

(i) It is an ancient obligate symbiosis that has existed from at least 250 million years ago that has been conserved concomitant with the evolution of land plants (Morton, 1990).

(ii) Most of the fungi appear to be obligately asexual, with only one unconfirmed exception reported (Tommerup and Sivasithamparam, 1990).

(iii) Sub-cellular morphological diversity of spores is unparalleled among the kingdom Fungi in organization, complexity and stability (Morton and Benny, 1990).

(iv) Organisms of a species are so widely dispersed that they can be found in all major continents (Morton and Bentivenga, 1994).

(v) Taxonomic groups do not correlate with ecological niches (Morton, 1990).

Mycorrhizal associations are potential factors in determining the diversity in ecosystems. They can probably modify the structure and functioning of a plant community in a complex and unpredictable way (Grime *et al.*, 1987). Any shift in the AM fungal population could have consequences for the composition of plant communities (survival, competition, floristic diversity) causing changes in the biology of natural ecosystems (Molina *et al.*, 1992).

The composition of plant communities may also affect that of fungal communities. Any factor (e.g. host-symbiont combinations) causing differential reproduction and survival of AM fungi (e.g. sporulation rates) will operate as a selective force on the composition of the soil population (Sanders and Fitter, 1992). Further, different abiotic factors like CO_2, pollutants may also provoke modifications in native plant communities thus modifying the mycorrhizal status of the ecosystems. Human activities like application of fertilizers, crop rotation and soil management may also alter the dynamics and diversity of the fungal community (Giovannetti and Gianinazzi-Pearson, 1994) and therefore, these mycorrhizal associations being dominant in agricultural/ horticultural cropping systems and tropical forestry have a lot of significance. Thus a knowledge of different factors influencing the population biology of AM fungi is essential in any attempt to use them in environmental conservation (Allen, 1991), biotechnology (Mulongoy *et al.*, 1992) or in sustainable agriculture/horticulture/forestry (Bethlenfalvay and Linderman, 1992).

Sustainability refers to productive performance of a system over time. It implies use of natural resources to meet the present needs without jeopardizing the future potential. The concept has an undefined time dimension. The magnitude of the time dimension depends on one's objectives, being shorter for economic factors and longer for concerns pertaining to environment, soil productivity, and land degradation. The shorter time dimension is generally less than a decade, while the longer time span may be up to five decades or more. The time dimension is also clearly addressed in the definition of sustainability adopted by the Technical Advisory Committee of the Consultative Group of International Agricultural Research: "Successful management of resources for agriculture to satisfy changing human needs while maintaining or enhancing the quality of the environment and conserving natural resources" (TAC, 1989). A sustainable system must be sustainable both ecologically and economically. From an agroecological perspective, sustainability may be defined as a measure of productivity over time per unit input of nonrenewable or a limiting resource.

Most soils of the tropics are of low inherent fertility. Crop yields are expectedly low, unless the nutrient status is enhanced by a regular or substantial addition of nutrients (fertilizers, manures, etc.). Soils of arid and semiarid regions are prone to production constraints imposed by physical and climatic processes, e.g., crusting, compaction, hard-setting, drought, erosion by wind and water, and high soil temperature. Productivity depends on input. Presently, agriculture in arid and semiarid tropics has low rates of energy input and low-to-moderate levels of traditional/subsistence farming. Much of the low-input or traditional agriculture of the developing world is not sustainable. The population pressure on the ecosystem in which these low-input systems are used results in severe land degradation. In view of the need to increase food, fibre and fuel production immediately and substantially and to ameliorate soil degradation, there is presently as enthusiastic interest in sustainable farming systems in the tropics. Some of the endeavors to achieve these in arid and semiarid ecosystems are conserving water, controlling erosion, use of improved cultivars and cropping systems, enhancing soil fertility, application of fertilizers as organic amendments, and increasing biological activity of soil flora and fauna. This approach reduces but does not eliminate the need for fertilizers and pesticides, and farm machinery and motorized equipment.

Currently there is considerable resistance against the use of chemical pesticides and fertilizers, because of their hazardous influence on the environment, and on soil, plant, animal, and human health. Hence, use of biofertilizers and biocontrol agents are recommended in practical agriculture/ horticulture/forestry. Nitrogen and phosphorus are two important plant nutrients. There are a large number of soil bacteria that are capable of fixing atmospheric nitrogen and making it available for plant growth. Bacteria like rhizobia, which live in association with roots of legumes, and the actinomycetes *Frankia,* which lives in association with casuarina, alder, and so forth, are examples of symbiotic nitrogen fixation. Bacteria like *Azotobacter* and *Azospirillum* and cyanobacteria like *Nostoc* and *Anabaena* are examples of organisms living freely in soil but capable of fixing atmospheric nitrogen. Phosphorus, which is the other major plant nutrient, is less mobile in soil solution. In most of the tropical soils, phosphorus is fixed and is in a form not readily available for plant growth. There are a number of fungi and bacteria that solubilize unavailable forms of phosphate and make it available for plant growth. There are certain fungi that form symbiotic association with the roots of plants and help in the uptake of phosphate from the labile pool.

Interest in AM fungi has reached a peak in recent years. The ability of these fungi to produce dramatic responses in plant growth is well documented. However, the application of this technology to commercial production of food, fibre or fuel has been minimal. One of the main reasons for this is the difficulty of inoculum production, the fungi being obligate symbionts (Jeffries, 1987). However this does not prevent using AM fungi in transplanted crops in which seedlings are raised in nursery beds or in polybags or root trainers and then planted in the field.

Need for AM fungal inoculation

Once the importance of AM symbiosis is recognized, a decision must be made as to whether the native population of AM fungi suffices as starting material or if there is a need to supplement the native species by external inoculation. If the native fungal population is adequate, then its efficacy in terms of plant growth improvement should be assessed. Generally spore counts are employed for assessing the native fungal population. This does not clearly indicate the population of AM fungi as sporulation depends on host and environmental factors. Alternatively, young fresh roots of native plants growing in the soil may be collected and stained for AM infection, but this is not always possible. Variations in extent of root colonization by a particular AM fungal species can depend on the host, fungal and environmental factors (Dodd and Jeffries, 1986) and an absence of infection at a particular time cannot indicate that the soil is devoid of AM fungi. A best indicator would be to grow the intended crop in the soil to which it would be transplanted and to evaluate the benefits for plant growth derived from the native AM fungi.

Some of these fungi are reported to cause no growth improvement in spite of maximum root colonization (Lovato *et al.*, 1992). Hence, inoculation with a fungus with high levels of colonization alone is not sufficient to harness maximum plant growth benefits. Further, AM fungi are known to have host preference, though not host specificity (Mosse, 1981). A fungal species and even isolates of the same species are known to vary markedly in the extent to which they colonize roots and in the benefits they confer to host plants (Rea and Tullio, 2005). This problem is further complicated by the fact that the host plant species (Plenchette *et al.*, 1983) and even cultivars of the same species (Graham and Syversten, 1985) respond differently to a specific fungus. This has led to selection of an efficient fungus for a particular host species (Reddy *et al.*, 1996). In all these cases it was observed that native AM fungal inoculum or native species were not efficient in enhancing the growth and nutrition of hosts to a great extent as compared to introduced AM inoculum. Hence, the need arises to screen and select an efficient fungus for each crop and to inoculate the substrate with such a fungus to get maximum benefits from this association.

Mycorrhizal dependency of plants

Plants differ greatly in their mycorrhizal dependence. Gerdemann (1975) defined Relative Mycorrhizal Dependency (RMD) as the degree to which a plant is dependent on the mycorrhizal condition to produce maximum growth or yield at a given level of soil fertility. The objective of this was to determine the extent of growth increase attributed to mycorrhizal condition. Plenchette *et al.* (1983) proposed a formula to calculate RMD of crop plants under field conditions by comparing plants in fumigated and unfumigated soils. This measures the extent of growth increase due to native endophytes and the calculated values are presented on a scale of 0 to 100%. There have been many reports that the introduction of mycorrhizal fungi can improve plant growth under unsterile conditions in spite of the presence of native endophytes (Jarstfer and Sylvia, 1992). Hence, Bagyaraj *et al.* (1988) proposed another formula which enables calculation of the Mycorrhizal Inoculation Effect (MIE) to assess the growth improvement brought about by inoculation with a mycorrhizal fungus in unsterile soil with indigenous AM fungi. MIE is very useful for the assessment of the extent to which introduced fungi compete with native endophytes to bring about a plant growth response. This information is of great value in practical agriculture, especially in developing countries, where farmers do not fumigate or sterilize either the nursery or the field.

Selection of efficient AM fungi

Where inoculation with selected AM fungi is required, it is essential to characterize the native AM fungal population first, as well as the soil characteristics, particularly soil P status which influences the efficacy of the fungus to a great extent. The selection process may then be carried out with host plants and inocula of various strains/species of AM fungi. Such experiments should be carried out under conditions which are nearest possible to the field conditions. Such pot experiments/micro plot experiments will be realistic and effective only when the exact host plant is used, as it is well-known that host plant species and even cultivars of the same species respond differently to a particular fungus. Further, the soil or substrate used must be the same or similar to the field soil conditions. This is also true with respect to the management practices like water, fertilizers and pesticide levels, as used in a sustainable agricultural system.

In the preliminary screening, generally where different species of AM fungi obtained from different sources or a culture collection are screened, performance of each species must be based on the plant growth and nutritional parameters apart from colonization levels. In the secondary screening, three or four fungal species which have performed well in the preliminary screening are compared with two to three predominant fungal isolates selected from the root zone soil of the host species under study. These fungi are again evaluated for their efficacy in terms of host plant growth and nutrition, alleviation of stress factors and ability to colonize and proliferate in the rhizosphere soil. This is essential because a fungus may heavily colonize the roots without considerably promoting plant growth. In such cases, where a plant species fails to significantly curtail the extent of root colonization by the mycorrhizal fungus, a depression in plant growth has been observed (Lovato *et al.*, 1992) and it has been attributed to the utilization of photosynthates by the fungus without compensation for better mineral nutrition of the host plant. The technique developed by Giovannetti and Mosse (1990) is most commonly used to evaluate the level of AM fungal colonization after staining the roots according to the procedure described by Kormanik *et al.* (1980). In this technique, arbuscules, the structures where exchanges between plant and fungus occur, cannot be evaluated. Trouvelot *et al.* (1986) developed another technique based on visual evaluation of the volume of the cortex occupied by AM fungi with a simple notation for richness in arbuscules. Similarly several other techniques, such as estimating total fungal biomass in the roots using non-vital staining (Koske and Gemma, 1989) and other staining techniques (Kormanik *et al.*, 1980), do not suffice as these procedures fail to differentiate between dead and alive fungal material. To combat these limitations, staining procedures based on physiological activities of fungi have been developed (Tisserant *et al.*, 1992). Hence, a combination of non-vital, vital and functional staining procedures can be reliable in an AM fungi selection program.

Appropriate technology for nursery raised crops

AM inoculum of suitably selected strains can be used for inoculation in the nursery bed. Growers only need to incorporate inoculum in the nursery beds or containers at the appropriate rate by hand. Seedlings thus raised will be colonized by the introduced fungus and then can be planted out in the field. In developed countries, seedlings are usually raised in fumigated soil or potting mix. Inoculation of such substrate with selected AM fungi for the crop/forest tree species is an easy and appropriate technology. Commercial AM inoculation by this method in the nursery production of citrus is applied in the USA (Palazzo *et al.*, 1994). Inoculation of the nursery with efficient AM fungi has been successful in the large-scale production of many horticultural plants in France (Gianinazzi *et al.*, 1990).

In many developing countries and sometimes in developed countries, seedlings are raised in unsterile soil or substrate. Inoculation with appropriate fungi in containers containing unsterilized potting mix was found to give vigorously growing seedlings of asparagus that performed well when planted out in the field (Powell *et al.*, 1985). The inoculation of unsterile nursery beds with efficient AM fungi enhances the growth of rootstocks of citrus and mango (Balakrishna Reddy and Bagyaraj, 1994). Further AM inoculation increased yield of chillies, tomato, capsicum and other vegetables (Bagyaraj, 2007). Increased growth of the seedlings of plantation crops like cashew was observed when inoculated with AM fungi (Lakshmipathy *et al.*, 2000). It was also observed that in AM inoculated plants there was an increased number of flowers, and the vase life of cut flowers of chrysanthemum and china asters was enhanced (Bagyaraj and Mallesh, 2000).

Use of AM fungi in rooting of cuttings, air layering and in overcoming transplant shock

Many horticultural plants are propagated through cuttings. Tropical root crops, such as cassava and sweet potato, are also propagated through cuttings. In such cases, rooting of the cuttings is very important. Enhanced rooting of cuttings through inoculation with AM fungi has been reported in apple (Plenchette *et al.*, 1981), cassava and sweet potato (Potty, 2004). More rooting in tamarind and cashew plants propagated through air layering was reported (Bagyaraj and Mallesh, 2000). Such plants not only withstood transplant shock but also established better when planted in field sites. AM-inoculated avocado plants withstood transplant shock better than uninoculated stock (Menge *et al.*, 1980).

Inoculum production

With AM fungi being obligate symbionts, there are many constraints in their large-scale commercial production and application. As attempts for the production of AM fungi in the artificial media have met with little or no success, the only method of production is in association with the host plant. Thus, at present the technique successful in mass production of AM inoculum is by pot culture. There are different types of AM inocula required for various purposes.

Different types of AM fungal inoculum have been developed such as spore inoculum, soil based inoculum, nutrient film technique, expended clay based inoculum, root based inoculum etc. These inocula have to be purchased by the farmer from an authentic commercial entrepreneur. But, sustainable/ organic farming always advocate cheaper sources of inputs and/or on-farm production of inputs. AM fungal inocula can be produced by the farmer himself. This method is described below.

On-farm production of AM inoculum by the farmers

The on-farm/on-site production of AM inoculum, close to the site of application using cheap local resources, can be the most appropriate method of mycorrhizal production and application at low cost (Barea and Jeffries, 1994). A simple method of producing AM inoculum on a farm level is described below:

Step 1: Land preparation

Clean the land of natural vegetation. Deep plough the soil and leave it for a month.

Step 2: Soil sterilization

Follow one of the methods of sterilizing soil

(a) Solarization

Spread polythene sheet over the wet soil;

(b) Fumigation

Inject fumigant (e.g., methyl bromide) in soil and cover with polyethylene sheet for four-five days;

(c) Heat sterilization

Burn agricultural waste material (e.g. straw) on the wet soil surface for three-four hours (Bagyaraj, 1992).

Step 3. Inoculation with AM fungus

Make 5-6 cm deep holes in the soil and place the starter culture (10-20g) of AM fungus in each hole. Sow the seeds of grass, sorghum, maize or any other suitable host.

Step 4. Harvest of inoculum

Cut the host plant after three-four months. Remove the soil along with the roots up to a depth of 15-20cm and use it as inoculum (after chopping roots into small bits) for inoculating the crop.

Application methods and time of application

It is important to standardize the dose and time of application, apart from the method of application, in order to harness the maximum benefit from the AM-host association. In a potential inoculum, every propagule is capable of colonizing the host root. However, in order to ensure a threshold level of colonization and quicken the process of colonization, a higher density of propagules should be present in the inoculum. Our experience has shown that about 150-200 infective propagules (determined by four-fold dilution) per g of substrate (sand: soil 1:1) will result in good colonization and establishment of the host. However, the number varies with other factors, such as their ability to tolerate soil (pH, nutrient status) and environmental factors (light intensity, temperature) and to proliferate. In horticultural/forest tree species, where planting is carried out at comparatively larger distances, high inoculum density is preferred, unlike in other agricultural crops where seedling density is high, hence roots have a greater probability of encountering the infective propagules.

Considering the time of application, it has been well demonstrated that earlier the inoculation, the greater the benefits to the host. Our experience reveals that inoculation at the time of sowing/planting will result in better colonization of the roots as they emerge, hence help in better establishment of the host. Experiments conducted with micropropagated banana and *Ficus benzamina* have revealed that the best time for inoculation would be just before planting out hardened plantlets (Shashikala *et al.*, 1999a, b; Ravolanirina *et al.*, 1989). However, Vidal *et al.* (1992) observed that inoculation of micropropagated avocado plantlets four weeks after acclimatization resulted in better establishment of these plantlets. Such varying views indicate that apart from identifying an efficient fungus for each host plant, there is a need to determine the optimum time of application for that particular host in order to maximize the benefits from AM fungi.

Methods of application of AM fungi generally include hand placement, placing below the seed material or in case of pot experiments, uniformly mixing the inoculum with the substrate. The importance of method of application of inoculum arises with the need to initiate colonization in the early stages of plant growth. Hence, from this view point, it is essential to place the inoculum close to the seed material such that the roots will come into contact with the fungal material as soon as they emerge. An experiment was conducted to standardize the method of inoculum application for fruit tree species raised in polybags. Polybags were filled with sand:soil mix in the ratio of 1:1. *Glomus mosseae* inoculum (15g) was inoculated into the polybags by different methods, like thoroughly mixing with the substrate, or placing as a layer at different depths, or placing as a band at different depths, or placing at a point at different depths. Papaya was the host used in this study. Based on the plant parameters, it was concluded that placing AM inoculum 8cm below the surface of the soil (7cm below the seed), at a point or as a layer, was the best method of inoculum placement for seedlings raised in polybags (Mamatha and Bagyaraj, 2000a). For seedlings raised on the nursery beds, placing the inoculum 2 cm below the seed was found to be the best approach (Mamatha and Bagyaraj, 2000b).

Production of seedlings inoculated with AM fungi

Plants multiplied by seedlings such as vegetables, fruits, forest, and ornamental plants can be inoculated with AM fungi due to the possibility of using a small amount of inoculum and to the easy handling, disinfecting and adjustment of the fertility of the substrate. The growth of the seedling in semi-controlled conditions promotes the establishment of inoculated AM fungi. Furthermore, the use of subsoil and organic material in the composition of the substrate, which have few or no AM fungal propagules, decrease the competition of the native AM fungi with the inoculated fungus. In many cases, there is also a disinfecting of the substrate to avoid pathogens and weeds killing the native AM fungi. This makes inoculation of the seedlings *via*ble and necessary when it is related to plants that present high mycorrhizal dependency or to seedlings that will be planted in areas that have few AM fungal propagules, such as those that are degraded or where there was a longtime culture of a non- AM species, such as *Pinus.*

There are many reports on nutritional improvement after inoculation of AM fungi and on the increase in the growth of seedlings, accelerating their formation and reducing the time and cost of nursery. There are, however, other less observed benefits promoted by the inoculation of seedlings. In the production of seedlings, both by conventional and cloning techniques, a difficult factor to control is the non-uniformity of plant growth. Even plants cloned *in vitro* show variation among the individuals in the formation of seedlings. This variability can be diminished with the inoculation of AM fungi. However, few works evaluate the uniformity of the seedlings, which should be considered to evaluate mycorrhizal seedlings, since it is an important characteristic that affects the production costs and the commercial value of the seedlings. Another benefit that seedling inoculation can bring is the increase in tolerance and even in protection against attacks by soil pathogens (Saggin Junior and da Silva, 2006). This happens due to the promotion of a better nutrition of plants with roots damaged by fungi or nematodes, and by direct or indirect antagonistic effects on the pathogens (Syl*via* and Chellemi, 2001).

Several technical aspects should be considered in the inoculation of seedlings with AM fungi, from the choice of the inoculum to the care that should be taken in the handling of the inoculated seedlings. Technology should be adapted to the local conditions of seedling production areas. The cost should be low to make the inoculation attractive to producers and nursery owners since, most times, the aim should not be the benefit of inoculation in seedling production, and one should search for the valuation of the produced seedling and the post-transplantation benefit.

Handling of mycorrhizal seedlings

When seedlings are mycorrhizal, besides the substrate and the fertilization, other techniques should be considered in the handling of seedlings, such as the use of pesticides, irrigation, and illumination. One of the most important factors in the handling of seedlings is the use of pesticides that can affect AM fungus, the formation, and the functioning of the symbiosis. The use of fumigating biocides, water vapour, or sunlight in the disinfestations of the substrate or of the seedbed, aiming at the elimination of phytopathogenic organisms, nematodes, and weeds, is common in the production of seedlings. These treatments greatly reduce the native AM fungal population, possibly resulting in blight and in symptoms of nutrient deficiency in plants with high mycorrhizal dependence, making it necessary to inoculate the seedlings.

The use of fungicides in seedlings and seeds is frequent, mainly to control soil -borne diseases and sometimes foliar diseases. Fungicides can severely affect AM fungi and the formation of mycorrhizae, or sometimes have a fungistatic effect and, surprisingly, can even foster the development of AM fungi (Trappe *et al.*, 1984). Among the fungicides that are very toxic to the AM fungi are Benomyl, Pentachloronitrobenzene, also called PCNB or Quintozene (Menge *et al.*, 1979) and Triadimefon (Bartschi, 1982). Benomyl is the fungicide whose action on most AM fungi has been studied, probably due to its great toxicity which affects all aspects of the development of the mycorrhizae. Since the use of Benomyl by nursery owners is common to control damping off of the plantlets, one should highlight the importance of not using this fungicide in seedlings that depend on mycorrhizae or in those inoculated with AM fungi. Fungicides such as Captan, Metalaxyl, Fosetyl-Al, and Thiram, have shown little or no depressive action on the AM fungi.

The use of herbicides in seedlings is less than other pesticides, because the control of weeds in the substrate is basically done through disinfestations and, when necessary, by manual pulling. However, the use of herbicides, such as Simazine and Trifluraline, in the formation of seedlings of certain species is common, such as in citrus. Grafting is done directly in the soil and only at the time of commercialization are the seedlings transferred to the containers. It is also common to use nonselective herbicides, such as Paraquat, to clean the nursery area. Trifluraline does not seem to affect AM fungi, while Simazine and Paraquat show a toxic effect on AM fungi. Therefore, due to the conflicting information available, one should avoid the use of herbicides in a nursery. The use of insecticides, acaricides and nematicides in seedlings is wide and their effects on AMF are less studied than other pesticides. Most tested insecticides and nematicides have little or no effect on the formation of spores and on root colonization. The specific interactions among active principles, AM fungal species, and plant species need to be better understood so that *via*ble control practices of pathogens in seedlings can be established without affecting mycorrhizal association.

Other factors that can affect AM fungi include irrigation and illumination, which are less often studied. The irrigation of mycorrhizal seedlings should be controlled to avoid anaerobic conditions, even if only for short periods of time, for the AM fungi demand high concentrations of O_2 for their development, physiological activity, and colonization. This care should be taken particularly in the acclimatization of micropropagated seedlings due to the plantlets' great need for humidity at this stage. This problem in acclimatization can be solved with the use of very porous substrates that promote fast drainage.

Illumination in the nursery indirectly affects the AM fungi through the effects on the photosynthetic rate of the host plant. The amount of photosynthates that move to the roots controls the infection of the roots by the AMF, through a mechanism not yet made clearly evident (Koide and Schreiner, 1992). Thus during the formation of seedlings its acclimatization to sunlight is necessary not only because of the adaptation to the conditions of the fields but also to benefit the establishment of the AMF, providing them with greater amount of photosynthates. In this aspect, the control of the seedling density by area and the selection of seedlings as to their size are also important practices in the handling of mycorrhized seedlings, for they influence illumination. The density of the seedlings should be particularly controlled if they are produced in containers, since they are determinant factors for the establishment of the density, the architecture of the seedling, the size of the leaves, and the rate of growth. The selection of seedlings in the nursery as to the uniformity of size, along with promoting perfect illumination, allows the elimination of plants with genetic problems and diseases.

Performance of mycorrhizal inoculation of seedlings in the nursery when planted in the field site

The application of AM fungi in the production of seedlings is promising as the beneficial effects have become evident in the initial stages of seedling production and transplantation to the fields, being able to increase productivity. Inoculated seedlings have superior growth and survival in the field over seedlings not inoculated with AM fungi. Many studies have shown that inoculation of nursery grown seedlings with selected mycorrhizal fungi enhances plant survival, growth, and establishment following out planting.

In annuals, seedlings raised in nursery beds or containers supplied with selected AM fungi and transplanted to the field have produced economic growth responses. This has been reported in agronomically important crops like chilli, finger millet, tobacco etc. (Bagyaraj, 2003) and horticultural crops such as tomato, capsicum, marigold etc. (Bagyaraj, 2007).

Further, these studies also brought out that application of phosphatic fertilizer can be reduced by nearly 50% (Table 1). In medicinal and aromatic plants AM fungal inoculation in the nursery not only increased the crop yield in the field but also that of the active ingredient (Arpana *et al.*, 2010). In *Solanum quitoense* inoculated seedlings established and grew vigorously where as uninoculated seedlings remained very small (Collazos *et al.*, 1986).

Table 1: Effect of different AM fungi and added phosphorus on fruit yield of chilli

Inoculation	Yield of chilli (kg/plot) Addition of P fertilizer		
	No Addition	Half the recommended level	Recommended level (75 kg/ha)
Uninoculated	0.27	0.37	0.43
Glomus fasciculatum	0.40	0.52	-
G. albidum	0.38	0.42	-
G. macrocarpum	0.32	0.40	-
G. caledonicum	0.37	0.41	-

There are only a few reports on the post-transplant effect of AM fungi inoculated seedlings of perennial plants like plantation crops, fruit trees and trees important in forestry. Coffee seedlings inoculated with AM fungi in the nursery exhibited higher survival rate and grew more vigorously in the field compared to uninoculated seedlings (Sieverding, 1991). The author felt that it may not only be due to improved plant nutrition but can be because of other positive effects like enhanced water uptake. In cashew we observed that seedlings raised with AM fungi in the nursery, exhibited higher level of grafting success and higher survival rate when planted in the field compared to seedlings which received no AM fungi in the nursery (Lakshmipathy *et al.*, 2004).

The most promising field for use of seedlings inoculated with AM fungi is in those seedlings that are meant for reforestation of degraded areas (Souza and Silva, 1996). In this case, the AM fungal inoculation is essential for the seedling to survive in the stressful environment and for reestablishment of the nutrient recycling process. The AM fungi enable leguminous plants in degraded soil to be better nourished so they can be able to fix atmospheric N_2, which is the main source of entrance of N in the system, increasing the accumulation of biomass and organic waste (Bethlenfalvay and Linderman, 1992). Besides that, the AM fungi are important so that the process of plant succession in the area to be replanted be activated, for they allow the entrance of plant species that are very dependent on this symbiosis and modify the competitive relations among plant species (Flores-Aylas *et al.*, 2003).

Inoculation of AM fungi along with other beneficial soil microorganisms

There are several reports on the interaction between AM fungi and other beneficial soil microorganisms like nitrogen fixers, phosphate solubilizers and PGPRs. These studies suggest that the interaction is synergistic with consequential benefit on plant growth. Recent studies have shown that inoculation with microbial consortia consisting of an efficient AM fungus together with a nitrogen fixer, P solubilizer and PGPR carefully screened and selected for a particular crop plant or forestry species is more beneficial than AM fungus alone in improving the growth, biomass and yield. The fundamental means of assessing the effectiveness of inoculating the nursery seedlings are their survival and growth in the field. Such studies have been done with ectomycorrhizal fungi mostly in the USA and Canada. The success of ectomycorrhizal fungal inoculation in improving seedling growth and survival after out planting is well recognized. However such studies with AM fungi alone or as microbial consortia are megre.

In the case of leguminous plant seedlings, joint inoculation of rhizobia and AM fungi is largely used for reforestation of degraded areas (Souza and Silva, 1996), and it can be used in fruit plants of Fabaceae, such as *Tamarindus indica* and *Inga edulis*. Introduction of other beneficial organisms, such as the rhizobacteria that promote plant growth (Germida and Walley, 1996; Lioussanne *et al.*, 2009) and are antagonistic to pathogens (Camprubi *et al.*, 1995), with the effect of fostering greater growth of plants and biological control, must be tried. In addition, phosphate - solubilizing microbes can be inoculated to increase the nutritional effect of AM fungi (Azcon-Aguilar and Barea, 1996).

In a recent study forest tree seedlings (*Acacia auriculiformis* and *Tectona grandis*) inoculated in the nursery with microbial consortia consisting of selected AM fungi and PGPRs were planted in degraded forests. Inoculated plants survived and grew much better compared to uninoculated plants 52 months after planting (Table 2).

Table 2: Response of *Acacia auriculiformis* and *Tectona grandis* to microbial inoculation in degraded forests at three locations 52 months after planting

Location	**Biovolume index [Plant height (cm) x Stem girth (mm)]**	
	Acacia auriculiformis	*Tectona grandis*
Mandya		
Uninoculated	10,390	746
Inoculated	20,479	3,556
Srirangapatna		
Uninoculated	11,661	944
Inoculated	17,889	1,818
Pandavapura		
Uninoculated	7,081	216
Inoculated	11,433	587

Conclusion

Many horticultural plants, e.g. tomato, onion and capsicum, some agronomic crops, e.g. tobacco and most of the tropical fruit trees and forest trees are first established in seedbeds or maintained during early development in nurseries before transplanting to the field. Ornamentals which are often important export products of tropical countries are raised in nursery containers.

The special circumstances of the pre-establishment of plants under "controlled" conditions allow a wide range of methods of incorporating the AM fungi inoculation technology into these production systems. Often, the seedbeds or potting substrates are disinfected (either with steam or with general biocides) to eradicate or to prevent soil-borne plant diseases and pests; furthermore, general biocides are often applied to avoid weed problems in nursery soil substrates. As said earlier, general biocides greatly reduce propagule density of indigenous AM fungi. Subsoils which have low pressure of soil-borne pathogens are also often used as potting material in nurseries; these soils contain little in the way of AM fungal populations. Some artificial potting media, like vermiculite, perlite or composted organic materials, do not have any AM fungi. There are many reports in literature that selected AM fungi (within all sources of inocula) have been successfully introduced to seedbeds and nursery potting materials.

The inoculation of seedlings with AM fungi is considered a safe and eco-friendly biotechnology, since these fungi occur naturally in the soils of most land ecosystems. Moreover, it is a low-cost technology, which makes its application very *via*ble. Further, it is essential to develop microbial consortia consisting of suitable AM fungi + PGPRs to inoculate a crop/forest plant in the nursery. This simple nursery technology will not only help to produce healthy, vigorously growing seedlings in the nursery but will also reduce transplant shock and ensure better survival, growth and productivity when out planted in the field. Finally, there is still a gap in the research and development framework, and technology transfer in this area. It is likely that inoculation with microbial consortia consisting of effective AM fungi + PGPRs will become an integral part of the nursery technology in the future. This simple technology will not only improve plant growth and productivity but will also reduce the usage of fertilizers and pesticides thus minimizing environmental pollution.

References

Abbott, L.K. and Robson, A.D (1977). The distribution and abundance of vesicular endophytes in some western Australian soils. *Aust. J. Bot.*, 25: 515-522.

Abbott, L.K. and Robson, A.D (1979). A quantitative study of the spores and anatomy of mycorrhizas formed of *Glomus,* with reference to its taxonomy. *Aust. J. Bot.,* 27: 363-375.

Abbott, L.K. and Robson, A.D (1981). Infectivity and effectiveness of five endomycorrhizal fungi: Competition with indigenous fungi in field soils. *Aust. J. Bot.,* 32: 621-630.

Allen, M.F. (1991). *The Ecology of Mycorrhizae.* Cambridge University Press, Cambridge, U.K.

Arpana, J., Bagyaraj, D.J., Prakasa Rao, E.V.S., Parameswaran, T.N. and Rahiman, B.A (2010). Evaluation of growth, nutrition and essential oil content with microbial consortia. *J. Soil Biol. Ecol.,* 30: 56-71.

Azcon-Aguilar, C and Barea, J.M. (1996). Arbuscular mycorrhizas and biological control of soil-borne plant pathogens - An overview of the mechanisms involved. *Mycorrhiza,* 6: 457-464.

Bagyaraj, D.J. (1992). Vesicular arbuscular mycorrhiza- Application in agriculture. In: *Methods in Microbiology* Vol. 24. (Eds.) J. R. Norris, D. J. Read and A. K. Varma, Academic Press, London: 359-374.

Bagyaraj, D.J. (2003). Mycorrhizal biotechnology for improved crop productivity. In: *Biotechnological Stratagies in Agro-Processing.* (Eds.) S.S. Marwaha and J.K. Arora, Asiatech. Pub. Inc., New Delhi: 80-84.

Bagyaraj, D.J. (2007). Arbuscular mycorrhizal fungi and their role in horticulture. In: *Recent Trends in Horticultural Biotechnology* (Eds.) R. Keshva Chandran, P.A. Nazeem, D. Girija, P.S. John and K.V. Peter, New India Pub. Agency, New Delhi: 53-58.

Bagyaraj, D.J. and Mallesh, B.C. (2000). Potentials for the use of VA mycorrhizain horticulture. In: *Glimpse in Plant Sciences.* (Eds.) K.R. Aneja, M.D Charaya, A. Aggarwal, D.K. Hans and S.A. Khan, Pragathi Prakashan, Meerut, India: 37-41.

Bagyaraj, D.J., Manjunath, A and Govinda Roa, Y.S (1988). Mycorrhizal inoculation effect on marigold, eggplant and citrus in an Indian soil. *J. Soil Biol. Ecol.,* 8: 98-103.

Balakrishna Reddy and Bagyaraj, D.J. (1994). Selection of efficient vesicular arbuscular mycorrhizal fungi for inoculating the mango root stock cultivar 'Nekkare'. *Scientia Hort.,* 59: 39-73.

Barea, J.M. and Jeffries, P. (1994). Arbuscular mycorrhizas in sustainable soil plant systems. In: *Mycorrhiza- Structure, Function, Molecular Biology and Biotechnology* (Eds.) A. Varma and B. Hock, Springer Verlag, Berlin, Germany : 521-560.

Bartschi H. (1982) Influence de quelques pesticides sur l'infection endomycorhizienne. In: *Mycorrhizae, an Integral Part of Plants: Biology and Perspectives for Their Use.* (Eds.) S. Gianinazzi, V. Gianinazzi-Pearson and A.Trouvelot, Paris: INRA: 259-266.

Bethlenfalvay, G.J. and Lindermann, R.G. (1992) *Mycorrhizae in Sustainable Agriculture.* ASA special Publication, Madison, WI, U.S.A.

Bidartando, M.L., Redecker, D., Hijri, L.,Wiemken, A., Bruns., T.D., Dominguez, L., Sersic, A., Leake, J.R. and Read, D.J. (2002). Epiparasitic plants specialized on arbuscular mycorrhizal fungi. *Nature,* 419: 389-392.

Blaszkowski, J. (1994). Comparative studies on the occurrence of the arbuscular fungi and mycorrhizae (Glomales) in cultivated and uncultivated soils of Poland. *Acta Mycol.*, 28: 93-140.

Borowicz, V. A. (2001). Do arbuscular mycorrhizal fungi alter plant-pathogen relations? *Ecology,* 82: 3057-3068.

Camprubi, A., Calvet, C. and Estaun, V. (1995). Growth enhancement of *Citrus reshni* after inoculation with *Glomus intraradices* and *Trichoderma aureoviride* and associated effects on microbial populations and enzyme activity in potting mixes. *Plant Soil,* 173: 233-238.

Caroline Machado and Bagyaraj, D.J. (1995). Mycorrhization bacteria and its influence on growth of cowpea. In: *Mycorrhizae, Biofertilizers for the Future.* (Eds.) A. Adholeya and Sujan Singh, Tata Energy Research Institute Pub. New Delhi: 192-196.

Collazos, S.H.H., Figueroa, B.P., Jordan, S.A., Patino, C.H. and Sieverding, E. (1986). Estudio preliminar sorbe micorriza vesiculo-arbuscular (MVA) en lulo (*Solanum quitoense* Lam.). *Acta Agron.*, 36: 34-36.

Dodd, J.C. and Jeffries, P. (1986). Early development of vesicular arbuscular mycorrhizas in autumn-sown cereals. *Soil. Biol. Biochem.,* 18: 149-154.

Doss, D.D. and Bagyaraj, D.J. (2001). Ecosystem dynamics of mycorrhizae. In: *Innovative Approaches of Microbiology* (Eds.) D.K. Maheshwari and R.C. Dubey, Bishen Singh and Mahendra Pal Singh Dehradun, India: 115-129.

Flores-Aylas W.W., Saggin-Jinior O.J., Siqueira J.O. and Davide, A.C. (2003). Efeito de *Glomus etunicatum* e fosforo no crescimento de species arboreas em semeadura direta. *Pesquisa Agropecuaria Brasileira,* 38: 257-266.

Frank, A.B (1985) Uber die out Wurzetsymbiose berohende ernahausing gewisser Baume durch utnerimdinche. *Berdent Bot. Gessel.,* 3: 128-145.

George, E., Marshner, H. and Jakobsen, I. (1995). Role of arbuscular mycorrhizal fungi in uptake of phosphorus and nitrogen from soil. *Crit. Rev. Biotechnol.,* 15: 253-270.

Gerdemann, J.W. (1975). Vesicular arbuscular mycorrhizae. In: *The Development and Function of Roots* (Eds.) Torrey, J.C and Clarkson D.T., Academic Press, New York, USA: 575-591.

Germida, J.J and Walley, F.L. (1996). Plant growth- promoting rhizobacteria alter rooting patterns and arbuscular mycorrhizal fungi colonization of field-grown spring wheat. *Biol. Fertil. Soils,* 1: 97-102.

Gianinazzi, S. and Gianinazzi-Pearson, V. (1986). Progress and headaches in endomycorrhiza biotechnology. *Symbiosis,* 2: 139-149.

Gianinazzi, S., Gianinazzi-Pearson, V. and Trouvelot, A. (1990), Potentialities and procedures for the use of endomycorrhizas with emphasis on high value crops. In: *Biotechnology of Fungi for Improving Plant Growth.* (Eds.) J.M. Whips and B. Lumsden, Cambridge University Press, Cambridge, U.K: 41-54.

Giovannetti, M and Mosse, B. (1990). An elevation of techniques for measuring vesicular arbuscular mycorrhizal infection in roots. *New Phytol.*, 84: 489-500.

Giovannetti, M. and Gianinnazzi-Pearson, V. (1994). Biodiversity in arbuscular mycorrhizal fungi. *Mycol. Res.*, 98: 705-715.

Graham, J.H. and Syversten, J.P. (1985). Host determinants of mycorrhizal dependency of citrus root stock seedlings. *New Phytol.*, 101: 667-676.

Grime, J. P., Mackey, J. M. L., Hiller, S. H. and Read, D. J. (1987). Floristic diversity in a model system using experimental microcosms. *Nature*, 328: 420-422.

Hall, I.R.S. (1977). Species and mycorrhizal infections of New Zealand Endogonaceae. *Trans. Br. Mycol. Soc.*, 68: 341-356.

Hayman, D.S. (1983). The physiology of vesicular arbuscular endomycorrhizal symbiosis. *Can. J. Bot.*, 61: 944-963.

Hirrel, M.C., Mehravaran, H. and Gerdemann, J.W. (1978). Vesicular arbuscular mycorrhizae in Chenopodiaceae and Cruciferaceae: do they occur? *Can. J. Bot.*, 56: 2813-2817.

Jarstfer, A.G. and Syl*via*, D.M. (1992). Inoculum production and inoculation strategies for vesicular arbuscular mycorrhizal fungi. In: *Soil Microbial Technologies: Application in Agriculture, Forestry and Environmental Management* (Ed.) B. Metting, Marcel Decker, New York, U.S.A: 349-377.

Jeffries, P. (1987). Use of mycorrhiza in agriculture . *CRC. Crit. Rev. Biotechnol.*, 5: 319-357.

Kendrick, B. and Berch, S. (1985). Mycorrhizae: Applications in agriculture and forestry. In: *Comprehensive Biotechnology*, Vol. 4 (Ed.) C.W. Robinson, Pergamon Press, Oxford: 109-150.

Khan, A.G. (1978). Vesicular arbuscular mycorrhizas in plants colonizing black wastes from bituminous coal mining in the Illawarra region of New South Wales. *New Phytol.*, 81: 53-63.

Koide, R.T and Schreiner, R.P. (1992). Regulation of the vesicular - arbuscular mycorrhizal symbiosis. *New Phytol.*, 27: 171-216.

Kormanik, P.P., Bryan, W.C. and Schultz. R.C. (1980). Procedure and equipment for staining large number of plant root samples for endomycorrhizal assay. *Cand. J. Microbiol.*, 26: 536-538.

Koske, R.E. and Gemma, J.N. (1989). A modified procedure for staining roots to detect VA mycorrhiza. *Mycol. Res.*, 92: 486-505.

Lakshmipathy, R., Bagyaraj, D.J. and Balakrishna, A.N. (2007). Can agricultural practices and land use patterns affect arbuscular mycorrhizal fungal population and diversity? In: *Fungi: Multifaceted Microbes*. (Eds.) B.N. Ganguli and S.K. Deshmukh, Anamaya Pub., New Delhi: 304-316.

Lakshmipathy, R., Balakrishna, A.N., Bagyaraj, D.J. and Kumar, D.P. (2000). Symbiotic response of cashew root stocks to different VA mycorrhizal fungi. *The Cashew,* 14: 20-24.

Lakshmipathy, R., Balakrishna, A.N., Bagyaraj, D.J., Sumana, D.A. and Kumar, D.P. (2004). Evaluation, grafting success and field establishment of cashew rootstock as influenced by VAM fungi. *Indian J. Exp. Biol.,* 42: 1132-1135.

Lioussanne, L., Beauregard, M., Hamel, C., Jolicoeur, M. and St-Arnaud, M. (2009). Interactions between arbuscular mycorrhizal fungi and soil microorganisms. In: *Advances in Mycorrhizal Science and Technology.* (Eds.) D. Khasa, Y. Piche and A.P Coughlan, NRC Research Press, Ottawa, Canada: 51-70.

Lovato, P., Guillemin, J.P. and Gianinazzi, S. (1992). Application of commercial arbuscular endomycorrhizal fungi inoculants to the establishment of micropropagated grapevine rootstock and pineapple plants. *Agron. J.,* 12: 873-880.

Mader, P., Edenhofer, S., Boller, T., Wiemken, A. and Niggli, U. (2000). Arbuscular mycorrhizae in a long term field trial comparing low input (organic biological) and high input (conventional) farming systems in a crop rotation. *Biol. Fertil. Soils,* 31: 150-156.

Mamatha, G. and Bagyaraj, D.J. (2000a). Effect of different methods of VAM inoculum application on plant growth and nutrient uptake of papaya raised in polybags. *J. Soil. Biol. Ecol.,* 20: 51-59.

Mamatha, G. and Bagyaraj, D.J. (2000b). Effect of different methods of VAM inoculum application on plant growth and nutrient uptake of tomato seedlings grown in raised nursery beds. *J. Soil. Biol. Ecol.,* 20: 71-77.

Meharg, A.A. and Cairney, J.W.G. (2000). Co-evaluation of mycorrhizal symbionts and their hosts to metal-contaminated environments. *Adv. Ecol. Res.,* 30: 69-112.

Menge, J.A., Johnson, E.L.V. and Minassian, V. (1979). Effect of heat treatment and three pesticides upon the growth and production of the mycorrhizal fungus *Glomus fasciculatum. New Phytol.,* 82: 473-480.

Menge, J.A., Larue, J., Labanauskas, C.K. and Johnson, E. (1980). The effect of two mycorrhizal fungi upon growth and nutrition of avocado seedlings grown with six fertilizer treatments. *J. Amer. Soc. Hort. Sci.,* 105: 400-404.

Miller, R.M. and Jastrow, J.D. (1992). The application of VA mycorrhizae to ecosystem restoration. In: *Mycorrhizal Functioning* (Ed.) M.F. Allen, Chapman and Hall Ltd., London, England.

Molina, R., Masscotte, H. and Trappe, J.M. (1992). Specificity phenomena in mycorrhizal symbioses: community - ecological consequences and potential implications. In: *Mycorrhizal Functioning* (Ed.) M.F. Allen, Chapman and Hall Ltd., New York, U.S.A.: 357-372.

Morton, J.B (1990). Species and clones of arbuscular mycorrhizal fungi (Glomales, Zygomycetes) and their role in macro and micro evolutionary process. *Mycotaxon,* 37: 493-515.

Morton, J.B and Benny, G.L. (1990). Revised classification of arbuscular mycorrhizal fungi (zygomycetes): a new order, Glomales, two new suborders, Glominaeae and Gigasporineae and two new families, Aculosporaceae and Gigasporaceae, with an emendation of Glomaceae, *Mycotaxon*, 37: 471-491.

Morton, J.B. and Bentivenga, S.P. (1994). Levels of diversity in endomycorrhizal fungi (Glomales, Zygomycetes) and their role in defining taxonomic and non-taxonomic groups. *Plant Soil*, 159: 47-59.

Mosse, B. (1977). Plant growth responses of vesicular-arbuscular mycorrhiza-IV. Soil given additional phosphorus. *New Phytol.*, 72: 127-136.

Mosse, B. (1981). *Vesicular Arbuscular Mycorrhizal Research for Tropical Agriculture*. Honolulu University of Haiwaii Press, HI, U.S.A.: 1-54.

Mulongoy, K., Gianinazzi, S., Roger, P.A and Dommergues, Y. (1992). Biofertilizers: Agronomic and environmental impacts and economics. In: *Biotechnology, Economic and Social Aspects; Issues for Developing Countries.* (Eds.) E.J De Silva, C. Rutledger and A. Sasson, UNESCO, Cambridge University Press, Cambridge, U.K.: 55-69.

Muthukumar, T. and Udaiyan, K. (2002). Growth and yield of cowpea as influenced by changes in arbuscular mycorrhiza in response to organic manuring. *J. Agron. Crop Sci.*, 188: 123-132.

Oehl, F., Sieverding, E., Ineichen, L., Mader, P., Boller, T. and Wiemken, A. (2003). Impact of land use intensity on the species diversity of arbuscular mycorrhizal fungi in agroecosystems of central Europe. *Appl. Environ. Microbiol.*, 69: 2816-2824.

Palazzo, D., Pommerening, B. and Vanadia, S. (1994). Effects of soil sterilization and vesicular arbuscular mycorrhiza on growth of sour orange (*Citrus aurantium* L.) seedlings. *Proc. Int. Soc. Citriculture.*, 2: 621-623.

Plenchette, C., Furlan, V. and Fortin, J.A. (1981). Growth stimulation of apple trees in unsterilized soil under field conditions with mycorrhizal inoculation. *Can. J. Bot.* 59: 2003-2008.

Plenchette, C., Fortin, J.A. and Furlan, V. (1983). Growth responses of several plant species to mycorrhiza in a soil of moderate P fertility. *Plant Soil.* 70: 199-209.

Potty, V.P. (2004). Arbuscular mycorrhizal fungi - a biofertilizer. Abstract of National Seminar on Horticulture Based Sustainable Income and Environmental Protection, 24-26 February 2004, Kohima, Nagaland, India: 12.

Powell, C.L., Bagyaraj, D.J., Clark, G.E and Cladwell, K.I. (1985). Inoculation with arbuscular mycorrhizal fungi in greenhouse production of asparagus seedlings. *N.Z. J. Agric. Res.* 28: 293-298.

Raj, J., Bagyaraj, D.J and Manjunath, A. (1981). Influence of soil inoculation with arbuscular mycorrhiza and a phosphate dissolving bacterium on plant growth and ^{32}P- uptake. *Soil Biol. Biochem.* 13: 105-108.

Ravolanirina, F., Gianinazzi, S., Trouvlet, A. and Carre, M. (1989). Production of endomycorrhizal explants of micropropagated grapevine rootstocks. *Agric. Ecosyst. Environ.* 29: 323-327.

Rea, E and Tullio, M. (2005). The agro-industrial production valorization: The possible role of arbuscular mycorrhizal fungi. In: *Microbial Biotechnology in Agriculture and Aquaculture* (Ed.) R.C. Ray, Science Publishers, Inc., Enfield, NH, USA. pp. 101-124.

Reddy, B., Bagyaraj, D.J. and Mallesha, B.C. (1996). Selection of efficient VA mycorrhizal fungi for acid lime. *Ind. J. Microbiol.* 36: 13-16.

Redhead, J.F. (1977). Endotrophic mycorrhizas in Nigeria: Species of the Endogonaceae and their distribution. *Trans. Br. Mycol. Soc.* 69: 275-280.

Saggin Junior, O.S. and da Silva, E.M.R. (2006). Production of seedlings inoculated with arbuscular mycorrhizal fungi and their performance after outplanting. In: *Handbook of Microbial Biofertilizers* (Ed. Rai, M.K.) International Book Distributing Co., Lucknow, India:. 354-386.

Sanders, I.R. and Fitters, A.H. (1992). Evidence for differential responses between host-fungus combinations of vesicular arbuscular mycorrhizas from a grassland. *Mycol. Res.* 415-455.

Schüßler, A. and Walker, C. (2010). http:// www. lrz.de/~schuessler/amphylo/.

Shashikala, B.N., Reddy, B.J.D. and Bagyaraj, D.J. (1999a). Influence of *Glomus mosseae* on the growth of micropropagated *Syngonium* and *Spathiphyllum* at varied levels of fertilizer phosphorus. *Crop. Res.* 18: 900.

Shashikala, B.N., Reddy, B.J.D. and Bagyaraj, D.J. (1999b). Response of micropropagated banana plantlets to *Glomus mosseae* at various levels of fertilizer phosphorus. *Ind. J. Exptl. Biol.* 37: 499-502.

Sieverding, E. (1991). *Vesicular-arbuscular mycorrhiza management in tropical agricultural systems.* TZ-Verlagsgesellscaft Rossdorf, Germany.

Souza F.A. and Silva E.M.R. (1996). Micorrizas arbusculares na revegetacao de areas degradades. In: *Aan cos em fundamentos e aplicacao de micorrhizas.* (Ed.) J.O Siquiera,

avras: Universidade Federal de Lavras/DCS e DCF. pp. 255-290.

Sparling, G.P. and Tinker, P.B. (1978). Mycorrhizal infection in Penning grassland. II. Effects of mycorrhizal infection on growth of some upland grasses on irradiated soil. *J. Appl. Ecol.* 15: 951-958.

Sylvia D.M. and Chellemi, D.O. (2001). Interactions among root-inhabiting fungi and their implications for biological control of root pathogens. *Adv. Agron.* 73: 1-33.

TAC Secretariat, 1989, *Sustainable Agricultural Production: Implication for International Agricultural Research,* Consultative Group on International Agricultural Research, Washington, D.C.

Tisserant, B., Gianinazzi-Pearson, V., Gianinazzi, S. and Gollute, A. (1992). In: Plant histochemical staining of fungal alkaline phosphatase activity for analysis of efficient endomycorrhizal infections. *Mycol. Res.* 97: 245-250.

Tommerup, I.C. and Sivasithamparam, K. (1990). Zygospores and asexual spores of *Gigaspora decipiens,* an arbuscular mycorrhizal fungus. *Mycol. Res.* 94: 897-900.

Trappe, J.M. (1977). Biogeography of hypogeous fungi: Trees, mammals and continental drift. In: Abstr. 2nd Int. Mycol. Congr. (Eds. Bigelow, H.E. and Simmons, E.G.) University of South Florida, Tampa, pp. 675.

Trappe, J.M., Molina. R. and Castellano, M. (1984). Reactions of mycorrhizal fungi and mycorrhiza formation to pesticides. *Ann. Rev. Phytopath.* 22:331-359.

Trouvelet, A., Kough, J.L. and Gianinazzi-Pearson, V. (1986). Measure du taux de mycorrhization VA d'un systeme radiculaire. Recherche de methods ayant une significant functionelle. In: *Physiological and Genetic Aspects of Mycorrhizae.* Proceedings 1st Eur. Symposium on Mycorrhizae. (Eds.) V. Gianinazzi-Pearson, and S.Gianinazzi, Instiut National de la Researche Agronomique. Paris, France.pp. 217-221.

Vidal, M.T., Azcon-Aguilar, C.K., Barea, J.M. and Pliego-Alfaro, F. (1992). Mycorrhizal inoculation enhances growth and development of micropropagated plants of avacado. *Hort. Sci.* 27: 785-787.

□□□

Microbial Diversity and Functions, 2012
© D.J. Bagyaraj, K.V.B.R. Tilak, H.K. Kehri (eds.), pp. 669-688
New India Publishing Agency, New Delhi (India)
E-mail : info@nipabooks.com; Website : www.nipabooks.com

Chapter 31

Use of Ectomycorrhizal Technology in Reclamation and Reforestation of Degraded Land

Bendangmenla, T. Ajungla, N.S. Jamir and G.D. Sharma

ABSTRACT

Ectomycorrhizal symbionts are important natural resources to revive degraded and deforested areas. Artificial inoculum of these mycobionts improves the growth of tree seedlings in nutrient poor soils due to the formation of extensive network of fungal mycelium responsible for the transfer of water and nutrients to the plants. Seedlings of tree species like, Pinus insularis, Quercus griffithii and Schima wallichii with ectomycorrhizae formed by Suillus bovinus, Boletus edulis and Scleroderma citrinum were studied in field condition. The study was conducted at two different stages of degraded sites. Results of the study indicated that seedlings with ectomycorrhizae were able to generate numerous lateral roots of greater length and thereby utilize water and nutrients more effectively than non- mycorrhizal ones The degree of mycorrhizal development varied in different host species with different fungal species. There was a strong relation between the colonization by mycorrhiza and the survival of seedlings. The study has suggested that mycorrhiza has a potential in reclamation of soil and reforestation of degraded land. S. bovinus., B. edulis and S. citrinum successfully colonized and promoted the growth of P. insulates Q. griffithii and S. wallichii respectively in reforestation programmes of degraded lands of Nagaland.

Keywords: Ectomycorrhizal technology, reclamation, reforestation, degraded lands.

Introduction

Ectomycorrhizas form symbiotic association between the terminal feeder roots of woody plant species and fungi. Within the root, the fungus ramifies between the outer cells forming a complex Hartig net and fungal sheath as an outer covering to provide a large surface area of contact between the fungus, soil and the host, allowing efficient transfer of metabolites (Taylor *et al.*, 2005).

The species richness and diversity of ectomycorrhizae is impressive in natural forests. Most of the terrestrial ecosystems are dominated by plants that require association with ectomycorrhizal fungi to achieve optimum productivity. According to Taylor *et al.* (2005), 7-8000 plant species may be capable of forming ectomycorrhizae. The great majority of ectomycorrhizal plants are woody perennials (Fitter and Moyersoen, 1996) although some sedges (*Kobresia* sp) and herbaceous *Polygonum* sp. also form ectomycorrhizae (Massiatte *et al.*, 1998). However, this minority of plant species is of enormous ecological and economical importance accounting for much of the world's wood production. Fungi capable of forming ectomycorrhizal associations are believed to be evolved on a number of occasions from a diverse range of saprophytic fungi (Hibbett *et al.*, 2000). The great majority (95%) of ectomycorrhizal fungi are homobasidiomycetes, the remaining being ascomycetes (4.8%) and a few zygomycetes within the genus *Endogone* (Molina *et al.*, 1992). Age of tree species and climatic conditions affect the succession of mycobionts on their roots (Rao *et al.*, 1996).

Ectomycorrhizal fungi play an important role in tree growth. The benefit of ectomycorrhiza for improving the growth of trees is root absorbing surface due to extensive network of fungal mycelium (Colinas *et al.*, 1994); binding of soil to create favourable soil structures (Borchers and Perry,1992); protection against pathogens (Mortier *et al.*, 1998) and facilitating below ground nutrient transfer among plants (Simard *et al.*, 1997a). Their beneficial role in production and supply of growth regulators, decreasing soil toxicity and increasing resistance to extreme soil temperature is also known (Heinrich *et al.*, 1989).

Status of wastelands in Nagaland

Rapid expansion of wastelands is taking place all over the world. According to Wastelands Atlas prepared by the Ministry of Rural Development, Government of India in collaboration with the National Remote Sensing Agency (NRSA), Hyderabad, the total area categorized as wastelands in Nagaland is 8.4 lakh hectares out of the total geographical area of 16.57 lakh hectares (Table 1, 2).

Table 1 : Nagaland- category-wise distribution and changes in wastelands

S	Wasteland Categories	2005-06	%	2003	%	Change	% diff
1	Land with Dense Scrub	972.55	5.87	1713.17	10.33	-740.62	-4.47
2	Land with Open Scrub	1011.02	6.10	59.18	0.36	951.84	5.74
3	Waterlogged and Marshy land- Permanent	0.00	0.00	0.60	0.00	-0.60	0.00
4	Waterlogged and Marshy land-Seasonal	0.00	0.00	1.03	0.01	-1.03	-0.01
5	Shifting cultivation area-Current Jhum	1239.09	7.47	1116.60	6.74	122.40	0.74
6	Shifting cultivation area-Abandoned-Jhum	1588.65	9.58	801.30	4.83	787.35	4.75
7	Under utilized/degraded notified forest land-Scrub dominated	0.00	0.00	8.19	0.05	-8.19	-0.05
8	Degraded pastures/ grazing land	0.00	0.00	0.34	0.00	-0.34	0.00
9	Sands-Riverine	0.00	0.00	0.55	0.00	-0.55	0.00
10	Barren rocky area	3.87	0.02	8.44	0.05	-4.57	-0.03
	Total	4815.18	29.04	3709.40	22.37	1105.78	6.67
	TGA	16579.00					

Table 2 : Category-wise Wastelands of Nagaland as per Wastelands Atlas of India, 2000

Category	Area in Ha.	Percentage of total
Land with or without scrub	1596.46	9.63
Shifting cultivation area	5224.65	31.51
Degraded forest land	1582.99	9.55
Total wasteland Area	**8404.10**	**50.69**

Source : Wastelands Maps prepared from Landsat Thematic Mapper/IRS LISS II/III Data, National Remote Sensing Agency(NRSA)

As evident from the table above, Jhum cultivation, commonly known as jhuming or Shifting Cultivation is a major cause of wastelands in Nagaland. This system of cultivation was ecologically *via*ble as long as population density was low and the fallow cycle was long enough to maintain the fertility of soil. However, with the increase in human population, the Jhum cycle has been shortened to 5 year or less. Frequent shifting from one land to another has affected the ecology of the region resulting in drastic loss of forest wealth, soil fertility, biodiversity and environment degradation. The area having Jhum cycle of 5 to 10 years is more vulnerable to weed invasion compared to Jhum cycle of 15 years (Singh *et al.,* 2003).

Ectomycorrhiza In Reclamation of Degraded Land

The use of ectomycorrhizal fungal inoculum is an important biological tool to rehabilitate degraded ecosystems. Mycorrhizal inoculation improves plant productivity especially in soil with low nutrient status (Ajungla *et al.,* 2003). The major advantage that a mycorrhizal association confers to plant and fungi is the enhanced supply of nutrients. The mycelia provide an increased surface area for nutrient uptake and improve nutrient acquisition of the host plants. In soil of low fertility, trees are more dependent on ectomycorrhizae for mineral nutrition (Smith and Read, 1997). The hyphae can penetrate small micro sites that are inaccessible to the much coarser plant roots. Active uptake of mobile nutrients such as phosphorus leads to the formation of nutrient depleted volumes of soil around roots and the mycorrhizal hyphae are able to bridge the annular spaces within soil, supplying nutrients from much distant soil.

The native ectomycorrhizal fungi *Suillus bovinus, Boletus edulis* and *Scleroderma citrinum* were selected and multiplied on MMN's medium. Seeds of native tree species *P. insularis, Q. griffithii* and *S .wallichii* were raised in nursery beds and transplanted in drgraded plots in randomized block design.. One month after transplantation, the selected isolates of fungi were inoculated (5mm diameter block) near the roots of the seedlings. Twenty five seedlings were maintained for each different fungus. No fungal inoculum was added to the control set of seedlings. Weeds and other debris were removed at regular intervals to reduce competition for nutrients and for the occurrence of pest and disease. The seedlings were irrigated as needed in order to avoid cross contamination. The effect of ectomycorrhizal fungi on the establishment and growth were assessed for one year at an interval of months after transplantation.

Assessment of Growth of Seedlings

Ectomycorrhizal fungi can contribute effectively up to 25% or more in root biomass of forests, thus contributing as a major structural component of the forest ecosystem (Manoharachary *et al.,* 2005). Studies have indicated that the ectomycorrhizal fungi possess great potential in forest productivity by adopting large scale inoculation practice (Ajungla *et al.,* 2001).

Majority of the work on inoculation with ectomycorrhizal fungi have been done in nurseries for production of bare root/container grown tree seedlings. Most of these studies were either conducted in glass-house or under controlled conditions which questions their suitability under field conditions where so many factors interact together. The studies carried out in laboratory condition on ectomycorrhizal fungi may show some differences in response to the natural environmental conditions due to complexity between soil, climate and biological components (Jha *et al.*, 1990). The selection of suitable ectomycorrhizal fungi for seedlings inoculations can enhance the success the forestation programmes (Harley and Smith, 1983). In order to gain maximum benefit from mycorrhizal inoculation, selection of fungal symbionts and field testing of mycorrhizal seedlings is an important strategy. The mycorrhizal association which enhances the growth and uptake of nutrient in tree need an assessment for its status in natural condition (Ajungla *et al.*, 2010).

Ectomycorrhizal fungi show strong physiological variability both within and between species; some species are better more adapted to certain ecological sites than do other species. Although ectomycorrhizal fungi are morphologically similar, but differ in their physiological functions (Allen, 1995). Artificial inoculations of most plant species with specific mycorrhizal fungus depend upon the efficiency of mycobionts to improve the growth of host under varied ecological conditions (Browning and Whitney, 1993). Marx and Cordell (1989) also noted that not all species of fungi that form ectomycorrhizae have equal benefits to their host under specific conditions and some are more effective than others.

In the present study, three ectomycorrhizal fungi induced differential benefits to the growth of the seedlings in terms of shoot height, needle/leaf length, root length, colar diameter and seedlings volume.

a. Ectomycorrhizal fungi and *Pinus insularis* (Endlicher)

Inoculation of ectomycorrhizal fungi consistently stimulated an increase in shoot height, needle length, seedlings volume, root collar diameter, lateral root length throughout the growing seasons as compared to uninoculated control. *S. bovinus* demonstrated its better compatibility with *P. insularis* by colonizing more of the root system than the other fungus and by mediating a far superior growth response (Table 3).

b. Ectomycorrhizal fungi and *Quercus griffithii* (Hoof. f.)

A positive response of inoculation with the three ectomycorrhizal fungi was also recorded in *Q. griffithii seedlings.* Comparatively, better growth characteristics were recorded with *B. edulis* and *Scleroderma citrinum* infected seedlings (Table 4).

c. Ectomycorrhizal fungi and *Schima wallichii* (Dc Korth)

S. wallichii seedlings were inoculated with the ectomycorrhizal fungi also had a significant effect on the growth of the seedlings. Mycorrhizal plants attained better growth than the mycorrhizal seedlings. However, there was no marked difference in the growth between the mycorrhizal seedlings. Out of the three ectomycorrhizal fungi, highest shoot height, root length, collar diameter and seedlings volume was recorded in seedlings with *B. edulis* (Table 4). A significant variation was found in all the growth parameters between mycorrhizal and non-mycorrhizal seedlings (Table 5, 6, 7).

Table 3 : Effect of ectomycorrhiza on the growth of *P. insularis* seedlings

Growth Parameters		Plot A				Plot B			
		M_4	M_8	M_{12}	M_{16}	M_4	M_8	M_{12}	M_{16}
Shoot height (cm)	M_0	11.2	15.8	35.6	72.2	11.3	16.1	36.4	77.6
	M_1	11.4	16.5	39.6	100.3	11.5	16.8	45.0	112.8
	M_2	11.6	16.9	45.5	125.3	11.7	17.4	50.0	133.0
	M_3	11.5	16.8	42.6	120.0	11.5	17.1	45.6	125.3
Needle length (cm)	M_0	5.1	7.9	17.4	20.0	5.2	8.3	17.4	23.1
	M_1	5.4	8.4	20.0	25.3	5.6	8.5	21.6	27.2
	M_2	5.6	8.5	21.8	27.8	5.7	8.9	25.8	27.2
	M_3	5.5	8.4	20.9	26.4	5.5	8.1	21.9	27.7
Root length (cm)	M_0	9.8	13.5	22.7	69.6	9.7	13.7	23.1	70.6
	M_1	10.4	16.4	27.2	78.0	10.5	16.6	31.7	778.4
	M_2	10.7	16.9	33.5	84.5	108	17.0	40.3	84.6
	M_3	10.6	16.7	33.1	81.4	107	16.8	35.2	79.7
Root colar diameter (cm)	M_0	0.18	0.22	0.27	0.40	0.17	0.23	0.28	0.43
	M_1	0.20	0.24	0.32	0.46	0.21	0.27	0.34	0.48
	M_2	0.22	0.26	0.34	0.50	0.23	0.29	0.36	0.53
	M_3	0.22	0.25	0.33	0.48	0.22	0.27	0.35	0.50
Seedlings volume (cm^3)	M_0	0.32	0.65	1.65	11.14	0.28	0.72	1.81	13.11
	M_1	0.42	0.94	2.78	16.50	0.46	1.21	3.66	18.06
	M_2	0.52	1.14	3.87	21.12	0.57	1.43	5.22	23.76
	M_3	0.51	1.04	3.60	18.75	0.52	1.22	4.31	19.92

Each value is the mean of 5 replication.

M_0 = Control

M_1 = *Scleroderma citrinum*

M_2 = *Suillus bovinus*

M_3 = *Boletus edulis*

Table 4 : Effect of ectomycorrhiza on the growth of *Q. griffithii* seedlings

Growth Parameters		Plot A				Plot B			
		M_4	M_8	M_{12}	M_{16}	M_4	M_8	M_{12}	M_{16}
Shoot height (cm)	M_0	8.2	11.8	33.3	94.0	8.2	12.1	32.5	89.9
	M_1	8.2	13.4	44.5	98.0	8.3	13.9	44.4	106.7
	M_2	8.4	13.2	37.0	82.8	8.3	13.5	49.0	104.1
	M_3	8.3	13.7	48.8	101.1	8.4	14.4	49.3	114.5
Leaf length (cm)	M_0	7.0	7.3	8.5	10.1	7.0	7.2	8.1	13.3
	M_1	7.2	7.4	8.4	11.3	7.0	7.6	14.4	16.9
	M_2	7.1	7.6	8.4	10.4	7.1	7.3	13.4	16.9
	M_3	7.2	7.6	10.4	11.8	7.1	7.6	13.7	17.2
Root length (cm)	M_0	6.6	10.4	14.1	24.0	6.5	10.6	14.4	24.3
	M_1	6.4	11.6	17.6	33.2	6.8	11.7	17.6	34.0
	M_2	6.7	11.4	17.0	29.9	6.7	11.5	16.5	30.4
	M_3	6.8	11.7	18.5	34.5	6.9	11.9	18.6	35.8
Root colar diameter (cm)	M_0	0.16	0.19	0.24	0.35	0.16	0.20	0.25	0.36
	M_1	0.17	0.23	0.31	0.44	0.17	0.24	0.31	0.46
	M_2	0.18	0.21	0.29	0.41	0.18	0.23	0.30	0.41
	M_3	0.16	0.24	0.33	0.47	0.18	0.26	0.33	0.49
Seedlings volume (cm^3)	M_0	0.17	0.37	0.88	2.94	0.17	0.42	0.90	3.15
	M_1	0.18	0.61	1.69	6.43	0.19	0.67	1.69	7.19
	M_2	0.22	0.50	1.43	5.03	0.22	0.61	1.48	5.11
	M_3	0.22	0.67	2.01	7.62	0.22	0.80	2.02	8.59

Each value is the mean of 5 replication.

M_0 = Control M_1 = *Scleroderma citrinum*

M_2 = *Suillus bovinus* M_3 = *Boletus edulis*

Table 5 : Effect of ectomycorrhiza on the growth of *S. wallichii* seedlings

Growth Parameters		Plot A				Plot B			
		M_4	M_8	M_{12}	M_{16}	M_4	M_8	M_{12}	M_{16}
Shoot height (cm)	M_0	7.4	11.8	39.5	102.3	7.5	11.8	33.6	110.6
	M_1	7.7	13.6	37.9	112.3	7.9	13.8	38.6	130.2
	M_2	7.5	13.0	34.1	107.1	7.7	13.6	35.0	109.0
	M_3	7.7	13.9	38.0	129.0	7.8	13.9	43.2	136.5
Leaf length (cm)	M_0	5.5	7.5	13.3	14.9	4.8	7.5	13.1	16.3
	M_1	6.9	10.3	16.8	18.5	5.3	10.6	16.8	19.6
	M_2	6.6	10.1	16.4	16.9	4.5	9.7	16.6	19.4
	M_3	6.8	10.4	16.5	19.5	5.3	10.7	16.9	20.0
Root length (cm)	M_0	7.0	11.2	14.6	28.1	7.1	11.3	14.5	28.3
	M_1	7.4	12.6	16.4	34.8	7.7	12.4	16.5	36.5
	M_2	7.3	12.4	16.4	35.5	7.4	12.5	16.4	36.5
	M_3	7.5	12.5	16.4	38.1	7.6	12.8	16.7	40.6
Root colar diameter (cm)	M_0	0.16	0.21	0.26	0.38	0.16	0.20	0.26	0.39
	M_1	0.18	0.24	0.32	0.50	0.19	0.26	0.32	0.52
	M_2	0.17	0.24	0.30	0.47	0.18	0.25	0.30	0.48
	M_3	0.19	0.24	0.34	0.53	0.19	0.26	0.34	0.55
Seedlings volume (cm^3)	M_0	0.18	0.49	0.99	4.06	0.18	0.45	0.98	4.31
	M_1	0.24	0.73	1.68	8.72	0.28	0.84	1.69	9.88
	M_2	0.21	0.71	1.46	7.86	0.23	0.78	1.48	8.42
	M_3	0.27	0.76	1.91	10.73	0.27	0.86	1.93	12.19

Each value is the mean of 5 replication.

M_0 = Control M_1 = *Scleroderma citrinum*

M_2 = *Suillus bovinus* M_3 = *Boletus edulis*

Table 6 : Analysis of variance (F) values of mycorrhizal and non-mycorrhizal seedlings.

	Source of variance	Variation between mycorrhizal and non - mycorrhizal seedlings
Pinus insularis	Shoot height	**162.90***
	Needle length	**136.54***
	Root length	**34.20****
	Root colar diameter	**53.30****
	Seedlings volume	**23.31****

Table 7 : Analysis of variance (F) values of mycorrhizal and non-mycorrhizal seedlings

	Source of variance	Variation between mycorrhizal and non - mycorrhizal seedlings
Quercus griffithii	Shoot height	**67.90***
	Leaf length	**67.60***
	Root length	**8.45****
	Root colar diameter	**45.09****
	Seedlings volume	**26.12****

Table 8 : Analysis of variance (F) values of mycorrhizal and non-mycorrhizal seedlings.

	Source of variance	Variation between mycorrhizal and non - mycorrhizal seedlings
Schima wallichii	Shoot height	**67.90***
	Leaf length	**67.60***
	Root length	**8.4****
	Root colar diameter	**45.1****
	Seedlings volume	**26.1****

* = Significant at $p < 0.05$ level ** = Significant at $p < 0.01$ level

Such differential response of the host plants to different fungal species can be attributed to the efficacy of specific mycorrhizal fungi in nutrient and water absorption and the amount of ectomycorrhizae formed during the growing seasons The results are supported by the findings of Tiwari and Mishra (1995) who reported that seedlings of *Pinus kesiya* inoculated with 4 different ectomycorrhizal fungi, among them *P. tinctorius* increased the plant height, colar diameter and root length as compared to non-inoculated seedlings. Singh and Devi (1998) also observed differential effects of three ectomycorrhizal fungi on the shoot, root or plant weight of *Pinus kesiya*. They found that ectomycorrhizal fungi like *Laccaria laccata* and *Scleroderma verrucosum* stimulated better growth of the seedlings. Rao *et al.* (1996) also observed differences in the growth of *P. kesiya* seedlings inoculated with 6 ectomycorrhizal fungi. They observed that the growth of seedlings in terms of shoot height, root length, root colar diameter was improved by *S. leteus* and *S. aurantium*. Ajungla *et al.* (2006) noted significant increase in Pine seedlings (*P. kesiya*) inoculated with ectomycorrhizal fungi as compared to non-mycorrhizal fungi. Similarly, Ajungla *et al.* (2010) also found that seedlings of *P. patula* infected with *Scleroderma citrinum* showed better enhanced growth performance than non-mycorrhizal ones.

Ectomycorrhiza and Survival of Seedlings In Waste Land Soil

Roots were harvested periodically and examined under binocular microscope for mycorrhizal colonization. Colonization of ectomycorrhizal fungi in the roots of the seedlings was visible between 40-60 days of inoculation. The degree of ectomycorrhizal formation differed in different species and also plant types (Table 8). The development of ectomycorrhizae was almost parallel to the growth of the seedlings. A positive correlation was found between the mycorrhizal infection and the various growth parameters (Table 10,11,12).

The ectomycorrhizal fungi had improved the growth of the seedlings. Survival and colonization showed a positive trend. There is a strong relation between percentage of colonization and percent of survival indicating that higher colonization of roots increased the survival of seedlings (Table 9). These findings are in confirmation with the studies of Umar Badshah (2006) who examined the growth and colonization in *Eucalyptus tinctorius*. Similarly, Bhat *et al.* (1997) also recorded highest survival of seedlings and colonization of roots with *Rhizopogon luteolus* followed by *P. tinctorius*. In addition, survival and growth of seedlings is also directly related to the way in which species can adjust their morphological and physiological characters to the environment (Rose, 2000).

Table 9: Development of ectomycorrhiza in less (Plot A) and more (Plot B) degraded land

		Plot A				Plot B			
		M_4	M_8	M_{12}	M_{16}	M_4	M_8	M_{12}	M_{16}
Pinus insularis	M_0	7	12	36	52	6	17	34	55
	M_1	43	60	80	92	40	60	81	93
	M_2	48	67	85	94	47	66	86	98
	M_3	46	63	82	92	44	63	85	96
Quercus griffithii	M_0	5	11	97	46	5	13	29	50
	M_1	33	50	77	89	39	55	79	90
	M_2	29	45	73	84	30	48	65	87
	M_3	39	58	80	90	41	60	80	92
Schima wallichii	M_0	5	10	23	44	6	13	24	42
	M_1	34	50	69	89	36	52	78	92
	M_2	31	47	63	84	27	50	66	87
	M_3	36	53	78	91	38	57	82	94

Each value is the mean of 5 replication.

M_0 = Control M_1 = *Scleroderma citrinum*

M_2 = *Suillus bovinus* M_3 = *Boletus edulis*

Table 10 : Survival of tree seedlings on less and more degraded land with ectomycorrhiza

		Plot A				Plot B			
		M_4	M_8	M_{12}	M_{16}	M_4	M_8	M_{12}	M_{16}
Pinus insularis	M_0	92	92	88	88	96	92	88	88
	M_1	96	92	92	92	100	96	96	96
	M_2	100	100	96	96	100	100	100	100
	M_3	96	96	96	92	96	96	96	96
Quercus griffithii	M_0	96	96	86	84	92	92	84	84
	M_1	96	92	92	88	96	96	92	92
	M_2	100	100	96	88	96	96	92	88
	M_3	96	96	92	92	92	92	92	92
Schima wallichii	M_0	92	92	84	80	96	96	96	92
	M_1	92	92	92	92	96	96	96	96
	M_2	100	100	92	92	96	96	92	92
	M_3	100	96	96	96	96	96	96	96

Each value is the mean of 5 replication.
M_0 = Control M_1 = *Scleroderma citrinum*
M_2 = *Suillus bovinus* M_3 = *Boletus edulis*

Table 11 : Correlation coefficient (r) between mycorrhizal infection (%) and growth parameters.

Pinus insularis	Mycorrhizal infection							
	Plot A				Plot B			
Parameters	M_0	M_1	M_2	M_3	M_0	M_1	M_2	M_3
Shoot height (cm)	NS	0.85	0.81	0.82	NS	0.84	0.87	0.82
Needle length (cm)	NS	0.94	0.93	0.94	NS	0.94	0.87	0.93
Root length (cm)	NS	0.84	0.83	0.85	NS	0.84	0.89	0.85
Root colar diameter (cm)	NS	0.91	0.88	0.86	NS	0.91	0.92	0.90
Seedlings volume (cm^3)	NS	0.80	0.75	0.77	NS	0.80	0.88	0.77

Table 12 : Correlation coefficient (r) between mycorrhizal infection (%) and growth of seedlings

Quercus griffithii	Mycorrhizal infection							
	Plot A				Plot B			
Parameters	M_0	M_1	M_2	M_3	M_0	M_1	M_2	M_3
Shoot height (cm)	NS	0.88	0.90	0.88	NS	0.86	0.91	0.84
Leaf length (cm)	NS	0.86	0.80	0.82	NS	0.84	0.82	0.82
Root length (cm)	NS	0.90	0.93	0.88	NS	0.81	0.94	0.88
Root colar diameter (cm)	NS	0.92	0.88	0.94	NS	0.92	0.92	0.93
Seedlings volume (cm^3)	NS	0.80	0.82	0.81	NS	0.88	0.87	0.81

Table 13 : Correlation coefficient (r) between mycorrhizal infection (%) and growth of tree seedlings

Schima wallichii	Mycorrhizal infection							
	Plot A				Plot B			
Parameters	M_0	M_1	M_2	M_3	M_0	M_1	M_2	M_3
Shoot height (cm)	NS	0.90	0.90	0.83	NS	0.85	0.91	0.83
Leaf length (cm)	NS	0.94	0.90	0.96	NS	0.97	0.82	0.98
Root length (cm)	NS	0.93	0.91	0.82	NS	0.87	0.94	0.83
Root colar diameter (cm)	NS	0.91	0.94	0.88	NS	0.90	0.92	0.90
Seedlings volume (cm^3)	NS	0.83	0.84	0.88	NS	0.83	0.87	0.76

Influence of Ectomycorrhizal Fungi on The Uptake of NPK

Results of the study suggest that the mycorrhizae formed by the three fungal species in association with the roots of host plants, play an important role in the cycling of macronutrients. In the three plant species studied, amount of different macronutrients was stimulated by different species of ectomycorrhizal fungi. It was noted that for each of tree species, best performance by a certain fungi in accumulating a given nutrient was not always observed for other nutrients (Figure 1 to 9).

Analysis of mineral nutrients in the mycorrhizal seedlings suggested that growth was greater in the mycorrhizal seedlings than the non-mycorrhizal ones which could be due to the increased uptake of nutrients especially of N and P. The greater root development in the seedlings treated with ectomycorrhizal fungi might have accounted for greater nutrient exploitation zone to the mycelium which ultimately resulted in better growth of seedlings. The cause of growth reduction in non-mycorrhizal maybe due to reduced root system, resulting in less exploitation of mineral nutrients. Kumar *et al.* (1991) also observed that the mycorrhizal seedlings have more extensive root system than non- mycorrhizal ones. The nutrient uptake capacity of a fungus was correlated with the amount of mycelium colonizing the soil. A significant variation was found in nitrogen content ($P<0.05$), phosphorous and potassium content ($P<0.01$) between the treated and control seedlings (Tables 13, 14, 15). There exists a correlation between the capability of the fungal species in

Table 14 : Analysis of variance (F) of mycorrhizal and non-mycorrhizal seedlings.

	Source of variance	Variation between mycorrhizal and non-mycorrhizal seedlings
Pinus insularis	Nitrogen	**6.53***
	Phosphorus	**18.0****
	Potassium	**72.0****

forming mycorrhizae with the roots of the seedlings and in stimulating the uptake of nutrients.

Table 15 : Analysis of variance (F) values of mycorrhizal and non-mycorrhizal seedlings.

	Source of variance	Variation between mycorrhizal and non-mycorrhizal seedlings
Quercus griffithii	Nitrogen	**5.50***
	Phosphorus	**29.8****
	Potassium	**91.60***

Table 16 : Analysis of variance (F) values of mycorrhizal and non-mycorrhizal seedlings

	Source of variance	Variation between mycorrhizal and non-mycorrhizal seedlings
Schima wallichii	Nitrogen	**5.62***
	Phosphorus	**11.2****
	Potassium	**85.9****

* = Significant at $p < 0.05$ level
** = Significant at $p < 0.01$ level

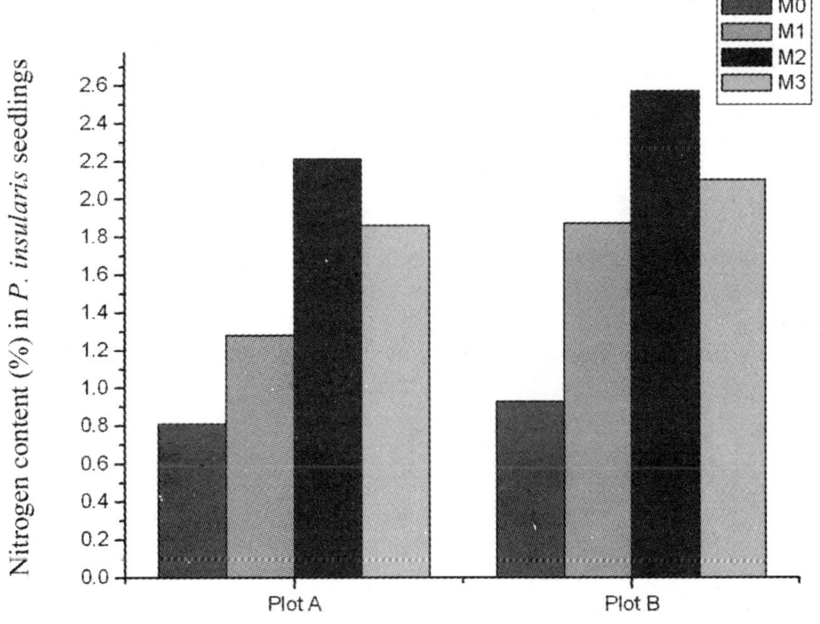

Figure 1

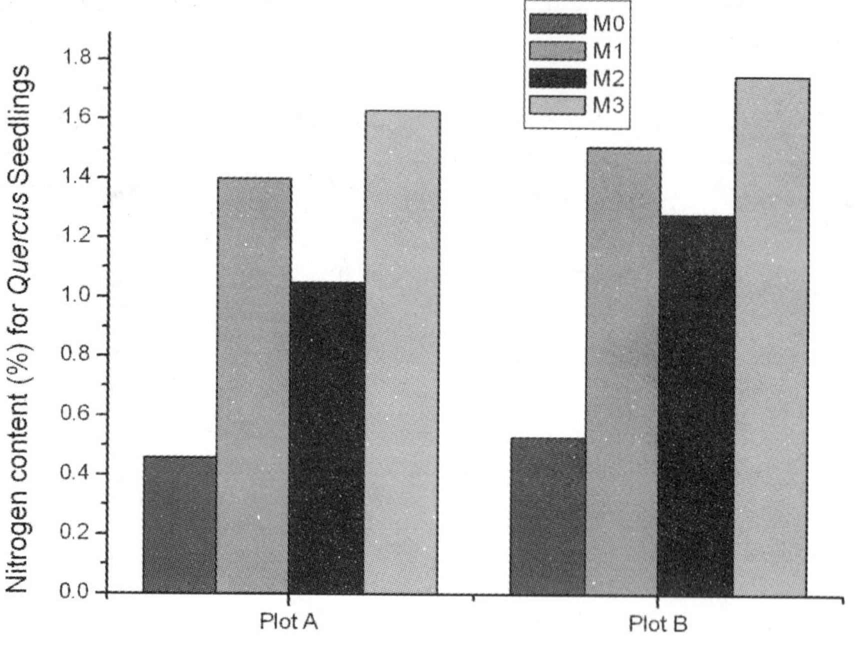

Figure 2

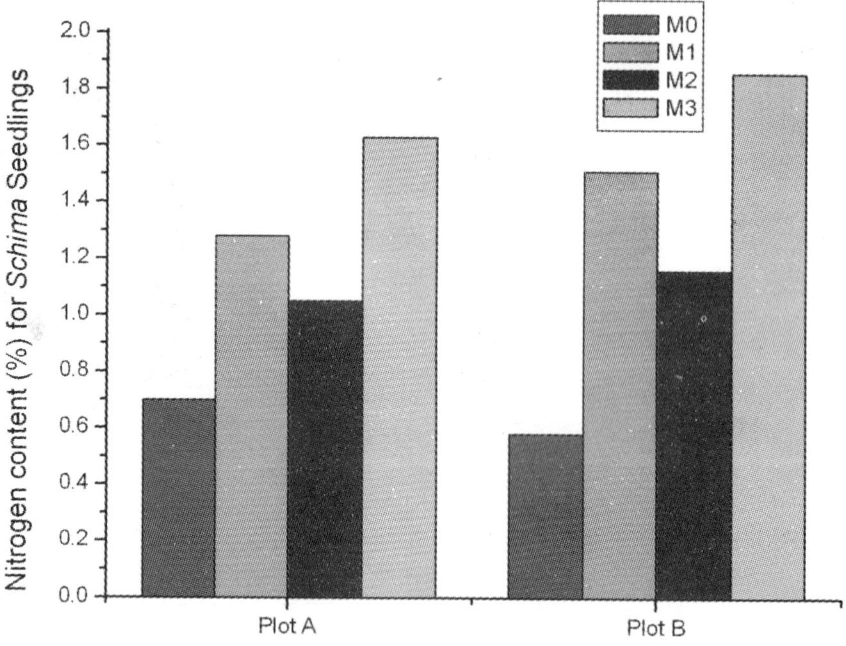

Figure 3

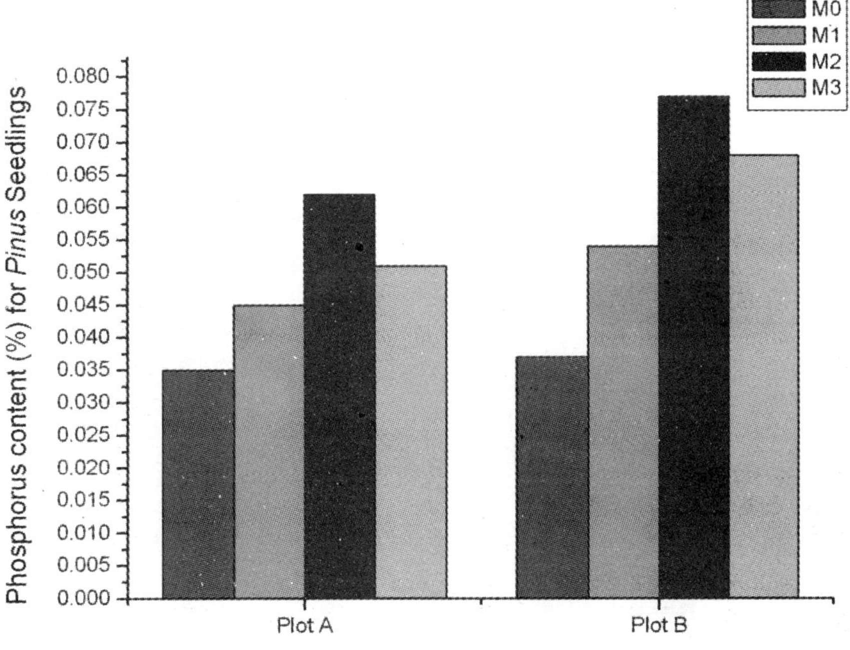

Figure 4

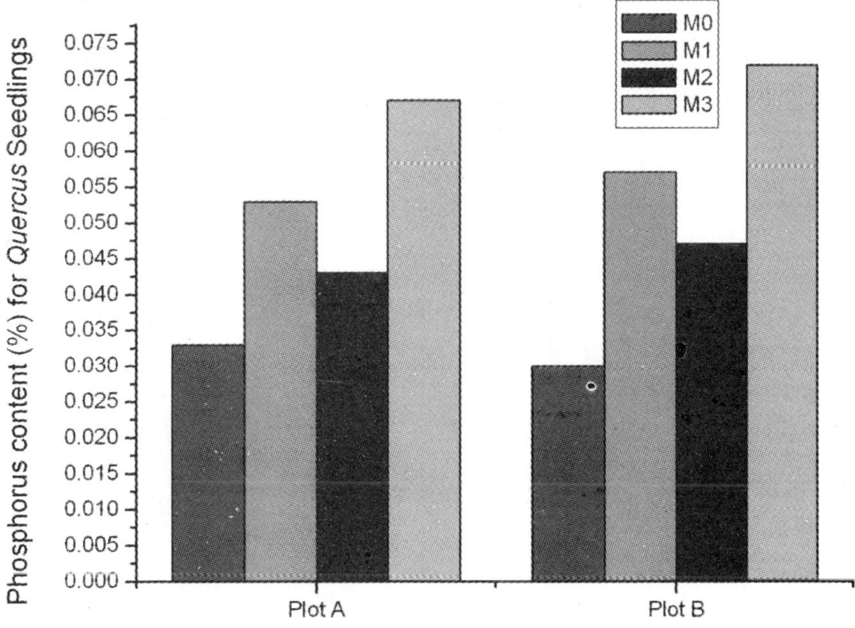

Figure 5

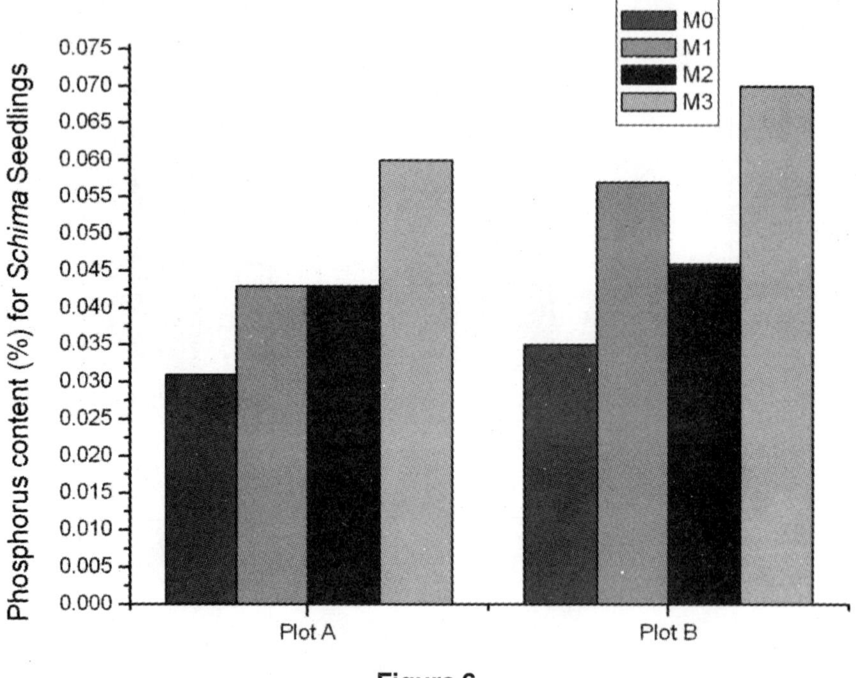

Figure 6

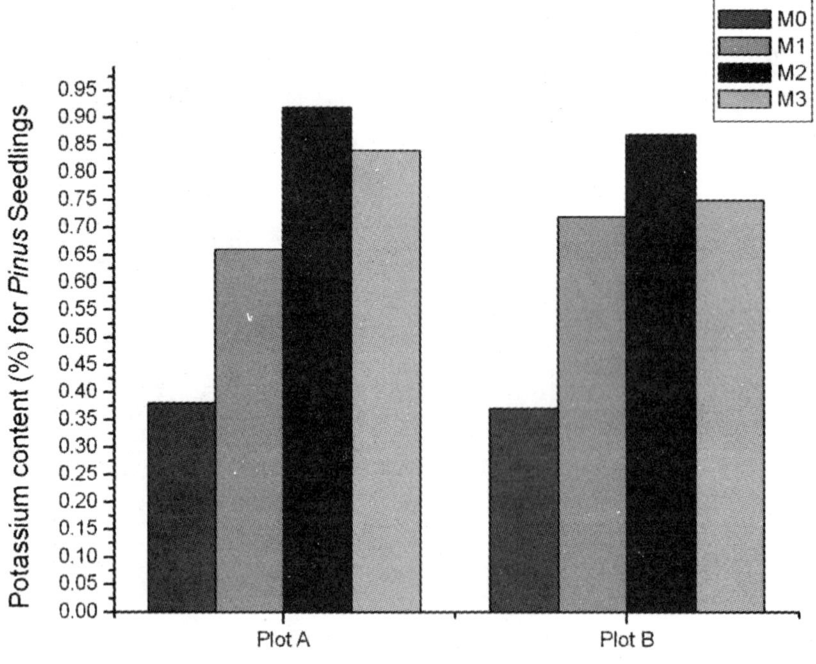

Figure 7

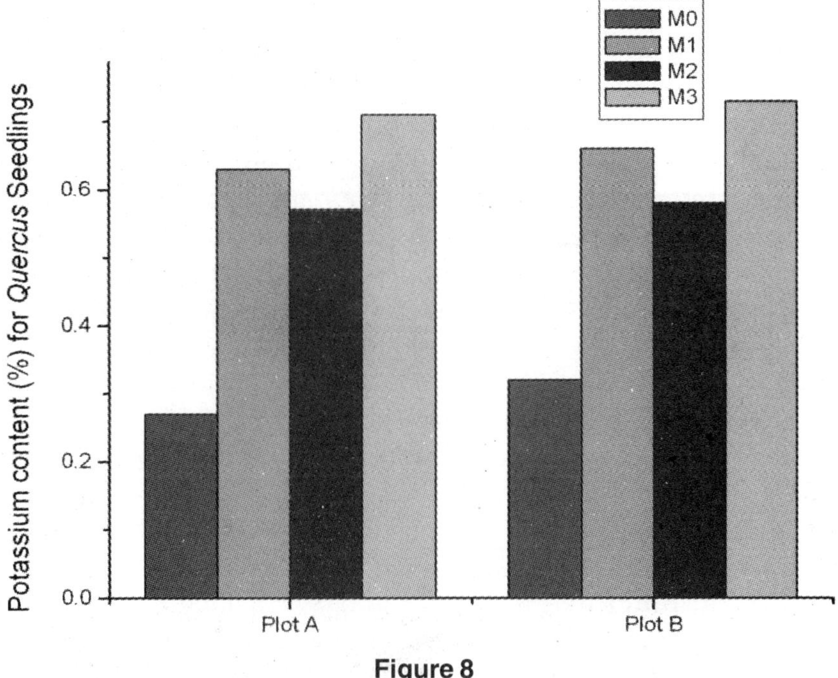

Figure 8

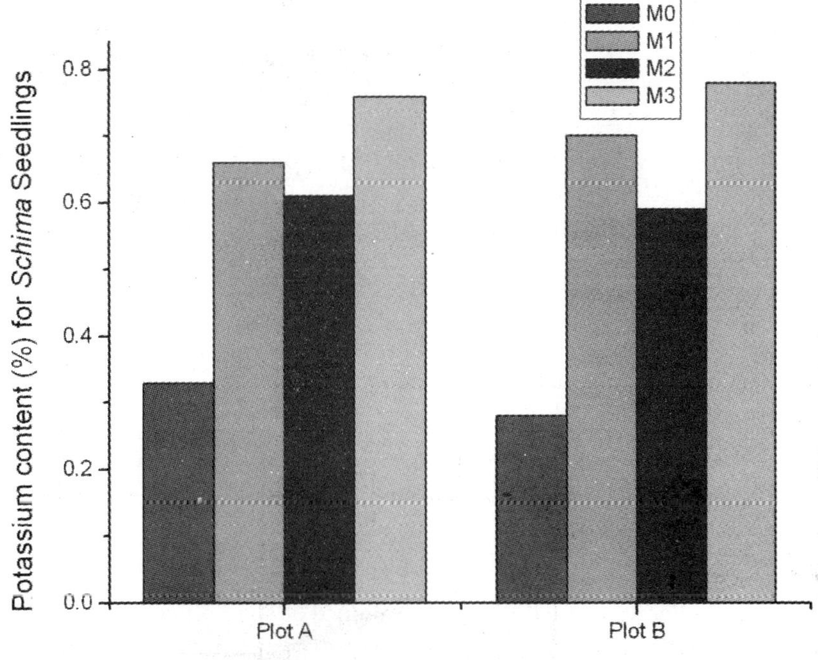

Figure 9

Conclusion

The problem of wasteland is a global one. Rehabilitation of wastelands is a difficult task.will not become rehabilitated within lifetime through the normal natural succession process. The planting of seedling with root system physiologically and ecologically adopted to accommodate the adversities of the planting sites can be an important biological tool in reclamation of wasteland. The species of ectomycorrhizal fungi used on various tree species in experimental inoculation programmes have shown a great promise for adopting large scale inoculation techniques in artificial rehabilitation of degraded land, which is a reliable, less time consuming technique and less expensive inoculum production technique. These field studies on adverse sites show that reclamation and forestation on such sites can be expedited by using seedlings tailored with ectomycorrhiza formed by fungi capable of growing under adverse conditions. Among the fungal species, *S. bovinus* and *B. edulis* are found to be efficient for artificial inoculation. It is possible to increase the growth of *P. insularis* seedlings with the inoculation of *S. bovinus*. Ectomycorrhizal fungi like *B. edulius* and *S. citrinum* can also be used successfully to colonize and promote the growth of *Q. griffithii* and *S. wallichii* seedlings in the degraded land areas.

References

Ajungla, T., Sharma, G.D. and Dkhar, M.S. (2001). Effect of heavy metal pollution on nutrient contents in ectomycorrhizal pine seedlings (*Pinus kesiya* Royle Ex. Gordon). *Vasundhara,* 6:1-4.

Ajungla, T., Sharma, G.D. and Dkhar, M.S. (2003). Heavy metal toxicity on dehydrogenase activity on rhizospheric soil of ectomycorrhizal pine seedlings in field condition. *J. Environ Biol.,* 24(4):461-463.

Ajungla, T., Sharma, G.D. and Dkhar, M.S. (2006). Uptake of phosphorus by ectomycorrhizal pine seedlings in the metal polluted soil. *J. Environ. Biol. and Conserv,* 11:15-18.

Ajungla, T., Imliyanger and Tzudir. (2010). Effects of ectomycorrhizal fungi on the growth performance of Pinus patula (Schiede ex Schlecht. & Cham). *Environ. Biol. And Conserv.,* 15:29-31.

Allen, E.B., Allen, M.F., Helm, D.J., Trappe, J.M., Molina, R. and Rincon, E. (1995). Patterns and regulation of mycorrhizal plant and fungal diversity. *Plant Soil,* 170:47-62.

Bhat, M., Jeyarajan, R. and Ramaraj, B. (1997). Bio-control of damping off of *Eucalyptus tereticornis* using ectomycorrhizae. *Ind. Forester,* 123(4):307-311.

Borchers, S.L. and Perry, D.A. (1992). The influence of soil texture and aggregation on carbon and nitrogen dynamics in Southwest Oregon forest and clear-cuts. *Can. J. For. Res.,* 21:198-305.

Browning, MHR and Whitney, RD. (1993). Infection of containerized jack pine and black spruce by *Laccaria* species and *Thelephora terrestris* and seedling survival and growth after out planting. *Can. J. For. Res.*, 23:330-333.

Colinas, C, Molina, R, Trappe, and Perry, D. (1994). Ectomycorrhizas and rhizosphere microorganisms of seedlings of *Pseudotsuga menziesii* (mirb.) Franco planted on a degraded site and inoculated with forest soil pretreated with selective biocides. *New Phytol.*, 127:529-537.

Fitter, A.H. and Moyersoen, B. (1996). Evolutionary trends in root-microbe symbioses. *Phil. Trans. Royal Soc. London. Series B-Biol. Sci.*, 351:1367-1375.

Harley, J.L. and Smith, S. E. (1983). *Mycorrhizal Symbiosis* , Academic press London, UK.

Heinrich, P.A., Muligen, D.R. and Patrick, J.F. (1989). The effect of ectomycorrhizal on the phosphorus and dry weight acquisition of Eucalyptus seedlings. *Plant Soil*, 108:147-149.

Hibbett, D.S., Gilber,t L.B. and Donaghue, M.J. (2000). Evolutionary instability of ectomycorrhizal symbiosis in basidiomycetes. *Nature*, 407:506-508.

http://dolr.nic.in/wasteland.2010. Nagaland–category–wise distribution and changes in wastelands.

http://dolr.nic.in/wasteland.2010. Category-wise Wastelands of Nagaland as per Wastelands Atlas of India, 2000.

Jha, B.N., Sharma, G.D. and Mishra, R.R. (1990). Effect of pH on the growth of ectomycorrhizal fungi *in vitro*. *Proc. Nat. Con. Mycorrhiza,* (Eds.) B.L. Jalali and H. Chand, Haryana Agric.University, Hisar: 66-67.

Kumar, R., Shukla, A.K., Sharma, G.D. and Mishra, R.R. (1991). Response of pine seedlings to fungicides. *Acta. Bot. Ind.*, 19:157-161.

Manoharachary, C., Sridhar, K., Singh, Reena, Adholeya, A., Suryanarayanan, T.S., Rawat, Seema and Johri, B.N. (2005). *Curr. Sci.*, 89(1):58-71.

Marx, D.H. and Cordell, C.E. (1989). In: *Biotechnology of fungi for improving plant growth* (Eds.) J.M. Whipps and R.D. Lumsden, Cambridge University Press, London

Massiatte, H. B., Melville, L.H., Peterson, R.L .and Luoma, D.L. (1998). Anatomical aspects of field ectomycorrhizas on *Polygonum viviparum* (Polygonaceae) and *Kobresia bellardii* (Cyperaceae). *Mycorrhiza*, 7:287-292.

Molina, R. and Trappe, J.M. (1992). Specificity phenomena in mycorrhizal symbiosis: Community ecological consequences and practical applications. In: *Mycorrhizal functioning* (Ed.) A. Allen, Chapman and Hall, New York.

Mortier, F., Le Tacon and Garbaye. (1998). Effect of inoculum type and inoculation dose on ectomycorrhizal development, root nurseries and growth of Douglas-fir seedlings inoculated with *Laccaria laccata* in a nursery. *Ann. Sci. For.*, 45:301-310.

Rao, C.S., Sharma, G.D. and Shukla, A.K. (1996). Ectomycorrhizal efficiency of various mycobionts with *Pinus kesiya* seedlings in forest and degraded soils. *Proc. Ind. Nat. Sci. Acad.B*, 62. 5:427-434.

Rose, S.A. (2000). *Seeds, saplings and gaps: Size matters. A study in the tropical rainforests of Guyana.* Ph.D. Thesis. Utrecht University, Utrecht, The Netherlands, Tropenbos Guyana Series 9, Tropenbos Guyana programme, Georgetown, Guyana.

Simard, S.W., Perry, D.A., Jones, M.D., Durall, D.M. and Molina, R. (1997a). Net transfer of carbon between tree species with shared ectomycorrhizal fungi. *Nature,* 388: 579-582.

Smith, S.E. and Read, D.J. (1997). *Mycorrhizal Symbiosis.* 2nd Edition. Academic Press, London.

Singh, J., Bora, I.P. and Baruah, A. (2003). Changes in the physio-chemical properties of soil under shifting cultivation with special reference to Karbi Anglong district of Assam. *Ind. J. Forest.,* 26(1):116-122.

Singh, M.S. and Devi, I.J. (1998). Effect of three ectomycorrhizal fungi on the growth of pine seedlings. *Vasundhara,* 3:51-53.

Taylor, A.E.S. and Alexander, I. (2005). The ectomycorrhizal symbiosis: life in the real world. *Mycologists,* 19:102-112.

Tiwari, S.C. and Mishra, R.R. (1995). Effects of *Boletus edulis, Laccaria laccata, Pisolithus* and *Rhizopogon luteolus* on the growth performance of *P. kesiya* (Royle Ex. Gordon) in N. E India. *Ind. J. Forest.,* 1(8):293-300.

Umar Badshah, N.K. and Naik, S.T. (2006). Growth and colonization in *Eucalyptus* tereticornis (hybrid) seedlings inoculated with different inoculum formulations of *Pisolithus tinctorius. Ind. Forester,* 575-580.

❑❑❑